Culinary Recipes

Culinary Recipes

Fifth Edition

Educational Task Force

JOHNSON & WALES UNIVERSITY College of Culinary Arts *Volume* III

Harborside Press
Providence, Rhode Island

Published by: Harborside Press
765 Allens Avenue
Providence, Rhode Island 02905

Senior developmental editors: Johnson & Wales Food/
Educational Task Force & Firebrand LLC
Cover design: Firebrand LLC
Interior design: Firebrand LLC
Production & Coordination: Firebrand LLC
Photography: YUM and Others
Photographer: Ronald Manville, YUM
Food stylist: James E. Griffin

ISBN: 1-890724-02-7

Printed in the United States of America

Contents

Message from Past Directors

As students of Culinary Arts, and as you go forth to pursue your professional goals, remember this phrase—*mise en place*. This is the magic that will guarantee your success . . . . if you live by it.

Socrates Z. Inonog, CCE, AAC
DSO, Executive Director
International Student Affairs

A superb meal does not happen by accident; it is the result of meticulous planning: purchasing, receiving, issuing, storing, coordinated teamwork, methodology, sanitary food handling, accurate timing, plating with artistic flair, and compatability with dining room service and table settings. And provided all of the above are accompanied by professional integrity.

So, keep on planning . . . don't cook and serve *on the cuff!!* Keep on planning.

Franz K. Lemoine
Founding Director
Johnson & Wales University
College of Culinary Arts

As the world's largest foodservice educator, we aspire to achieve prominence among the future gastronomes of the world with a clear commitment to culinary arts. We combine the classical tradition of the past with the unique demands of today to prepare you for solid careers in foodservice.

As a Johnson & Wales student, you are laying your *groundwork for greatness* through formalized education. The core of our educational philosophy is sharply focused on hands-on career education, which is so vital in providing you with the opportunity for an improved quality of life and contribution to your community.

You are a member of a very unique and proud team—the College of Culinary Arts—which has a moral obligation to see you succeed. With membership privileges also come responsibility—your commitment to work hard, to utilize the resources provided by the University, and always to take pride in yourself. This is the formula for success. Remember: *When hard-learned skills and natural talent meet opportunity, achievement is not only possible, it's inevitable.*

John J. Bowen
Executive Vice President
Johnson & Wales University

Foreword

Culinary: Recipes is a culmination of ideas arranged in a systematic format to be easily understood and followed.

A recipe is not a scientific rule that cannot be altered, modified, or improved.

Cooking was and will always be a creative art. A chef's responsibility is to satisfy one of life's pleasures through knowledge and artistic talent.

Presently we are faced with challenges that were not encountered in the past. The chef today must be able to meet the needs of ever-changing consumers, ones with different habits, beliefs, cultures, and lifestyles.

Other challenges facing the chef today are technical and agricultural advances; for example, production of plant and animal products, delivery, and marketing of those products. Every day new vegetables and fruits are available, and we are challenged to create dishes from them.

A chef's experience and skills are only as good as a recipe. You need both to produce a quality dish.

R. M. Nograd, CMC
Dean of Practicum Education
Johnson & Wales University

Preface

The Johnson & Wales University, College of Culinary Arts, Volume I—*Culinary: Fundamentals,* Volume II—*Culinary: Service,* and Volume III—*Culinary: Recipes,* represent the University's commitment to excellence in education. The new edition will enable the college to be on the cutting edge as a trendsetter, introducing new ideas, new procedures, and technologies to our students and to the foodservice industry.

Each volume will be used as a reference tool from which faculty and students can teach and learn. These textbooks will emphasize theory and practical applications in nutrition, sanitation, cost control, and marketing.

Volume III, *Culinary: Recipes,* contains over 800 recipes. The recipes contained in this volume were selected in order to teach the classical as well as nutritional value of food. The recipes also show and demonstrate the range of the simple to the more complex methods of preparation and presentation.

It is important to understand that there are numerous ways of writing a recipe, but consistency is the key to a quality product.

Introduction to Hazard Analysis Critical Control Point (HACCP)

The Hazard Analysis Critical Control Point (HACCP) was established by the Pillsbury Company in 1971 for the National Aeronautics and Space Administration (NASA). The primary purpose of HACCP is to ensure that food served to the customer is safe.

All foods in a foodservice operation pass through a number of critical areas including receiving, storage, preparation, cooking, holding, service, reheating, and cooling. At each step, contamination is a possibility. It is imperative that the foodservice operator identify the critical control point as foods flow through the operation.

Each of the food sections in this text includes a discussion indicating how to apply HACCP in each particular food section.

Food and Drug Administration Internal Cooking Temperatures

Beef	140° F
Fish	140° F
Lamb	140° F
Liver	140° F
Veal	145° F
Game (e.g., venison)	155° F
Ground beef	155° F
Pork	155° F
Poultry	165° F
Stuffed pasta	165° F
All stuffed meats	165° F

- Cold protein salads; pasta salads and all ingredients: 40° F.
- Microwave cooking: heat an additional 25° F or higher than conventional oven products cooking temperatures. For example, in a conventional oven, cook a chicken casserole to 165° F; in a microwave oven, cook a chicken casserole to 190° F.
- Shell eggs, broken and prepared for immediate service to consumer: 145° F or above.
- Temperature danger zone: 40° F to 140° F.
- Internal reheating temperature: 165° F.
- Internal cooking temperature for a milk- or cream-based soup (e.g., New England clam chowder, cream of broccoli): 165° F.
- Internal cooking temperature for stews and seafood chowders: 165° F.
- Internal holding temperature: 140° F.

- Roasts can be cooked to an internal temperature of 130° F **if** held at that temperature for 121 minutes.
- Roasts can be cooked to an internal temperature of 140° F **if** held at that temperature for 12 minutes.

Anaphylaxis

A customer's health and safety is a very important priority. Anaphylactic (allergic) reactions can be life-threatening. The Food Allergy Network, 4744 Holly Avenue, Fairfax, VA 22030-5647 has indicated the most common food-borne allergens. The ingredients which most often cause allergic reactions are noted throughout the text in the marginal notes. The food item may cause a life-threatening reaction if eaten by individuals with food allergies. If a reaction does occur, 911 should be called *immediately*.

Educational Task Force

Acknowledgments

The three volumes of Culinary—I: Fundamentals; II: Service; *and* III: Recipes—*have been written by people who truly love culinary arts. I wish to thank all the faculty, administration, and friends of the University for their support and participation in this tremendous undertaking. Their collective work represents their dedication to educating students who will keep the flame of culinary inspiration alive for future generations.*

Thomas L. Wright

Vice President
Culinary Education

Special thanks to the following individuals for their tireless efforts and dedication in producing these textbooks:

Pauline Allsworth
Linda Beaulieu
Dr. Barbara Bennett
Lynn Dieterich
James Griffin
Karl Guggenmos
Edward Korry
Victoria A. McCrae
Paul J. McVety
Robert Nograd
Pamela Peters
Jacquelyn B. Scott
Christine Stamm
Michel Vienne
Bradley Ware

Educational Task Force

Carolyn Buster
Martha Crawford
Elaine Cwynar
Mary Ann DeAngelis
John Dion
Jean-Jacques Dietrich
Meridith Ford
Karl Guggenmos
Frederick Haddad
Lars Johannson
Edward Korry
Robert Nograd
Pamela Peters
Patrick Reed
Janet Rouslin
Christine Stamm
Frank Terranova
William Travis
Bradley Ware

Providence Administration, Faculty and Support Staff

Thomas L. Wright, M.S., Vice President of Culinary Education

Jean-Michel Vienne, C.C.P., C.E.P., C.A.P., Dean, College of Culinary Arts

Dorothy Jeanne Allen, M.S., Associate Professor; A.S., B.S., M.S., Johnson & Wales University

Pauline Allsworth, Office Manager

Frank Andreozzi, B.S., Assistant Professor; B.S., Providence College

Charles Armstrong, A.O.S., Instructor; A.O.S., Culinary Institute of America

Soren Arnoldi, Danish Master Chef, Associate Instructor; Falke Hotel, Tivoli Gardens Wivex, Palace Hotel, Copenhagen, Apprenticeship

John Aukstolis, A.S., Instructor; A.S., Johnson & Wales University

Adrian Barber, A.O.S., Associate Instructor; A.O.S., Culinary Institute of America

Claudia Berube, A.S., Instructor; A.S., Johnson & Wales University

Steven Browchuk, M.A., Certified T.I.P.S. Trainer, Associate Professor; B.A., Roger Williams College; M.A., University of Sorbonne; M.A., Middlebury College

Victor Calise, Associate Instructor

Carl Calvert, B.S., Instructor; A.O.S., B.S., Johnson & Wales University

Gerianne Chapman, M.B.A., Associate Professor; A.O.S., B.S., Johnson & Wales University; B.A., George Washington University; M.B.A., University of Rhode Island

John S. Chiaro, M.S., C.E.C., C.C.E., Associate Professor; B.A., Rhode Island College; M.S., Johnson & Wales University

Cynthia Coston, A.S., Instructor; A.S., Schoolcraft College

Laurie Coughlin, Administrative Assistant

Martha Crawford, B.S., C.W.P.C., Instructor; A.O.S., Culinary Institute of America; B.S., University of Michigan

Elaine R. Cwynar, B.A., Associate Instructor; A.S., Johnson & Wales University; B.A., University of Connecticut

William J. Day, M.S., C.F.E., Associate Professor and Director of Continuing Education; B.S., Bryant College; M.S., Johnson & Wales University

Mary Ann DeAngelis, M.S., Assistant Professor; B.S., M.S., University of Rhode Island

Richard DeMaria, B.S., Instructor; B.S., University of Rhode Island

Jean-Luc Derron, Associate Instructor; Hotel Schwanen Switzerland; steinli Trade School, Switzerland, Apprenticeship; Certification, Department of Labor and Trade, Switzerland; Confiserie Bachmann, Switzerland, Apprenticeship

Lynn Dieterich, Coordinator Faculty Support Services

Jean Jacques Dietrich, M.A., Senior Instructor; A.S., New York City Technological College; B.A., Hunter College; M.A., Johnson & Wales University

John R. Dion, B.S,., C.E.C., C.C.E., Associate Instructor; A.O.S., Culinary Institute of America; B.S., Johnson & Wales University

Rene R. Dionne, Director of Corporate Relations/Purchasing

Reginald B. Dow, A.O.S., Storeroom Manager; A.O.S., Culinary Institute of America

Kevin Duffy, B.S., Instructor; B.S., Johnson & Wales University

Thomas Dunn, B.S., Instructor; A.O.S., B.S., Johnson & Wales University

Roger Dwyer, B.A., Instructor; B.A., George Washington University

Neil Fernandes, B.S., Storeroom Office Manager; B.S. Johnson & Wales University

Paula Figoni, M.B.A., Instructor; B.S., University of Massachusetts; M.S., University of California; M.B.A., Simmons College Graduate School of Management

Ernest Fleury, M.S., Associate Professor, A.O.S., Johnson & Wales University; A.S., Community College of Rhode Island; B.S., M.S., Johnson & Wales University

Meridith Ford, B.S., Instructor; A.O.S., B.S., Johnson & Wales University

James Fuchs, A.O.S., Instructor; A.O.S., Johnson & Wales University

Nancy Garnett-Thomas, M.S., R.D., L.D.N., Associate

Professor; A.O.S., Culinary Institute of America; B.A., Colby College; M.S., University of Rhode Island

William Gormley, B.S., Instructor; A.O.S., B.S., Johnson & Wales University

James Griffin, M.S., C.W.C., C.C.E., Associate Dean & Associate Professor; A.O.S., B.S., M.S., Johnson & Wales University

Frederick Haddad, A.O.S., C.E.C., C.C.E., Associate Instructor; A.O.S., Culinary Institute of America

Rainer Hienerwadel, B.S., Instructor; A.O.S., B.S., Johnson & Wales University

J. Jeffrey Howard, B.A., Instructor; B.A., University of Massachusetts

Lars E. Johansson, C.P.C., C.C.E., Director, International Baking & Pastry Institute

Steven Kalble, A.S., Instructor; A.S., Johnson & Wales University

Linda Kender, B.S., Associate Instructor; A.S., B.S., Johnson & Wales University

Edward Korry, M.A., Assistant Professor; B.A., University of Chicago; M.A., University of Cairo

C. Arthur Lander, B.S.; Instructor; B.S., Johnson & Wales University

Kelly Lawton, Administrative Assistant

Hector Lipa, B.S., C.E.C., C.C.E., Associate Instructor; B.S., University of St. Augustine, the Philippines

Laird Livingston, A.O.S., C.E.C., C.C.E., Associate Instructor; A.O.S., Culinary Institute of America

Michael D. Marra, M.Ed., Associate Professor; B.A., M.Ed., Providence College

Susan Desmond-Marshall, M.S., Associate Professor; B.S., University of Maine; M.S., Johnson & Wales University

Victoria A. McCrae, Assistant to the Vice President

Diane McGarvey, B.S., Instructor; A.O.S., B.S., Johnson & Wales University

Jack McKenna, B.S., C.E.C., C.C.E., C.C.P., Director of Special Projects

Paul J. McVety, M.Ed., Assistant Dean and Associate Professor; A.S., B.S., Johnson & Wales University; M.Ed., Providence College

Michael Moskwa, M.Ed., Assistant Professor; B.A., University of Rhode Island; M.Ed., Northeastern University

Sean O'Hara, M.S., Certified T.I.P.S. Trainer, Instructor; A.O.S., B.S., M.S., Johnson & Wales University

George O'Palenick, M.S., C.E.C., C.C.E., Associate Professor; A.O.S., Culinary Institute of America; A.S., Jamestown Community College; B.S., M.S., Johnson & Wales University

Robert Pekar, B.S., Associate Instructor; A.O.S., Culinary Institute of America; A.S., Manchester Community College; B.S., Johnson & Wales University

Pamela Peters, A.O.S.,C.E.C., C.C.E., Director of Culinary Education; A.O.S., Culinary Institute of America

David Petrone, B.S., Associate Instructor; A.O.S., B.S., Johnson & Wales University

Felicia Pritchett, M.S., Associate Professor; A.O.S., B.S., M.S., Johnson & Wales University

Thomas J. Provost, Instructor

Ronda Robotham, B.S., Instructor; B.S., Johnson & Wales University

Robert Ross, B.S., Associate Instructor; A.S., B.S., Johnson & Wales University

Janet Rouslin, B.S., Instructor; B.S., University of Maine

Cynthia Salvato, A.S., C.E.P.C., Instructor; A.S., Johnson & Wales University

Stephen Scaife, B.S., C.E.C., C.C.E., Associate Instructor; A.O.S., Culinary Institute of America, B.S., Johnson & Wales University

Gerhard Schmid, Associate Instructor; European Apprenticeship, Germany

Louis Serra, B.S., C.E.C., Instructor; A.O.S., B.S, Johnson & Wales University

Christine Stamm, M.S., C.W.C., Associate Professor; A.O.S., B.S., M.S., Johnson & Wales University

Laura Schwenk, Administrative Assistant

Adela Tancayo-Sannella, Certified T.I.P.S. Trainer, Associate Instructor

Mary Ellen Tanzi, B.A., Instructor; B.A., Rhode Island College

Frank Terranova, B.S., C.E.C., C.C.E., Associate Instructor; B.S., Johnson & Wales University

Segundo Torres, B.S., Associate Instructor; B.S., Johnson & Wales University

Helene Houde-Trzcinski, M.S., Instructor; B.S., M.S., Johnson & Wales University

Peter Vaillancourt, B.S., Instructor; B.S., Roger Williams College

Paul VanLandingham, Ed.D, C.E.C., IMP, CFBE, C.C.E., Professor; A.O.S, Culinary Institute of America; B.S., Roger Williams College; M.A., Anna Maria College; Ed.D., Nova University

Suzanne Vieira, M.S., R.D., L.D.N., Department Chair, Foodservice Academic Studies; Associate Professor; B.S., Framingham State College; M.S., University of Rhode Island

Bradley Ware, M.Ed., C.C.C., C.C.E., Associate Professor; A.S., Johnson & Wales University; B.S., Michigan State University; M.Ed., Providence College; C.A.G.S., Salve Regina University

Gary Welling, A.O.S., Instructor; A.O.S., Johnson & Wales University

Robin Wheeler, Receptionist

Ed Wilroy, B.A., Continuing Education Coordinator; A.O.S., Johnson Wales University; B.A., Auburn University

Kenneth Wollenberg, B.S., Associate Instructor; A.O.S., B.S., Johnson & Wales University

Robert Zielinski, A.S., Instructor; A.S., Johnson & Wales University

Branch Campuses Administration and Faculty

CHARLESTON

Karl Guggenmos, B.S., C.E.C., G.C.M.C., Director of Culinary Education

Diane Aghapour, B.S., Instructor

Patricia Agnew, M.E., Assistant Professor

Donna Blanchard, B.A., Instructor

Robert Bradham, Instructor

Matthew Broussard, C.W.C., Instructor

Jan Holly Callaway, Instructor

Wanda Crooper, B.S., C.C.E., C.W.P.C., Associate Instructor

James Dom, M.S., Associate Professor

Armin Gronert, G.C.M.P.C., Associate Instructor

Kathy Hawkins, Instructor

David Hendrieksen, B.S., C.C.E., C.C.C., Associate Instructor

Andrew Hoxie, M.A., Assistant Professor

John Kacaia, C.E.C., Instructor

Michael Koons, A.O.S., C.E.C., C.C.E., Associate Instructor

Audrey McKnight, A.O.S., Instructor

Mary McLellan, M.S., Adjunct

Marcel Massenet, C.E.P.C., Associate Instructor

Stephen Nogle, A.A.S., C.E.C., C.C.E., Associate Instructor

Daniel Polasek, Instructor

Frances Ponsford, B.S., Instructor

Lloyd Regier, Ph.D., Adjunct

Victor Smurro, B.S., C.C.C., Associate Instructor

Susan Wigley M.Ed., C.C.E., C.W.C., Associate Professor

NORFOLK

Robert Nograd, Acting Director

Fran Adams, M.S., Instructor; B.S., M.S., Old Dominion University

Guy Allstock, III, M.S., Storeroom Instructor; B.S., M.S., Johnson & Wales University

Christian Barney, B.A., Associate Instructor; B.A., Old Dominion University

Ed Batten, A.O.S., Instructor; A.O.S., Johnson & Wales University

Susan Batten, C.E.C., C.C.E., Culinary Technology Degree, Associate Instructor; Culinary Technology Degree, Asheville Buncombe Technical Institute

Bettina Blank, M.S., Instructor; B.S., Grand Valley State University; M.S., Boston University

Dedra Butts, B.S., Instructor; B.S., Johnson & Wales University

Tim Cameron, M.A., C.E.C., Associate Professor; B.A., Milligan College; M.A., Old Dominion

Donna Curtis, B.A., Instructor; B.A., Northern Michigan University; Reading Specialist Degree, Memphis State University.

Art M. Elvins, A.A.S., C.E.C., Associate Instructor; A.A.S., Johnson & Wales University

Kristen Fletcher, R.D., M.S., Instructor; B.S., M.S., Virginia Polytechnic Institute

Scarlett Holmes-Paul, M.A., Instructor; B.S. Western Michigan University; M.A. Eastern Michigan University

Joan Hysell, M.Ed., Instructor; B.P.S., SUNY Institute of Technology at Utica/Rome; M.Ed., Ohio University

John Keating, M.S., Oenology Instructor; B.S., Georgetown University; M.S., George Washington University

Lisa Kendall, M.A., Instructor; B.A., State University of New York; M.A., Old Dominion University

Greg Kopanski, M.S., Instructor; B.S., New York University; M.S., Old Dominion University.

Jerry Lanuzza, B.S., Instructor; B.S., Johnson & Wales University

Peter Lehmuller, B.A., Instructor; A.O.S., Culinary Institute of America; B.A., State University of New York, Albany

Alex Leuzzi, M.S., Associate Instructor; B.S., Pikesville College; M.S., Fairleigh Dickinson University

Melanie Loney, M.S., Associate Instructor; B.S., M.S., Old Dominion University

Mary Matthews, M.S., Instructor; B.S., M.S., Old Dominion University

Carrie Moranha, A.A.S., Dining Room /Beverage Instructor; A.A.S. Johnson & Wales University

Maureen Nixon, M.A., Instructor; B.A., North Carolina State University; M.A., Norfolk State University.

Shelly Owens, B.A., Baking & Pastry Instructor; B.A., Townson State University

Patrick Reed, A.O.S, C.C.C., C.C.E., Associate Instructor; A.O.S., Culinary Institute of America

Gregory Retz, B.S., Instructor; A.A.S., Johnson & Wales University; B.S., Virginia Polytechnic

Steven Sadowski, C.E.C., A.O.S., Associate Instructor; A.O.S., Johnson & Wales University

Bonita Startt, M.S ., Instructor; B.S., M. S., Old Dominion University

Fred Tiess, A.A.S., Instructor; A.A.S., State University of New York, Poughkeepsie; A.O.S., Culinary Institute of America

NORTH MIAMI

Donato Becce, Instructor; Diploma di Qualifica, Instituto Professionale, Alberghiero di Stato, Italy

Kenneth Beyer, B.B.A., Instructor; A.B.A., Nichols College; B.B.A., University of Miami

Drue Brandenburg, B.S., C.C.E., C.E.C., Instructor; A.O.S., Culinary Institute of America; B.S., Oklahoma State University

Dennis Daugherty, M.Ed., Instructor; B.S., University of Maryland; M.Ed., Pennsylvania State University

Melvin Davis, B.A., Instructor; B.A. University of Maryland

Alberto Diaz, English Master Pastry Chef, Instructor

Claus Esrstling, C.E.C., Instructor

John Goldfarb, B.S., Instructor; A.O.S., Culinary Institute of America; B.S., Florida International University

John Harrison, B.S., Instructor; A.O.S., Culinary Institute of America; B.S., University of New Haven

James Hensley, Instructor

Giles Hezard, Instructor; Certification of Professional Aptitide - College D'Enseignement Technique Masculin, Audincourt, France

Alan Lazar, B.A., Instructor; B.A. Monmouth College

Lucille Ligas, M.Ed., Assistant Professor, Indiana University of Pennsylvania; B.S. Ed. Indiana University of Pennsylvania

Charles Miltenberger, C.E.P.C., Instructor

Betty Murphy, M.S.Ed. Instructor; B.S.Ed. Eastern Illinois University; M.S.Ed., University of Guam

Larry Rice, M.S., Instructor; A.S., Johnson & Wales University; B.S., Florida International University; M.S., Florida International University

Mark Testa, Ph.D., Associate Professor; A.A.S., State University of New York at Farmingdale; B.P.S., New York Institute of Technology; M.A.L.S. State University of New York at Stony Brook; Ph.D., Barry University

Todd Tonova, M. S., Instructor; A.O.S., Culinary Institute of America; B.S., Florida International University; M.S., Florida International University

Karen Woolley, B.S., Instructor; A.O.S., Culinary Institute of America; B.S., Florida State University

VAIL

Todd M. Rymer, M.S., Director; B.A., New College; M.S., Florida International University

Paul Ferzacca, A.O.S., Instructor; A.O.S. Kendall College

David Hite, A.S., Instructor; A.S. Johnson & Wales University

Robert Kuster, Instructor; Diploma, Swiss Hotel School, Lucerne; Diploma, Trade School, Cook's Apprenticeship, Lucerne; Diploma, Institute Stavia, Estavater Le-Lac

Katie Mazzia; B.S. R.D., Instructor; R.D., Saint Joseph's Health Center; B.S., Ohio State

Paul Reeves, B.S., Instructor; B.S., Saint Cloud State University

David B. Sanchez, A.O.S., Instructor; A.O.S., Johnson & Wales University

Culinary Advisory Council

Scott Armendinger, Editor, Journal Publications, Rockland, ME

Michael P. Berry, Vice President of Food Operations and Concept Development, Walt Disney World, Orlando, FL

Edward Bracebridge, Chef Instructor, Blackstone Valley Tech, Upton, MA

Gerry Fernandez, Technical Service Specialist, General Mills, Inc., Minneapolis, MN

John D. Folse, C.E.C., A.A.C., Owner, Executive Chef, Chef John Folse & Company, Donaldsonville, LA

Ira L. Kaplan, President, Servolift/Eastern Corp., Boston, MA

Gustav Mauler, VP, Food & Beverage, Treasure Island Hotel, Las Vegas, NV

Franz Meier, President, MW Associates, Columbia, SC

Roland Mesnier, Executive Pastry Chef, The White House, Washington, DC

Stanley Nicas, Chef/Owner, Castle Restaurant, Leicester, MA

Robert J. Nyman, President, The Nyman Group, Livingston, NJ

Johnny Rivers, Food & Beverage Manager/Executive Chef, Thyme & Associates,

Joseph Schmidt, Owner, Joseph Schmidt Confections, San Francisco, CA

Martin Yan, President, Yan Can Cook, Inc., Foster City, CA

Johnson & Wales University *Distinguished Visiting Chefs 1979–1997*

1 Dr. Jean Joaquin
2 Garry Reich
3 Dr. Hans J. Bueschkens
4 Michael Bourdin
5 Christian Inden
6 Casey Sinkeldam
7 John Kempf
8 Bernard S. Urban
9 Marcel Paniel
10 Lutz Olkiewicz
11 Dr. Joel Robuchon
12 Ray Marshall
13 Francis Hinault
14 Wally Uhl
15 Gunther Heiland
16 Dr. Pierre Franey
17 Jean-Jacques Dietrich
18 Uri Guttmann
19 William Spry
20 Dr. Stanley Nicas
21 Dr. Paul Elbling
22 Angelo Paracucchi
23 Albert Kellner
24 Hans K. Roth
25 Gerhard Daniel
26 Jacques Noe
27 Andre Rene
28 Dr. Anton Mosimann
29 Dr. Roger Verge
30 Gerhard Schmid
31 Karl Ronaszeki

32 Jacques Pepin
33 Klauss Friedenreich
34 Arno Schmidt
35 Lucien Vannier
36 Dr. Wolfgang Bierer
37 Dr. John L. Bandera
38 Albert Marty
39 Dr. Siegfried Schaber
40 Dr. Michael Minor
41 Raimund Hofmeister
42 Henry Haller
43 Dr. Noel Cullen
44 Dr. Carolyn Buster
45 Dr. Madeleine Kamman
46 Udo Nechutnys
47 Andrea Hellrigl
48 George Karousos
49 Warren LeRuth
50 Rene Mettler
51 Dr. Johnny Rivers
52 Milos Cihelka
53 Dr. Louis Szathmary
54 Philippe Laurier
55 Dr. Hans J. Schadler
56 Franz Klampfer
57 Jean-Pierre Dubray
58 Neil Connolly
59 Joachim Caula
60 Dr. Emeril LaGasse†
61 Dr. Roland Mesnier
62 Bernard Dance
63 Hartmut Handke
64 James Hughes†
65 Paul Bocuse
66 Dr. Martin Yan
67 Marcel Desaulniers
68 Heinz H. Veith
69 Benno Eigenmann
70 Johanne Killeen & George Germon
71 Dr. John D. Folse
72 Dr. Christian Rassinoux
73 Dr. Gustav E. Mauler
74 Dr. Keith Keogh
75 Clayton Folkners
76 Kenneth Wade
77 Dr. Roland E. Schaeffer
78 Dr. William Gallagher
79 Van P. Atkins
80 Hiroshi Noguchi
81 Jasper White
82 Albert Kumin
83 Alfonso Contrisciani†
84 Dr. Victor Gielisse
85 Reimund D. Pitz
86 Daniel Bruce†
87 Antoine Schaefers
88 Michael Ty
89 Phil Learned
90 Joseph Schmidt
91 John Halligan
92 Willy O. Rossel
93 John J. Vyhnanek
94 Roberto Gerometta
95 Robert A. Trainor†
96 Ewald & Susan Notter
97 Joseph Amendola
98 David Paul Johnson
99 Thomas Pedersen
100 André Soltner
101 Christian Clayton†
102 Konstantinos Exarchos
103 Christian Chemin
104 Lars Johansson
105 Paul O'Connell†

SPECIAL FRIENDS

John J. Bowen
Joseph P. Delaney
Socrates Inonog
Franz K. Lemoine

† Alumni
Deceased

Partial List of Companies Associated with Johnson & Wales University

Adam's Mark Hotels and Resorts
Allied Domecq Retailing
American General Hospitality, Inc.
AmeriClean Systems, Inc.
Angelica Uniform Group
Antigua Hotel Association
Aramark Services, Inc.
Automatic Sales, Inc.
AVTECH Industries
Bacardi & Company, Ltd.
Bacon Construction Company
Balfour Foundation
Banfi Vintners
Basic American Frozen Foods
Bertoill, USA, Inc.
Boston Chicken, Inc.
Boston Park Plaza Hotel
Braman Motors
Brinker International
Bristol Hotel Company
Bugaboo Creek Steakhouse
Bushiri Hotel Aruba
Campbell Food Service Company
Carlson Companies, Inc.
Carnival Cruise Lines
Cartier, Inc.
Celebrity Cruise Lines
Choice Hotels
Citizens Financial Group
Cleveland Range, Inc.
Club Corporation International
Comstock-Castle Stove Company
Concord Hospitality
Cookshack
Cookson America, Inc.
Coors Brewing Company
Crabtree McGrath Associates
Daka Restaurants, Inc.
Darden Restaurants
Deer Valley Resort

Denny's Restaurants
Dial Corporation
Digital Equipment Corporation
DiLeonardo International
Doral Arrowwood
E.A. Tosi & Sons Company, Inc.
E-H Enterprises
Ecolab, Inc.
Edison Electric Institute
Edwards Super Food Store
EGR International
Electric Cooking Council
Eurest Dining Service
F. Dick
Felchlin, Inc.
Feinstein Foundation
Flik International Corporation
Forbes
Friendly Ice Cream Corporation
Frymaster
G.S. Distributors
Garland Commercial Industries
Gavin Sales Company
General Mills
Godfather's Pizza, Inc.
The Golden Corral Corporation
Grand Western Brands, Inc.
Grisanti, Inc.
Groen, a Dover Industries Co.
Hallsmith-Sysco Food Services
Harman Management Corporation
Harris-Teeter, Inc.
Harvard University
Hasbro, Inc.
Hatch-Jennings, Inc.
HERO
Hiram Walker & Sons, Inc.
Hilton Hotels
Hobart Corporation
Houlihan's Restaurant Group
Houston's Restaurants
Hyatt Hotels Corporation
Ice-O-Matic
Ikon
Intercontinental Hotels
International Metro Industries
Interstate Hotels
Keating of Chicago, Inc.
Kiawah Island Resorts
Kraft Foods, Inc.
Lackman Food Service
Le Meridien Hotel Boston
L.J. Minor Corporation
Legal Sea Foods, Inc.
Loews Hotels
Longhorn Steaks, Inc.
Lyford Cay Foundation, Inc.
Manor Care Health Services
Market Forge Company
Marriott International, Inc.
Max Felchlin, Inc.
Massachusetts Electric Company
McCormick & Company, Inc.
Moet & Chandon
Morris Nathanson Design
Motel 6
MTS Seating
Nabisco Brands, Inc.
Narragansett Electric Company
National Votech Educators
National Banner Company, Inc.
National Prepared Foods Assoc.
National Student Organization
Nestle USA, Inc.
New England Electric System
New World Development Company
Norwegian Seafood Council
Opryland Hotel
Paramount Restaurant Supply
PepsiCo, Inc.
Pillsbury Corp.
The Proctor & Gamble Co.
Providence Beverage
Prudential Insurance Company
Quadlux
The Quaker Oats Company
Radisson Hospitality Worldwide
Ralph Calise Fruit & Produce
Red Lion Hotels
Renaissance Hotels & Resorts
Restaurant Data Concepts
Rhode Island Distributing Company
Rhode Island Foundation
Rich Products Corporation
The Ritz-Carlton Hotel
Robert Mondavi Winery
Robot Coupe
Ruth's Chris Steak House
Saunders Hotel Group
Joseph Schmidt Confections
Schott Corporation
Select Restaurants, Inc.
Servolift/Eastern Corp.
Sharp Electronic Corporation
Somat Corporation
Southern Foods
State of Rhode Island, Department of Education
Stonehard
Sun International
Sunrise Assisted Living
Swiss American Imports, Ltd.
Swiss Chalet Fine Foods
Sysco Corporation
TACO, Inc.
Taco Bell
Tasca Ford Sales, Inc.
Tekmatex, Inc.
The Delfield Company
The Waldorf-Astoria
Thermodyne Foodservice Products
Toastmaster
Tufts University
Tyson Foods, Inc.
U.S.D.A./Bell Associates
United States Army
United States Navy
University of Connecticut
Vail Associates
Vulcan Hart Corporation
Walt Disney World
Wells Manufacturing Company
Wyatt Corporation
Wyndham Hotels & Resorts

The Techniques

Assembling:

1. Collect and prepare all of the ingredients to the formula.
2. Clear the area for assembly.
3. Fit the pieces together according to formula instructions or instructor's guidelines.

Baking:

1. Preheat the oven.
2. Position the item appropriately in the oven.
3. Check for appropriate firmness and/or color.

Blending:

1. Combine the dry ingredients on low speed.
2. Add the softened fat(s) and liquid(s).
3. Mix the ingredients on low speed.
4. Increase the speed gradually.

Blooming:

Gelatin sheets or leaves:

1. Fan the sheets out.
2. Cover the sheets in liquid.
3. Sheets are bloomed when softened.

Granular gelatin:

1. Sprinkle the gelatin.
2. Gelatin is ready when it is cream of wheat consistency.

Boil: (at sea level)

1. Bring the cooking liquid to a rapid boil.
2. Stir the contents, and cook the food product throughout.
3. Serve hot.

Braise:

1. Heat the braising pan to the proper temperature.
2. Sear and brown the food product to a golden color.
3. Degrease and deglaze.
4. Cook the food product in two-thirds liquid until fork-tender.

Brushing:

1. Use a pastry brush.
2. Lightly apply the glaze.

Caramelizing:

Wet method:

1. Use an extremely clean pot.
2. Place the sugar and water on high heat.
3. Never stir the mixture once the sugar begins to dissolve.
4. Once caramelized, shock in ice water.

Chopping:

1. Use a sharp knife.
2. Hold the food product properly.
3. Cut with a quick downward motion.

Coating:

1. Use a coating screen, with a sheet pan underneath.
2. Ensure that the product is the correct temperature.
3. Coat the product using an appropriately-sized utensil.

Combing:

1. Prepare the item with the appropriate amount of icing.
2. Drag a clean comb across the surface.

Combining:

1. Prepare the components to be combined.
2. Add one to the other, using the appropriate mixing method (if needed).

Cooking:

1. Choose the appropriate heat application: baking, boiling, simmering, etc.
2. Prepare the formula according to instructions.
3. Cook according to instructions.

Covering:

1. Prepare the product to be covered with a sticky layer of buttercream, food gel, or other medium that is called for in the formula.
2. Roll out the covering material.
3. Cover the product with the covering material.
4. Smooth the covered product with a bowl scraper to remove wrinkles and air bubbles.

Creaming:

1. Soften the fats on low speed.
2. Add the sugar(s) and cream; increase the speed slowly.
3. Add the eggs one at a time; scrape the bowl frequently.
4. Add the dry ingredients in stages.

Crystallizing:

1. Boil the product for 30 seconds to 2 minutes.
2. Drain the water; cook until tender and translucent.
3. Strain the sugar syrup.
4. Dredge the product; dry on a screen.

Cutting:

1. Use a sharp knife to cut to the directed size.

Decorating:

1. Follow the formula's instructions or the instructor's guidelines to appropriately decorate each cake or pastry.

Deep-Fry:

1. Heat the frying liquid to the proper temperature.
2. Submerge the food product completely.
3. Fry the product until it is cooked throughout.

Dipping:

1. Prepare the product to the proper dipping temperature.
2. Carefully submerge the product.
3. Dry on parchment paper or a screen.

Dredging:

1. Coat the food product.
2. Sprinkle or toss the product in an appropriate dredging application.

Filling:

1. Cut open the food product.
2. Carefully spread the filling using an icing spatula.
3. Carefully pipe the filling using a pastry bag.

Flambéing:

1. Heat the food product and liqueur in a shallow pan.
2. The pan needs to be very hot.
3. Carefully tilt the pan into the open flame; ignite the liqueur.

Folding:

Do steps 1, 2, and 3 in one continuous motion.

1. Run a bowl scraper under the mixture, across the bottom of the bowl.
2. Turn the bowl counterclockwise.
3. Bring the bottom mixture to the top.

Freezing:

1. Prepare the product.
2. Place the product in the freezing cabinet for the appropriate length of time.

Frying:

1. Heat the frying liquid to the appropriate temperature.
2. Place the food product into the hot liquid.
3. Cook the product, turning frequently, until golden brown and tender.

Grill/Broil:

1. Clean and heat the grill/broiler.
2. To prevent sticking, brush the food product with oil.

Heating:

1. Prepare the food product according to the formula's instructions.
2. Choose the appropriate method of heating (on the range or stove top, in the oven, etc.)
3. Apply the product to the heat.

Icing:

1. Use a clean icing spatula.
2. Work quickly and neatly.

Kneading:

1. Prepare the kneeding surface with the appropriate medium (flour, cornstarch, etc.).
2. Press and form the dough into a mass using soft, determined strokes.
3. Continue kneading until appropriate consistency and/or temperature is achieved.

Laminating:

1. Allow a proper time to rest dough.
2. Roll the dough out to a ½-inch to ¾-inch thickness.
3. Evenly spread the fat.
4. Allow a proper time for the dough to rest.
5. Refrigerate for several hours.

Melting:

1. Prepare the food product to be melted.
2. Place the food product in an appropriately sized pot over direct heat or over a double boiler.
3. Stir frequently or occasionally, depending on the delicacy of the product, until melted.

OR

1. Place the product on a sheet pan or in a bowl, and place in a low oven until melted.

Mixing:

1. Follow the proper mixing procedure: creaming, blending, whipping, or combination.

Molding:

1. Wash your hands.
2. Clear the work area.
3. Carefully shape the substance.

Napper:

1. Stir the sauce frequently.
2. Frequently check the consistency of the sauce.

En Papillote:

1. Properly prepare the parchment paper.
2. Cook the food product until the packet is puffed and brown.

Peeling:

1. Use a clean paring knife or peeler.
2. Do not peel over an unsanitary surface.

Piping:

With bag:

1. Use a bag with a disposable tip; cut the bag at 45-degree angle.
2. Fill to no more than half full.
3. Burp the bag.

With cone:

1. Cut and fold the piping cone to the appropriate size.
2. Fill the cone with a small amount.
3. Fold the ends to form a triangle.
4. Pipe the desired designs.

Poaching:

1. Bring the liquid to a boil; then reduce to a simmer.
2. Submerge and anchor the product.
3. Do not overcook the product.

Portioning:

1. Mark the product for portioning, using a ruler, if necessary.
3. Cut, spoon, or scoop the product with the appropriate-sized utensil.

Pouring:

1. Place the product in an appropriate container for pouring: a pitcher or large ladle.
2. Pour the product into desired containers or over another product.

Proofing:

1. Set a proof box to the proper temperature.
2. Do not over- or underproof the dough.

Pulling:

1. Pay attention; do not get distracted.
2. Sugar must be at an even temperature.
3. Work with clean equipment and hands.

Puréeing:

1. Do not overfill the food processor.
2. First pulse the food processor.
3. Turn food processor to maximum speed to puree food.

Reducing:

1. Bring the sauce to a boil; then reduce to a simmer.
2. Stir often; reduce to the desired consistency.

Ribboning:

1. Use a high speed on the mixer.
2. Do not overwhip the egg yolks.

Roast:

1. Sear the food product, and brown evenly.
2. Elevate the food product in a roasting pan.
3. Determine doneness, and consider carryover cooking.
4. Let the food product rest before carving.

Rolling:

1. Prepare the rolling surface by dusting with the appropriate medium (flour, cornstarch, etc.).
2. Use the appropriate style pin (stick pin or ball bearing pin) to roll the dough to desired thickness; rotate the dough during rolling to prevent sticking.

Rubbing:

1. Use a pastry cutter to keep the fat in large pieces.
2. Add the liquid in stages.

Sauté:

1. Heat the sauté pan to the appropriate temperature.
2. Evenly brown the food product.
3. For a sauce, pour off any excess oil, reheat, and deglaze.

Scalding:

1. Heat the liquid on high heat.
2. Do not boil the liquid.

Shallow-Fry:

1. Heat the cooking medium to the proper temperature.
2. Cook the food product throughout.
3. Season, and serve hot.

Shaping:

1. Prepare the medium to be shaped.
2. Prepare the surface area for shaping.
3. Mold medium into desired shapes according to the instructor's directions.

Simmer and Poach:

1. Heat the cooking liquid to the proper temperature.
2. Submerge the food product completely.
3. Keep the cooked product moist and warm.

Simmering:

1. Place the prepared product in an appropriate-sized pot.

3. Bring the product to a boil, then reduce the heat to allow the product to barely boil.
4. Cook until desired doneness is achieved.

Slicing:

1. Prepare the product for cutting; clean and clear the work area.
2. Slice the product using the "claw" grasp or the rocking motion.

Slow Baking:

1. Use an appropriate baking dish.
2. Use hot water in the pan.
3. Replenish the water when needed.

Soaking:

1. Place the item(s) to be soaked in a large bowl or appropriate container.
2. Pour water or other liquid over the items to be soaked.
3. Allow to sit until desired saturation or softening is achieved.

Spiking:

1. Use chocolate that is appropriately tempered.
2. Drag the truffles across a screen.
3. Use three to four clean swift strokes.

Spreading:

1. Using an icing spatula or off-set spatula, smooth the icing or other spreading medium over the surface area.

Steam: (Traditional)

1. Place a rack over a pot of water.
2. Prevent steam vapors from escaping.
3. Shock or cook the food product throughout.

Stew:

1. Sear, sauté, sweat, or blanch the main food product.
2. Deglaze the pan, if desired.
3. Cover the food product with simmering liquid.
4. Remove the bouquet garni.

Stir-Fry:

1. Heat the oil in a wok until hot but not smoking.
2. Keep the food in constant motion; use the entire cooking surface.

Stretching:

1. Cover the surface area with a clean pastry cloth.
2. First roll the dough with pin.
3. Use the back of your hands to stretch the dough.

Stuffing:

1. Place the stuffing inside the cavity of the food product using a gloved hand or a piping bag.
2. Be sure to fill the cavity completely.

Tempering:

1. Whisk the eggs vigorously while ladling hot liquid.

Thickening:

1. Mix a small amount of sugar with the starches.
2. Create a slurry.
3. Whisk vigorously until thickened and translucent.

Washing:

1. Brush or spray the product to the desired saturation.
2. Do not under- or oversaturate.

Whipping:

1. Hold the whip at a 55-degree angle.
2. Create circles, using a circular motion.
3. The circular motion needs to be perpendicular to the bowl.

1

Stocks and Sauces

Stocks and Sauces

The world is nothing without life, and all that lives takes nourishment.

Thus wrote the earliest gourmet and connoisseur, Jean-Anthelme Brillat-Savarin. Here at Johnson & Wales University, young men and women receive the foundation of culinary excellence. Because each nation has its own particular style of cuisine, the culinary arts and their culinarians know no boundaries; and as the size of the world of today has been reduced through international travel, culinarians have truly become ambassadors of goodwill. It is they, through their gastronomic creations, who contribute so much toward an atmosphere of warm hospitality when one is far from home. They are the custodians of the culinary future. They are the true culinary ambassadors of their school, Johnson & Wales University, and their country, the United States of America.

Hans J. Bueschkens, CEC
Distinguished Visiting Chef
January 20–22, 1980
Former President, World Association of Cooks Societies

The HACCP Process—Stocks

The following recipe illustrates how stocks flow through the HACCP process.

Brown Stock

Receiving

- Bones, frozen—temperature at or below 0° F (−17.8° C) with no signs of thawing; packaging intact
- Bones, fresh—free of slime, and odor; packaging intact
- Tomato paste—check cans for swelling, leakage, flawed seals, rust, and dents
- Vegetables (carrots, celery, onions, and leeks)—packaging intact; no cross contamination from other foods on the truck; no signs of insect or rodent activity
- Herbs and spices (black peppercorns, bay leaves, parsley stems, and thyme leaves)—packaging intact

Storage

- Bones, frozen—Store in freezer at or below 0° F (−17.8° C).
- Bones, fresh—Store under refrigeration, with a product temperature not to exceed 40° F (4.4° C). Store below already cooked foods.
- Vegetables (carrots, celery, onions, and leeks)—Store under refrigeration, with a product temperature not to exceed 40° F (4.4° C). Store above and away from raw, potentially hazardous foods.
- Tomato paste—Store in dry storage at 50° F (10° C) with a relative humidity of 50% to 60%.
- Herbs and spices (black peppercorns, bay leaves, and thyme leaves)—Store in dry storage at 50° F (10° C) with a relative humidity of 50% to 60%. Keep dry.

Thawing Bones

- Thaw under refrigerated storage. Air temperature of refrigerated unit should be 38° F (3.3° C) or lower.

Preparation and Cooking

- Preheat the oven to 425° F.
- Peel the outer layer of the carrots, and wash.
- Trim the celery, and wash.
- Peel the outer layer of the onions, and wash.
- Trim the leeks, and wash.
- Chop vegetables on a clean and sanitized cutting board with clean and sanitized utensils.

- Split the bones (if needed) on a clean and sanitized cutting board with clean and sanitized utensils.
- Place the bones in a roasting pan, and roast in the oven until well browned.
- Remove the pan from the oven, spread the tomato paste lightly over the bones. Return the pan to the oven for an additional few minutes, being careful not to burn the tomato product.
- Remove the bones from the oven, drain the excess grease, and reserve. Place the bones in a suitable stockpot or steam kettle with a spigot.
- Add the cold liquid to a height of 8 inches above the bones.
- Apply heat, and bring to a simmer.
- Using the same pan as for roasting the bones, add a small quantity of fat, and place the pan on a hot flat-top.
- When the fat is hot, add the mirepoix, and sauté until browned.
- Skim the stock of impurities and fat.
- After skimming, add the browned mirepoix and bouquet garni to the stock.
- Deglaze the roasting pan with some of the hot stock, and return to the stock.
- Simmer until the proper richness is achieved, approximately 6 hours.

Cooling

- Put the sauce in shallow pans at a product depth of 2 inches or less.
- Place the pans with product into an ice bath, immersed to level of product within pan. Stir frequently, and lower the temperature to 40° F (4.4° C) or less within 4 hours.
- Cover loosely with plastic wrap when cooled.
- Label with the date, time, and name of the product.
- Store on the upper shelf of a refrigerated unit at 40° F (4.4° C) or lower.

Reheating

- Reheat to 165° F (73.9° C) or higher in 2 hours.

The HACCP Process—Sauces

The following recipe illustrates how sauces flow through the HACCP process.

Béchamel Sauce

Receiving

- Milk—should have a sweetish taste. Container should be intact. Milk should be delivered at 40° F (4.4° C) or lower.
- Butter—should have a sweet flavor, firm texture, and uniform color. Butter should be free of specks and mold, and the packaging should be intact. Butter should be delivered at 40° F (4.4° C) or lower.
- Flour—packaging intact
- Vegetables (onions)—packaging intact; no cross-contamination from other foods on the truck; no signs of insect or rodent activity
- Spices (salt, pepper, and nutmeg)—packaging intact

Storage

- Milk—Store under refrigeration at 40° F (4.4° C) or lower. Store above and away from raw, potentially hazardous foods.
- Butter—Store under refrigeration at 40° F (4.4° C) or lower. Store above and away from raw, potentially hazardous foods.
- Flour—Store in dry storage at 50° F (10° C) with a relative humidity of 50% to 60%.
- Vegetables (onions)—Store under refrigeration, with a product temperature not to exceed 40° F (4.4° C). Store above and away from raw, potentially hazardous foods.

- Spices (salt, pepper, and nutmeg)—Store in dry storage at 50° F (10° C) with a relative humidity of 50% to 60%. Keep dry.

Preparation and Cooking

- Peel the outer layer of the onions, wash, and insert cloves into both ends of onions
- In a saucepan, heat the milk with the onion clouté, and simmer for 10 minutes.
- In another saucepan, heat the clarified butter over moderate heat.
- Gradually whisk in sifted flour to make a roux. Using a wooden spoon, mix the roux thoroughly, cooking it approximately 5 to 6 minutes. The roux will have a slightly nutty aroma when finished. Do not brown. Remove it from the heat, and cool slightly.
- Remove the onion clouté from the milk.
- Gradually whisk hot milk into the roux. Bring to a boil. Reduce to a simmer, and skim the impurities. Simmer for 20 minutes, or until the proper flavor and consistency are achieved.
- Season with salt, white pepper, and nutmeg.
- Strain through a fine China cap into a suitable container.

Holding and Service

- Serve immediately.
- Hold at 140° F (60° C) or higher for less than 2 hours and stir frequently.

Cooling

- Put the sauce in shallow pans at product depth of 2 inches or less.
- Place the pans with the product into an ice bath, immersed to the level of the product within pan. Stir frequently, and lower the temperature to 40° F (4.4° C) or less within 6 hours.
- Cover loosely with plastic wrap when cooled.
- Label with the date, time, and name of product.
- Store on the upper shelf of a refrigerated unit at 40° F (4.4° C) or lower.

Reheating

- Reheat to 165° F (73.9° C) or higher in 2 hours.
- Reheat only the quantity needed.

Nutritional Notes

Traditional sauces can be a significant source of sodium, fat, and cholesterol. Fortunately, it is relatively simple to modify most sauce recipes to lower fat, saturated fat, cholesterol, and sodium and at the same time maintain the characteristic flavor of the sauce. Some sauces do not adapt well to such modifications (e.g., béarnaise); in such cases, it is best to either use small amounts of the sauce or substitute with a lower-fat and lower-cholesterol alternative.

Many chefs throughout the United States are using a number of the following suggestions with great success and satisfaction:

- Enhance flavor with fresh herbs, spices, and wines, and reduce the amount of salt used. Good-quality ingredients minimize the need for salt and other high-sodium flavor enhancers, such as MSG.
- Remove all visible fat by chilling the stock to solidify the fat or simply spooning it off. Finish the fat removal process by floating absorbent paper on the surface of the stock.
- Reduce the amount of fat added in preparation of the sauce. For example, brown onions by roasting, or use tomato paste or caramelized sugar to darken a sauce, when appropriate. Also, to add flavor, use Canadian bacon in place of regular bacon.
- Rather than using a roux, which is high in fat, thicken sauces with arrowroot, cornstarch, modified food starch (e.g., instant flour), grated raw potato, or bread crumbs.

- Concentrated vegetable purée or puréed cooked legumes also thicken as well as add flavor, fiber, vitamins, and minerals. Roast or slow-bake vegetables before puréeing to develop a deep, rich flavor and experience better thickening. Puréed vegetables and legumes need to be stabilized with a starch if used as the primary thickening agent.
- Use yogurt in place of sour cream. Yogurt works best as a thickener if it is drained in a filter and stabilized with cornstarch before using. It is best to warm yogurt slightly before adding it to a hot sauce.
- Replace heavy cream with evaporated skim milk or dry milk solids. Once again, thicken with a starch if a thicker sauce is desired. When thickening with low-fat dairy products or puréed vegetables or legumes, add them near the end of the cooking process and do not allow the sauce to boil.
- Reduced stocks can be used as simple, low-fat, low-salt, and full-flavored sauces.
- Dried butter or sour cream crystals can be used to impart the flavor of butter or sour cream without adding fat. These products, however, do contain salt, a component that must be factored into the recipe.

Brown Stock

Beef, Veal, Lamb, Fowl, and Game

YIELD: 12 GALLONS. SERVING SIZE: AS NEEDED.

COOKING TECHNIQUES:
Roast, Simmer

Roast:
1. Sear the food product, and brown evenly.
2. Elevate the food product in a roasting pan.
3. Determine doneness, and consider carryover cooking.
4. Let the food product rest before carving.

Simmer and Poach:
1. Heat the cooking liquid to the proper temperature.
2. Submerge the food product completely.
3. Keep the cooked product moist and warm.

GLOSSARY:
Mirepoix: roughly chopped vegetables
Bouquet garni: bouquet of herbs and spices
Dépouiller: to skim impurities/grease
Chinois: cone-shaped strainer

HACCP:
Cool to 40° F or lower.

HAZARDOUS FOOD:
Bones

NUTRITION:
Calories: 17.9
Fat: .395 g
Protein: .54 g

CHEF NOTES:
1. The best bones to create a flavorful and gelatinous stock are a combination of beef and veal bones.
2. Cooking time of stocks will vary according to the type and size of the bones.

INGREDIENTS:

50 pounds	Bones
	Oil, if needed
1 No. 10 can (6 pounds, 5 ounces)	Tomato paste
15 gallons	Cold liquid

Mirepoix:

5 pounds	Leeks (white part only), washed, trimmed, and reserved separately
5 pounds	Carrots, washed and peeled
5 pounds	Celery, washed and trimmed
5 pounds	Onions, peeled

Bouquet Garni: (2 each as below)

2 tablespoons	Whole black peppercorns
4 each	Bay leaves
3 ounces	Parsley stems
2 tablespoons	Dried thyme leaves

METHOD OF PREPARATION:

1. Preheat the oven to 425° F.
2. Split bones (if needed), and rinse thoroughly in cold water, removing all blood, which causes discoloration.
3. Place the bones in a roasting pan, and brush with oil, if needed. Roast in the oven until well browned.
4. Remove the pan from the oven, and spread the tomato paste lightly over the bones. This will result in additional color and flavor in the stock. Return the pan to the oven for an additional few minutes, being careful not to burn the tomato product.
5. Remove the bones from the oven, drain, and reserve the grease. Place the bones in a suitable stockpot or steam-jacketed kettle with a spigot. Add the leeks.

6. Add the cold liquid to a height of 8 inches above the bones.
7. Apply heat, and heat to a simmer.
8. Using the same pan as for roasting the bones, add a small quantity of fat, and place on a hot flat-top.
9. When the fat is hot, add the remaining mirepoix, and sauté until browned.
10. **Dépóuiller** the stock; then add the browned mirepoix and both bouquet garni.
11. Deglaze the roasting pan with some of the hot stock, and add back into the stock.
12. Simmer until the proper richness is achieved, approximately 6 hours for beef or veal bones, and continue to dépouiller as needed.
13. Strain the stock through a **chinois** into another stockpot. Return to a boil, and reduce to the desired strength; then transfer to a suitable container, and place in a cooling sink or chill blaster. Cool to an internal temperature of 40° F or lower. Label, date, and refrigerate.
14. Reheat to a boil.

Court Boullion

YIELD: 1 QUART. SERVING SIZE: AS NEEDED.

INGREDIENTS:

1½ quarts	Water
2 ounces	Lemon juice
2 ounces	Dry white wine

Mirepoix:

4 ounces	Onions, peeled
4 ounces	Leeks (white part only), washed and trimmed
2 ounces	Celery, washed and trimmed

Bouquet Garni:

1 teaspoon	Whole black peppercorns
2 each	Bay leaves
3 ounces	Parsley stems
1 teaspoon	Dried thyme leaves

METHOD OF PREPARATION:

1. Place all of the ingredients into the poaching pan, heat to a boil, and **dépouiller** as needed; then reduce to a gentle simmer for 30 minutes. Strain through a **chinois.**
2. Use immediately, or hold at 140° F. Cool to an internal temperature of 40° F or lower (within 6 hours). Label, date, and refrigerate.

COOKING TECHNIQUE:

Simmer

Simmer and Poach:

1. Heat the cooking liquid to the proper temperature.
2. Submerge the food product completely.
3. Keep the cooked product moist and warm.

GLOSSARY:

Mirepoix: roughly chopped vegetables
Bouquet garni: bouquet of herbs and spices
Dépouiller: to skim impurities/grease
Chinois: cone-shaped strainer

HACCP:

Hold at 140° F.
Cool to 40° F or lower.

NUTRITION:

Calories: 13.4
Fat: .097 g
Protein: .416 g

CHEF NOTES:

1. There are many variations of court-bouillons, all containing an acid and a mirepoix with herbs and spices. Classically, two types of court-bouillons were made: one with white wine and one with red wine.
2. Traditionally, court-bouillon was a poaching liquid for fish and seafood. The liquid was then used to produce a sauce.

Fish Stock

YIELD: 3 GALLONS. SERVING SIZE: AS NEEDED FOR USE.

INGREDIENTS:

12 pounds	Lean white fish bones and heads
2 ounces	Vegetable oil

Mirepoix:

1½ pounds	Onions, peeled
1 pound	Celery, washed and trimmed
1 pound	Leeks (white part only), washed and trimmed
1 quart	Dry white wine
3 gallons	Water, cold

Bouquet Garni:

1 tablespoon	Whole white peppercorns
2 each	Bay leaves
3 ounces	Parsley stems
2 teaspoons	Dried thyme leaves

METHOD OF PREPARATION:

1. Clean and remove all blood clots and gills from the head. Split the bones, and wash thoroughly in cold water.
2. Heat the oil in a stockpot, add the mirepoix and fish bones, cover, and let **sweat** 10 to 15 minutes until the fish bones turns opaque.
3. Deglaze with white wine, and let reduce by half; then add the water and bouquet garni.
4. Gently simmer until the proper flavor is achieved; **dépouiller** occasionally.
5. Strain the stock through a chinois into another stockpot. Return to a boil, reduce to the desired strength; then transfer to a suitable container, and place in a cooling sink or blast chiller. Cool to an internal temperature of 40° F or lower. Label, date, and refrigerate.
6. Reheat to a boil.

COOKING TECHNIQUE:

Simmer

Simmer and Poach:

1. Heat the cooking liquid to the proper temperature.
2. Submerge the food product completely.
3. Keep the cooked product moist and warm.

GLOSSARY:

Mirepoix: roughly chopped vegetables
Bouquet garni: bouquet of herbs and spices
Sweat: to sauté under a cover
Dépouiller: to skim impurities/grease

HACCP:

Cool to 40° F or lower.

HAZARDOUS FOOD:

Fish bones and heads

NUTRITION:

Calories: 17.1
Fat: .606 g
Protein: .204 g

CHEF NOTES:

1. The best fish bones are from sole, haddock, whiting, and turbot. All blood clots and gills should always be removed from the head.
2. Fish bones from salmon, mackerel, and bluefish are too dark in color, too strong in flavor, and, especially, too oily for stocks.

Vegetable Stock

YIELD: 1 GALLON. SERVING SIZE: AS NEEDED.

COOKING TECHNIQUE:
Simmer

Simmer and Poach:
1. Heat the cooking liquid to the proper temperature.
2. Submerge the food product completely.
3. Keep the cooked product moist and warm.

GLOSSARY:
Mirepoix: roughly chopped vegetables
Bouquet garni: bouquet of herbs and spices
Sweat: to sauté under a cover
Deglaze: to add liquid to hot pan

HACCP:
Cool to 40° F or lower.

NUTRITION:
Calories: 43.2
Fat: 1.81 g
Protein: .624 g

INGREDIENTS:

4 tablespoons	Vegetable oil
6 cloves	Garlic, peeled and mashed
1 pint	White wine
1 gallon	Water, cold

Mirepoix:

1½ pounds	Onions, peeled
8 ounces	Leeks (white part only), washed, trimmed, split, and then re-washed
8 ounces	Carrots, washed and peeled, then re-washed
8 ounces	Celery, washed and trimmed

Bouquet Garni:

1 tablespoon	Whole white peppercorns
3 each	Bay leaves
3 ounces	Parsley stems
1 teaspoon	Thyme leaves, dried

METHOD OF PREPARATION:

1. Heat the oil in a stockpot, add the mirepoix and garlic, cover, and **sweat.**
2. **Deglaze** with white wine, let reduce by half, and then add water.
3. Add the bouquet garni, and simmer until the proper flavor is achieved.
4. Strain the stock through a chinois into another stockpot. Return to a boil, and reduce to the desired strength; then transfer to a suitable container, and place in cooling sink or blast chiller. Cool to an internal temperature of 40° F or lower. Label, date, and refrigerate.
5. Reheat to a boil.

White Stock

Chicken, Beef, Veal, Fowl, and Game

YIELD: 12 GALLONS. SERVING SIZE: AS NEEDED FOR USE.

INGREDIENTS:

50 pounds	Bones
15 gallons	Cold liquid

Mirepoix:

5 pounds	Leeks (white part only), washed, trimmed, and reserved separately
5 pounds	Carrots, washed and peeled
5 pounds	Celery, washed and trimmed
5 pounds	Onions, peeled

Bouquet Garni: (prepare 2)

2 tablespoons	Whole black peppercorns
4 each	Bay leaves
3 ounces	Parsley stems
2 tablespoons	Thyme leaves, dried

METHOD OF PREPARATION:

1. Split the bones (if needed), and rinse thoroughly in cold water, removing all blood and fat, which causes discoloration.
2. Place the bones in a suitable stockpot or steam-jacketed kettle with a spigot.
3. Add the cold liquid to a height of 8 inches above the bones.
4. Apply heat, and heat to a simmer; **dépouiller.** Add the leeks.
5. After approximately 1 hour, add the rest of the mirepoix (uncooked) and both bouquet garni.
6. Simmer until the proper richness is achieved, approximately 4 hours, and continue to dépouiller, as needed.
7. Strain the stock through a **chinois** into another stockpot. Return to a boil and reduce to desired strength, then transfer to a suitable container, and place in cooling sink or blast chiller. Cool to an internal temperature of 40° F or lower. Label, date, and refrigerate.
8. Reheat to a boil.

COOKING TECHNIQUE:

Simmer

Simmer and Poach:

1. Heat the cooking liquid to the proper temperature.
2. Submerge the food product completely.
3. Keep the cooked product moist and warm.

GLOSSARY:

Mirepoix: roughly chopped vegetables
Bouquet garni: bouquet of herbs and spices
Dépouiller: to skim impurities/grease
Chinois: cone-shaped strainer

HACCP:

Cool to 40° F or lower.

HAZARDOUS FOOD:

Bones

NUTRITION:

Calories: 19
Fat: .101 g
Protein: .546 g

CHEF NOTES:

1. The most flavorful bones to use are from the ribs, loin, neck, shank, and major joint knuckles. The bones must be fresh.
2. The cooking time of stocks will vary according to the type and size of the bones.

Apple Mint Sauce

Serve this sauce with lamb, pork, or poultry.

COOKING TECHNIQUE:
Not applicable

NUTRITION:
Calories: 18
Fat: 0 g
Protein: 0 g

YIELD: ABOUT 1 QUART. SERVING SIZE: TWO OUNCES.

INGREDIENTS:

1 pound	Granny Smith apples, unpeeled, cored, and thinly sliced
2 ounces	Water
2 tablespoons	Lemon juice
Pinch	Cinnamon
	Honey, to taste
2 tablespoons	Fresh mint, chopped

METHOD OF PREPARATION:

Simmer the apples, water, lemon juice, and cinnamon over low heat for 15 minutes, until the apples become tender. Purée the apple mixture and honey in a food processor. Stir in the chopped mint. Serve warm or chilled.

Barbecue Sauce

YIELD: 1 GALLON. SERVING SIZE: NOT APPLICABLE.

COOKING TECHNIQUE:

Simmer

Simmer and Poach:

1. Heat the cooking liquid to the proper temperature.
2. Submerge the food product completely.
3. Keep the cooked product moist and warm.

GLOSSARY:

Chinois: cone-shaped strainer

HACCP:

Cool to 40° F within 6 hours.

NUTRITION: per oz

Calories: 35
Fat: 1.19 g
Protein: .426 g

INGREDIENTS:

4 ounces	Clarified butter
1 pound	Onions, peeled and diced
6 cloves	Garlic, peeled and crushed
2 quarts	Prepared spicy barbecue sauce (preferably Open Pit)
12 ounces	Ketchup
6 ounces	Red wine vinegar
12 ounces	Light brown sugar
6 ounces	Freshly squeezed lemon juice
4 ounces	Dijon mustard
3 ounces	Worcestershire sauce
1 teaspoon	Thai chili sauce
2 tablespoons	Soy sauce
2 teaspoons	Fish sauce
2 each	Bay leaves
1 quart	Water, or as needed
	Salt and freshly ground black pepper, to taste

METHOD OF PREPARATION:

1. Melt the butter in a saucepan, add the onions, and sauté until translucent.
2. Add all of the remaining ingredients and bring to a boil. Reduce heat and simmer for 30 minutes, or until thickness is achieved.
3. Adjust seasonings, and remove from the heat. Strain through a **chinois.** Cool to an internal temperature of 40° F within 6 hours. Label, date, and refrigerate.

CHEF NOTE:

There are many variations of barbecue sauce. Prepared mustard can be used instead of Dijon. The amount of acid utilized varies by personal taste or how the sauce will be used. Chili sauce; add as desired.

Béarnaise Sauce

YIELD: 1 QUART. SERVING SIZE: TWO OUNCES.

INGREDIENTS:

5 ounces	Pasteurized egg yolks
24 ounces	Clarified butter, heated to 165° F
1 tablespoon	Freshly chopped tarragon leaves
1 tablespoon	Fresh parsley, washed, excess moisture removed, and chopped
	Salt, to taste

Reduction:

1 tablespoon	Tarragon leaves, dried
6 ounces	Dry white wine
1 tablespoon	White vinegar
6 each	Black peppercorns, crushed
2 each	Bay leaves

METHOD OF PREPARATION:

1. In a saucepan, place all of the ingredients for the **reduction,** and reduce to half of the volume. Cool; then strain the liquid into the egg yolks in a stainless steel bowl.
2. Over a double boiler, whip the egg yolk mixture until the eggs start to ribbon. Then remove from the heat.
3. Slowly add the hot clarified butter to the egg mixture, whipping constantly.
4. Adjust the seasonings, then fold in the fresh tarragon and parsley. Serve immediately, or hold warm for a maximum of 30 minutes.

COOKING TECHNIQUE:

Simmer

Simmer and Poach:

1. Heat the cooking liquid to the proper temperature.
2. Submerge the food product completely.
3. Keep the cooked product moist and warm.

GLOSSARY:

Reduction: evaporation of liquid by boiling

HAZARDOUS FOOD:

Egg yolks

NUTRITION:

Calories: 347
Fat: 37.3 g
Protein: 2.08 g

CHEF NOTES:

1. This sauce is served with broiled meats, fish, and egg dishes.
2. To create sauce Choron, add 8 ounces of tomato sauce. Serve with grilled meats and fish.

Béchamel Sauce

YIELD: 1 GALLON. SERVING SIZE: TWO OUNCES.

INGREDIENTS:

4 quarts	Milk
1 each	Onion **clouté,** peeled and cut in half
6 ounces	Clarified butter
6 ounces	All-purpose flour, sifted
	Salt and ground white pepper, to taste
	Nutmeg, to taste

METHOD OF PREPARATION:

1. In a saucepan, heat the milk with the onion clouté, and simmer for 10 minutes.
2. In another saucepan, heat the clarified butter over moderate heat.
3. Gradually add flour to make a blonde roux. Using a wooden spoon, mix the roux thoroughly, and cook it approximately 5 to 6 minutes. Remove from the heat, and cool slightly.
4. Remove the onion clouté from the milk.
5. Gradually add the hot milk to the roux, whisking constantly. Heat to a boil; reduce to a simmer. Simmer for 20 minutes or until the proper flavor and consistency are achieved.
6. Season to taste.
7. Strain through a fine **chinois** into a suitable container. Hold at 140° F, or cool to an internal temperature of 40° F or lower within 6 hours. Label, date, and refrigerate.
8. Reheat to 165° F.

COOKING TECHNIQUE:

Simmer

Simmer and Poach:

1. Heat the cooking liquid to the proper temperature.
2. Submerge the food product completely.
3. Keep the cooked product moist and warm.

GLOSSARY:

Clouté: studied with cloves
Chinois: cone-shaped strainer
Bain-marie: hot-water bath

HACCP:

Hold at 140° F or cool to an internal temperature of 40° F or lower.

HAZARDOUS FOOD:

Milk

NUTRITION:

Calories: 67.2
Fat: 4.23 g
Protein: 2.34 g

CHEF NOTES:

1. Béchamel sauce is a basic white cream sauce consisting simply of thickened, seasoned milk. Béchamel is often used as a binding agent or to make compound sauces.
2. The sauce is ready when the proper thickness has been achieved and the "floury" taste is cooked away.
3. To prevent a dried surface (skin) from forming while holding the sauce in a **bain-marie,** cover the surface with plastic wrap.

Beurre Blanc

YIELD: 1 QUART. SERVING SIZE: TWO OUNCES.

COOKING TECHNIQUE:
Simmer

Simmer and Poach:
1. Heat the cooking liquid to the proper temperature.
2. Submerge the food product completely.
3. Keep the cooked product moist and warm.

GLOSSARY:
Chinois: cone-shaped strainer

HACCP:
Hold at 140° F or above.

HAZARDOUS FOOD:
Heavy cream

NUTRITION:
Calories: 188
Fat: 19.9 g
Protein: .4 g

INGREDIENTS:

1 pint	Dry white wine
8 ounces	Heavy cream
2 ounces	Freshly squeezed lemon juice
2 ounces	Shallots, peeled and diced
1½ pounds	Sweet butter, cut into small pieces
	Salt and ground white pepper, to taste
1 tablespoon	Fresh parsley, washed, excess moisture removed, and chopped

METHOD OF PREPARATION:

1. In a saucepan over medium heat, combine the wine, heavy cream, lemon juice, and shallots. Heat to a boil; then lower the heat, and simmer until the liquid reduces by two thirds.
2. Remove from the heat, and slowly whisk in the butter until all is totally incorporated.
3. Season to taste; then add parsley, and serve. Hold at 140° F or above for a maximum of 10 minutes.

CHEF NOTES:
1. This butter and wine sauce can accompany any meat or fish dish.
2. The fresh parsley can be replaced with other herbs, such as tarragon, watercress, dill, or any other fine herbs, to create a new butter and wine sauce.
3. Sauce can be strained through a **chinois** before seasoning and adding parsley.

Beurre Rouge

YIELD: 1 QUART. SERVING SIZE: TWO OUNCES.

COOKING TECHNIQUE:

Simmer

Simmer and Poach:

1. Heat the cooking liquid to the proper temperature.
2. Submerge the food product completely.
3. Keep the cooked product moist and warm.

GLOSSARY:

Bain-marie: hot-water bath

HACCP:

Hold at 140° F or above.

NUTRITION:

Calories: 165
Fat: 17.2 g
Protein: .261 g

INGREDIENTS:

1 pint	Dry red wine
4 ounces	Balsamic vinegar
2 ounces	Shallots, peeled and diced
1½ pounds	Sweet butter, cut in small pieces
	Salt and ground white pepper, to taste
1 tablespoon	Fresh parsley, washed, excess moisture removed, and chopped

METHOD OF PREPARATION:

1. In a saucepan, over medium heat, combine the wine, vinegar, and shallots. Simmer until the liquid is reduced by two thirds.
2. Remove from the heat, and gradually whisk in butter until it is fully incorporated.
3. Season to taste; then add the parsley. Serve immediately, or hold at 140° F or above. (See chef note 3.)

CHEF NOTES:

1. This sauce can be served with lightly grilled meats or fish.
2. There are many variations to this sauce. For example, heavy cream can be used in the reduction to replace the vinegar. Seasonings can vary to include garlic and red or black pepper.
3. Butter sauces will separate if held in a **bain-marie** for more than 10 to 15 minutes.

Brown Specialty Butter Sauce

YIELD: 1 QUART. SERVING SIZE: TWO OUNCES.

COOKING TECHNIQUE:
Simmer

Simmer and Poach:
1. Heat the cooking liquid to the proper temperature.
2. Submerge the food product completely.
3. Keep the cooked product moist and warm.

NUTRITION:
Calories: 210
Fat: 23 g
Protein: .295 g

INGREDIENTS:

8 ounces	Demi-glacé, heated to a boil (see page 26)
4 ounces	Dry white wine
2 ounces	Freshly squeezed lemon juice
1 pound	Unsalted butter, cold but slightly softened
	Salt and freshly ground black pepper, to taste
	Cayenne pepper, to taste

METHOD OF PREPARATION:

1. In a heavy sauté pan over medium heat, combine the demi-glacé, white wine, and lemon juice. Heat to a simmer, then slowly add the butter, piece by piece, stirring constantly.
2. Season, to taste, with salt, pepper, and cayenne pepper.
3. Serve immediately. This sauce is not recommended for holding.

CHEF NOTES:
1. The sauce will separate if allowed to stand.
2. If the sauce separates, pour off the butter from the top, reheat the butter to 165° F, and then whisk the butter back into the sauce.
3. Brown specialty butter sauce can be served with smoked lamb.

Cardinal Sauce

YIELD: 1 QUART. SERVING SIZE: TWO OUNCES.

INGREDIENTS:

2 ounces	Clarified butter
2 ounces	All-purpose flour
1 ounce	Vegetable oil
2 pounds	Lobster shells, crushed
4 ounces	Brandy or cognac
2 ounces	Tomato paste
1 quart	Lobster or fish stock, heated to a boil (see page 10)
3 ounces	Dry sherry
¼ teaspoon	Cayenne pepper
	Salt, to taste
4 ounces	Heavy cream

Mirepoix:

8 ounces	Onions, peeled
4 ounces	Celery, washed and trimmed
4 ounces	Carrots, washed and peeled and rewashed

METHOD OF PREPARATION:

1. In a large saucepan, heat the oil, and sear the lobster shells. When they are bright red, add the mirepoix, and lightly sauté until the vegetables are golden brown.
2. **Flambé** the vegetables and lobster shells with the brandy or cognac. Add the tomato paste and then the lobster or fish stock. Let simmer for 30 minutes.
3. In a saucepan, heat the butter, and add the flour. Cook 5 to 6 minutes to create a blond roux.
4. Ladle some of the stock into the roux and dissolve; add the sherry, bring to a boil; reduce to a simmer, and cook an additional 10 minutes; **dépouiller** as needed. Season with cayenne pepper and salt.
5. Strain the sauce through a **chinois** into a suitable container.
6. Temper the heavy cream, and add it to the strained sauce.
7. Serve immediately, or hold at 140° F or above.

COOKING TECHNIQUE:

Simmer

Simmer and Poach:

1. Heat the cooking liquid to the proper temperature.
2. Submerge the food product completely.
3. Keep the cooked product moist and warm.

GLOSSARY:

Mirepoix: roughly chopped vegetables
Flambé: to flame
Dépouiller: to skim impurities/grease
Chinois: cone-shaped strainer

HACCP:

Hold at 140° F or above.

HAZARDOUS FOOD:

Lobster shells

NUTRITION:

Calories: 105
Fat: 6.28 g
Protein: 6.67 g

Colbert Butter

YIELD: 20 OUNCES. SERVING SIZE: TWO OUNCES.

COOKING TECHNIQUE:
Not applicable

HACCP:
Freeze or chill to 40° F or below.

NUTRITION:
Calories: 329
Fat: 36.8 g
Protein: .416 g

INGREDIENTS:

1 pound	Butter, softened and whipped
2 ounces	Fresh parsley, washed, excess moisture removed, and chopped
1 ounce	Fresh lemon juice
	Ground white pepper, to taste
½ teaspoon	Fresh tarragon leaves, minced
2 ounces	Meat glaze, heated to a boil and cooled

METHOD OF PREPARATION:

1. In a mixing bowl, combine all of the ingredients, and mix well.
2. Pipe the butter onto parchment paper, and roll into 1-inch-thick cylindrical shapes.
3. Freeze or chill to 40° F or below.
4. For service, slice the butter into 1-ounce medallions by dipping a knife into hot water.
5. Hold the medallions on ice.

Creole Butter Sauce

YIELD: 2 QUART. SERVING SIZE: TWO OUNCES.

COOKING TECHNIQUES:
Boil, Simmer

Boil: (at sea level)
1. Bring the cooking liquid to a rapid boil.
2. Stir the contents, and cook the food product throughout.
3. Serve hot.

Simmer and Poach:
1. Heat the cooking liquid to the proper temperature.
2. Submerge the food product completely.
3. Keep the cooked product moist and warm.

GLOSSARY:
Blanch: to parboil

HACCP:
Hold at 140° F.

HAZARDOUS FOOD:
Crabmeat

NUTRITION:
Calories: 234
Fat: 23.2 g
Protein: 3.44 g

CHEF NOTES:
1. To create a Cajun butter sauce, use crawfish tails instead of lump crabmeat.
2. This sauce can be served with broiled, grilled, or pan-fried fish and shellfish.

INGREDIENTS:

1 pint	Dry white wine
2 ounces	Freshly squeezed lemon juice
1 pound	Tomatoes, washed, cored, **blanched,** peeled, seeded, and diced
2 cloves	Garlic, peeled and mashed into a purée
2 each	Scallions, washed, trimmed, and cut ¼-inch thick
4 ounces	Tomato purée
1 teaspoon	Hot sauce
⅛ teaspoon	Cayenne pepper
2 pounds	Butter
1 pound	Lump crabmeat, flaked, removing any shell fragments

METHOD OF PREPARATION

1. In a saucepan, over moderate heat, combine the wine and lemon juice. Heat to a boil, and reduce to half.
2. Add the tomatoes, garlic, scallions, tomato purée, and seasonings, and simmer 20 minutes.
3. In a sauté pan, heat 2 ounces of the butter, and sauté the crabmeat until heated throughout. Hold covered at 140° F.
4. Cut the remaining butter into small pieces.
5. Remove the saucepan from heat, and gradually whisk in the butter until it is fully incorporated.
6. Add the reserved crabmeat, and blend together.
7. Adjust the seasoning, to taste. Hold at 140° F for a maximum of 10 to 15 minutes.

Creole Sauce

YIELD: 1 GALLON. SERVING SIZE: TWO OUNCES.

COOKING TECHNIQUES:
Sauté, Simmer

Sauté:
1. Heat the sauté pan to the appropriate temperature.
2. Evenly brown the food product.
3. For a sauce, pour off any excess oil, reheat, and deglaze.

Simmer and Poach:
1. Heat the cooking liquid to the proper temperature.
2. Submerge the food product completely.
3. Keep the cooked product moist and warm.

GLOSSARY:
Macédoine: ¼-inch dice
Blanch: to parboil

HACCP:
Hold at 140° F.

NUTRITION:
Calories: 25.3
Fat: 1.33 g
Protein: .584 g

INGREDIENTS:

3 ounces	Clarified butter or oil
4 ounces	Scallions, washed, trimmed, and sliced ¼-inch wide
4 cloves	Garlic, peeled and mashed to a purée
1 teaspoon	Curry powder
½ teaspoon	Chili powder
⅛ teaspoon	Hot sauce
8 ounces	Tomato purée
8 ounces	Vegetable stock, heated to a boil (see page 11)
2 pints	Demi-glacé, heated to a boil (see page 26)
	Salt and freshly ground black pepper, to taste

Macédoine:

1 pound	Onions, peeled
8 ounces	Carrots, washed and peeled
8 ounces	Celery, washed and trimmed
1½ pounds	Green peppers, washed and seeded
2 pounds	Tomatoes, washed, cored, **blanched,** peeled, and seeded

METHOD OF PREPARATION:

1. In a large saucepan, heat the butter or oil, and sauté the onions until they are translucent.
2. Add the remaining macédoine, scallions, and garlic, and lightly sauté until golden brown. Add the curry, chili powder, and hot sauce, and blend well.
3. Add the tomato purée, white stock, and demi-glacé. Let the sauce simmer for 20 minutes.
4. Season, to taste, and serve immediately, or hold at 140° F.

CHEF NOTE:
Creole sauce can be served with many meals. The type of stock or basic sauce used would denote the type of meat or fish with which the sauce would be served.

Cumberland Sauce

YIELD: 1 QUART. SERVING SIZE: TWO OUNCES.

COOKING TECHNIQUES:

Boil, Simmer

Boil: (at sea level)

1. Bring the cooking liquid to a rapid boil.
2. Stir the contents, and cook the food product throughout.
3. Serve hot.

Simmer and Poach:

1. Heat the cooking liquid to the proper temperature.
2. Submerge the food product completely.
3. Keep the cooked product moist and warm.

HACCP:

Chill to 40° F or below.

NUTRITION:

Calories: 76.8
Fat: .066 g
Protein: .318 g

INGREDIENTS:

3 ounces	Shallots, peeled and diced
6 ounces	Freshly squeezed orange juice
2 ounces	Freshly squeezed lemon juice
8 ounces	Currant jelly
1 teaspoon	Ginger, ground
½ teaspoon	Mustard
1 tablespoon	Arrowroot
2 tablespoons	White wine
12 ounces	Port wine
Zest of 2	Oranges
Zest of 1	Lemon

METHOD OF PREPARATION

1. Combine the shallots, juices, and currant jelly. Heat to a boil; reduce the heat, and simmer for 15 minutes, whisking to blend in the jelly.
2. Add the ginger and dry mustard, and continue to simmer for 5 minutes.
3. Dissolve the arrowroot in the white wine, and thicken the sauce. Simmmer for 15 minutes; then add the port wine. Return to a boil; then add the zests. Chill to 40° F or below within 6 hours.

CHEF NOTE:

Cumberland sauce is served cold with pâtés, terrines, smoked meats, and game.

Curry Sauce

YIELD: 1 GALLON. SERVING SIZE: TWO OUNCES.

COOKING TECHNIQUES:
Sauté, Simmer

Sauté:
1. Heat the sauté pan to the appropriate temperature.
2. Evenly brown the food product.
3. For a sauce, pour off any excess oil, reheat, and deglaze.

Simmer and Poach:
1. Heat the cooking liquid to the proper temperature.
2. Submerge the food product completely.
3. Keep the cooked product moist and warm.

GLOSSARY:
Chinois: cone-shaped strainer

HACCP:
Hold at 140° F.
Cool to 40° F or lower.

NUTRITION:
Calories: 82.5
Fat: 6.57 g
Protein: 1.82 g

CHEF NOTES:
1. The term *curry* comes from the southern Indian word *Kari,* meaning "sauce." Curry powder is an integral ingredient in all curries.
2. The type of stock used depends on the type of meat or fish served.

INGREDIENTS:

8 ounces	Clarified butter
1 ounce	Parsley stems, washed and chopped
2 tablespoons	Fresh thyme leaves
2 each	Bay leaves
2 tablespoons	Madras curry powder
	Salt and ground white pepper, to taste
6 ounces	Flour
1 quart	Coconut milk, heated to a boil
2 quarts	Chicken stock, heated to a boil (see **chef notes** 2 on page 7)
12 pounds	Onions, peeled, and diced
1 pound	Celery stalks, washed, and diced
6 ounces	Tart apples, washed, peeled, cored, and diced

METHOD OF PREPARATION:

1. In a saucepan, heat the butter, and sauté the onions, celery, and apples until the onions are translucent.
2. Add the seasonings, and stir.
3. Dust with flour, and cook on low heat, stirring constantly for 3 to 5 minutes.
4. Add the coconut milk and stock, and heat to a boil; reduce the heat, and simmer for 15 minutes.
5. Adjust the seasonings. Strain through a **chinois,** and hold at 140° F.
6. Cool to 40° F or lower within 6 hours. Label, date, and refrigerate.
7. Reheat to a boil.

Demi-Glacé

YIELD: 1 GALLON. SERVING SIZE: TWO OUNCES.

INGREDIENTS:

1 gallon	Espagnole sauce, heated to a boil (see page 27)
1 gallon	Brown stock, heated to a boil (see page 7)
8 ounces	Madeira wine
	Salt and freshly ground black pepper, to taste

Bouquet Garni:

1 tablespoon	Whole black peppercorns
2 each	Bay leaves
3 ounces	Parsley stems
½ teaspoon	Thyme leaves, dried
1 teaspoon	Tarragon, dried

METHOD OF PREPARATION:

1. In a saucepan, combine the Espagnole sauce and brown stock, and heat to a boil; reduce to a simmer.
2. **Dépouiller** as needed. Add the bouquet garni and Madeira wine.
3. Simmer until reduced to 50 % of the original volume, and season to taste.
4. Strain through a **chinois** into a suitable container. Hold at 140° F or cool to an internal temperature of 40° F or lower within 6 hours. Label, date, and refrigerate.
5. Reheat to 165° F or higher.

COOKING TECHNIQUE:

Simmer

Simmer and Poach:

1. Heat the cooking liquid to the proper temperature.
2. Submerge the food product completely.
3. Keep the cooked product moist and warm.

GLOSSARY:

Bouquet garni: bouquet of herbs and spices
Dépouiller: to skim impurities/grease
Chinois: cone-shaped strainer

HACCP:

Hold at 140° F or cool to an internal temperature of 40° F or lower.

NUTRITION:

Calories: 3.52
Fat: .018 g
Protein: .065 g

CHEF NOTES:

1. Demi-glacé is the most refined basic brown sauce. The long hours of simmering and the reduction process create the final results of a richer flavor, darker color, improved luster, and more palatable sauce.
2. The demi-glacé is used as a base sauce for brown compound sauces.
3. Madeira is a wine fortified with brandy; therefore, reduction is not needed. It can be omitted if the demi-glacé is used for another type of wine sauce.

Espagnole Sauce

YIELD: 1 GALLON. SERVING SIZE: AS NEEDED

INGREDIENTS:

4 quarts	Brown beef stock, heated to boil (see page 7)
10 ounces	Brown roux
8 ounces	Slab bacon, diced into ½-inch pieces
8 ounces	Tomato purée
	Salt and ground white pepper, to taste

Mirepoix:

8 ounces	Onions, peeled
4 ounces	Carrots, washed and peeled
4 ounces	Celery, washed and trimmed

Bouquet Garni:

1 tablespoon	Whole, black peppercorns
3 each	Bay leaves
3 ounces	Parsley stems
1 teaspoon	Dried thyme leaves

METHOD OF PREPARATION:

1. In a saucepan, gradually whisk the hot brown stock into a brown roux until it is dissolved completely.
2. Heat to a boil; reduce to a simmer, and **dépouiller** as needed.
3. In a sauté pan, render the slab bacon; then remove the bacon. Add the mirepoix, and sauté until browned. Drain the excess fat.
4. Add the tomato purée, and simmer for about 5 minutes to cook out the acidity. Add this mixture to the stock/roux mixture.
5. Add the bouquet garni, continue to simmer, and dépouiller as needed until the desired flavor and consistency are achieved. Adjust the seasonings as necessary.
6. Strain through a **chinois** into a suitable container. Hold at 140° F, or cool to an internal temperature of 40° F or lower within 6 hours. Label, date, and refrigerate.
7. Reheat to 165° F or higher.

COOKING TECHNIQUES:

Boil, Simmer

Boil: (at sea level)

1. Bring the cooking liquid to a rapid boil.
2. Stir the contents, and cook the food product throughout.
3. Serve hot.

Simmer and Poach:

1. Heat the cooking liquid to the proper temperature.
2. Submerge the food product completely.
3. Keep the cooked product moist and warm.

GLOSSARY:

Mirepoix: roughly chopped vegetables
Bouquet garni: bouquet of herbs and spices
Dépouiller: to skim impurities/grease
Chinois: cone-shaped strainer

HACCP:

Hold at 140° F or cool to an internal temperature of 40° F or lower.
Reheat to 165° F or higher.

NUTRITION:

Calories: 47.1
Fat: 4.32 g
Protein: 1.23 g

CHEF NOTE:

Espagnole sauce is a basic brown sauce with the added flavor of the accessory elements.

Grated Fresh Horseradish

YIELD: 1 QUART. SERVING SIZE: AS NEEDED.

COOKING TECHNIQUE:
Not applicable

HACCP:
Hold refrigerated at 40° F or below.

NUTRITION:
Calories: 50.6
Fat: .091 g
Protein: .59 g

INGREDIENTS:

2 pounds (2 pieces)	Fresh horseradish, soaked in cold water overnight, then peeled and grated
2 ounces	White vinegar
6 ounces	Granulated sugar

METHOD OF PREPARATION:

1. Combine all of the ingredients, and mix well.
2. Allow to stand at room temperature for 1 hour before service. Hold refrigerated at 40° F or below until needed.

CHEF NOTE:
This sauce can be served with roast prime rib of beef or boiled corned beef.

Green Peppercorn Sauce

YIELD: 1 QUART. SERVING SIZE: TWO OUNCES.

INGREDIENTS:

1 pint	Dry white wine
1 pint	Heavy cream
2 ounces	Shallots, peeled and diced
1 ounce	Green peppercorns, drained and rinsed of brine
1½ pounds	Butter, cut into small pieces
	Salt, to taste

METHOD OF PREPARATION:

1. In a saucepan over medium heat combine the wine, cream, shallots, and peppercorns, and heat to a boil. Reduce the heat, and simmer until reduced by half.
2. Remove from the heat, and gradually whisk in the butter until it is fully incorporated into the sauce.
3. Add salt, if needed, and serve, or hold for service at 140° F or above.

COOKING TECHNIQUES:
Boil, Simmer

Boil: (at sea level)
1. Bring the cooking liquid to a rapid boil.
2. Stir the contents, and cook the food product throughout.
3. Serve hot.

Simmer and Poach:
1. Heat the cooking liquid to the proper temperature.
2. Submerge the food product completely.
3. Keep the cooked product moist and warm.

HACCP:
Hold at 140° F or above.

HAZARDOUS FOOD:
Heavy cream

NUTRITION:
Calories: 432
Fat: 45.6 g
Protein: 1.54 g

CHEF NOTES:
1. This sauce can be used with poached salmon, grilled meats, and fish.
2. To fortify sauce, add 4 ounces of glacé de viande to make the sauce richer.

Hollandaise Sauce

YIELD: 1 QUART. SERVING SIZE: TWO OUNCES.

INGREDIENTS:

5 ounces	Pasteurized egg yolks
24 ounces	Clarified butter, heated to 165° F
	Salt and ground white pepper, to taste

Reduction:

5 ounces	Dry white wine
2 tablespoons	White vinegar
4 each	Whole black peppercorns, crushed
1 each	Bay leaf

METHOD OF PREPARATION:

1. In a saucepan, combine all of the ingredients for the **reduction,** and heat to a boil. Reduce to half of the volume, strain, and cool.
2. In a bowl, whisk the cooled reduction into the egg yolks.
3. Whisk the egg yolk mixture over a double boiler, cooking until the eggs start to ribbon. Remove the eggs from heat.
4. Slowly add the hot clarified butter to the eggs, whisking constantly.
5. Adjust the seasonings, and serve immediately.

COOKING TECHNIQUES:

Boil, Steam

Boil: (at sea level)

1. Bring the cooking liquid to a rapid boil.
2. Stir the contents, and cook the food product throughout.
3. Serve hot.

Steam: (Traditional)

1. Place a rack over a pot of water.
2. Prevent steam vapors from escaping.
3. Shock or cook the food product throughout.

GLOSSARY:

Reduction: evaporation of liquid by boiling

HACCP:

Heat to 165° F.

HAZARDOUS FOOD:

Pasteurized egg yolks

NUTRITION:

Calories: 386
Fat: 42.4 g
Protein: 1.53 g

CHEF NOTES:

1. Hollandaise sauce can be served with poached fish, boiled vegetables, and many other dishes.
2. If the sauce breaks, whisk ice water into the mixture, or start with a small base of egg yolks and butter, and then incorporate the broken sauce.

Horseradish Sauce (cold)

COOKING TECHNIQUE:
Not applicable

HACCP:
Hold chilled at 40° F or below.

HAZARDOUS FOOD:
Heavy cream

NUTRITION:
Calories: 82.8
Fat: 6.98 g
Protein: .797 g

YIELD: 1 QUART. SERVING SIZE: AS NEEDED.

INGREDIENTS:

8 ounces	Prepared horseradish, drained
2 tablespoons	White wine vinegar
1 ounce	Granulated sugar
1 teaspoon	Dry mustard
	Salt and ground white pepper, to taste
10 ounces	Heavy cream

METHOD OF PREPARATION:

1. Combine all of the ingredients except the cream, and mix well.
2. Whip the heavy cream until stiff.
3. With a rubber spatula fold the horseradish mixture into the cream.
4. Hold chilled at 40° F or below, until needed for service.

CHEF NOTE:
This sauce is served with roast beef, or cold meats.

Hungarian Sauce

YIELD: 1 GALLON. SERVING SIZE: TWO OUNCES.

COOKING TECHNIQUES:
Sauté, Simmer

Sauté:
1. Heat the sauté pan to the appropriate temperature.
2. Evenly brown the food product.
3. For a sauce, pour off any excess oil, reheat, and deglaze.

Simmer and Poach:
1. Heat the cooking liquid to the proper temperature.
2. Submerge the food product completely.
3. Keep the cooked product moist and warm.

GLOSSARY:
Deglaze: to add liquid to hot pan grease
Temper: to equalize two extreme temperatures

HACCP:
Hold to 140° F.
Cool to 40° F or below.

HAZARDOUS FOOD:
Sour cream
Heavy cream

NUTRITION:
Calories: 49.2
Fat: 3.97 g
Protein: 1.19 g

INGREDIENTS:

3 ounces	Clarified butter or oil
1 pound	Onions, peeled and thinly sliced
1 pound	Green bell peppers, washed, seeded, and cut into rings
2 each	Red bell peppers, washed, seeded, and cut into rings
24 ounces	Fresh tomatoes, washed, cored, blanched, seeded, and chopped
2 tablespoons	Sweet Hungarian paprika
1 tablespoon	Hot Hungarian paprika
2 ounces	Dry sherry
2 quarts	Chicken velouté, heated to a boil (see page 68)
4 ounces	White chicken stock, heated to a boil (see page 12)
8 ounces	Heavy cream
8 ounces	Sour cream
	Salt and ground white pepper, to taste

METHOD OF PREPARATION:

1. In a large sauté pan, heat the butter, and sauté the onions until they are translucent.
2. Add the peppers, tomatoes, and paprika. Sauté until the peppers are translucent.
3. **Deglaze** the pan with the sherry, and let the alcohol evaporate.
4. Incorporate the velouté and white stock. Heat to a boil; reduce the heat, and then simmer for 20 minutes, or until the proper flavor is achieved.
5. Remove from the heat, **temper** the heavy and sour creams, and add to the sauce. Season to taste, and serve immediately, or hold at 140° F. Cool to 40° F or below within 6 hours.
6. Reheat to a boil.

CHEF NOTE:
Hungarians are known for their use of paprika and sour cream, and their recipes usually contain a hot and spicy flavoring.

Lemon Caper Sauce

YIELD: 1 QUART. SERVING SIZE: TWO OUNCES.

INGREDIENTS:

1 quart	Fish **velouté,** heated to a boil (see page 68)
3 each	Anchovy fillets, mashed into a paste
	Ground white pepper, to taste
1 tablespoon	Fresh parsley, washed, excess moisture removed, and chopped

Reduction:

1 tablespoon	Capers
Zest of 1	Lemon
3 ounces	Dry white wine or dry vermouth

METHOD OF PREPARATION:

1. In a saucepan, combine the capers, lemon zest, and wine or vermouth. Heat to a boil, and reduce by two thirds.
2. Combine the reduction with the fish **velouté;** then blend in the anchovy paste.
3. Season, to taste, with white pepper, and add the parsley. Serve immediately, or hold at 140° F or above.

COOKING TECHNIQUE:

Simmer

Simmer and Poach:

1. Heat the cooking liquid to the proper temperature.
2. Submerge the food product completely.
3. Keep the cooked product moist and warm.

GLOSSARY:

Reduction: evaporation of liquid by boiling
Velouté: velvety-textured sauce or soup

HACCP:

Hold at 140° F or above.

NUTRITION:

Calories: 61.9
Fat: 3.88 g
Protein: 2.62 g

CHEF NOTES:

1. Capers and anchovies are high in salt, which is why this recipe does not require additional salt.
2. This sauce can be served with poached fish.

Marinara Sauce

YIELD: 1 GALLON. SERVING SIZE: TWO OUNCES.

COOKING TECHNIQUE:

Simmer

Simmer and Poach:

1. Heat the cooking liquid to the proper temperature.
2. Submerge the food product completely.
3. Keep the cooked product moist and warm.

GLOSSARY:

Blanch: to par cook
Sweat: to sauté under a cover

HACCP:

Hold at 140° F, or cool to an internal temperature of 40° F or lower.

NUTRITION:

Calories: 46.6
Fat: 3.63 g
Protein: .735 g

CHEF NOTE:

Served with Italian pasta dishes.

INGREDIENTS:

8 ounces	Olive oil
12 cloves	Fresh garlic, peeled and minced
10 pounds	Fresh plum tomatoes, washed, cored, **blanched,** peeled, seeded, and chopped (reserve the juice)
6 ounces	Basil leaves, washed and finely chopped
2 ounces	Fresh parsley, washed, excess moisture removed, and chopped
	Salt and freshly ground pepper, to taste

METHOD OF PREPARATION:

1. In a sauté pan, heat the olive oil, add the garlic, and **sweat** for 1 minute. (Do not brown.)
2. Add the diced tomatoes with the juice, and cook over medium heat until thickened.
3. Add the herbs, and season to taste; then remove from the heat. Hold at 140° F, or cool to an internal temperature of 40° F or lower within 6 hours. Label, date, and refrigerate.
4. Reheat to a boil.

Mayonnaise

YIELD: 1 GALLON. SERVING SIZE: NOT APPLICABLE.

COOKING TECHNIQUE:
Not applicable

HACCP:
Hold below 40° F.

HAZARDOUS FOOD:
Egg yolks

NUTRITION:
Calories: 193
Fat: 21.2 g
Protein: .464 g

INGREDIENTS:

12 ounces	Pasteurized egg yolks
4 ounces	Dijon mustard
	Salt, to taste
4 ounces	Sugar
3 ounces	White vinegar
3 quarts	Vegetable oil
4 ounces	Lemon juice

METHOD OF PREPARATION:

1. In a mixing bowl, combine the egg yolks, mustard, salt, sugar, and vinegar.
2. Whip the mixture to combine; then slowly add the oil in a thin, steady stream, and continue to whip, allowing air to be incorporated as the oil flows.
3. After all of the oil is incorporated, add the lemon juice.
4. Adjust the seasoning as needed.
5. Label, date, and refrigerate.

CHEF NOTES:

1. Dry mustard, Worcestershire sauce, and Tabasco sauce can be added, to taste, for additional flavor.
2. The holding temperature must be below 40° F.
3. For a lighter color add a few tablespoons of hot water.

Meat Glaze (Glacé de Viande)

COOKING TECHNIQUE:

Simmer

Simmer and Poach:

1. Heat the cooking liquid to the proper temperature.
2. Submerge the food product completely.
3. Keep the cooked product moist and warm.

GLOSSARY:

Reduction: evaporation of a liquid by boiling
Bain-marie: hot-water bath
Chinois: cone-shaped strainer

NUTRITION: (per 4 oz.)

Calories: 40
Fat: 2.1 g
Protein: 1.5 g

YIELD: 1 QUART. SERVING SIZE: AS NEEDED.

INGREDIENTS:

Reduction:

10 quarts	Brown beef stock

METHOD OF PREPARATION:

1. In a 12-quart stockpot, simmer the beef stock to **reduce.**
2. Keep a ladle in a **bain-marie** of cold water next to the stockpot. Skim the stock every 10 to 15 minutes.
3. When the stock has reduced by half (about 3 to 4 hours), strain it through a **chinois** with three layers of damp cheesecloth into a 5-quart stockpot.
4. Simmer until 2 quarts remain, and skim.
5. When the glaze is reduced to one tenth of its original volume (1 quart), it should have a honey-like consistency. Strain through a fine chinois (no cheesecloth; it will stick) into a stainless steel bain-marie.
6. Cool, and cover tightly. It should have a rubbery texture.
7. It can be kept in the refrigerator for 1 month or more.

Mornay Sauce

YIELD: 1 GALLON. SERVING SIZE: TWO OUNCES.

INGREDIENTS:

3½ quarts	Béchamel sauce, heated to 140° F and strained through a chinois (see page 16)
8 ounces	Gruyère cheese, grated
8 ounces	Parmesan cheese, grated
8 ounces	Heavy cream, heated to 140° F
	Salt and ground white pepper, to taste

METHOD OF PREPARATION:

1. In a saucepan, combine the béchamel sauce and cheeses; then simmer over low heat until the cheeses are melted and the mixture is smooth.
2. **Temper** the cream, add to the sauce, and season, to taste. Serve immediately, or hold at 140° F or above.

COOKING TECHNIQUE:

Simmer

Simmer and Poach:

1. Heat the cooking liquid to the proper temperature.
2. Submerge the food product completely.
3. Keep the cooked product moist and warm.

GLOSSARY:

Temper: to equalize two extreme temperatures

HACCP:

Hold at 140° F or above.

HAZARDOUS FOOD:

Heavy cream

NUTRITION:

Calories: 106
Fat: 8.92 g
Protein: 3.32 g

CHEF NOTE:

Mornay sauce is a cheese sauce used for coating crêpes, vegetables, potatoes, or pasta dishes. It also can be used with poached fish or eggs. If used for browning (au gratine) add 2 to 4 egg yolks per one quart of sauce.

Mustard Herb Sauce

YIELD: ABOUT 24 OUNCES. SERVING SIZE: TWO OUNCES.

COOKING TECHNIQUE:
Not applicable

HAZARDOUS FOOD:
Hard-cooked eggs

NUTRITION:
Calories: 246
Fat: 26 g
Protein: 2.8 g

INGREDIENTS:

5 each	Hard-cooked egg yolks, mashed
4 ounces	Dijon-style mustard
6 tablespoons	Red wine vinegar
10 to 12 ounces	Vegetable oil
	White pepper, to taste
5 each	Hard-cooked egg whites, sieved or finely minced
2 tablespoons	Shallots, minced
1 tablespoon	Fresh dill, chopped
1 tablespoon	Fresh basil, chopped
1 tablespoon	Fresh coriander, chopped

METHOD OF PREPARATION:

Completely blend the egg yolks, mustard, and vinegar in a bowl. In a slow, steady stream, pour in the oil, constantly whipping it as you would in making mayonnaise. When the mixture emulsifies, add the remaining ingredients, and chill.

CHEF NOTE:
This sauce is a good accompaniment for hot or cold poultry and beef dishes.

Oriental Peanut Sauce

YIELD: 1 QUART. SERVING SIZE: TWO OUNCES.

INGREDIENTS:

8 ounces	Raw peanuts or peanut butter
1 clove	Garlic, peeled and mashed to a purée
12 ounces	Coconut milk
10 ounces	White chicken stock, heated to a boil (see page 12)
2 teaspoons	Soy sauce
2 teaspoons	Granulated sugar
1 teaspoon	Crushed red pepper flakes, or to taste
	Salt, to taste

METHOD OF PREPARATION:

1. If using fresh peanuts, grind to a paste.
2. In a wok or saucepan, over moderate heat, combine all of the ingredients. Heat to a boil, stirring constantly; reduce the heat, and simmer until the sauce is thick.
3. Adjust the seasonings, to taste. Serve immediately, or hold at 140° F.

COOKING TECHNIQUE:

Simmer

Simmer and Poach:

1. Heat the cooking liquid to the proper temperature.
2. Submerge the food product completely.
3. Keep the cooked product moist and warm.

HACCP:

Hold at 140° F.

NUTRITION:

Calories: 130
Fat: 11.6 g
Protein: 4.6 g

CHEF NOTES:

1. This recipe originated in Indonesia. Plain yogurt can be substituted for the coconut milk.
2. This recipe can be used for grilled fish, chicken, or pork.

Parsley Sauce

YIELD: 1 PINT. SERVING SIZE: ONE AND ONE-HALF OUNCES.

COOKING TECHNIQUE:
Sauté

Sauté:

1. Heat the sauté pan to the appropriate temperature.
2. Evenly brown the food product.
3. For a sauce, pour off any excess oil, reheat, and deglaze.

INGREDIENTS:

10 ounces	Salted butter, heated to 165° F
1½ ounces	Freshly squeezed lemon juice
Zest of 1	Lemon
2 ounces	Fresh dill, washed, stemmed, and chopped
3 ounces	Fresh parsley, washed, excess moisture removed, and chopped fine

METHOD OF PREPARATION:

Just before service, combine the butter with the remaining ingredients, and heat to 165° F.

Peach Sauce

YIELD: 1 QUART. SERVING SIZE: TWO OUNCES.

COOKING TECHNIQUE:

Simmer

Simmer and Poach:

1. Heat the cooking liquid to the proper temperature.
2. Submerge the food product completely.
3. Keep the cooked product moist and warm.

NUTRITION:

Calories: 43.6
Fat: .638 g
Protein: .726 g

INGREDIENTS:

1 ounce	Brown sugar
3 ounces	Peach preserves
1 ounce	Peach liqueur
1 quart	Brown stock, heated to boil (see page 7)
1½ ounces	Cornstarch
2 ounces	Currant jelly
½ ounce	Lemon juice
	Salt, to taste

METHOD OF PREPARATION:

1. In a saucepan, caramelize the sugar.
2. Add the preserves, liqueur, and stock to the sugar, and heat to a boil. Simmer for a few minutes. Dilute the cornstarch in the cold stock (water). Stir into the sauce, and continue to simmer for 5 more minutes.
3. Add the currant jelly and lemon juice, and season, to taste. Hold at 140° F.

CHEF NOTE:

Reserve 3 ounces of brown stock to dilute cornstarch.

Piquant Mustard Dill Sauce

COOKING TECHNIQUE:
Not applicable

NUTRITION:
Calories: 423
Fat: 43.9 g
Protein: 1.08 g

YIELD: 1 PINT. SERVING SIZE: TWO OUNCES.

INGREDIENTS:

2 small cloves	Garlic, minced or pressed
4 ounces	Dijon-style mustard
2 ounces	Sugar, or
1 ounce	Honey
½ teaspoon	Ground white pepper
Pinch	Salt
1 ounce	Red wine vinegar
8 ounces	Vegetable oil
6 tablespoons	Fresh dill, finely chopped

METHOD OF PREPARATION:

Blend all of the ingredients together, except the oil and dill. Slowly whip in the oil as you would for mayonnaise. Stir in the dill and chill.

CHEF NOTE:
This sauce, which is traditionally served with gravad lox, also goes well with other fish.

Poivrade Sauce

YIELD: 10 SERVINGS. SERVING SIZE: SIX OUNCES.

COOKING TECHNIQUES:
Sauté, Simmer

Sauté:
1. Heat the sauté pan to the appropriate temperature.
2. Evenly brown the food product.
3. For a sauce, pour off any excess oil, reheat, and deglaze.

Simmer and Poach:
1. Heat the cooking liquid to the proper temperature.
2. Submerge the food product completely.
3. Keep the cooked product moist and warm.

GLOSSARY:
Chinois: cone-shaped strainer

INGREDIENTS:

1½ ounces	Butter
4 ounces	Onions, peeled, and diced
10 ounces	White wine
4 ounces	Red wine vinegar
6 each	Black peppercorns
1 sprig	Fresh parsley, washed
½ teaspoon	Thyme
2 each	Bay leaves
2 ounces	Madeira wine
16 ounces	Demi-glacé (see page 26)
	Salt and freshly ground black pepper, to taste

METHOD OF PREPARATION:

1. Sauté the onions in the butter until they are translucent.
2. Add the wine and the vinegar, peppercorns, parsley, thyme, bay leaves, Madeira, and demi-glacé.
3. Simmer for 35 minutes, or until reduced by one third. Season with salt and pepper.
4. Strain through a **chinois** mousseline before use.

Raisin Sauce

YIELD: 1 QUART. SERVING SIZE: TWO OUNCES.

INGREDIENTS:

8 ounces	Port wine
4 ounces	Yellow raisins
4 ounces	Dark raisins
16 ounces	Brown stock, heated to a boil (see page 7)
6 ounces	Demi-glacé, heated to a boil (see page 26)
1 tablespoon	Granulated sugar
2 ounces	Red wine vinegar
8 ounces	Butter, cut into small pieces

METHOD OF PREPARATION:

1. Combine the port wine and raisins, allowing them to soak for 20 minutes. Place this mixture over low heat, and simmer (do not boil) for 3 minutes.
2. Add the stock, demi-glacé, sugar, and vinegar, continuing to **reduce** slowly for 12 to 15 minutes, or until a syrupy consistency is achieved.
3. Strain the sauce from the raisins into another saucepan, reserving the raisins. Heat the sauce to a boil; remove from the heat, and gradually whisk in the butter until fully incorporated. Return the raisins to the sauce, and serve. Hold at 140° F or above.

COOKING TECHNIQUES:

Simmer, Boil

Simmer and Poach:

1. Heat the cooking liquid to the proper temperature.
2. Submerge the food product completely.
3. Keep the cooked product moist and warm.

Boil: (at sea level)

1. Bring the cooking liquid to a rapid boil.
2. Stir the contents, and cook the food product throughout.
3. Serve hot.

GLOSSARY:

Reduce: to evaporate liquid by boiling

HACCP:

Hold at 140° F or above.

NUTRITION:

Calories: 72.1
Fat: .084 g
Protein: .606 g

CHEF NOTE:

This sauce is served with roast duck, braised brisket of beef, and meat balls.

Ravigote Sauce (Verte)

YIELD: 1 GALLON. SERVING SIZE: TWO OUNCES.

COOKING TECHNIQUE:
Not applicable

GLOSSARY:
Blanch: to par cook

HACCP:
Chill at 40° F or lower until service.

NUTRITION:
Calories: 152
Fat: 38.6 g
Protein: 1.1 g

INGREDIENTS:

3 ½ quarts	Mayonnaise (see page 35)
3 ounces	Fresh parsley, washed, excess moisture removed, and finely chopped
4 ounces	Fresh chives, washed and finely chopped
3 ounces	Fresh chervil leaves, washed and finely chopped
12 ounces	Fresh spinach leaves, **blanched** and finely chopped
2 ounces	Freshly squeezed lemon juice, or to taste
	Salt, to taste

METHOD OF PREPARATION:

Mix all of the ingredients together, and chill at 40° F or lower until service time.

CHEF NOTES:
1. This sauce is served cold.
2. This sauce can be served as a dressing for salads or with cold meats, or used as a dip for crudités.

Red Pepper Cream Sauce

YIELD: 1 QUART. SERVING SIZE: TWO OUNCES.

INGREDIENTS:

3 ounces	Olive oil
8 each	Red bell peppers, washed, seeded, and chopped
4 ounces	Shallots, peeled, and diced
9 ounces	White wine
3 ounces	Dry vermouth
12 ounces	Heavy cream
8 ounces	Butter, cut into small pieces
	Salt and ground white pepper, to taste

METHOD OF PREPARATION:

1. In a saucepan, heat the oil, and sauté the peppers until tender; then purée.
2. Combine the shallots, wine, and vermouth; simmer until the shallots are translucent and the liquid is **reduced** by half.
3. Add the red pepper purée to the shallot mixture, and reduce by half again.
4. Add the cream, and reduce until the sauce coats the back of a spoon.
5. Remove from the heat, and gradually whisk in the butter until it is fully incorporated.
6. Strain through a **chinois** into a suitable container. Serve immediately, or hold at 140° F.

COOKING TECHNIQUES:
Sauté, Simmer

Sauté:
1. Heat the sauté pan to the appropriate temperature.
2. Evenly brown the food product.
3. For a sauce, pour off any excess oil, reheat, and deglaze.

Simmer and Poach:
1. Heat the cooking liquid to the proper temperature.
2. Submerge the food product completely.
3. Keep the cooked product moist and warm.

GLOSSARY:
Reduction: evaporation of liquid by boiling
Chinois: cone-shaped strainer

HACCP:
Hold at 140° F.

HAZARDOUS FOOD:
Heavy cream

NUTRITION:
Calories: 127
Fat: 12.4 g
Protein: .551 g

Rémoulade Sauce

YIELD: 1 GALLON. SERVING SIZE: TWO OUNCES.

COOKING TECHNIQUE:
Not applicable

HACCP:
Chill to 40° F or below.

HAZARDOUS FOOD:
Mayonnaise

NUTRITION:
Calories: 311
Fat: 33.3 g
Protein: 1.51 g

INGREDIENTS:

3 quarts	Mayonnaise (see page 35)
4 ounces	Anchovies, drained and mashed into a paste
1 pound	Gherkins, diced
4 ounces	Capers, drained and rinsed
3 ounces	Fresh parsley, washed, excess moisture removed, and chopped
2 tablespoons	Fresh-chopped chervil leaves

METHOD OF PREPARATION:

Place the mayonnaise in a bowl, add the remaining ingredients, and mix thoroughly. Chill to 40° F or below, until service.

CHEF NOTE:
This sauce can be served with fried fish, seafood, grilled entrées, or marinated vegetables, or used as a dip for crudités.

COOKING TECHNIQUE:

Roast

Roast:

1. Sear the food product, and brown evenly.
2. Elevate the food product in a roasting pan.
3. Determine doneness, and consider carryover cooking.
4. Let the food product rest before carving.

HACCP:

Hold refrigerated at 40° F or lower.

HAZARDOUS FOOD:

Egg yolks

NUTRITION:

Calories: 208
Fat: 17.8
Protein: 2.67 g

Roasted Garlic and Sun-Dried Tomato Purée

YIELD: APPROXIMATELY 20 OUNCES. SERVING SIZE: AS NEEDED.

INGREDIENTS:

3 heads	Garlic
4 ounces	Olive oil
6 ounces	Sun-dried tomatoes, packed in oil, excess oil drained and reserved
2 ounces	Freshly squeezed lemon juice
1 ounce	Balsamic vinegar
1 ½ ounces	Pasteurized egg yolks
½ teaspoon	Cayenne pepper flakes, crushed, or to taste
	Salt and freshly ground black pepper, to taste

METHOD OF PREPARATION:

1. Preheat the oven to 375° F.
2. Wrap the garlic heads individually in aluminum foil, place on a baking tin, and roast until soft.
3. Squeeze the garlic pulp from the heads, and discard the peelings.
4. In a food processor, combine the garlic pulp, sun-dried tomatoes, lemon juice, vinegar, egg yolks, and seasonings. Pulse to blend together; then, with the motor running, slowly, add the remaining olive oil and the reserved tomato oil. Hold refrigerated at 40° F or lower. Bring to room temperature for service.

Rosemary-Orange Mayonnaise

YIELD: 20 OUNCES. SERVING SIZE: TWO OUNCES.

COOKING TECHNIQUE:
Not applicable

HACCP:
Hold refrigerated at 40° F or lower.

NUTRITION:
Calories: 354
Fat: 36.8 g
Protein: .934 g

INGREDIENTS:

16 ounces	Mayonnaise (see page 35)
3 ounces	Frozen orange concentrate, thawed
1 ounce	Freshly squeezed lemon juice
½ teaspoon	Worcestershire sauce
1½ ounces	Fresh rosemary, washed, stems removed, and chopped
	Salt and ground white pepper, to taste
	Tabasco, to taste
Zest of 1	Orange, chopped

METHOD OF PREPARATION:

1. Combine the mayonnaise and orange concentrate, and whisk together.
2. Add the remaining ingredients, and blend together.
3. Taste, and adjust the seasonings as required. Hold at 40° F or lower.

CHEF NOTE:
If rosemary needles are too dry, use ground rosemary.

Salsa Picante con Tomatillo

YIELD: 1 QUART. SERVING SIZE: TWO OUNCES.

COOKING TECHNIQUE:
Simmer

Simmer and Poach:
1. Heat the cooking liquid to the proper temperature.
2. Submerge the food product completely.
3. Keep the cooked product moist and warm.

GLOSSARY:
Brunoise: ⅛-inch dice

HACCP:
Refrigerate at 40° F or below.
Heat to 165° F.

NUTRITION:
Calories: 19.9
Fat: .463 g
Protein: .609 g

INGREDIENTS:

24 ounces	Tomatillos, husks removed, cored, and diced
8 ounces	Onions, peeled and **brunoise**
3 cloves	Garlic, peeled and mashed into a purée
2 tablespoons	Finely chopped fresh cilantro
1 each	Hot green chili, charred over flame, peeled, seeded, and brunoise
1 teaspoon	Granulated sugar
1	Lime, for juice
	Salt to taste

METHOD OF PREPARATION:

1. Combine all of the ingredients; add juice of lime, then refrigerate at 40° F or below for 6 to 8 hours.
2. Just before serving, heat to 165° F. Season with salt.

Sauce Allemande

YIELD: 1 GALLON. SERVING SIZE: TWO OUNCES.

COOKING TECHNIQUE:
Simmer

Simmer and Poach:
1. Heat the cooking liquid to the proper temperature.
2. Submerge the food product completely.
3. Keep the cooked product moist and warm.

GLOSSARY:
Velouté: velvety-textured sauce or soup
Temper: to equalize two extreme temperatures

HACCP:
Hold at 140° F, or chill to 40° F or lower.

HAZARDOUS FOOD:
Pasteurized egg yolks
Heavy cream

NUTRITION:
Calories: 77.8
Fat: 5.76 g
Protein: 1.66 g

INGREDIENTS:

3½ quarts	Veal **velouté** (see page 68)
4 ounces	Pasteurized egg yolks
8 ounces	Heavy cream
2 ounces	Lemon juice
	Salt and white pepper to taste

METHOD OF PREPARATION:

1. In a saucepan, heat the velouté to a boil.
2. Combine egg yolks and heavy cream in a mixing bowl, **temper,** and add to the velouté.
3. Whisk in the lemon juice; season with salt and white pepper.
4. Hold at 140° F, or cool to 40° F or lower within 6 hours. Label, date, and refrigerate.

Sauce Bercy

YIELD: 1 QUART. SERVING SIZE: TWO OUNCES.

INGREDIENTS:

1 ounce	Clarified butter
2 ounces	Shallots, peeled, and diced
8 ounces	Dry white wine
8 ounces	Fish stock, heated to a boil (see page 10)
1 ounce	Freshly squeezed lemon juice
1 pint	Thick fish velouté, heated to a boil (see page 68)
4 ounces	Heavy cream
	Salt and ground white pepper, to taste
2 ounces	Fresh parsley, washed, excess moisture removed, and chopped

METHOD OF PREPARATION:

1. In a saucepan, melt the butter, and lightly sauté the shallots. Add the white wine, fish stock, and lemon juice. Heat to a boil, and reduce the liquid by half.
2. Add the fish velouté, heat to a boil, and **dépouiller** as needed. Simmer for 15 minutes, or until the proper flavor is achieved.
3. **Temper** the heavy cream, and add it to the sauce. Season, to taste, and serve garnished with parsley.

COOKING TECHNIQUE:
Simmer

Simmer and Poach:
1. Heat the cooking liquid to the proper temperature.
2. Submerge the food product completely.
3. Keep the cooked product moist and warm.

GLOSSARY:
Dépouiller: to skim impurities/grease
Temper: to equalize two extreme temperatures

HAZARDOUS FOOD:
Heavy cream

NUTRITION:
Calories: 70.3
Fat: 5.22 g
Protein: .551 g

CHEF NOTE:
Sauce Bercy is used for poached fish.

Sauce Bigarade

YIELD: 1 QUART. SERVING SIZE: TWO OUNCES.

COOKING TECHNIQUE:
Simmer

Simmer and Poach:
1. Heat the cooking liquid to the proper temperature.
2. Submerge the food product completely.
3. Keep the cooked product moist and warm.

GLOSSARY:
Reduction: evaporation of liquid by boiling
Flambé: to flame
Deglaze: to add liquid to hot pan

HACCP:
Hold at 140° F.

NUTRITION:
Calories: 78.4
Fat: 3.82 g
Protein: .393 g

INGREDIENTS:

1½ pint	Demi-glacé, heated to a boil (see page 26)
1½ ounces	Butter
3 ounces	Grand Marnier or Curaçao
Zest of 1	Orange, minced

Reduction:

1 ounce	Clarified butter
2 ounces	Shallots, peeled and minced
1 tablespoon	Sugar
1 ounce	Brandy or cognac
3 ounces	Dry red wine
1 ounce	Raspberry vinegar or red wine
6 ounces	Fresh orange juice
4 ounces	Tomato purée

METHOD OF PREPARATION:

1. In a saucepan, heat 1 ounce of butter, and sauté the shallots. Add sugar, and lightly caramelize.
2. Add brandy or cognac, and **flambé. Deglaze** with dry red wine and vinegar, and reduce by two thirds.
3. Add the orange juice, and reduce by half.
4. Add the tomato purée, and cook for several minutes to remove the acidity.
5. Add the demi-glacé to the reduction, and simmer until the proper flavor is reached.
6. Remove from the heat, and whisk in 1½ ounces of butter; then add the Grand Marnier or Curaçao and orange zest. Serve immediately, or hold at 140° F.

CHEF NOTES:
1. The objective of this sauce is to find a proper balance of different flavors: sweet, sour, and savory. Serve it with roast duck, roast pork, or baked ham.
2. If the demi-glacé is tomato flavored, omit tomato purée.

Sauce Bordelaise

YIELD: 1 QUART. SERVING SIZE: TWO OUNCES.

INGREDIENTS:

1 quart	Demi-glacé, heated to a boil (see page 26)
	Salt and white pepper, to taste
4 ounces	Bone marrow, poached, drained, and diced

Reduction:

1 ounce	Clarified butter
1 ounce	Shallots, peeled, diced
8 ounces	Red Bordeaux wine
½ teaspoon	Thyme
2 each	Bay leaves

METHOD OF PREPARATION:

1. In a sauté pan, melt the butter, add the shallots, and sauté. **Deglaze** the pan with the wine, add the thyme and bay leaves, and reduce by two thirds.
2. Add the demi-glacé, and let simmer for 20 minutes. When the proper flavor is reached, season, to taste, and remove from the heat. Strain, and hold at 140° F.
3. Just before serving, add the bone marrow.

COOKING TECHNIQUES:

Simmer, Poach

Simmer and Poach:

1. Heat the cooking liquid to the proper temperature.
2. Submerge the food product completely.
3. Keep the cooked product moist and warm.

GLOSSARY:

Reduction: evaporation of liquid by boiling

Deglaze: to add liquid to hot pan

HACCP:

Hold at 140° F.

HAZARDOUS FOOD:

Bone marrow

NUTRITION:

Calories: 60.9
Fat: 5.17 g
Protein: .142 g

CHEF NOTES:

1. Sauce Bordelaise requires the use of red Bordeaux wine.
2. Sauce Bordelaise can be served with grilled or sautéed beef.

Sauce Bourguignonne

YIELD: 1 QUART. SERVING SIZE: TWO OUNCES.

COOKING TECHNIQUE:
Simmer

Simmer and Poach:
1. Heat the cooking liquid to the proper temperature.
2. Submerge the food product completely.
3. Keep the cooked product moist and warm.

GLOSSARY:
Reduction: evaporation of liquid by boiling
Bouquet garni: bouquet of herbs and spices

HACCP:
Hold at 140° F.

NUTRITION:
Calories: 31.5
Fat: 1.66 g
Protein: .336 g

INGREDIENTS:

Reduction:

1 ounce	Clarified butter
2 ounces	Shallots, peeled, and diced
4 ounces	White mushrooms, washed and sliced
8 ounces	Red Burgundy wine
1 quart	Demi-glacé, heated to a boil (see page 26)
	Salt and freshly ground black pepper, to taste

Bouquet Garni:

5 each	Whole black peppercorns
2 each	Bay leaves
3 each	Parsley stems
½ teaspoon	Dried thyme leaves

METHOD OF PREPARATION:

1. In a sauté pan, melt the butter, and lightly sauté the shallots and mushrooms; deglaze the pan with the wine, and reduce by half.
2. Add the demi-glacé and bouquet garni, and simmer for 10 minutes. When the proper flavor is achieved, remove bouquet garni, season to taste, and remove from the heat. Hold at 140° F.

CHEF NOTES:
1. Sauce Bourguignonne requires the use of red Burgundy wine flavored with mushroom essence.
2. Sauce Bourguignonne can be served with sautéed meat or can accompany stewed meats.

Sauce Chasseur

YIELD: 1 QUART. SERVING SIZE: TWO OUNCES.

COOKING TECHNIQUE:

Simmer

Simmer and Poach:

1. Heat the cooking liquid to the proper temperature.
2. Submerge the food product completely.
3. Keep the cooked product moist and warm.

GLOSSARY:

Reduction: evaporation of liquid by boiling

Deglaze: to add liquid to hot pan

HACCP:

Hold at 140° F or above.

NUTRITION:

Calories: 42.6
Fat: 32.6 g
Protein: .437 g

INGREDIENTS:

Reduction:

2 ounces	Clarified butter
2 ounces	Shallots, peeled, and diced
4 ounces	Mushrooms, cleaned and sliced
4 ounces	Dry white wine
½ teaspoon	Freshly chopped chervil leaves
1 teaspoon	Freshly chopped tarragon leaves
1½ pints	Demi-glacé, heated to a boil (see page 26)
4 ounces	Tomato sauce, heated to a boil (see page 76)
	Salt and ground white pepper, to taste

METHOD OF PREPARATION:

1. In a sauté pan, heat the butter, and lightly sauté the shallots and mushrooms. **Deglaze** the pan with the white wine, and reduce by half.
2. Add the herbs to the reduction with the demi-glacé and tomato sauce. Heat to a boil; reduce the heat, and simmer for 20 minutes.
3. When the proper flavor is achieved, season, to taste, and remove from the heat. Serve immediately, or hold at 140° F or above.

CHEF NOTES:

1. This sauce also can be made by omitting the tomato sauce and replacing it with 4 ounces of chopped tomato, which is sautéed with the mushrooms and shallots.
2. Sauce Chasseur can be served with sautéed or grilled beef and chicken.

Sauce Diable

YIELD: 1 QUART. SERVING SIZE: TWO OUNCES.

COOKING TECHNIQUE:
Simmer

Simmer and Poach:
1. Heat the cooking liquid to the proper temperature.
2. Submerge the food product completely.
3. Keep the cooked product moist and warm.

GLOSSARY:
Deglaze: to add liquid to hot pan
Dépouiller: to skim impurities/grease

HACCP:
Hold at 140° F or above.

NUTRITION:
Calories: 29.3
Fat: 1.64 g
Protein: .205 g

INGREDIENTS:

1 quart	Demi-glacé, heated to a boil and strained (see page 26)
⅛ teaspoon	Cayenne pepper
	Salt, to taste
½ teaspoon	Freshly ground black pepper

Reduction:

1 ounce	Clarified butter
2 ounces	Shallots, peeled, and diced
8 ounces	White wine
1 tablespoon	White vinegar
¼ teaspoon	Dried thyme
2 each	Bay leaves

METHOD OF PREPARATION:

1. In a sauté pan, melt the butter, and lightly sauté the shallots.
2. **Deglaze** the pan with the wine and vinegar.
3. Add the thyme and bay leaves, and reduce by half.
4. Add the demi-glacé, heat to a boil, and **dépouiller** as needed. Reduce the heat and simmer for 20 minutes.
5. Add the cayenne pepper, salt, and black pepper. Remove from the heat, and strain to taste.
6. Serve immediately, or hold at 140° F or above.

CHEF NOTE:
Sauce Diable is served with broiled, grilled, or roasted chicken or sautéed **emincé** of veal.

Sauce Forestière

YIELD: 1 QUART. SERVING SIZE: TWO OUNCES.

INGREDIENTS:

Reduction:

4 ounces	Mushrooms
2 ounces	Clarified butter
2 ounces	Shallots
2 ounces	Red Burgundy wine
1 quart	Demi-glacé, heat to a boil (see page 26)
	Salt and freshly ground white pepper, to taste
2 ounces	Additional clarified butter to sauté sliced mushrooms

METHOD OF PREPARATION:

1. Remove the stems from the caps of the mushrooms, reserving separately. Chop the stems, and slice the caps.
2. In a sauté pan, heat the butter, and lightly sauté the shallots and mushroom stems. **Deglaze** with wine, and reduce by half.
3. Add the demi-glacé, and heat to a boil; reduce to a simmer; cook for 20 minutes, and season, to taste.
4. When the proper flavor is achieved, remove from the heat, and strain.
5. Sauté the sliced mushroom caps in the additional butter, and add to the sauce.

COOKING TECHNIQUE:

Simmer

Simmer and Poach:

1. Heat the cooking liquid to the proper temperature.
2. Submerge the food product completely.
3. Keep the cooked product moist and warm.

GLOSSARY:

Reduction: evaporation of liquid by boiling
Deglaze: to add liquid to hot pan

NUTRITION:

Calories: 66.6
Fat: 6.43 g
Protein: .333 g

CHEF NOTES:

1. This sauce is served with sautéed veal and chicken or braised beef and veal.
2. When serving Forestière sauce with white meats, use white Burgundy wine in the reduction.

Sauce Maltaise

COOKING TECHNIQUE:

Mix

NUTRITION:

Calories: 159
Fat: 15.2 g
Protein: 1.88 g

YIELD: 1 QUART. SERVING SIZE: TWO OUNCES.

INGREDIENTS:

1 quart	Hollandaise sauce (see page 30)
Juice of 1	Blood orange, strained
Zest of 1	Blood orange, steeped in hot water

METHOD OF PREPARATION:

Fold and incorporate the juice and zest of a blood orange into the hollandaise sauce just before service.

CHEF NOTE:

Serve the sauce with poached fish.

Sauce Mousseline

YIELD: 1 QUART. SERVING SIZE: TWO OUNCES.

COOKING TECHNIQUE:
Not applicable

HACCP:
Heat and hold at 140° F.

HAZARDOUS FOOD:
Heavy cream

NUTRITION:
Calories: 157
Fat: 15.6 g
Protein: 1.71 g

INGREDIENTS:

24 ounces	Hollandaise sauce, heated to 140° F (see page 30)
8 ounces	Heavy cream, whipped

METHOD OF PREPARATION:

Fold and incorporate the cream into the hollandaise sauce just before service. Hold at 140° F.

CHEF NOTE:
This sauce can be served with fish, seafood, and vegetables.

Sauce Newburg

YIELD: 1 QUART. SERVING SIZE: TWO OUNCES.

INGREDIENTS:

2 ounces	Clarified butter
8 ounces	Onion, peeled, and chopped
4 ounces	Celery, washed, trimmed, and chopped
4 ounces	Carrots, washed, peeled, and chopped
2 pounds	Lobster and shrimp shells, crushed
1 teaspoon	Spanish paprika
6 ounces	Dry sherry
2 ounces	Tomato purée
1 quart	Fish velouté, heated to a boil (see page 68)
4 ounces	Heavy cream
	Salt, to taste
⅛ teaspoon	Cayenne pepper

METHOD OF PREPARATION:

1. In a saucepan, melt the butter, and sauté the vegetables until the onions are translucent. Add the shells and the paprika, cook until the shells turn red.
2. Deglaze the pan with the sherry, and cook to evaporate the alcohol.
3. Add the tomato purée.
4. Add the velouté, and heat to a boil; reduce the heat, and simmer for 20 minutes; **dépouiller** as needed. Strain the sauce through a **chinois.**
5. **Temper** the heavy cream, and add it to the sauce; then add the remaining seasonings.
6. Serve immediately, or hold at 140° F.

COOKING TECHNIQUE:
Simmer

Simmer and Poach:
1. Heat the cooking liquid to the proper temperature.
2. Submerge the food product completely.
3. Keep the cooked product moist and warm.

GLOSSARY:
Dépouiller: to skim impurities/grease
Chinois: cone-shaped strainer
Temper: to equalize two extreme temperatures

HACCP:
Hold at 140° F.

HAZARDOUS FOOD:
Lobster and shrimp shells
Heavy cream

NUTRITION:
Calories: 126
Fat: 9.67 g
Protein: 2.49 g

CHEF NOTE:
This sauce is served with poached fish or fish casserole.

Sauce Normande

YIELD: 1 QUART. SERVING SIZE: TWO OUNCES.

COOKING TECHNIQUE:
Simmer

Simmer and Poach:
1. Heat the cooking liquid to the proper temperature.
2. Submerge the food product completely.
3. Keep the cooked product moist and warm.

GLOSSARY:
Velouté: velvety-textured sauce or soup
Temper: to equalize two extreme temperatures

HAZARDOUS FOOD:
Heavy cream

NUTRITION:
Calories: 119
Fat: 10.9 g
Protein: 2.37 g

INGREDIENTS:

1 quart	Fish **velouté,** heated to a boil (see page 68)
8 ounces	Fish stock (see page 10)
2 ounces	Mushroom essence
	Salt and ground white pepper, to taste
2 ounces	Heavy cream
4 ounces	Butter

METHOD OF PREPARATION:

1. Combine the velouté, fish stock, and mushroom essence, and heat to a boil; reduce the heat, and simmer until the proper consistency is achieved.
2. Season, to taste.
3. **Temper** the heavy cream, and add it to the sauce. Whisk in the butter piece by piece. Once the butter is incorporated, serve immediately.

CHEF NOTE:
This sauce originated from the region of Normandy, France, which is known for its fresh fish, crustaceans, and dairy products.

Sauce Paloise

YIELD: 1 QUART. SERVING SIZE: TWO OUNCES.

INGREDIENTS:

2 tablespoons	Mint leaves, **chiffonade**
1 teaspoon	Mint essence
1 quart	Hollandaise sauce, heated to 140° F (see page 30)

METHOD OF PREPARATION:

Fold and incorporate the mint leaves and essence into the Hollandaise sauce just before service. Hold at 140/dg F.

COOKING TECHNIQUE:

Not applicable

GLOSSARY:

Chiffonade: ribbon cut of leafy greens

HACCP:

Heat to and hold at 140° F.

NUTRITION:

Calories: 176
Fat: 17.1 g
Protein: 2.08 g

CHEF NOTE:

This sauce can be served with fish, seafood, and vegetables.

Sauce Périgourdine

YIELD: 1 QUART. SERVING SIZE: TWO OUNCES.

COOKING TECHNIQUE:

Simmer

Simmer and Poach:

1. Heat the cooking liquid to the proper temperature.
2. Submerge the food product completely.
3. Keep the cooked product moist and warm.

GLOSSARY:

Temper: to equalize two extreme temperatures

HACCP:

Hold at 140° F or above.

HAZARDOUS FOOD:

Heavy cream

NUTRITION:

Calories: 33
Fat: 1.18 g
Protein: .712 g

INGREDIENTS:

12 ounces	Madeira wine
2 ounces	Truffles, diced
24 ounces	Demi-glacé, heated to a boil, diced (see page 26)
	Salt and freshly ground black pepper, to taste
2 ounces	Foie gras
1 ounce	Heavy cream

METHOD OF PREPARATION:

1. In a sauce pan, combine the wine and truffles, and simmer, reducing by half.
2. Add the demi-glacé, and heat to a boil. Reduce the heat, and simmer for 20 minutes. Season, to taste.
3. In a mixing bowl, combine the foie gras and heavy cream.
4. **Temper** the foie gras/cream mixture, and add to the sauce.
5. Serve immediately, or hold at 140° F or above.

CHEF NOTES:

1. This sauce is from Périgueux, the capital of the Dordogne region of France.
2. Périgord is a province within the Dordogne region where the black truffles are found.
3. *À la Périgourdine* means "in the style of Périgord" and is in reference to the use of truffles and foie gras.

Sauce Rachel

COOKING TECHNIQUE:
Not applicable

HACCP:
Heat to and hold at 140° F.

NUTRITION:
Calories: 134
Fat: 12.8 g
Protein: 1.65 g

YIELD: 1 QUART. SERVING SIZE: TWO OUNCES.

INGREDIENTS:

3 ounces	Tomato sauce, heated to a boil (see page 76)
5 ounces	Demi-glacé, heated to a boil (see page 26)
24 ounces	Hollandaise sauce, heated to 140° F (see page 30)

METHOD OF PREPARATION:

Fold and incorporate the tomato sauce and demi-glacé into the Hollandaise sauce just before service. Hold at 140° F.

CHEF NOTE:
This sauce can be served with fish, seafood, and vegetables.

Sauce Soubise

YIELD: 1 QUART. SERVING SIZE: TWO OUNCES.

INGREDIENTS:

1 quart	Béchamel sauce (see page 16)
2 ounces	Clarified butter
1 pound	Onions, peeled, and thinly sliced
2 ounces	Heavy cream
	Salt and ground white pepper, to taste

METHOD OF PREPARATION:

1. In a saucepan, heat the béchamel sauce to a simmer.
2. In another saucepan, melt the butter, and **sweat** the onions until they are translucent. Purée the onions in a food processor, and add to the béchamel sauce. Simmer for an additional 10 minutes.
3. **Temper** the heavy cream, add to the béchamel sauce, adjust the seasonings, and serve.
4. Hold at 140° F or above.

COOKING TECHNIQUE:

Simmer

Simmer and Poach:

1. Heat the cooking liquid to the proper temperature.
2. Submerge the food product completely.
3. Keep the cooked product moist and warm.

GLOSSARY:

Sweat: to sauté under a cover
Temper: to equalize two extreme temperatures

HACCP:

Hold at 140° F or above.

HAZARDOUS FOOD:

Heavy cream

NUTRITION:

Calories: 123
Fat: 10.7 g
Protein: 1.24 g

CHEF NOTE:

Soubise purée is onions and rice, cooked and puréed. Used also as thickening for soup or sauce.

Sauce Suprême

YIELD: 1 GALLON. SERVING SIZE: TWO OUNCES.

INGREDIENTS:

1 gallon	White chicken stock, heated to a boil (see page 12)
6 ounces	Butter
6 ounces	Flour
8 ounces	Heavy cream
	Salt, and white pepper to taste

METHOD OF PREPARATION:

1. In a saucepan or stockpot, heat the chicken stock to a boil. Reduce to a simmer.
2. Melt the butter, and add the flour. Mix until the starch is completely cooked, about 5 to 7 minutes. Do not brown.
3. **Temper** the roux with some of the stock.
4. Return the roux to the stock, whisking quickly to prevent lumps, and cook for 10 minutes. Strain through a **chinois** into another saucepan or stockpot.
5. Temper the heavy cream with some **velouté.** Add the tempered cream to the velouté.
6. Adjust the seasonings and consistency, if needed.

COOKING TECHNIQUE:

Simmer

Simmer and Poach:

1. Heat the cooking liquid to the proper temperature.
2. Submerge the food product completely.
3. Keep the cooked product moist and warm.

GLOSSARY:

Temper: to equalize two extreme temperatures
Chinois: cone-shaped strainer
Velouté: velvety-textured sauce or soup

HAZARDOUS FOOD:

Heavy cream

NUTRITION:

Calories: 48
Fat: 3.5 g
Protein: 1.6 g

Sauce Velouté

YIELD: 1 GALLON. SERVING SIZE: TWO OUNCES.

INGREDIENTS:

8 ounces	Clarified butter
8 ounces	All-purpose flour, sifted
5 quarts	White (chicken, fish, veal, or vegetable) stock, heated to a boil (see page 12)
	Salt and ground white pepper, to taste

METHOD OF PREPARATION:

1. In a saucepan, heat the clarified butter over moderate heat. Gradually stir in sifted flour to make a blonde roux.
2. Using a wooden spoon, mix the roux thoroughly, and cook 5 to 6 minutes. The roux will be blonde with a light hazelnut odor.
3. Gradually whisk the stock into the roux. Heat the sauce to a boil; reduce the heat, and simmer for 20 minutes, or until the proper flavor and consistency are achieved. **Dépouiller** as necessary.
4. Season, to taste.
5. Strain through a fine **chinois** into a suitable container. Hold at 140° F, or cool to an internal temperature of 40° F or lower within 6 hours. Label, date, and refrigerate.
6. Reheat to a boil.

COOKING TECHNIQUE:

Simmer

Simmer and Poach:

1. Heat the cooking liquid to the proper temperature.
2. Submerge the food product completely.
3. Keep the cooked product moist and warm.

GLOSSARY:

Dépouiller: to skim impurities/grease
Chinois: cone-shaped strainer

HACCP:

Hold at 140° F or cool to an internal temperature of 40° F or lower.

NUTRITION:

Calories: 67.7
Fat: 4.49 g
Protein: 1.37 g

CHEF NOTE:

This velouté sauce is a base for different compound sauces.

Sauce Zingara

YIELD: 1 QUART. SERVING SIZE: TWO OUNCES.

COOKING TECHNIQUE:
Simmer

Simmer and Poach:
1. Heat the cooking liquid to the proper temperature.
2. Submerge the food product completely.
3. Keep the cooked product moist and warm.

HACCP:
Hold at 140° F or above.

NUTRITION:
Calories: 52.8
Fat: 4.68 g
Protein: 1.58 g

INGREDIENTS:

2 ounces	Clarified butter
2 ounces	White mushrooms, cleaned and sliced
2 ounces	Smoked ham, cut into thin strips
2 ounces	Cooked beef tongue, cut into thin strips
2 ounces	Dill pickle, cut into thin strips
1 ounce	Tomato purée
1½ pints	Demi-glacé, heated to a boil (see page 26)
⅛ teaspoon	Cayenne pepper

METHOD OF PREPARATION:
1. In a sauté pan, melt the butter, and sauté the mushrooms. Add the ham, tongue, and dill pickle.
2. Add the tomato purée and demi-glacé.
3. Heat to a boil, and simmer for about 20 minutes, until the proper flavor is achieved. Add the cayenne pepper to taste and season as needed. Serve immediately, or hold at 140° F or above.

CHEF NOTE:
This sauce can be served with grilled beef, steaks, roast veal, and braised beef roulades.

Scotch Honey Sauce

YIELD: 1 QUART. SERVING SIZE: TWO OUNCES.

COOKING TECHNIQUE:

Simmer

Simmer and Poach:

1. Heat the cooking liquid to the proper temperature.
2. Submerge the food product completely.
3. Keep the cooked product moist and warm.

HACCP:

Hold at 140° F or above.

NUTRITION:

Calories: 261
Fat: 8.68 g
Protein: .328 g

INGREDIENTS:

18 ounces	Honey
8 ounces	Drambuie (see chef note)
6 ounces	Butter, cut into ½-ounce pieces
4 each	Cloves

METHOD OF PREPARATION:

1. In a saucepan, combine the honey and Drambuie. Heat until the honey thins out; then add the cloves. Simmer for 5 minutes.
2. Slowly whisk the butter into the honey. Remove the cloves, and serve immediately, or hold at 140° F or above.

CHEF NOTE:

Drambuie is a golden, Scotch-based liqueur that is sweetened with heather honey and flavored with herbs.

Sweet and Sour Sauce (basic) No. 1

YIELD: 1 QUART. SERVING SIZE: TWO OUNCES.

COOKING TECHNIQUE:
Boil

Boil: (at sea level)
1. Bring the cooking liquid to a rapid boil.
2. Stir the contents, and cook the food product throughout.
3. Serve hot.

HACCP:
Hold at 140° F or above.

NUTRITION:
Calories: 67.9
Fat: 0.22 g
Protein: .419 g

INGREDIENTS:

8 ounces	White vinegar
3 ounces	Soy sauce
6 ounces	Granulated sugar
	Salt and ground white pepper, to taste
10 ounces	Pineapple juice
1½ ounces	Cornstarch
3 ounces	Cold water

METHOD OF PREPARATION:

1. In a saucepan, combine all of the ingredients, except the cornstarch and water, and heat to a boil. Cook until the sugar is dissolved.
2. Dissolve the cornstarch in the water.
3. Whisk the cornstarch mixture into the sauce to thicken it. Boil for an additional 5 minutes; then remove from the heat. Hold at 140° F or above.

Sweet and Sour Sauce No. 2

YIELD: 2 QUARTS. SERVING SIZE: 2 OUNCES.

COOKING TECHNIQUE:

Boil

Boil: (at sea level)

1. Bring the cooking liquid to a rapid boil.
2. Stir the contents, and cook the food product throughout.
3. Serve hot.

HACCP:

Hold at 140° F or above.

NUTRITION:

Calories: 31.9
Fat: .093 g
Protein: .253g

INGREDIENTS:

3 ounces	Sugar
12 ounces	Palm vinegar (or substitute rice wine vinegar)
12 ounces	Pineapple juice
4 ounces	Ketchup
½ ounce	Soy sauce
1 tablespoon	Worcestershire sauce
¼ teaspoon	Hot sauce
2 tablespoons	Cornstarch, dissolved in 4 tablespoons of cold water
1 pound	Canned pineapple chunks
1 (4-ounce)	Green bell pepper, washed, seeded, and cut into strips
1 (4-ounce)	Red bell pepper, washed, seeded, and cut into strips
6 each	Scallions, washed, trimmed, and cut into 3-inch pieces
	Salt, to taste

METHOD OF PREPARATION:

1. In a wok, combine the first eight ingredients, and heat the mixture to a boil. Cook for 2 minutes.
2. Add the remaining ingredients, return to a boil, and cook for 2 more minutes, or until the desired consistency is achieved. Serve immediately, or hold at 140° F or above.

Sweet and Sour Sauce No. 3

YIELD: 1 QUART. SERVING SIZE: TWO OUNCES.

COOKING TECHNIQUES:

Boil, Simmer

Boil: (at sea level)

1. Bring the cooking liquid to a rapid boil.
2. Stir the contents, and cook the food product throughout.
3. Serve hot.

Simmer and Poach:

1. Heat the cooking liquid to the proper temperature.
2. Submerge the food product completely.
3. Keep the cooked product moist and warm.

GLOSSARY:

Reduction: evaporation of liquid by boiling

NUTRITION:

Calories: 118
Fat: 4.76 g
Protein: 2.03 g

INGREDIENTS:

16 ounces	White duck stock, heated to a boil (see page 12)
8 ounces	Honey
8 ounces	Demi-glacé, heated to a boil (see page 26)
4 ounces	Pine nuts, roasted
1½ ounces	Cornstarch
1½ ounces	Red wine vinegar
	Salt and ground white pepper, to taste

METHOD OF PREPARATION:

1. Fortify the duck stock by boiling to **reduce** by half.
2. Add the honey, demi-glacé, and pine nuts.
3. Return to a boil.
4. Combine the cornstarch and vinegar, and add gradually to thicken the sauce to the desired consistency. Cook for 10 minutes, taste, and season.

Tartar Sauce

YIELD: 1 GALLON. SERVING SIZE: TWO OUNCES.

COOKING TECHNIQUE:
Not applicable

HACCP:
Refrigerate at 40° F or lower.

HAZARDOUS FOOD:
Mayonnaise

NUTRITION:
Calories: 349
Fat: 38.6 g
Protein: .741 g

INGREDIENTS:

3½ quarts	Mayonnaise (see page 35)
8 ounces	Dill pickles, diced
8 ounces	Onions, peeled and diced
1 ounce	Capers, drained and rinsed
1 ounce	Fresh chervil leaves, finely chopped
2 ounces	Fresh tarragon, finely chopped
1 ounce	Fresh parsley leaves, washed, excess moisture removed, and chopped
2 ounces	Freshly squeezed lemon juice

METHOD OF PREPARATION:

Mix all of the ingredients together in a bowl, and refrigerate at 40° F or lower until served.

CHEF NOTES:

1. Serve this sauce cold. It can be served with fried fish, seafood, grilled entrées, or vegetables.
2. As an alternative, for a Southern-type preparation, add 1 tablespoon of Cajun spice, or to taste, and mix well.
3. The original tartar sauce is made of hardboiled egg yolk-based mayonnaise.
4. Onions can be replaced with chopped chives.

Teriyaki Sauce

YIELD: 1 QUART. SERVING SIZE: TWO OUNCES.

COOKING TECHNIQUE:
Boil

Boil: (at sea level)

1. Bring the cooking liquid to a rapid boil.
2. Stir the contents, and cook the food product throughout.
3. Serve hot.

GLOSSARY:

Flambé: to flame
Chinois: cone-shaped strainer

HACCP:

Chill to 40° F or below.

NUTRITION:

Calories: 40.3
Fat: .416 g
Protein: 1.83 g

INGREDIENTS:

12 ounces	Mirin (see chef note)
12 ounces	Soy sauce
12 ounces	White chicken stock, heated to a boil (see page 12)
2 tablespoons	Onion, finely minced
1 teaspoon	Ginger root, finely minced
4 cloves	Garlic, peeled and mashed into a purée

METHOD OF PREPARATION:

1. Heat the mirin in a wok or saucepan over moderate heat. **Flambé,** and allow the alcohol to burn off.
2. Add the soy sauce, stock, onion, ginger root, and garlic to the wok. Heat to a boil; then strain through a **chinois.**
3. Cool to 40° F or below within 6 hours, and refrigerate until needed.

CHEF NOTE:

Mirin is a low-alcohol, sweet, golden wine made from glutinous rice. Essential to the Japanese cook, mirin adds sweetness and flavor to a variety of dishes, sauces, and glazes. It is available in all oriental markets and the gourmet section of some supermarkets. Mirin also is known simply as rice wine.

Tomato Sauce

YIELD: 1 GALLON. SERVING SIZE: AS NEEDED.

COOKING TECHNIQUES:
Sauté, Simmer

Sauté:
1. Heat the sauté pan to the appropriate temperature.
2. Evenly brown the food product.
3. For a sauce, pour off any excess oil, reheat, and deglaze.

Simmer and Poach:
1. Heat the cooking liquid to the proper temperature.
2. Submerge the food product completely.
3. Keep the cooked product moist and warm.

GLOSSARY:
Sweat: to sauté under a cover
Chinois: cone-shaped strainer

HACCP:
Hold at 140° F, or cool immediately to 40° F.

NUTRITION:
Calories: 22.4
Fat: 1.03 g
Protein: .633 g

CHEF NOTES:
1. This is a basic tomato sauce. If a specific taste is desired, simply adjust the quantities of the ingredients accordingly.
2. This tomato sauce must be cooked until the acidity and bitterness of the tomato products are removed. The heat source should be controlled carefully, so that the sauce is not scorched or burned. The best pot in which to cook tomato sauce is one made of stainless steel. Tomato sauce or products should never be cooked or stored in aluminum pots.
3. Basic tomato sauce has many variations. Some variations include roast pork or veal bones. Others include a mirepoix.
4. An alternative tomato sauce is marinara sauce (see page 34).
5. Salt pork or bacon fat can replace the oil.

INGREDIENTS:

2 ounces	Olive oil
1 pound	Onions, peeled and diced
8 to 10 cloves	Fresh garlic, peeled and mashed into a paste
3 each	Bay leaves
2 tablespoons	Basil leaves, dried
½ teaspoon	Red pepper flakes, crushed
4 ounces	Tomato paste
1 #10 can (6 pounds)	Tomatoes, crushed
3 quarts	Vegetable stock or water, heated to a boil (see page 11)
	Salt and freshly ground black pepper, to taste

METHOD OF PREPARATION:

1. Heat the oil in a heavy pot, add the onions and garlic, and **sweat** until the onions are translucent.
2. Add the remaining ingredients, and heat to a boil; reduce to a simmer.
3. Simmer for approximately 45 minutes, stirring, and dépouiller as needed.
4. Remove from the heat, and strain through a **chinois** into a suitable container. Taste, and adjust the seasoning as needed.
5. Hold at 140° F, or cool immediately to 40° F; cover and refrigerate.

Vermouth Sauce

YIELD: 20 OUNCES SERVING SIZE: TWO OUNCES.

INGREDIENTS:

2 ounces	Olive oil
4 ounces	Scallions, washed, trimmed, and cut into ¼-inch slices
4 ounces	Dry vermouth
4 ounces	Tomato purée
2 ounces	Fish stock, concentrated (see page 11)
8 ounces	Heavy cream
4 ounces	Chilled butter, cut into ½-ounce pieces
3 ounces	Fresh parsley, washed, excess moisture removed, and chopped
	Salt and ground white pepper, to taste

METHOD OF PREPARATION:

1. In a sauté pan, heat the olive oil, and sauté the scallions until wilted. Add the vermouth, heat to a boil, and **reduce** by half.
2. Add the tomato purée and fish stock, and simmer for 5 minutes.
3. **Temper** the cream, and add to the sauce; continue to simmer until reduced to a desired consistency.
4. Piece by piece, whisk the butter into the sauce; add the parsley, and season to taste. Hold at 140° F or above.

COOKING TECHNIQUES:

Sauté, Boil, Simmer

Sauté:

1. Heat the sauté pan to the appropriate temperature.
2. Evenly brown the food product.
3. For a sauce, pour off any excess oil, reheat, and deglaze.

Boil: (at sea level)

1. Bring the cooking liquid to a rapid boil.
2. Stir the contents, and cook the food product throughout.
3. Serve hot.

Simmer and Poach:

1. Heat the cooking liquid to the proper temperature.
2. Submerge the food product completely.
3. Keep the cooked product moist and warm.

GLOSSARY:

Reduction: evaporation of liquid by boiling
Temper: to equalize two extreme temperatures

HACCP:

Hold at 140° F or above.

HAZARDOUS FOOD:

Heavy cream

NUTRITION:

Calories: 135
Fat: 13.1 g
Protein: .972 g

White Horseradish Sauce

YIELD: 1 QUART SERVING SIZE: TWO OUNCES.

INGREDIENTS:

4 ounces	Fresh horseradish, peeled, washed, and grated
3 ounces	White wine
1½ pints	Béchamel sauce, heated to 165° F, strained (see page 16)
1¼ teaspoons	Cayenne pepper
	Salt, to taste
4 ounces	Heavy cream

METHOD OF PREPARATION:

1. Cover the horseradish with wine, and refrigerate at 40° F or below overnight.
2. In a saucepan, heat the grated horseradish and white wine, and reduce by simmering until nearly dry.
3. Add the béchamel sauce to the horseradish, and simmer for 5 minutes. Add the cayenne pepper and salt, to taste.
4. **Temper** the heavy cream, and add to the sauces. Serve immediately, or hold at 140° F or above.

COOKING TECHNIQUE:

Simmer

Simmer and Poach:

1. Heat the cooking liquid to the proper temperature.
2. Submerge the food product completely.
3. Keep the cooked product moist and warm.

GLOSSARY:

Temper: to equalize two extreme temperatures

HACCP:

Refrigerate at 40° F or below.
Hold at 140° F or above.

HAZARDOUS FOOD:

Heavy cream

NUTRITION:

Calories: 85.7
Fat: 7.37 g
Protein: .879 g

CHEF NOTES:

1. Preparation should begin at least 1 day in advance.
2. If the horseradish is soaked in cold water overnight before use, the sharpness will be removed, and it will become more crisp, which will enable easier grating.
3. This sauce is English in origin, traditionally served with boiled beef brisket or short ribs of beef. In some European countries, this sauce is flavored with vinegar and sugar.

White Tarragon Sauce

YIELD: 1 QUART   SERVING SIZE: TWO OUNCES.

INGREDIENTS:

8 ounces	Dry white wine
2 ounces	Shallots, peeled and diced
8 ounces	Chicken velouté, heated to a boil (see page 68)
1 tablespoon	Fresh tarragon, chopped
	Salt and ground white pepper, to taste
8 ounces	Heavy cream

METHOD OF PREPARATION:

1. In a saucepan, combine the white wine and shallots. Heat to a boil, and reduce the liquid by half.
2. Add the chicken velouté, and heat to a boil; reduce the heat to a simmer, and cook for 10 minutes. Strain the sauce through a **chinois** into another saucepan.
3. Add the fresh tarragon, and let the sauce simmer an additional 5 minutes. Season, to taste.
4. **Temper** the heavy cream, and add it to the sauce. Serve immediately, or hold at 140° F or above.

COOKING TECHNIQUE:

Simmer

Simmer and Poach:

1. Heat the cooking liquid to the proper temperature.
2. Submerge the food product completely.
3. Keep the cooked product moist and warm.

GLOSSARY:

Chinois: cone-shaped strainer
Temper: to equalize two extreme temperatures

HACCP:

Hold at 140° F or above.

HAZARDOUS FOOD:

Heavy cream

NUTRITION:

Calories: 26.4
Fat: .989 g
Protein: .39 g

CHEF NOTES:

1. Many variations of this sauce can be made; for example, chicken stock can be used in place of velouté or in the reduction process. The pan can be deglazed with white wine and heavy cream, reduced, and then garnished with fresh tarragon.
2. Fresh tarragon is an herb with a licorice or anisette flavor. It can accompany any white meat.

Yogurt Horseradish Sauce

YIELD: 1 GALLON SERVING SIZE: TWO OUNCES.

COOKING TECHNIQUE:
Not applicable

HACCP:
Refrigerate at 40° F or below

HAZARDOUS FOOD:
Yogurt

NUTRITION:
Calories: 31.2
Fat: 1.21 g
Protein: 1.56 g

INGREDIENTS:

5 pounds	Yogurt
1 pound	Prepared horseradish
2 ounces	Freshly squeezed lemon juice
3 ounces	Fresh parsley, washed, excess moisture removed, and chopped

METHOD OF PREPARATION:

Combine all of the ingredients in a stainless steel bowl, and blend until smooth. Label, date, and refrigerate at 40° F or below until served.

CHEF NOTE:
This sauce can be served with brisket, corned beef, or prime rib.

2

Appetizers

2

Appetizers

An ounce of caviar rolled in smoked salmon will produce mumbles of ecstasy.

Andre Launay
Caviare and After

As the eyes of the culinary world shift toward the United States and its chefs, we have a tremendous responsibility not only to embrace culinary education but also to share the results of such education. We are the new chefs of the world! Our products and techniques are creating excitement in kitchens worldwide. As custodians of our culinary future, we must continue to create unique and contemporary plates but never forget our classical heritage. It is the knowledge of these kitchen basics that allows us to evolve and create. Here at Johnson & Wales University, all men and women seeking a future in culinary arts are ensured the basic knowledge necessary for a successful career. Learn these basics well, and realize what great opportunity awaits one who wears the toque with pride and professionalism. Most of all, remember—You are the new chefs of the world!

John D. Folse, CEC, AAC
Distinguished Visiting Chef
March 24–26, 1991
Owner/Executive Chef, A Taste of Louisiana

The HACCP Process

The following recipe illustrates how appetizers flow through the HACCP process.

Asparagus Quiche

Receiving

- Vegetables (asparagus) fresh—Check for signs of spoilage or pest infestation.
- Vegetables (asparagus) frozen—temperature at or below 0° F (17.8° C) with no signs of thawing
- Cheese, swiss or cheddar—Check to see that each type has its characteristic flavor and texture, as well as uniform color. Cheese should be delivered at 40° F (4.4° C) or lower.
- Pie shells—packaging intact
- Whole shell eggs—Shells should not be cracked or dirty. Containers should be intact. Whole shell eggs should be delivered at 40° F (4.4° C) or lower.
- Milk, whole—should have a sweetish taste. Container should be intact. Milk should be delivered at 40° F (4.4° C) or lower.
- Spices (salt and pepper)—packaging intact

Storage

- Vegetables (asparagus) fresh—Store under refrigerated storage, with a product temperature not to exceed 40° F (4.4° C). Store above and away from raw, potentially hazardous foods.
- Vegetables (asparagus) frozen—store in freezer at or below 0° F (−17.8° C).
- Cheese—Store under refrigerated storage, with a product temperature not to exceed 40° F (4.4° C). Store above and away from raw, potentially hazardous foods.
- Pie shells—Store in freezer at 0° F (−17.8° C) until needed.
- Eggs—Store under refrigeration at 40° F (4.4° C) or lower. Store above and away from raw, potentially hazardous foods.
- Milk—Store under refrigeration at 40° F (4.4° C) or lower. Store above and away from raw, potentially hazardous foods.
- Spices (salt and pepper)—Store in dry storage at 50° F (10° C) with a relative humidity of 50% to 60%. Keep dry.

Thawing Asparagus

- Thaw the package of asparagus under refrigeration at 40° F (4.4° C) or lower.

Preparation and Cooking

- Clean and wash the fresh asparagus, or thaw the frozen asparagus.
- Chop the asparagus on a clean and sanitized cutting board with clean and sanitized utensils.
- Grate the cheese, and add to the bottom of the pie shells.
- Portion the asparagus over the cheese.
- Beat the eggs slightly; then add the milk, salt, and pepper.
- Portion the mixture over the asparagus.
- Bake in a 375° F oven about 45 minutes. Let stand for 10 minutes before cutting.
- Bake the quiche to internal temperature of 145° F (62.8° C) or higher.

Holding and Service

- Hold at 140° F (60° C) or higher.
- Serve immediately.

Note: Any product that contains eggs must be disposed of at the end of the day.

Nutritional Notes

Whether they are hot or cold, hors d'oeuvre or canapés, appetizers should appeal to the eye as well as the palate, serving to enhance the appetite, not drown it. Foods high in fat, such as full-fat cheeses, sour cream, and puff pastry, may be replaced with lower-fat alternatives, both to prevent coating the palate and drowning the appetite and to provide selections in line with current nutritional guidelines. Crudités served with dips made from nonfat yogurt, nonfat cottage cheese, or low-fat ricotta cheese as the base in place of the traditional sour cream, cream cheese, or cream can provide an exciting blend of colors, textures, flavors, and shapes while meeting the nutritional goals of lowering fat and increasing fiber.

Other recommendations for lowering fat, saturated fat, and salt while increasing fiber in appetizers include the following:

- *Use legume-based dips or fillings, such as hummus or vegetarian chili.*
- *Grill, bake, broil, or steam rather than fry. Use lean meats in place of those high in fat. For example, serve baked marinated chicken fingers in place of fried chicken wings, or marinated vegetables in place of fried vegetables.*
- *Replace full-fat dairy ingredients with nonfat or low-fat varieties. For example, use nonfat yogurt that has been drained in a filter, or blended nonfat cottage cheese with a bit of lemon juice added, in place of sour cream.*
- *Offer fruit or low-sodium vegetable juice cocktails.*
- *Reduce fat in recipes and replace saturated fat, such as butter, partly or completely with unsaturated fat, such as olive oil or canola oil.*
- *Reduce salt called for in recipes and limit use of high-salt ingredients such as MSG, caviar, and salted anchovies.*
- *Prepare forcemeats in-house, using less fat and leaner cuts. Offer low-fat fish sausages.*
- *Replace high-fat pastry with layers of filo dough sprayed lightly with unsaturated oil.*
- *Use whole-grain doughs and breads whenever possible.*

Baked Eggs

YIELD: 10 SERVINGS. SERVING SIZE: ONE EGG.

COOKING TECHNIQUES:

Sauté, Steam

Sauté:

1. Heat the sauté pan to the appropriate temperature.
2. Evenly brown the food product.
3. For a sauce, pour off any excess oil, reheat, and deglaze.

Steam: (Traditional)

1. Place a rack over a pot of water.
2. Prevent steam vapors from escaping.
3. Shock or cook the food product throughout.

HAZARDOUS FOOD:

Eggs

NUTRITION:

Calories: 234
Fat: 16.6 g
Protein: 9.3 g

INGREDIENTS:

4 ounces	Olive oil
12 cloves	Garlic, peeled and finely minced
6 pounds	Canned, crushed tomatoes, drained
½ teaspoon	Ground cumin
½ teaspoon	Spanish paprika
¼ teaspoon	Cayenne pepper
	Salt, to taste
10 each	Eggs
3 ounces	Fresh parsley, washed, excess moisture removed, and chopped

METHOD OF PREPARATION:

1. In a sauté pan, heat the olive oil, add the garlic, and sauté 1 minute.
2. Add the drained tomatoes, and sauté on low heat for 10 minutes to cook the tomatoes.
3. Add the seasonings, and continue to sauté for 5 more minutes.
4. Transfer the tomatoes into a half hotel pan. Using a spoon, hollow out 10 evenly divided spaces. Break the eggs one at a time into a cup, and transfer into each space.
5. Cover the pan tightly with aluminum foil so that no steam escapes, and finish cooking on the stove for about 3 to 5 minutes. Whites should be firm, but yolks should remain soft.
6. Serve immediately on a preheated plate, garnished with parsley sprinkled over the sauce.

CHEF NOTE:

This is a Spanish-style preparation that was inherited from the Moors who occupied Spain for hundreds of years.

Bouchée à la Reine

YIELD: 10 SERVINGS. SERVING SIZE: ONE BOUCHÉE.

INGREDIENTS:

2 ounces	Butter
3 ounces	Shallots, peeled and **brunoise**
15 ounces	Chicken breast, poached and diced
10 ounces	Mushrooms, washed, trimmed, and quartered
20 ounces	Chicken **velouté,** seasoned, heated to a boil (see page 68)
	Salt and ground white pepper, to taste
10 each	Bouchée (puff pastry shells), prebaked (see page 87)

METHOD OF PREPARATION:

1. Preheat the oven to 275° F.
2. In a sauté pan, melt the butter, and sauté the shallots until translucent.
3. Add the chicken breast, and sauté for 1 to 2 minutes; then add the mushrooms.
4. Continue to sauté until the mushrooms are soft. Add the velouté, and heat to 165° F. Season, to taste. Hold at 140° F or above.
5. Warm the puff pastry shells in the oven.
6. To serve, place the pastry shell on a preheated plate, and ladle à la Reine into the shell, allowing enough to overflow onto the plate.

COOKING TECHNIQUES:

Bake, Sauté, Simmer

Bake:

1. Preheat the oven.
2. Place the food product on the appropriate rack.

Sauté:

1. Heat the sauté pan to the appropriate temperature.
2. Evenly brown the food product.
3. For a sauce, pour off any excess oil, reheat, and deglaze.

Simmer and Poach:

1. Heat the cooking liquid to the proper temperature.
2. Submerge the food product completely.
3. Keep the cooked product moist and warm.

GLOSSARY:

Brunoise: ⅛-inch dice
Velouté: velvety-textured sauce or soup

HAZARDOUS FOOD:

Chicken breast

NUTRITION:

Calories: 186
Fat: 10.7 g
Protein: 15 g

CHEF NOTES:

1. A bouchée is a small savory puff pastry shell that may be filled with a variety of foods. The word *bouchée* means "a mouthful."
2. À la Reine is named after Louis XVI's wife and Queen Marie Antoniette and refers to using chicken in some form.

Puff Pastry Shells (Bouchée)

YIELD: 10 BOUCHÉES. SERVING SIZE: ONE EACH.

COOKING TECHNIQUE:

Bake

Bake:

1. Preheat the oven.
2. Place the food product on the appropriate rack.

HAZARDOUS FOOD:

Egg whites
Egg yolks

NUTRITION:

Calories: 253
Fats: 17.5 g
Protein: 3.4 g

INGREDIENTS:

1 pound	Puff pastry, fresh or frozen
	Egg whites, as needed, lightly beaten
	Egg yolks, as needed, lightly beaten

METHOD OF PREPARATION:

1. Preheat the oven to 450° F.
2. If using fresh dough, roll to ⅓-inch thickness, and/or if using frozen dough, stick two sheets together with egg white, and roll lightly. The thickness will be close to the thickness of the fresh dough.
3. Cut two rounds of dough with a round-straight or fluted 3-inch cutter.
4. With a 2-inch, round-straight cutter, cut out the center of one round only. Brush the ring with egg yolk.
5. Place the full round of dough on a sheet pan lined with parchment paper. Place the cut round of dough on top of the full circle, egg yolk-side down, and press gently to stick both pieces together.
6. Poke the center of the bottom piece with a fork.
7. Place a form 3 inches high in each corner of the sheet pan. Place an inverted glazing screen on top.
8. Bake in the oven for 7 minutes; lower the temperature to 425° F, and bake for 8 more minutes, or until cooked throughout and lightly browned.

Guidelines:

1. Roll out the dough in four directions, not just one, or the dough will rise unevenly.
2. The dough should be cool when working it.
3. Apply the egg yolk carefully. Do not allow drippings on the side of the dough, or it will not rise.
4. Do not grease the sheet pan. Dampen the pan with water before placing the dough on it, or have a small pan of water in the oven. Steam helps the dough to rise during baking.
5. Chill the bouchée at 40° F or below for 10 minutes before baking.
6. Do not open the oven during the first 5 minutes of baking.
7. Smaller bouchées will require less baking time.
8. Store the bouchées in a cool, dry place or freeze.

CHEF NOTES:

1. After working with the dough, it should rest before baking.
2. If deeper bouchée is needed, the thickness of the rim must be twice the thickness of the base.
3. Before placing pan in the oven, place sheet pan paper on top of the bouchée. It will prevent the bouchée from collapsing.

Cheese and Spinach Bourekas

YIELD: 10 SERVINGS. SERVING SIZE: THREE OUNCES.

INGREDIENTS:

2 ounces	Olive oil
4 ounces	Onions, peeled and dice
1 pound	Spinach, washed, stemmed, and drained of excess moisture
	Oil, as needed
1 pound	Feta cheese, crumbled
12 sheets	Phyllo dough, thawed

METHOD OF PREPARATION:

1. Preheat the oven to 425° F.
2. In a sauté pan, heat the oil, add the onions, and sauté until translucent. Add the spinach, and sauté until the spinach is wilted.
3. Remove, and transfer the mixture to a bowl; cool slightly. Add the feta cheese, and mix well.
4. Fill the bottom of a hotel pan with ½ inch of oil. Dip the phyllo dough, three sheets at a time, in the oil, and place them flat on a clean towel.
5. Arrange six leaves together on the towel, and add the spinach mixture along the length of the dough, like a sausage.
6. Roll the phyllo dough, with the aid of the towel, into a roulade, and transfer to a parchment-lined sheet pan, with the seam-side down. Repeat to make a second roulade.
7. Thirty minutes before service, place in the oven, and bake the bourekas for approximately 20 minutes, or until the dough is golden brown. Hold at 140° F for 10 minutes. To serve, slice each roulade into five portions, place a portion on a preheated plate, and serve immediately. This item cannot be held in a **bain-marie.**

COOKING TECHNIQUES:

Sauté, Bake

Sauté:

1. Heat the sauté pan to the appropriate temperature.
2. Evenly brown the food product.
3. For a sauce, pour off any excess oil, reheat, and deglaze.

Bake:

1. Preheat the oven.
2. Place the food product on the appropriate rack.

GLOSSARY:

Bain-marie: hot-water bath

HACCP:

Hold at 140° F.

NUTRITION:

Calories: 259
Fat: 23.5 g
Protein: 8.01 g

CHEF NOTE:

This recipe is a national appetizer in Greece.

Duxelle Stuffed Mushrooms

(Champignons Farcis à la Duxelle)

YIELD: 10 SERVINGS. SERVING SIZE: FOUR MUSHROOMS.

INGREDIENTS:

	Clarified butter, as needed
4 ounces	Onions, peeled and diced
4 ounces	Shallots, peeled and diced
1½ pounds	Small mushrooms, cleaned, stemmed, trimmed, and diced
8 ounces	White wine
3 ounces	Fresh parsley, washed, excess moisture removed, and chopped
	Salt and freshly ground black pepper, to taste
40 each	Large mushrooms, cleaned and stems removed

METHOD OF PREPARATION:

1. In a large, heavy-bottomed sauté pan, heat the butter, and sauté the onions and shallots. Add the mushrooms, and continue to sauté for 4 to 5 minutes. Then add the white wine.
2. **Reduce** the mushroom and wine mixture until it reaches a dry consistency. When dry, add the parsley, and season, to taste. Cool slightly.
3. Using a pastry bag with a plain tube, pipe the mushroom mixture into the large mushroom caps. Place the mushrooms on a baking tray, and bake until lightly browned.

COOKING TECHNIQUES:

Sauté, Simmer, Bake

Sauté:

1. Heat the sauté pan to the appropriate temperature.
2. Evenly brown the food product.
3. For a sauce, pour off any excess oil, reheat, and deglaze.

Simmer and Poach:

1. Heat the cooking liquid to the proper temperature.
2. Submerge the food product completely.
3. Keep the cooked product moist and warm.

Bake:

1. Preheat the oven.
2. Place the food product on the appropriate rack.

GLOSSARY:

Reduction: evaporation of liquid by boiling

NUTRITION:

Calories: 65.6
Fat: .685 g
Protein: 3.63 g

CHEF NOTES:

1. If the mushroom mixture is too moist, add some bread crumbs.
2. Crabmeat or shrimp may be added for variety.
3. For large quantities, bind with seasoned béchamel sauce.

COOKING TECHNIQUE:
Sauté

Sauté:

1. Heat the sauté pan to the appropriate temperature.
2. Evenly brown the food product.
3. For a sauce, pour off any excess oil, reheat, and deglaze.

GLOSSARY:

Brunoise: ⅛-inch dice
Concassé: peeled, seeded, roughly chopped

HACCP:

Chill to 40° F or below.

NUTRITION:

Calories: 309
Fat: 27.2 g
Protein: 2.93 g

CHEF NOTE:

This dish is served as a cold appetizer, or it can be used to stuff tomatoes or as a part of an antipasto dish.

Eggplant and Vegetable Relish

(Caponata)

YIELD: 10 SERVINGS. SERVING SIZE: THREE OUNCES.

INGREDIENTS:

2 pounds	Eggplant, peeled and diced into ½-inch cubes
	Salt, as needed
4 ounces	Olive oil
6 ounces	Onions, peeled and diced
6 ounces	Tomato **concassé**
3 ounces	Golden raisins
3 ounces	Pitted black olives, sliced
3 ounces	Pitted green olives, sliced
1 ounce	Capers, rinsed of brine
1 ounce	Pine nuts
4 ounces	Celery stalks, washed, trimmed, and **brunoise**
1 ounce	Red wine vinegar
	Salt and freshly ground black pepper, to taste

METHOD OF PREPARATION:

1. Place the eggplant in a nonreactive bowl. Mix with salt, press down, and set aside.
2. In a sauté pan, heat 2 ounces of the oil; add the onions, and sauté until translucent. Add the tomatoes, and continue to sauté for 2 more minutes; then add the raisins, olives, capers, and pine nuts. Sauté for 2 more minutes, and set aside.
3. Rinse and drain the eggplant, and squeeze out the excess moisture.
4. In a separate sauté pan, heat the remaining olive oil. Add the eggplant, and sauté until browned; then add the celery, and sauté until tender.
5. Add the vinegar and tomato mixture, and mix well.
6. Continue to cook until all of the liquid evaporates.
7. Adjust the seasoning, remove from the heat, and chill to 40° F or below. Serve at room temperature.

Egg Rolls

YIELD: 10 SERVINGS. SERVING SIZE: TWO EGG ROLLS.

INGREDIENTS:

1 ounce	Soybean oil
1 pound	Ground pork
4 ounces	Mushrooms, washed, stems trimmed, and thinly sliced
1 pound	Cabbage, washed, cored, and cut **chiffonade**
1 pound	Celery, washed, trimmed, and diced
8 ounces	Soy sauce
2 ounces	Sake
1 ounce	Sugar
1 pound	Bean sprouts, washed and drained
20 each	Egg rolls, 6-inch wrappers
2 each	Eggs, lightly beaten
1 quart	Peanut oil
3 ounces	Mustard sauce (see chef notes)

METHOD OF PREPARATION:

1. In a wok or heavy pan heat the soybean oil over high heat. Add the ground pork, and stir-fry for 2 minutes.
2. Add the mushrooms, cabbage, celery, soy sauce, sake, and sugar, and stir-fry for another minute. Drain the excess liquid, and transfer the contents of the wok to a bowl.
3. Add the bean sprouts to the mixture, and chill in the refrigerator at 40° F or below.
4. For each egg roll, shape approximately 2 ounces of filling into a 4-inch by 1-inch cylinder. Place the filling diagonally across the center of a wrapper. Lift the lower triangular flap over the filling, and tuck the point underneath on the far side, leaving the upper point of the wrapper exposed. Bring each of the two small end flaps, one at a time, up to the top of the enclosed filling, and press the points down firmly. Brush the upper and exposed triangle of dough with lightly beaten egg. Roll the wrapper into a neat package. The beaten egg will seal the edges and keep the wrapper intact.
5. Place the filled egg rolls on a plate, and cover with a dry kitchen towel.
6. To fry, heat the peanut oil in a wok over moderate heat. Bring the temperature to about 375° F, and deep-fry the egg rolls for 3 to 4 minutes, or until golden brown and crisp. Place on absorbent paper to drain.
7. Serve on a preheated plate, with mustard sauce on the side.

COOKING TECHNIQUES:

Stir-Fry, Deep-Fry

Stir-Fry:

1. Heat the oil in a wok until hot but not smoking.
2. Keep the food in constant motion; use the entire cooking surface.

Deep-Fry:

1. Heat the frying liquid to the proper temperature.
2. Submerge the food product completely.
3. Fry the product until it is cooked throughout.

GLOSSARY:

Chiffonade: ribbons of leafy greens

HACCP:

Refrigerate at 40° F or below
Hold at 140° F or above.

HAZARDOUS FOOD:

Ground pork
Eggs

NUTRITION:

Calories: 592
Fat: 48 g
Protein: 18.2 g

CHEF NOTES:

1. Mustard sauce is Chinese dry mustard and cold water combined to a thick consistency.
2. Sweet and sour sauce no. 1 or no. 2 can also be served with egg rolls (see pages 71 and 72). Hold at 140° F or above.
3. Fry egg rolls to order.

Semolina Dumplings over Leaf Spinach, Beurre Blanc

(Noques de Semoule sur Epinard)

YIELD: 10 SERVINGS. SERVING SIZE: FOUR OUNCES.

COOKING TECHNIQUE:
Sauté

Sauté:
1. Heat the sauté pan to the appropriate temperature.
2. Evenly brown the food product.
3. For a sauce, pour off any excess oil, reheat, and deglaze.

HAZARDOUS FOOD:
Milk
Egg yolks

NUTRITION:
Calories: 420
Fat: 33 g
Protein: 9.2 g

INGREDIENTS:

1 quart	Milk
3 ounces	Butter
8 ounces	Semolina
2 each	Egg yolk
	Salt and nutmeg, to taste
12 ounces	Beurre blanc (see page 17)
6 ounces	Clarified butter
2 pounds	Spinach, washed, reserved, and drained

METHOD OF PREPARATION:

1. Add the butter to the milk, and heat to a full boil.
2. Pour semolina into the boiling milk.
3. Cook over low heat for 10 minutes, scraping the bottom of the pot with a wood spatula to avoid burning.
4. Take the mixture off of the fire, and incorporate the two egg yolks.
5. Cool on a buttered half-sheet pan, and refrigerate.
6. Cut into round shapes, and sauté in 2 ounces of clarified butter to a golden color.
7. Sauté the spinach in the remaining clarified butter, and season.
8. Spoon the sautéed leaf spinach into the center of the plate. Place the sautéed dumplings over the spinach. Drizzle beurre blanc around the dumplings.

Meat Pies with Mango Relish

(Empanadas con Mangoes)

YIELD: 10 SERVINGS. SERVING SIZE: ONE EACH.

COOKING TECHNIQUES:
Sauté, Bake

Sauté:
1. Heat the sauté pan to the appropriate temperature.
2. Evenly brown the food product.
3. For a sauce, pour off any excess oil, reheat, and deglaze.

Bake:
1. Preheat the oven.
2. Place the food product on the appropriate rack.

GLOSSARY:
Render: to melt fat
Brunoise: ⅛-inch dice

HACCP:
Cook to an internal temperature of 165° F or above; hold at 140° F.

HAZARDOUS FOOD:
Ground pork
Eggs

NUTRITION:
Calories: 1014
Fat: 67.6 g
Protein: 37 g

CHEF NOTE:
These meat pies are staples of Mexico, but popular all over Central and South America.

INGREDIENTS:

3 pounds	Frozen puff pastry
4 each	Eggs, lightly beaten
5 each	Hard-boiled eggs, cut in half
2 each	Eggs beaten with 1 tablespoon of water
	Mango relish

Stuffing:

2 ounces	**Rendered** bacon fat
2 pounds	Ground pork
2 ounces	Onions, peeled and diced
4 cloves	Garlic, peeled and finely minced
2 ounces	Pitted green olives, coarsely chopped
2 ounces	Smoked ham, **brunoise**
½ teaspoon	Cumin
½ teaspoon	Chili powder
	Salt, to taste

METHOD OF PREPARATION:

1. Heat the bacon fat, add the pork, and sauté until the meat crumbles.
2. Add the onions and garlic, and sauté until the meat is cooked throughout, or to an internal temperature of 165° F. Remove, and cool to room temperature.
3. Preheat the oven to 425° F.
4. Roll out the puff pastry to ¼-inch thickness, and cut into 6-inch rounds.
5. Combine the cooled meat with the olives, ham, and seasonings. Add the eggs, and mix well.
6. Place one scoop, approximately 3 ounces, of stuffing on half of the puff pastry round; add half a hard-boiled egg, and press the egg down into the filling.
7. Moisten the edges of the pastry, and fold over the stuffing; then press to close the edges.

8. Place on a baking pan, brush with the eggwash, and let rest for 20 minutes.
9. Bake in the oven until golden brown and the pie is baked throughout, or to an internal temperature of 165° F or above.
10. Serve immediately, or hold at 140° F.

Mussels, Mariner's Style

(Moules Marinières)

YIELD: 10 SERVINGS. SERVING SIZE: EIGHT TO TWELVE OUNCES.

COOKING TECHNIQUE:

Simmer

Simmer and Poach:

1. Heat the cooking liquid to the proper temperature.
2. Submerge the food product completely.
3. Keep the cooked product moist and warm.

GLOSSARY:

Velouté: velvety-textured sauce or soup

HAZARDOUS FOOD:

Mussels

NUTRITION:

Calories: 290
Fat: 16 g
Protein: 24.1 g

INGREDIENTS:

4 quarts	Fresh mussels, scrubbed, debearded, and washed
6 ounces	Chablis wine
3 ounces	Shallots, peeled and finely chopped
	Salt and pepper, to taste
5 ounces	Butter
2½ tablespoons	Fresh parsley, washed, excess moisture removed, and chopped

METHOD OF PREPARATION:

Place all the ingredients (except the parsley) in a 3-gallon stew pot. Cook on high heat for 6 to 8 minutes or until all mussels are open. Stir frequently with a spoon. Sprinkle on the chopped parsley just before serving. Serve in soup plates, with some cooking liquid ladled over the mussels.

CHEF NOTE:

Farm-grown mussels are generally meatier, more flavorful, cleaner, and less sandy (gritty) than those caught in the ocean. Classic "Moules Marinières" is a preparation of cultured mussels that is cooked in white wine, shallots, and parsley. The cooking liquid is then thickened with cream. Once cooked, half of the shell is removed from each mussel, and the mussel **velouté** is poured over the mussel in the shell.

Oysters Rockefeller

(Huitres Florentine)

YIELD: 10 SERVINGS. SERVING SIZE: FIVE OYSTERS.

COOKING TECHNIQUE:
Bake

Bake:
1. Preheat the oven.
2. Place the food product on the appropriate rack.

GLOSSARY:
Bain-marie: hot-water bath

HACCP:
Hold at 140° F.

HAZARDOUS FOOD:
Oysters

NUTRITION:
Calories: 221
Fat: 15.4 g
Protein: 8.18 g

INGREDIENTS:

50 each	Oysters, cleaned and shucked; remove and discard top shell
2 pounds	Spinach filling (see below)
1 quart	Hollandaise sauce, heated to 140° F (see page 30)
5 each	Lemons
4 ounces	Kale, washed and trimmed
10 each	Cherry tomatoes, cut in half

METHOD OF PREPARATION:

1. Preheat the oven to 350° F.
2. On a sheet pan, lay out the oysters. Top each oyster with spinach filling.
3. Bake the oysters until they become firm, about 8 minutes. Remove.
4. Cover the oysters with hollandaise sauce, and glaze to order under the broiler or salamander.
5. Cut the lemons into crowns for garnish.
6. To serve, arrange the oysters on a preheated plate; then garnish with a lemon crown, kale, and a cherry tomato. Serve immediately, or hold at 140° F.

Spinach Filling

INGREDIENTS:

2 ounces	Clarified butter
4 cloves	Garlic, peeled and finely minced
2 pounds	Frozen spinach, thawed, chopped, and excess moisture removed
4 ounces	Béchamel sauce (see page 16)
½ teaspoon	Nutmeg
1½ ounces	Pernod
	Salt and freshly ground black pepper, to taste

METHOD OF PREPARATION:

1. In a sauté pan, heat the clarified butter, and sauté the garlic.
2. Add the chopped spinach, béchamel sauce, nutmeg, and Pernod. Sauté until heated throughout and well mixed.
3. Season, to taste. Remove from the heat, and cool quickly to reserve color and texture.

CHEF NOTE:
Oysters can be held for a few minutes in a **bain-marie** but are best served immediately.

COOKING TECHNIQUE:
Boil

Boil: (at sea level)

1. Bring the cooking liquid to a rapid boil.
2. Stir the contents, and cook the food product throughout.
3. Serve hot.

HACCP:
Hold at 140° F or above.

HAZARDOUS FOOD:
Sour cream
Eggs
Egg yolk

NUTRITION:
Calories: 554
Fat: 29.4 g
Protein: 13.8 g

Polish Dumplings with Sour Cream

(Pierogi)

YIELD: 10 SERVINGS. SERVING SIZE: FOUR OUNCES.

INGREDIENTS:

Dough:

1 pint	Sour cream
1½ pounds	All-purpose flour
1 ounce	Butter, melted
2 each	Eggs
1 each	Egg yolk
2 teaspoons	Salt
2 teaspoons	Vegetable oil
	Flour, as needed
	Salted water, as needed
3 ounces	Clarified butter
6 ounces	Sour cream, whisked until smooth

Filling:

8 ounces	Farmer's cheese
½ ounce	Butter, melted
1 each	Egg, cracked and lightly beaten
1 ounce	Granulated sugar
½ ounce	Freshly squeezed lemon juice

METHOD OF PREPARATION:

1. In a large bowl, combine the ingredients for the dough mix, and knead until soft. Divide the dough in half, place in a covered bowl, and allow to rest for 10 minutes.
2. Combine the ingredients for the filling, and mix well.
3. On a lightly floured surface, roll each half of the dough into a ¼-inch-thick round. Using a 3-inch circle cutter, cut the dough.
4. Place 1 tablespoon of filling, off-center, onto each circle of dough, and then fold the dough over the filling, pressing edges together to seal. Reserve on a lightly floured tray, covered.

5. In a saucepot, heat the salted water to a boil; add the pierogi, and cook until tender, or about 5 to 8 minutes. Drain on absorbent paper.
6. To serve, in a sauté pan, heat the butter, add the pierogi, and sauté until heated throughout. Add the sour cream, and toss until heated and well coated.
7. Serve immediately, or hold at 140° F or above.

Lobster Quiche

(Quiche de Homard)

YIELD: 10 SERVINGS. SERVING SIZE: ONE QUICHE.

INGREDIENTS:

	Pie crust (see page 808)
2 (1-pound)	Lobsters, boiled, meat removed, and chopped
20 each	Asparagus tips separated from stems, peeled, tied, and cooked in salted water, drained
5 each	Eggs
18 ounces	Heavy cream
1 teaspoon	Old Bay seasoning

METHOD OF PREPARATION:

1. Roll the dough ⅛-inch thick. Line a small, fluted, round 3- to 3½-inch mold. (Let the dough slightly overlap the edge of the mold.)
2. Place the lobster meat and two asparagus tips in each mold.
3. Mix the cream, eggs, and seasoning in a mixing bowl, and fill each mold with the mixture.
4. Bake in a 425° F oven for 15 to 18 minutes, or until golden brown and solid.
5. Unmold to serve.

COOKING TECHNIQUE:

Bake

Bake:

1. Preheat the oven.
2. Place the food product on the appropriate rack.

HAZARDOUS FOOD:

Lobster
Eggs
Heavy cream

NUTRITION:

Calories: 312
Fat: 25.3 g
Protein: 9.65 g

CHEF NOTES:

1. Other filling ingredients can be used, such as crabmeat, mussels, crayfish, shrimp, or oysters.
2. Keep the asparagus stems for future use.

Snails with Wild Mushrooms

(Escargots aux Cèpes)

YIELD: 10 SERVINGS. SERVING SIZE: FOUR SNAILS.

COOKING TECHNIQUE:
Simmer

Simmer and Poach:
1. Heat the cooking liquid to the proper temperature.
2. Submerge the food product completely.
3. Keep the cooked product moist and warm.

HACCP:
Hold at 140° F for above.

HAZARDOUS FOOD:
Heavy cream

NUTRITION:
Calories: 301
Fat: 13.6 g
Protein: 9.56 g

INGREDIENTS:

40 each	Canned escargots, drained
1 pint	Dry red wine
1 ounce	Dried cèpes
1 pint	Port wine
4 ounces	Meat glaze, heated to a boil (see page 36)
1 pint	Heavy cream
1 ounce	Freshly chopped parsley, excess moisture removed
6 cloves	Garlic, peeled and finely minced
	Salt and ground white pepper, to taste
3 ounces	Croutons

METHOD OF PREPARATION:

1. In a skillet, combine the stock from the escargots and the red wine. Heat to a boil. Then reduce the heat and simmer.
2. Meanwhile, soak the cépes in the port wine until they soften, then strain.
3. Add the strained port wine to the reduction. Simmer until 2 tablespoons of liquid remain and add the meat glaze.
4. Temper the cream, and add to the pan along with the parsley and garlic. Simmer, tossing frequently. Cook until the liquid reduces to the proper thickness. Season, to taste. Add the escargots and cépes.
5. Serve immediately, or hold at 140° F or above.
6. To serve, place four escargots in a preheated ramekin. Spoon the sauce and mushrooms over, and garnish with croutons.

CHEF NOTE:
Cèpes is a type of mushroom that is seldom found fresh in the United States. The dried type is an excellent product, if used properly. Other mushrooms, such as morels or chanterelles, also will give good flavor.

Stuffed Artichoke Bottoms with Smoked Salmon

(Artichaut au Saumon Gratiné)

YIELD: 10 SERVINGS. SERVING SIZE: ONE ARTICHOKE.

COOKING TECHNIQUE:

Bake

Bake:

1. Preheat the oven.
2. Place the food product on the appropriate rack.

HAZARDOUS FOOD:

Smoked salmon

NUTRITION:

Calories: 166
Fat: 8.32 g
Protein: 10.5 g

INGREDIENTS:

10 each	Artichoke bottoms, fresh or canned
1 each	Lemon, cut in half
4 ounces	Mushroom duxelles (see under beef wellington, page 248)
5 ounces	Smoked salmon, trimmed and cut into ten slices
6 ounces	Mornay sauce (see page 37)
¼ teaspoon	Nutmeg
2 ounces	Swiss cheese, grated
3 ounces	Watercress, washed and excess moisture removed

METHOD OF PREPARATION:

1. Preheat the oven to 375° F.
2. If using fresh artichokes, add the lemon, and cook in boiling water for 20 to 30 minutes until tender. Cool in cold running water, and remove the leaves (can be served with a vinaigrette). Trim the bottom neatly. If using canned artichokes, cut a small piece off of the bottom to stabilize the bottom.
3. Place 1 teaspoon of duxelle on each artichoke. Place one slice of smoked salmon on top, and coat with mornay sauce.
4. Place the artichokes on a sheet pan. Sprinkle the nutmeg and cheese over each one.
5. Bake for 6 to 8 minutes, or until cheese melts and is golden brown.
6. To serve, place on a small preheated plate, and garnish with watercress.

CHEF NOTE:

Artichokes are deemed tender when a leaf can be pulled out easily.

COOKING TECHNIQUES:
Sauté

HACCP:
Hold warm at 140° F.

HAZARDOUS FOOD:
Eggs
Milk
Crabmeat
Heavy cream

NUTRITION:
Calories: 326
Fat: 22.9 g
Protein: 13.5 g

Stuffed Crêpes with Crabmeat

(Les Crêpes au Crabe Bénédictine)

YIELD: 10 SERVINGS. SERVING SIZE: ONE CRÊPE.

INGREDIENTS:

5 ounces	All-purpose flour
2 each	Eggs
2 ounces	Butter, melted
10 ounces	Milk
	Salt and ground white pepper, to taste

Filling:

2 ounces	Shallots, peeled and minced
2 ounces	Clarified butter
6 ounces	Snow crabmeat
1 tablespoon	Dijon mustard
1 pint	Béchamel sauce (thick) (see page 16)
	Salt and ground white pepper, to taste

Glacage:

8 ounces	Hollandaise sauce (see page 30)
4 ounces	Heavy cream, whipped
4 ounces	Parmesan cheese

METHOD OF PREPARATION:

1. Sift the flour into a mixing bowl, and make a well in the center. Add the eggs one at a time, beating well after each addition.
2. Add the melted butter and milk. Blend thoroughly. Let the batter rest 30 minutes.
3. In a saucepan, heat the butter, and sauté the shallots until they are translucent.
4. Add the crabmeat and Dijon mustard, and blend in the béchamel sauce.
5. Season with the salt and white pepper. Remove from the heat, and let cool.
6. In a nonstick pan, make the crêpes (6 inches or 7 inches in diameter).
7. Lay the crêpes out on a cutting board. Add the filling, and roll. Fold in the ends. Transfer to a half-sheet pan, and place in a 300° F oven to heat.
8. Fold the whipped cream into the hollandaise sauce, and blend gently.
9. Ladle the sauce over the crêpes. Sprinkle with Parmesan cheese, and glaze under the salamander. (Be careful, because they can burn easily.) Remove, and hold warm at 140° F for service.

Stuffed Mushrooms

YIELD: 10 SERVINGS. SERVING SIZE: THREE EACH.

INGREDIENTS:

30 each	Medium-sized mushrooms, cleaned, stems removed
2 ounces	Olive oil
8 ounces	Onions, peeled and diced
4 ounces	Marsala wine
8 ounces	Gruyère cheese, grated
4 ounces	Parmesan cheese, grated
4 ounces	Bread crumbs
4 ounces	Heavy cream
½ teaspoon	Ground nutmeg
	Salt and ground white pepper, to taste
3 ounces	Fresh parsley, washed, excess moisture removed, and chopped

METHOD OF PREPARATION:

1. Preheat the oven to 400° F.
2. Chop the mushroom stems.
3. In a sauté pan, heat the oil, add the onions, and sauté until translucent. Add the mushroom stems with the wine, and sauté until tender.
4. Combine the sautéed mushroom stems with 2 ounces of the Gruyère cheese, Parmesan cheese, bread crumbs, cream, seasonings, and parsley.
5. Using a pastry bag, pipe the mushroom mixture into the mushroom caps. Place the mushrooms on a baking tray.
6. Sprinkle the remaining Gruyère cheese on each cap. Bake in the oven until the mushrooms are tender and the cheese is lightly browned. Serve immediately, or hold at 140° F.

COOKING TECHNIQUES:

Sauté, Bake

Sauté:

1. Heat the sauté pan to the appropriate temperature.
2. Evenly brown the food product.
3. For a sauce, pour off any excess oil, reheat, and deglaze.

Bake:

1. Preheat the oven.
2. Place the food product on the appropriate rack.

GLOSSARY:

Bain-marie: hot-water bath

HACCP:

Hold at 140° F.

HAZARDOUS FOOD:

Heavy cream

NUTRITION:

Calories: 471
Fat: 39.5 g
Protein: 14.9 g

CHEF NOTE:

Mushrooms will lose their firmness and darken if held too long in a **bain-marie.**

3

Soups

3

Soups

Soup, it is to dinner what a portico or a peristyle is to a building; that is to say that it is not only the first part of it, but it must be devised in such a manner as to set the tone of the whole banquet, in the same way as the overture of an opera announces the subject of the work.

Grinod de la Reyniere
French gastronome (1758–1838)

The HACCP Process

The following recipe illustrates how soups flow through the HACCP process.

Cream of Broccoli Soup

Receiving

- Butter—should have a sweet flavor, firm texture, and uniform color. Butter should be free of specks and mold, and the packaging should be intact. Butter should be delivered at 40° F (4.4° C) or lower
- Oil—packaging intact
- Flour—packaging intact
- Vegetables (broccoli, celery, and onions)—packaging intact; no cross-contamination from other foods on the truck; no signs of insect or rodent activity
- Milk—should have a sweetish taste. Container should be intact. Milk should be delivered at 40° F (4.4° C) or lower.
- Spices (salt, pepper, and nutmeg)—packaging intact

Storage

- Butter—Store under refrigeration at 40° F (4.4° C) or lower. Store above and away from raw, potentially hazardous foods.
- Oil—Store in dry storage at 50° F (10° C) with a relative humidity of 50% to 60%.
- Flour—Store in dry storage at 50° F (10° C) with a relative humidity of 50% to 60%.
- Vegetables (broccoli, celery)—Store under refrigeration with a product temperature not to exceed 40° F (4.4° C). Store above and away from raw, potentially hazardous foods.
- Milk—Store under refrigeration at 40° F (4.4° C) or lower. Store above and away from raw, potentially hazardous foods.
- Spices (salt, pepper, onions, and nutmeg)—Store in dry storage at 50° F (10° C) with a relative humidity of 50% to 60%. Keep dry.

Preparation and Cooking

- Peel the outer layer of the onions, and wash.
- Trim the celery, and wash.
- Separate the broccoli florets and stems. Wash the broccoli.
- Chop the vegetables on a clean and sanitized cutting board with clean and sanitized utensils.
- In a stockpot, heat the clarified butter or oil. Add the white mirepoix and broccoli stems, and sweat until the onions are translucent. Add the flour; then cook for 3 to 5 minutes to create a thickening agent.
- Bring the milk to a simmer, add to the vegetables, bring to a slight boil; reduce, and simmer for about 30 minutes, skimming as needed. Let the soup simmer until the broccoli stems are tender and the proper flavor is achieved. Strain the stock. Purée the vegetables in a food processor. Combine the stock and vegetables in a stockpot, and bring back to a boil. Keep the soup at an internal temperature of 165° F (73.9° C) or higher.
- In boiling chicken stock or water, blanch the broccoli florets, and then shock them under cold, running water. Drain and reheat.
- Adjust the seasonings, and serve hot. Garnish with hot broccoli florets.

Holding and Service

- Serve immediately.
- Hold at 140° F (60° C) or higher for less than 2 hours, and stir frequently.

Cooling

- Put the soup in shallow pans at a product depth of 2 inches or less.
- Place the pans with the product into an ice bath, immersed to the level of the product within the pan. Stir frequently, and lower the temperature to 40° F (4.4° C) or less within 6 hours.
- Cover loosely with plastic wrap when cooled.
- Label with the date, time, and name of the product.
- Store on the upper shelf of a refrigerated unit at 40° F (4.4° C) or lower.

Reheating

- Reheat to 165° F (73.9° C) or higher in 2 hours.

Nutritional Notes

Traditional soups can be a significant source of sodium, fat, and cholesterol. Fortunately, it is relatively simple to modify most soup recipes to lower fat, saturated fat, cholesterol, and sodium and at the same time maintain the characteristic flavor of the soup.

Many chefs throughout the United States are using a number of the following suggestions with great success and satisfaction:

- *Enhance the flavor with fresh herbs, spices, and wines, and reduce the amount of salt used. Good-quality ingredients minimize the need for salt and other high-sodium flavor enhancers, such as MSG.*
- *Remove all visible fat by chilling the soup to solidify the fat or simply spooning it off. Finish the fat removal process by floating absorbent paper on the surface of the soup.*
- *Reduce the amount of fat added in the preparation of the soup. For example, brown onions by roasting, or use tomato paste or caramelized sugar to darken a soup, when appropriate. Also, use Canadian bacon in place of regular bacon to add flavor.*
- *Rather than using a roux, which is high in fat, thicken soups with arrowroot, cornstarch, modified food starch (e.g., instant flour), grated raw potato, or bread crumbs. Cooking grains or pasta in hearty soups serves to thicken them.*
- *Concentrated vegetable purée or puréed cooked legumes also thicken as well as add flavor, fiber, vitamins, and minerals. Roast or slow-bake vegetables before puréeing to develop a deep, rich flavor and experience better thickening. Puréed vegetables and legumes need to be stabilized with a starch if used as the primary thickening agent.*
- *Use yogurt in place of sour cream. Yogurt works best as a thickener if it is drained in a filter and stabilized with cornstarch before using. It is best to warm yogurt slightly before adding it to a hot soup.*
- *Replace heavy cream with evaporated skim milk or dry milk solids. Once again, thicken with a starch if a thicker soup is desired. When thickening with low-fat dairy products or puréed vegetables or legumes, add them near the end of the cooking process and do not allow the soup to boil.*
- *Increase fiber and flavor by using whole-grain products whenever pasta or grains are called for in a recipe. Garnish soups with vegetables or croutons made of toasted whole-grain bread cubes.*
- *Dried butter or sour cream crystals can be used to impart the flavor of butter or sour cream without adding fat. These products, however, do contain salt, which must be factored into the recipe.*

Beef Bouillon Celestine

YIELD: 50 SERVINGS. SERVING SIZE: EIGHT OUNCES.

INGREDIENTS:

5 pounds	Beef bones (not marrow)
5 pounds	Neck meat
15 quarts	White beef stock, cold (see page 12)

Bouquet de Marmite:

30 ounces	Carrots, washed and peeled
15 stalks	Celery, trimmed and washed
5 each	Large onion, split in half
5 cloves	Garlic
5 sprigs	Thyme
5 each	Bay leaf
15 each	Black peppercorns, crushed
	Salt, to taste

METHOD OF PREPARATION:

1. Wash the bones, and place in a pot. Add water to cover bones. Heat to a boil. Remove the foam, drain, and discard the water. Wash the bones again.
2. Place the washed bones with the neck meat in a **marmite.** Add the cold beef stock. Heat to a boil; reduce the heat, and simmer. When the foam and fat come to the surface, **dépouiller.**
3. Add the **bouquet de marmite,** and simmer until the meat is tender, approximately 2 hours. When done, strain through a **chinois.**
4. Season the bouillon, and hold at 140° F or higher.
5. Discard everything except the meat, keeping it for further use (e.g., forcemeat or ragout).

COOKING TECHNIQUE:

Boil

Boil: (at sea level)

1. Bring the cooking liquid to a rapid boil.
2. Stir the contents, and cook the food product throughout.
3. Serve hot.

GLOSSARY:

Bouquet de marmite: vegetable roots, herbs and spices
Marmite: stockpot
Dépouiller: to skim impurities/grease
Chinois: cone-shaped strainer

HACCP:

Hold at 140° F or higher.

HAZARDOUS FOOD:

Beef bones and neck meat

NUTRITION:

Calories: 308
Fat: 16.3 g
Protein: 16.2 g

CHEF NOTE:

Celestine denotes the use of julienne of crêpes. (See next page for the celestine recipe.)

Celestine

GLOSSARY:

Julienne: matchstick strips

HAZARDOUS FOOD:

Eggs

INGREDIENTS:

20 ounces	Flour
5 each	Eggs
2½ ounces	Oil
½ teaspoon	Salt
20 ounces	Water
15 ounces	Freshly chopped parsley

METHOD OF PREPARATION:

1. Mix all of the ingredients into a smooth batter, and rest it for about 15 minutes.
2. Fry the thin crêpes on both sides.
3. Roll the crêpes, split in length and cut **julienned.**
4. Heat the celestine in a little bouillon, and place in a preheated soup cup. Pour the bouillon over the celestine, and serve with chopped parsley as garnish.

Beef Broth

YIELD: 50 SERVINGS. SERVING SIZE: EIGHT OUNCES.

INGREDIENTS:

5 gallons	Brown beef stock, heated to a boil (see page 7), fortified with beef base
	Salt and ground white pepper, to taste
6 ounces	Freshly chopped parsley, excess moisture removed

Bouquet Garni:

1 tablespoon	Black peppercorns, crushed
6 each	Bay leaves
4 ounces	Parsley stems
1½ teaspoons	Dried thyme leaves

Mirepoix:

2 pounds	Onions, peeled
13 ounces	Leeks (white parts only), washed and trimmed
13 ounces	Carrots, washed and peeled
4 stalks	Celery, washed and trimmed

METHOD OF PREPARATION:

1. In a stockpot, add the beef stock, bouquet garni, and mirepoix. Heat to a boil. Reduce the heat, and simmer for half an hour, or until the proper flavor is achieved.
2. Strain the broth through a **chinois** lined with four or five layers of wet cheesecloth into a soup insert.
3. Taste, and adjust the seasonings. Serve in a preheated cup, garnished with parsley, or hold at 140° F.
4. Cool to an internal temperature of 40° F or lower within 6 hours.
5. Reheat to a boil, and hold at 140° F.

COOKING TECHNIQUES:

Boil, Simmer

Boil: (at sea level)

1. Bring the cooking liquid to a rapid boil.
2. Stir the contents, and cook the food product throughout.
3. Serve hot.

Simmer and Poach:

1. Heat the cooking liquid to the proper temperature.
2. Submerge the food product completely.
3. Keep the cooked product moist and warm.

GLOSSARY:

Bouquet garni: herbs and spices
Mirepoix: roughly chopped vegetables
Chinois: cone-shaped strainer

HACCP:

Hold at 140° F.
Cool to 40° F or lower.

NUTRITION:

Calories: 156
Fat: 6.61 g
Protein: 5.76 g

CHEF NOTES:

1. This broth is used as the base for a consommé.
2. Beef broth can be served garnished with blanched, julienned vegetables, grains, or pasta.

Beef Consommé

YIELD: 50 SERVINGS. SERVING SIZE: EIGHT OUNCES.

INGREDIENTS:

5 gallons	Cold brown beef stock or beef broth strong (see page 7)
10 each	Egg whites, slightly whipped
3 pounds	Ground beef, lean
16 each	Black peppercorns
6 each	Bay leaves
3 ounces	Parlsey stems
1½ teaspoons	Thyme leaves

Mirepoix:

12 ounces	Onion, peeled, cut **brunoise**
2 pounds	Carrots, washed, peeled, cut brunoise
4 stalks	Celery, washed, trimmed, cut brunoise
2 pints	Tomato purée

METHOD OF PREPARATION:

1. In a mixing bowl, combine the lean ground beef, mirepoix, tomato purée, herbs, spices, salt, and white pepper to taste. Mix the egg whites and meat mixture until blended. Refrigerate.
2. In a **marmite,** blend the cold beef stock with the above clarifying ingredients.
3. Place on moderate heat. Carefully watch the clarifying ingredients to make sure they do not scorch. Stir occasionally, until a raft forms. Then stop stirring.
4. Simmer the soup for 1½ hours or to the desired strength, making sure the raft does not break or sink. Remove the first cup of consommé through the spigot, and discard.
5. In a **chinois** lined with four to five layers of wet cheesecloth, slowly strain the liquid into a soup insert, separating the clarifying ingredients from the liquid. Hold at 140° F.
6. Adjust the seasonings. Remove all of the fat from the consommé, and serve very hot with the appropriate garnish.
7. Cool to an internal temperature of 40° F or lower within 6 hours.
8. Reheat to 165° F.

COOKING TECHNIQUES:

Simmer, Boil

Simmer and Poach:

1. Heat the cooking liquid to the proper temperature.
2. Submerge the food product completely.
3. Keep the cooked product moist and warm.

Boil: (at sea level)

1. Bring the cooking liquid to a rapid boil.
2. Stir the contents, and cook the food product throughout.
3. Serve hot.

GLOSSARY:

Mirepoix: roughly chopped vegetables
Brunoise: ⅛-inch dice
Marmite: stockpot
Chinois: fine cone-shaped strainer

HACCP:

Hold at 140° F.
Chill to 40° F.

HAZARDOUS FOODS:

Egg whites
Ground beef

NUTRITION:

Calories: 116
Fat: 4.34 g
Protein: 13.8 g

CHEF NOTES:

1. There are many types of consommé. To create a chicken consommé, add ground chicken and use cold chicken stock. To create a fish consommé, use cold fish stock and lean white fish, omit the carrots and black peppercorns, use white peppercorns, and replace the onions with leeks; the tomato purée is optional. To create a vegetable consommé, use the vegetable stock, increase the egg whites, and replace the onions with leeks.
2. Consommé should always have an appropriate garnish that will be served with it in the consommé cup. The consommé will be identified by the name of the garnish. For example, consommé celestine is consommé served with a julienne of crêpes.
3. If the stock is gelatinous, allow it to liquify before using.

Black Bean Soup

YIELD: 10 SERVINGS.   SERVING SIZE: EIGHT OUNCES.

INGREDIENTS:

1 pound	Black beans, sorted, rinsed, and soaked in cold water overnight, then drained
2 ounces	Clarified butter or oil
4 ounces	Bacon, cut into ¼-inch dice
2½ quarts	Brown beef stock, heated to a boil (see page 7)
4 cloves	Garlic, peeled and mashed into a purée
1 each	Smoked ham hocks
2 each	Bay leaves
8 ounces	Andouille sausage, cut into ¼-inch dice
	Hot sauce, to taste
	Salt and freshly ground black pepper, to taste
4 each	Scallions, washed, trimmed, and sliced thin
2 ounces	Fresh parsley, washed, excess moisture removed, and chopped

Macédoine:

6 ounces	Onions, peeled
4 ounces	Carrots, washed and peeled
4 ounces	Green bell peppers, washed and seeded

METHOD OF PREPARATION:

1. Place the drained, soaked black beans in a stockpot; cover with cold water, and simmer until tender, then drain.
2. In another stockpot, heat the oil or butter, and **render** the bacon. Add the macédoine of vegetables, and sauté until the onions are translucent. Add the beef stock, garlic, ham hocks, and bay leaves. Heat to a boil; reduce the heat, and simmer for 1 hour.
3. Add the beans and andouille sausage, and continue to simmer until beans are cooked and the ham hocks are tender. Season, to taste.
4. Remove the ham hocks from the soup, and cool slightly. Remove the meat from the bones, and cut into a ¼-inch dice. Return the meat to the soup.
5. Serve immediately in preheated cups, garnished with scallions and parsley, or hold at 140° F or above. Cool to 40° F or lower within 6 hours. Reheat to 165° F.

COOKING TECHNIQUE:

Sauté, Simmer

Sauté:

1. Heat the sauté pan to the appropriate temperature.
2. Evenly brown the food product.
3. For a sauce, pour off any excess oil, reheat, and deglaze.

Simmer and Poach:

1. Heat the cooking liquid to the proper temperature.
2. Submerge the food product completely.
3. Keep the cooked product moist and warm.

GLOSSARY:

Macédoine: ¼-inch dice
Render: to melt fat

HACCP:

Hold at 140° F or above.
Cool to 40° F or lower within 6 hours.
Reheat to 165° F.

NUTRITION:

Calories: 331
Fat: 13.6 g
Protein: 18.6 g

CHEF NOTES:

1. Preparation must be started 1 day in advance to allow beans to soak.
2. Use caution when seasoning with salt, because the ham hocks, bacon, and sausage will add salt to the soup. To reduce the fat in this recipe, use lean ham and a ham bone in place of the ham hocks.
3. Alternative beans, such as red kidney beans, can be substituted in this recipe, but the name will change to red bean soup.

Bouillon Milanaise

YIELD: 10 SERVINGS. SERVING SIZE: EIGHT OUNCES.

INGREDIENTS:

3 quarts	White chicken stock (see page 12)
1 (2-pound)	Whole chicken, washed
10 ounces	Mushrooms, washed, stems trimmed, sliced
4 ounces	Tomato paste
5 ounces	Ham cut **julienne**
5 ounces	Spaghetti, broken into 1-inch pieces (optional)
	Salt and freshly ground black pepper, to taste
10 ounces	Parmesan cheese, grated

METHOD OF PREPARATION:

1. In a stockpot, combine the chicken and chicken stock. Bring to a boil. Then reduce and simmer for 40 minutes, or until the chicken is fork-tender. After 30 minutes, add the mushrooms.
2. Remove the chicken, remove the skin, debone, and dice.
3. Return the chicken meat to the boiling chicken stock. Add the tomato paste and ham. Add spaghetti, if desired. Reduce the heat, and simmer for 5 more minutes, or until the pasta is cooked. Season, to taste.
4. Serve immediately, or hold at 140° F. Cool to 40° F or below within 6 hours.
5. To serve, ladle the soup into preheated bouillon cups, and garnish with cheese.

COOKING TECHNIQUE:

Simmer

Simmer and Poach:

1. Heat the cooking liquid to the proper temperature.
2. Submerge the food product completely.
3. Keep the cooked product moist and warm.

GLOSSARY:

Julienne: matchstick strips

HACCP:

Hold at 140° F or above.
Cool to 40° F or under.

HAZARDOUS FOOD:

Chicken

NUTRITION:

Calories: 313
Fat: 15.2 g
Protein: 33.1 g

CHEF NOTE:

Cook pasta separately and add to each cup when soup is served.

Cannellini Bean and Pasta Soup

(Pasta e Fagioli)

YIELD: 10 SERVINGS. SERVING SIZE: EIGHT OUNCES.

INGREDIENTS:

10 ounces	Dried cannellini beans, sorted, washed
3 quarts	Vegetable stock, (see page 11)
2 ounces	Olive oil
2 ounces	Salt pork diced **brunoise**
4 cloves	Garlic, peeled and finely minced
2 ounces	Fresh parsley, washed, excess moisture removed, and chopped
4 ounces	Celery stalks, washed, trimmed, and diced brunoise
1 pound	Tomato **concassé**
½ teaspoon	Oregano, dried
¼ teaspoon	Marjoram, dried
½ teaspoon	Ground white pepper
	Salt, to taste
10 ounces	Rigatoni pasta, cooked **al dente**
3 ounces	Parmesan cheese, grated

METHOD OF PREPARATION:

1. In a stockpot, soak the beans in vegetable stock overnight; refrigerate at 40° F or below.
2. Heat the beans to a boil in the soaking liquid; reduce the heat, and simmer until tender. Add additional stock if needed.
3. When done, remove half of the beans, and purée in a food processor; then return to the pot.
4. In a sauté pan, heat the oil, add the salt pork, sautéing until brown. Add the garlic, parsley, celery, and tomato concassé, and continue to sauté for 10 more minutes; then add the mixture to the soup.
5. Add the herbs, and season, to taste. Add the pasta, return to a boil; remove from the heat, and hold at 140° F or above. To serve, ladle into preheated cups or soup bowls, and sprinkle with cheese.

COOKING TECHNIQUES:

Sauté, Boil, Simmer

Sauté:

1. Heat the sauté pan to the appropriate temperature.
2. Evenly brown the food product.
3. For a sauce, pour off any excess oil, reheat, and deglaze.

Boil: (at sea level)

1. Bring the cooking liquid to a rapid boil.
2. Stir the contents, and cook the food product throughout.
3. Serve hot.

Simmer and Poach:

1. Heat the cooking liquid to the proper temperature.
2. Submerge the food product completely.
3. Keep the cooked product moist and warm.

GLOSSARY:

Brunoise: ⅛-inch dice
Concassé: peeled, seeded, roughly chopped
Al dente: to the bite

HACCP:

Refrigerate at 40° F or below.
Hold at 140° F or above.

NUTRITION:

Calories: 406
Fat: 18.9 g
Protein: 18.7 g

CHEF NOTE:

Start the preparation 1 day in advance to allow the beans to soak.

Cheddar Tomato Soup

YIELD: 50 SERVINGS. SERVING SIZE: EIGHT OUNCES.

COOKING TECHNIQUES:
Boil, Simmer

Boil: (at sea level)
1. Bring the cooking liquid to a rapid boil.
2. Stir the contents, and cook the food product throughout.
3. Serve hot.

Simmer and Poach:
1. Heat the cooking liquid to the proper temperature.
2. Submerge the food product completely.
3. Keep the cooked product moist and warm.

GLOSSARY:
Temper: to equalize two extreme temperatures

HAZARDOUS FOOD:
Heavy cream

NUTRITION:
Calories: 120
Fat: 8.53 g
Protein: 6.34 g

INGREDIENTS:

1 (#10 can)	Tomatoes, chopped
1 pound	Onion, chopped
8 ounces	Celery, washed, diced
2 gallons	White chicken stock (see page 12)
1 quart	Heavy cream
1 teaspoon	Basil, ground
	Salt, white pepper, seasoned, to taste
16 ounces	Cheddar cheese, shredded

METHOD OF PREPARATION:

1. Combine the tomatoes, onions, and celery in the food processor, and blend until coarsely chopped.
2. Pour into a large saucepan, and add the chicken stock. Heat to a boil over medium heat. Simmer, uncovered, for 20 minutes.
3. **Temper** the heavy cream, and add to the tomato soup.
4. Add the basil, salt, and pepper, to taste. Simmer over low heat for 5 minutes.
5. Add the cheddar cheese, and stir constantly until the cheese melts. Serve immediately.

Chicken and Leek Soup

(Cock-a-Leekie)

YIELD: 10 SERVINGS. SERVING SIZE: EIGHT OUNCES.

INGREDIENTS:

1½ quarts	White chicken stock (see page 12)
20 ounces	Chicken thighs
10 ounces	Leeks (white part only), washed, trimmed, and cut on a diagonal into ¼-inch slices
5 ounces	Barley, washed and drained
	Salt and ground white pepper, to taste
3 ounces	Fresh parsley, washed, excess moisture removed, and chopped

METHOD OF PREPARATION:

1. In a **marmite,** combine the stock and chicken, and heat to a boil, **dépouiller** as needed.
2. Add the leeks and barley, reduce the heat, and simmer until the barley and chicken pieces are tender, approximately 45 minutes.
3. Remove the chicken from the pot. Remove the fat, bones, and skin. Discard. Dice the chicken meat and add back to the rest of the ingredients.
4. Season, to taste, and serve immediately in a preheated soup cups, or hold at 140° F or higher.
5. Garnish each portion with parsley.

COOKING TECHNIQUES:

Boil, Simmer

Boil: (at sea level)

1. Bring the cooking liquid to a rapid boil.
2. Stir the contents, and cook the food product throughout.
3. Serve hot.

Simmer and Poach:

1. Heat the cooking liquid to the proper temperature.
2. Submerge the food product completely.
3. Keep the cooked product moist and warm.

GLOSSARY:

Marmite: stockpot
Dépouiller: to skim impurities/grease

HACCP:

Hold at 140° F or higher.

HAZARDOUS FOOD:

Chicken thighs

NUTRITION:

Calories: 165
Fat: 3.75 g
Protein: 15.1 g

Chicken Soup with Poached Eggs

(Zuppa Pavese)

YIELD: 10 SERVINGS. SERVING SIZE: EIGHT OUNCES.

COOKING TECHNIQUE:

Simmer

Simmer and Poach:

1. Heat the cooking liquid to the proper temperature.
2. Submerge the food product completely.
3. Keep the cooked product moist and warm.

HACCP:

Hold at 140° F.

HAZARDOUS FOOD:

Eggs

NUTRITION:

Calories: 283
Fat: 13.8 g
Protein: 14.9 g

INGREDIENTS:

2½ quarts	White chicken stock (see page 12)
3 cloves	Garlic, peeled and finely minced
10 ounces	Canned tomatoes, drained, peeled, seeded, and chopped
10 ounces	Leeks (white part only), washed, trimmed, and thinly sliced
	Salt and ground white pepper, to taste
10 each	Poached eggs (see page 709)
10 each	Round croutons
3 ounces	Parmesan cheese, freshly grated
3 ounces	Fresh parsley, washed, excess water removed, and chopped

METHOD OF PREPARATION:

1. In a stockpot, heat the chicken stock to a boil; season, and fortify with chicken demi-glacé or soup base.
2. Add the garlic, tomatoes, and leeks, and simmer for 30 minutes. Taste, and season. Hold at 140° F.
3. Just before service, reheat the poached eggs in some of the hot soup.
4. To serve, place a warm poached egg in a preheated soup cup, and ladle hot soup over egg. Place a crouton on top of the soup, and sprinkle with cheese and parsley.

CHEF NOTES:

1. *Pavese* refers to the area of northern Italy where this style of soup is popular.
2. Do not add salt before soup base. Taste, then adjust seasoning.

Chicken Velouté with Smoked Tongue

(Velouté Agnés Sorel)

YIELD: 10 SERVINGS. SERVING SIZE: EIGHT OUNCES.

INGREDIENTS:

2½ quarts	White chicken stock (see page 12)
4 ounces	Butter
1 pound	Mushrooms, washed, stems removed, and diced; slice and reserve caps
4 ounces	All-purpose flour
10 ounces	Boneless chicken breast, skin removed
4 ounces	Smoked beef tongue, cut into thin strips
	Salt and ground white pepper, to taste
3 ounces	Freshly chopped parsley, excess moisture removed

METHOD OF PREPARATION:

1. Place beef tongue to boil, and cook until fork-tender. Approx. 2 hours.
2. Place chicken stock to boil, add chicken breast and poach until tender. Remove and set to cool.
3. In a sauce pan, melt butter; add mushroom stems and sauté until all moisture evaporates. Add flour and cook until the raw odor of flour turns into hazelnut aroma. Add some stock to dilute roux, then add roux to stock and cook on low temperature 15–20 minutes; then strain, and return on stove or into steam or trunion kettle.
4. Cut chicken breast and tounge into thin strips, add to the velouté. Add sliced mushrooms. Simmer for 10 minutes, taste and season. Hold above 140° F until service. Serve in a preheated bouillon cup with chopped parsley.

COOKING TECHNIQUE:

Simmer

Simmer and Poach:

1. Heat the cooking liquid to the proper temperature.
2. Submerge the food product completely.
3. Keep the cooked product moist and warm.

HACCP:

Hold at 140° F.
Cool to an internal temperature of 40° F.
Reheat to 165° F.

HAZARDOUS FOOD:

Chicken breast

NUTRITION:

Calories: 228
Fat: 12.3 g
Protein: 16.5 g

CHEF NOTES:

1. Mistress of the French king Charles VII, Agnés Sorel was a celebrated cook, who gave her name to several dishes.
2. Tongue must be placed to cook early. It takes 2 hours or more to tenderize.

Consommé with Cheese Croutons

(Consommé aux Diablotins)

YIELD: 10 SERVINGS. SERVING SIZE: EIGHT OUNCES.

INGREDIENTS:

2½ quarts	Beef consommé, heated to a boil and held at 140° F or above (see page 112)
Diablotin:	
4 ounces	Parmesan cheese, grated
10 each	French bread, round croutons (see page 906)

METHOD OF PREPARATION:

1. Preheat the broiler or salamander.
2. Spread the grated Parmesan cheese on toast, and broil until lightly browned.
3. To serve, ladle the consommé into a preheated bouillon cup, and serve a crouton on the side.

COOKING TECHNIQUES:

Boil, Broil

Boil: (at sea level)

1. Bring the cooking liquid to a rapid boil.
2. Stir the contents, and cook the food product throughout.
3. Serve hot.

Grill/Broil:

1. Clean and heat the grill/boiler.
2. To prevent sticking, brush the food product with oil.

HACCP:

Hold at 140° F or above.

NUTRITION:

Calories: 175
Fat: 4.45 g
Protein: 12.9 g

Consommé Julienne

YIELD: 10 SERVINGS. SERVING SIZE: EIGHT OUNCES.

INGREDIENTS:

2½ quarts	Beef consommé, seasoned (see page 112)
8 ounces	Carrots, washed, peeled, and cut **julienne**
4 ounces	Celeriac, washed, peeled, and cut julienne
4 ounces	Leek (white part only), cleaned, and cut julienne
1½ ounces	Fresh parsley, washed, excess moisture removed, and chopped
	Salt and ground white pepper, to taste

METHOD OF PREPARATION:

1. Heat the consommé to a boil, season, and hold warm at 140° F.
2. Cook all vegetables **al dente** in the seasoned liquid; drain, mix, and hold warm at 140° F.
3. To serve, place the julienne into a warm bouillon cup, pour in the consommé, and sprinkle parsley on top.

COOKING TECHNIQUE:

Simmer

Simmer and Poach:

1. Heat the cooking liquid to the proper temperature.
2. Submerge the food product completely.
3. Keep the cooked product moist and warm.

GLOSSARY:

Julienne: matchstick strips
Al dente: to the bite

HACCP:

Hold warm at 140° F.

NUTRITION:

Calories: 53.1
Fat: .178 g
Protein: 5.97 g

CHEF NOTE:

Any noncoloring root vegetable can be used, e.g., turnip, celeriac, kohlrabi, or parsnip.

Consommé Madriléne

YIELD: 10 SERVINGS. SERVING SIZE: EIGHT OUNCES.

INGREDIENTS:

2½ quarts	Beef consommé, flavored and seasoned (see page 112)
10 ounces	Tomato **concassé**
1½ ounces	Fresh parsley, washed, excess moisture removed, and chopped

METHOD OF PREPARATION:

1. Heat the consommé to a boil.
2. Heat the tomato concassé.
3. To serve, place the concassé into a warm bouillon cup, pour in the consommé, and sprinkle parsley on top.

COOKING TECHNIQUE:

Simmer

Simmer and Poach:

1. Heat the cooking liquid to the proper temperature.
2. Submerge the food product completely.
3. Keep the cooked product moist and warm.

GLOSSARY:

Concassé: peeled, seeded, roughly chopped

HACCP:

Heat to 165° F or higher.

NUTRITION:

Calories: 75.1
Fat: .328 g
Protein: 6.82 g

CHEF NOTES:

1. This consommé originally is served cold, and jellied.
2. When the clarification is prepared, 3 ounces of tomato pulp is added to each quart of liquid.
3. In some places tapioca is added for enriching the consommé.

Consommé with Profiteroles

YIELD: 10 SERVINGS. SERVING SIZE: EIGHT OUNCES.

INGREDIENTS:

2½ quarts	Beef consommé, heated to a boil (see page 112)
8 ounces	Pâté à choux no. 2 (see page 838)

METHOD OF PREPARATION:

1. Preheat the oven to 400° F.
2. Fill a pastry bag with a ¼-inch plain tube, and fill with pâté à choux batter.
3. On parchment paper, line half of the sheet pan. Pipe small mounds of batter, approximately 40 to 50.
4. Bake in the oven until golden brown and very light.
5. Serve the consommé in preheated bouillon cups, and float four to five profiteroles on top.
6. Hold the consommé at 140° F or above. Hold the profiteroles in a dry, closed container.

COOKING TECHNIQUES:

Simmer, Bake

Simmer and Poach:

1. Heat the cooking liquid to the proper temperature.
2. Submerge the food product completely.
3. Keep the cooked product moist and warm.

Bake:

1. Preheat the oven.
2. Place the food product on the appropriate rack.

HACCP:

Hold consommé at 140° F or above.

NUTRITION:

Calories: 59.6
Fat: 2.06 g
Protein: 6.1 g

CHEF NOTES:

1. The pâté à choux batter can be seasoned further with the addition of grated cheese and freshly ground pepper.
2. It can be filled with purée of foie gras and the name changed to *consommé strasbourgoise.*

Consommé with Veal Dumplings

(Consommé au Quenelle de Veau)

YIELD: 10 SERVINGS. SERVING SIZE: EIGHT OUNCES.

COOKING TECHNIQUES:
Simmer, Poach

Simmer and Poach:
1. Heat the cooking liquid to the proper temperature.
2. Submerge the food product completely.
3. Keep the cooked product moist and warm.

HACCP:
Hold in stock at 140° F or above.

HAZARDOUS FOODS:
Veal shoulder meat
Heavy cream
Eggs

NUTRITION:
Calories: 246
Fat: 11.4 g
Protein: 20.7 g

INGREDIENTS:

2½ quarts	Beef consommé, heated to a boil and seasoned (see page 112)

Veal Dumplings:

1 pound	Veal shoulder meat, fat removed and ground
9 ounces	Panada, chilled, see below
4 ounces	Heavy cream
4 each	Eggs
	Salt and ground white pepper, to taste
	Ginger ground to taste
	Veal stock, as needed for poaching, heated to a boil (see page 12)

METHOD OF PREPARATION:

1. Purée the meat in a food processor. Add the chilled panada (broken into small pieces), cream with egg mixture, and pass through a wire mesh.
2. Season, to taste, and chill in the refrigerator overnight at 40° F or below.
3. Using 2 large teaspoons, form the meat into football-shaped dumplings. Test the mixture by poaching a few dumplings in seasoned veal stock. If the poached dumplings are too firm, add a little ice water. If too soft, add some chilled egg white. Hold the dumplings in the stock at 140° F or above.
4. To serve, place three dumplings in a preheated boullion cup, and ladle in hot consommé.

Panada

INGREDIENTS:

8 ounces	Water
2 ounces	Butter
5½ ounces	All-purpose flour

METHOD OF PREPARATION:

1. Heat the water and butter to a boil in a saucepan. Simmer until the butter is melted.
2. Add the flour all at once. Cook until a ball forms, peeling away from the side of the pan.
3. Chill at 40° F or below. Cover to prevent a crust from forming.

Corn Chowder

YIELD: 10 SERVINGS. SERVING SIZE: EIGHT OUNCES.

INGREDIENTS:

2 ounces	Salt pork, rind removed and diced
4 ounces	Clarified butter
6 ounces	Onions, peeled and diced **brunoise**
8 ounces	Celery, washed and diced brunoise
4 ounces	All-purpose flour
2 quarts	White chicken stock, heated to a boil (see page 12)
8 ounces	Potatoes, washed, peeled, and diced
1 pound	Fresh or frozen corn kernels
	Salt and ground white pepper, to taste
¼ teaspoon	Ground nutmeg
8 ounces	Heavy or light cream
2 ounces	Chopped fresh parsley

METHOD OF PREPARATION:

1. In a large stockpot, render the salt pork with a small amount of butter. When rendered, remove from the pan, add the remaining butter, and sauté the onions and celery.
2. When the onions are translucent, dust with flour, and cook for 5 minutes; gradually whisk in the hot chicken stock. Add the corn kernels. Heat to a boil. Reduce the heat, and simmer for about 30 minutes, or until the proper flavor is achieved.
3. In a separate pan, starting with cold, salted water, cook the potatoes until tender. When tender, add to the soup.
4. **Temper** the cream, and add to the chowder. Hold at 140° F. Adjust the seasonings. Garnish with fresh chopped parsley.
5. Cool to an internal temperature of 40° F or lower within 6 hours.
6. Reheat to a boil.

COOKING TECHNIQUES:

Sauté, Simmer

Sauté:

1. Heat the sauté pan to the appropriate temperature.
2. Evenly brown the food product.
3. For a sauce, pour off any excess oil, reheat, and deglaze.

Simmer and Poach:

1. Heat the cooking liquid to the proper temperature.
2. Submerge the food product completely.
3. Keep the cooked product moist and warm.

GLOSSARY:

Brunoise: ⅛-inch dice
Temper: to equalize two extreme temperatures

HACCP:

Hold at 140° F.
Cool to 40° F or below.

HAZARDOUS FOOD:

Heavy or light cream

NUTRITION:

Calories: 402
Fat: 28.2 g
Protein: 6.56 g

CHEF NOTES:

1. Chicken stock may be replaced with vegetable stock (see page 11).
2. The addition of creamed corn will intensify the flavor, and thicken the chowder.

Crab and Greens Soup

(Callaloo)

YIELD: 10 SERVINGS. SERVING SIZE: EIGHT OUNCES.

INGREDIENTS:

3 ounces	Vegetable oil
3 ounces	Onions, peeled and diced **brunoise**
5 cloves	Garlic, peeled and finely minced
1 pound	Callaloo or Swiss chard greens cut **chiffonade**
8 ounces	Coconut milk
2 quarts	White chicken stock (see page 12)
10 ounces	Crabmeat, cleaned
	Salt and ground black pepper, to taste
¼ teaspoon	Hot sauce

METHOD OF PREPARATION:

1. In a sauté pan, heat the oil; add the onions and garlic, and sauté until the onions are translucent. Add the chiffonade of greens, and sauté 1 or 2 more minutes until they are wilted.
2. Add the coconut milk and chicken stock, and heat to a boil; reduce the heat, and simmer 5 more minutes.
3. Add the crabmeat, and continue to simmer until heated throughout.
4. Season, to taste, and add the hot sauce.
5. Serve immediately in preheated cups, or hold at 140° F or above.

COOKING TECHNIQUES:

Sauté, Simmer

Sauté:

1. Heat the sauté pan to the appropriate temperature.
2. Evenly brown the food product.
3. For a sauce, pour off any excess oil, reheat, and deglaze.

Simmer and Poach:

1. Heat the cooking liquid to the proper temperature.
2. Submerge the food product completely.
3. Keep the cooked product moist and warm.

GLOSSARY:

Brunoise: ⅛-inch dice
Chiffonade: ribbons of leafy greens

HACCP:

Hold at 140° F or above.

NUTRITION:

Calories: 189
Fat: 13.6 g
Protein: 8.17 g

CHEF NOTE:

This soup is a specialty of the island of Barbados.

Cream of Asparagus Soup

(Crème Argenteuil)

YIELD: 10 SERVINGS. SERVING SIZE: EIGHT OUNCES.

COOKING TECHNIQUES:

Sauté, Simmer

Sauté:

1. Heat the sauté pan to the appropriate temperature.
2. Evenly brown the food product.
3. For a sauce, pour off any excess oil, reheat, and deglaze.

Simmer and Poach:

1. Heat the cooking liquid to the proper temperature.
2. Submerge the food product completely.
3. Keep the cooked product moist and warm.

GLOSSARY:

Brunoise: ⅛-inch dice
Chinois: cone-shaped strainer
Temper: to equalize two extreme temperatures

HACCP:

Hold at 140° F or above.

HAZARDOUS FOOD:

Heavy cream

NUTRITION:

Calories: 244
Fat: 15.1 g
Protein: 9.03 g

INGREDIENTS:

4 ounces	Butter
4 ounces	Shallots, peeled and diced **brunoise**
4 ounces	Celery stalks, washed, trimmed, and diced brunoise
2 pounds	Asparagus, tips removed and reserved, stems peeled, chopped
6 ounces	All-purpose flour
2½ quarts	Vegetable stock, heated to a boil (see page 11)
	Salt and ground white pepper, to taste
6 ounces	Heavy cream
2 ounces	Clarified butter

METHOD OF PREPARATION:

1. In a stockpot, melt the butter, and sauté the shallots until they are translucent. Add the celery and the asparagus stems, and continue to sauté for 10 minutes.
2. Dust with flour, and cook for a few minutes more to remove the raw flavor from the flour.
3. Pour the boiling vegetable stock over the vegetables, and cook until the asparagus stems are tender.
4. In a food mill, purée the soup. Strain the purée through a **chinois,** and return to a saucepan. Heat to a boil, and season, to taste.
5. **Temper** the heavy cream, and add to the soup. Heat to a boil. Hold at 140° F or above.
6. In a sauté pan, heat the remaining butter, and sauté the asparagus tips.
7. To serve, place the tips in a preheated soup cup, and ladle in the soup.

CHEF NOTE:

Traditionally, this soup is prepared with white asparagus and is named for a village in northern France where the best asparagus is grown.

Cream of Avocado Soup

COOKING TECHNIQUE:
Not applicable

HACCP:
Chill to 40° F or lower.

HAZARDOUS FOOD:
Sour cream

NUTRITION:
Calories: 433
Fat: 35.9 g
Protein: 8.9 g

YIELD: 10 SERVINGS. **SERVING SIZE:** EIGHT OUNCES.

INGREDIENTS:

1½ quarts	Béchamel sauce, chilled to 40° F or lower (see page 16)
6 each	Fresh ripe avocados, washed and peeled, pit removed
1 ounce	Fresh mint leaves, washed and coarsely chopped
8 ounces	Sour cream
1 ounce	Freshly squeezed lime juice
	Salt and ground white pepper, to taste
	Vegetable stock (chilled), as needed (see page 11)
	Fresh mint sprigs for garnish

METHOD OF PREPARATION:

1. In a food processor bowl, combine the béchamel and avocados, and purée until smooth. Add mint, and continue to process until totally incorporated.
2. In a large bowl, whisk the sour cream until smooth. Add the lime juice to the cream, and whisk until incorporated.
3. Fold the avocado purée into the cream. Season, to taste, with salt and pepper, and thin to the proper consistency with the stock. Chill to 40° F or lower, until needed for service.
4. To serve, ladle the soup into chilled cups or stemmed glasses, and garnish with mint sprigs.

Cream of Carrot Soup

(Crème Crecy)

YIELD: 10 SERVINGS. SERVING SIZE: EIGHT OUNCES.

INGREDIENTS:

4 ounces	Butter
4 ounces	Onions, peeled and diced **brunoise**
20 ounces	Carrots, washed, peeled, and thinly sliced
3 ounces	All-purpose flour
2½ quarts	Vegetable stock, heated to a boil (see page 11)
5 ounces	Rice
2 each	Egg yolks
2 ounces	Heavy cream
	Salt and ground white pepper, to taste

METHOD OF PREPARATION:

1. In a stockpot, melt the butter, add onions, and sauté until translucent. Add the carrots, and sauté for 5 minutes.
2. Dust the carrots and onions with the flour, and cook for 3 minutes. Add boiling vegetable stock, and heat to a boil.
3. Add the rice, reduce the heat, and simmer the soup until the carrots and rice are very tender.
4. Puree the soup in a food processor, strain through a **chinois,** and return to the stove.
5. Mix the egg yolks and cream, **temper,** and add to the soup; hold at 140° F or above.
6. Season the soup, to taste. Serve in a preheated cup.

COOKING TECHNIQUES:
Sauté, Simmer

Sauté:
1. Heat the sauté pan to the appropriate temperature.
2. Evenly brown the food product.
3. For a sauce, pour off any excess oil, reheat, and deglaze.

Simmer and Poach:
1. Heat the cooking liquid to the proper temperature.
2. Submerge the food product completely.
3. Keep the cooked product moist and warm.

GLOSSARY:
Brunoise: ⅛-inch dice
Chinois: cone-shaped strainer
Temper: to equalize two extreme temperatures

HACCP:
Hold at 140° F or above.

HAZARDOUS FOODS:
Egg yolks
Heavy cream

NUTRITION:
Calories: 202
Fat: 12.4 g
Protein: 5.39 g

Cream of Cauliflower Soup

(Crème Du Barry)

YIELD: 10 SERVINGS. SERVING SIZE: EIGHT OUNCES.

INGREDIENTS:

4 ounces	Clarified butter
Mirepoix:	
4 ounces	Onions, peeled
2 ounces	Celery, washed and trimmed
2½ lbs	Cauliflower, fresh, washed, cored, flowerets separated and reserved, stems peeled, and diced **brunoise**
4 ounces	All-purpose flour
2½ quarts	White chicken stock (see page 12)
	Salt and ground white pepper, to taste
⅛ teaspoon	Nutmeg, ground
8 ounces	Heavy cream
8 ounces	White chicken stock or vegetable stock, heated to a boil

METHOD OF PREPARATION:

1. In a stockpot, heat the clarified butter. Add the mirepoix and cauliflower stems. **Sweat** onions until translucent. Add the flour, and cook for 3 to 5 minutes to create a thickening agent, stirring constantly.
2. Heat the chicken stock to a simmer, and add to the vegetables. Heat to a slight boil. Then reduce the heat, and simmer for about 30 minutes. Let the soup simmer until the cauliflower is tender and the proper flavor is achieved. Strain off the liquid through a **chinois;** then purée the vegetables in a food processor. Combine the liquid and vegetables in a stockpot. Heat to a boil.
3. Temper the heavy cream, and add to the soup. Heat to 165° F but do not return to a boil.
4. Blanch the cauliflower flowerets in the steamer. Shock them under cold, running water. Drain thoroughly and reheat. Hold at 140° F or above.
5. Adjust the seasonings, and serve hot, in a preheated soup cup. Garnish with hot cauliflower flowerets.
6. Cool to 40° F or lower within 6 hours.
7. Reheat to 165° F.

COOKING TECHNIQUE:

Simmer

Simmer and Poach:

1. Heat the cooking liquid to the proper temperature.
2. Submerge the food product completely.
3. Keep the cooked product moist and warm.

GLOSSARY:

Mirepoix: roughly chopped vegetables
Brunoise: ⅛-inch dice
Sweat: to sauté under a cover
Chinois: cone-shaped strainer

HACCP:

Hold at 140° F.
Cool to 40° F or below.
Reheat to 165° F.

HAZARDOUS FOODS:

Milk
Heavy cream

NUTRITION:

Calories: 240
Fat: 18.4 g
Protein: 6.77 g

Cream of Chicken Princesse

(Crème de Volaille, Princesse)

YIELD: 10 SERVINGS. SERVING SIZE: EIGHT OUNCES.

COOKING TECHNIQUE:

Simmer

Simmer and Poach:

1. Heat the cooking liquid to the proper temperature.
2. Submerge the food product completely.
3. Keep the cooked product moist and warm.

GLOSSARY:

Temper: to equalize two extreme temperatures
Velouté: velvety-textured sauce or soup

HAZARDOUS FOODS:

Chicken breast
Heavy cream

NUTRITION:

Calories: 198
Fat: 10.8 g
Protein: 13.6 g

INGREDIENTS:

10 ounces	Chicken breast, deboned and skinned
2½ quarts	Chicken stock (see page 12)
4 ounces	Butter, clarified
4 ounces	All-purpose flour
4 ounces	Heavy cream
5 ounces	Asparagus tips, washed
	Salt and ground white pepper, to taste
1½ ounces	Fresh parsley, washed, excess moisture removed, and finely chopped

METHOD OF PREPARATION:

1. Poach the chicken breast in the chicken stock. When tender, remove and let cool. Cut the chicken in ¼-inch dice. Reserve.
2. In a saucepan, melt the clarified butter. Add the flour, cook for 5 to 7 minutes to create a blonde roux, remove from the heat, and set aside.
3. Gradually add chicken stock into the roux, and heat to a second boil. Reduce the heat, and simmer for about 30 minutes. Simmer until the proper flavor is achieved, and strain.
4. Blanch the asparagus tips in the chicken stock, cool, and reserve.
5. **Temper** the heavy cream, and add to the **velouté.** Adjust the seasonings to taste.
6. At the time of service, warm the chicken and asparagus. Serve the velouté hot in a preheated soup cup and garnish with diced chicken and asparagus tips.

CHEF NOTE:

Princesse denotes the use of asparagus. The classical method of service is to place two thin slices of poached chicken and two small asparagus tips in a hot cup and add hot chicken velouté.

Fish Chowder

YIELD: 50 SERVINGS. **SERVING SIZE:** FIVE OUNCES.

INGREDIENTS:	
8 ounces	Salt pork, rind removed, cut, and diced
2 pounds	Onions, peeled, dice brunoise
1 pound	Bread flour
2 gallons	Fish stock (see page 10), heated to a boil
2 pounds	Potatoes, washed, peeled, diced
5 pounds	Cod fillets, rinsed
	Salt and ground white pepper, to taste
2 pints	Heavy cream
6 ounces	Freshly chopped parsley, excess moisture removed

METHOD OF PREPARATION:

1. In a large stockpot, **render** the salt pork. Add the onions, and sauté until translucent.
2. Add the flour to create a roux. Cook over medium heat for 5 to 7 minutes.
3. Gradually whisk in the hot fish stock. Blend until smooth. Heat to a boil. Reduce to a simmer.
4. Add the diced potatoes, and simmer.
5. When the potatoes are three-fourths cooked, add the diced cod, and finish cooking.
6. Season lightly with salt and ground white pepper.
7. Just before serving, **temper** the heavy cream, and add to soup. Hold at 140° F. Serve in preheated soup cup.
8. Garnish each preheated soup cup with chopped parsley, and serve.
9. Cool to an internal temperature of 40° F or lower within 6 hours.
10. Reheat to 165° F.

COOKING TECHNIQUES:

Sauté, Simmer

Sauté:

1. Heat the sauté pan to the appropriate temperature.
2. Evenly brown the food product.
3. For a sauce, pour off any excess oil, reheat, and deglaze.

GLOSSARY:

Render: to melt fat
Temper: to equalize two extreme temperatures

HACCP:

Hold at 140° F.
Chill to 40° F.

HAZARDOUS FOODS:

Cold fillets
Heavy cream

NUTRITION:

Calories: 196
Fat: 10.1 g
Protein: 10.3 g

CHEF NOTE:

To avoid cross-contamination, the whole cod fillets may be placed in the fish stock and broken down naturally.

French Onion Soup

(Soupe à l'Oignon Gratinée)

YIELD: 10 SERVINGS. SERVING SIZE: EIGHT OUNCES.

COOKING TECHNIQUES:

Sauté, Simmer

Sauté:

1. Heat the sauté pan to the appropriate temperature.
2. Evenly brown the food product.
3. For a sauce, pour off any excess oil, reheat, and deglaze.

Simmer and Poach:

1. Heat the cooking liquid to the proper temperature.
2. Submerge the food product completely.
3. Keep the cooked product moist and warm.

GLOSSARY:

Julienne: matchstick strips
Deglaze: to add liquid to a hot pan

NUTRITION:

Calories: 260
Fat: 18.5 g
Protein: 9.8 g

INGREDIENTS:

4 ounces	Clarified butter or oil
1½ pounds	Onions, peeled and cut **julienne**
⅛ ounce	Thyme leaves, dried
4 ounces	Sherry
1 quart	Beef stock (see page 12)
1½ quarts	Chicken stock (see page 12)
	Salt and ground white pepper, to taste
	Croutons, as needed, toasted
5 ounces	Gruyère cheese, grated

METHOD OF PREPARATION:

1. In a small stockpot, heat the oil or butter until very hot. Add the onions, and sauté until they are caramelized.
2. **Deglaze** with sherry, and simmer until the alcohol evaporates.
3. Season with thyme, and add the beef and chicken stock.
4. Let simmer for about 1 hour, or until the proper flavor is achieved.
5. Pour the soup in a crock or bowl. Top with a crouton and grated cheese, and brown under a salamander or in an oven. Serve when the top is golden brown.

CHEF NOTE:

There are variations of onion soup. For example, dust sautéd onions with flour, cook for 3 minutes, add Madeira or Port wine, then add stock or water. Cook and serve with dry slices of French bread floating in the soup.

Garlic and Egg Soup

YIELD: 10 SERVINGS. SERVING SIZE: EIGHT OUNCES.

COOKING TECHNIQUES:
Sauté, Simmer

Sauté:
1. Heat the sauté pan to the appropriate temperature.
2. Evenly brown the food product.
3. For a sauce, pour off any excess oil, reheat, and deglaze.

Simmer and Poach:
1. Heat the cooking liquid to the proper temperature.
2. Submerge the food product completely.
3. Keep the cooked product moist and warm.

GLOSSARY:
Concassé: peeled, seeded, roughly chopped
Temper: to equalize two extreme temperatures

HACCP:
Hold at 140° F.

HAZARDOUS FOOD:
Eggs

NUTRITION:
Calories: 196
Fat: 12.5 g
Protein: 5.55 g

INGREDIENTS:

3 ounces	Olive oil
	Croutons
8 cloves	Garlic, peeled and finely minced
½ teaspoon	Spanish paprika
2½ quarts	White chicken stock, heated to a boil (see page 12)
10 ounces	Tomato **concassé**
	Salt and ground white pepper, to taste
5 each	Eggs

METHOD OF PREPARATION:

1. In a saucepan, heat the olive oil.
2. Add the garlic, and sauté until it begins to brown; then sprinkle with paprika, and add the croutons.
3. Pour the boiling stock over the croutons, add the tomato concassé, and simmer until the bread almost falls apart.
4. Season to taste.
5. Break the eggs into a bowl, beat lightly with a whisk, **temper,** and pour into the soup. Simmer just 1 minute more. Serve immediately in preheated cups, or hold at 140° F.

CHEF NOTES:
1. Do not allow the soup to return to a boil after adding the eggs, because the eggs will curdle.
2. This is a Spanish-style preparation.

SOUPS

Grilled Corn Soup with Chili Cream

YIELD: 10 SERVINGS. SERVING SIZE: EIGHT OUNCES.

INGREDIENTS:

5 cloves	Garlic, peeled
1 ounce	Oil
5 ears	Fresh corn on the cob, husks and silks removed
2 quarts	Vegetable stock (see page 11)
2 each	Jalapeño chilies, preferably red, washed, seeded, and finely minced
5 ounces	Heavy cream
	Salt, to taste
	Chili cream

Mirepoix:

5 ounces	Carrots, washed and peeled
5 ounces	Celery stalks, washed and trimmed
5 ounces	Onions, peeled

METHOD OF PREPARATION:

1. Preheat the oven to 400° F.
2. In a small sauté pan, place the garlic in oil, cover with aluminum foil, and roast in the oven for 15 minutes.
3. Brush the corn with garlic oil, and grill, turning frequently, until golden brown.
4. In an appropriate sized **marmite,** combine the stock, mirepoix of vegetables, and jalapeño chilies. Bring to a boil; reduce the heat, and simmer for 20 to 30 minutes.
5. Remove the corn kernels from the cob, add them to the marmite, and continue to simmer until all of the vegetables are very tender.
6. In a food processor, purée the soup. Do not strain. Bring back to a boil, and reduce to desired consistency.
7. **Temper** the heavy cream, and then add it to the soup. Season, to taste, and serve immediately in preheated cups, garnished with chili cream, or hold at 140° F or higher.

COOKING TECHNIQUES:

Roast, Grill, Simmer

Roast:

1. Sear the food product, and brown evenly.
2. Elevate the food product in a roasting pan.
3. Determine doneness, and consider carryover cooking.
4. Let the food product rest before carving.

Grill/Broil:

1. Clean and heat the grill/boiler.
2. To prevent sticking, brush the food product with oil.

Simmer and Poach:

1. Heat the cooking liquid to the proper temperature.
2. Submerge the food product completely.
3. Keep the cooked product moist and warm.

GLOSSARY:

Mirepoix: roughly chopped vegetables
Marmite: stockpot
Temper: to equalize two extreme temperatures

HACCP:

Hold at 140° F.

HAZARDOUS FOOD:

Heavy cream

NUTRITION:

Calories: 143
Fat: 9.05 g
Protein: 2.88 g

Hawaiian Seafood Chowder with Pesto Crostini

YIELD: 10 SERVINGS. SERVING SIZE: EIGHT OUNCES.

INGREDIENTS:

2 ounces	Olive oil
8 cloves	Garlic, peeled and finely minced
2 ounces	Fresh cilantro, washed
1 pint	White wine
2 quarts	Clam juice
20 each	Littleneck clams, scrubbed
20 (1-ounce)	Scallops, muscle removed
20 each	Mussels, debearded and scrubbed
20 (16 to 20 count)	Shrimp, washed, peeled, and deveined
10 ounces	Ono, cut into 1-inch pieces
4 ounces	Pancetta, cut into ¼-inch dice
1 pint	Vegetable stock, heated to a boil (see page 11)
1 bunch	Fresh chives, washed, dried, and cut into 3-inch lengths
	Salt and ground white pepper, to taste
10 each	Pesto crostini (see page 138)

Mirepoix:

8 ounces	Onions, peeled
6 ounces	Carrots, washed and peeled
6 ounces	Celery stalks, washed and trimmed

METHOD OF PREPARATION:

1. In a large stockpot, heat 1 ounce of the olive oil with the garlic, and sauté until the garlic turns golden. Add the cilantro, wine, and clam juice. Bring the liquid to a boil; reduce the heat to a simmer.
2. Add the shellfish to the broth, and poach until the shells open and/or the shrimp and scallops are just firm. Immediately remove, and spread out on

COOKING TECHNIQUES:

Sauté, Poach, Broil

Sauté:

1. Heat the sauté pan to the appropriate temperature.
2. Evenly brown the food product.
3. For a sauce, pour off any excess oil, reheat, and deglaze.

Simmer and Poach:

1. Heat the cooking liquid to the proper temperature.
2. Submerge the food product completely.
3. Keep the cooked product moist and warm.

Grill/Broil:

1. Clean and heat the grill/boiler.
2. To prevent sticking, brush the food product with oil.

GLOSSARY:

Mirepoix: roughly chopped vegetables
Chinois: cone-shaped strainer

HACCP:

Hold at 40° F or lower.
Hold at 140° F or higher.

HAZARDOUS FOODS:

Clams
Scallops
Mussels
Shrimp
Ono

NUTRITION:

Calories: 507
Fat: 18.6 g
Protein: 45 g

CHEF NOTE:

Seabass or catfish can be substituted for the ono, a yellowtail tuna.

a tray to cool. Remove the top shells from the clams and mussels, and loosen the mussel within each shell. Hold at 40° F or lower until needed for service.

3. Cook the ono in the broth, remove, and cool; then refrigerate to 40° F or below.
4. In a separate stockpot, strain the broth through a cheesecloth-lined **chinois,** and reserve. Heat the remaining olive oil. Add the pancetta, and sauté for 2 to 3 minutes. Add the mirepoix, and continue to sauté until the vegetables begin to soften.
5. Add the vegetable stock and strained broth. Return to a boil; reduce the heat, and simmer for 15 to 20 minutes, or until vegetables are tender. Increase heat, and reduce the liquid by half. Season, to taste.
6. For service, hold 1 pint of broth in a shallow hotel pan at 140° F. Hold the remaining broth in a separate container at 140° F or higher.
7. Separate the shellfish and fish into portions, allowing two clams, two scallops, two mussels, two shrimp, and one piece of ono for each.
8. Reheat the seafood in the broth, and arrange in a preheated bowl. Ladle the hot soup over, and serve immediately, garnished with chives and a crostini.

Pesto Crostini

INGREDIENTS:

10 slices	French baguette bread slices, cut on a diagonal, ⅓-inch thick
10 sprigs	Fresh cilantro sprigs, washed

Pesto:

4 ounces	Sesame oil
4 ounces	Dry-roasted peanuts
4 ounces	Cilantro leaves, washed
1 tablespoon	Fermented black beans
1 tablespoon	Shredded kaffir lime leaves
	Salt to taste

METHOD OF PREPARATION:

1. In a food processor, combine all of the pesto ingredients, and blend to a paste.
2. Toast one side of each crouton; then spread the untoasted side with pesto. Place the croutons under a broiler or salamander until the pesto is bubbling. Garnish each with a cilantro sprig.

Hot and Sour Soup

YIELD: 10 SERVINGS. SERVING SIZE: EIGHT OUNCES.

COOKING TECHNIQUE:
Simmer

Simmer and Poach:
1. Heat the cooking liquid to the proper temperature.
2. Submerge the food product completely.
3. Keep the cooked product moist and warm.

GLOSSARY:
Julienne: matchstick strips

HACCP:
Hold at 140° F or above.

HAZARDOUS FOODS:
Lean pork
Eggs

NUTRITION:
Calories: 229
Fat: 11.1 g
Protein: 15.1 g

INGREDIENTS:

2 quarts	Chicken stock (see page 12)
6 ounces	Lean pork, cooked and cut **julienne**
4 ounces	Straw mushrooms, washed, trimmed, and sliced
6 ounces	Bean curd, shredded
4 ounces	Bamboo shoots
2 ounces	Dried wooden ears (Chinese black mushrooms), soaked in warm water for 30 minutes, drained, and cut julienne
2 ounces	Soy sauce
1 ounce	White vinegar
1½ teaspoons	Salt
1 teaspoon	Ground white pepper
3 ounces	Cornstarch, dissolved in 6 ounces of cold water
5 each	Eggs, beaten lightly
2 ounces	Sesame seed oil
5 each	Scallions, washed, trimmed, and thinly sliced

METHOD OF PREPARATION:

1. In a stockpot or wok, heat the chicken stock to a boil; then add the pork, mushrooms, bean curd, bamboo shoots, and wooden ears. Simmer for 2 minutes; then add the soy sauce, white vinegar, salt, and pepper.
2. Thicken the soup to a light syrupy consistency with the cornstarch mixture.
3. Just before service, pour the eggs through a colander into the simmering soup. Stir gently until the eggs are cooked. Hold at 140° F or above.
4. To serve, ladle the soup into preheated soup cups, and garnish with sesame seed oil and sliced green onions for flavor.

CHEF NOTE:
This is a Chinese recipe from the Szechwan region, known for its hot and oily foods.

Hungarian Goulash Soup

YIELD: 10 SERVINGS. SERVING SIZE: EIGHT OUNCES.

INGREDIENTS:

1 ounce	Vegetable oil
4 ounces	Salt pork, rind removed, and diced **brunoise**
6 ounces	Onions, peeled and diced brunoise
1 pound	Beef chuck, trimmed of fat and silverskin, cut into ½-ounce cubes
4 ounces	Green bell peppers, washed, seeded, and diced brunoise
6 ounces	Tomatoes, washed, cored, **blanched,** peeled, seeded, and diced brunoise
½ ounce	Hungarian paprika
2 quarts	Brown beef stock, heated to a boil (see page 7)
8 ounces	Potatoes, washed, peeled, and diced
2 cloves	Garlic, peeled and minced to a purée
¼ teaspoon	Caraway seeds, bruised
1 each	Lemon, zest only, blanched
	Hot sauce, to taste
	Spätzels, as needed (see page 482)
	Salt and freshly ground black pepper, to taste

METHOD OF PREPARATION:

1. In a large stockpot, heat the oil, and render the salt pork. When rendered, remove the salt pork from the pan, and sauté the beef cubes until they are golden brown. Then, remove the beef, and reserve.
2. Add the onions, and sauté until translucent. Then, add the green peppers, tomatoes, and garlic, and sauté for 5 minutes more.
3. Return the meat to the pot; add the paprika, and **sweat** for about 5 minutes.
4. Add the beef stock and heat to a boil; reduce the heat to a simmer.
5. In a separate pot, starting with cold water, cook the potatoes until they are three-fourths done. Drain; then add to the goulash soup. Simmer the soup until the meat and potatoes are completely tender.
6. Season with caraway seeds, salt, black pepper, and hot sauce.
7. To serve, place spätzel into preheated cups, and ladle soup over them. Garnish with lemon zest.

COOKING TECHNIQUE:

Simmer

Simmer and Poach:

1. Heat the cooking liquid to the proper temperature.
2. Submerge the food product completely.
3. Keep the cooked product moist and warm.

GLOSSARY:

Brunoise: ⅛-inch dice
Blanch: to par cook
Sweat: to sauté under a cover

NUTRITION:

Calories: 368
Fat: 22.3 g
Protein: 19.6 g

CHEF NOTE:

A goulash is a Hungarian-style soup or stew flavored with Hungarian paprika and is traditionally hot and spicy. This dish is commonly served with spätzel, which are small flour dumplings.

Italian Fish Soup

YIELD: 50 SERVINGS. SERVING SIZE: EIGHT OUNCES.

COOKING TECHNIQUE:
Simmer

Simmer and Poach:
1. Heat the cooking liquid to the proper temperature.
2. Submerge the food product completely.
3. Keep the cooked product moist and warm.

GLOSSARY:
Brunoise: ⅛-inch dice
Concassé: peeled, seeded, roughly chopped
Deglaze: to add liquid to hot pan

HAZARDOUS FOOD:
Cod fillets

NUTRITION:
Calories: 198
Fat: 6 g
Protein: 13 g

INGREDIENTS:

6 ounces	Olive oil
3 pounds	Onions, peeled and diced **brunoise**
6 stalks	Celery, washed, trimmed, and diced brunoise
1 pound	Leeks (white part only), peeled and chopped fine
3 heads	Garlic, peeled and mashed to a purée
3 tablespoons	Oregano
3 tablespoons	Marjoram
3 tablespoons	Basil
1 pint	Dry white wine
6 (#2½ cans)	Tomatoes, drained and **concassé**
3 gallons	Fish stock (see page 10)
6 pounds	Cod fillets, diced
	Salt and freshly ground black pepper, to taste
2 ounces	Fresh parsley, washed, stems removed, excess moisture removed, and chopped
	Garlic toast, as needed

METHOD OF PREPARATION:

1. Heat the olive oil in a stockpot. Sauté the onions, celery, and leeks until the onions are translucent. Add the garlic and seasonings. **Deglaze** with white wine, and reduce by half.
2. Add the crushed tomatoes and fish stock. Heat to the first boil. Simmer for 10 to 15 minutes until the proper flavor is achieved.
3. Just before service, add the cod fillets, and cook for 10 minutes until the fish is white and flaky. Adjust the seasonings.
4. To serve, place the fish soup in a preheated soup cup. Serve with garlic toast, and garnish with chopped parsley.

CHEF NOTE:
In Southern Italy this soup is prepared with fish heads and pieces of fish with bones. It is served either as a meal or as an addition to a pasta dish.

Italian Tomato and Bread Soup

YIELD: 10 SERVINGS. SERVING SIZE: EIGHT OUNCES.

INGREDIENTS:

4 ounces	Olive oil
8 ounces	Onion, peeled and diced **brunoise**
8 cloves	Garlic, peeled and mashed to a purée
5 pounds	Ripe tomatoes, cored, **blanched,** peeled, seeded,and coarsely chopped, reserving juices
4 ounces	Fresh basil or marjoram leaves, washed and coarsely chopped
1½ quarts	White chicken or vegetable stock, heated to a boil (see pages 11 and 12)
	Salt and freshly ground black pepper, to taste
1 pound	Day-old sourdough bread, trimmed of crust and cut into 1-inch cubes
4 ounces	Parmesan cheese, freshly grated

METHOD OF PREPARATION:

1. In stockpot, heat the oil; then add the onions, and sauté until translucent, or about 10 minutes. Add the garlic, and sauté for another 2 to 3 minutes.
2. Add the tomatoes with the reserved juice, the herbs, and the stock, and season, to taste. Heat the liquid to a boil.
3. Reduce the heat, partially cover the stockpot, and simmer for 30 to 40 minutes.
4. Adjust the seasonings to taste, and hold at a minimum of 140° F.
5. For service, place eight to 10 bread cubes in a preheated bowl, and ladle in the soup. Garnish with Parmesan cheese.

COOKING TECHNIQUES:

Sauté, Simmer

Sauté:

1. Heat the sauté pan to the appropriate temperature.
2. Evenly brown the food product.
3. For a sauce, pour off any excess oil, reheat, and deglaze.

Simmer and Poach:

1. Heat the cooking liquid to the proper temperature.
2. Submerge the food product completely.
3. Keep the cooked product moist and warm.

GLOSSARY:

Brunoise: ⅛-inch dice
Blanch: to par cook

HACCP:

Hold at 140° F.

NUTRITION:

Calories: 62.7
Fat: 2.87 g
Protein: 2.16 g

CHEF NOTES:

1. The Italian name for this soup is "Pappa al Pomodoro."
2. The bread cubes can be toasted in olive oil and garlic for added flavor and texture.
3. This soup also is commonly served at room temperature.

Lentil Soup

YIELD: 50 SERVINGS. SERVING SIZE: EIGHT OUNCES.

INGREDIENTS:

5 pounds	Lentils, cleaned
4 gallons	Brown beef stock (see page 7)
7 ounces	Oil or butter
7 ounces	Bacon, diced
3 each	Ham hocks
3 pounds	Sausage links, mild, diced small
6 each	Bay leaves
12 ounces	Freshly chopped parsley, excess moisture removed
	Salt and ground black pepper, to taste

Mirepoix:

1 pound	Onions, peeled and diced **brunoise**
10 ounces	Carrots, washed, peeled, and diced brunoise
3 stalks	Celery, washed, trimmed, and diced brunoise
9 cloves	Garlic, peeled and mashed

METHOD OF PREPARATION:

1. In a stockpot, render the bacon with a small amount of oil. Add the sausage and the mirepoix. Sauté until the onions are translucent.
2. Add the beef stock, ham hocks, bay leaves, and lentils. Heat to the first boil. Reduce to a simmer.
3. Simmer for about 2 hours, or until the lentils and the meat from the ham hocks are tender.
4. Remove the meat from the ham hocks, dice brunoise, and return to the soup. Remove the soup from the heat, and hold at 140° F. Adjust the seasonings.
5. Place in preheated soup cups, and serve.
6. Cool to an internal temperature of 40° F or lower within 6 hours.
7. Reheat at 165° F.

COOKING TECHNIQUES:

Sauté, Boil, Simmer

Sauté:

1. Heat the sauté pan to the appropriate temperature.
2. Evenly brown the food product.
3. For a sauce, pour off any excess oil, reheat, and deglaze.

Boil: (at sea level)

1. Bring the cooking liquid to a rapid boil.
2. Stir the contents, and cook the food product throughout.
3. Serve hot.

Simmer and Poach:

1. Heat the cooking liquid to the proper temperature.
2. Submerge the food product completely.
3. Keep the cooked product moist and warm.

GLOSSARY:

Mirepoix: roughly chopped vegetables
Brunoise: ⅛-inch dice

HACCP:

Hold at 140° F.
Chill to 40° F.

HAZARDOUS FOODS:

Sausage links
Bacon
Ham

NUTRITION:

Calories: 465
Fat: 22.2 g
Protein: 33.1 g

CHEF NOTES:

1. Use caution when seasoning with salt because the ham hocks, bacon, and sausage will add salt to the soup. To reduce the fat in this recipe, use lean ham and a ham bone in place of the ham hocks.
2. This recipe has many variations and can be garnished with any type of vegetable. The ingredients depend on the country in which it is made.

Little Neck Clam Soup

YIELD: 10 SERVINGS. SERVING SIZE: EIGHT OUNCES.

INGREDIENTS:

4 ounces	Olive oil
1 pound	Onions, peeled and diced **brunoise**
8 cloves	Garlic, peeled and mashed into a purée
½ tablespoon	Oregano, dried
½ tablespoon	Marjoram, dried
½ tablespoon	Basil, dried
	Salt and freshly ground black pepper, to taste
8 ounces	Dry white wine
1 pound	Canned tomatoes, drained, seeded, and roughly chopped
3 quarts	Clam juice or fish stock, heated to a boil (see page 10)
50 each	Little neck clams, cleaned and soaked in cold water for 1 hour, held at 40° F or below until needed
10 each	Round croutons, as needed
2 each	Limes, washed and sliced ¼-inch thick

METHOD OF PREPARATION:

1. In a stockpot, heat the olive oil, and sauté the onions until they are translucent. Add the garlic and seasonings; then **deglaze** the pan with the wine. Reduce the liquid by half.
2. Add the tomatoes, clam juice, or stock, and heat to a boil. Simmer for about 10 to 15 minutes, or until the proper flavor is achieved.
3. Just before service, add the little neck clams and cook for 15 minutes, or until the shells open. Discard the unopened shells.
4. To serve, place the clams in a preheated soup cup, add a crouton, and ladle hot soup over the crouton. Garnish with a slice of lime.

COOKING TECHNIQUE:

Simmer

Simmer and Poach:

1. Heat the cooking liquid to the proper temperature.
2. Submerge the food product completely.
3. Keep the cooked product moist and warm.

GLOSSARY:

Brunoise: ⅛-inch dice
Deglaze: to add liquid to a hot pan

HACCP:

Hold at 40° F or below.
Hold soup at 140° F or above.

HAZARDOUS FOOD:

Clams

NUTRITION:

Calories: 276
Fat: 15.1 g
Protein: 8.83 g

CHEF NOTES:

1. This soup is a Mediterranean regional dish, and there are many variations. Little neck clams are used because they are tender and can be served either in or out of the shells.
2. If the soup is to be held, remove the clams and chill to 40° F or below. Hold the soup at 140° F, and reheat the clams to order in a small saucepan with some of the soup.

Lübeck Shrimp Soup

YIELD: 10 SERVINGS. **SERVING SIZE:** EIGHT OUNCES.

INGREDIENTS:

2 ounces	Butter
12 ounces (30 to 40 count)	Shrimp, peeled and deveined. Save shells.
4 ounces	Onion, peeled and diced **brunoise**
3 ounces	Celery stalks, washed, trimmed, and diced brunoise
2 ounces	Carrot, washed, peeled, and diced brunoise
1 clove	Garlic, peeled and mashed into a purée
2 tablespoons	All-purpose flour
1½ quart	Fish stock, heated to a boil (see page 10)
8 ounces	Dry white wine
2 each (only)	Juniper berries
½ teaspoon	Thyme leaves, dried
	Salt and freshly ground black pepper, to taste
8 ounces	Half-and-half or heavy cream
1 ounce	Blanched almonds, finely ground
1 tablespoon	Freshly squeezed lemon juice
1 ounce	Fresh dill leaves, washed and finely minced

METHOD OF PREPARATION:

1. In a saucepot, melt the butter, add the shrimp, and sauté until just firm, approximately 3 to 5 minutes. Remove the shrimp, and cool; then cover, and refrigerate at 40° F or lower until needed for service.
2. Add shells to the pot, sauté for 10 minutes; then add the onions, celery, carrots, and garlic. Continue to sauté until the onion becomes translucent.
3. Sprinkle the flour over the vegetables and shells; then stir and cook another 5 minutes.
4. Add the stock, wine, and seasonings; heat to a boil, cover loosely, reduce the heat, and simmer for 30 minutes.
5. Strain the soup through a **chinois** into another saucepan, pressing down on the shells and vegetables with a ladle to remove all of the liquid.
6. Return the soup to a boil. Temper the half-and-half or cream, and add it to the soup. Stir in the almonds, and hold at 140° F or higher.
7. Just before service, add the reserved cooked shrimp, lemon juice, and dill. Ladle into preheated cups.

COOKING TECHNIQUE:
Sauté

Sauté:
1. Heat the sauté pan to the appropriate temperature.
2. Evenly brown the food product.
3. For a sauce, pour off any excess oil, reheat, and deglaze.

GLOSSARY:
Brunoise: ⅛-inch dice
Chinois: cone-shaped strainer

HACCP:
Refrigerate at 40° F or lower.
Hold at 140° F or higher.

HAZARDOUS FOODS:
Shrimp & shells
Half-and-half or heavy cream

NUTRITION:
Calories: 208
Fat: 15.7 g
Protein: 7.09 g

CHEF NOTES:
1. For a more intense shrimp flavor, half of the cooked shrimp can be puréed with the half-and-half in a food processor.
2. This soup was created at a restaurant in the port of Lübeck, which is near Hamburg, Germany.

Manhattan Clam Chowder

YIELD: 10 SERVINGS. SERVING SIZE: EIGHT OUNCES.

INGREDIENTS:

10 each	Quahog clams, washed
2½ quarts	Cold water
4 ounces	Dry white wine
1 teaspoon	Fresh thyme leaves, divided
1 each	Bay leaf
3 ounces	Bacon, diced **brunoise**
1 ounce	Vegetable oil
4 ounces	Onion, peeled and diced brunoise
2 ounces	Celery stalks, washed, trimmed, and diced brunoise
2 cloves	Garlic, peeled and crushed
12 ounces	Canned whole tomatoes, drained, seeded, and diced brunoise
2 ounces	Tomato purée
1 pound	Potatoes, washed, peeled, and diced brunoise
	Salt, to taste
	Cayenne pepper, to taste

METHOD OF PREPARATION:

1. In a stockpot combine the clams with the water, wine, ¼ teaspoon of the thyme, and the bay leaf. Cover, and simmer for 10 to 15 minutes, or until the clams open.
2. Remove the pot from the heat, and let the stock rest for 10 minutes. Strain the liquid through a **chinois** lined with four or five layers of cheesecloth, and reserve the broth, holding at 140° F. Remove the clam meat from the shells, chop, and reserve in a bowl.
3. In another stockpot, **render** the bacon with the oil. Remove bacon from the pot; add the onions and celery. Sauté until the onion is translucent; then add the garlic, tomatoes, and tomato purée, and cook another 5 minutes.
4. Add the reserved broth, and heat the liquid to a boil. Reduce the heat to a simmer, and allow to cook for 30 minutes.
5. In a separate pot, cook the potatoes in salted water until tender. Drain, and add to the soup.
6. Add the reserved clam meat and the remaining thyme; season, to taste. Serve in preheated cups or bowls.
7. Hold at 140° F; cool to 40° F or below within 6 hours. Reheat to 165° F.

COOKING TECHNIQUES:

Sauté, Simmer

Sauté:

1. Heat the sauté pan to the appropriate temperature.
2. Evenly brown the food product.
3. For a sauce, pour off any excess oil, reheat, and deglaze.

Simmer and Poach:

1. Heat the cooking liquid to the proper temperature.
2. Submerge the food product completely.
3. Keep the cooked product moist and warm.

GLOSSARY:

Brunoise: ⅛-inch dice
Chinois: cone-shaped strainer
Render: to melt fat

HACCP:

Hold at 140° F, and cool to 40° F or below.
Reheat to 165° F.

HAZARDOUS FOOD:

Clams

NUTRITION:

Calories: 151
Fat: 7.31 g
Protein: 6.62 g

CHEF NOTE:

1. A chowder is a thick, chunky seafood soup, of which clam chowder is the most well known. The name comes from the French *chaudière,* a caldron in which fishermen made their stews fresh from the sea. New England–style is made with milk or cream; Manhattan-style is made with tomatoes. Chowder can contain any of several varieties of seafood and vegetables. The term also is used to describe any thick, rich soup containing chunks of food (e.g., corn chowder).
2. It is at the chef's discretion to cook potatoes separately. Starch released during cooking will add to the thickness of the chowder.

Meatball Soup

(Giouvarlakia)

YIELD: 10 SERVINGS. SERVING SIZE: EIGHT OUNCES.

INGREDIENTS:

2 pounds	Beef neck or chuck meat
2 pounds	Beef knuckles, with bones
5 quarts	Brown beef stock, heated to a boil (see page 7)
	Salt and ground white pepper, to taste

Bouquet Garni: (2 each)

1 ounce	Parsley stems
2 each	Bay leaves
1 teaspoon	Whole black peppercorns
1 sprig	Fresh thyme

Meatballs:

1 pound	Ground beef
5 ounces	Long-grained rice
6 ounces	Onions, peeled and diced **brunoise**
½ teaspoon	Freshly chopped mint
1 tablespoon	Fresh parsley, washed, excess moisture removed, and chopped
2 each	Eggs, cracked and lightly beaten
	Salt and freshly ground black pepper, to taste

METHOD OF PREPARATION:

1. In a large stockpot, combine the beef with the bones, stock, and bouquet garni. Heat the liquid to a boil; reduce the heat, and simmer for 2 hours; **dépouiller** as necessary.
2. Strain the bouillon through a fine **chinois,** and reserve the meat for other purposes. Return the bouillon to a boil, adjust the seasoning, and hold at 140° F or above.
3. To prepare the meatballs, place the ground beef in a mixing bowl. Add the remaining ingredients, and mix well.
4. Shape into ½-ounce balls.
5. In a saucepan, heat half of the bouillon to a boil, add the meatballs, reduce the heat, and simmer for 20 to 25 minutes, or until the rice is cooked. Hold at 140° F or above in bouillon.
6. To serve, place five to six meatballs in a preheated cup, and ladle in the reserved bouillon.

COOKING TECHNIQUES:

Boil, Simmer

Boil: (at sea level)

1. Bring the cooking liquid to a rapid boil.
2. Stir the contents, and cook the food product throughout.
3. Serve hot.

Simmer and Poach:

1. Heat the cooking liquid to the proper temperature.
2. Submerge the food product completely.
3. Keep the cooked product moist and warm.

GLOSSARY:

Bouquet garni: bouquet of herbs and spices
Brunoise: ⅛-inch dice
Dépouiller: to skim impurities/grease
Chinois: cone-shaped strainer

HACCP:

Hold at 140° F or above.

HAZARDOUS FOODS:

Beef neck or chuck meat
Beef knuckles
Ground beef
Eggs

NUTRITION:

Calories: 522
Fat: 30.5 g
Protein: 37.6 g

CHEF NOTES:

1. This soup is of Greek origin, and it is popular in the Balkans.
2. To save time and money, the beef stock can be fortified with soup base.

Mexican Corn Soup

(Sopa de Elote)

YIELD: 10 SERVINGS. SERVING SIZE: EIGHT OUNCES.

INGREDIENTS:

2½ pounds	Frozen or canned corn
1 quart	Chicken stock, heated to a boil (see page 12)
3 ounces	Butter
10 ounces	Scallions, washed and thinly sliced
3 ounces	Green chili peppers, seeded and chopped
1 quart	Milk, heated to a boil
	Salt, to taste
5 ounces	Sour cream
3 ounces	Pimientos, diced **brunoise**

METHOD OF PREPARATION:

1. Place the corn and stock in a food processor, and purée.
2. Melt the butter in a sauté pan, add the scallions, and sauté until translucent. Add the chili peppers, and continue to sauté for 2 more minutes.
3. In a saucepan, combine the corn mixture with the scallions and peppers, and heat the liquid to a boil.
4. Add the milk, and cook for 5 more minutes; season with salt, and hold at 140° F.
5. To serve, ladle the soup into preheated cups, garnish with a spoonful of sour cream, and sprinkle with diced pimientos.

COOKING TECHNIQUE:

Simmer

Simmer and Poach:

1. Heat the cooking liquid to the proper temperature.
2. Submerge the food product completely.
3. Keep the cooked product moist and warm.

GLOSSARY:

Brunoise: ⅛-inch dice

HACCP:

Hold at 140° F.

HAZARDOUS FOODS:

Milk
Sour cream

NUTRITION:

Calories: 273
Fat: 14.5 g
Protein: 7.96 g

CHEF NOTE:

In Mexico this soup is made with just water. The corn is placed to boil in water; when tender, it is mashed, discarding the outer skin of the kernels.

Minestrone Soup

YIELD: 50 SERVINGS. SERVING SIZE: EIGHT OUNCES.

INGREDIENTS:

3 pounds	White beans, sorted, washed, soaked in cold water overnight, and drained
2 pounds	White cabbage, outer leaves removed; cored, and chopped
3 each	Smoked ham hocks
3 gallons	White chicken stock, heated to a boil (see page 12)
1 pound	Salt pork, rind removed, medium diced
6 ounces	Oil
1 pound	Onions, peeled and diced **brunoise**
4 stalks	Celery, washed, peeled, trimmed, and diced brunoise
1 pound	Carrots, washed, peeled, and diced brunoise
6 (#2½ cans)	Whole tomatoes, drained, peeled, seeded, and chopped rough
9 cloves	Garlic, peeled and mashed
1 pound	Potatoes, washed, peeled, and diced
1 pound	Ditalini pasta
3 ounces	Fresh basil, washed and chopped
	Salt and ground white pepper, to taste
1 pound	Parmesan cheese, grated

METHOD OF PREPARATION:

1. Starting with cold water, cook the white beans with the cabbage and ham hock until they are tender. Remove the ham hocks, debone, and dice. Then add the beans and diced ham hock to the soup.
2. In a large stockpot, **render** the salt pork in a small amount of oil. When the salt pork is rendered, sauté the vegetables. Add the chicken stock and crushed tomatoes. Heat to a boil. Simmer.
3. Starting with cold water, separately cook the potatoes until they are al dente; drain, and add to the soup.
4. Cook the ditalini in boiling, salted water until al dente. Shock under cold running water. Drain, set aside, and lightly oil.
5. Let the soup simmer for about 30 minutes until the proper flavor is achieved. Adjust the seasonings with fresh basil, salt, and pepper.
6. Place the pasta in the bottom of a preheated soup cup. Pour the hot soup over the pasta; top with ½-teaspoon of Parmesan cheese, and serve.

COOKING TECHNIQUES:

Boil, Simmer, Sauté

Boil: (at sea level)

1. Bring the cooking liquid to a rapid boil.
2. Stir the contents, and cook the food product throughout.
3. Serve hot.

Simmer and Poach:

1. Heat the cooking liquid to the proper temperature.
2. Submerge the food product completely.
3. Keep the cooked product moist and warm.

Sauté:

1. Heat the sauté pan to the appropriate temperature.
2. Evenly brown the food product.
3. For a sauce, pour off any excess oil, reheat, and deglaze.

GLOSSARY:

Brunoise: ⅛-inch dice
Render: to melt fat

NUTRITION:

Calories: 422
Fat: 25.5 g
Protein: 15.6 g

CHEF NOTE:

This soup is a meal in some parts of Italy and is often made with different ingredients. A variety of dried beans and pastas can be used. The region of Italy determines whether the soup contains meat.

Mushroom Veloute

YIELD: 50 SERVINGS. SERVING SIZE: EIGHT OUNCES.

COOKING TECHNIQUES:

Sauté, Simmer

Sauté:

1. Heat the sauté pan to the appropriate temperature.
2. Evenly brown the food product.
3. For a sauce, pour off any excess oil, reheat, and deglaze.

Simmer and Poach:

1. Heat the cooking liquid to the proper temperature.
2. Submerge the food product completely.
3. Keep the cooked product moist and warm.

GLOSSARY:

Brunoise: ⅛-inch dice

NUTRITION:

Calories: 259
Fat: 21.4 g
Protein: 6.61 g

INGREDIENTS:

1 pound	Butter
1 pound	Shallots, peeled and diced **brunoise**
1½ pounds	Celery, washed, trimmed, and diced brunoise
5 pounds	Mushrooms, washed and sliced; reserve some for garnish
1¾ pounds	White roux
2½ gallons	Vegetable stock (see page 11), heated to a boil
	Salt and ground white pepper, to taste

METHOD OF PREPARATION:

1. Place 1 pound of butter in a suitable pot to melt.
2. Add the shallots, sauté lightly. Add the celery and mushrooms. Sauté for 10 minutes.
3. Add boiling vegetable stock, and heat to a boil. Reduce the heat, and simmer until the vegetables are tender.
4. Dilute the roux with some of the hot liquid, and add to the soup. Continue to simmer for 10 more minutes.
5. Purée the soup in a robot coupe, and return it to the stove top. Heat to a boil. Season, to taste. Hold at 140° F.
6. Sauté the reserved mushroom slices in the remaining ½ pound of butter.
7. To serve, place the mushrooms in a preheated soup cup, ladle the soup over, and serve hot.

CHEF NOTE:

Mushrooms can be stemmed, chopped, the caps sliced and sautéed separately, and returned to the puréed soup. This will enhance the flavor of the velouté.

Mussel Soup

(Mussel Brose)

YIELD: 10 SERVINGS. SERVING SIZE: EIGHT OUNCES.

INGREDIENTS:

30 each	Mussels, debearded and washed under running cold water
4 ounces	Leeks (white part only), washed, trimmed, and roughly chopped
4 ounces	Celery, washed, trimmed and chopped
1 ounce	Parsley leaves, washed
8 ounces	Apple cider
2 ounces	Butter, melted
2 ounces	All-purpose flour
2 quarts	Milk, heated to a boil
4 ounces	Heavy cream
	Salt and freshly ground black pepper, to taste
½ teaspoon	Nutmeg, ground

METHOD OF PREPARATION:

1. In a stainless steel pot, combine the mussels, leeks, celery, parsley, and cider, and heat the liquid to a boil.
2. Reduce the heat, and let simmer 10 minutes, shaking the pot, allowing the mussels to steam thoroughly until opened.
3. Remove the open mussels, and set aside. Remove the mussels from the shells, and hold covered at 140° F or higher.
4. Strain the liquid from the mussels through a double cheesecloth. Melt the butter in a sauté pan, and add the flour to make a roux blonde; then add the boiling milk, and cook, stirring constantly for 5 to 6 minutes.
5. Add the strained liquid to the cream sauce to create the **velouté.**
6. Heat the mixture to a boil. **Temper** the cream, and add to the soup; then reduce the heat, and simmer for 2 minutes. Add the seasoning to taste, and serve immediately or hold at 140° F or higher until service.
7. For service, place three mussels into a preheated soup cup, and ladle in the soup.

COOKING TECHNIQUES:

Boil, Simmer

Boil: (at sea level)

1. Bring the cooking liquid to a rapid boil.
2. Stir the contents, and cook the food product throughout.
3. Serve hot.

Simmer and Poach:

1. Heat the cooking liquid to the proper temperature.
2. Submerge the food product completely.
3. Keep the cooked product moist and warm.

GLOSSARY:

Velouté: velvety-textured sauce or soup
Temper: to equalize two extreme temperatures

HACCP:

Hold covered at 140° F or higher.

HAZARDOUS FOODS:

Mussels
Milk
Heavy cream

NUTRITION:

Calories: 380
Fat: 18.9 g
Protein: 27.5 g

CHEF NOTE:

For additional flavor, a teaspoon of minced garlic can be added just before service.

Oriental Chicken Curry Soup with Noodles and Vegetables

(Ga-li-tong-mein)

YIELD: 10 SERVINGS. SERVING SIZE: EIGHT OUNCES.

COOKING TECHNIQUE:
Stir-Fry

Stir-Fry:
1. Heat the oil in a wok until hot but not smoking.
2. Keep the food in constant motion; use the entire cooking surface.

GLOSSARY:
Julienne: matchstick strips
Blanch: to par cook
Al dente: to the bite

HAZARDOUS FOOD:
Chicken

NUTRITION:
Calories: 224
Fat: 9.21 g
Protein: 20.2 g

INGREDIENTS:

	Peanut oil, as needed
1 pound	Boneless chicken, skin removed, diced
4 ounces	Onions, peeled and cut **julienne**
1 teaspoon	Ginger root, peeled and cut julienne
3 quarts	White chicken stock (see page 12)
4 ounces	Cellophane noodles, soaked in hot water for 15 minutes, drained, and cut into 2-inch long pieces
2 each	Black mushrooms, soaked in warm water for 30 minutes, drained, and julienned
6 ounces	Chinese cabbage (bok choy), washed, trimmed, and cut julienne
2 ounces	Coconut milk
¼ teaspoon	Curry powder
	Salt and ground white pepper, to taste
½ teaspoon	Granulated sugar
2 ounces	Snow peas, washed, **blanched**, and cut julienne
3 ounces	Scallions, washed, trimmed, and cut into ¼-inch slices

METHOD OF PREPARATION:

1. In a wok or large sauté pan, heat the oil, and stir-fry the chicken until golden brown. Remove the chicken and reserve. Add the onions and ginger root to the wok or pan and sauté until onions are translucent. Add chicken stock, and heat to a boil.

CHEF NOTE:
This is a regional soup of Thailand.

2. Add the cellophane noodles, and continue to cook until the noodles are **al dente.** Add the black mushrooms; return the chicken, Chinese cabbage, and the coconut milk, and simmer for another 10 minutes.
3. Adjust the seasonings with curry powder, salt, white pepper, and sugar.
4. Pour into preheated soup cups, and garnish with the snow peas and sliced green onions.

Petite Marmite

YIELD: 10 SERVINGS. SERVING SIZE: EIGHT OUNCES.

INGREDIENTS:

3 quarts	Brown beef stock (see page 7)
1½ pounds	Beef, chuck
1 pound	Chicken breast, boneless, skinless
1 ounce	Butter, clarified or oil
5 ounces	Leek whites, washed, trimmed, split, washed (again), and cut **julienne**
5 ounces	Carrots, washed, peeled, and cut julienne
5 ounces	Celery, washed, peeled, and cut julienne
5 ounces	Turnips, washed, peeled, and cut julienne
	Seasoned, to taste
1½ ounces	Parsley, fresh, washed, excess moisture removed, and chopped

METHOD OF PREPARATION:

1. In a stockpot, pour the beef stock; cover and place on an open flame to boil. When boiling, add the beef chuck, and simmer for 90 minutes, covered.
2. Add the chicken breasts to the stockpot, and simmer an additional 20 minutes or until tender. Remove the beef chuck and chicken breasts, and julienne.
3. In a large sauté pan, sweat the vegetables until they are translucent. Deglaze with a ladle of beef stock, and add back to the stockpot, along with the beef and chicken, and simmer for 15 minutes.
4. Adjust the seasonings, garnish with chopped parsley, and serve.
5. Cool to an internal temperature of 40° F or lower within 6 hours.
6. Reheat to a boil.

COOKING TECHNIQUE:

Simmer

Simmer and Poach:

1. Heat the cooking liquid to the proper temperature.
2. Submerge the food product completely.
3. Keep the cooked product moist and warm.

GLOSSARY:

Julienne: matchstick strips

HACCP:

Cool to internal temperature of 40° F or lower.

HAZARDOUS FOOD:

Beef
Chicken breast

NUTRITION:

Calories: 132
Fat: 5.8 g
Protein: 17.6 g

CHEF NOTES:

1. *Petite* in French means "small," and *marmite* denotes "stockpot" or "soup pot." Traditionally, this soup was served in a small china marmite. In the past, poached beef marrow was served with this soup.
2. In classical service, the vegetables and meats are placed in the preheated bouillon cup and clear bouillon or consommé is poured over. It is then garnished with chopped parsley.

Pheasant and Lentil Soup

YIELD: 10 SERVINGS. SERVING SIZE: EIGHT OUNCES.

INGREDIENTS:

1 two-pound	Pheasant, giblets removed, rinsed in cold water, reserving liver for other usage
2 each	Bay leaves
1 medium-sized	Onion, **clouté**
	Salt and freshly ground black pepper, to taste
	Cold water as needed
2 ounces	Clarified butter, melted
8 ounces	Bacon, cut into ½-inch dice
4 ounces	Celery stalks, washed, trimmed, and diced **brunoise**
1 pound	Lentils, sorted and washed
16 ounces	Dry red wine
2 ounces	Tomato paste
8 ounces	Medium dry sherry
1½ ounces	Balsamic vinegar
1 tablespoon	Dijon mustard
¼ teaspoon	Grated nutmeg
2 ounces	Fresh chives, washed and snipped

Mirepoix:

8 ounces	Onion, peeled
4 ounces	Carrots, washed and peeled
4 ounces	Leeks, washed and trimmed (use white part only)

METHOD OF PREPARATION:

1. In a suitable pot, combine the pheasant, bay leaves, onion clouté, and mirepoix. Season to taste, and add water to 3 inches above the pheasant.
2. Heat the liquid to a boil; then reduce the heat, and simmer uncovered until the pheasant is fork-tender.
3. Remove the pheasant to a tray, and allow to cool until easily handled.
4. Return the stock to a boil, and reduce to 3 quarts. Strain through a chinois, and reserve the stock, discarding the solids.

COOKING TECHNIQUES:

Simmer, Sauté

Simmer and Poach:

1. Heat the cooking liquid to the proper temperature.
2. Submerge the food product completely.
3. Keep the cooked product moist and warm.

Sauté:

1. Heat the sauté pan to the appropriate temperature.
2. Evenly brown the food product.
3. For a sauce, pour off any excess oil, reheat, and deglaze.

GLOSSARY:

Clouté: studded with cloves
Mirepoix: roughly chopped vegetables
Brunoise: ⅛-inch dice

HACCP:

Hold at 140° F or higher.
Chill to 40° F or below.

HAZARDOUS FOOD:

Pheasant

NUTRITION:

Calories: 538
Fat: 20 g
Protein: 43.3 g

CHEF NOTES:

1. Preparation can be started 1 day in advance to cook pheasant and reduce the stock.
2. Additional brunoise of onion and carrot can be added with the celery.
3. For a heavier texture, remove half of the lentils once they are tender, and purée; return the purée to the soup.

5. In another pot, heat the butter, add the bacon, and sauté until the pieces are crisp. Remove the pieces of bacon with a slotted spoon, and reserve for garnish.
6. Add the celery, and sauté over medium heat for approximately 5 minutes. Add the lentils, and stir to coat with fat.
7. Add the reserved stock, wine, and tomato paste, and heat the liquid to a boil. Reduce the heat, and simmer until the lentils are tender, or about 30 minutes.
8. When the lentils are cooked, add the sherry, vinegar, mustard, and nutmeg, and adjust the seasoning to taste. Return to a boil, then remove from heat. Hold at 140° F or higher. Chill to 40° F or below within 6 hours. Reheat to a boil.
9. Skin and debone the pheasant. Julienne the meat, discarding the skin and bones, and add the meat to the soup.
10. For service, ladle the soup into a preheated cup, and garnish with reserved crisp bacon and fresh chives.

Polish Mushroom and Barley Soup

YIELD: 10 SERVINGS. SERVING SIZE: SIX OUNCES.

INGREDIENTS:

1 ounce	Dried mushrooms, soaked in warm water for 30 minutes
2½ quarts	Brown beef stock, heated to a boil (see page 7)
1 pint	Water
8 ounces	Pearl barley
3 ounces	Butter
4 ounces	Carrots, washed, peeled, and diced **brunoise**
8 ounces	Potatoes, washed, peeled, and diced brunoise
2 ounces	Celery stalks, washed, trimmed, and diced brunoise
4 ounces	Fresh or frozen green beans, **blanched** and cut into ¼-inch pieces
½ ounce	Fresh parsley, washed, excess moisture removed, and chopped
	Salt and freshly ground black pepper, to taste
4 ounces	Sour cream

METHOD OF PREPARATION:

1. Drain the mushrooms, reserving liquid, and cut into ½-inch pieces.
2. In a saucepan, combine the stock and water, reserving the mushroom liquid. Heat to a boil, and add the barley; cover and simmer until the barley is **al dente.**
3. In a sauté pan, heat the butter, and add the carrots, potatoes, and celery; sauté until al dente. Add to the barley, and simmer for 30 minutes or until the vegetables are tender.
4. Add the green beans, and continue to cook for 5 minutes. Season, to taste. Serve immediately or hold at 140° F or above.
5. To serve, ladle the soup into a preheated cup, and garnish with sour cream.

COOKING TECHNIQUES:

Simmer, Sauté

Simmer and Poach:

1. Heat the cooking liquid to the proper temperature.
2. Submerge the food product completely.
3. Keep the cooked product moist and warm.

Sauté:

1. Heat the sauté pan to the appropriate temperature.
2. Evenly brown the food product.
3. For a sauce, pour off any excess oil, reheat, and deglaze.

GLOSSARY:

Brunoise: ⅛-inch dice
Blanch: to par cook
Al dente: to the bite

HACCP:

Hold at 140° F or above.

HAZARDOUS FOOD:

Sour cream

NUTRITION:

Calories: 219
Fat: 10.4 g
Protein: 4.73 g

CHEF NOTE:

In season, use fresh green beans.

Polynesian Chicken and Vegetable Soup

(Tinolang Manok)

YIELD: 10 SERVINGS. SERVING SIZE: EIGHT OUNCES.

COOKING TECHNIQUES:
Stir-Fry, Simmer

Stir-Fry:
1. Heat the oil in a wok until hot but not smoking.
2. Keep the food in constant motion; use the entire cooking surface.

Simmer and Poach:
1. Heat the cooking liquid to the proper temperature.
2. Submerge the food product completely.
3. Keep the cooked product moist and warm.

GLOSSARY:
Julienne: matchstick strips
Fork-tender: without resistance

HACCP:
Hold at 140° F or above.

HAZARDOUS FOOD:
Chicken

NUTRITION:
Calories: 225
Fat: 12.1 g
Protein: 22.1 g

INGREDIENTS:

2 tablespoons	Soybean oil
1 ounce	Ginger root, peeled and sliced into ⅛-inch pieces
2 cloves	Garlic, peeled and thinly sliced
12 ounces	Onions, peeled and cut **julienne**
1 each	Whole chicken, washed and cut into pieces
4 tablespoons	Fish sauce (see chef note)
2 quarts	White chicken stock, heated to a boil (see page 12)
1 pound	Winter melon, washed, peeled, seeded, and cut into 1-inch cubes
2 bunches	Bok choy, washed, trimmed, and cut into 2-inch sections
	Salt and ground white pepper, to taste
10 each	Scallion fans, washed

METHOD OF PREPARATION:

1. In a wok, heat the oil over moderate heat.
2. Add the ginger and garlic, and stir-fry until lightly brown.
3. Add the onions, and cook until translucent.
4. Add the chicken and lightly brown; then add the fish sauce and chicken stock, and simmer until the chicken is **fork-tender.**
5. Add the melon and bok choy, and simmer until the melon is tender but not overcooked, approximately 5 minutes.
6. Taste, and adjust the seasonings.
7. Serve in preheated soup cups or bowls, garnished with scallion fans, or hold at 140° F or above.

CHEF NOTE:
Fish sauce is known as *Patis* and is available in supermarkets or Oriental markets.

Purée of Celery Soup

YIELD: 50 SERVINGS. SERVING SIZE: EIGHT OUNCES.

INGREDIENTS:

6 ounces	Butter
3 pounds	Celery, washed, trimmed, and diced **brunoise**
1 pound	Onion, peeled and diced brunoise
6 cloves	Garlic, peeled and mashed
7 pounds	Potatoes, washed, peeled, and cut, into pieces
3 gallons	Vegetable stock (see pages 11), heated to a boil
	Salt and ground white pepper, to taste
½ teaspoon	Ground nutmeg

METHOD OF PREPARATION:

1. In a stockpot, heat the butter, and sauté the celery and onions until translucent. Add the garlic.
2. Add the stock and the potatoes. Heat to a boil. Reduce to a simmer.
3. Simmer the soup for 1 hour until the potatoes are tender.
4. When the potatoes and celery are tender, purée the soup in a robot coupe. Hold at 140° F.
5. Season to taste, and serve in a preheated soup cup.
6. Cool to an internal temperature of 40° F or lower within 6 hours.
7. Reheat to 165° F.

COOKING TECHNIQUES:

Sauté, Boil, Simmer

Sauté:

1. Heat the sauté pan to the appropriate temperature.
2. Evenly brown the food product.
3. For a sauce, pour off any excess oil, reheat, and deglaze.

Boil: (at sea level)

1. Bring the cooking liquid to a rapid boil.
2. Stir the contents, and cook the food product throughout.
3. Serve hot.

Simmer and Poach:

1. Heat the cooking liquid to the proper temperature.
2. Submerge the food product completely.
3. Keep the cooked product moist and warm.

GLOSSARY:

Brunoise: ⅛-inch dice

HACCP:

Hold at 140° F.
Chill to 40° F.

NUTRITION:

Calories: 128
Fat: 4.74 g
Protein: 5.85 g

CHEF NOTE:

For a smoother textured soup, add some roux or velouté.

COOKING TECHNIQUE:
Simmer

Simmer and Poach:
1. Heat the cooking liquid to the proper temperature.
2. Submerge the food product completely.
3. Keep the cooked product moist and warm.

NUTRITION:
Calories: 139
Fat: 4.87 g
Protein: 6.28 g

Purée of Potato Leek Soup

YIELD: 50 SERVINGS. SERVING SIZE: EIGHT OUNCES.

INGREDIENTS:

6 ounces	Butter, clarified
2 pounds	Leeks (use only the white part), washed, trimmed, split, and rough chopped into small pieces
6 cloves	Garlic, peeled and minced
7 pounds	Potatoes, peeled, washed, rough chopped into small pieces
3 gallons	Vegetable stock (see page 11)
	Salt, white pepper
½ teaspoon	Nutmeg
1 pound	Leeks, whites (garnish), washed, trimmed, split, and cross-cut

METHOD OF PREPARATION:

1. In a stockpot, heat the clarified butter or oil, and lightly sauté the leeks. Add the vegetable stock and potatoes, and heat to the first boil. Reduce to a simmer.
2. Simmer the soup until the potatoes are tender.
3. When the potatoes are tender, strain, and pass the mixture through a food mill or robot coupe.
4. Place the soup in a stockpot. Heat to a boil. Simmer to the desired consistency. Adjust the seasonings.
5. In a separate saucepan, poach the julienne of leeks in the vegetable stock. Add to the soup as a garnish.

Purée of Split Pea Soup

(Potage St. Germain)

YIELD: 10 SERVINGS. SERVING SIZE: EIGHT OUNCES.

INGREDIENTS:

4 ounces	Smoked bacon, diced small
1 ounce	Oil
4 cloves	Garlic, peeled and mashed
3 quarts	Vegetable stock, cold (see page 11)
1 pound	Split peas, sorted, washed, soaked, and drained
2 each	Ham hocks
	Salt and ground white pepper, to taste
	Croutons, as needed, diced, seasoned, and toasted with clarified butter

Mirepoix:

4 ounces	Onions, peeled
4 ounces	Carrots, washed and peeled
4 ounces	Celery, washed and trimmed

METHOD OF PREPARATION:

1. In a stockpot, **render** the bacon in oil. Add the mirepoix, and sauté until the onions are translucent.
2. Add the cold stock, ham hocks, and split peas. Heat to the first boil; then reduce to a simmer.
3. Simmer and **dépouiller** if necessary for about 2 hours or until the split peas and the meat from the ham hocks are tender.
4. Remove the ham hocks from the soup; debone and dice the meat, and strain the liquid. Purée the vegetables in a food processor. Combine the stock and puréed vegetables in the stockpot. Reheat to a boil. Garnish with the diced meat from the ham hocks. Adjust the seasonings.
5. Place in preheated soup cups, and serve with toasted croutons.

COOKING TECHNIQUE:

Simmer

Simmer and Poach:

1. Heat the cooking liquid to the proper temperature.
2. Submerge the food product completely.
3. Keep the cooked product moist and warm.

GLOSSARY:

Mirepoix: roughly chopped vegetables
Render: to melt fat
Dépouiller: to skim impurities/grease

NUTRITION:

Calories: 328
Fat: 10.2 g
Protein: 21.7 g

CHEF NOTES:

1. If the soup is kept in a bain-marie, the puréed peas will settle to the bottom. Mix well before serving.
2. To avoid the above, use brown roux to thicken the soup.

Purée of Tuscan Bean Soup

YIELD: 10 SERVINGS. SERVING SIZE: EIGHT OUNCES.

INGREDIENTS:

2 ounces	Bacon, ⅛-inch dice
Mirepoix:	
4 ounces	Onion, peeled
2 ounces	Carrots, washed and peeled
2 ounces	Celery stalk, washed and trimmed
6 cloves	Garlic, peeled and mashed into a purée
1½ quarts	White chicken or vegetable stock, heated to a boil (see pages 11 or 12)
1 pound	Smoked ham hocks
1 pound	Dried white cannellini beans, sorted, washed, and soaked in cold water overnight, then drained
2 pounds	Tomatoes, cored, blanched, peeled, seeded, and coarsely chopped
	Salt and ground white pepper, to taste
	Chopped fresh marjoram or basil leaves, to taste
	Croutons, as neeeded

METHOD OF PREPARATION:

1. In a medium stockpot, render the bacon. Add mirepoix and garlic, and sauté until the onions are translucent.
2. Add the stock and ham hocks. Heat to a boil; reduce the heat, and simmer for 45 minutes.
3. Add the beans, and continue to simmer for about 2 hours or until the white beans and the meat from the ham hocks are tender.
4. Remove the ham hocks from the soup; remove the meat from the bones, and cut into ¼-inch dice. Strain most of the liquid off; remove and purée half of the beans. Return the liquid, beans, and bean purée to the pot. Add the tomatoes and diced ham, and season to taste. Return the soup to a boil; reduce the heat and simmer 15 minutes.
5. Add the fresh herbs, and serve immediately in a preheated bowl, and garnish with croutons, or hold at 140° F or higher. Cool to 40° F within 6 hours.
6. Reheat to a boil.

COOKING TECHNIQUES:

Sauté, Simmer

Sauté:

1. Heat the sauté pan to the appropriate temperature.
2. Evenly brown the food product.
3. For a sauce, pour off any excess oil, reheat, and deglaze.

Simmer and Poach:

1. Heat the cooking liquid to the proper temperature.
2. Submerge the food product completely.
3. Keep the cooked product moist and warm.

GLOSSARY:

Mirepoix: roughly cut vegetables

HACCP:

Hold at 140° F or higher.
Cool to 40° F within 6 hours.

NUTRITION:

Calories: 380
Fat: 14.1 g
Protein: 28.7 g

CHEF NOTES:

1. Use caution when seasoning with salt, because the ham hocks will add salt to the soup. To reduce the fat in this recipe, use lean ham and a ham bone in place of the ham hocks.
2. This recipe has many variations. Any type of white beans may be substituted.
3. Diced proscuitto can be used instead of ham hocks and, in this case, should be added when the beans are half cooked.
4. Fresh fennel can be used instead of celery.

Rice and Sausage Soup, Venice Style

YIELD: 50 SERVINGS. SERVING SIZE: SIX OUNCES.

INGREDIENTS:

10 ounces	Pancetta (Italian fat-back bacon)
4 pounds	Italian sausage
3½ pounds	Onions, peeled and diced
3½ pounds	Turnips, washed, peeled, and diced
3½ ounces	Olive oil
3 gallons	White chicken stock (see page 12), heated to a boil
3 ounces	Freshly chopped parsley, excess moisture removed
5 ounces	Parmesan cheese, grated
	Salt and ground black pepper, to taste
½ pound	Arborio rice

METHOD OF PREPARATION:

1. Preheat the oven to 350° F.
2. Place the Italian sausage on a sheet pan, and cook in the oven until the meat is approximately half cooked. Remove, and slice 1-inch thick. Hold at 140° F.
3. Slice the pancetta, and dice it into ⅛-inch pieces. Place it in a small stockpot with the olive oil, and **render** until lightly browned. Remove the pancetta. Add the onions, and sauté until translucent.
4. Add the sausage and turnips. Continue to sauté for 5 minutes.
5. Strain off the fat. Return the pancetta to the stockpot, and add the chicken stock. Heat to a boil. Simmer.
6. Add the parsley, cheese, salt, and pepper.
7. Add the rice, stir, and simmer for 25 minutes or until the rice is tender. Hold at 140° F. Garnish with grated cheese.
8. Cool to an internal temperature of 40° F or lower within 6 hours.
9. Reheat to 165° F.

COOKING TECHNIQUES:

Sauté, Boil, Simmer

Sauté:

1. Heat the sauté pan to the appropriate temperature.
2. Evenly brown the food product.
3. For a sauce, pour off any excess oil, reheat, and deglaze.

Boil: (at sea level)

1. Bring the cooking liquid to a rapid boil.
2. Stir the contents, and cook the food product throughout.
3. Serve hot.

Simmer and Poach:

1. Heat the cooking liquid to the proper temperature.
2. Submerge the food product completely.
3. Keep the cooked product moist and warm.

HACCP:

Hold at 140° F.
Chill to 40° F.

HAZARDOUS FOOD:

Italian sausage

NUTRITION:

Calories: 308
Fat: 21 g
Protein: 12.2g

Roasted Garlic, Sun-Dried Tomatoes, and Cilantro Soup

YIELD: 10 SERVINGS. SERVING SIZE: EIGHT OUNCES.

COOKING TECHNIQUES:

Roast, Simmer

Roast:

1. Sear the food product, and brown evenly.
2. Elevate the food product in a roasting pan.
3. Determine doneness, and consider carryover cooking.
4. Let the food product rest before carving.

Simmer and Poach:

1. Heat the cooking liquid to the proper temperature.
2. Submerge the food product completely.
3. Keep the cooked product moist and warm.

HACCP:

Hold at 140° F or above.

HAZARDOUS FOOD:

Heavy cream

NUTRITION:

Calories: 321
Fat: 21.5 g
Protein: 5.53 g

INGREDIENTS:

4 heads (6 ounces)	Garlic
8 ounces (2 each)	Spanish onions
4 ounces	Spanish olive oil
3 quarts	White chicken or veal stock, heated to a boil (see page 12)
	Salt and freshly ground pepper, to taste
3 ounces	Diced, sun-dried tomatoes, drained from oil (reserve oil)
8 ounces	Heavy cream
1 each	Lemon, zest only, minced
10 each	Thin slices sour dough bread or corn bread, brushed with reserved oil from tomatoes, and toasted
3 ounces	Cilantro leaves, chopped

METHOD OF PREPARATION:

1. Preheat the oven to 350° F.
2. Roll the onions and garlic in olive oil, and wrap each individually in aluminum foil. Reserve the remaining oil. Place on a baking tray, and roast until tender. This will take approximately 1½ hours for the onions and 45 minutes for the garlic.
3. Remove from the oven, and allow to cool slightly. Peel the onions, and squeeze out the purée from garlic, discarding the skins.
4. Purée the onions and garlic with the remaining olive oil.
5. In a medium-sized saucepan, add the purée to the boiling stock. Return the mixture to a boil, then lower heat, and simmer until reduced by one third.
6. Add the tomatoes, and season, to taste. Then add cream, and continue to simmer until desired consistency is achieved. Hold at 140° F or above.
7. Just before service, add the lemon zest.
8. To serve, place one toasted bread slice in the bottom of a preheated soup cup. Ladle the soup over the bread, and serve generously garnished with cilantro.

CHEF NOTES:

1. Sun-dried tomatoes need cooking time; otherwise they remain chewey.
2. Onion and garlic can be adjusted to guests' taste.

Scotch Barley Soup

YIELD: 50 SERVINGS. SERVING SIZE: EIGHT OUNCES.

INGREDIENTS:

3 pounds	Lamb meat, cut into ¼-inch dice
8 ounces	Clarified butter

Mirepoix:

8 ounces	Onions, peeled and diced brunoise
8 ounces	Leek whites, washed, split, and diced brunoise
3 stalks	Celery, washed, trimmed, and diced brunoise
8 ounces	Carrots, washed, peeled, and diced brunoise
8 ounces	Turnips, washed, peeled, and diced brunoise
3 gallons	Lamb stock or brown stock (see page 7), heated to a boil
2 pounds	Barley, washed in cold running water, may be precooked
	Salt and ground black pepper, to taste

METHOD OF PREPARATION:

1. In a large stockpot, sauté the mirepoix in the clarified butter. Add the diced lamb, and sweat, covered, for about 5 minutes or until the vegetables are translucent.
2. Add the stock; heat to the first boil, and reduce to a simmer.
3. Add the barley, and simmer until the barley and meat are fully cooked and tender. Hold at 140° F.
4. Adjust the seasonings, and serve in preheated soup cups.
5. Cool to an internal temperature of 40° F or lower within 6 hours.
6. Reheat to 165° F.

COOKING TECHNIQUES:

Sauté, Boil, Simmer

Sauté:

1. Heat the sauté pan to the appropriate temperature.
2. Evenly brown the food product.
3. For a sauce, pour off any excess oil, reheat, and deglaze.

Boil: (at sea level)

1. Bring the cooking liquid to a rapid boil.
2. Stir the contents, and cook the food product throughout.
3. Serve hot.

Simmer and Poach:

1. Heat the cooking liquid to the proper temperature.
2. Submerge the food product completely.
3. Keep the cooked product moist and warm.

GLOSSARY:

Mirepoix: roughly chopped vegetables

HACCP:

Hold at 140° F.
Chill to 40° F.

HAZARDOUS FOOD:

Lamb meat

NUTRITION:

Calories: 227
Fat: 9.45 g
Protein: 13 g

CHEF NOTE:

Scotch barley soup is native to Scotland, and the main ingredient is lamb meat. Barley is a cereal or grain used in soups, stews, and porridge. In certain Eastern European countries it is served as a starch. This soup also can be served as a main meal; the barley is used to extend the meat.

She-Crab Soup

YIELD: 10 SERVINGS. SERVING SIZE: EIGHT OUNCES.

INGREDIENTS:

3 ounces	Butter
3 ounces	Onions, diced **brunoise**
2 ounces	All-purpose flour
3 cloves	Garlic, finely minced
2 quarts	Seafood or fish stock, heated to a boil (see page 10)
2 each	Bay leaves
	Salt and ground white pepper, to taste
10 ounces	Crab meat, cleaned and shredded
5 ounces	Heavy cream

METHOD OF PREPARATION:

1. In a suitable **marmite,** melt the butter, add the onions, and sauté until translucent, but do not brown.
2. Add the flour, and make a blonde roux.
3. Add the garlic to the blonde roux; then add the stock and bay leaves. Reduce the heat, and simmer 10 minutes.
4. Remove the bay leaves, and season to taste; then add the crab meat, and continue to simmer 5 more minutes.
5. **Temper** the cream, and add to the soup.
6. Serve immediately in preheated cups, or hold at 140° F.

COOKING TECHNIQUE:

Simmer

Simmer and Poach:

1. Heat the cooking liquid to the proper temperature.
2. Submerge the food product completely.
3. Keep the cooked product moist and warm.

GLOSSARY:

Brunoise: ⅛-inch dice
Marmite: stockpot
Temper: to equalize two extreme temperatures

HACCP:

Hold at 140° F.

HAZARDOUS FOODS:

Crabmeat
Heavy cream

NUTRITION:

Calories: 174
Fat: 13 g
Protein: 7.6 g

CHEF NOTE:

Optional ingredients that can be added to the soup are sherry, pimientos, diced green bell peppers, and roe. This soup is very popular in South Carolina.

Southern Vegetable Soup

YIELD: 10 SERVINGS. SERVING SIZE: EIGHT OUNCES.

INGREDIENTS:

2 ounces	Salt pork, cut into small dice
10 ounces	Beef, bottom round, cut into small cubes
8 ounces	Canned peeled tomatoes, drained, seeded, and chopped
3½ quarts	Beef stock, heated to a boil (see page 12)
2 ounces	Frozen green beans
2 ounces	Red beans, cooked
4 ounces	Onions, peeled and diced **brunoise**
3 ounces	Celery stalks, washed, trimmed, and diced brunoise
6 ounces	Green cabbage, washed, cored and **chiffonade**
3 ounces	Carrots, washed, peeled, and diced brunoise
2 ounces	Frozen corn kernels
2 ounces	Frozen okra, sliced
2 ounces	Zucchini, washed, trimmed, and cut in ½-inch dice
	Salt and freshly ground black pepper, to taste

METHOD OF PREPARATION:

1. In a large **marmite,** place the salt pork, and render the fat, stirring frequently until browned. Add the beef, reduce the heat, and sauté until browned.
2. Add the tomatoes, and sauté for another 2 minutes.
3. Add the boiling stock, and simmer until the meat is **al dente.**
4. Add all other ingredients, and continue to simmer until vegetables are tender. Season, to taste, and serve immediately in preheated cups, or hold at 140° F or higher.

COOKING TECHNIQUES:

Sauté, Boil, Simmer

Sauté:

1. Heat the sauté pan to the appropriate temperature.
2. Evenly brown the food product.
3. For a sauce, pour off any excess oil, reheat, and deglaze.

Boil: (at sea level)

1. Bring the cooking liquid to a rapid boil.
2. Stir the contents, and cook the food product throughout.
3. Serve hot.

Simmer and Poach:

1. Heat the cooking liquid to the proper temperature.
2. Submerge the food product completely.
3. Keep the cooked product moist and warm.

GLOSSARY:

Brunoise: ⅛-inch dice
Chiffonade: ribbons of leafy greens
Marmite: stockpot
Al dente: to the bite

HACCP:

Hold at 140° F.

HAZARDOUS FOOD:

Beef

NUTRITION:

Calories: 180
Fat: 11.8 g
Protein: 8.7 g

SOUPS

Spanish Vegetable Soup

(Catalan)

YIELD: 10 SERVINGS. SERVING SIZE: EIGHT OUNCES.

COOKING TECHNIQUE:
Simmer

Simmer and Poach:
1. Heat the cooking liquid to the proper temperature.
2. Submerge the food product completely.
3. Keep the cooked product moist and warm.

GLOSSARY:
Brunoise: ⅛-inch dice
Al dente: to the bite

HACCP:
Hold at 140° F or above.

NUTRITION:
Calories: 247
Fat: 9.13 g
Protein: 8.91 g

INGREDIENTS:

3 quarts	Beef stock, heated to a boil (see page 12)
6 ounces	Onions, peeled and diced **brunoise**
6 ounces	Carrots, washed, peeled, and diced brunoise
6 ounces	Turnips, washed, peeled, and diced brunoise
3 ounces	Celery, washed and diced brunoise
1 pound	Potatoes, washed, peeled, and diced
2 cloves	Garlic, peeled and mashed to a purée
2 each	Bay leaves
6 ounces	Long-grained rice, soaked in cold water for 1 hour
8 ounces	Cured chorizo sausage, skin removed, and cut in ½-inch dice
1 pinch	Ground saffron
	Salt and ground white pepper, to taste

METHOD OF PREPARATION:

1. In a large stockpot, combine the beef stock and vegetables with the garlic and bay leaves. Heat the liquid to a boil; reduce the heat, and simmer until the vegetables are **al dente.**
2. Drain the rice, and add to the stockpot. Add the chorizo and saffron; continue to simmer for 15 to 20 minutes more or until the rice is tender.
3. Season, to taste, and serve immediately in a preheated cup or bowl, or hold at 140° F or above.

CHEF NOTE:
Spain is noted for hearty, delicious soups. This soup is from Catalonia and may be served as a meal with crusty bread.

Spinach Soup

YIELD: 10 SERVINGS. SERVING SIZE: EIGHT OUNCES.

INGREDIENTS:

2 ounces	Clarified butter
2 pounds	Fresh spinach, washed, stemmed, and roughly chopped
4 ounces	Blonde roux
2 quarts	Brown beef stock, heated to a boil (see page 7)
	Salt and ground white pepper, to taste
1 tablespoon	Lemon juice
2½ ounces	Pasteurized egg yolks
5 ounces	Feta cheese
5 ounces	Croutons

METHOD OF PREPARATION:

1. In a saucepan, heat the butter, and **sweat** the spinach for 5 minutes.
2. In a separate saucepan, **temper** the roux with beef stock, and then gradually whisk in the remaining hot beef stock. Blend well, and heat to a boil; then combine with the spinach.
3. Simmer for 30 minutes. Season, to taste, and add lemon juice.
4. Temper the egg yolks, and add to the soup. Serve immediately, or hold at 140° F or above.
5. For service, place ½ ounce of feta cheese in a preheated cup; add the croutons, and ladle the soup in the cup.

COOKING TECHNIQUE:

Simmer

Simmer and Poach:

1. Heat the cooking liquid to the proper temperature.
2. Submerge the food product completely.
3. Keep the cooked product moist and warm.

GLOSSARY:

Sweat: sauté under a cover
Temper: to equalize two extreme temperatures

HACCP:

Hold at 140° F or above.

HAZARDOUS FOOD:

Egg yolks

NUTRITION:

Calories: 274
Fat: 18.1 g
Protein: 9.56 g

CHEF NOTE:

This soup will lose color if it is held too long in heat.

SOUPS

Vegetable Soup, Asti Style

YIELD: 50 SERVINGS. SERVING SIZE: EIGHT OUNCES.

INGREDIENTS:

2 pounds	White beans sorted, washed and soaked overnight in cold water
3 gallons	White chicken stock (see page 12), heated to a boil
6 ounces	Salt pork, rind removed, diced medium
	Oil as needed
12 ounces	Onions, peeled and diced **brunoise**
4 stalks	Celery, washed, trimmed, and diced brunoise
12 ounces	Carrots, washed, peeled, and diced
2 pounds	White cabbage, cored and cut **chiffonade**
6 cloves	Garlic, peeled and mashed
12 ounces	Potatoes, washed, peeled, and diced brunoise
12 ounces	Pasta, ditalini
6 ounces	Freshly chopped basil, excess moisture removed
	Salt and ground white pepper, to taste
1 pound	Parmesan cheese, grated

METHOD OF PREPARATION:

1. In a large stockpot, **render** the salt pork in a small amount of oil. When the salt pork is rendered, remove it from the pot; then add the vegetables, and sauté. Next, add the chicken stock. Heat to a boil. Simmer.
2. Starting with cold water in a separate pot, cook the white beans until they are tender. Strain, and add to the soup.
3. Starting with cold water in a separate saucepan, cook the potatoes until they are tender. Strain, and add to the soup.
4. In boiling, salted water with oil, cook the ditalini until it is al dente. Shock it under cold running water; drain and oil it slightly, and set it aside.
5. Simmer the soup for about 30 minutes until the proper flavor is achieved. Adjust the seasonings with fresh basil, salt, and pepper. Hold at 140° F.
6. Place the pasta in the bottom of a preheated soup cup. Pour the hot soup over it. Top with ½ teaspoon grated Parmesan cheese, and serve.
7. Cool to an internal temperature of 40° F or lower within 6 hours.
8. Reheat to 165° F.

COOKING TECHNIQUES:

Sauté, Boil, Simmer

Sauté:

1. Heat the sauté pan to the appropriate temperature.
2. Evenly brown the food product.
3. For a sauce, pour off any excess oil, reheat, and deglaze.

Boil: (at sea level)

1. Bring the cooking liquid to a rapid boil.
2. Stir the contents, and cook the food product throughout.
3. Serve hot.

Simmer and Poach:

1. Heat the cooking liquid to the proper temperature.
2. Submerge the food product completely.
3. Keep the cooked product moist and warm.

GLOSSARY:

Brunoise: ⅛-inch dice
Chiffonade: ribbons of leafy greens
Render: to melt fat

HACCP:

Hold at 140° F.
Chill to 40° F.

NUTRITION:

Calories: 253
Fat: 11.5 g
Protein: 11.6 g

CHEF NOTE:

This recipe is basically a white minestrone soup. Asti is in the northern region of Italy, by the Piedmont area, and not well known for the use of tomatoes. Asti, however, is known for the famous sparkling wine, Asti Spumante.

Velouté of Vegetable Soup

YIELD: 50 SERVINGS. SERVING SIZE: EIGHT OUNCES.

COOKING TECHNIQUE:

Simmer

Simmer and Poach:

1. Heat the cooking liquid to the proper temperature.
2. Submerge the food product completely.
3. Keep the cooked product moist and warm.

GLOSSARY:

Mirepoix: roughly chopped vegetables
Concassé: peeled, seeded, and roughly chopped
Sweat: to sauté under cover
Singer: to dust with flour

HACCP:

Hold at 140° F.
Chill to 40° F.

NUTRITION:

Calories: 206
Fat: 10.3 g
Protein: 7.35 g

INGREDIENTS:

1 pound	Clarified butter
1 pound	All-purpose flour
3 pounds	Tomatoes, **concassé**
12 cloves	Garlic, peeled and mashed
12 ounces	Tomato purée
3¼ gallons	Vegetable stock (see page 11), heated to a boil
	Salt and ground white pepper, to taste
3 teaspoons	Thyme leaves
2 teaspoons	Basil
2 teaspoons	Marjoram
1 teaspoon	Fennel seeds
6 ounces	Freshly chopped parsley, excess moisture removed
1 pound	Brown rice, cooked

Mirepoix:

1½ pounds	Onions, peeled
¾ pound	Leek whites, washed, trimmed, and split
4 stalks	Celery, washed and trimmed
¾ pound	Carrots, washed and peeled

METHOD OF PREPARATION:

1. In a stockpot, melt the clarified butter. Add the mirepoix, and **sweat** until the onions are translucent. **Singer;** then cook for 5 minutes on low heat to create a thickening agent. Add the tomatoes, and cook for an additional 2 minutes. Set aside.
2. Heat the vegetable stock to a boil, and add mirepoix; whisk well to eliminate lumps of the flour. Heat to a second boil. Add the tomato purée, herbs, and spices. Reduce the heat, and simmer for about 30 minutes, or until the carrots are tender and the proper flavor is achieved. Add the cooked brown rice. Hold at 140° F.
3. Strain and purée the vegetable; return puréed vegetable to the liquid. Bring to a boil. Add the cooked rice. Hold at 140° F.
4. Adjust the seasonings, and serve hot in preheated soup cups.
5. Cool to an internal temperature of 40° F or lower within 6 hours.
6. Reheat to 165° F.

Velouté Chartreuse

COOKING TECHNIQUE:
Simmer

Simmer and Poach:
1. Heat the cooking liquid to the proper temperature.
2. Submerge the food product completely.
3. Keep the cooked product moist and warm.

HACCP:
Hold at 140° F or higher.

HAZARDOUS FOOD:
Heavy cream

NUTRITION:
Calories: 304
Fat: 25 g
Protein: 7.8 g

YIELD: 10 SERVINGS. SERVING SIZE: EIGHT OUNCES.

INGREDIENTS:

4 ounces	Butter, clarified
4 ounces	Flour, all-purpose
3 quarts	White chicken stock (see page 12)
3 each	Tomatoes, washed, cored, peeled, seeded, and diced
4 ounces	Spinach, washed, dried, and cut large
4 ounces	Mushrooms, washed and sliced
2 ounces	Butter
1 teaspoon	Chervil
4 ounces	Foie gras (goose liver pâté)
10 ounces	Heavy cream
	Salt and pepper to taste

METHOD OF PREPARATION:

1. In a skillet, heat 2 ounces of butter, and sauté the mushrooms. Reserve.
2. In a 1-gallon stockpot, place the clarified butter; heat, add flour, stir, and make a white roux. Cook for 3 minutes. Add the chicken stock; stir, and heat to a boil. Simmer for 20 minutes.
3. Add the tomatoes, spinach, mushrooms, and chervil, and simmer for 15 minutes.
4. Dilute the foie gras in heavy cream, and add the foie gras to the pot. Heat, and hold at 140° F. Adjust the seasoning with salt and pepper. Mix well; then serve.

Velouté of Potato Soup

YIELD: 50 SERVINGS. SERVING SIZE: EIGHT OUNCES.

INGREDIENTS:

1 pound	Butter
1 pound	Shallots, peeled, and diced **brunoise**
1½ pound	Celery, washed, trimmed, and diced brunoise
5 pounds	Potato, washed, peeled, and diced
1¾ pound	White roux
2½ gallons	Vegetable stock (see page 11), heated to a boil
1 tsp.	Nutmeg
	Salt and ground white pepper, to taste

METHOD OF PREPARATION:

1. Place butter into a stock pot to melt.
2. Add the shallots; sauté lightly, and then add the celery and potatoes. Sauté for 10 minutes.
3. Add the boiling vegetable stock, and reheat to a boil. Reduce the heat, and simmer until the potatoes are tender.
4. Dilute the roux with some of the hot liquid; then add to the soup, and continue simmering for 10 more minutes.
5. Purée the soup in a robot coupe; return it to the stove, and reheat to a boil. Season, to taste, and hold at 140° F.
6. Serve the soup in a preheated soup cup.

COOKING TECHNIQUE:
Simmer

Simmer and Poach:
1. Heat the cooking liquid to the proper temperature.
2. Submerge the food product completely.
3. Keep the cooked product moist and warm.

GLOSSARY:
Brunoise: ⅛-inch dice

HACCP:
Hold at 140° F.

NUTRITION:
Calories: 224
Fat: 13.9 g
Protein: 6.48 g

Velouté of Tomato Soup

YIELD: 50 SERVINGS. SERVING SIZE: EIGHT OUNCES.

INGREDIENTS:

1¼ pint	Butter, clarified
13 ounces	Onion, peeled and diced **brunoise**
13 ounces	Carrots, washed, peeled, and diced brunoise
4 stalks	Celery, washed, trimmed, and diced brunoise
3 teaspoons	Garlic, peeled, and minced
13 ounces	All-purpose flour
1 #10 can	Tomatoes, drained, pulp puréed in robot coupe
6¼ tablespoons	Tomato paste
6¼ quarts	White veal stock (see page 12), heated to a boil
1 quart	Heavy cream, heated to a boil
6 each	Bay leaves
	Salt and ground white pepper, to taste

METHOD OF PREPARATION:

1. In a stockpot, heat the clarified butter. Add the carrots, onions, and celery, and then sauté. Stir, cooking, until the onions are translucent. Add the garlic, and stir. (Do not burn the garlic.)
2. **Singer** and cook roux on low heat. Add the tomato paste, and cook for 3 minutes.
3. Add the stock slowly, and stir vigorously. Heat to a boil, and add the bay leaves. Simmer.
4. Add the canned tomatoes, and cook for 1¼ hours, stirring occasionally. Then remove the bay leaves.
5. When cooked, pass the soup through a **chinois** fin, and purée the solid ingredients in a robot coupe.
6. Put everything but the bay leaves back into the stockpot; and heat to a boil. Dilute the warm cream, stir, and add to the velouté. Hold at 140° F.
7. Adjust the seasonings as needed.
8. Cool to an internal temperature of 40° F or lower within 6 hours.
9. Reheat to 165° F.

COOKING TECHNIQUES:
Sauté, Boil

Sauté:
1. Heat the sauté pan to the appropriate temperature.
2. Evenly brown the food product.
3. For a sauce, pour off any excess oil, reheat, and deglaze.

Boil: (at sea level)
1. Bring the cooking liquid to a rapid boil.
2. Stir the contents, and cook the food product throughout.
3. Serve hot.

GLOSSARY:
Brunoise: ⅛-inch dice
Singer: to dust with flour
Chinois: medium cone-shaped strainer

HACCP:
Hold at 140° F.
Chill to 40° F.

HAZARDOUS FOOD:
Heavy cream

NUTRITION:
Calories: 258
Fat: 20 g
Protein: 3.94 g

CHEF NOTE:
In quantity food production, add cream only to the quantity of soup being served. Do not cream all of the soup at one time.

Velouté of Mushrooms

(Velouté au Girolle)

YIELD: 10 SERVINGS. SERVING SIZE: EIGHT OUNCES.

INGREDIENTS:

6 ounces	Butter
10 ounces	Girolles (chanterelles), cleaned and carefully sliced
10 ounces	Mushrooms, cleaned and puréed in a robot coupe
4 ounces	All-purpose flour
2½ quarts	White chicken stock, heated to a boil (see page 12)
8 ounces	Heavy cream
	Salt and ground white pepper, to taste

METHOD OF PREPARATION:

1. Sauté the mushrooms slowly; let them remain in the butter for 3 to 4 minutes.
2. Add the flour; stir, and make a roux blonde. Cook for 3 minutes. Add the stock; simmer for 30 minutes, and stir.
3. In 2 ounces of butter, sauté the girolles for 5 minutes.
4. Add the sautéed girolles to soup and simmer for 5 minutes. **Temper** the cream, and add to the soup. Season with salt and pepper. Taste, and adjust the seasonings, as necessary.

COOKING TECHNIQUES:

Sauté, Simmer

Sauté:

1. Heat the sauté pan to the appropriate temperature.
2. Evenly brown the food product.
3. For a sauce, pour off any excess oil, reheat, and deglaze.

Simmer and Poach:

1. Heat the cooking liquid to the proper temperature.
2. Submerge the food product completely.
3. Keep the cooked product moist and warm.

GLOSSARY:

Temper: to equalize two extreme temperatures

HAZARDOUS FOOD:

Heavy cream

NUTRITION:

Calories: 279
Fat: 17.5 g
Protein: 10.1 g

CHEF NOTE:

If the girolles are canned, make sure to add the liquid to the soup.

White Kidney Bean Soup

(Crème Bretonne)

YIELD: 10 SERVINGS. SERVING SIZE: EIGHT OUNCES.

INGREDIENTS:

4 ounces	Butter, melted
1 ounce	Onions, peeled and diced **brunoise**
1 pound	Leeks (white part only), washed, split, rewashed, and chopped
2½ quarts	White chicken stock, heated to a boil (see page 12)
1 pound	Dried white kidney beans, sorted, washed, and soaked overnight in cold water, then drained
2 tablespoons	Tomato paste
	Salt and ground white pepper, to taste
8 ounces	Heavy cream
5 ounces	Croutons

METHOD OF PREPARATION:

1. In a stockpot, heat the butter, and sauté the onions and leeks until translucent. Add the chicken stock, beans, and tomato paste. Heat the liquid to a boil; reduce the heat, and simmer until the beans are tender. Add more liquid, if necessary.
2. Strain the liquid, and reserve. Purée the vegetables and beans in a food processor.
3. Return puréed beans to the liquid. Heat to a boil; reduce heat to a simmer.
4. Temper the heavy cream and add to the simmering soup. Taste and adjust seasoning.
5. Serve immediately, or hold at 140° F or above. Cool to 40° F or below within 6 hours.
6. To serve, ladle the soup into a preheated soup cup or bowl. Garnish with croutons.

COOKING TECHNIQUE:

Simmer

Simmer and Poach:

1. Heat the cooking liquid to the proper temperature.
2. Submerge the food product completely.
3. Keep the cooked product moist and warm.

GLOSSARY:

Brunoise: ⅛-inch dice
Temper: to equalize two extreme temperatures

HACCP:

Hold at 140° F or above, and cool to 40° F or under.

HAZARDOUS FOOD:

Heavy cream

NUTRITION:

Calories: 397
Fat: 18.4 g
Protein: 17.2 g

SOUPS

Wisconsin Cheddar Cheese Soup

YIELD: 10 SERVINGS. SERVING SIZE: EIGHT OUNCES.

INGREDIENTS:

4 ounces	Clarified butter
1 teaspoon	Paprika
3 ounces	Bread flour
1 quart	White chicken stock, heated to a boil (see page 12)
1 pound	Wisconsin cheddar cheese, finely grated
1 quart	Milk, heated to a boil
	Salt, to taste
	Hot sauce, to taste

Brunoise:

4 ounces	Carrots, washed and peeled
4 ounces	Celery, washed and peeled
4 ounces	Green pepper, roasted, peeled, and seeded
4 ounces	Red pepper, washed, dried, roasted or blistered, peeled, and seeded
4 ounces	Onions, peeled

METHOD OF PREPARATION:

1. In a stockpot or saucepot, heat the clarified butter, and sauté the vegetables. Add the paprika and lightly dust with flour. Cook for 5 to 7 minutes to create a thickening agent.
2. Add the hot chicken stock, and heat to a boil. Reduce to a simmer. Cook for 20 to 30 minutes or until the vegetables are tender. Gradually add the grated cheese and milk, and stir constantly until combined.
3. Add the seasonings, and serve in preheated soup cups, or hold at 140° F.
4. Cool to an internal temperature of 40° F or lower within 6 hours.
5. Reheat to 165° F.

COOKING TECHNIQUE:

Simmer

Simmer and Poach:

1. Heat the cooking liquid to the proper temperature.
2. Submerge the food product completely.
3. Keep the cooked product moist and warm.

GLOSSARY:

Brunoise: ⅛-inch dice
Temper: to equalize two extreme temperatures

HACCP:

Hold at 140° F.
Cool to an internal temperature of 40° F or lower.
Reheat to 165° F.

HAZARDOUS FOOD:

Milk

NUTRITION:

Calories: 392
Fat: 29.2 g
Protein: 16.3 g

CHEF NOTE:

Add the cheese and milk carefully, so that the soup does not scorch (burn).

Zuppa Millefanti

YIELD: 10 SERVINGS. SERVING SIZE: EIGHT OUNCES.

COOKING TECHNIQUE:
Simmer

Simmer and Poach:
1. Heat the cooking liquid to the proper temperature.
2. Submerge the food product completely.
3. Keep the cooked product moist and warm.

HAZARDOUS FOOD:
Eggs

NUTRITION:
Calories: 131
Fat: 5.78 g
Protein: 10 g

INGREDIENTS:

2 quarts	Beef broth, heated to a boil (see page 111)
	Salt and freshly ground black pepper, to taste
4 ounces	Fresh bread crumbs
4 ounces	Parmesan cheese, grated
3 each	Eggs
2 ounces	Fresh parsley, washed, excess moisture removed, and chopped

METHOD OF PREPARATION:

1. In a saucepot, heat the broth to a boil, and season, to taste.
2. Combine the bread crumbs, cheese, eggs, and parsley, and mix well.
3. Add the mixture to the boiling broth; do not stir, but allow the mixture to float to the surface of the soup.
4. After 1 minute, whisk vigorously to create different forms.
5. Serve immediately in preheated consommé cups.

4

Salads and Dressings

Salads and Dressings

Lettuce, like conversation, requires a good deal of oil, to avoid friction, and keep the company smooth.

Charles Dudly Warner
Writer, 1871

To be a cook is to understand that food preparation is more than a profession. It is a discipline: a discipline that requires practice of its craft; a discipline that requires the continuing respect for basic fundamentals of the art and science of food and its preparation. It is a discipline that requires the daily study of its heritage, its present climate, and its possible future trends. It is a discipline in which we share differences and commonality with the world, but it is one in which one person can have a great impact and make a difference. Most of all, it is a discipline and a profession that not only builds character, but also demonstrates it on a daily basis.

Here at Johnson & Wales, you will learn the fundamentals of preparation and the science of food. In your hearts you will find the character and discipline to be a chef.

Keith Keogh, CEC, AAC
Distinguished Visiting Chef
October 27–29, 1991
National President
American Culinary Federation

The HACCP Process

The following recipe illustrates how salads flow through the HACCP process.

Caesar Salad

Receiving

- Vegetables (garlic and romaine)—packaging intact; no cross-contamination from other foods on the truck; no signs of insect or rodent activity
- Spices (salt and pepper)—packaging intact
- Anchovies—packaging intact
- Mustard, Dijon—packaging intact
- Olive oil—packaging intact
- Vinegar, wine—packaging intact
- Lemon juice—packaging intact
- Pasteurized eggs—container intact; temperature at 40° F (4.4° C) or lower
- Croutons—packaging intact
- Cheese, Parmesan—Check to see that each type has its characteristic flavor and texture, as well as uniform color. Cheese should be delivered at 40° F (4.4° C) or lower.

Storage

- Vegetables (garlic and romaine)—Store under refrigeration, with a product temperature not to exceed 40° F (4.4° C). Store above and away from raw, potentially hazardous foods.
- Spices (salt and pepper)—Store in dry storage at 50° F (10° C) with a relative humidity of 50% to 60%. Keep dry.
- Anchovies—Store in dry storage at 50° F (10° C) with a relative humidity of 50% to 60%.
- Mustard, Dijon—Store in dry storage at 50° F (10° C) with a relative humidity of 50% to 60%.
- Olive oil—Store in dry storage at 50° F (10° C) with a relative humidity of 50% to 60%.
- Vinegar, wine—Store in dry storage at 50° F (10° C) with a relative humidity of 50% to 60%.
- Lemon juice—Store in dry storage at 50° F (10° C) with a relative humidity of 50% to 60%.
- Pasteurized eggs—Store under refrigeration at 40° F (4.4° C) or lower. Store above and away from raw, potentially hazardous foods.
- Croutons—Store in dry storage at 50° F (10° C) with a relative humidity of 50% to 60%.
- Cheese, Parmesan—Store under refrigeration, with a product temperature not to exceed 40° F (4.4° C). Store above and away from raw, potentially hazardous foods.

Preparation and Cooking

- Season the bowl with the garlic clove, and use the salt as an abrasive. Remove the salt and garlic when the bowl is well seasoned.
- Add the anchovy, mashed with a fork. Move the anchovy to one side of the bowl, and add the pasteurized egg and mustard.
- Blend the oil into the egg, slowly and steadily, forming an emulsion.
- Add the wine vinegar, lemon juice, and pepper. Mix well.
- Add the romaine, and roll it into all of the previous ingredients by rotating the service spoon and fork from the back to the front of the bowl until the lettuce is fully coated.
- Add the croutons, and roll as directed in the preceding step.
- Add the cheese, and roll. Serve immediately.

Holding and Service

- Serve immediately.

Note

- Use pasteurized eggs when no cooking is done.
- All of the ingredients should be chilled at 40° F (4.4° C) before preparation takes place.

Nutritional Notes

Most salad ingredients are a concentrated source of nutrients. Some of the ingredients commonly added, however, are high in fat and, therefore, calories. Just a few of these ingredients can nutritionally ruin a perfectly good salad. The following is a partial list of common salad ingredients that should be omitted or used sparingly:

- *Salad dressing—4 tablespoons add about 300 calories and 32 grams of fat.*
- *Sunflower seeds—1 tablespoon adds 52 calories and 9 grams of fat.*
- *Bacon—1 slice adds 36 calories and 3 grams of fat.*
- *Cheese—1 ounce adds 110 calories and about 9 grams of fat.*
- *Oil—1 teaspoon adds 40 calories and 4.7 grams of fat.*

The following are suggestions for lowering fat, saturated fat, and cholesterol in salads:

- *Use baked, unbuttered croutons, made from whole-grain bread, to increase fiber and nutrition.*
- *Use low-fat or nonfat cheeses.*
- *Use commercial low-fat or nonfat mayonnaise. (It is difficult to make low-fat or nonfat mayonnaise in the kitchen.)*
- *Make salad dressings by using mono- and polyunsaturated oils, such as olive, canola, sesame seed, corn, safflower, soybean, or walnut. All oils have the same amount of fat and calories, but they differ in the amount of saturated and unsaturated fats they contain.*
- *Use flavorful oils such as olive, sesame seed, or walnut. The more intense the flavor of the oil, the less oil is needed, thereby reducing total fat.*
- *Vary the flavor by using vinegars such as raspberry, apricot, or balsamic, and infuse oils and vinegars with herbs. Try using fruit juices in the dressing. These robust flavors also can take the place of salt.*
- *Use egg whites in place of whole eggs. When making emulsified salad dressings, use egg substitutes in place of yolks, or reduce the number of yolks used.*
- *When serving the salad, leave high-fat dressings on the side, so that the customer can control the amount added to the salad.*

Main-dish salads can be meals full of high fiber, vitamins, minerals, protein, and complex carbohydrates if the ingredients are selected carefully. For example, fresh, crisp vegetables can be complemented by seafood, low-fat poultry, or lean meats. Additionally, dressings can be nonfat or low-fat commercial varieties or exotic house dressings made according to the aforementioned guidelines.

American Salad

(Salade Amèricaine)

YIELD: 10 SERVINGS. SERVING SIZE: FOUR TO FIVE OUNCES.

COOKING TECHNIQUE:
Not applicable

GLOSSARY:
Julienne: matchstick strips

NUTRITION:
Calories: 402
Fat: 29.4 g
Protein: 5.48 g

INGREDIENTS:

1 head	Romaine
5 each	Tomatoes, washed, peeled, seeded, and sliced into ¼-inch pieces
2 pounds	New potatoes, washed, poached and sliced into ¼-inch pieces
1 pound	Celery, washed and cut **julienned**
1 each	Onion (medium), peeled and sliced thin
3 each	Eggs, hard-boiled
20 ounces	Vinaigrette (see page 230)

METHOD OF PREPARATION:

1. Line a serving plate with the salad greens.
2. Arrange the slices of tomatoes and potatoes over the greens.
3. Peel the hard-boiled eggs, and slice them with an egg slicer. Arrange the egg slices over the tomatoes and potatoes.
4. Sprinkle the julienne of celery and onions rings on top of the eggs.
5. Flavor with vinaigrette.

CHEF NOTES:

1. Boil the potatoes a day in advance in moderately salted water.
2. Always choose uniform size potatoes.

Beet and Onion Salad

YIELD: 10 SERVINGS. SERVING SIZE: SIX OUNCES.

COOKING TECHNIQUE:
Simmer

Simmer and Poach:
1. Heat the cooking liquid to the proper temperature.
2. Submerge the food product completely.
3. Keep the cooked product moist and warm.

HACCP:
Hold at 40° F or below.

NUTRITION:
Calories: 228
Fat: 11.9 g
Protein: 3.44 g

INGREDIENTS:

3 pounds	Fresh beets, washed and trimmed, leaving 1-inch of stems and roots
	Cold water, as needed
1 ounce	Red wine vinegar
8 ounces	Onions, peeled, sliced, and separated into rings
2 heads	Bibb lettuce
2 ounces	Fresh parsley, washed, excess moisture removed, and chopped

Dressing:

1 ounce	Mild mustard
2 ounces	Red wine vinegar
4 ounces	Granulated sugar
1 teaspoon	Salt
2 ounces	Dry red wine
4 ounces	Vegetable oil
4 ounces	Beet cooking water
	White pepper, to taste

METHOD OF PREPARATION:

1. In a saucepan, place the beets; add water to cover, and vinegar. Heat the liquid to a boil; reduce the heat to a simmer, and cook until the beets are tender. Drain the liquid, reserving 4 ounces for the dressing. Shock the beets in an ice bath, and drain. Remove the skin, trim the ends, and cut into ¼-inch slices.
2. In a separate bowl, combine all of the dressing ingredients, and vigorously whisk together.
3. Add the beets and onions, and mix gently. Hold at 40° F or below until service.
4. To serve, place the salad greens on chilled salad plates, top with the beet salad, and garnish with parsley.

Caesar Salad

YIELD: 10 SERVINGS. SERVING SIZE: THREE OUNCES.

COOKING TECHNIQUE:
Not applicable

Glossary:
Emulsion: incorporation of two liquids

HACCP:
All ingredients must be chilled to 40° F or less.

HAZARDOUS FOOD:
Egg yolks

NUTRITION:
Calories: 175
Fat: 12.5 g
Protein: 7.64 g

INGREDIENTS:

6 large cloves	Garlic, peeled and cut in half
1 teaspoon	Salt
5 each	Anchovy fillets
2 tablespoons	Dijon mustard
2 ounces	Olive oil
1 tablespoon	Red wine vinegar
2 ounces	Freshly squeezed lemon juice
5 each	Eggs, boiled 3 minutes and white discarded
	Freshly ground black pepper, to taste
2 heads (2 pounds)	Romaine, washed, dried, and torn into bite-sized pieces
	Croutons
3 ounces	Grated Parmesan cheese

METHOD OF PREPARATION:

1. Season the bowl with the garlic cloves, using the salt as an abrasive. Remove the salt and garlic when the bowl is well seasoned.
2. Add the anchovies, and mash with a fork. Move the anchovies to one side of the bowl. Add the egg yolks and mustard.
3. Blend the oil into the egg, slowly and steadily, forming an **emulsion.**
4. Add the wine vinegar, lemon juice, and pepper. Mix well.
5. Add the romaine, and toss lettuce. Toss all of the above by rotating the service spoon and fork from the back to the front of the bowl until the lettuce is fully coated.
6. Add the croutons, and toss as in step 5.
7. Add the cheese, and toss again. Serve immediately.

CHEF NOTES:

1. When adding lemon juice, supplement with five dashes of Worcestershire sauce and three dashes of Tabasco sauce.
2. When adding lemon juice, supplement with five dashes of soy sauce.
3. Blend the oil and lemon juice, and emulsify; then add dry mustard instead of Dijon.
4. Egg yolks may be omitted due to the concern over salmonella poisoning.
5. When increasing Caesar salad to a larger amount, do not increase the amount of garlic, anchovies, and mustard proportionately. That is, if you double the recipe, use less than twice the amount of garlic, anchovies, and mustard.
6. All ingredients must be chilled to 40° F or less.

Calamari Salad

YIELD: 50 SERVINGS. SERVING SIZE: FOUR OUNCES.

COOKING TECHNIQUE:
Blanch

Glossary:
Blanch: to cook
Brunoise: finely diced

HAZARDOUS FOOD:
Squid rings

NUTRITION:
Calories: 147
Fat: 13.8 g
Protein: 1.25 g

INGREDIENTS:

8 pounds	Squid rings
1 gallon	Court-bouillon
1 pound	Pepperoncini, diced **brunoise**
1 pound	Green peppers, washed, cored, and diced brunoise
1 pound	Red peppers, washed, cored, and diced brunoise
8 ounces	Celery, washed and diced brunoise
8 ounces	Red onions, peeled and diced brunoise
	Salad greens, as needed
	Fresh parsley, as needed

Dressing:

1½ pints	Olive oil
½ pint	White vinegar
4 ounces	Granulated sugar
2 ounces	Fresh basil, washed and roughly chopped
2 ounces	Fresh oregano, washed and roughly chopped
½ teaspoon	Marjoram
3 cloves	Garlic, peeled and mashed into paste
¼ teaspoon	Crushed red pepper
	Salt and black pepper, to taste

METHOD OF PREPARATION:

1. Heat the court-bouillon to a boil. Add the squid rings, and return the bouillon to a boil.
2. Cook for 2 to 3 minutes. Drain the squid. Slightly chill.
3. In a large bowl, combine all of the ingredients. Marinate for at least 2 hours.
4. Serve on top of greens, and garnish with fresh parsley.

CHEF NOTE:
Calamari salad is served throughout the southern Mediterranean region.

California Cobb Salad

YIELD: 10 SERVINGS. SERVING SIZE: FOUR OUNCES.

INGREDIENTS:

1 pound	Bacon, laid out on a sheet pan
1 pound	Savoy cabbage, washed, cored, and cut **chiffonade**
24 ounces	French dressing (see page 227)
1 head	Romaine lettuce, washed, trimmed, and cut chiffonade
3 each	Ripe tomatoes, washed, cored, blanched, peeled, seeded, and diced
4 each	Avocados, peeled, diced **brunoise,** and tossed with 1 ounce of lemon juice
1 pound	Smoked turkey breast, diced brunoise
10 ounce	Bleu cheese, crumbled

METHOD OF PREPARATION:

1. Preheat the oven to 350° F.
2. Bake the bacon until brown and crispy; then remove, and drain on absorbent paper. Crumble or cut bacon into ½-inch pieces.
3. Marinate the savoy cabbage in 6 ounces of the dressing. Hold at 40° F or lower until service.
4. Place the romaine lettuce on a chilled plate. Arrange the vegetables, turkey, and cheese into long strips over the top of the romaine lettuce, leaving the center free.
5. For service, drain the excess dressing, and arrange the cabbage down the center of salad. Serve immediately, offering additional dressing on the side.

COOKING TECHNIQUE:

Bake

Bake:

1. Preheat the oven.
2. Place the food product on the appropriate rack.

GLOSSARY:

Chiffonade: ribbons of leafy greens
Brunoise: ⅛-inch dice

HACCP:

Hold at 40° F or lower.

NUTRITION:

Calories: 855
Fat: 71.8 g
Protein: 33.5 g

CHEF NOTES:

1. The ingredients in cobb salad vary, but avocado is a standard item.
2. All ingredients must be chilled to 40° F or lower.

COOKING TECHNIQUE:
Blanch

Glossary:
Blanch: to parboil

NUTRITION:
Calories: 131
Fat: 8.56 g
Protein: 2.2 g

Carrot Salad with Asparagus

(Bagatelle)

YIELD: 10 SERVINGS. SERVING SIZE: THREE OUNCES.

INGREDIENTS:

20 each	Asparagus, peeled
2 pounds	Carrots, washed, peeled, and grated
1 head	Romaine lettuce, washed and cored
10 ounces	Mushrooms, washed and sliced
10 ounces	Vinaigrette (see page 230)

METHOD OF PREPARATION:

1. **Blanch** the asparagus, shock it in cold water.
2. Grate the carrots by using a grater or a robot coupe with the proper attachment.
3. Line the salad plate with the romaine lettuce.
4. Arrange the grated carrots on each plate (about 3 ounces on each). Then decorate each plate with four asparagus and mushrooms.
5. Season with the vinaigrette.

Chef's Salad

YIELD: 25 SERVINGS.   SERVING SIZE: EIGHT OUNCES.

INGREDIENTS:

2 heads	Red leaf lettuce, washed, cored, and cut into bite-sized pieces
1 head	Romaine lettuce
3 pounds	Ham, cut into thin strips
3 pounds	Turkey breast, cut into thin strips
3 pounds	Cheddar cheese
13 each	Eggs, hard-cooked, peeled and cut into quarters
6 each	Tomatoes, cut into wedges
50 each	Black olives
25 each	Scallions, cut into fans (2 per scallion)

METHOD OF PREPARATION:

1. Combine the cut red leaf and romaine lettuce in a large bowl. Mix thoroughly.
2. Fill separate bowls or plates with the cut lettuce, and arrange 2 ounces each of meat and cheese, radiating from the center.
3. Place two egg quarters and two tomato wedges on each salad.
4. Garnish with two black olives and two scallion fans.
5. The chef's choice of dressing may be served either on top or on the side of the salad.

COOKING TECHNIQUE:

Not applicable

HAZARDOUS FOODS:

Ham
Turkey breast
Eggs

NUTRITION:

Calories: 424
Fat: 24.6 g
Protein: 44.4 g

CHEF NOTES:

1. There are many variations of chef's salad. The meats and cheeses used depend on the person preparing the salad. The choice of dressing is selected by the guest or customer.
2. This salad is usually served as a main course.
3. It has a similarity to Italian antipasto.

Chicken and Walnut Salad

YIELD: 10 SERVINGS. SERVING SIZE: AS NEEDED.

COOKING TECHNIQUE:
Poach

Simmer and Poach:

1. Heat the cooking liquid to the proper temperature.
2. Submerge the food product completely.
3. Keep the cooked product moist and warm.

HAZARDOUS FOODS:

Chicken breast
Yogurt

NUTRITION:

Calories: 370
Fat: 28.5 g
Protein: 20.9 g

INGREDIENTS:

	White chicken stock to cover, heated to a boil (see page 12)
1 pound	Boneless, skinless chicken breast, cut in half
6 ounces	Shelled and chopped walnuts
1 pound	Mixed green salad
1 each	Red delicious apple (unpeeled), washed, cored, and seeded
1 each	Green delicious apple (unpeeled), washed, cored, and seeded
1 each	Lemon, juice reserved
6 ounces	Cheddar cheese, cut into small cubes

Dressing:

4 ounces	Yogurt
2 tablespoons	Red wine vinegar
1 tablespoon	Dijon mustard
4 ounces	Olive oil
1 tablespoon	Finely chopped walnuts
	Salt and pepper, to taste

METHOD OF PREPARATION:

1. Add the chicken breast to the chicken stock. Simmer until the chicken is fully cooked and firm to the touch. Drain, cool, and dice small.
2. Wash the mixed greens (iceberg, romaine, chicory, escarole, bibb lettuce, endive, watercress, spinach, radicchio—use whatever is available). Pat dry in a clean towel. Set aside.
3. Slice the apples, pour lemon juice over them, and toss.
4. Combine all of the ingredients for the dressing in a mixing bowl.
5. Mix the chicken breast with the walnuts, apples, and cheddar cheese cubes. Add the dressing, and toss well. Refrigerate for 1 hour.
6. Arrange the mixed greens on a cold plate, and place the chicken mixture on top.

CHEF NOTES:

1. Salad can be garnished with scallion flowers, radishes, or fresh watercress.
2. This salad can be made by using leftover roast or poached chicken.
3. For large quantity production, diced boiled potatoes can be added.
4. For additional flavor, diced dill cucumbers can be incorporated.

Chinese Vegetable Salad

YIELD: 10 SERVINGS. SERVING SIZE: THREE OUNCES.

INGREDIENTS:

8 ounces	Chinese cabbage, washed, trimmed, and cut **chiffonade**
4 ounces	Fresh bean sprouts, rinsed and drained
1 pound	Fresh bamboo shoots, **blanched** and cut **julienned**
8 ounces	Canned lychees, drained

Dressing:

3 cloves	Garlic, peeled and minced
½ teaspoon	Ginger root, peeled and grated
1 ounce	Dark soy sauce
1 ounce	Light soy sauce
1½ ounces	Rice wine vinegar
1 ounce	Mirin (available in Oriental markets)
1 ounce	Peanut oil
½ ounce	Sesame oil
1 ounce	Toasted sesame seeds

METHOD OF PREPARATION:

1. Mix together the garlic, ginger root, soy sauces, vinegar, and mirin. Whisk together for 5 minutes.
2. Gradually add the oils and sesame seeds, whisking constantly. Cover and chill at 40° F or below.
3. To serve, arrange the remaining salad ingredients on cold plates, and drizzle with dressing.

COOKING TECHNIQUE:
Not applicable

GLOSSARY:
Chiffonade: ribbons of leafy greens
Blanch: to parboil
Julienne: matchstick strips

HACCP:
Chill at 40° F or below.

NUTRITION:
Calories: 191
Fat: 15.9 g
Protein: 2.74 g

CHEF NOTE:
If canned bamboo shoots are used, no blanching is necessary.

Curly Endive with Peppers

YIELD: 10 SERVINGS. SERVING SIZE: FOUR OUNCES.

INGREDIENTS:

1 head	Red leaf lettuce
1 head	Curly endive (chicory), cleaned, washed, and torn into pieces
5 each	Green bell peppers
5 each	Red bell peppers

Dressing:

6 ounces	Olive oil
1 ounce	Freshly squeezed lemon juice
2 each	Chili peppers, washed, seeded, and diced **brunoise**
2 ounces	Shallots, peeled and diced brunoise
2 ounces	Cilantro leaves, washed and coarsely chopped
	Salt and ground black pepper, to taste

METHOD OF PREPARATION:

1. Place the endive and red leaf lettuce in a large bowl on ice.
2. In a 450° F oven or over a grill, char the peppers until the skin is blistered. Place the peppers in a paper bag to rest for 10 minutes.
3. Peel the peppers, de-seed them, and cut them into strips.
4. Combine all of the ingredients for the dressing, and whisk together. Refrigerate at 40° F or below until needed.
5. To serve, arrange the endive and lettuce on a chilled salad plate, arrange the peppers over the salad mix, and drizzle with dressing to order.

COOKING TECHNIQUES:
Roast or Grill

Roast:
1. Sear the food product, and brown evenly.
2. Elevate the food product in a roasting pan.
3. Determine doneness, and consider carryover cooking.
4. Let the food product rest before carving.

Grill/Broil:
1. Clean and heat the grill/broiler.
2. To prevent sticking, brush the food product with oil.

GLOSSARY:
Brunoise: 1/8-inch dice

HACCP:
Refrigerate at 40° F or below.

NUTRITION:
Calories: 182
Fat: 17.3 g
Protein: 1.4 g

COOKING TECHNIQUE:
Not applicable

GLOSSARY:
Julienne: matchstick strips

HACCP:
Hold at 40° F or below.

NUTRITION:
Calories: 49.5
Fat: .451 g
Protein: 2.32 g

Daikon Radish and Tomato Salad

(Ensaladang Labanos at Kamatis)

YIELD: 10 SERVINGS. SERVING SIZE: THREE OUNCES.

INGREDIENTS:

3 each	Daikon radishes, washed, peeled, and shredded on a large-hole grater
1 ounce	Salt
5 each	Scallions, washed and cut **julienne**
5 each	Tomatoes, washed, cored, and cut into wedges
2 heads	Romaine lettuce, washed and torn into bite-sized pieces
1 pint	Palm vinegar (or substitute rice wine vinegar)
¼ teaspoon	Ground white pepper
	Salt, to taste
2 each	Carrots, washed, peeled, and cut julienne

METHOD OF PREPARATION:

1. Place the daikon in a bowl. Add salt, and mix well; then allow to stand for 15 to 20 minutes.
2. Rinse the daikon in cold water; drain and squeeze out the liquid.
3. Add all of the other ingredients except the carrots; mix well, and let stand in a cool place for 15 minutes.
4. Arrange the salad on a chilled plate, and decorate with shredded carrots.
5. Serve immediately, or hold at 40° F or below.

Florida Conch Salad

COOKING TECHNIQUE:
Not applicable

HAZARDOUS FOOD:
Conch meat

NUTRITION:
Calories: 353
Fat: 14.5 g
Protein: 35.6 g

YIELD: 50 SERVINGS. SERVING SIZE: SIX OUNCES.

INGREDIENTS:

16 pounds	Conch meat, sliced paper-thin
1½ pounds	Pepperoncini, thinly sliced
1½ pounds	Onions, peeled and thinly sliced
1½ pounds	Green peppers, washed, seeded, and thinly sliced
1½ pounds	Red peppers, washed, seeded, and thinly sliced
1½ pounds	Celery, washed and diced small
4 heads	Romaine lettuce, washed
50 each	Fresh parsley sprigs, washed

Dressing:

1½ pints	Olive oil
½ pint	White vinegar
4 ounces	Granulated sugar
2 ounces	Fresh basil, washed and roughly chopped
2 ounces	Fresh oregano, washed and roughly chopped
½ teaspoon	Marjoram
3 cloves	Garlic, peeled and mashed into a paste
¼ teaspoon	Crushed red pepper
	Salt and black pepper, to taste

METHOD OF PREPARATION:

1. In a large bowl, combine all of the ingredients (including dressing).
2. Marinate overnight.
3. Before serving, mix well, taste, and adjust seasoning.
4. Serve on top of the romaine lettuce, and garnish with fresh parsley sprig.

German Potato Salad

YIELD: 50 SERVINGS. SERVING SIZE: SIX OUNCES.

INGREDIENTS:

2 each	Bay leaves
5 each	White peppercorns, crushed
12 pounds	Red bliss potatoes, washed
3 pounds	Onions, peeled and cut **julienne**
3 ounces	Fresh parsley, washed and finely chopped

Dressing:

12 ounces	Cider vinegar
1 pint	Vegetable oil
6 ounces	Bacon fat, **rendered**
1 pound	Granulated sugar
2 ounces	Salt
	White pepper, to taste

METHOD OF PREPARATION:

1. Wrap the bay leaves and peppercorns in cheesecloth, and tie with string to make a sachet.
2. In a large stockpot, add the potatoes, sachet, and cold water, and heat to a boil. Cook the potatoes until tender; drain cool on sheet pan, and slice ¼-inch thick slices.
3. In a mixing bowl, mix all of the ingredients for the dressing; whip vigorously, and combine with the onions and cooked potatoes.
4. Serve on chilled salad plates, and garnish with diced, cooked bacon and chopped parsley.
5. For best result, prepare this salad one day in advance.

COOKING TECHNIQUE:

Boil

Boil: (at sea level)

1. Bring the cooking liquid to a rapid boil.
2. Stir the contents, and cook the food product throughout.
3. Serve hot.

GLOSSARY:

Render: to melt fat
Julienne: matchstick strips

NUTRITION:

Calories: 259
Fat: 12.6 g
Protein: 2.51 g

Globe Artichoke with Olive Oil and Garlic

YIELD: 10 SERVINGS. SERVING SIZE: ONE ARTICHOKE.

COOKING TECHNIQUE:
Simmer

Simmer and Poach:
1. Heat the cooking liquid to the proper temperature.
2. Submerge the food product completely.
3. Keep the cooked product moist and warm.

GLOSSARY:
Brunoise: ⅛-inch dice

NUTRITION:
Calories: 267
Fat: 16 g
Protein: 6.19 g

INGREDIENTS:

10 medium-sized	Artichokes
4 ounces	Fresh lemon juice
1 pound	Onions, peeled and diced **brunoise**
1 pound	Carrots, peeled and diced brunoise
6 cloves	Garlic, peeled and diced brunoise
¼ teaspoon	Dried thyme leaves
	Salt and freshly ground black pepper, to taste
4 ounces	Olive oil
8 ounces	Dry white wine
1 quart	Acidulated water
10 ounces	Green or red salad leaves
3 ounces	Fresh Italian parsley (flat leaved), washed, excess moisture removed, and chopped
	Red wine vinaigrette, as needed (see page 230)

METHOD OF PREPARATION:

1. Remove the stems and tough bottom leaves from the artichokes. Cut about 1½ inches off of the tops. Using scissors, trim the points from the remaining leaves.
2. Place the artichokes in a large bowl or hotel pan, add lemon juice, and cover with cold water to prevent discoloration.
3. Combine all of the remaining ingredients, except the salad leaves, parsley, and vinaigrette, in a large pot. Add the artichokes and the acidulated water.
4. Partially cover the pot, and cook over medium heat until tender, about 1 hour. If necessary, add more water during the cooking process.
5. When cooked, remove the artichokes from the pan, and invert on a baking tray to drain. Allow to cool to room temperature. Reserve the vegetables and 1½ pints of the cooking liquid.
6. To serve, place a salad leaf on a plate, and set one artichoke in the center. Spoon some of the vegetables and reserved cooking liquid over the artichoke, and garnish with parsley. Offer vinaigrette on the side.

CHEF NOTES:
1. Artichokes are deemed tender when a leaf can be pulled out easily.
2. Holding the artichokes in acidulated water prevents discoloration.
3. If budget allows, use extra virgin olive oil.

Greek Salad

YIELD: 10 SERVINGS. SERVING SIZE: FOUR OUNCES.

COOKING TECHNIQUE:
Not applicable

GLOSSARY:
Chiffonade: ribbons of leafy greens
Blanch: to par cook

HACCP:
Hold at 40° F or below.

HAZARDOUS FOOD:
Sardines
Anchovy fillets

NUTRITION:
Calories: 330
Fat: 30.6 g
Protein: 7.52 g

INGREDIENTS:

8 ounces	Olive oil
2 ounces	Red wine vinegar
1 head	Romaine lettuce, cored, washed, drained, and cut **chiffonade**
8 ounces	Tomatoes, washed, cored, **blanched,** peeled, and cut into wedges
1 pound	Small cucumbers, washed, peeled, and cut into ¼-inch round slices
8 ounces	Green bell peppers, washed, seeded, and cut into rings
8 ounces	Small-sized onions, peeled and sliced into rings
8 ounces	Feta cheese, crumbled
30 each	Calamata olives
10 each	Salted sardines or anchovy fillets

METHOD OF PREPARATION:

1. Combine the oil and vinegar, and set aside as the dressing.
2. On chilled plates, construct the salad in the following order: lettuce, tomatoes, cucumbers, peppers, onions; drizzle with dressing, and garnish with cheese, olives, and sardines or anchovies. Hold at 40° F or below.

Grilled Chicken Salad

YIELD: 10 SERVINGS. SERVING SIZE: SIX OUNCES.

COOKING TECHNIQUE:
Grill

Grill/Broil:
1. Clean and heat the grill/broiler.
2. To prevent sticking, brush the food product with oil.

GLOSSARY:
Blanch: to parboil
Chiffonade: ribbons of leafy greens

HAZARDOUS FOOD:
Chicken breasts

NUTRITION:
Calories: 438
Fat: 29.4 g
Protein: 35 g

INGREDIENTS:

3 ounces	Lime juice
8 ounces	Olive oil
¼ ounce	Chili paste
1 clove	Garlic, peeled and minced
3½ pounds	Boneless, skinless chicken breasts
3 each	Red peppers, washed, seeded, and diced
3 cloves	Garlic, peeled and minced
1 ounce	Cilantro, washed and chopped
4 ounces	Balsamic vinegar
1 ounce	Scallions, finely chopped
2 each	Large tomatoes, washed, cored, **blanched,** peeled, seeded, and chopped
1 large head	Romaine lettuce, cut **chiffonade**
2 each	Avocados, washed, peeled, and cut into thin slices. Add lemon juice to prevent discoloration.
2 each	Limes, cut into wedges
	Seasoned, to taste

METHOD OF PREPARATION:

1. In a bowl, combine the lime juice, olive oil, chili paste, and garlic. Add the chicken, and toss to coat. Marinate in the refrigerator for 1 hour.
2. Grill the chicken breast for 4 to 5 minutes on each side or until golden brown and cooked throughout. Let cool for 10 minutes, and cut into strips. Place the chicken in a bowl, and add the peppers, cilantro, garlic, scallions, and vinegar. Toss gently. Season, as needed, with salt and pepper.
3. To assemble, place the lettuce on a plate or a platter. Sprinkle with chopped tomatoes. Neatly place the chicken on top. Garnish with avocado fans and lime wedges.

Grilled Tuna, Tomatoes, and Fusilli Pasta Salad

YIELD: 50 SERVINGS. SERVING SIZE: FIVE OUNCES.

COOKING TECHNIQUE:
Grill

Grill/Broil:
1. Clean and heat the grill/broiler.
2. To prevent sticking, brush the food product with oil.

GLOSSARY:
Julienne: matchstick strips
Al dente: to the bite

HAZARDOUS FOOD:
Tuna steak

NUTRITION:
Calories: 275
Fat: 16.4 g
Protein: 14.1 g

INGREDIENTS:

1½ pints	Olive oil
5 pounds	Fresh tuna steak
5 pounds	Fusilli pasta
2 gallons	Water
1 pound	Sun-dried tomatoes cut **julienne**
6 cloves	Garlic, peeled and crushed
2 ounces	Italian parsley, washed and chopped
2 ounces	Fresh basil, washed and chopped
12 ounces	Balsamic vinegar
	Salt and white pepper, to taste

METHOD OF PREPARATION:

1. Brush the tuna steaks with 2 tablespoons of olive oil. Prepare the grill. When the grill is hot, grill the tuna steaks until firm.
2. In a pot of boiling salted water, cook the pasta until **al dente,** about 5 to 7 minutes. Drain and cool.
3. In a sauté pan, heat 2 ounces of olive oil over medium heat. Add the garlic, and stir for 1 minute; then add the sun-dried tomatoes. Remove from the heat, and cool.
4. Place the pasta and tuna, sliced in 1 inch pieces, in a bowl; add the remaining ingredients and mix to incorporate. Adjust the seasonings, and serve.

CHEF NOTE:
Grilled tuna salad is a favorite of the Mediterranean area, predominantly southern Italy.

Grilled Vegetable Salad

YIELD: 10 SERVINGS. SERVING SIZE: FOUR OUNCES.

INGREDIENTS:

10 each	Ripe tomatoes, washed and cored
5 each	Red bell peppers, washed and seeded
5 each	Yellow bell peppers, washed and seeded
3 each	Zucchini, washed and trimmed
3 each	Yellow squash, washed and trimmed
	Vegetable or olive oil, as needed
	Salad greens, as available, washed and cut **chiffonade**
10 sprigs	Fresh oregano

Dressing:

12 ounces	Olive oil
4 ounces	Freshly squeezed lemon juice
4 cloves	Garlic, peeled and mashed
2 ounces	Fresh oregano leaves, washed and roughly chopped
1 pound	Onions, peeled and cut **julienne**
	Salt and freshly ground black pepper, to taste

METHOD OF PREPARATION:

1. In a bowl, combine all of the ingredients for the dressing, and whisk together. Refrigerate at 40° F or lower, until needed.
2. Cut tomatoes into ½-inch slices and the peppers into ½-inch wide strips.
3. Slice the zucchini and yellow squash on the bias into ½-inch wide pieces.
4. Preheat the broiler or grill.
5. Lightly coat each vegetable slice with oil, and grill until **al dente.** Transfer to sheet pans, and refrigerate to 40° F or lower, to cool. For service, arrange the salad greens on a chilled plate. Arrange slices of each vegetable over the greens, and ladle dressing over the salad. Garnish with an oregano sprig. Hold at 40° F or lower.

COOKING TECHNIQUE:
Grill

Grill/Broil:
1. Clean and heat the grill/broiler.
2. To prevent sticking, brush the food product with oil.

GLOSSARY:
Julienne: matchstick strips
Chiffonade: ribbons of leafy greens
Al dente: to the bite

HACCP:
Chill to 40° F.

NUTRITION:
Calories: 373
Fat: 33.4 g
Protein: 3.41 g

CHEF NOTES:
1. All ingredients must be chilled to 40° F or lower.
2. If salads are to be held, do not add dressing until the last minute.

Hot Wilted-Spinach Salad

YIELD: 10 SERVINGS. SERVING SIZE: FOUR OUNCES.

INGREDIENTS:

2 pounds	Fresh spinach leaves, washed, dried, and stems removed
8 ounces	Bacon, cut into ½-inch pieces
2 ounces	Brown sugar
2 ounces	Olive oil
1 ounce	Wine vinegar
1 tablespoon	Lemon juice
	Salt and freshly ground black pepper, to taste

METHOD OF PREPARATION:

1. Tear the spinach into bite-sized pieces, and place in a bowl.
2. In a sauté pan, place the bacon; sauté until it begins to brown and most of the fat is **rendered.**
3. Add the sugar, and melt.
4. When the bacon is fully cooked and the sugar is melted, add the oil, vinegar, and lemon juice, and season, to taste.
5. Mix well; then pour over the spinach leaves, and toss. Serve immediately.

COOKING TECHNIQUE:

Sauté

Sauté:

1. Heat the sauté pan to the appropriate temperature.
2. Evenly brown the food product.
3. For a sauce, pour off any excess oil, reheat, and deglaze.

GLOSSARY:

Render: to melt fat

NUTRITION:

Calories: 135
Fat: 9.68 g
Protein: 4.91 g

CHEF NOTES:

1. In addition to spinach, ripe olives, onions, cooked eggs, or sliced mushrooms can be added.
2. The dressing can be supplemented with Worcestershire sauce or garlic.
3. Flaming brandy can be poured over the spinach.
4. White sugar can be used instead of brown sugar.

Iceberg Lettuce

COOKING TECHNIQUE:
Not applicable

NUTRITION:
Calories: 264
Fat: 26 g
Protein: 3.04 g

YIELD: 10 SERVINGS. SERVING SIZE: ONE-SIXTH WEDGE.

INGREDIENTS:

2 heads	Iceberg lettuce, cleaned, left whole
5 each	Tomatoes, washed, cored, and peeled
1½ ounces	Fresh parsley, washed, excess moisture removed, and chopped

Dressing—Roquefort

1 pint	French dressing (see page 227)
4 ounces	Roquefort (blue) cheese, crumbled

METHOD OF PREPARATION:

1. Place the french dressing in a blender. Add the cheese, and blend to a smooth mixture.
2. To serve, cut the iceberg lettuce into five wedges each.
3. Slice the peeled tomatoes into eight wedges each.
4. Place a salad wedge on a chilled 7-inch plate, and pour dressing on the salad.
5. Decorate with four wedges of tomatoes; sprinkle chopped parsley on the entire plate.

Italian Pepper Salad

YIELD: 50 SERVINGS. SERVING SIZE: SIX OUNCES.

COOKING TECHNIQUE:
Boil

Boil: (at sea level)

1. Bring the cooking liquid to a rapid boil.
2. Stir the contents, and cook the food product throughout.
3. Serve hot.

GLOSSARY:

Al dente: to the bite

NUTRITION:

Calories: 186
Fat: 14.6 g
Protein: 1 g

INGREDIENTS:

50 each	Banana or cubanele peppers
1 gallon	Water, to boil the peppers
1 pound	Granulated sugar
2 ounces	Salt
5 each	Bay leaves
1 teaspoon	Crushed peppercorns
5 cloves	Garlic, peeled and mashed
1 quart	Rémoulade sauce (see page 48)
1 quart	Ravigote sauce (see page 46)

METHOD OF PREPARATION:

1. In a large stockpot, add the seasonings to the water, and heat to a boil. Then boil peppers until they are **al dente.** When done, transfer the peppers into hotel pans. Pour cooking liquid to cover peppers, chill, and marinate overnight.
2. Before service, drain peppers; place on chilled salad plate. Serve with remoulade sauce or ravigote sauce.
3. For better presentation, diced pimientos can be added.

Lettuce Salad

YIELD: 10 SERVINGS. SERVING SIZE: FOUR OUNCES.

COOKING TECHNIQUE:

Not applicable

GLOSSARY:

Brunoise: ⅛-inch dice
Emulsion: incorporation of two liquids

HACCP:

Hold at 40° F or below.

NUTRITION:

Calories: 127
Fat: 12.8 g
Protein: 1.2 g

INGREDIENTS:

3 heads	Boston lettuce, trimmed, washed, drained, and gently torn into pieces
8 ounces	Cherry tomatoes, washed and drained
5 ounces	Green olives, split in half lengthwise

Dressing:

4 ounces	Olive oil
1½ ounces	Red wine vinegar
2 ounces	Onions, peeled and diced **brunoise**
	Salt and ground white pepper, to taste

METHOD OF PREPARATION:

1. Arrange the lettuce on chilled salad plates, and decorate with tomatoes and olives.
2. Hold at 40° F or below until ready for service.
3. In a container with a lid, combine all of the dressing ingredients; cover and shake until the mixture **emulsifies.** Hold at 40° F or below.
4. For service, present the salad, and offer a dressing tableside.

Lobster Salad Parisienne

YIELD: 10 SERVINGS. SERVING SIZE: ONE LOBSTER (APPROXIMATELY ONE AND ONE-FOURTH POUNDS)

INGREDIENTS:

10 each	Live lobsters
1 pound	Carrots, washed, peeled, and diced
8 ounces	Celery, peeled and diced
8 ounces	Frozen peas
8 ounces	Red pepper or pimiento, diced
1 pint	Mayonnaise (see page 35)
10 ounces	Salad greens in season
	Salt, to taste

METHOD OF PREPARATION:

1. In a large stockpot with boiling water, boil the lobsters for 7–8 minutes. Then shock the lobsters in ice water.
2. When chilled, remove the claws, deshell, and reserve meat for garnish. Cut the lobsters in half, lengthwise. Carefully remove the meat from the shells, and reserve for garnish. Remove the tail meat, and dice. Discard the tamale and sac. Clean and save the shell.
3. In a large mixing bowl, mix together the diced vegetables, peas, mayonnaise, and the diced lobster meat.
4. When the salad is thoroughly mixed and seasoned, fill the shells, and place on a large platter with greens. Garnish with the claw meat, and serve chilled.

COOKING TECHNIQUE:

Boil

Boil: (at sea level)

1. Bring the cooking liquid to a rapid boil.
2. Stir the contents, and cook the food product throughout.
3. Serve hot.

HAZARDOUS FOODS:

Lobster
Mayonnaise

NUTRITION:

Calories: 498
Fat: 36.8 g
Protein: 30.7 g

CHEF NOTE:

This recipe originated in Paris. Lobster Salad Parisienne is served as a main course and usually is garnished with marinated vegetables, sliced cucumbers, tomatoes, and hard-cooked eggs.

Salad Lorette

YIELD: 10 SERVINGS. SERVING SIZE: FOUR OUNCES.

INGREDIENTS:

1 head	Romaine lettuce, washed, trimmed, and cut **chiffonade**
3 each	Belgian endive, washed and cut **julienne**
5 each	Beets, washed, cooked, peeled, and cut julienne
20 each	Walnut halves
	Fresh cream dressing (see page 226)

METHOD OF PREPARATION:

Arrange the vegetables in bouquets on chilled salad plates. Place two walnut halves on top, and serve with fresh cream dressing. Serve cold.

COOKING TECHNIQUE:

Not applicable

GLOSSARY:

Chiffonade: ribbons of leafy greens
Julienne: matchstick strips

NUTRITION:

Calories: 32.2
Fat: .258 g
Protein: 1.64 g

Marinated Mushroom Salad

YIELD: 10 SERVINGS. SERVING SIZE: THREE OUNCES.

COOKING TECHNIQUE:
Not applicable

NUTRITION:
Calories: 269
Fat: 23.3 g
Protein: 4.1 g

INGREDIENTS:

2 pounds	Button mushrooms, cleaned, sliced thin, or selected small buttons that are left whole
8 ounces	Red bell pepper, washed, seeded, and diced into ¼-inch pieces
1 ounce	Lemon juice
3 heads	Romaine lettuce, washed and cut or shredded into fine strips
8 ounces	Green peas

Dressing:

8 ounces	Olive oil
2 ounces	Lemon juice
1 ounce	Granulated sugar
1½ ounces	Fresh basil, washed and chopped
1½ ounces	Fresh oregano, washed and chopped
¼ teaspoon	Marjoram
	Salt and cracked black pepper, to taste

METHOD OF PREPARATION:

1. Combine the mushrooms with the diced red peppers in a mixing bowl with 1 ounce of lemon juice, and refrigerate.
2. In a mixing bowl, vigorously mix the olive oil with the rest of the lemon juice, sugar, basil, oregano, marjoram, salt, and pepper, to create the dressing. Then blend with the mushrooms and peppers.
3. Chill the salad plates, and line with shredded lettuce. Arrange the mushroom salad on top of the lettuce, and garnish with the green peas.

CHEF NOTE:
This salad can be garnished with diced pimiento or diced red bell pepper and parsley.

Marinated Shrimp and Citrus Salad

YIELD: 10 SERVINGS. SERVING SIZE: FOUR OUNCES.

INGREDIENTS:

	Court-bouillon, as needed (see page 9)
3 pounds (16–20 count)	Shrimp, washed in cold water
2 each	Grapefruits, washed, peeled, and sectioned
4 each	Oranges, washed, peeled, and sectioned
4 ounces	Red onions, peeled, and cut **julienne**
2 ounces	Celery stalks, washed and cut **brunoise**
10 leaves	Salad greens, washed, drained, and trimmed
	Lemon juice, as needed
4 each	Pears, washed, peeled, cored, sliced, and coated with lemon juice
2 ounces	Pimientos, rinsed and cut julienne

Dressing:

8 ounces	Vegetable oil
4 ounces	Cider vinegar
2 ounces	Granulated sugar
1 teaspoon	Dry mustard
1 teaspoon	Onions, peeled and finely minced
½ tablespoon	Celery seed
	Salt and ground white pepper, to taste

METHOD OF PREPARATION:

1. In a large pot, heat the court-bouillon to a boil, and poach the shrimp for 3 to 5 minutes or until just firm. Shock in cold water, peel, and devein.
2. In a large mixing bowl, combine all of the ingredients for the dressing, and vigorously whisk together.
3. In a separate mixing bowl, combine the cooked shrimp, citrus fruits, onions, and celery. Pour the dressing over, toss together, and cover. Marinate refrigerated at 40° F or lower for about 1 hour.
4. For service, arrange the salad greens on chilled plates, and spoon the shrimp salad onto the center of each. Garnish with pear slices and pimiento strips. Serve immediately, or hold refrigerated at 40° F or lower.

COOKING TECHNIQUE:

Poach

Simmer and Poach:

1. Heat the cooking liquid to the proper temperature.
2. Submerge the food product completely.
3. Keep the cooked product moist and warm.

GLOSSARY:

Julienne: matchstick strips
Brunoise: ⅛-inch dice

HACCP:

Refrigerate at 40° F.

HAZARDOUS FOOD:

Shrimp

NUTRITION:

Calories: 404
Fat: 24.4 g
Protein: 22.1 g

CHEF NOTES:

1. This salad can be served as a first course or as a main course.
2. All ingredients must be chilled to 40° F or lower.

Mediterranean Salad

YIELD: 10 SERVINGS. SERVING SIZE: SIX OUNCES.

COOKING TECHNIQUE:

Not applicable

GLOSSARY:

Blanch: to parcook
Brunnoise: ⅛-inch dice
Chiffonade: ribbons of leafy greens

NUTRITION:

Calories: 280
Fat: 25.7 g
Protein: 2.55 g

INGREDIENTS:

1 pound	Globe tomatoes, washed, cored, **blanched,** peeled, seeded, and diced
½ pound	Cucumbers, washed, peeled, seeded and diced
½ pound	Red peppers, washed, skin removed, seeded and diced
4 ounces	Spanish onions, peeled and diced **brunoise**
½ pound	Pitted niçoise olives, sliced
½ pound	Zucchini (unpeeled), washed and diced
½ pound	Yellow squash (unpeeled), washed and diced
½ ounce	Capers, drained and rinsed
1 head	Romaine lettuce, washed, cut **chiffonade,** for garnish

Dressing:

8 ounces	Olive oil
4 ounces	Red wine vinegar
1 ounce	Granulated sugar
¼ ounce	Fresh basil, washed and roughly chopped
¼ ounce	Fresh oregano, washed and roughly chopped
⅛ teaspoon	Marjoram
1 clove	Garlic, peeled and mashed into a paste
	Salt and black pepper, to taste

METHOD OF PREPARATION:

1. Combine all of the ingredients for the dressing in a bowl. Whip vigorously.
2. Place all diced vegetables and capers in a large mixing bowl. Blend with the vinaigrette dressing, and marinate for 1 hour.
3. Serve on a chilled salad plate. Garnish with chiffonade of romaine lettuce.

Mexican Corn Salad

YIELD: 10 SERVINGS. SERVING SIZE: SIX OUNCES.

COOKING TECHNIQUE:

Blanch

GLOSSARY:

Blanch: to parboil
Brunoise: ⅛-inch dice

NUTRITION:

Calories: 565
Fat: 41 g
Protein: 12.4 g

INGREDIENTS:

10 pounds	Corn kernels, frozen
2 pounds	Red peppers, washed, seeded, and diced
2 pounds	Green peppers, washed, seeded, and diced
6 each	Chili peppers, washed, seeded, and diced
3 pounds	Globe tomatoes, washed, cored, **blanched,** peeled, seeded, and chopped
2 pounds	Red onions, peeled and diced **brunoise**
2 pounds	Zucchini, washed and diced

Dressing:

12 ounces	Vegetable oil
4 ounces	Lemon juice
2 ounces	White vinegar
¼ ounce	Fresh oregano, washed and chopped
¼ ounce	Fresh cilantro, washed and chopped
½ teaspoon	Ground coriander, toasted in a sauté pan
½ teaspoon	Ground cumin, toasted in sauté pan
	Salt, to taste
¼ teaspoon	Hot pepper sauce (e.g., Tabasco)

Garnishes:

3 pounds	Monterey Jack cheese, diced
2 pounds	Black olives
4 ounces	Scallions, cut into flowers
	Salad greens as needed

METHOD OF PREPARATION:

1. Combine all of the ingredients for the dressing in a bowl. Whip vigorously.
2. Combine the corn kernels with the diced vegetables in a different bowl. Marinate with the dressing in the refrigerator for 1 hour.
3. Place the corn salad on top of greens on chilled salad plates. Garnish with Monterey Jack cheese, black olives, and scallion flowers.

CHEF NOTE:

This recipe is a Mexican-American salad served in the Southwest as either an appetizer or an accompaniment. Some variations of this recipe can be called corn salsa.

COOKING TECHNIQUE:

Not applicable

HAZARDOUS FOOD:

Eggs

NUTRITION:

Calories: 85.8
Fat: 4.07 g
Protein: 3.99 g

Mimosa Salad

YIELD: 10 SERVINGS. SERVING SIZE: FOUR OUNCES.

INGREDIENTS:

2 heads	Iceberg lettuce, trimmed, washed, and torn in bite-sized pieces
5 each	Eggs, hard-boiled, peeled, and grated
3 ounces	Freshly chopped parsley, excess moisture removed
	Salt and ground white pepper, to taste
10 ounces	Basic French dressing (see page 227)

METHOD OF PREPARATION:

1. Place the lettuce in the ice water, and refrigerate at 40° F or below until needed for service.
2. Mix the grated eggs with the parsley, and season to taste.
3. Drain and dry the lettuce leaves. Place in a bowl and mix with the French dressing.
4. Place salad on chilled salad plate.
5. Sprinkle the egg mixture over the salad and serve.

CHEF NOTE:

The name *Mimosa* refers to the garnish of hard boiled egg mixed with parsley.

Mixed Greens with Wild Mushrooms

YIELD: 10 SERVINGS. SERVING SIZE: FOUR OUNCES.

INGREDIENTS:

1 head	Boston lettuce, washed, drained, and gently torn in pieces
1 bunch	Escarole, washed, drained, and gently torn in pieces
3 heads	Radicchio, leaves separated, washed, drained, and gently torn into pieces
1 head	Romaine, leaves separated, washed, drained, and gently torn into pieces
2 ounces	Olive oil
1 pound	Wild mushrooms, cleaned and trimmed
20 each	Cherry tomatoes, washed and dried
	Salt and freshly ground black pepper, to taste
1 pint	Vinaigrette dressing (see page 230)

METHOD OF PREPARATION:

1. In a large bowl, combine all of the salad greens, toss together, cover with a wet paper towel, and refrigerate at 40° F or lower until needed.
2. Just before service, in a sauté pan, heat the olive oil, and add the mushrooms. Sauté for 3 to 5 minutes; then add the cherry tomatoes, and continue to sauté until the tomatoes are warmed throughout. Season, to taste.
3. To serve, arrange the salad leaves on chilled 7-inch plates; ladle the warm mushrooms with the tomatoes into the center of the salads, and offer the vinaigrette dressing tableside.

COOKING TECHNIQUE:

Sauté

Sauté:

1. Heat the sauté pan to the appropriate temperature.
2. Evenly brown the food product.
3. For a sauce, pour off any excess oil, reheat, and deglaze.

HACCP:

Refrigerate at 40° F or lower.

NUTRITION:

Calories: 503
Fat: 49.3
Protein: 3.82 g

CHEF NOTE:

If the mushrooms being used are large, cut into halves, quarters, or slices.

Niçoise Salad

YIELD: 50 SERVINGS. SERVING SIZE: SIX OUNCES.

INGREDIENTS:

5 pounds	Red bliss potatoes, washed
5 pounds	Green beans, washed, ends cut
9 each	Tomatoes, cut in wedges
6 pounds	Canned white tuna, flaked
13 each	Eggs, hard-cooked, cut in quarters
50 each	Black olives, to garnish
5 heads	Boston lettuce, washed and used for base

Dressing:

1 pint	Wine vinegar
3 pints	Olive oil
	Salt and white pepper, to taste
6 cloves	Garlic, peeled and crushed

METHOD OF PREPARATION:

1. Wash the potatoes, and place in a stockpot with water. Heat to a boil, and cook until tender.
2. Drain the potatoes; cool and slice.
3. In another pot, heat salted water to a boil. Add the green beans, and cook until **al dente.** Drain, and shock in ice water.
4. Vigorously whip all of the ingredients for the dressing until they are temporarily emulsified.
5. Arrange the potatoes, green beans, tomatoes, tuna, eggs, and black olives on top of the Boston lettuce so that they are radiating from the center. Top with the dressing before serving.

COOKING TECHNIQUE:

Boil

Boil: (at sea level)

1. Bring the cooking liquid to a rapid boil.
2. Stir the contents, and cook the food product throughout.

GLOSSARY:

Al dente: to the bite

HAZARDOUS FOOD:

Eggs

NUTRITION:

Calories: 389
Fat: 30.5 g
Protein: 18.1 g

CHEF NOTES:

1. Use the oil from the tuna as part of the dressing.
2. The original niçoise has no potatoes and uses anchovy fillets for garnish.

Phillipino Mixed-Fruit Salad

(Flordeliz Salad)

YIELD: 10 SERVINGS. SERVING SIZE: THREE OUNCES.

COOKING TECHNIQUE:
Not applicable

HACCP:
Hold refrigerated at 40° F or below.

HAZARDOUS FOOD:
Cream cheese

NUTRITION:
Calories: 387
Fat: 21 g
Protein: 7.43 g

INGREDIENTS:

14 ounces	Cream cheese
10 ounces	Canned condensed milk
1 #2½ can (27 ounces)	Fruit cocktail, drained
1 #303 can (1 pound)	Peaches, drained and diced
1 #2 can (19 ounces)	Pineapple tidbits
1 #3 can (1 pound)	Mandarin oranges, drained
2 each (8 ounces)	Apples, peeled, cored, and diced
1 head	Boston lettuce, washed, cored, leaves separated and wiped dry
20 each	Halved walnuts
2 ounces	Shredded coconut
10 each	Cherries, with stems

METHOD OF PREPARATION:

1. In a large bowl, blend together the cream cheese and condensed milk. Add the canned fruits and apples, and mix well. Cover and refrigerate at 40° F or below for at least 1 hour.
2. To serve, line a chilled salad plate with lettuce. Scoop the salad on the center of the lettuce, and garnish with walnuts and coconut. Place a cherry on top, and serve, or hold refrigerated at 40° F or below.

COOKING TECHNIQUE:

Not applicable

HAZARDOUS FOOD:

Anchovies

HACCP:

Hold chilled at 40° F or lower.

NUTRITION:

Calories: 173
Fat: 15.1 g
Protein: 4.72 g

Portuguese Salad with Vinaigrette Dressing

YIELD: 10 SERVINGS. SERVING SIZE: THREE OUNCES.

INGREDIENTS:

2 heads	Boston lettuce, washed, cored, and leaves separated
4 ounces	Fresh watercress, washed and trimmed
8 ounces	Fresh tomatoes, washed, cored, blanched, peeled, and cut into wedges
4 ounces	Red onion, peeled and thinly sliced
30 each	Anchovy-stuffed green olives
	Vinaigrette dressing (see page 230)
2 ounces	Fresh parsley, washed, excess moisture removed, and chopped

METHOD OF PREPARATION:

1. In a large bowl, combine all of the ingredients except the dressing and parsley, and toss together. Hold chilled at 40° F or lower.
2. Just before service, toss the salad with the dressing, divide among chilled plates, and garnish with parsley.

CHEF NOTE:

If the Boston leaves are too large, gently tear them into smaller pieces.

Red Cabbage Salad

YIELD: 10 SERVINGS. SERVING SIZE: THREE OUNCES.

COOKING TECHNIQUE:
Not applicable

GLOSSARY:
Julienne: matchstick strips
Acidulated: with acid added

HACCP:
Hold at 40° F or below.

NUTRITION:
Calories: 154
Fat: .771 g
Protein: 2.82 g

INGREDIENTS:

2 heads	Red cabbage, outer leaves removed, core cut out, washed, and shredded
1 teaspoon	Caraway seed, crushed
5 each	Roma apples, washed, peeled, cored, and cut **julienne**

Marinade:

1 quart	Boiling water
4 ounces	Red wine vinegar
6 ounces	Granulated sugar
1 teaspoon	Salt

METHOD OF PREPARATION:

1. Place the shredded cabbage in a suitable noncorrosive bowl.
2. In a saucepan, combine the marinade ingredients, and heat to a boil for 1 minute.
3. Pour the boiling marinade over the cabbage, and toss. Marinate overnight, holding at 40° F or below.
4. Drain the marinade, and discard. Add the caraway seed and apples, and mix well.

CHEF NOTES:

1. Preparation should be started 1 day in advance to allow time for the cabbage to marinate.
2. If apples are julienned ahead, hold in **acidulated** water.
3. This salad can be mixed with mayonnaise or sour cream.

COOKING TECHNIQUE:

Not applicable

HACCP:

Chill to 40° F or lower.

HAZARDOUS FOODS:

Salmon
Eggs
Caviar

NUTRITION:

Calories: 209
Fat: 7.75 g
Protein: 19.8 g

Smoked Salmon Tartare with Black Caviar

YIELD: 10 SERVINGS. SERVING SIZE: THREE OUNCES.

INGREDIENTS:

1½ pounds	Smoked salmon, finely minced
3 each	Hard-boiled eggs, peeled and grated
1½ ounces	Shallots, peeled and finely minced
	Freshly ground black pepper, to taste
10 ounces	Fresh watercress sprigs, washed
10 slices	Thin dark rye bread slices
2 ounces	Black caviar (salmon roe)
3 each	Lemons, washed and cut into wedges

METHOD OF PREPARATION:

1. In a mixing bowl, combine the salmon, eggs, shallots, and pepper. Mix on medium speed until the eggs are no longer visible. Cover and chill at 40° F or lower until ready for service.
2. Arrange the watercress on the salad plates.
3. Pipe the salmon tartare through a star tube onto the bread slices, and set on top of the watercress.
4. Garnish the tartare with salmon roe, and garnish the plate with lemon wedges.

CHEF NOTE:

Watercress can be dressed with a light vinaigrette, if desired.

Spinach Salad

YIELD: 10 SERVINGS. SERVING SIZE: THREE OUNCES.

COOKING TECHNIQUE:

Not applicable

NUTRITION:

Calories: 258
Fat: 21 g
Protein: 9.41 g

INGREDIENTS:

4 pounds	Spinach, stemmed, washed, and drained, stems removed
8 cloves	Garlic, peeled and mashed into a purée
20 ounces	Vinaigrette dressing (see page 230)
5 ounces	Romano cheese, grated
1 ounce	Paprika

METHOD OF PREPARATION:

1. Place spinach leaves on chilled salad plates. Hold at 40° F or below.
2. Mix the garlic into the dressing. Hold chilled at 40° F or below.
3. To serve, ladle the dressing over the salad, and sprinkle with cheese and paprika.

CHEF NOTE:

Spinach will wilt if held too long before service.

Tabbouleh Salad

YIELD: 30 SERVINGS. SERVING SIZE: SIX OUNCES.

COOKING TECHNIQUE:
Blanch

GLOSSARY:
Blanch: to parboil

NUTRITION:
Calories: 216
Fat: 14.3
Protein: 3.86 g

INGREDIENTS:

2 pounds	Burgul wheat
½ gallon	Water or vegetable stock, heated to a boil (see page 11)
10 each	Cucumbers, peeled, seeded, and diced
2 pounds	Red onions, peeled and diced
2 pounds	Red peppers, washed, cored, and diced small
24 each	Tomatoes, washed, **blanched,** peeled, seeded, and diced
8 ounces	Scallions, washed and sliced
6 ounces	Fresh (flat leafed) parsley, washed, excess moisture removed, and chopped
¾ pound	Fresh mint, chopped
2 heads	Romaine lettuce

Dressing:

1½ pints	Olive oil
6 ounces	Lemon juice
7 cloves	Garlic, mashed into a paste
	Salt and pepper, to taste

METHOD OF PREPARATION:

1. Rehydrate the burgul wheat in boiling water or stock. Then place it in a large mixing bowl, and allow it to cool.
2. Reserve some of the diced cucumbers, red and green peppers, and herbs for garnish. Mix the rest with the burgul wheat and dressing. Portion onto the greens, and garnish with tomatoes, scallions, and fresh mint.

CHEF NOTE:
Tabbouleh salad is a favorite in the Middle East and certain areas of the Mediterranean. It is most commonly prepared with only olive oil, lemon juice, and fresh mint.

Tomato and Mozzarella Salad

COOKING TECHNIQUE:
Not applicable

GLOSSARY:
Chiffonade: ribbons of leafy greens

HACCP:
Hold at 40° F or below.

NUTRITION:
Calories: 162
Fat: 11 g
Protein: 9.43 g

YIELD: 10 SERVINGS. SERVING SIZE: THREE OUNCES.

INGREDIENTS:

1 head	Romaine lettuce, washed, trimmed, and cut **chiffonade**
2 pounds	Medium-sized, firm tomatoes, washed, cored, and cut into ⅛″ slices
10 ounces	Mozzarella cheese, cut into ⅛″ slices
1 ounce	Basil leaves, stems removed, washed, and cut chiffonade
2 ounces	Olive oil
5 each	Lemons, washed and cut in half

METHOD OF PREPARATION: 1

1. On chilled salad plates, place a bed of chiffonade of romaine.
2. Divide the slices of tomatoes over the chiffonade.
3. Place the mozzarella in a small bowl, add the basil and oil, and mix well. Hold at 40° F or below.
4. To order, spoon the mozzarella mixture over the salads, and garnish with a half lemon.

METHOD OF PREPARATION: 2

1. Slice the mozzarella same size as the tomatoes.
2. Place all lettuce leaves on chilled salad plate.
3. Arrange tomato and mozzarella slices in overlapping method on lettuce.
4. Sprinkle the chiffonade of basil on top, and drizzle with olive oil.

Tortellini Salad

YIELD: 50 SERVINGS. SERVING SIZE: SIX OUNCES.

INGREDIENTS:

8 pounds	Cheese tortellini
	Oil, as needed
4 pounds	Broccoli flowerettes, washed, stems cut
2 pounds	Carrots, washed, peeled, and diced small
2 pounds	Celery, washed and diced small
1½ pounds	Red bell peppers, washed, seeded, and cut **julienne**
1½ pounds	Green bell peppers, washed, seeded, and cut julienne
1½ pounds	Fresh mushrooms, washed and sliced
4 heads	Red leaf lettuce

Dressing:

2½ quarts	Olive oil
1½ quarts	Red wine vinegar
4 ounces	Granulated sugar
2 ounces	Fresh basil, washed and roughly chopped
2 ounces	Fresh oregano, washed and roughly chopped
½ teaspoon	Marjoram
3 cloves	Garlic, peeled and mashed into a paste
	Salt and black pepper, to taste

METHOD OF PREPARATION:

1. In boiling salted water, cook the tortellini until **al dente.** Rinse under cold running water, drain, and place in a large bowl. Toss with a small amount of oil.
2. Combine all of the ingredients for the dressing in a mixing bowl, and whip vigorously.
3. Mix the dressing with the vegetables and tortellini right before plating.
4. Serve on chilled salad plates, and garnish with greens.

COOKING TECHNIQUE:

Boil

Boil: (at sea level)

1. Bring the cooking liquid to a rapid boil.
2. Stir the contents, and cook the food product throughout.

GLOSSARY:

Julienne: matchstick strips
Al dente: to the bite

NUTRITION:

Calories: 551
Fat: 49.9 g
Protein: 9 g

Tossed Salad

COOKING TECHNIQUE:
Blanch

GLOSSARY:
Blanch: to parboil

NUTRITION:
Calories: 66.8
Fat: 2.75 g
Protein: 2.85 g

YIELD: 10 SERVINGS. SERVING SIZE: SIX OUNCES.

INGREDIENTS:

1 head	Red leaf lettuce (outer leaves removed), washed, leaves torn
2 heads	Green oakleaf lettuce (outer leaves removed), washed, leaves torn
5 each	Globe tomatoes, washed, cored, **blanched,** peeled, and cut into wedges
½ pound	Carrots, washed, peeled, trimmed, and sliced thin
½ pound	Cucumbers, washed, peeled, seeded, and sliced
1 ounce	Chives, washed, chopped fine
6 ounces	Pitted black olives
1 pint	Vinaigrette dressing (see page 230)

METHOD OF PREPARATION:

1. Mise en place ingredients for salad, and refrigerate.
2. Mise en place dressing (vinaigrette preferably).
3. At the time of service and when filling each order, toss the lettuce in a bowl with the vinaigrette until lightly but evenly coated. Portion the lettuce onto each chilled salad plate.
4. Decorate each plate of lettuce with tomato wedges, carrots, cucumbers, chives, and olives.
5. Serve immediately.

Waldorf Salad

YIELD: 50 SERVINGS. SERVING SIZE: SIX OUNCES.

COOKING TECHNIQUE:
Not applicable

HAZARDOUS FOODS:
Mayonnaise
Heavy cream

NUTRITION:
Calories: 435
Fat: 34.9 g
Protein: 5.71 g

INGREDIENTS:

12 pounds	Red and green apples, washed, cored, and diced
4 ounces	Lemon juice, to treat the apples
6 pounds	Celery, washed and diced small
2 pounds	Walnuts, roughly chopped
	Salad greens, as needed for serving, washed and trimmed

Chantilly Dressing:

3 pints	Mayonnaise (see page 35)
16 ounces	Heavy cream
8 ounces	Granulated sugar

METHOD OF PREPARATION:

1. Place the diced apples in a bowl, and treat them with lemon juice so they do not turn brown. Add the celery and walnuts.
2. Mix ingredients together for the dressing. Just before service, blend the dressing together with the other ingredients.
3. Serve in a large bowl or on chilled salad plates on lettuce greens. Garnish with grapes.

CHEF NOTE:
This salad calls for celeriac. Celery is used only if celeriac is not available.

COOKING TECHNIQUE:
Not applicable

NUTRITION:
Calories: 97
Fat: 2.16 g
Protein: 9.11 g

Watercress Salad

(Salade à la Cressonnière)

YIELD: 10 SERVINGS. SERVING SIZE: FOUR OUNCES.

INGREDIENTS:

2 pounds	Watercress, washed, trimmed, and held on ice until service
8 ounces	New potatoes, boiled, peeled, then sliced
20 ounces	Yogurt dressing (see below)
20 each	Cherry tomatoes, washed, dried, and cut in half
10 each	Radishes, washed, trimmed, and carved into flowers
2 each	Hard-boiled eggs, peeled, and grated
1 tablespoon	Semi-hot paprika
2 ounces	Freshly chopped parsley, excess moisture removed

Yogurt Dressing:

1 pint	Low-fat yogurt
5 ounces	Nonfat cream cheese
1 tablespoon	Lemon juice
3 drops	Hot sauce
1 tablespoon	Worcestershire sauce
	Salt and ground white pepper, to taste

METHOD OF PREPARATION:

1. In a bowl, combine all of the ingredients; mix well and allow to stand 30 minutes before using. Hold at 40° F or below.
2. For service, drain and arrange the watercress on chilled salad plates. Add the potatoes, and drizzle with dressing.
3. Garnish with tomatoes, radish flower, and eggs.
4. Sprinkle chives, and add paprika in straight lines from top down between the tomato halves. Sprinkle with parsley, and serve immediately.

Floridian Fruit Vinaigrette

YIELD: 1½ QUARTS. SERVING SIZE: TWO OUNCES.

COOKING TECHNIQUE:
Not applicable

HACCP:
Refrigerate at 40° F or lower.

NUTRITION:
Calories: 187
Fat: .46 g
Protein: 1.33 g

INGREDIENTS:

5 each	Oranges, peeled and sectioned
2 each	Apples, peeled, cored, and thinly sliced
8 ounces	Strawberries, washed, stemmed, and cut in half (if large)
2 each	Lemons, peeled and sectioned
8 ounces	Red seedless grapes, washed and stemmed
8 ounces	Granulated sugar
2 ounces	Cointreau

METHOD OF PREPARATION:

1. Place all of the ingredients, except the liqueur, into the bowl of a food processor, and purée.
2. Transfer the purée into a glass or crock container, add the Cointreau, and mix well. Refrigerate at 40° F or lower at least 12 hours or overnight.

CHEF NOTE:
Preparation should begin 1 day ahead to allow acid flavor to develop.

Fresh Cream Dressing

COOKING TECHNIQUE:
Not applicable

HAZARDOUS FOOD:
Heavy cream

NUTRITION:
Calories: 61
Fat: 5.68 g
Protein: .568 g

YIELD: 10 OUNCES. SERVING SIZE: TWO OUNCES.

INGREDIENTS:

8 ounces	Heavy cream
2½ ounces	Lemon juice
½ teaspoon	Salt
½ teaspoon	Dry mustard
½ teaspoon	Ground black pepper

METHODS OF PREPARATION:

Place all of the ingredients in a blender or small mixer, and whip just before service. Hold at 40° F or below.

French Dressing

YIELD: 16 SERVINGS. SERVING SIZE: TWO OUNCES.

COOKING TECHNIQUE:
Not applicable

NUTRITION:
Calories: 360
Fat: 40.5 g
Protein: 0 g

INGREDIENTS:

8 ounces	Red wine vinegar
24 ounces	Vegetable or olive oil
	Salt and white pepper, to taste

METHOD OF PREPARATION:

Whip all of the ingredients vigorously until they are temporarily emulsified.

Italian Dressing

YIELD: 64 SERVINGS. SERVING SIZE: TWO OUNCES.

COOKING TECHNIQUE:

Not applicable

NUTRITION:

Calories: 325
Fat: 35.4 g
Protein: 0 g

INGREDIENTS:

2½ quarts	Olive oil
1½ quarts	Red wine vinegar
4 ounces	Granulated sugar
2 ounces	Fresh basil, washed and roughly chopped
2 ounces	Fresh oregano, washed and roughly chopped
½ teaspoon	Marjoram
3 cloves	Garlic, peeled and mashed into a paste
	Salt and black pepper, to taste
4 ounces	Shallots, peeled and minced

METHOD OF PREPARATION:

Combine all of the ingredients in a mixing bowl. Whip vigorously, and and serve immediately. (This dressing is a temporary emulsion and must be mixed before each use.)

CHEF NOTE:

A temporary emulsion will hold longer if the oil is mixed gradually into the acid.

Socrates Dressing

YIELD: 64 SERVINGS. SERVING SIZE: TWO OUNCES.

COOKING TECHNIQUE:
Not applicable

GLOSSARY:
Brunoise: finely diced

HAZARDOUS FOODS:
Mayonaise
Eggs

NUTRITION:
Calories: 318
Fat: 34.3 g
Protein: 1.04 g

INGREDIENTS:

3 quarts	Mayonnaise (see page 35)
4 ounces	White vinegar
2 ounces	Worcestershire sauce
1 ounce	Mustard powder
½ ounce	White pepper
1 ounce	Oregano
¼ ounce	Cayenne pepper
½ ounce	Tarragon
½ ounce	Marjoram
½ ounce	Dill seed
½ ounce	Chervil
½ ounce	Cumin
1 ounce	Salt
1 ounce	Sugar
3 cloves	Garlic, peeled and mashed
8 ounces	Gherkins diced **brunoise**
6 each	Eggs, hard-cooked, peeled, and diced brunoise
6 ounces	Onions, peeled and diced brunoise
2 ounces	Fresh parsley, washed, finely chopped, and excess moisture removed

METHOD OF PREPARATION:

1. In a mixing bowl, combine the mayonnaise, vinegar, and Worcestershire sauce. Blend thoroughly; then add seasonings.
2. Add the gherkins, eggs, onions, and parsley to the mayonnaise. Blend until well incorporated.
3. Refrigerate for 2 days to allow the flavors of the spices and herbs to be enhanced.
4. Serve with a tossed or mixed green salad.

Vinaigrette Dressing

YIELD: 1 CUP.

HACCP:

Hold chilled at 40° F or lower.

NUTRITION:

Calories: 190
Fat: 21.3 g
Protein: .048 g

INGREDIENTS:

2 cloves	Garlic, peeled and mashed into a purée
2 ounces	Sherry vinegar
6 ounces	Extra virgin olive oil
	Salt and freshly ground black pepper, to taste

METHOD OF PREPARATION:

1. Combine the garlic and sherry vinegar in a bowl.
2. Add the olive oil in a fine stream, and whisk the ingredients together.
3. Hold at 40° F or lower.

CHEF NOTE:

The dressing is a temporary emulsion. It must periodically be vigorously rewhipped.

5

Entrées

5

Entrées

Pheasants snared in Colchis in Africa game-birds,
These are the rarities that have to be chased,
White goose and duckling,
Gaudy in their gay plumes,
Are left to the populace, not to our taste.

Petronius
The Satyricon, No. 92

Cooking is a means of expressing oneself. It is an art form that is similar to the work of the painter, the sculptor, or the musician, and, like such artists, for a chef to succeed, it is fundamental to understand the basics—Picasso would not have been so well praised had he not gone through his classical period. The chef has to evolve from this and learn by experiment, harmonizing only the best ingredients and produce with good taste to compose his own personal works of art. It is also a question of love, not only for oneself, but also for others.

After the correct training of 8 to 10 years, at least, to learn your craft, and 3 to 5 years to find your own identity, only then, with a team of cuisiniers to work under you, with you, and to do the utmost for you, can you expect to begin the real task of Chef de Cuisine or Chef Patron.

La Grande Cuisine Francaise is still alive, and when it is executed with knowledge, the right reasoning, and spirit, it is always fashionable and can inspire talented chefs all over the world.

Michel Bourdin, CMA-COMN
Distinguished Visiting Chef
March 9–March 11, 1980
Chef de Cuisine
The Connaught Hotel, London
President de l'Academie Culinaire de France
(Filiale de Gde. Bretagne)

The HACCP Process—Beef

The following recipe illustrates how beef flows through the HACCP process.

Prime Rib of Beef

Receiving

- Beef rib roast 1 #109—packaging intact. Beef should be delivered at 40° F (4.4° C) or lower.
- Spices (salt, pepper, and garlic powder)—packaging intact
- Vegetables (onions, celery, and carrots)—packaging intact; no cross-contamination from other foods on the truck; no signs of insect or rodent activity
- Brown stock (already made)

Storage

- Beef rib roast—Store under refrigeration, with a product temperature not to exceed 40° F (4.4° C). Store below already cooked foods.
- Spices (salt, pepper, and garlic powder)—Store in dry storage at 50° F (10° C) with a relative humidity of 50% to 60%. Keep dry.
- Vegetables (onion, celery, and carrots)—Store under refrigeration, with a product temperature not to exceed 40° F (4.4° C). Store above and away from raw, potentially hazardous foods.
- Brown stock—Store under refrigeration, with a product temperature not to exceed 40° F (4.4° C). Store above and away from raw, potentially hazardous foods.

Preparation and Cooking

- Peel the outer layer of onions, and wash.
- Trim and wash the celery.
- Peel and wash the carrots.
- Chop the vegetables on a clean and sanitized cutting board with clean and sanitized utensils.

- Season the roast under the fat cap, and then place it on top of the cut mirepoix in a roasting pan.
- Place the rib roast in an oven preheated to 400° F, and reduce to 325° F. Cook until a temperature of 130° F (54.4° C) is reached in the center of the meat. The Food and Drug Administration (FDA) recommends that beef be cooked to an internal temperature of 145° F (62.8° C).
- Remove the rib roast from the oven.
- Remove the grease from the pan, deglaze the pan with the stock, adjust the seasonings, and strain the juice.

Holding and Service

- Hold at 140° F (60° C) or higher.
- Let the meat rest before carving. Slice the meat to order, across the grain. Serve with the juice.

Cooling

- Place the beef in a chill blaster. Lower the temperature to 40° F (4.4° C) or less within 6 hours.
- Cover loosely with plastic wrap when cooled.
- Label with the date, time, and name of product.
- Store on the upper shelf of a refrigerator unit at 40° F (4.4° C) or lower.

Reheating

- Reheat at 165° F (73.9° C) or higher in 2 hours.

The HACCP Process—Lamb

The following recipe illustrates how lamb flows through the HACCP process.

Rack of Lamb with Parsley

Receiving

- Rack of lamb—packaging intact. Lamb should be delivered at 40° F (4.40° C) or lower.
- Butter—should have a sweet flavor, firm texture, and uniform color. Butter should be free of specks and mold, and the packaging should be intact. Butter should be delivered at 40° F (4.4° C) or lower.
- Vegetables (garlic)—packaging intact; no cross-contamination from other foods on the truck; no signs of insect or rodent activity.
- Herbs (parsley)—no visible signs of discoloration
- Bread crumbs—packaging intact
- Spices (salt and pepper)—packaging intact

Storage

- Lamb—Store under refrigeration, with a product temperature not to exceed 40° F (4.4° C). Store below already cooked foods.
- Butter—Store under refrigeration, with a product temperature not to exceed 40° F (4.4° C). Store above and away from raw, potentially hazardous foods.
- Vegetables (garlic)—Store under refrigeration, with a product temperature not to exceed 40° F (4.4° C). Store above and away from raw, potentially hazardous foods.
- Herbs (parsley)—Store under refrigeration at 40° F (4.4° C) or lower.
- Bread crumbs—Store in dry storage at 50° F (10° C) with a relative humidity of 50% to 60%.
- Spices (salt and pepper)—Store in dry storage at 50° F (10° C) with a relative humidity of 50% to 60%.

Preparation and Cooking

- Peel and mince the garlic.
- Wash and dry the parsley.
- Mince the garlic and chop the parsley on a clean and sanitized cutting board with clean and sanitized utensils.

- Brown the rack of lamb in a skillet with heated butter.
- Place the lamb in a roasting pan in a 450° F oven for 30 minutes. Then remove the meat from the oven, turn it, season with salt and pepper, sprinkle the bread crumb mixture over rack of lamb, and press down with a spoon or a spatula.
- Return the lamb to the oven, and cook for 15 more minutes, until it reaches a medium-rare temperature. The FDA recommends that lamb be cooked to an internal temperature of 145° F (62.8° C).

Holding and Service

- Hold at 140° F (60° C) or higher.
- Serve with proper garnishes, and carve tableside.

The HACCP Process—Pork

The following recipe illustrates how pork flows through the HACCP process.

Pork Cacciatore

Receiving

- Pork loin, boneless—Fat should be firm and white, and the lean portions light pink. Pork should be delivered at 40° F (4.4° C) or lower.
- Olive oil—packaging intact.
- Vegetables (garlic, onions, green peppers, and mushrooms)—packaging intact; no cross-contamination from other foods on the truck; no signs of insect or rodent activity
- Tomato sauce—Check cans for swelling, leakage, flawed seals, rust, or dents
- Herbs and spices (oregano, rosemary, marjoram, black pepper, and salt)—packaging intact

Storage

- Pork loin—Store under refrigeration, with a product temperature not to exceed 40° F (4.4° C). Store below already cooked foods.
- Olive oil—Store in dry storage at 50° F (10° C) with a relative humidity of 50% to 60%.
- Vegetables (garlic, onions, green peppers, and mushrooms)—Store under refrigeration, with a product temperature not to exceed 40° F (4.4° C). Store above and away from raw, potentially hazardous foods.
- Tomato sauce—Store in dry storage at 50° F (10° C) with a relative humidity of 50% to 60%.
- Herbs and spices (oregano, rosemary, marjoram, black pepper, and salt)—Store in dry storage at 50° F (10° C) with a relative humidity of 50% to 60%.

Preparation and Cooking

- Peel and mince the garlic.
- Peel the outer layer of onions, and wash.
- Seed and wash green peppers.
- Wash the mushrooms.
- Mince the garlic, and cut onions, green peppers, and mushrooms on a clean and sanitized cutting board with clean and sanitized utensils.
- Slice the pork loin into portions.
- Heat the olive oil in a large sauté pan. Add the meat, and brown the pork slices well on all sides. Remove the meat, and reserve.
- Sauté the onions and green peppers, stirring often to prevent burning. Add the garlic and then the mushrooms, and sauté until the vegetables are tender. Add the pork chops and spices, and sweat for a few moments.
- Heat the tomato sauce, and add it to the meat and vegetables. Simmer for 45 minutes until the pork is tender. Adjust the seasonings. The internal cooking temperature of pork should be 155° F (68° C).

Holding and Service

- Hold at 140° F (60° C) or higher.
- Serve immediately.

Cooling

- Put the pork cacciatore in shallow pans at a product depth of 2 inches or less.
- Place the pans with the product into an ice bath, immersed to the level of the product within the pan. Stir frequently, and lower the temperatures to 40° F (4.4° C) or less within 4 hours.

Reheating

- Reheat to 165° F (73.9° C) or higher in 2 hours.

The HACCP Process—Poultry

The following recipe illustrates how poultry flows through the HACCP process.

Chicken à la Kiev

Receiving

- Chicken breasts, whole, boneless—Check for firm flesh. There should be no sign of odor or stickiness. Chicken should be delivered at 40° F (4.4° C) or lower.
- Butter—should have a sweet flavor, firm texture, and uniform color. Butter should be free of specks and mold, and the packaging should be intact. Butter should be delivered at 40° F (4.4° C) or lower.
- Vegetables (garlic)—packaging intact; no cross-contamination from other foods on the truck; no signs of insect or rodent activity
- Herbs (chives)—no visible signs of discoloration
- Spices (salt and pepper)—packaging intact
- Flour—packaging intact
- Bread crumbs—packaging intact
- Eggs, whole—shells not cracked or dirty. Container should be intact. Whole shell eggs should be delivered at 40° F (4.4° C) or lower.
- Milk—should have a sweetish taste. Container should be intact. Milk should be delivered at 40° F (4.4° C) or lower.
- Lemons—fine-textured skin, heavy for size, and uniform color

Storage

- Chicken—Store under refrigeration, with a product temperature not to exceed 40° F (4.4° C). Store below already cooked foods.
- Butter—Store under refrigeration at 40° F (4.4° C) or lower. Store above and away from raw, potentially hazardous foods.
- Vegetables (onion, celery, and carrots)—Store under refrigeration, with a product temperature not to exceed 40° F (4.4° C). Store above and away from raw, potentially hazardous foods.
- Herbs (chives)—Store under refrigeration at 40° F (4.4° C) or lower.
- Spices (salt and pepper)—Store in dry storage at 50° F (10° C) with a relative humidity of 50% to 60%. Keep dry.
- Flour—Store in dry storage at 50° F (10° C) with a relative humidity of 50% to 60%.
- Bread crumbs—Store in dry storage at 50° F (10° C) with a relative humidity of 50% to 60%.
- Eggs, whole—Store under refrigeration at 40° F (4.4° C) or lower. Store above and away from raw, potentially hazardous foods.
- Milk—Store under refrigeration at 40° F (4.4° C) or lower. Store above and away from raw, potentially hazardous foods.
- Lemons—Store under refrigeration at 40° F (4.4° C) or lower.

Preparation and Cooking

- Peel and mince the garlic.
- Wash, dry, and slice the chives.
- Mince the garlic and slice the chives on a clean and sanitized cutting board with clean and sanitized utensils.

- Place the butter in a mixing machine, using the paddle, and mix at low speed until the butter turns smooth (but not too soft).
- Add the minced garlic and chives, and season with salt and pepper.
- Flatten the boneless chicken breasts, skin-side down, using a meat mallet.
- In the center of each breast, place a good tablespoon of butter, roll up, and fold the ends. Place in the refrigerator until slightly firm.
- Dredge each breast through flour, an egg wash, and bread crumbs. Pat off the excess crumbs.
- Preheat a deep-fat fryer to 350° F. Brown the chicken lightly; remove, drain on absorbent paper, and transfer to a sheet pan.
- Preheat the oven to 350° F, and finish baking chicken in the oven for 15 to 20 minutes. Cook to an internal temperature of 165° F (73.9° C).
- Using a paring knife, cut a 1-inch slit in the top of each breast.

Holding and Service

- Hold at 140° F (60° C) or higher.
- Garnish each breast with a lemon slice, and serve immediately.

The HACCP Process—Seafood

The following recipe illustrates how seafood flows through the HACCP process.

Baked Lobster

Receiving

- Lobsters—Shells should be hard and heavy. There should be movement, and the tails should curl under when the lobsters are picked up.
- Vegetables (shallots)—packaging intact; no cross-contamination from other foods on the truck; no signs of insect or rodent activity
- Herbs (parsley)—no visible signs of discoloration
- Butter—should have a sweet flavor, firm texture, and uniform color. Butter should be free of specks and mold, and the packaging should be intact. Butter should be delivered at 40° F (4.4° C) or lower.
- Spices (salt and pepper)—packaging intact
- Tabasco sauce—packaging intact
- Bread crumbs—packaging intact
- Lemons—fine-textured skin, heavy for size, and uniform color

Storage

- Lobsters—Store under refrigeration at 40° F (4.4° C) or lower.
- Vegetables (shallots)—Store under refrigeration, with a product temperature not to exceed 40° F (4.4° C). Store above and away from raw, potentially hazardous foods.
- Herbs (parsley)—Store under refrigeration at 40° F (4.4° C) or lower.
- Butter—Store under refrigeration at 40° F (4.4° C) or lower. Store above and away from raw, potentially hazardous foods.
- Spices (salt and pepper)—Store in dry storage at 50° F (10° C) with a relative humidity of 50% to 60%. Keep dry.
- Tabasco sauce—Store in dry storage at 50° F (10° C) with a relative humidity of 50% to 60%.
- Bread crumbs—Store in dry storage at 50° F (10° C) with a relative humidity of 50% to 60%.
- Lemons—Store under refrigeration at 40° F (4.4° C) or lower.

Preparation and Cooking

- Peel and dice the shallots.
- Wash, dry, and chop the parsley.
- Peel the shallots and dice the parsley on a clean

sanitized cutting board with clean and sanitized utensils.

- In a skillet, heat 2 ounces of butter, and sauté the shallots for 2 to 3 minutes, stirring occasionally.
- In a mixing bowl, mix the parsley, bread crumbs, salt, pepper, Tabasco sauce, and shallots. Mix well.
- Place the lobster shell-side down on a sheet pan. Fill the body with the bread mixture, and sprinkle some of the bread mixture over the tail meat. Spoon some melted butter over the bread crumbs.
- Bake at 400° F for 16 to 20 minutes until the bread crumbs are well browned. The internal temperature should be 145° F (62.8° C).

Holding and Service

- Serve immediately.

The HACCP Process—Veal

The following recipe illustrates how veal flows through the HACCP process.

Veal Franconian

Receiving

- Veal, top round cutlets—Fat should be firm and white, and the lean portion light pink. Veal should be delivered at 40° F (4.4° C) or lower.
- Flour—packaging intact
- Eggs, whole—shells not cracked or dirty. Container should be intact. Whole shell eggs should be delivered at 40° F (4.4° C) or lower.
- Lemons—fine-textured skin, heavy for size, and uniform color
- Milk—should have a sweetish taste. Container should be intact. Milk should be delivered at 40° F (4.4° C) or lower.
- Bread crumbs—packaging intact
- Spices (salt and pepper)—packaging intact

Storage

- Veal—Store under refrigeration, with a product temperature not to exceed 40° F (4.4° C). Store below already cooked foods.
- Flour—Store in dry storage at 50° F (10° C) with a relative humidity of 50% to 60%.
- Eggs, whole—Store under refrigeration at 40° F (4.4° C) or lower. Store above and away from raw, potentially hazardous foods.
- Lemons—Store under refrigeration at 40° F (4.4° C) or lower.
- Milk—Store under refrigeration at 40° F (4.4° C) or lower. Store above and away from raw, potentially hazardous foods.
- Bread crumbs—Store in dry storage at 50° F (10° C) with a relative humidity of 50% to 60%.
- Spices (salt and pepper)—Store in dry storage at 50° F (10° C) with a relative humidity of 50% to 60%. Keep dry.

Preparation and Cooking

- Trim the fat and sinew. On a clean and sanitized cutting board with clean and sanitized utensils, pound the veal between plastic wrap until it is ½-inch to ¾-inch thick.
- Dredge the veal pieces in the flour.
- Blend the eggs, salt, pepper, and milk, and then dip the veal pieces in the mixture.
- Roll the veal pieces in the bread crumbs, pressing the crumbs evenly onto the veal.
- Arrange the veal pieces on a parchment-lined baking sheet. Cover and refrigerate until the crumbs set (about 2 hours). Store under refrigeration at 40° F (4.4° C) or lower.
- Pan-fry the veal in vegetable oil until golden brown (about 5 to 8 minutes). Cook to an internal temperature of 145° F (62.8° C).

Holding and Service

- Hold at 140° F (60° C) or higher.
- Garnish each cutlet with a lemon slice, and serve immediately.

Nutritional Notes

While entrées are good sources of many nutrients, they also can be a primary source of fat, saturated fat, cholesterol, and calories. The following are some suggestions for reducing fat, saturated fat, and cholesterol in meat, poultry, and seafood entrées:

- *Use lean meats.* Prime grades of meat are higher in fat. Meats with lower fat content are recommended when they are practical for the intended use.

Beef: Choose round, tenderloin, flank, or ground that is 90% or more fat-free.
Pork: Choose round (leg) and tenderloin.
Lamb: Choose leg and sirloin chop.
Veal: Choose shoulder, rib, loin, or leg. (Note that veal is higher in cholesterol than are other meats.)

- *Offer entrées featuring wild game.* Wild game is very low in fat, whereas domesticated game, such as duck or goose, is high in fat. Skin should be removed from wild game before eating.
- *Offer entrées featuring chicken and turkey.* If it is prepared separately, poultry can be cooked with the skin on and then removed before eating. If the poultry is part of a mixed dish, however, the skin should be removed before cooking. Ground turkey and chicken should be skinless and without added fat.
- *Feature seafood items.* Most seafood is low in fat. Fish that are high in fat contain polyunsaturated oils and are low in saturated fats.
- *Reduce portion size.* Limit portions of cooked meat, poultry, and seafood to 3 or 4 ounces.
- *Use low-fat cooking methods.* Baking, broiling, steaming, en papillote, poaching, simmering, blanching, and microwaving limit the amount of added fat required to prepare an entrée.
- *Stir-fry or sauté.* Use a small amount of oil, or use a well-seasoned pan and omit the oil.
- *Cook meats on a rack.* Cooking meat or poultry on a rack allows the fat to drain.

Baked Meat Loaf

YIELD: 50 SERVINGS. SERVING SIZE: FIVE OUNCES.

INGREDIENTS:

13 pounds	Ground beef
2 pounds	White bread, chopped rough
2 pints	Milk
10 ounces	Oil
2 pounds	Onions, peeled and diced **brunoise**
2 heads	Garlic, peeled and minced
13 ounces	Parmesan cheese
13 each	Eggs
2 ounces	Freshly chopped parsley, washed, trimmed, and excess moisture removed
	Salt, to taste
1 teaspoon	Ground black pepper
13 ounces	Tomato purée
13 ounces	Oil

METHOD OF PREPARATION:

1. Preheat oven to 350° F.
2. Place the meat in a mixing bowl. Soak the white bread in the milk and squeeze out any excess liquid. Add to the meat.
3. Sauté the onions in oil until they are translucent, and then cool. Add to the meat.
4. Add the Parmesan cheese, eggs, parsley, and seasonings to the meat.
5. Mix all the items well to incorporate all the ingredients. Make a small patty for testing the seasoning. Fry and taste. Adjust seasoning.
6. Shape the meat into four loaves. Place them in hotel pans, and cover them with a mixture of the tomato purée and oil.
7. Refrigerate for 15 minutes. Bake in the oven for 45 minutes to 1 hour, or until an internal temperature of 155° F is reached.
8. Allow the loaves to set before slicing.
9. Slice loaves into 1-inch-thick pieces, and serve with a desired sauce (either brown or tomato, heated to a boil).
10. Hold at 140° F.

COOKING TECHNIQUE:

Bake

Bake:

1. Preheat the oven.
2. Place the food product on the appropriate rack.

GLOSSARY:

Brunoise: ⅛-inch dice

HACCP:

Hold at 140° F.

HAZARDOUS FOODS:

Ground beef
Eggs
Milk

NUTRITION:

Calories: 376
Fat: 22 g
Protein: 30 g

CHEF NOTES:

1. A meat loaf often is used for utilization purposes, that is, to use ingredients on hand to avoid waste. Therefore, ground pork, veal, ground turkey, or a combination may be substituted for the ground beef.
2. For a smoother and finer texture, grind the meat again after all ingredients are incorporated.
3. Test for flavor by frying a small pattie before shaping into loaf.

Baked Salisbury Steak

YIELD: 50 SERVINGS. SERVING SIZE: SIX OUNCES.

INGREDIENTS:

8 ounces	Butter, clarified
2 pounds	Onions, peeled and diced **brunoise**
1 pound	Green bell peppers, washed, seeded, and diced brunoise
1 pound	Red bell peppers, washed, seeded, and diced brunoise
10 pounds	Ground beef
10 pounds	Ground pork
6 ounces	Freshly chopped parsley, excess moisture removed
20 each	Whole eggs
2 teaspoon	Ground black pepper
	Salt, to taste

METHOD OF PREPARATION:

1. Preheat the oven to 350° F.
2. Sauté the onions and peppers in the clarified butter, and cool to internal temperature below 40° F.
3. Place the ground meats in a bowl. Add the onions and peppers.
4. Add the remaining ingredients, using the eggs as needed. Adjust the seasonings as needed.
5. Thoroughly mix all of the ingredients in 20-quart mixer equipped with a flat paddle. Shape into 8-ounce oval steaks, and place on an oiled sheet pan or wire rack.
6. Bake in the oven until the internal temperature of 150° F is reached.
7. When done, remove from the oven. Serve on a preheated dinner plate with an appropriate sauce.

COOKING TECHNIQUE:

Bake

Bake:

1. Preheat the oven.
2. Place the food product on the appropriate rack.

GLOSSARY:

Brunoise: ⅛-inch dice

HACCP:

Hold at 140° F.

HAZARDOUS FOODS:

Ground beef
Ground pork
Whole eggs

NUTRITION:

Calories: 612
Fat: 44.4 g
Protein: 47.2 g

CHEF NOTES:

1. Salisbury steak can be served with many sauces. The most traditional are brown mushroom sauce and tomato sauce.
2. Ground veal or ground turkey meat may be used in the place of ground pork.

Barbecued Beef with BBQ Sauce

YIELD: 50 PORTIONS. SERVING SIZE: FOUR OUNCES.

COOKING TECHNIQUE:
Sauté

Sauté:
1. Heat the sauté pan to the appropriate temperature.
2. Evenly brown the food product.
3. For a sauce, pour off any excess oil, reheat, and deglaze.

GLOSSARY:
Baste: to brush with sauce

HAZARDOUS FOOD:
Shaved steak

NUTRITION:
Calories: 324
Fat: 12.1 g
Protein: 34.2 g

INGREDIENTS:

13 pounds	Shaved steak

Barbecue Sauce:

1½ pounds	Onion, peeled and minced
6 each	Garlic cloves, minced
4 tablespoons	Butter
1 quart	Ketchup
1 quart	Bourbon
4 tablespoons	Maple syrup
4 tablespoons	Worcestershire sauce
2 teaspoons	Mustard powder
2 teaspoons	Salt
2 teaspoons	Chili powder

METHOD OF PREPARATION:

1. Preheat tilting skillet to 400° F.
2. Using rubber gloves, separate the shaved steak.
3. Spray the tilting skillet with nonstick spray.
4. Add the shaved steak, and sauté quickly. Do not overcook.
5. **Baste** with the barbecue sauce. Hold over 140° F, and serve.

Barbecue Sauce

METHOD OF PREPARATION:

1. Sauté the onions, garlic, and butter in a heavy saucepan.
2. Add the remaining ingredients. Heat to a boil. Simmer for 10 minutes. Use to baste meats for grilling.

CHEF NOTE:
Simmer 10 to 15 minutes longer to reduce to a thicker sauce for serving with cooked meat. This is especially good for ribs or pork chops.

Beef Burgundy

YIELD: 50 SERVINGS. SERVING SIZE: EIGHT OUNCES.

INGREDIENTS:

20 pounds	Bottom round beef
5 ounces	Oil
2 pounds	Onions, peeled and diced **brunoise**
1 teaspoon	Marjoram
1 teaspoon	Thyme
1 teaspoon	Spanish paprika
1 pint	Red Burgundy wine
3 quarts	Demi-glace, heated to a boil (see page 26)
1 pound	Smoked bacon, ¼-inch dice
2 pounds	Pearl onions, peeled
5 pounds	Mushrooms, washed, cut into quarters
	Salt and freshly ground black pepper, to taste
15 pounds	Buttered egg noodles (see page 473) or
15 pounds	Rice pilaf (see page 517)

METHOD OF PREPARATION:

1. Trim and peel the fat from the beef. Cut the meat ½-inch dice.
2. Sear the beef cubes in hot oil until lightly browned. Do not overcrowd the pan. Remove and reserve; pour off any excess fat, add the diced onions, and sauté until the onions are translucent. Add the marjoram, thyme, and paprika to the meat, and blend.
3. **Deglaze** the pan with the red Burgundy wine, reduce by half, add the demi-glace, and cover the pan with parchment paper and aluminum foil.
4. Simmer until the meat is fork-tender.
5. Render the diced bacon in a sauté pan. Add peeled onions and mushrooms, and sauté. Add to the stew.
6. Remove the stew from the heat, **dépouiller,** hold at 140° F, adjust the seasonings, and serve with egg noodles or rice pilaf.

COOKING TECHNIQUES:

Sauté, Stew

Sauté:

1. Heat the sauté pan to the appropriate temperature.
2. Evenly brown the food product.
3. For a sauce, pour off any excess oil, reheat, and deglaze.

Stew:

1. Sear, sauté, sweat, or blanch the main food product.
2. Deglaze the pan, if desired.
3. Cover the food product with simmering liquid.
4. Remove the bouquet garni.

GLOSSARY:

Brunoise: ⅛-inch dice
Deglaze: to add liquid to hot pan
Dépouiller: to skim impurities/grease

HACCP:

Heat to boil.
Internal temperature of 140° F.
Hold at 140° F.

HAZARDOUS FOOD:

Beef

NUTRITION:

Calories: 896
Fat: 31.4 g
Protein: 73.3 g

Beef Fajita

YIELD: 50 SERVINGS. **SERVING SIZE:** SIX OUNCES.

INGREDIENTS:

14 pounds	Top round beef
6 ounces	Oil
50 each	Flour tortillas, warm
8 each	Large tomatoes, **concassé**
6 ounces	Scallions, washed and sliced
3 pounds	Cheddar cheese, shredded
1½ pounds	Sour cream

Marinade:

1 pint	Fresh lime juice
1 quart	Tomato juice
2 teaspoons	Hot sauce
4 cloves	Garlic, peeled and minced
6 each	Jalapeño peppers, seeded and diced
3 each	Red onions, peeled and diced
6 each	Green peppers, washed, seeded, and diced
1 tablespoon	Ground cumin
2 tablespoons	Chili powder
	Salt, to taste
3 ounces	Fresh cilantro, washed leaves chopped

METHOD OF PREPARATION:

1. Trim and remove the excess fat and silverskin from the beef. Cut the meat into thin strips.
2. Combine the marinade ingredients, and pour over the meat. Marinate for 1 to 2 hours in the refrigerator.
3. In a hot sauté pan, sauté the meat in oil, and add the red onions and green peppers. Cook the beef to an internal temperature of 140° F. Drain any excess fat. Sauté until the onions are translucent; then add the ground cumin, chili powder, salt, and cilantro. Adjust the seasonings, and place mixture in a hotel pan. Keep hot until needed.
4. To prepare individual portions, warm the tortillas on a sheet pan, either in a 350° F oven or lightly under the broiler. Do not dry out tortillas. Place 4 ounces of the meat mixture on a tortilla, and top with chopped tomato, scallions, cheese, and sour cream. Roll the tortilla, seam-side down, around the filling, tucking in the ends. Place the rolled tortillas in hotel pans, and hold at 140° F.

COOKING TECHNIQUE:

Sauté

Sauté:

1. Heat the sauté pan to the appropriate temperature.
2. Evenly brown the food product.
3. For a sauce, pour off any excess oil, reheat, and deglaze.

GLOSSARY:

Concassé: peeled, seeded, roughly chopped

HACCP:

Hold at 140° F.

HAZARDOUS FOODS:

Top round of beef
Sour cream

NUTRITION:

Calories: 534
Fat: 24 g
Protein: 41.5 g

CHEF NOTES:

1. There are many variations of beef fajita. This recipe is a regional dish prepared throughout South America. The region determines the ingredients used and the degree of spiciness.
2. Do not prepare meat in advance. It drys out and gets tough when kept in the bain-marie for extended time.

Beef Pot Pie

YIELD: 50 SERVINGS. SERVING SIZE: EIGHT OUNCES.

COOKING TECHNIQUES:

Stew, Bake

Stew:

1. Sear, sauté, sweat, or blanch the main food product.
2. Deglaze the pan, if desired.
3. Cover the food product with simmering liquid.
4. Remove the bouquet garni.

Bake:

1. Preheat the oven.
2. Place the food product on the appropriate rack.

GLOSSARY:

Macédoine: ¼-inch dice
Sear: to brown quickly
Al dente: to the bite

HACCP:

Hold at 140° F.

HAZARDOUS FOOD:

Beef

NUTRITION:

Calories: 664
Fat: 35 g
Protein: 48.7 g

INGREDIENTS:

12 ounces	Oil
15 pounds	Beef bottom round, fat and connective tissue removed, cut into 1-ounce cubes

Macédoine:

2 pounds	Onions, peeled and diced
3 pounds	Carrots, washed, peeled, and diced
3 pounds	Celery stalks, washed, peeled, and diced
1 (#10 can)	Crushed tomatoes, chopped
1 gallon	Demi-glace (see page 26)
1 teaspoon	Dried thyme leaves
4 pounds	Potatoes, washed, peeled and scooped parisienne
	Salt and ground white pepper, to taste

METHOD OF PREPARATION:

1. Heat the oil in a braising pan. Add the meat, and **sear** to form a brown crust on the beef cubes. Remove the meat, and drain the fat. Sauté the onions in same utensil as heat until they are translucent. Add the carrots and celery.
2. Add the crushed tomatoes, browned meat cubes, demi-glace, and thyme. Heat to a boil and simmer until the meat is fork-tender. Hold at 140° F.
3. In a saucepan, place the potatoes in cold, salted water, and cook until **al dente.** Drain in a colander, rinse under cold water. Add the potatoes to the stew, and simmer for an additional 10 minutes. Do not overcook potatoes.
4. Degrease the excess fat from the top of the stew. Season, to taste.
5. Transfer the stew into 2-inch deep hotel pan, and top with a rolled out pie crust (see page 808).
6. Egg wash the pie crust, and bake at 350° F until brown and crust is baked.

Pie Crust

YIELD: 6 POUNDS. SERVING SIZE: NOT APPLICABLE.

INGREDIENTS:

1 ounce	Salt
3 pounds	Pastry flour
2 pounds	Butter or shortening
1 pound	Cold water
4 each	Eggs for egg wash

METHOD OF PREPARATION:

1. Sift the salt and flour in a mixing bowl.
2. Break the shortening or butter into small pieces, and flake into the fat and flour.
3. Make a well in the fat and flour mixture, add cold water to the well, and mix the dough until a ball is formed. Then cover the dough and refrigerate.
4. Roll the dough (on a floured table or pastry cloth) until ⅛-inch thick. Cut the dough into the shape of the casserole. Place on a parchment-papered tray, and brush with the egg wash. Bake in a 375° F oven for 12 minutes, or until golden brown. Cool and decorate the pot pies, as needed.

COOKING TECHNIQUE:

Bake

Bake:

1. Preheat the oven.
2. Place the food product on the appropriate rack.

HAZARDOUS FOOD:

Eggs

NUTRITION:

See Beef Pot Pie

CHEF NOTES:

1. There are many variations of beef pot pie. Any garden-fresh or frozen vegetables may be used to add richness to the stew (e.g., green peas or mushrooms).
2. Commercial frozen pie dough sheets or puff pastry may be purchased.

Beef Tournedos Chasseur

YIELD: 10 SERVINGS. SERVING SIZE: SIX OUNCES.

INGREDIENTS:

4 pounds	Beef tenderloin, peeled
16 ounces	Sauce chasseur (see page 57)
20 each	Mushroom caps, large, fluted
20 each	Croutons, round, toasted

METHOD OF PREPARATION:

1. Cut the beef tenderloin into 3-ounce medallions, two per portion.
2. Either sauté or broil the medallions, to order, cooking until the desired internal temperature is reached, but to a minimum internal temperature of 140° F.
3. Sauté the mushroom caps.
4. Place the medallions on the croutons, and **nappé** with sauce chasseur. Garnish with mushroom caps.

COOKING TECHNIQUES:

Broil, Sauté

Grill/Broil:

1. Clean and heat the grill/broiler.
2. To prevent sticking, brush the food product with oil.

Sauté:

1. Heat the sauté pan to the appropriate temperature.
2. Evenly brown the food product.
3. For a sauce, pour off any excess oil, reheat, and deglaze.

GLOSSARY:

Nappé: coated

HACCP:

Cook to a minimum temperature of 140° F.

HAZARDOUS FOOD:

Beef tenderloin

NUTRITION:

Calories: 723
Fat: 51 g
Protein: 47 g

CHEF NOTE:

Tournedos can be served with a variety of sauces or compound butter.

Beef Wellington

(Filet de Boeuf Wellington)

YIELD: 10 SERVINGS. SERVING SIZE: SIX OUNCES.

INGREDIENTS:

2 ounces	Clarified butter
2 ounces	Tomato puree
	Salt and freshly ground black pepper, to taste
4 pounds	Beef tenderloin, trimmed of all fat and silverskin
3 tablespoons	Vegetable oil
2 pounds	Puff pastry
12 ounces	Prepared goose liver, puréed
2 each	Eggs, for egg wash
20 ounces	Sauce Périgourdine, heated to a boil (see page 64) (hold at 140° F.)

Mushroom Duxelle:

1 pound	Mushrooms, washed
4 ounces	Shallots, peeled and diced **brunoise**
6 ounces	Onions, peeled and diced brunoise

METHOD OF PREPARATION:

1. Preheat the oven to 400° F.
2. Using a robo-coupe, purée the mushrooms.
3. In a sauté pan, melt the butter, and sauté the shallots and onions. Add the mushrooms. Add the tomato purée, and season to taste. Cook until the mixture is reduced to a fine, dry paste.
4. Season the beef, and **truss.** In a large skillet, heat the oil, and sear the meat on all sides. Cool in a refrigerator at 40° F or below.
5. Roll out the pastry to ¼-inch thickness.
6. Mix duxelle, and season.
7. Remove the string from the meat. Cover the entire tenderloin with the goose liver and the mushroom duxelle mixture.
8. Place the meat on the dough, fold the pastry around the meat, and seal. Place on a baking pan lined with parchment paper.
9. Brush the surface with the egg wash, and decorate with strips of dough.
10. Pierce the dough on top to let steam escape. Bake in the oven for 35 to 40 minutes, or until golden brown on top and bottom. The internal temperature should be 140° F.
11. Remove and allow to rest at 140° F for 25 minutes.
12. To serve, ladle the sauce onto a preheated plate. Slice the Wellington at least ¾ inches thick, and place a slice on top of the sauce.

COOKING TECHNIQUES:
Sauté, Bake

Sauté:
1. Heat the sauté pan to the appropriate temperature.
2. Evenly brown the food product.
3. For a sauce, pour off any excess oil, reheat, and deglaze.

Bake:
1. Preheat the oven.
2. Place the food product on the appropriate rack.

GLOSSARY:
Truss: to tie or secure
Brunoise: finely diced

HAZARDOUS FOODS:
Beef tenderloin
Goose liver
Eggs

NUTRITION:
Calories: 783
Fat: 48.7 g
Protein: 38.5 g

CHEF NOTES:
1. Traditionally, Beef Wellington is served with Sauce Périgourdine, which is flavored with Madeira wine and truffles.
2. Beef Wellington can be made without Duxelle, using chopped truffles instead, with brioche dough instead of puff pastry, and Madeira sauce instead of Périgourdine.
3. Beef Wellington can be served on a platter, carved tableside, and garnished with a bouquet of vegetables and noisettes or chateau potatoes.

Braised Short Ribs of Beef

YIELD: 50 SERVINGS. SERVING SIZE: TWO RIBS.

INGREDIENTS:

20 pounds	Beef short ribs
	Seasoned flour, as needed
10 ounces	Oil
1¼ pint	Red wine
1¼ pint	Tomato pureé
2½ tablespoons	Marjoram
2½ tablespoons	Basil
5 quarts	Demi-glace, heated to a boil (see page 26)
	Salt and ground white pepper, to taste
2½ teaspoons	Horseradish, prepared
	Brown beef stock (see page 7), heated to a boil
8 ounces	Freshly chopped parsley, excess moisture removed

Mirepoix:

2½ pounds	Onions, peeled and diced
2½ pounds	Carrots, washed, peeled, and chopped
2½ pounds	Celery, washed, trimmed, and chopped
1¼ heads	Garlic, peeled and crushed

METHOD OF PREPARATION:

1. Preheat oven to 350° F.
2. **Dredge** the short ribs in the seasoned flour, and shake off any excess.
3. In a braising pan, heat the oil, and **sear** the ribs until they are light brown. Remove the ribs from the pan, and reserve.
4. In the same oil, sauté the **mirepoix** until lightly browned. Add the wine, tomato pureé, and demi-glace with the herbs and seasonings. Heat to a boil. If the sauce is too thick, add beef stock as needed. Return the ribs to the pan.
5. Cover with parchment paper and foil, and braise in the oven until the meat is **fork-tender.**
6. When the ribs are done, remove them from the braising pan and hold at 140° F. Degrease the sauce, strain, flavor with horseradish, and adjust the seasonings. Serve it over the short ribs on a preheated dinner plate. Garnish with fresh, chopped parsley.

COOKING TECHNIQUE:
Braise

Braise:
1. Heat the braising pan to the proper temperature.
2. Sear and brown the food product to a golden color.
3. Degrease and deglaze.
4. Cook the food product in two thirds liquid until fork-tender.

GLOSSARY:
Mirepoix: roughly chopped vegetables
Dredge: to coat with flour
Sear: to brown quickly
Fork-tender: without resistance

HACCP:
Hold at 140° F.

HAZARDOUS FOOD:
Beef short ribs

NUTRITION:
Calories: 375
Fat: 20.2 g
Protein: 29.8 g

Braised Steak with Onions

YIELD: 10 SERVINGS.   SERVING SIZE: SIX OUNCES.

INGREDIENTS:

4 pounds	Beef, bottom round, cut into 6-ounce slices
5 ounces	Oil or fat
	Seasoned flour, as needed
1 quart	Demi-glace, heated to a boil (see page 26)
2 pounds	Onions, peeled and cut **julienne**
1 tablespoon	Hungarian sweet paprika
	Salt and freshly ground black pepper, to taste
3 ounces	Fresh parsley, washed, excess moisture removed, and chopped

METHOD OF PREPARATION:

1. Preheat the oven to 350° F.
2. Pound the meat with a heavy mallet to tenderize.
3. In a sauté pan, heat 3 ounces of the oil or fat.
4. Dredge the steaks in seasoned flour, shake off the excess, and **sear** on both sides. **Shingle** into a baking pan.
5. Pour the demi-glace over the steaks, and seal the pan with aluminum foil.
6. Braise the steaks in oven until tender, about 45 minutes. Hold at 140° F or above.
7. In another sauté pan, heat the remaining oil or fat, and sauté onions until they become translucent. Add the paprika, and sauté 1 minute more, or until the fat turns red. Season to taste, and hold at 140° F.
8. To serve, place the steak on a preheated dinner plate, and top with onions. Ladle sauce around the steak, and sprinkle chopped parsley over the onions.

COOKING TECHNIQUE:
Braise

Braise:
1. Heat the braising pan to the proper temperature.
2. Sear and brown the food product to a golden color.
3. Degrease and deglaze.
4. Cook the food product until fork-tender.

GLOSSARY:
Julienne: matchstick strips
Sear: to brown quickly
Shingle: to overlap

HACCP:
Hold at 140° F or above.

HAZARDOUS FOOD:
Sirloin butt

NUTRITION:
Calories: 593
Fat: 29.4 g
Protein: 57.8 g

Broiled Beef Kebabs

YIELD: 50 SERVINGS. SERVING SIZE: SIX OUNCES.

INGREDIENTS:

20 pounds	Top sirloin of beef
8 each	Red bell peppers, washed, seeded, and cut into uniform size (1 inch)
8 each	Green bell peppers, washed, seeded, and cut into uniform size (1 inch)
5 each	Large onions, peeled and cut into uniform size (1 inch)
50 each	Mushroom caps, washed, stems cut

Marinade:

8 ounces	Soy sauce
8 ounces	Oil
8 ounces	Honey
8 ounces	Hoisin sauce
2 tablespoons	Hot pepper sauce
1 ounce	Fresh ginger, peeled and grated
8 cloves	Garlic, peeled and minced
	Freshly ground black pepper, as needed

METHOD OF PREPARATION:

1. Trim and remove the excess fat from the top beef sirloin, and then slice it lengthwise, ½-inch thick. Cut across the grain, into cubes (1 inch).
2. In a mixing bowl, mix together the ingredients listed to create a marinade. Pour the marinade over the beef, and marinate in a refrigerator for 2 to 4 hours.
3. Soak the wooden skewers in water (to prevent burning). Then alternate, sliding pieces of the beef and the vegetables onto the skewers.
4. Broil the beef kebabs to order. When serving, remove the wooden skewers.

COOKING TECHNIQUE:

Broil

Grill/Broil:

1. Clean and heat the grill/broiler.
2. To prevent sticking, brush the food product with oil.

HACCP:

Cook to an internal temperature of 140° F.

HAZARDOUS FOOD:

Top sirloin

NUTRITION:

Calories: 448
Fat: 19.3 g
Protein: 56.6 g

CHEF NOTES:

1. Beef kebabs can be served with many variations of rice pilaf.
2. The FDA recommends that beef be cooked to an internal temperature of 140° F.
3. This is an oriental recipe. The marinade can be modified according to the geographic region in which it is served.

Chateaubriand

YIELD: 10 SERVINGS. SERVING SIZE: SIX OUNCES.

INGREDIENTS:

2 each (4 pounds)	Beef tenderloin, center cut, trimmed and silverskin removed
	Vegetable oil, as needed
	Salt and freshly ground black pepper, to taste

METHOD OF PREPARATION:

1. Divide the beef tenderloin into five 12-ounce cuts. Reserve the trimmings for future use.
2. Rub the beef with oil, and season generously with salt and pepper.
3. Preheat the broiler, and broil the beef until medium-rare, or to an internal temperature of 140° F. Hold at 140° F for 3 minutes.
4. Slice on an angle before serving (three slices per portion).
5. To serve, arrange slices overlapping on a preheated dinner plate. **Nappé** with a sauce.

COOKING TECHNIQUE:

Broil

Grill/Broil:

1. Clean and heat the grill/broiler.
2. To prevent sticking, brush the food product with oil.

GLOSSARY:

Nappé: coated

HACCP:

Cook to an internal temperature of 130° F.

HAZARDOUS FOOD:

Beef tenderloin

Nutrition:

Calories: 401
Fat: 29.5 g
Protein: 32.1 g

CHEF NOTES:

1. The classic Chateaubriand is presented on a platter and served tableside. Chateaubriand is to be served with a bouquetiere of vegetables, broiled tomatoes, and chateau potatoes.
2. Sauces such as Périgourdine, béarnaise, or choron are among those that would complement this dish.

Chili Con Carne

YIELD: 50 SERVINGS. SERVING SIZE: EIGHT OUNCES.

INGREDIENTS:

2 pounds	Red beans, sorted, washed, and soaked overnight
10 ounces	Oil
1 pound	Onions, peeled and diced **brunoise**
1 pound	Green bell peppers, washed, seeded, and diced brunoise
1 head	Garlic, peeled and minced
15 pounds	Ground beef
5 pounds	Tomatoes, washed, cored, blanched, peeled, seeded, and rough chopped
2 pints	Tomato purée
2 ounces	Chili powder
2 ounces	Cumin, ground
	Crushed red pepper, to taste
	Salt and freshly ground black pepper, to taste

METHOD OF PREPARATION:

1. In a saucepan, cover the red beans with 1 gallon of cold water. Heat to a boil. Simmer until tender, then drain, and keep warm.
2. In a braising pan, sauté the onions in hot oil. When the onions are translucent, add the green peppers and garlic. Sauté lightly.
3. Add the ground beef. Sauté until half cooked. Drain the excess fat. Add the tomatoes, tomato purée, and seasonings. Continue to sauté to cook out the acidity of the tomato product until meat is fully done.
4. Add the cooked red beans. Let the chili simmer until the desired flavor is reached. Adjust the seasonings. Serve.
5. This dish can be served with brown or white rice and garnished with sour cream, cheese, and chopped onions.

COOKING TECHNIQUES:

Boil, Sauté, Simmer

Boil: (at sea level)

1. Bring the cooking liquid to a rapid boil.
2. Stir the contents, and cook the food product throughout.
3. Serve hot.

Sauté:

1. Heat the sauté pan to the appropriate temperature.
2. Evenly brown the food product.
3. For a sauce, pour off any excess oil, reheat, and deglaze.

Simmer and Poach:

1. Heat the cooking liquid to the proper temperature.
2. Submerge the food product completely.
3. Keep the cooked product moist and warm.

GLOSSARY:

Brunoise: ⅛-inch dice

HAZARDOUS FOOD:

Ground beef

NUTRITION:

Calories: 404
Fat: 21.7 g
Protein: 33.5 g

CHEF NOTES:

1. This dish has many variations and was once considered a poor person's meal. In some areas, the meat is diced, and in others no beans are added.
2. In American regional cuisine, this recipe is made with chili peppers that are roasted, peeled, chopped, and added to the meat. There are many variations of dried beans that can be used in this recipe.
3. *Chili con carne* is Spanish for "chili with meat." This dish is a melange of diced or ground beef and chili peppers or chili powder (or both). It originated in the Lone Star State, and Texans, who commonly refer to it as "a bowl of red," consider it a crime to add beans to the mixture. In many parts of the United States, however, beans are requisite, and the dish is called "chili con carne with beans."
4. *Devil's chili* denotes "hot."
5. Remove the meat and meat products to serve the chili as a meatless dish.

Fresh Glazed Beef Brisket

YIELD: 50 SERVINGS. SERVING SIZE: SIX OUNCES.

COOKING TECHNIQUES:
Simmer, Braise

Simmer and Poach:
1. Heat the cooking liquid to the proper temperature.
2. Submerge the food product completely.
3. Keep the cooked product moist and warm.

Braise:
1. Heat the braising pan to the proper temperature.
2. Sear and brown the food product to a golden color.
3. Degrease and deglaze.
4. Cook the food product in two thirds liquid until fork-tender.

GLOSSARY:
Bouquet garni: herbs and spices
Fork-tender: without resistance

HAZARDOUS FOOD:
Brisket of beef

NUTRITION:
Calories: 458
Fat: 30 g
Protein: 25 g

INGREDIENTS:

25 pounds	Brisket of beef, trimmed
3 teaspoons	Pickling spice
	Water, as needed

Bouquet Garni: (3 large bundles)

15 each	Black peppercorns
6 each	Bay leaves
9 ounces	Parsley stems
3 teaspoons	Thyme leaves

Glaze:

2½ pints	Honey
5 tablespoons	Dry mustard
15 ounces	Brown sugar
1½ teaspoons	Cloves, ground
½ pint	White wine
½ pint	Brown beef stock (see page 7) heated to a boil
½ pint	Brown corn syrup

METHOD OF PREPARATION:

1. Preheat the oven to 350° F.
2. Place the brisket in cold water. Heat to a boil. Add the bouquet garni and the pickling spice. Reduce to a simmer, and cook until **fork-tender.**
3. Place all of the ingredients for the glaze in a saucepan, and simmer for 10 minutes until a syrupy consistency is reached.
4. Remove the brisket from the pot, and transfer it to a braising pan. Pour the glaze over the meat, and braise in the oven for 15 to 20 minutes.
5. Remove from the oven and let rest 15 or 20 minutes. Slice the meat into portions on the bias, across the grain, and lightly brush the slices with the glaze. Serve on preheated dinner plates.

Grilled Rib-Eye Steak with Oysters and Béarnaise Sauce

YIELD: 10 SERVINGS. SERVING SIZE: EIGHT OUNCES.

INGREDIENTS:

10 (6-ounce)	Rib-eye steaks
	Vegetable oil, as needed
	Salt and freshly ground black pepper, to taste
20 each	Shucked oysters, sautéed in butter, to order
20 ounces	Béarnaise sauce, heated (see page 15)

METHOD OF PREPARATION:

1. Brush the steaks with oil, and grill to order (e.g., rare, medium-rare, medium, medium-well, well).
2. Season each lightly with salt and black pepper.
3. For service, place the steak on a preheated dinner plate, place two sautéed oysters on the steak, and **nappé** with béarnaise sauce.

COOKING TECHNIQUES:

Grill, Sauté

Grill/Broil:

1. Clean and heat the grill/broiler.
2. To prevent sticking, brush the food product with oil.

Sauté:

1. Heat the sauté pan to the appropriate temperature.
2. Evenly brown the food product.
3. For a sauce, pour off any excess oil, reheat, and deglaze.

GLOSSARY:

Nappé: coated

HAZARDOUS FOODS:

Steak
Oysters

NUTRITION:

Calories: 850
Fat: 66.9 g
Protein: 51 g

CHEF NOTES:

1. Hold the steaks and oysters in a cooler at 40° F or lower, and remove only when the order has to be prepared.
2. Sauté the oysters to order while grilling the steak.

Hungarian Goulash

YIELD: 50 SERVINGS. SERVING SIZE: FOUR TO FIVE OUNCES.

COOKING TECHNIQUE:
Stew

Stew:
1. Sear, sauté, sweat, or blanch the main food product.
2. Deglaze the pan, if desired.
3. Cover the food product with simmering liquid.
4. Remove the bouquet garni.

GLOSSARY:
Brunoise: ⅛-inch dice
Sear: brown quickly

HACCP:
Hold at 140° F.

HAZARDOUS FOOD:
Bottom round of beef

NUTRITION:
Calories: 308
Fat: 14.1 g
Protein: 36.9 g

INGREDIENTS:

12½ pounds	Bottom round of beef, ½-inch dice
6 ounces	Oil
5 pounds	Onions, peeled and diced **brunoise**
3 tablespoons	Spanish paprika
3 teaspoons	Hungarian paprika
1½ pounds	Tomato purée
3 quarts	Brown demi-glace, heated to a boil (see page 26)
3 teaspoons	Caraway seeds, crushed
6 cloves	Garlic, peeled and minced
3 each	Lemons, zested
	Salt and ground black pepper, to taste

METHOD OF PREPARATION:

1. Wash the cubed bottom round of beef in cold water.
2. Heat the oil, add the meat, and **sear.** Remove and reserve. Pour off any excess fat. Add the onions. Cook until the onions are translucent.
3. Add the meat, paprika, tomato purée, and demi-glace. Blend well. Cover with parchment paper and aluminum foil. Let simmer until the meat is tender.
4. When the meat is tender, remove from the heat. Hold at 140° F. Add the crushed caraway seeds, minced garlic, and shredded lemon zest. Season, to taste, and serve.

CHEF NOTES:
1. Goulash is a spicy, heavy, souplike Hungarian dish. The preparation depends on its region. *Goulash* means "shepherd."
2. Hungarian goulash is traditionally served with dumplings. Most commonly, the eastern European variation of spätzel is used.

Japanese Beef and Vegetables

(Sukiyaki)

YIELD: 10 SERVINGS. SERVING SIZE: TEN OUNCES.

INGREDIENTS:

	Peanut oil, as needed
2½ pounds	Top round of beef, trimmed and cut into ⅛-inch thick strips
4 ounces	Soy sauce
3 tablespoons	Granulated sugar
10 each	Scallions, washed, trimmed, and cut into 1-inch strips
10 ounces	Onions, peeled and sliced ½-inch thick
10 ounces	Mushrooms, washed, stems trimmed, and thinly sliced
1 pound	Tofu
1 pound	Fresh spinach, washed, stemmed, and cut **chiffonade**
1 pound	Bamboo shoots, washed and cut julienne
10 ounces	Mirin (available in Oriental markets)
1½ pound	Shirataki noodles, cooked, cooled, and cut into thirds

METHOD OF PREPARATION:

1. In a wok or heavy sauté pan, heat a small amount of the peanut oil. Add the beef strips, and stir-fry quickly, just to brown the meat. Season with soy sauce and sugar; then remove and set aside.
2. Clean the wok or pan, then heat some more peanut oil. Add the scallions, onions, mushrooms, tofu, spinach, and bamboo shoots. Stir-fry quickly for about 2 minutes. Add the mirin and noodles, and return the cooked meat to the mixture. Stir-fry an additional minute, or just until hot and well incorporated.
3. Remove from wok or pan, and serve on a preheated dinner plate.

COOKING TECHNIQUE:

Stir-Fry

Stir-Fry:

1. Heat the oil in a wok until hot but not smoking.
2. Keep the food in constant motion; use the entire cooking surface.

GLOSSARY:

Chiffonade: ribbons of leafy greens
Bain-marie: hot-water bath

HAZARDOUS FOOD:

Beef

NUTRITION:

Calories: 440
Fat: 10.8 g
Protein: 47.6 g

CHEF NOTE:

This dish will lose color and texture if held in a **bain-marie.**

London Broil

YIELD: 50 SERVINGS. SERVING SIZE: FIVE OUNCES.

INGREDIENTS:

17 pounds	Beef flank steaks

Marinade:

2 pints	Oil
2 pints	Red wine
1 head	Garlic, peeled and chopped
2 teaspoons	Salt
1 teaspoon	Ground black pepper
12 ounces	Onions, peeled and diced **brunoise**
5 tablespoons	Dijon mustard
	Mushroom sauce (see Sauce Forestière, page 58) as needed, heated to 165° F

METHOD OF PREPARATION:

1. Peel the silverskin and trim the fat from the meat.
2. Mix all of the remaining ingredients together for the marinade. Marinate the flank steaks for 24 hours. Turn the steaks every few hours.
3. Remove the steaks from the marinade, and pat dry with paper towels.
4. Broil or grill the steaks **à la minute,** or until the internal temperature of 140° F is reached and the steak is medium-rare. Hold at 140° F.
5. Slice thin, against the grain, on the bias.
6. Place the meat on a preheated dinner plate, and serve with mushroom sauce.

COOKING TECHNIQUE:
Grill or Broil

Grill/Broil:
1. Clean and heat the grill/broiler.
2. To prevent sticking, brush the food product with oil.

GLOSSARY:
Brunoise: ⅛-inch dice
À la minute: cooked to order

HACCP:
Hold at 140° F.

HAZARDOUS FOOD:
Beef flank steaks

NUTRITION:
Calories: 500
Fat: 33.5 g
Protein: 42.5 g

Meat Balls in Tomato Sauce

YIELD: 50 SERVINGS. SERVING SIZE: FOUR OUNCES.

INGREDIENTS:

2 ounces	Oil
12 pounds	Ground beef
1½ pounds	Onions, peeled and diced **brunoise**
3 pounds	White bread, cut into large cubes
	Milk, as needed
12 each	Whole eggs
3 ounces	Freshly chopped parsley, excess moisture removed
1 head	Garlic, peeled and minced
1 teaspoon	Ground black pepper
	Salt, to taste
2 gallons	Tomato sauce, heated to a boil (see page 76)

METHOD OF PREPARATION:

1. Preheat the oven to 350° F.
2. Heat the oil in a saute pan, and sauté the onions until translucent. Remove from the heat, and cool.
3. Place the ground beef in a mixing bowl, and add the sautéed onions.
4. Soak the white bread in the milk; squeeze out any excess milk, and add the bread to the meat.
5. Add the eggs, parsley, and seasonings to the meat. Place all ingredients in a 20-quart mixer bowl. Using a mixer paddle, thoroughly mix all of the ingredients. Shape into 2-ounce round balls, and place on a sheet pan.
6. Bake in the oven for 15 minutes, or until the meatballs reach an internal temperature of 155° F. When done, transfer the meatballs to a hotel pan.
7. Pour the tomato sauce on the meat balls, and cover with parchment paper and foil. Bake for an additional 30 minutes. Hold at 140° F.
8. Serve the meatballs with any type of pasta.

COOKING TECHNIQUES:
Sauté, Bake

Sauté:
1. Heat the sauté pan to the appropriate temperature.
2. Evenly brown the food product.
3. For a sauce, pour off any excess oil, reheat, and deglaze.

Bake:
1. Preheat the oven.
2. Place the food product on the appropriate rack.

GLOSSARY:
Brunoise: ⅛-inch dice

HACCP:
Hold at 140° F.

HAZARDOUS FOODS:
Ground beef
Milk
Whole eggs

NUTRITION:
Calories: 165
Fat: 4.6 g
Protein: 7.58 g

CHEF NOTE:
1. For best results, pass the beef through a grinder a second time.
2. For efficient production, use a 2-ounce scoop to portion the mixture before rolling by hand.

Meat Lasagna

YIELD: 50 SERVINGS. SERVING SIZE: EIGHT OUNCES.

COOKING TECHNIQUES:

Sauté, Boil, Bake

Sauté:

1. Heat the sauté pan to the appropriate temperature.
2. Evenly brown the food product.
3. For a sauce, pour off any excess oil, reheat, and deglaze.

Boil: (at sea level)

1. Bring the cooking liquid to a rapid boil.
2. Stir the contents, and cook the food product throughout.
3. Serve hot.

Bake:

1. Preheat the oven.
2. Place the food product on the appropriate rack.

GLOSSARY:

Brunoise: ⅛-inch dice

HACCP:

Hold at 140° F.

HAZARDOUS FOODS:

Ricotta cheese
Eggs

NUTRITION:

Calories: 672
Fat: 28.1 g
Protein: 46.5 g

INGREDIENTS:

Meat Filling:

	Oil, as needed
15 ounces	Onions, washed, peeled, and diced **brunoise**
7 pounds	Ground beef
5 ounces	Tomato purée
	Salt and ground black pepper, to taste

Cheese Filling:

5 pounds	Ricotta cheese
2 pounds	Mozzarella cheese
2 bunches	Fresh chives, washed and chopped
13 each	Eggs
2 gallons	Tomato sauce, cooled to 40° F (see page 76)
7 pounds	Lasagna sheets, frozen
2 pounds	Mozzarella cheese, sliced
2 pounds	Parmesan cheese, grated

METHOD OF PREPARATION:

1. Preheat oven to 350° F.
2. Heat the boil in a sauté pan, and sauté the onions until they are translucent. Add the ground beef, and cook until it crumbles. Drain off any excess grease. Add the tomato purée, salt, and black pepper. Set the mixture aside, and refrigerate.
3. In a mixing bowl, mix the ricotta and mozzarella cheeses with the chives and eggs. Then refrigerate.
4. Cook the lasagna sheets in boiling, salted water with oil. Test for proper doneness. Shock under cold running water. Lightly oil the noodles, laying them out separately on a sheet pan, with plastic wrap separating the noodles.
5. To assemble the lasagna, place some tomato sauce in the bottom of a lightly oiled hotel pan, then alternate layers of noodles, meat, cheese mixture, and sauce. Finish with an upper layer of noodles and mozzarella cheese.
6. Bake in oven for 40 to 45 minutes, and hold at 140° F.
7. Remove from the oven, and let rest for 15 minutes. Before serving, sprinkle with Parmesan cheese. Cut into 8-ounce square portions, and serve.

New England Boiled Dinner

YIELD: 10 SERVINGS. SERVING SIZE: SIX OUNCES (MEAT ONLY).

INGREDIENTS:

6 pounds	Corned beef
1 gallon	Cold water, or to cover
2 ounces	Pickling spices
1 head	White cabbage (medium-sized), washed, cored, and cut into wedges
5 each	Carrots, washed, peeled, and **tournéed**
5 each	Turnips, washed, peeled, and tournéed
10 each	Potatoes, washed, peeled, and tournéed
10 each	Onions (small-sized), peeled
	Salt and ground white pepper, to taste
1 pound	Butter, heated to 165° F
3 ounces	Fresh parsley, washed, excess moisture removed, and chopped
10 ounces	White horseradish sauce (see page 78) or mustard

METHOD OF PREPARATION:

1. Place the beef in a large stock pot, and cover with water. Add pickling spices, and heat the liquid to a boil. Skim the surface; then reduce the heat, and simmer covered until the meat is **fork-tender.** This will require 2½ to 3 hours' cooking time.
2. Remove the beef, and hold covered at 140° F. Strain the broth and **dépouiller.**
3. In cold water or cold brown beef stock, boil the carrots, potatoes, and turnips, using separate cooking vessels for each item. Cook each vegetable until tender; then drain and season, to taste. Hold at 140° F.
4. Trim any excess fat from the beef, and slice thin against the grain on a bias. Arrange in a hotel pan, and cover with some of the hot broth. Hold at 140° F for service.
5. To serve, place the cabbage wedges on a preheated dinner plate. Lay the sliced beef over the cabbage, and arrange the remaining vegetables around the edge. Drizzle with hot butter, and garnish with parsley. Offer the sauce separately.

COOKING TECHNIQUES:
Simmer, Boil

Simmer and Poach:
1. Heat the cooking liquid to the proper temperature.
2. Submerge the food product completely.
3. Keep the cooked product moist and warm.

Boil: (at sea level)
1. Bring the cooking liquid to a rapid boil.
2. Stir the contents, and cook the food product throughout.
3. Serve hot.

GLOSSARY:
Tournéed: trimmed to a large olive shape
Fork-tender: without resistance
Dépouiller: to skim impurities/grease

HACCP:
Hold at 140° F.

HAZARDOUS FOOD:
Corned beef

NUTRITION:
Calories: 768
Fat: 50 g
Protein: 32 g

CHEF NOTES:
1. By separately cooking the vegetables in the broth, each can be easily cooked to the desired tenderness. Vegetables also could be prepared in a steamer, then allowed to rest in hot broth to absorb the flavor.
2. Although this dish is considered a "boiled" dinner, it is important that the liquid in which the beef is cooked is maintained at a simmer to avoid toughening the meat.

New Mexico Stewed Steak

YIELD: 10 SERVINGS. SERVING SIZE: SIX OUNCES.

INGREDIENTS:

4 pounds	Beef sirloin, trimmed and cut into 10 steaks
2 ounces	Vegetable oil, heated
12 ounces	Onions, peeled and thinly sliced
12 ounces	Canned peeled tomatoes, drained (reserve juice), seeded, and chopped
8 ounces	Pitted black olives
3 ounces	Fresh parsley, washed, excess moisture removed, and chopped

Marinade:

4 ounces	Red wine
8 ounces	White wine
3 each	Anaheim chili peppers, washed, seeded, and sliced
5 cloves	Garlic, peeled and minced
2 ounces	Brown sugar
1 teaspoon	Salt

METHOD OF PREPARATION:

1. Combine the ingredients for marinade, and mix well. Add the steaks, and turn to coat each well. Refrigerate at 40° F or lower for at least 1 hour.
2. Remove the steaks from the marinade, and pat each dry. Reserve the marinade.
3. In a large sauté pan, heat the oil, and when almost smoking, **sear** the meat on both sides, browning lightly. Remove and arrange the steaks overlapping in a **braisière.**
4. In the same sauté pan as the steaks were seared, place the onions, and sauté until translucent. Add the tomatoes, and continue to sauté for 3 more minutes.
5. Add the marinade, bring to a boil, and then pour the mixture over the steaks. Add the olives, cover tightly with aluminum foil, and stew until **fork-tender,** approximately 30 minutes.
6. To serve, place a steak on a preheated dinner plate, and **nappé** with pan sauce. Garnish with parsley. Hold at 140° F or above.

COOKING TECHNIQUES:

Sauté, Stew

Sauté:

1. Heat the sauté pan to the appropriate temperature.
2. Evenly brown the food product.
3. For a sauce, pour off any excess oil, reheat, and deglaze.

Stew:

1. Sear, sauté, sweat, or blanch the main food product.
2. Deglaze the pan, if desired.
3. Cover the food product with simmering liquid.
4. Remove the bouquet garni.

GLOSSARY:

Sear: to brown quickly
Braisière: braising or stewing pan
Fork-tender: without resistance
Nappé: coated

HACCP:

Chill to 40° F.
Hold at 140° F.

NUTRITION:

Calories: 542
Fat: 32.9 g
Protein: 38.4 g

CHEF NOTE:

For institutional food service, the use of beef top round is recommended.

Old-Fashioned Beef Stew

YIELD: 50 SERVINGS. SERVING SIZE: EIGHT OUNCES.

INGREDIENTS:

20 pounds	Beef, bottom round
40 ounces	Oil
5 pounds	Potatoes, washed, peeled, and cut into ½-inch pieces
5 pounds	Onions, peeled
2½ teaspoons	Dried thyme
5 #10 cans	Crushed tomatoes
5 quarts	Brown sauce, heated to 165° F
3¾ pounds	Carrots, washed, peeled, and diced into ¼-inch pieces
3¾ pounds	Celery, washed, trimmed, and cut into ¼-inch pieces
20 ounces	Frozen green peas
20 ounces	Mushrooms, washed and sliced
5 ounces	Clarified butter
	Salt and freshly ground black pepper, to taste
15 ounces	Fresh chopped parsley

METHOD OF PREPARATION:

1. In a saucepan, place the potatoes in cold, salted water. Cook until tender. Drain in a colander, and rinse in cold water.
2. Heat the oil in a braising pan. Sear the meat to a golden brown on all sides. Remove, and reserve the meat. (Brown the meat in two stages if necessary, to prevent sweating.)
3. In the same pan used for the meat, sauté the onions until they are translucent. Add the carrots and celery to the onions, and sauté for 4 minutes.
4. Add the crushed tomatoes, brown sauce, and thyme. Bring to a boil, and skim as needed.
5. Return the meat to the pan, and simmer until the meat is 90% done. Add the cooked potatoes, and simmer an additional 10 minutes.
6. Skim the grease from the stew, and adjust the seasonings, to taste.
7. Heat the butter in a sauté pan. Sauté the mushrooms for 2 minutes. Add the peas, and heat thoroughly; then add the mushrooms and peas to the stew. Hold at 140° F.
8. Serve the stew hot, garnished with chopped parsley.

COOKING TECHNIQUE:

Saute

Sauté:

1. Heat the sauté pan to the appropriate temperature.
2. Evenly brown the food product.
3. For a sauce, pour off any excess oil, reheat, and deglaze.

HACCP:

Hold at 140° F.

HAZARDOUS FOOD:

Beef, bottom round

NUTRITION:

Calories: 901
Fat: 47.7 g
Protein: 68.4 g

CHEF NOTE:

Beef bottom round is very tough and it takes two or more hours of cooking to tenderize. Internal temperature does not apply to this type of meat.

Oriental Barbecued Short Ribs

YIELD: 50 SERVINGS. **SERVING SIZE:** FIVE OUNCES OR ONE RIB.

COOKING TECHNIQUES:
Broil, Bake

Grill/Broil:
1. Clean and heat the grill/broiler.
2. To prevent sticking, brush the food product with oil.

Bake:
1. Preheat the oven.
2. Place the food product on the appropriate rack.

HACCP:
Hold at 165° F.

HAZARDOUS FOOD:
Beef short ribs

NUTRITION:
Calories: 289
Fat: 19.1 g
Protein: 19 g

INGREDIENTS:

50 (5-ounce)	Beef short ribs
1 head	Garlic, peeled and minced
3 quarts	Soy sauce
1 pint	Sesame oil
5 tablespoons	Sesame seeds
4 ounces	White vinegar
8 ounces	Sugar
2 ounces	Hot mustard, dry
50 each	Scallions, cut into scallion fans, for garnish

METHOD OF PREPARATION:

1. Preheat the oven to 375° F.
2. Trim the excess fat off the short ribs. Score the meat, cutting every ½-inch, almost to the bone.
3. Combine all the ingredients (except the scallions) to create a marinade. Pour the marinade over the ribs, and marinate overnight.
4. Remove the meat from the marinade, and grill it over charcoals or broil it under a broiler. Broil for about 15 minutes, or until well-browned. Start with the meat-side first, and finish with the bones toward the heat. Place the broiled meat into hotel pans.
5. Continue cooking the short ribs in the oven until the meat on the bone is fork-tender.
6. Garnish the short ribs with scallion fans, and serve.

CHEF NOTE:
Hold ribs above 165° F.

Oriental Pepper Steak

YIELD: 10 SERVINGS. SERVING SIZE: SIX OUNCES.

INGREDIENTS:

2½ pounds	Beef flank steaks, excess fat and silverskin removed
1½ ounces	Mirin (available in Oriental markets)
1 tablespoon	Granulated sugar
1½ ounces	Soy sauce
2 ounces	Cornstarch
12 ounces	Peanut oil
12 ounces	Onions, peeled and cut **julienne**
1 pound	Green bell peppers, washed, seeded, and cut julienne
1 pound	Red bell peppers, washed, seeded, and cut julienne
1½ teaspoons	Ginger root, peeled and minced
2 teaspoons	Hot sauce

METHOD OF PREPARATION:

1. Cut the beef lengthwise into strips 1½-inch wide; then crosswise into ¼-inch slices.
2. Mix the mirin, sugar, soy sauce, and cornstarch in a large bowl. Add the beef slices, and coat thoroughly. Marinate for 20 minutes
3. Heat a wok or heavy skillet over high heat for 30 seconds. Add 4 ounces of the oil, and heat for 30 seconds more. Reduce the heat, add the meat, and stir-fry until the meat is lightly browned; then remove the meat, and reserve.
4. Add 8 ounces more oil, or as needed, to the wok. Heat for 30 seconds, or until smoking. Add the onions, peppers, and ginger, and stir-fry over medium heat for about 2 minutes, or until the peppers are cooked.
5. Add the meat to the vegetables, mix, and add the hot sauce.

COOKING TECHNIQUE:

Stir-Fry

Stir-Fry:

1. Heat the oil in a wok until hot but not smoking.
2. Keep the food in constant motion; use the entire cooking surface.

GLOSSARY:

Julienne: matchstick strips

HAZARDOUS FOOD:

Beef

NUTRITION:

Calories: 499
Fat: 33.3 g
Protein: 32.2 g

CHEF NOTE:

Pepper steak must be prepared á la minute.

Roasted Beef Tenderloin with Tarragon Sauce

(Filet de Boeuf Rôti à l'Estragon)

YIELD: 10 SERVINGS. SERVING SIZE: FIVE OUNCES.

INGREDIENTS:

1 (3½ to 4 pounds)	Beef tenderloin, trimmed and silverskin removed
10 slices	Bacon
4 ounces	Tarragon vinegar
4 ounces	Water
1 pint	White wine
3 ounces	Shallots, peeled and diced **brunoise**
1 teaspoon	Crushed peppercorns
1 pound	Mushrooms, cleaned, trimmed, and sliced
2 ounces	Butter
8 ounces	Heavy cream
	Salt and freshly ground black pepper, to taste
1½ ounces	Fresh tarragon leaves, washed, and chopped

METHOD OF PREPARATION:

1. Preheat the oven to 400° F. Combine the vinegar and water. Bard the beef with the bacon, and **truss** with butcher's twine.
2. Place the tenderloin in a roasting pan, and roast in the oven until well browned but still rare, basting often with the vinegar mixture. (The internal temperature should be 140° F.) Remove, and hold at 140° F.
3. Pour the cooking liquid into a saucepan, and add the wine, shallots, and peppercorns. Boil to reduce by half. **Dépouiller,** strain through a **chinois,** and set aside.
4. Place butter in a sauce pan, heat, add mushrooms, and sauté 2 minutes. Add the chopped tarragon and sauté half a minute more.
5. Add the reduction to the strained sauce. Heat to a simmer.
6. **Temper** the heavy cream with some of the sauce and add to the simmering sauce. Simmer to slightly thicken, taste, and adjust seasoning.
7. To serve, slice the beef on an angle, allowing three slices per serving. Arrange the slices on a preheated dinner plate, and **nappé** with sauce.

COOKING TECHNIQUES:

Roast, Simmer

Roast:

1. Sear the food product, and brown evenly.
2. Elevate the food product in a roasting pan.
3. Determine doneness, and consider carryover cooking.
4. Let the food product rest before carving.

Simmer and Poach:

1. Heat the cooking liquid to the proper temperature.
2. Submerge the food product completely.
3. Keep the cooked product moist and warm.

GLOSSARY:

Brunoise: ⅛-inch dice
Truss: to tie or secure
Dépouiller: to skim impurities/grease
Chinois: cone-shaped strainer
Temper: to equalize two extreme temperatures
Nappé: coated

HACCP:

Hold at 140° F or above.

HAZARDOUS FOODS:

Beef tenderloin
Heavy cream

NUTRITION:

Calories: 365
Fat: 25.9 g
Protein: 25.5 g

Roast Rib-Eye of Beef with Yorkshire Pudding and Horseradish Sauce

YIELD: 10 SERVINGS. SERVING SIZE: FIVE OUNCES.

INGREDIENTS:

	Salt and freshly ground black pepper, to taste
½ teaspoon	Garlic powder
1 (5-pound)	Boneless rib-eye of beef, trimmed
4 ounces	Meat glaze, heated to a boil (see page 36)
1 quart	Brown stock, heated to a boil (see page 7)
	White horseradish sauce, hot (see page 78)
	Yorkshire pudding (see page 268)

Mirepoix:

8 ounces	Onions, peeled
8 ounces	Carrots, washed and peeled
8 ounces	Celery stalks, washed and trimmed

METHOD OF PREPARATION:

1. Preheat the oven to 300° F.
2. Mix together the salt, pepper, and garlic powder. Rub the seasonings into the meat.
3. Place the **mirepoix** in a **plaque à rôtir,** and set the meat on top.
4. Place the pan in the oven, and roast to an internal temperature of 135° F to 140° F, which will require 2 to 2½ hours. Remove the roast from the pan, and hold at 140° F.
5. Place the plaque à rôtir on the stove, and brown the mirepoix. **Dépouiller** the fat, and reserve for the Yorkshire pudding. Add the meat glaze and brown stock, and heat to a boil.
6. Strain the sauce into a suitable container, and hold at 140° F until service.
7. For service, slice the roast, and **nappé** with sauce. Offer with Yorkshire pudding and horseradish sauce.

COOKING TECHNIQUE:

Roast

Roast:

1. Sear the food product, and brown evenly.
2. Elevate the food product in a roasting pan.
3. Determine doneness, and consider carryover cooking.
4. Let the food product rest before carving.

GLOSSARY:

Mirepoix: roughly chopped vegetables
Plaque à rôtir: roasting pan
Dépouiller: to skim impurities/grease
Nappé: coated

HACCP:

Heat to an internal temperature of 135° F to 140° F.
Hold at 140° F.

HAZARDOUS FOOD:

Beef

NUTRITION:

Calories: 514
Fat: 30.4 g
Protein: 49.7 g

CHEF NOTE:

Traditionaly the rib is placed on a rack above the pudding to allow the roast dripping into the pudding.

Yorkshire Pudding

YIELD: 10 SERVINGS. SERVING SIZE: THREE OUNCES.

COOKING TECHNIQUE:
Bake

Bake:
1. Preheat the oven.
2. Place the food product on the appropriate rack.

HACCP:
Preheat the oven to 475° F.

HAZARDOUS FOODS:
Eggs
Milk

NUTRITION:
Calories: 514
Fat: 30.4 g
Protein: 49.7 g

INGREDIENTS:

4 each	Eggs
12 ounces	All-purpose flour
1 pint	Milk
	Salt, to taste
½ teaspoon	Nutmeg, ground
2 ounces	Reserved roasting fat or bacon fat

METHOD OF PREPARATION:

1. Preheat the oven to 475° F.
2. Combine all of the ingredients except the fat in a bowl, and mix into a smooth batter.
3. In a 2-inch half hotel pan, heat the fat. Pour the batter into the hot fat in the pan, and place in the oven. Bake until done, approximately 20 to 30 minutes, or until when pierced with a skewer, it comes out dry.
4. Using a sharp knife, divide into 10 servings. Serve with the roast beef.

Roast Sirloin of Beef

(Contrefilet Rôti)

YIELD: 10 SERVINGS. SERVING SIZE: FIVE OUNCES.

INGREDIENTS:

1 teaspoon	Freshly ground black pepper
1½ teaspoons	Salt
1 (5-pound)	Sirloin strip, excess fat removed
1 quart	Demi-glace, heated to a boil and seasoned (see page 26)

Mirepoix:

2 each (8 ounces)	Onions, peeled
2 each (6 ounces)	Carrots, washed and peeled
3 stalks (3 ounces)	Celery, washed and trimmed
3 cloves	Garlic, peeled

METHOD OF PREPARATION:

1. Preheat the oven to 375° F.
2. Rub the meat with the salt and pepper.
3. In a suitable plaque à rôtir, place the meat on top of the **mirepoix.**
4. Roast the meat in the oven for 10 minutes; then reduce temperature to 300° F. Baste the meat with the demi-glace.
5. Continue to roast to an internal temperature of 140° F, basting often to avoid creation of a hard crust on the meat.
6. Remove the meat, and hold at 140° F for 12 to 15 minutes before slicing.
7. Place the plaque a rôtir on the stove. Add any remaining demi-glace, and cook over medium heat for 5 minutes. Adjust the seasoning. Strain the liquid through a **chinois** into a suitable container. Hold at 140° F or above. Skim off excess fat before serving.
8. To serve, thinly slice the meat to order, allowing three slices per serving, and **nappé** with demi-glace sauce.

COOKING TECHNIQUE:

Roast

Roast:

1. Sear the food product, and brown evenly.
2. Elevate the food product in a roasting pan.
3. Determine doneness, and consider carryover cooking.
4. Let the food product rest before carving.

GLOSSARY:

Mirepoix: roughly chopped vegetables
Chinois: cone-shaped strainer
Nappé: coated

HACCP:

Roast to an internal temperature of 140° F.
Hold at 140° F.

HAZARDOUS FOOD:

Sirloin strip

NUTRITION:

Calories: 301
Fat: 11.6 g
Protein: 35.7 g

Roast Top Round of Beef

YIELD: 50 SERVINGS. SERVING SIZE: SIX OUNCES.

INGREDIENTS:

20 pounds	Top round of beef
	Salt and ground black pepper, as needed for meat seasoning
	Garlic powder and mustard, as needed for meat seasoning
2½ quarts	Brown beef stock (see page 7), heated to a boil

Mirepoix:

2 pounds	Onions, peeled and diced
2 pounds	Celery, washed, peeled, and diced
1½ pounds	Carrots, washed, peeled, and diced

METHOD OF PREPARATION:

1. Preheat the oven to 350° F.
2. Trim the top round, and remove part of the thick fat.
3. Rub the seasonings into the meat.
4. Place the mirepoix in the bottom of a roasting pan with the meat on top. Roast until an internal temperature of 140° F is reached. Hold at 140° F.
5. Remove the roast from the pan, degrease the drippings, and **deglaze** the pan with beef stock. Heat to a boil, and **reduce** to the proper consistency.
6. Strain the sauce, and adjust the seasonings.
7. Let the meat rest for 20 minutes, and split it in half in line with the grain.
8. Slice approximately 6-ounce portions against the grain, and arrange them in a hotel pan. Hold at 140° F.
9. Serve on a preheated dinner plate with the seasoned, strained sauce.

COOKING TECHNIQUE:
Roast

Roast:
1. Sear the food product, and brown evenly.
2. Elevate the food product in a roasting pan.
3. Determine doneness, and consider carryover cooking.
4. Let the food product rest before carving.

GLOSSARY:
Mirepoix: roughly chopped vegetables
Deglaze: to add liquid to a hot pan
Reduce: to evaporate liquid by boiling

HACCP:
Hold at 140° F.

NUTRITION:
Calories: 346
Fat: 9 g
Protein: 58.2 g

Sauerbraten

YIELD: 10 SERVINGS. SERVING SIZE: SIX OUNCES.

COOKING TECHNIQUE:
Braise

Braise:
1. Heat the braising pan to the proper temperature.
2. Sear and brown the food product to a golden color.
3. Degrease and deglaze.
4. Cook the food product in two thirds liquid until fork-tender.

GLOSSARY:
Julienne: matchstick strips
Fork-tender: without resistance
Chinois: cone-shaped strainer

HACCP:
Hold at 140° F or higher.

NUTRITION:
Calories: 700
Fat: 26.5 g
Protein: 87.2 g

CHEF NOTE:
Preparation needs to begin 3 to 4 days in advance.

INGREDIENTS:

6 pounds	Bottom round of beef, trimmed of fatty silverskin, split, and tied with butcher's twine
	Vegetable oil, as needed
6 ounces	Red wine
3 ounces	Gingersnap crumbs

Marinade:

4 ounces	Red wine vinegar
12 ounces	Red wine
8 ounces	Onion, peeled and cut **julienne**
4 ounces	Carrots, washed, peeled and cut julienne
2 cloves	Garlic, peeled and chopped
1 ounce	Brown sugar
2 each	Bay leaves
1 teaspoon	Black peppercorns, crushed
1 teaspoon	Salt

METHOD OF PREPARATION:

1. Place the meat in a deep hotel pan. Combine the ingredients for the marinade, and pour over the meat. Cover and refrigerate for 3 to 4 days, turning the meat in the marinade each day.
2. Preheat the oven to 300° F.
3. Remove the meat, and pat dry.
4. In a braising pan, heat the oil until almost smoking, and add the meat. Sear on all sides until brown.
5. Add the marinade to the braising pan; cover and braise in the oven for 2 to 3 hours, or until the meat is **fork-tender.**
6. Remove the meat, and hold covered at 140° F or higher. Strain 1 quart of the liquid through a **chinois** into a saucepan. Heat to a boil, dépouiller, and reduce the liquid by one fourth.
7. Add the wine, and simmer for 10 minutes. Stir in the gingersnap crumbs, and simmer for 3 minutes. Let stand for 5 minutes to allow the crumbs to be absorbed. Hold at 140° F.
8. To serve, slice the meat across the grain, and nappé with the sauce.

Sautéed Beef with Mushrooms and Onions

YIELD: 50 SERVINGS. SERVING SIZE: SIX OUNCES.

INGREDIENTS:

15 pounds	Top round of beef, fat and silverskin removed, cut into strips
5 ounces	Oil
2½ pounds	Onions, peeled and cut **julienne**
2½ pounds	Mushrooms, cleaned and sliced
20 ounces	Burgundy wine
5 pints	Beef demi-glace (see page 26)
3 ounces	Fresh parsley, washed and finely chopped
	Salt and freshly ground black pepper, to taste

METHOD OF PREPARATION:

1. Heat the oil in a sauté pan, add the meat, and sauté over moderately high heat. Remove the meat when it is lightly browned.
2. In the same pan, sauté the onions and mushrooms until the onions are translucent and the mushrooms are browned. Then remove the vegetables from the pan.
3. Remove any excess fat, **deglaze** the pan with the Burgundy wine, and add the meat, vegetables, and demi-glace. Adjust the seasonings. Garnish with chopped parsley, and serve.

COOKING TECHNIQUES:

Sauté, Stew

Sauté:

1. Heat the sauté pan to the appropriate temperature.
2. Evenly brown the food product.
3. For a sauce, pour off any excess oil, reheat, and deglaze.

Stew:

1. Sear, sauté, sweat, or blanch the main food product.
2. Deglaze the pan, if desired.
3. Cover the food product with simmering liquid.
4. Remove the bouquet garni.

GLOSSARY:

Julienne: matchstick strips
Deglaze: to add liquid to hot pan

HAZARDOUS FOOD:

Beef

NUTRITION:

Calories: 296
Fat: 9.53 g
Protein: 44 g

Steak Diane

(Filet de Boeuf Diane)

YIELD: 10 SERVINGS. SERVING SIZE: FIVE OUNCES.

INGREDIENTS:

5 ounces	Shallots, peeled and diced **brunoise**
5 ounces	Mushrooms, cleaned and sliced
2 ounces	Dijon mustard
1 ounce	Worcestershire sauce
4 ounces	Demi-glace, seasoned (see page 26)
8 ounces	Heavy cream
	Salt and freshly ground black pepper, to taste
4 pounds	Tenderloin of beef, cleaned, trimmed, and sliced into 20 slices
10 sprigs	Fresh parsley
2 ounces	Butter
2 ounces	Oil
1 clove	Garlic, peeled and minced

METHOD OF PREPARATION:

1. Sauté the shallots and mushrooms. Add the mustard, Worcestershire sauce, and demi-glace. Heat to a boil. Add the heavy cream, and season to taste. Reduce to medium viscosity.
2. Sauté the fillets to order (two slices per serving); place on a preheated dinner plate. **Nappé** with sauce, and garnish with parsley sprigs.

COOKING TECHNIQUE:
Sauté

Sauté:
1. Heat the sauté pan to the appropriate temperature.
2. Evenly brown the food product.
3. For a sauce, pour off any excess oil, reheat, and deglaze.

GLOSSARY:
Brunoise: ⅛-inch dice
Nappé: coated

HAZARDOUS FOODS:
Heavy cream
Tenderloin of beef

NUTRITION:
Calories: 387
Fat: 24.4 g
Protein: 33.5 g

CHEF NOTE:
Steak Diane originally was always prepared tableside.

Stuffed Green Peppers

YIELD: 25 SERVINGS. SERVING SIZE: ONE EACH.

COOKING TECHNIQUES:

Bake

Bake:

1. Preheat the oven.
2. Place the food product on the appropriate rack.

GLOSSARY:

Brunoise: ⅛-inch dice
Al dente: to the bite
Blanch: to parbroil

HACCP:

Hold at 140° F.

HAZARDOUS FOODS:

Ground beef
Whole eggs

NUTRITION:

Calories: 336
Fat: 18.1 g
Protein: 20.1 g

INGREDIENTS:

50 each	Green bell peppers, medium, washed, and seeded
2 gallons	Tomato sauce (see page 76)

Stuffing:

7 pounds	Ground beef
5 ounces	Oil
15 ounces	Onions, peeled and diced **brunoise**
3 pounds	Rice, cooked **al dente,** cooled completely (see page 517)
1 head	Garlic, peeled and minced
1 teaspoon	Ground black pepper
1 teaspoon	Salt
15 each	Whole eggs

METHOD OF PREPARATION:

1. Preheat the oven to 350° F.
2. In a saucepan, place the peppers in salted, boiling water. **Blanch,** drain, and shock in an ice bath. Reserve.
3. Sauté the onions in oil until they are translucent; then cool.
4. Place all of the stuffing ingredients in a mixing bowl. Mix together, adjust the seasonings, and stuff the peppers with approximately four ounces of stuffing each.
5. Place some tomato sauce on the bottom of the pan; arrange the stuffed peppers on top of the sauce, and cover with remaining tomato sauce.
6. Cover the pan with parchment paper and foil. Bake in the oven until stuffing is done and the peppers are tender.
7. Hold at 140° F. Adjust the seasonings, and serve two peppers per portion on preheated dinner plates.

CHEF NOTE:

The best pepper for stuffing is the cubanelle, also known as a banana or Italian pepper.

Taco Bar: Beef for Tacos

YIELD: 50 SERVINGS. SERVING SIZE: SIX OUNCES.

INGREDIENTS:

50 each	Flour tortillas/taco shells

Beef Filling:

2 pounds	Onion, diced
8 ounces	Jalapeño peppers, diced **brunoise**
8 ounces	Sweet red pepper, diced brunoise
1 quart	Olive oil
7 pounds	Ground beef, lean
2 tablespoons	Chili powder

Toppings:

2 pounds	Globe tomatoes, chopped
2 pounds	Monterey Jack cheese, grated
½ gallon	Picante sauce or salsa
32 ounces	Olives, sliced
32 ounces	Red onion, diced
32 ounces	Scallions, sliced
2 quarts	Sour cream
3 heads	Iceberg lettuce, chopped
2 (#10 cans)	Refried beans
1 (#10 can)	Guacamole
32 ounces	Jalapeño peppers, sliced

METHOD OF PREPARATION:

1. Sauté the onion in olive oil until translucent. Add the Jalapeño peppers and red peppers. Add the ground beef, and sauté until well done.
2. Season with chili powder. Drain the ground beef mixture, and dispose of the fat properly. Reserve the meat. Hold at 140° F.
3. Set up the taco bar with the toppings.

COOKING TECHNIQUE:

Sauté

Sauté:

1. Heat the sauté pan to the appropriate temperature.
2. Evenly brown the food product.
3. For a sauce, pour off any excess oil, reheat, and deglaze.

GLOSSARY:

Brunoise: ⅛-inch dice

HAZARDOUS FOODS:

Sour cream
Cheese
Ground beef

HACCP:

Hold meat above 140° F. Keep bar below 40° F.

NUTRITION:

Calories: 587
Fat: 39.3 g
Protein: 26.7 g

Yankee Pot Roast

YIELD: 50 SERVINGS. SERVING SIZE: FIVE OUNCES.

COOKING TECHNIQUE:
Braise

Braise:
1. Heat the braising pan to the proper temperature.
2. Sear and brown the food product to a golden color.
3. Degrease and deglaze.
4. Cook the food product in two thirds liquid until fork-tender.

GLOSSARY:
Mirepoix: roughly chopped vegetables
Brunoise: ⅛-inch dice
Truss: to tie or secure
Fork-tender: without resistance
Dépouiller: to skim impurities/grease
Nappé: coated

HACCP:
Hold at 140° F.

HAZARDOUS FOOD:
Beef bottom round

NUTRITION:
Calories: 388
Fat: 17 g
Protein: 44.5 g

CHEF NOTE:
Recommended to thicken the sauce with cornstarch 1.5 ounce per 1 quart of sauce.

INGREDIENTS:

20 pounds	Beef bottom round
6 ounces	Oil
8 ounces	Tomato purée
3 cloves	Garlic, peeled and minced
1 teaspoon	Marjoram
1 teaspoon	Thyme
1 teaspoon	Basil
1 teaspoon	Spanish paprika
1 pint	Dry red wine
½ gallon	Brown beef stock (see page 7), heated to a boil
½ gallon	Demi-glace (see page 26), heated to a boil

Mirepoix:

2 pounds	Onions, peeled and diced **brunoise**
1 pound	Celery, washed and diced brunoise
1 pound	Carrots, washed, peeled, and diced brunoise

METHOD OF PREPARATION:

1. Preheat the oven to 350° F.
2. Trim the bottom round, remove any excess fat and connective tissue, and **truss** the roast with butcher twine.
3. Mix the demi-glace and beef stock, and heat to a boil.
4. Heat the oil in a braising pan over medium-high heat, sear and brown the meat on all sides, and then remove.
5. Add the mirepoix and garlic to the braising pan, and brown lightly. Reduce the heat. Add the tomato purée, herbs, spices, and wine. Add the boiling demi-glace and season. Place the browned meat in the sauce and braise until **fork-tender,** approximately 2½ to 3 hours.
6. Remove the meat from the pan, and allow to rest for 30 minutes. Hold at 140° F.
7. Strain and **dépouiller** the sauce. Adjust the seasonings and consistency, if necessary. Hold at 140° F for service.
8. For service, remove the string from the meat, and slice the meat thinly against the grain. Place three slices on a preheated dinner plate, and **nappé** with the sauce.

Baked Southwest Eggs

YIELD: 10 SERVINGS. SERVING SIZE: SIX OUNCES.

COOKING TECHNIQUES:
Sauté, Bake

Sauté:
1. Heat the sauté pan to the appropriate temperature.
2. Evenly brown the food product.
3. For a sauce, pour off any excess oil, reheat, and deglaze.

Bake:
1. Preheat the oven.
2. Place the food product on the appropriate rack.

GLOSSARY:
Julienne: matchstick strips
Al dente: to the bite
Render: to melt fat

HACCP:
Hold at 140° F.

HAZARDOUS FOODS:
Heavy cream
Eggs

NUTRITION:
Calories: 640
Fat: 50.5 g
Protein: 34.9 g

CHEF NOTE:
Lobster mushrooms add great color and texture to this dish. Other available mushrooms can be substituted.

INGREDIENTS:

2 ounces	Olive oil
6 ounces	Red bell pepper, washed, seeded, and cut **julienne**
4 ounces	Poblano pepper, washed, seeded, and cut julienne
6 ounces	Lobster or shiitake mushrooms, cleaned and trimmed
1 pound	Mexican chorizo sausage
8 ounces	Heavy cream
24 each	Eggs, lightly beaten
	Salt and freshly ground black pepper, to taste
1 pound	Fresh spinach leaves, washed, stemmed, and steamed; excess moisture removed, then chopped
2 ounces	Sun-dried tomatoes (packed in oil), drained and rough cut
8 ounces	Monterey Jack cheese with Jalapeños, grated
2 ounces	Fresh cilantro leaves, washed and coarsely chopped
	Pico de gallo (see page 671)
10 each	Cilantro sprigs, as needed

METHOD OF PREPARATION:

1. Preheat oven to 450° F.
2. In a sauté pan, heat the olive oil until it is almost smoking. Add the peppers, and sauté, tossing frequently, until just **al dente.** Remove from the sauté pan, and allow to cool on a tray.
3. Add the mushrooms, and sauté until al dente. Cool on the tray with the peppers.
4. Remove the casings from the chorizo. Crumble the chorizo, and, in a medium-sized pan, fry it until it is cooked and the fat is **rendered.** Reserve the fat separately from the chorizo.
5. Add the cream to the eggs, whisk together, and season lightly.
6. Place a heavy, cast-iron or porcelain-lined baking dish in the oven to heat. When hot, add 2 ounces of the rendered chorizo fat, and, when it is hot, pour in half of the egg mixture. Add half of the spinach, mushrooms, peppers, and tomatoes.
7. Place in the oven, and bake until just set, or for about 20 minutes.

8. When the top is set, remove from the oven, add the chorizo and cheese, and then pour in the remaining egg mixture. Add the balance of the spinach, mushrooms, peppers, and tomatoes, and return to the oven.
9. Bake until the eggs are firm, or for about another 20 minutes. Remove from the oven, and allow to stand 15 minutes at 140° F before service.
10. For service, place one portion of eggs on a preheated plate, and surround with pico de gallo. Garnish with fresh cilantro sprigs.

Basque Omelet

YIELD: 10 SERVINGS. SERVING SIZE: EIGHT OUNCES.

INGREDIENTS:

1 recipe	Basic omelet (see page 701)
6 ounces	Spanish olive oil
1½ pounds	Onions, peeled and ½-inch diced
1 pound	Ham, diced
1½ pounds	Ripe tomatoes, washed, cored, **blanched,** peeled, seeded, and diced
12 ounces	Canned pimientos, drained (reserve liquid) and cut **julienne**
	Salt and freshly ground black pepper, to taste

METHOD OF PREPARATION:

1. Prepare the egg mixture for the basic omelet, and hold at 40° F or below.
2. In a sauté pan, heat the oil; then add the onions, and sauté until they become translucent.
3. Add the ham, tomatoes, pimientos and reserved pimiento liquid, sautéing over medium heat until the liquid evaporates, or for about 15 minutes.
4. Season, to taste, remove from the heat, and hold at 140° F.
5. Prepare the omelets as per the recipe, filling each with some of the onion mixture just before rolling and removing from the pan.
6. For service, turn the omelets onto a preheated plate, and serve immediately.

COOKING TECHNIQUES:
Shallow-Fry, Sauté

Shallow-Fry:
1. Heat the cooking medium to the proper temperature.
2. Cook the food product throughout.
3. Season, and serve hot.

Sauté:
1. Heat the sauté pan to the appropriate temperature.
2. Evenly brown the food product.
3. For a sauce, pour off any excess oil, reheat, and deglaze.

GLOSSARY:
Blanched: to parboil
Julienne: matchstick strips

HACCP:
Hold at 40° F or below.
Hold at 140° F.

HAZARDOUS FOOD:
Eggs

NUTRITION:
Calories: 514
Fat: 39 g
Protein: 28.3 g

CHEF NOTES:
1. In Spain, this omelet would be made with Serrano or Bayonne ham.
2. The Basque onion mixture also can be used with basic scrambled eggs (see page 715). Heat a portion of 2 ounces of onion mixture in butter or olive oil, add 6 ounces of egg mixture, and stir until just set.
3. The Spanish name for this dish is *Piperade.*

California-Style Ranch Eggs

(Huevos Rancheros)

YIELD: 10 SERVINGS.   SERVING SIZE: TWO EGGS.

INGREDIENTS:

10 each	Corn tortillas
20 each	Poached eggs (see page 709)
5 each	Ripe avocados, washed, peeled, thinly sliced lengthwise, and drizzled with lemon juice
10 ounces	Monterey Jack cheese, grated

Mexican Tomato Sauce:

1 ounce	Olive oil
8 ounces	Onions, peeled and diced **brunoise**
1 quart	Tomato juice
2 ounces	Tomato purée
3 each	Jalapeño peppers, washed, charred, peeled, seeded, and finely minced
1½ teaspoons	Cumin, ground
½ teaspoon	Oregano, ground
	Salt and ground white pepper, to taste

METHOD OF PREPARATION:

1. In a medium-sized saucepan, heat the oil, add the onions, and sauté until the onions become translucent.
2. Add the remaining ingredients for the sauce, and bring the mixture to a boil. Reduce the heat, and simmer for 20 to 30 minutes, or until the desired consistency is achieved. Remove from the heat, and hold at 140° F.
3. Preheat the oven to 350° F.
4. Wrap the tortillas in aluminum foil, and warm in the oven.
5. Have the eggs ready, and re-warm them in a water bath held at 140° F or higher, or prepare the eggs to order.
6. For service, place a corn tortilla in the center of a preheated plate. Arrange the slices of half an avocado around the outer edge of the tortilla. Place two poached eggs in the center, and **nappé** with sauce.
7. Sprinkle cheese over the dish, and place it in the oven until the cheese is melted. Serve immediately.

COOKING TECHNIQUES:

Sauté, Poach

Sauté:

1. Heat the sauté pan to the appropriate temperature.
2. Evenly brown the food product.
3. For a sauce, pour off any excess oil, reheat, and deglaze.

Simmer and Poach:

1. Heat the cooking liquid to the proper temperature.
2. Submerge the food product completely.
3. Keep the cooked product moist and warm.

GLOSSARY:

Brunoise: ⅛-inch dice
Nappé: coated

HACCP:

Hold at 140° F.

HAZARDOUS FOOD:

Eggs

NUTRITION:

Calories: 538
Fat: 37.8 g
Protein: 24.3 g

CHEF NOTES:

1. Flour tortillas could be used as an alternative.
2. Cheddar or queso fresca can also be substituted for the Monterey Jack.
3. Avocado changes color as a result of oxidation.

Crabmeat Benedict

YIELD: 10 SERVINGS. SERVING SIZE: TWO EGGS.

INGREDIENTS:

5 ounces	Butter
20 ounces	Crabmeat, shell particles removed and meat shredded
10 each	English muffins, lightly toasted and kept warm
20 each	Poached eggs (see page 709)
20 ounces	Mornay sauce, heated to 165° F and held at 140° F (see page 37)

METHOD OF PREPARATION:

1. In a sauté pan, melt the butter, add the crabmeat, and sauté until heated throughout.
2. To serve, place muffin halves, toasted-side up, on preheated plate. Top each with a portion of crabmeat. Place one poached egg on each, and **nappé** with sauce.

COOKING TECHNIQUES:

Poach, Sauté

Simmer and Poach:

1. Heat the cooking liquid to the proper temperature.
2. Submerge the food product completely.
3. Keep the cooked product moist and warm.

Sauté:

1. Heat the sauté pan to the appropriate temperature.
2. Evenly brown the food product.
3. For a sauce, pour off any excess oil, reheat, and deglaze.

GLOSSARY:

Nappé: coated

HACCP:

Heat to 165° F.
Hold at 140° F.

HAZARDOUS FOODS:

Crabmeat
Eggs

NUTRITION:

Calories: 910
Fat: 57.3 g
Protein: 49.9 g

CHEF NOTES:

1. The crabmeat can be replaced with smoked whitefish or smoked salmon.
2. Hollandaise sauce can be used in place of mornay sauce.

(Gemüse) Vegetable Ragout with Poached Eggs

YIELD: 10 SERVINGS. SERVING SIZE: SIX OUNCES.

COOKING TECHNIQUES:
Sauté, Braise, Poach

Sauté:
1. Heat the sauté pan to the appropriate temperature.
2. Evenly brown the food product.
3. For a sauce, pour off any excess oil, reheat, and deglaze.

Braise:
1. Heat the braising pan to the proper temperature.
2. Sear and brown the food product to a golden color.
3. Degrease and deglaze.
4. Cook the food product in two thirds liquid until fork-tender.

Simmer and Poach:
1. Heat the cooking liquid to the proper temperature.
2. Submerge the food product completely.
3. Keep the cooked product moist and warm.

GLOSSARY:
Brunoise: ⅛-inch dice
Blanch: to par cook
Bouquet garni: bouquet of herbs and spices
Nappé: coated

HACCP:
Hold at 140° F or above.

HAZARDOUS FOOD:
Eggs

NUTRITION:
Calories: 428
Fat: 32 g
Protein: 8.81 g

CHEF NOTES:
1. Eggs can be pre-poached and then reheated to order in 140° F water.
2. The vegetable mixture is the Basque preparation, known as *Piperade*.
3. *Gemüse* is *vegetable* in Germany.

INGREDIENTS:

6 ounces	Olive oil
12 ounces	Onions, peeled and diced **brunoise**
6 ounces	Eggplant, washed, peeled, and diced brunoise
4 each	Red bell peppers, washed, roasted, peeled, seeded, and diced brunoise
	Salt and freshly ground black pepper, to taste
1 pound	Ripe tomatoes, washed, cored, **blanched,** peeled, seeded, and diced brunoise
4 ounces	Clarified butter
1 ounce	Freshly squeezed lemon juice
1 pound	Mushrooms, cleaned, stems trimmed, and sliced
1 bottle	Red wine (28 ounces) (preferably Côtes de Rhone)
2 cloves	Garlic, peeled and minced
10 each	Eggs
1 ounce	Cornstarch, dissolved in 1 ounce of cold water

Bouquet Garni:

1 ounce	Parsley stems
1 each	Bay leaf
2 to 3 each	Sprigs of fresh thyme

METHOD OF PREPARATION:

1. Preheat the oven to 350° F.
2. Heat the olive oil in a large sautoir. Add the onions, and sauté until the onions become translucent. Add the eggplant and peppers, and sauté another 6 to 8 minutes.
3. Add the tomatoes, and season to taste. Cover tightly and braise in the oven until the vegetables are tender, or about 30 minutes. Hold at 140° F or above.

4. Melt the butter in another sauté pan, and when the butter begins to brown, add the mushrooms. Season, to taste, and sauté over high heat until the mushrooms are browned. Reserve at room temperature.
5. In a shallow saucepan, heat the wine to a boil. Add the bouquet garni, reduce the heat, and simmer for 5 minutes.
6. Break the eggs into individual cups, and slide each into the wine. Poach until the whites are firm but the yolks remain soft. (See the preparation for basic poached eggs, page 709.) Remove and hold in cool water. Reheat for service in 140° F or higher water.
7. Strain the poaching liquid into another saucepan. Return to a boil, and reduce by one half.
8. Dissolve the cornstarch in cold water, and whisk the mixture into the boiling wine until the desired consistency is achieved; season, to taste, then add the reserved mushrooms. Hold the sauce at 140° F or higher.
9. To serve, place a portion of vegetables on a preheated dinner plate. Place a poached egg in the center, and **nappé** with sauce.

Scotch Woodcock

YIELD: 10 SERVINGS. SERVING SIZE: TWO EACH.

INGREDIENTS:

5 ounces	Anchovy paste
20 slices	White bread, toasted
10 portions	Scrambled eggs (see page 715)
20 each	Anchovy fillets
	Salt and ground black pepper, to taste

METHOD OF PREPARATION:

1. Spread the anchovy paste on the toast, and cut each slice in half on a diagonal.
2. Arrange the toast on a preheated plate, with one complete slice positioned in the center.
3. Prepare the eggs, and top the center slice of toast with the scrambled eggs. Place an anchovy fillet over the eggs, and serve immediately.

COOKING TECHNIQUE:

Sauté

Sauté:

1. Heat the sauté pan to the appropriate temperature.
2. Evenly brown the food product.
3. For a sauce, pour off any excess oil, reheat, and deglaze.

HAZARDOUS FOOD:

Scrambled eggs

NUTRITION:

Calories: 725
Fat: 47.6 g
Protein: 33.6 g

CHEF NOTES:

1. Carefully season the eggs with salt, because the anchovies are salty.
2. For eye appeal, paprika can be sprinkled on the eggs.

Braised Pheasant with Stuffed Prunes

YIELD: 10 SERVING. SERVING SIZE: ONE FOURTH PHEASANT.

COOKING TECHNIQUES:
Braise, Boil

Braise:
1. Heat the braising pan to the proper temperature.
2. Sear and brown the food product to a golden color.
3. Degrease and deglaze.
4. Cook the food product in two thirds liquid until fork-tender.

Boil: (at sea level)
1. Bring the cooking liquid to a rapid boil.
2. Stir the contents, and cook the food product throughout.
3. Serve hot.

GLOSSARY:
Mirepoix: roughly chopped vegetables
Truss: to tie or secure
Sear: to brown quickly
Fork-tender: without resistance
Dépouiller: to skim impurities/grease

HACCP:
Hold at 140° F or above.

HAZARDOUS FOOD:
Pheasant

NUTRITION:
Calories: 655
Fat: 18.5 g
Protein: 90.5 g

CHEF NOTE:
If needed, use cornstarch or arrowroot diluted in cold liquid to thicken sauce.

INGREDIENTS:

3 (4-pound)	Pheasants, giblets removed, rinsed and dried
	Salt and ground white pepper, to taste
2 ounces	Vegetable oil
½ teaspoon	Dried marjoram
½ teaspoon	Dried basil
½ teaspoon	Paprika
	Zest of two oranges
3 ounces	Cranberry jelly
8 ounces	Prune juice
1½ quarts	Brown game stock, heated to a boil (see page 7)
	Salt and freshly ground black pepper, to taste
30 (three per serving)	Dried prunes

Mirepoix:

6 ounces	Carrots, washed and peeled
6 ounces	Celery, washed and trimmed
6 ounces	Onions, peeled

METHOD OF PREPARATION:

1. Preheat the oven to 400° F.
2. Season the cavities of the pheasants, and **truss.**
3. In a braising pan, heat the oil, and **sear** the pheasants on all sides until golden. Remove, and drain the excess fat from the pan.
4. Add the mirepoix to the pan, and sauté.
5. Add the spices, orange zest, and mix. Add the jelly, juice, and stock.
6. Heat to a boil, season with salt and pepper, and then add the pheasants. Cover tightly and place in oven; cook until the meat is **fork-tender,** approximately 1½ hours.
7. Remove the pheasants from the braising pan, and allow to cool slightly.
8. Strain some of the sauce over the prunes; cover and simmer for 15 minutes. Hold at 140° F.

9. Reduce the remaining sauce by boiling to a desired consistency; then strain and **dépouiller.** Hold at 140° F or above.
10. Split the pheasants into quarters, and debone; then cover with sauce, and hold at 140° F or above.
11. Make a stuffing from the trimmings. Split the prunes, stuff with stuffing, and bake. Serve as a garnish with the pheasant.

Braised Pheasant with Scotch Honey Sauce

YIELD: 10 SERVINGS. SERVING SIZE: ONE FOURTH PHEASANT.

INGREDIENTS:

3 each	Pheasants, giblets removed, and rinsed in cold water
	Salt and freshly ground black pepper, to taste
2 ounces	Vegetable oil
½ teaspoon	Dried marjoram
½ teaspoon	Dried basil
½ teaspoon	Sweet paprika
1 each	Orange, zest only
2 quarts	Brown stock, heated to a boil (see page 7)
	Scotch honey sauce (see page 71)

Mirepoix:

8 ounces	Carrots, washed and peeled
8 ounces	Celery stalks, washed and trimmed
8 ounces	Onions, peeled

METHOD OF PREPARATION:

1. Preheat the oven to 400° F.
2. Season the cavities of the pheasants, and **truss** the birds.
3. In a braising pan, heat the oil, and sear the pheasants on all sides until golden. Remove the pheasants, and set aside.
4. Add the mirepoix to the pan, and sauté until the onions turn translucent.
5. Add the spices, orange zest, and stock.
6. Heat the liquid to a boil, and return the pheasants to the pan. Cover tightly, place in the oven, and braise until the meat is **fork-tender,** approximately 1½ hours.
7. Remove the pheasants from the braising pan, and hold at 140° F for 15 minutes.
8. Cut the pheasants into quarters and partially debone. Hold covered at 140° F until needed for service.

COOKING TECHNIQUES:

Sauté, Braise

Sauté:

1. Heat the sauté pan to the appropriate temperature.
2. Evenly brown the food product.
3. For a sauce, pour off any excess oil, reheat, and deglaze.

Braise:

1. Heat the braising pan to the proper temperature.
2. Sear and brown the food product to a golden color.
3. Degrease and deglaze.
4. Cook the food product in two thirds liquid until fork-tender.

GLOSSARY:

Mirepoix: roughly chopped vegetables
Truss: to tie or secure
Fork-tender: without resistance

HACCP:

Hold at 140° F or higher.

HAZARDOUS FOOD:

Pheasant

NUTRITION:

Calories: 681
Fat: 37.5 g
Protein: 56.1 g

9. Place the braising pan over high heat, and reduce the liquid to 8 ounces; then strain the juice.
10. Add the reduced juice to the sauce, and serve immediately or hold at 140° F or higher.
11. To serve, place a portion of pheasant on a preheated dinner plate, and nappé with sauce.

Braised Rabbit with Pearl Onions

YIELD: 10 SERVINGS. SERVING SIZE: SIX OUNCES.

INGREDIENTS:

3 each	Rabbits, washed and dried
3 ounces	Vegetable oil
	Seasoned flour for **dredging**
1½ quarts	Veal demi-glacé, heated to a boil (see page 26)
1½ pounds	Pearl onions, **blanched** and peeled
2 tablespoons	Granulated sugar
2 tablespoons	Clarified butter
1 each	Orange, zest only

Marinade:

4 ounces	Carrots, washed, peeled, and diced
4 ounces	Onions, peeled and diced
4 ounces	Celery stalk, washed, trimmed, and diced
1 clove	Garlic, peeled and minced
½ teaspoon	Black peppercorns, crushed
¼ teaspoon	Salt
¼ teaspoon	Juniper berries
1 each	Bay leaf
2 each	Whole cloves
½ teaspoon	Dried thyme
1 pint	Red wine
2 ounces	White vinegar

METHOD OF PREPARATION:

1. Cut the rabbits into eighths, and place in a deep hotel pan.
2. Combine all of the ingredients for the marinade, and mix well.
3. Pour the marinade over the rabbits; cover and refrigerate for 24 hours.
4. Preheat the oven to 375° F.
5. Drain and reserve the marinade from the rabbit. Pat the rabbit pieces dry.
6. In a sauté pan, heat the oil, dredge the rabbit pieces in flour, shaking off excess, and sear pieces on all sides.
7. Drain any excess oil from the pan. Transfer the rabbit pieces to a clean

COOKING TECHNIQUES:

Braise, Sauté

Braise:

1. Heat the braising pan to the proper temperature.
2. Sear and brown the food product to a golden color.
3. Degrease and deglaze.
4. Cook the food product in two thirds liquid until fork-tender.

Sauté:

1. Heat the sauté pan to the appropriate temperature.
2. Evenly brown the food product.
3. For a sauce, pour off any excess oil, reheat, and deglaze.

GLOSSARY:

Dredge: to coat with flour
Blanch: to par cook
Fork-tender: without resistance

HACCP:

Hold at 140° F.

HAZARDOUS FOOD:

Rabbit

NUTRITION:

Calories: 467
Fat: 23.8 g
Protein: 32.2 g

CHEF NOTE:

Preparation should begin 1 day ahead to allow time to marinate the rabbit.

braising pan; then add the reserved marinade and demi-glace. Bring the sauce to a boil; cover the pan and place in the oven to braise until the meat becomes **fork-tender.** This will take approximately 1 hour. Turn the meat several times during the cooking time.

8. Meanwhile, in a sauté pan, over low heat, melt the sugar. Add the onions and toss; then add the butter. Sauté for 5 minutes or until lightly colored. Add the orange zest.
9. When tender, remove the rabbit pieces, and hold covered at 140° F.
10. Strain the sauce, return to a boil, and reduce to a desired consistency. Add the onions and adjust the seasoning. Return the rabbit pieces to the sauce, and hold at 140° F.
11. To serve, arrange the pieces of rabbit on a preheated dinner plate, and nappé with the sauce.

Braised Squab with Chocolate

YIELD: 10 SERVINGS. SERVING SIZE: ONE EACH.

INGREDIENTS:

10 (8-ounce)	Squabs, rinsed, dried, and **trussed**
4 ounces	Spanish olive oil
8 ounces	Onion, peeled and thinly sliced
12 ounces	Canned tomatoes, drained (reserve liquids) and chopped
12 cloves	Garlic, peeled and minced
6 whole	Cloves
2 each	Bay leaves
16 ounces	Dry white wine
2 ounces	Sherry vinegar
	Salt and freshly ground pepper, to taste
3 ounces	Semisweet chocolate, grated

METHOD OF PREPARATION:

1. Preheat the oven to 350° F.
2. In sauté pan, heat the oil until it is smoking, and **sear** the squab on all sides.
3. Place the squabs in hotel pan. Add the remaining ingredients, except the chocolate.
4. Cover tightly, and place in the oven. Braise until fork-tender, or about 45 to 60 minutes. The internal temperature must be a minimum of 165° F. Remove the squab, and hold covered at 140° F.
5. Place the braising pan over medium heat, and add the chocolate. Stir until the chocolate is melted.
6. Purée the sauce, and then strain through a **chinois.** Hold at 140° F.
7. To serve, split and partially debone squab. Arrange the halves on a preheated dinner plate, and **nappé** with the sauce.

COOKING TECHNIQUE:
Braise

Braise:
1. Heat the braising pan to the proper temperature.
2. Sear and brown the food product to a golden color.
3. Degrease and deglaze.
4. Cook the food product in two thirds liquid until fork-tender.

GLOSSARY:
Truss: to tie or secure
Sear: to brown quickly
Chinois: cone-shaped strainer
Nappé: coated

HACCP:
The internal temperature must be a minimum of 165° F.
Hold at 140° F.

HAZARDOUS FOOD:
Squab

NUTRITION:
Calories: 867
Fat: 68.4 g
Protein: 43.4 g

Prosciutto-Stuffed Quail

YIELD: 10 SERVINGS. SERVING SIZE: ONE EACH.

COOKING TECHNIQUES:
Sauté, Bake

Sauté:
1. Heat the sauté pan to the appropriate temperature.
2. Evenly brown the food product.
3. For a sauce, pour off any excess oil, reheat, and deglaze.

Bake:
1. Preheat the oven.
2. Place the food product on the appropriate rack.

GLOSSARY:
Brunoise: ⅛-inch dice
Deglaze: to add liquid to hot pan
Nappé: coated

HACCP:
Bake to an internal temperature of 165° F, and hold at 140° F.

HAZARDOUS FOOD:
Quail

NUTRITION:
Calories: 314
Fat: 16 g
Protein: 25.1 g

INGREDIENTS:

10 each	Quail, partially deboned
2 ounces	Olive oil
4 ounces	Marsala wine
1 pint	Demi-glace, heated to a boil and seasoned (see page 26)

Stuffing:

5 ounces	Prosciutto ham, diced **brunoise**
5 ounces	Apple, washed, peeled, cored, and diced brunoise
2 ounces	Pistachio nuts, shelled and peeled
2 ounces	Dried dates, pitted and chopped
1 ounce	Bread crumbs
1 ounce	Butter, melted
¼ teaspoon	Ginger, ground
¼ teaspoon	Thyme, ground
¼ teaspoon	Ground white pepper
	Salt, to taste

METHOD OF PREPARATION:

1. Preheat the oven to 325° F.
2. Combine ingredients for stuffing, and mix well.
3. Rinse the quail, and pat dry. Fill with the stuffing mixture. Fold the wings under the birds.
4. In a sauté pan, heat the olive oil until it is almost smoking, and sear the birds on all sides. Then, transfer the birds to a hotel pan.
5. Discard the fat from the pan, and **deglaze** with Marsala wine. Cook to reduce by half.
6. Add the demi-glace; return to a boil, and then pour over the quail.
7. Bake in the oven for 10 minutes or until tender, or to an internal temperature of 165° F, and hold at 140° F.
8. To serve, place the quail on preheated dinner plates, and **nappé** with sauce.

Quail in Potato Nest with Cherry Sauce

(Caille en Nids de Pommes de Terre aux Cerises)

YIELD: 10 SERVINGS. SERVING SIZE: ONE EACH.

INGREDIENTS:

5 each	Granny Smith apples, peeled, cored, and diced **brunoise**
3 ounces	Walnuts, roughly chopped
	Salt and ground white pepper, to taste
10 each	Quail, partially deboned
10 each	Black cherries, pitted
10 each	Potato nests (see page 510)

Sauce:

1 ounce	Brandy
4 ounces	Port wine
	Zest of 1 orange
2 ounces	Cherry jam
20 each	Black cherries, pitted
¼ teaspoon	Thyme
¼ teaspoon	Marjoram
20 ounces	Chicken demi-glace, heated to a boil and seasoned (see page 26)

METHOD OF PREPARATION:

1. Preheat oven to 425° F.
2. Mix the diced apples with walnuts, and season.
3. Stuff the cavities of the quail with the apples and walnut mix. Place a cherry into each cavity, and **truss.**
4. Place the stuffed quail in a roasting pan. Roast in the hot oven for about 10 minutes, or until lightly browned.
5. Reduce the temperature to 350° F, and roast for an additional 5 minutes. Remove from the pan. Hold covered at 140° F until service.
6. To prepare the sauce, **deglaze** the roasting pan with brandy and wine. Add all of the other ingredients. Heat to a simmer, and continue to cook for an additional 5 minutes. If needed, add 2 ounces of beurre manié.
7. To serve, set a potato nest on a preheated dinner plate, and place a quail in the nest. Offer the sauce tableside.

COOKING TECHNIQUES:

Roast, Simmer

Roast:

1. Sear the food product, and brown evenly.
2. Elevate the food product in a roasting pan.
3. Determine doneness, and consider carryover cooking.
4. Let the food product rest before carving.

Simmer and Poach:

1. Heat the cooking liquid to the proper temperature.
2. Submerge the food product completely.
3. Keep the cooked product moist and warm.

GLOSSARY:

Brunoise: ⅛-inch dice
Truss: to tie or secure
Deglaze: to add liquid to hot pan

HACCP:

Hold at 140° F.

HAZARDOUS FOOD:

Quail

Nutrition:

Calories: 572
Fat: 29.9 g
Protein: 25.7 g

Roast Partridge, Dijon Style

(Perdreau Rôti Dijonnaise)

YIELD: 10 SERVINGS. SERVING SIZE: ONE-HALF PARTRIDGE.

INGREDIENTS:

5 each	Whole partridges
12 ounces	Dijon mustard, 3 tablespoons reserved
20 slices	Bacon
6 ounces	Game fumet
4 ounces	Cognac
1 pint	Heavy cream
	Salt and pepper, to taste
10 slices	Toast (white bread), cut into diagonals

METHOD OF PREPARATION:

1. Trim the partridges, and spread the mustard over each bird. Bard each with three or four slices of bacon, and **truss.**
2. Place the birds in a roasting pan. Roast in an oven at 425° F for 20 to 25 minutes.
3. Remove the birds from the roasting pan, and hold warm at 140° F. Take 2 ounces of the fat from the roasting pan, and place in a skillet. Add the fumet, stir, and heat to a boil. Add the cognac, and cook for 3 minutes.
4. Add the cream and 3 tablespoons of Dijon mustard. Stir, simmer, and **reduce** the sauce slowly. Season with salt and pepper. Remove the string and barding from the birds.
5. Cut the partridges in half. Set each half on top of two toast diagonals, and spoon the sauce on top.

COOKING TECHNIQUE:

Roast

Roast:

1. Sear the food product, and brown evenly.
2. Elevate the food product in a roasting pan.
3. Determine doneness, and consider carryover cooking.
4. Let the food product rest before carving.

GLOSSARY:

Truss: to tie or secure
Reduction: evaporation of liquid by boiling

HACCP:

Hold warm at 140° F.

HAZARDOUS FOODS:

Partridge
Game fumet
Heavy cream

NUTRITION:

Calories: 493
Fat: 23.6 g
Protein: 44.4 g

COOKING TECHNIQUES:

Sauté, Simmer

Sauté:

1. Heat the sauté pan to the appropriate temperature.
2. Evenly brown the food product.
3. For a sauce, pour off any excess oil, reheat, and deglaze.

Simmer and Poach:

1. Heat the cooking liquid to the proper temperature.
2. Submerge the food product completely.
3. Keep the cooked product moist and warm.

GLOSSARY:

Flambé: to flame
Nappé: coated

HAZARDOUS FOOD:

Venison saddle

NUTRITION:

Calories: 350
Fat: 13.4 g
Protein: 25.1 g

Venison Medallions with Peppercorn Sauce

(Noisettes de Chevreuil Poivrade)

YIELD: 10 SERVINGS. SERVING SIZE: SIX OUNCES.

INGREDIENTS:

3 ounces	Oil
1 each	Venison, saddle, trimmed, loins removed, cut into 20 small noisettes
	Salt and freshly gound black pepper, to taste
3 ounces	Cognac or brandy
12 ounces	Poivrade sauce, heated to a boil (see page 43)
3 ounces	Clarified butter
2 each	Grapefruits, peeled and sectioned (at least 20 segments)
10 slices	White bread, toasted, with two rounds cut from each

METHOD OF PREPARATION:

1. In a sauté pan, heat the oil. Quickly sauté the noisettes until browned, turning once. Season, to taste. Hold the noisettes on a sheet pan in the oven at 140° F. Drain off the fat.
2. Add the cognac or brandy, and **flambé.** Add the poivrade sauce. Cover and simmer over low heat for 4 to 6 minutes. Hold at 140° F or above.
3. To serve, place two toast rounds on a preheated dinner plate, and place a noisette on each. **Nappé** with sauce, and garnish each with a grapefruit segment.

Crown Roast of Lamb with Mint

YIELD: 10 SERVINGS.   SERVING SIZE: SIX OUNCES (BONE IN).

COOKING TECHNIQUE:

Roast

Roast:

1. Sear the food product, and brown evenly.
2. Elevate the food product in a roasting pan.
3. Determine doneness, and consider carryover cooking.
4. Let the food product rest before carving.

GLOSSARY:

Chinois: cone-shaped strainer

HACCP:

Cook to an internal temperature of 165° F.
Hold at 140° F.

HAZARDOUS FOOD:

Rack of lamb

NUTRITION:

Calories: 326
Fat: 20.5 g
Protein: 23.9 g

INGREDIENTS:

4 (8-bone) (1½ pounds)	Racks of lamb, prepared in a crown shape (see chef note 3)

Marinade:

12 cloves	Garlic, peeled and minced
1 tablespoon	Chopped fresh mint leaves
2 tablespoons	Fresh parsley, washed, excess moisture removed, and chopped
	Salt and freshly ground black pepper, to taste
	Piri piri, to taste (see page 672)
2 teaspoons	Spanish hot paprika
3 ounces	Spanish olive oil
2 cups	Dry white wine
20 ounces	Brown lamb stock, heated to a boil (see page 7)
2 ounces	Fresh mint leaves
1 teaspoon	Sugar

METHOD OF PREPARATION:

1. Place the crown roast in a half hotel pan or roasting pan.
2. Combine the ingredients for the marinade, and pour over the roast. Marinate for 24 hours, spooning the marinade over the roast at several intervals, so the flavors permeate the meat.
3. Preheat the oven to 425° F.
4. Remove the crown roast from the marinade (reserve the marinade), and place it in a roasting pan. Place it in the oven, and roast until medium rare, or to an internal temperature of 130° F. Remove and hold at 140° F for at least 20 minutes before slicing.
5. Add the marinade to the lamb stock, and heat to a boil. Reduce the heat, and simmer for 30 minutes. Strain through a fine **chinois.**
6. Sprinkle sugar over the mint leaves, and then mince the leaves. Add the mint to the strained stock, and return to a boil. Hold at 140° F.
7. To serve, slice the roast between the bones. Place two chops on a preheated dinner plate, and ladle the mint sauce over the meat.

CHEF NOTES:

1. Preparation should begin 1 day in advance to allow the roast to marinate.
2. The use of sugar in chopping mint prevents bruising of the leaves, which causes bitterness.
3. Crown roast is formed by tying the loins into a circle; with the Frenched rib, the bones are standing upright. The loin of the meat is in the center.

Grilled Lamb Chops with Curried Carrot and Rosemary Sauce

YIELD: 10 SERVINGS. SERVING SIZE: SIX OUNCES.

INGREDIENTS:

20 (3-ounce), or	**Frenched** lamb chops, or
10 (6-ounce)	Double-cut lamb chops, one bone removed
	Salt and freshly ground black pepper, to taste
	Vegetable oil, as needed

Sauce:

4 pounds	Carrots, peeled and juice extracted, or
1 quart	Fresh carrot juice (available at health food stores)
1 pint	Lamb stock, heated to a boil (see page 7)
1 ounce	Fresh shallots, peeled and diced **brunoise**
1 large clove	Garlic, peeled and mashed into a purée
½ teaspoon	Madras curry powder
12 (3-inch)	Fresh rosemary sprigs
2 ounces	Butter, cut in pieces and chilled
	Salt and freshly ground black pepper, to taste

METHOD OF PREPARATION:

1. Prepare the grill, or heat broiler to maximum.
2. Season the lamb chops on each side, being particularly generous with the pepper, and coat each with oil. Refrigerate at 40° F or lower for at least 30 minutes
3. In a saucepan, place the carrot juice, and bring to a boil. Reduce the heat, and simmer until the quantity is reduced by half.
4. Add the lamb stock, shallots, garlic, curry powder, and two (only) rosemary sprigs. Bring the liquid to a boil; reduce the heat, and simmer for 15 minutes, to allow the flavors to mingle.
5. Strain the sauce through a **chinois** into another saucepan. Return to the heat, and whisk in the butter. Hold at 140° F or higher.
6. Grill the lamb chops over high heat, turning once, to an internal temperature of 145° F. Serve immediately.
7. For service, arrange two chops (or one if double) on a preheated dinner plate, crossing the Frenched bones. **Nappé** with sauce, and garnish with a rosemary sprig.

COOKING TECHNIQUES:

Grill, Broil, Simmer

Grill/Broil:

1. Clean and heat the grill/broiler.
2. To prevent sticking, brush the food product with oil.

Simmer and Poach:

1. Heat the cooking liquid to the proper temperature.
2. Submerge the food product completely.
3. Keep the cooked product moist and warm.

GLOSSARY:

Brunoise: ⅛-inch dice
Chinois: cone-shaped strainer
Frenched: rib bones cleaned
Nappé: coated

HACCP:

Refrigerate at 40° F or lower.
Hold at 140° F or higher.

HAZARDOUS FOOD:

Lamb

NUTRITION:

Calories: 550
Fat: 36.1 g
Protein: 36 g

CHEF NOTE:

Lamb chops will have the best flavor if charred on the outside and medium-rare on the inside.

Lamb in Egg Sauce

YIELD: 10 SERVINGS. SERVING SIZE: SIX OUNCES.

COOKING TECHNIQUES:

Sauté, Roast

Sauté:

1. Heat the sauté pan to the appropriate temperature.
2. Evenly brown the food product.
3. For a sauce, pour off any excess oil, reheat, and deglaze.

Roast:

1. Sear the food product, and brown evenly.
2. Elevate the food product in a roasting pan.
3. Determine doneness, and consider carryover cooking.
4. Let the food product rest before carving.

GLOSSARY:

Sear: to brown quickly
Fork-tender: without resistance
Deglaze: to add liquid to hot pan
Chinois: cone-shaped strainer

HACCP:

Hold at 140° F.

HAZARDOUS FOODS:

Lamb chops
Egg yolks
Heavy cream

NUTRITION:

Calories: 562
Fat: 36.9 g
Protein: 50.8 g

INGREDIENTS:

4 ounces	Vegetable oil
4 cloves	Garlic, peeled and finely minced
3 ounces	Pine nuts, shelled and peeled
10 ounces	Fresh spinach leaves, washed, stemmed, blanched, excess moisture removed, and roughly chopped
	Salt and freshly ground black pepper, to taste
6 ounces	Prosciutto, thinly sliced into 20 slices
20 (3-ounce)	Lamb chops, boned, Frenched, and a pocket cut in the eye of the chop
6 ounces	Dry white wine
1½ ounces	Egg yolks
2 ounces	Parmesan cheese, grated
4 ounces	Heavy cream
1 tablespoon	Freshly squeezed lemon juice
2 ounces	Fresh parsley, washed, excess moisture removed, and chopped

METHOD OF PREPARATION:

1. Preheat the oven to 350° F.
2. In a sauté pan, heat half of the oil; add the garlic and pine nuts, and sauté until lightly browned. Add the spinach, and sauté an additional 3 to 5 minutes. Season, to taste, then remove and cool to room temperature.
3. Divide the spinach mixture among the prosciutto slices, and roll each.
4. Season the pocket of the chops, and stuff with a prosciutto roll, pressing lightly.
5. Heat the remaining oil, and **sear** the chops on both sides. Transfer to a baking pan, and roast in the oven until **fork-tender.** Hold at 140° F.
6. Discard the oil from the sauté pan, **deglaze** with the wine, and reduce to half the liquid. Strain through a **chinois,** and hold at 140° F.
7. Just before service, place the egg yolks in a stainless steel bowl, and whisk lightly. Add the cheese, heavy cream, lemon juice, and parsley, and whisk all together. Slowly whisk in the reduced wine. Heat, and hold at 140° F.
8. To serve, arrange two chops on a preheated dinner plate, and nappé with sauce.

CHEF NOTE:

The sauce cannot be held for more than 15 minutes, or the egg yolks will coagulate.

Lamb Kebabs with Zucchini

YIELD: 10 SERVINGS. SERVING SIZE: FIVE OUNCES.

INGREDIENTS:

4 pounds	Boneless leg of lamb
3 each	Small-sized zucchini, washed and roughly cut into 1-inch pieces
10 each	Mushrooms, washed and stemmed
1 each	Large onion, peeled and cut into 1-inch pieces

Marinade:

16 ounces	Dry red wine
12 ounces	Olive oil
½ tablespoon	Dried oregano
1 teaspoon	Salt
	Freshly ground black pepper, to taste
4 cloves	Garlic, peeled and finely minced

METHOD OF PREPARATION:

1. Remove all of the fat and connective tissue from the lamb, and cut it into 1-ounce cubes.
2. In a large bowl, combine the marinade ingredients, and mix well. Add the lamb, and toss to coat all pieces with the herbs and spices, and marinate at 40° F or below for at least 30 minutes.
3. Drain the marinade from the meat, and reserve.
4. Prepare the skewers, alternating vegetables with lamb cubes. Place on metal trays, and hold at 40° F or below until needed for service.
5. Preheat the grill or broiler.
6. To order, grill the kebabs, turning and brushing with the marinade; cook to an internal temperature of 140° F. Serve immediately on preheated dinner plates.

COOKING TECHNIQUES:

Grill or Broil

Grill/Broil:

1. Clean and heat the grill/broiler.
2. To prevent sticking, brush the food product with oil.

HACCP:

Hold at 40° F or below.
Cook to an internal temperature of 140° F.

HAZARDOUS FOOD:

Boneless leg of lamb

NUTRITION:

Calories: 686
Fat: 46.7 g
Protein: 52.8 g

CHEF NOTES:

1. In the southern part of Greece, this kebab is served with lemon wedges. The same dish can be made with eggplant, in which case it would be called *souvlakia me melitzanes*.
2. If using wooden skewers, soak skewers in water to prevent them from burning.

Lamb Stew with Peas

(Cordrero con Guisantes)

YIELD: 10 SERVINGS. SERVING SIZE: EIGHT OUNCES.

INGREDIENTS:

4 pounds	Shoulder of lamb, cubed
3 ounces	Olive oil
8 ounces	Onions, peeled and diced **brunoise**
4 each	Red peppers, washed, cored, and diced
8 ounces	Tomato **concassé**
4 cloves	Garlic, peeled and minced
2 each	Chili peppers, washed, seeded, and chopped
1 tablespoon	Spanish paprika
½ teaspoon	Saffron
1 pound	Fresh or frozen green peas
	Salt and ground black pepper, to taste

METHOD OF PREPARATION:

1. Brown the cubed lamb, and place in a large casserole.
2. Add oil, if needed, in the same vessel in which the meat was browned, and sauté the onions until lightly browned.
3. Add the red peppers, and sauté for 2 minutes. Add the tomato concassé, garlic, and chili peppers, and sauté for 5 minutes. Add the paprika and saffron.
4. Pour on the meat, season, and mix well.
5. Add some water or demi-glace. Add the peas, and cover with a lid.
6. Cook on low heat, or bake in a 320° F oven until the meat is tender.

COOKING TECHNIQUES:

Sauté, Stew

Sauté:

1. Heat the sauté pan to the appropriate temperature.
2. Evenly brown the food product.
3. For a sauce, pour off any excess oil, reheat, and deglaze.

Stew:

1. Sear, sauté, sweat, or blanch the main food product.
2. Deglaze the pan, if desired.
3. Cover the food product with simmering liquid.
4. Remove the bouquet garni.

GLOSSARY:

Brunoise: ⅛-inch dice
Concassé: peeled, seeded, and roughly chopped

HAZARDOUS FOOD:

Lamb

NUTRITION:

Calories: 478
Fat: 33.2
Protein: 31.3

CHEF NOTES:

1. This dish should be served with very little or no sauce; liquid should be given in very small quantities.
2. This dish also can be prepared with olives, fava beans, or chick peas instead of green peas.

Rack of Lamb with Parsley

(Carré d'Agneau Persillade)

YIELD: 10 SERVINGS. SERVING SIZE: TWO RIBS.

INGREDIENTS:

6 pounds	Rack of lamb, trimmed and Frenched
6 ounces	Clarified butter
1 head	Garlic, peeled and finely chopped
1 ounce	Fresh parsley, washed, excess moisture removed, and chopped
10 ounces	Bread crumbs, mix with garlic and parsley
	Salt and pepper, to taste

METHOD OF PREPARATION:

1. Brown the rack of lamb in a skillet with heated butter.
2. Place the lamb in a roasting pan in a 450° F oven for 5 minutes. Remove the meat from the oven, turn it, and season with salt and pepper. Sprinkle the bread crumb mixture over the side of the rack, and press it down with a spoon or a spatula.
3. Return the lamb to the oven, and cook for 10 more minutes, until it reaches a medium-rare temperature of 140° F. Carve tableside. Serve with proper garnishes.

COOKING TECHNIQUE:
Roast

Roast:
1. Sear the food product, and brown evenly.
2. Elevate the food product in a roasting pan.
3. Determine doneness, and consider carryover cooking.
4. Let the food product rest before carving.

HAZARDOUS FOOD:
Rack of Lamb

NUTRITION:
Calories: 609
Fat: 30.5 g
Protein: 58.3 g

CHEF NOTE:
Fried parsley is recommended as a garnish.

Roast Leg of Lamb with Tarragon

(Gigot d'Agneau Rôti à l'Estragon)

YIELD: 10 SERVINGS. SERVING SIZE: SIX OUNCES.

INGREDIENTS:

1 each	Leg of lamb (approximately 5 pounds boneless)
	Salt and freshly ground black pepper, to taste
½ teaspoon	Tarragon, ground
3 ounces	Oil, heated
1 quart	Lamb demi-glace, seasoned
1 ounce	Cornstarch
3 ounces	Tarragon leaves
1 head	Garlic, peeled and left whole
5 stems	Fresh tarragon

Mirepoix:

8 ounces	Onions, peeled
8 ounces	Carrots, washed and peeled
8 ounces	Celery, cleaned and washed

METHOD OF PREPARATION:

1. Season the lamb with salt, pepper, and ground tarragon. Truss leg with butcher's twine.
2. **Sear** and brown in hot oil, place on mirepoix in roasting pan, and roast in a 325° F oven until the internal temperature reaches 140° F. Baste with demi-glace every 15 to 20 minutes. Set aside the meat when it is done.
3. Place the roasting pan on the stove, add the lamb demi-glace, and heat to a simmer.
4. Let cook for 10 minutes, and strain.
5. Degrease the sauce, heat to a boil, thicken with diluted cornstarch, and let simmer for 6 minutes. Taste, and adjust the seasonings.
6. Remove 10 tarragon leaves for garnishing the portioned lamb; chop the remaining leaves, and add to the sauce.
7. To serve, slice on a bias, three slices per portion. Place on heated dinner plate, **nappé** with sauce, and garnish with a tarragon leaf.

COOKING TECHNIQUE:

Roast

Roast:

1. Sear the food product, and brown evenly.
2. Elevate the food product in a roasting pan.
3. Determine doneness, and consider carryover cooking.
4. Let the food product rest before carving.

GLOSSARY:

Mirepoix: roughly chopped vegetables
Sear: to brown quickly
Nappé: coated

HACCP:

Cook to an internal temperature of 140° F.

HAZARDOUS FOOD:

Leg of lamb

NUTRITION:

Calories: 460
Fat: 18.7 g
Protein: 47.8 g

Roast Saddle of Lamb

(Selle d'Agneau Richelieu)

YIELD: 10 SERVINGS. SERVING SIZE: FIVE OUNCES.

INGREDIENTS:

2 each	Lamb saddles, approximately 2 pounds each
	Salt and freshly ground black pepper, to taste
	Fresh rosemary, to taste
2 pounds	Trimmings and lamb bones, for making a stock
1 quart	Brown lamb stock, heated to a boil (see page 7)
1 tablespoon	Cornstarch, dissolved in 2 tablespoons cold water
	Bread crumbs, as needed
	Clarified butter, as needed
4 ounces	Dijon mustard

Mirepoix:

4 ounces	Carrots, washed and peeled
4 ounces	Celery, washed and trimmed
4 ounces	Onions, peeled

METHOD OF PREPARATION:

1. Trim the lamb saddles, reserving trimmings. Rub the salt, black pepper, and rosemary into the saddles, and place in a roasting pan.
2. Preheat the oven to 425° F. Place the mirepoix, lamb trimmings, and bones around the saddles in the roasting pan.
3. Place the pan in the oven for 15 minutes; reduce temperature to 375° F, and continue to roast for an additional 25 minutes. When done, the internal temperature should be 140 to 145° F. Remove the lamb, and set it aside.
4. Place the roasting pan over medium heat, and **deglaze** the pan with the lamb stock. Simmer until the desired taste is achieved; then strain through a **chinois.**
5. Return the stock to a boil, and thicken with the cornstarch mixture. Season to taste, and reserve.
6. In a sauté pan, heat the butter, and sauté the bread crumbs until golden.
7. Remove the lamb meat from the bones, slice it on an angle, and place it back on the bones. Spread the Dijon mustard over the meat, and coat the saddles with the bread crumbs. Warm in the oven for 3 to 4 minutes. Serve immediately.

COOKING TECHNIQUE:
Roast

Roast:
1. Sear the food product, and brown evenly.
2. Elevate the food product in a roasting pan.
3. Determine doneness, and consider carryover cooking.
4. Let the food product rest before carving.

GLOSSARY:
Mirepoix: roughly chopped vegetables
Deglaze: to add liquid to hot pan
Chinois: cone-shaped strainer

HACCP:
The internal temperature should be 120 to 125° F.

HAZARDOUS FOOD:
Lamb

NUTRITION:
Calories: 374
Fat: 18.4 g
Protein: 30.7 g

CHEF NOTE:
The name *Richelieu* generally refers to a garnish for large cuts of meats, comprising stuffed tomatoes, duxelle, chateau potatoes, and braised lettuce.

Roast Shoulder of Lamb

YIELD: 50 SERVINGS. SERVING SIZE: SIX OUNCES.

INGREDIENTS:

25 pounds	Boneless lamb shoulder
	Salt, to taste
2½ tablespoons	Rosemary leaves, dried
2½ tablespoons	Black peppercorns, crushed
2 heads	Garlic, peeled and roughly chopped
1¼ pints	White wine
1¼ gallons	Brown lamb stock, heated to a boil (see page 7)

Mirepoix:

1 pound	Onions, peeled and washed
2 pounds	Carrots, peeled and washed
2 pounds	Celery stalks, peeled, washed, and roughly cut on a bias

METHOD OF PREPARATION:

1. Preheat the oven to 375° F.
2. Trim the excess fat.
3. Mix the seasonings together, and rub them into the meat. Roll and tie the lamb.
4. Combine the mirepoix in a roasting pan. Add the garlic, and place the lamb on top. Roast in the oven for 30 minutes. Reduce the heat to 325° F, and continue to roast to an internal temperature of 130° F. Turn the meat in the roasting pan during the roasting period, to roast evenly. The roasting time will be approximately 1½ hours.
5. When done, remove the meat, and hold at 140° F. Allow it to rest for 15 minutes before slicing.
6. Place the roasting pan on the stove, and caramelize the vegetables. Remove the excess fat, and **deglaze** the pan with wine. Reduce to half volume. Add the lamb stock, reduce the heat, and simmer until the desired consistency is achieved. Strain, and adjust the seasonings. Hold at 140° F.
7. To serve, slice the meat across the grain. Arrange slices on a preheated dinner plate, and **nappé** with sauce.

COOKING TECHNIQUE:

Roast

Roast:

1. Sear the food product, and brown evenly.
2. Elevate the food product in a roasting pan.
3. Determine doneness, and consider carryover cooking.
4. Let the food product rest before carving.

GLOSSARY:

Mirepoix: roughly chopped vegetables
Deglaze: to add liquid to hot pan
Nappé: coated

HACCP:

Hold at 140° F.

HAZARDOUS FOOD:

Lamb

NUTRITION:

Calories: 405
Fat: 20 g
Protein: 47 g

CHEF NOTE:

The FDA recommends that lamb be cooked to an internal temperature of 140° F.

Shepherd's Pie

YIELD: 50 SERVINGS. SERVING SIZE: FIVE OUNCES.

INGREDIENTS:

10 ounces	Oil
2 pounds	Onions, peeled and diced **brunoise**
12 pounds	Ground lamb
	Salt and ground black pepper, to taste
1 teaspoon	Rosemary, ground
15 each	Whole eggs
10 pounds	Duchess potatoes (see page 479)

METHOD OF PREPARATION:

1. Preheat the oven to 350° F.
2. Heat the oil in a sauté pan, and sauté the onions.
3. Add the ground lamb to the sauté pan, and sauté. Cook to an internal temperature of 155° F. Drain off the excess fat. Season with the salt, pepper, and rosemary. Allow the meat to cool slightly.
4. Add 10 eggs to the meat, and incorporate.
5. On the bottom of hotel pans, spread the duchess potato mix ⅓-inch thick, and let set for 15 minutes.
6. Place a layer of the sautéed meat mixture on top of the potatoes, approximately 1-inch thick.
7. Pipe the rest of the potatoes on top of the meat in each pan, using a star tube with a pastry bag.
8. Brush the top of the potatoes with five beaten eggs, and bake for 30 minutes, or until golden brown.
9. Hold at 140° F.

COOKING TECHNIQUES:

Sauté, Bake

Sauté:

1. Heat the sauté pan to the appropriate temperature.
2. Evenly brown the food product.
3. For a sauce, pour off any excess oil, reheat, and deglaze.

Bake:

1. Preheat the oven.
2. Place the food product on the appropriate rack.

GLOSSARY:

Brunoise: ⅛-inch dice

HACCP:

Hold at 140° F.

HAZARDOUS FOODS:

Ground lamb
Eggs

NUTRITION:

Calories: 454
Fat: 29.1 g
Protein: 30.7 g

CHEF NOTES:

1. There are many variations of shepherd's pie.
2. The pie can be served with tomato sauce or lamb demi-glace. Pour 1 ounce of hot sauce on a preheated dinner plate, and place the pie in the center of the sauce. (Sauce should never be poured on top of the pie.)
3. For individual portions, this entrée may be served in a casserole dish.

Spinach-Stuffed Leg of Lamb

(Gigot d'Agneau Florentine)

YIELD: 10 SERVINGS. SERVING SIZE: FIVE OUNCES.

INGREDIENTS:

1½ pounds	Lean ground lamb
	Salt and ground white pepper, to taste
¼ teaspoon	Rosemary, ground
2 each	Hard-cooked egg whites, chopped
1 ounce	Red bell pepper, washed and diced
3 each	Whole eggs
2 ounces	Butter
4 ounces	Onions, peeled and finely chopped
1 pound	Fresh spinach, washed, stemmed, and drained well
1 clove	Garlic, peeled and finely minced
	Bread crumbs, optional
5 pounds	Leg of lamb, boneless
4 ounces	White wine
1 quart	White lamb stock, heated to a boil (see page 12)

METHOD OF PREPARATION:

1. Preheat the oven to 375° F.
2. Place the ground lamb in a bowl. Add the salt, white pepper, rosemary, egg whites, red pepper, and two of the eggs, and incorporate.
3. In a sauté pan, heat the butter, and sauté the onions until translucent. Add spinach and continue to sauté until wilted.
4. Add the remaining egg, additional seasoning, and garlic. If the spinach is too moist, sifted bread crumbs may be incorporated.
5. Stuff the leg of lamb with one layer of ground lamb and one layer of spinach, and **truss.**
6. Place the lamb in a roasting pan, and roast in the oven for 30 minutes, or until the lamb reaches an internal temperature of 140° F. Hold the lamb at 140° F for 30 minutes before slicing.
7. Place the roasting pan on the stove, and heat until the meat juices in the pan are caramelized. Pour off the excess fat. **Deglaze** the pan with the wine. Add the lamb stock, and simmer for 10 minutes. Adjust the seasonings, and strain through a **chinois.** Hold at 140° F or above.
8. To serve, remove the strings from the lamb, and slice. Arrange two slices on a preheated dinner plate, and **nappé** with sauce.

COOKING TECHNIQUE:

Roast

Roast:

1. Sear the food product, and brown evenly.
2. Elevate the food product in a roasting pan.
3. Determine doneness, and consider carryover cooking.
4. Let the food product rest before carving.

GLOSSARY:

Truss: to tie or secure
Deglaze: to add liquid to hot pan
Chinois: cone-shaped strainer
Nappé: coated

HACCP:

Cook to an internal temperature of 140° F.

HAZARDOUS FOODS:

Ground lamb
Eggs
Leg of lamb

NUTRITION:

Calories: 505
Fat: 25.8 g
Protein: 54.4 g

Baked Ham with Raisin Sauce

YIELD: 50 SERVINGS. SERVING SIZE: FOUR OUNCES.

INGREDIENTS:

15 pounds	Ham, smoked, shankless, and trimmed
	Whole cloves, inserted into ham, as needed
4 ounces	Dijon mustard
7 ounces	Honey
4 ounces	Brown sugar
½ ounce	Nutmeg
2 quarts	Raisin sauce, heated to a boil

METHOD OF PREPARATION:

1. Score the surface of the ham into diamond shapes. Place a whole clove at the center of each diamond.
2. Mix the mustard, honey, brown sugar, and nutmeg in a mixing bowl.
3. Place the ham on a rack in a roasting pan. Bake in a 350° F oven. While baking, brush the honey glaze on the ham frequently, until the ham is golden brown and has reached an internal temperature of 155° F, approximately 1 hour. Hold at 140° F.
4. Remove the ham from the oven, and let it rest for 30 minutes before carving. Slice 4-ounce portions, and serve with the raisin sauce on warm plates.

COOKING TECHNIQUE:

Bake

Bake:

1. Preheat the oven.
2. Place the food product on the appropriate rack.

HACCP:

Hold at 140° F.

NUTRITION:

Calories: 254
Fat: 8.19 g
Protein: 34.5 g

CHEF NOTE:

Deck or conventional ovens work best. If using convection oven, deglaze the ham during the last 15 minutes of baking.

Barbecued Beer-Cured Pork Ribs

YIELD: 10 SERVINGS. SERVING SIZE: SIXTEEN OUNCES (BONE-IN).

INGREDIENTS:

10 pounds	Pork spareribs, portioned to 16-ounce servings
3 quarts	Beer
	Water, to cover
	Salt, to taste
2 tablespoons	Black peppercorns, crushed
1 head	Garlic, thinly sliced
2 ounces	Fresh rosemary leaves, washed and coarsely chopped
2 ounces	Fresh tarragon leaves, washed and chopped
2 ounces	Fresh thyme leaves, washed and chopped
1 tablespoon	Oregano leaves, dried
1 tablespoon	Basil leaves, dried
10 ounces	Honey
2 quarts	Barbecue sauce (see page 14)
	Fresh herb sprigs, to garnish

METHOD OF PREPARATION:

1. In a large pot, place the spareribs; then add all of the remaining ingredients, except the honey and barbecue sauce.
2. Bring the liquid to a boil; reduce the heat and simmer until the ribs are **fork-tender.** This will take approximately 1½ hours.
3. Once tender, remove the ribs from the liquid, and brush, while hot, with honey on both sides.
4. If possible, allow the ribs to cool to room temperature.
5. Grill or broil the ribs, brushing generously with sauce. Hold at 140° F or higher.
6. Serve, cut in portions, on preheated dinner plates. Garnish with additional fresh herbs.

COOKING TECHNIQUES:

Simmer, Grill/Broil

Simmer and Poach:

1. Heat the cooking liquid to the proper temperature.
2. Submerge the food product completely.
3. Keep the cooked product moist and warm.

Grill/Broil:

1. Clean and heat the grill/broiler.
2. To prevent sticking, brush the food product with oil.

GLOSSARY:

Fork-tender: without resistance

HACCP:

Hold at 140° F.

HAZARDOUS FOOD:

Pork spareribs

NUTRITION:

Calories: 851
Fat: 42 g
Protein: 42 g

CHEF NOTES:

1. Allowing the ribs to cool permits the honey to penetrate.
2. Ribs need frequent turning during the grilling or broiling process due to the sugar from the honey which may otherwise burn.

Broiled Ham Slices

YIELD: 50 SERVINGS. SERVING SIZE: AS NEEDED.

INGREDIENTS:

12½ pounds	Baked Virginia ham, fat trimmed

METHOD OF PREPARATION:

1. Slice the ham into 4-ounce portions.
2. Store in a holding pan, and keep refrigerated.
3. Broil on both sides, and hold at 140° F. Remove, and transfer to hotel pans for service.

COOKING TECHNIQUE:

Broil

Grill/Broil:

1. Clean and heat the grill/broiler.
2. To prevent sticking, brush the food product with oil.

HACCP:

Hold at 140° F.

NUTRITION:

Calories: 66.7
Fat: 2.6 g
Protein: 10.1 g

Calabrian Roast Pork

YIELD: 10 SERVINGS. SERVING SIZE: FIVE OUNCES.

COOKING TECHNIQUES:
Roast, Shallow-Fry

Roast:
1. Sear the food product, and brown evenly.
2. Elevate the food product in a roasting pan.
3. Determine doneness, and consider carryover cooking.
4. Let the food product rest before carving.

Shallow-Fry:
1. Heat the cooking medium to the proper temperature.
2. Cook the food product throughout.
3. Season, and serve hot.

GLOSSARY:
Chinois: cone-shaped strainer
Charred: grilled or broiled until skin breaks

HACCP:
Hold at 140° F or above.

HAZARDOUS FOOD:
Pork

NUTRITION:
Calories: 393
Fat: 19.9 g
Protein: 38.5 g

INGREDIENTS:

5 pounds	Pork shoulder, deboned
1 quart	Veal demi-glace, heated to a boil (see page 26)
2 sprigs	Fresh thyme, washed
5 cloves	Garlic, peeled and finely minced
3 ounces	Olive oil
12 ounces	Zucchini, washed, peeled, and sliced into same length as peppers
1 pound	Red bell peppers, washed, **charred,** peeled, seeded, and cut into 1½-inch strips
1 pound	Green bell peppers, washed, charred, peeled, seeded, and cut into 1½-inch strips
	Salt and freshly ground black pepper, to taste

METHOD OF PREPARATION:

1. Preheat the oven to 375° F.
2. Trim and tie the meat; then place in a roasting pan.
3. Heat the demi-glace to a boil, add the thyme and garlic, and boil for 1 minute.
4. Pour the demi-glace over the meat, and roast in the oven for 20 minutes. Reduce the heat to 325° F, and continue to roast until an internal temperature of 175° F is achieved.
5. Remove the meat from the pan, and hold at 140° F or above.
6. Strain the sauce through a **chinois.**
7. In a sauté pan, heat the olive oil, and fry the zucchini slices, but do not brown. Add the peppers, and warm throughout. Season to taste, and hold at 140° F or above.
8. To serve, slice the meat, allowing three slices per serving. Ladle the sauce onto a preheated dinner plate, and arrange the meat slices, overlapping, on top of the sauce.
9. Garnish the pork with the peppers and zucchini, arranging them in the order of the colors of the Italian flag; green, white, and red.

Chinese Barbecued Pork Spareribs

YIELD: 10 SERVINGS. SERVING SIZE: EIGHT OUNCES.

COOKING TECHNIQUE:
Roast

Roast:
1. Sear the food product, and brown evenly.
2. Elevate the food product in a roasting pan.
3. Determine doneness, and consider carryover cooking.
4. Let the food product rest before carving.

HAZARDOUS FOOD:
Pork spareribs

NUTRITION:
Calories: 831
Fat: 52.5 g
Protein: 51.7 g

INGREDIENTS:

10 pounds	Pork spareribs (uncut), excess fat and breastbone removed
24 ounces	Prepared plum sauce
10 each	Scallion fans

Marinade:

5 ounces	Soy sauce
2½ ounces	Honey
2½ ounces	Prepared Hoisin sauce
2½ ounces	Wine vinegar
1½ ounces	Mirin
1¼ teaspoons	Red pepper flakes, soaked in warm water for 10 to 15 minutes
3 ounces	Granulated sugar
	Red food coloring, as needed

METHOD OF PREPARATION:

1. Combine all of the marinade ingredients, and pour over the ribs in a hotel pan. Marinate in the refrigerator for a minimum of 5 hours or overnight, turning the ribs several times while marinating.
2. Preheat the oven to 375° F.
3. Place the ribs on a wire rack inserted into a roasting pan. Roast for 45 minutes; then raise the temperature to 450° F, and roast for another 15 minutes, or until golden brown. Brush the ribs with the plum sauce, and remove from the oven. Hold at 140° F or above.
4. To serve, arrange three to four ribs on a preheated dinner plate, and garnish with the scallion fans.

Dijon Pork Cutlet

YIELD: 50 SERVINGS. SERVING SIZE: TWO- TO THREE-OUNCE CUTLETS.

COOKING TECHNIQUES:
Sauté, Bake

Sauté:
1. Heat the sauté pan to the appropriate temperature.
2. Evenly brown the food product.
3. For a sauce, pour off any excess oil, reheat, and deglaze.

Bake:
1. Preheat the oven.
2. Place the food product on the appropriate rack.

GLOSSARY:
Dredge: to coat with flour
Shingle: to overlap

HACCP:
Hold at 165° F.

HAZARDOUS FOOD:
Pork cutlets

NUTRITION:
Calories: 418
Fat: 20 g
Protein: 38.8 g

INGREDIENTS:

100 each	Pork cutlets, 3 ounces each, tenderized
	Seasoned flour, as needed
1 pint	Salad oil
8 ounces	Dijon mustard
3 quarts	Demi-glace, heated to a boil

METHOD OF PREPARATION:

1. Preheat the oven to 300° F.
2. **Dredge** the cutlets in the seasoned flour, and shake off any excess.
3. Sauté both sides of the cutlets in a tilting skillet.
4. **Shingle** the sautéed cutlets into a 2-inch hotel pan.
5. Incorporate the mustard into the demi-glace. Heat to a boil. Adjust the seasonings.
6. Pour the sauce over the cutlets two-thirds high.
7. Bake in the oven for 10 minutes.
8. Hold at 165° F.

Hungarian Stuffed Cabbage Rolls

YIELD: 10 SERVINGS. SERVING SIZE: FIVE OUNCES.

INGREDIENTS:

1 ounce	Rendered fat from bacon
4 cloves	Garlic, peeled and finely minced
12 ounces	Onions, peeled and diced **brunoise**
8 ounces	Rice
3 each	Ham hocks, smoked and boiled in 2 quarts of water
3 tablespoons	Hungarian paprika
	Salt and freshly ground black pepper, to taste
1 pound	Ground pork
2 each	Eggs
1 head	Green cabbage, washed, cored, **blanched** or steamed
2 pounds	Canned sauerkraut, rinsed under cold water and drained
8 ounces	Smoked pork butt, trimmed of fat and cut **julienne**

METHOD OF PREPARATION:

1. Preheat the oven to 350° F.
2. In a sauté pan, heat the fat, add the garlic and onions, and sauté until tender. Add the rice, half of ham hock, stock, half of the paprika, and season, to taste. Cover and simmer for 10 minutes; then remove and set aside to cool.
3. In a bowl, combine the ground pork, cooled rice mixture, and eggs, and season to taste. Mix well.
4. Separate 20 cabbage leaves, cut off thick vein, fill them with the stuffing, and roll them up.
5. Line the bottom of a 2-inch-deep hotel pan with half of the drained sauerkraut, and arrange the cabbage rolls on top. Top the remaining sauerkraut with the smoked pork, and add the remainder of the ham hock stock. Cover tightly, and braise in the oven for 1 hour, or until cooked throughout. Hold at 140° F.
6. Add the remaining paprika to the boiling demi-glace, and hold at 140° F.
7. To serve, arrange two cabbage rolls on a preheated dinner plate, and top with the sauerkraut.

COOKING TECHNIQUES:

Sauté, Braise

Sauté:

1. Heat the sauté pan to the appropriate temperature.
2. Evenly brown the food product.
3. For a sauce, pour off any excess oil, reheat, and deglaze.

Braise:

1. Heat the braising pan to the proper temperature.
2. Sear and brown the food product to a golden color.
3. Degrease and deglaze.
4. Cook the food product in two thirds liquid until fork-tender.

GLOSSARY:

Brunoise: ⅛-inch dice
Blanch: to parboil
Julienne: matchstick strips

HACCP:

Hold at 140° F.

HAZARDOUS FOODS:

Pork
Eggs

NUTRITION:

Calories: 428
Fat: 17.6 g
Protein: 27.4 g

CHEF NOTE:

Stuffed cabbage is one of the national foods of Hungary. It is always prepared very fatty and served with boiled potatoes.

COOKING TECHNIQUE:

Braise

Braise:

1. Heat the braising pan to the proper temperature.
2. Sear and brown the food product to a golden color.
3. Degrease and deglaze.
4. Cook the food product in two thirds liquid until fork-tender.

GLOSSARY:

Deglaze: to add liquid to hot pan
Fork-tender: without resistance

HAZARDOUS FOOD:

Pork chops

NUTRITION:

Calories: 345
Fat: 17.6 g
Protein: 40.9 g

Pork Chops Braised in White Wine

YIELD: 50 SERVINGS. SERVING SIZE: FIVE OUNCES.

INGREDIENTS:

1½ tablespoons	Sage leaves, dried, crumbled
1½ tablespoons	Rosemary leaves, dried, crumbled
2 tablespoons	Garlic, peeled and minced
3 tablespoons	Salt
1½ tablespoons	Freshly ground black pepper
4 ounces	Clarified butter
4 ounces	Oil
50 (5-ounce)	Pork chops, fat trimmed
1 quart	Dry white wine
1 quart	White chicken stock, hot (see page 12)
2 ounces	Freshly parsley, washed and finely chopped

METHOD OF PREPARATION:

1. Combine the sage, rosemary, garlic, salt, and black pepper. Press into both sides of the pork chops.
2. Heat a tilting skillet, and add the butter and oil. Place the chops in the hot fat, and brown on each side. Remove and place overlapping in hotel pans.
3. Pour off the excess fat from the skillet, and **deglaze** with the wine. Heat to a boil, and reduce by half. Add the chicken stock and season. Pour stock on chops to cover two thirds, and braise in oven until **fork-tender.**
4. Remove the chops from the pan, and keep hot. Heat the sauce to a boil, skimming as needed.
5. Adjust the seasonings, and strain the sauce; stir in the parsley, and pour the sauce over the pork chops just before service.

Pork Chops, Normande Style

(Côtes de Porc à la Normande)

YIELD: 10 SERVINGS. SERVING SIZE: FIVE OUNCES.

INGREDIENTS:

10 (5-ounce)	Pork chops, trimmed of fat
	Salt and ground white pepper, to taste
5 each	Red apples, washed, cored, and sliced
5 each	Yellow apples, washed, cored, and sliced
4 ounces	Apple cider
2 ounces	Heavy cream
3 ounces	Butter
3 ounces	Oil

METHOD OF PREPARATION:

1. Preheat the oven to 325° F.
2. In a sauté pan, heat the oil and butter. Season the pork chops with salt and pepper, and **sear** until well browned, about 5 to 6 minutes. Remove the chops, and reserve the fat.
3. Coat the bottom of the baking dish or roasting pan with some of the reserved fat. Place the red apple slices on the bottom, and arrange the pork chops on top. Fan with half of the yellow apple slices, and add the apple cider.
4. Cover and bake in the oven for 20 to 25 minutes, or until **fork-tender.**
5. Transfer the chops to a hotel pan, and with a skimmer, transfer the apples. Hold covered at 140° F or above.
6. Heat the liquid in the roasting pan to a boil. **Temper** the cream, and whisk the pan juices; simmer until the desired consistency is achieved. Strain through a **chinois** into a suitable container, and hold at 140° F or above.
7. To serve, arrange the chops on a preheated dinner plate. **Nappé** with sauce, and garnish with the apple slices.

COOKING TECHNIQUES:
Sauté, Bake

Sauté:
1. Heat the sauté pan to the appropriate temperature.
2. Evenly brown the food product.
3. For a sauce, pour off any excess oil, reheat, and deglaze.

Bake:
1. Preheat the oven.
2. Place the food product on the appropriate rack.

GLOSSARY:
Sear: to brown quickly
Fork-tender: without resistance
Temper: to equalize two extreme temperatures
Chinois: cone-shaped strainer
Nappé: coated

HAZARDOUS FOOD:
Pork chops

NUTRITION:
Calories: 321
Fat: 20.8 g
Protein: 16 g

Pork Chops Provençale

YIELD: 50 SERVINGS. SERVING SIZE: THREE OUNCES OF MEAT.

INGREDIENTS:

16 pounds	Pork chops, washed and dried
4 tablespoons	Olive oil
12 cloves	Garlic, peeled and minced
5 each	Onions, large, peeled, and cut **julienne**
2 pounds	Black olives, sliced
4 pounds	Mushrooms, cleaned, cut into quarters
4 pounds	Tomatoes, washed and diced
6 ounces	Anchovies, diced (optional)
3 each	Eggplant, washed, peeled, and diced
4 teaspoons	Ground black pepper
	Salt, as needed

METHOD OF PREPARATION:

1. Preheat the oven to 350° F.
2. Mark the pork chops on the broiler or grill, and place in a hotel pan.
3. Sauté the onions. Add the garlic, eggplant, mushrooms, tomatoes, anchovies, and olives and season.
4. Top the pork chops with the sautéed vegetables.
5. Bake in the oven until fork-tender.
6. Season, to taste, and serve on preheated dinner plates.

COOKING TECHNIQUES:

Sauté, Simmer, Bake

Sauté:

1. Heat the sauté pan to the appropriate temperature.
2. Evenly brown the food product.
3. For a sauce, pour off any excess oil, reheat, and deglaze.

Simmer and Poach:

1. Heat the cooking liquid to the proper temperature.
2. Submerge the food product completely.
3. Keep the cooked product moist and warm.

Bake:

1. Preheat the oven.
2. Place the food product on the appropriate rack.

GLOSSARY:

Julienne: matchstick strips

HAZARDOUS FOOD:

Pork chops

NUTRITION:

Calories: 252
Fat: 11.3 g
Protein: 29.7 g

CHEF NOTE:

Anchovies and olives are salted.

Pork Chow Mein

YIELD: 10 SERVINGS. SERVING SIZE: SIX OUNCES.

COOKING TECHNIQUE:

Sauté

Sauté:

1. Heat the sauté pan to the appropriate temperature.
2. Evenly brown the food product.
3. For a sauce, pour off any excess oil, reheat, and deglaze.

GLOSSARY:

Julienne: matchstick strips

HACCP:

Hold at 140° F or above.

HAZARDOUS FOOD:

Pork butt

NUTRITION:

Calories: 599
Fat: 37.7 g
Protein: 36.2 g

INGREDIENTS:

6 ounces	Oil
4 pounds	Boneless, skinless pork butt, fat and connective tissues removed and sliced into thin strips
12 ounces	Onions, peeled and cut **julienne**
8 ounces	Celery, washed and cut on a bias
8 ounces	Mushrooms, cleaned and sliced
1 pound	Canned bamboo shoots, drained
1 pound	Canned water chestnuts, drained
1 pint	Chicken stock, heated to a boil (see page 12)
2 ounces	Soy sauce
	Salt and ground white pepper, to taste
1 tablespoon	Cornstarch, dissolved in 4 tablespoons of cold water
1 pound	Canned bean sprouts, drained and rinsed in cold water
1 pound	Canned fried noodles, spread on a sheet pan and heated before serving

METHOD OF PREPARATION:

1. In a wok or large sauté pan, heat 2 ounces of the oil. Add the pork, and sauté until lightly browned; then remove, and reserve the meat.
2. Clean the wok or sauté pan; then add the remaining 4 ounces of oil, and heat.
3. Add the onions, celery, and mushrooms, and sauté until the onions are translucent. Add the bamboo shoots and water chestnuts, and continue to sauté for 2 more minutes.
4. Add the chicken stock, and heat to a boil; then reduce the heat, and simmer for 10 more minutes. Add the soy sauce, and season, to taste.
5. Thicken the stock with the cornstarch mixture; then add the bean sprouts, and return the pork to the pan.
6. Heat the mixture back to a boil; then reduce the heat, and simmer for 2 minutes. Adjust the seasonings and consistency, as needed.
7. Serve immediately on preheated dinner plate over noodles, or hold at 140° F or above.

Pork Egg Rolls

YIELD: 50 SERVINGS. SERVING SIZE: 2 ROLLS.

COOKING TECHNIQUES:
Stir-Fry, Deep-Fry

Stir-Fry:
1. Heat the oil in a wok until hot but not smoking.
2. Keep the food in constant motion; use the entire cooking surface.

Deep-Fry:
1. Heat the frying liquid to the proper temperature.
2. Submerge the food product completely.
3. Fry the product until it is cooked throughout.

GLOSSARY:
Chiffonade: ribbons of leafy greens
Brunoise: ⅛-inch dice
Saké: rice wine

HAZARDOUS FOOD:
Ground pork

NUTRITION:
Calories: 282
Fat: 17.4 g
Protein: 12 g

CHEF NOTES:
1. This recipe also can be served with sweet and sour sauce.
2. Individually quick-frozen egg rolls, with and without meat, may be purchased for quantity food production.

INGREDIENTS:

5 ounces	Soybean oil
5 pounds	Ground pork
1¼ pounds	Mushrooms cleaned and thinly sliced
5 pounds	Cabbage washed, cored, outer leaves removed, and cut **chiffonade**
3 ounces	Celery, washed, trimmed and diced **brunoise**
1¼ quarts	Soy sauce
10 ounces	**Saké**
5 ounces	Sugar
5 ounces	Bean sprouts, washed
10 packages	Egg roll wrappers
10 each	Eggs, lightly beaten
15 ounces	Mustard sauce (Chinese mustard and cold water, combined to a thick consistency)
	Frying oil, as needed

METHOD OF PREPARATION:

1. Add the soybean oil to a wok over high heat. Add the ground pork, and stir-fry for 2 minutes.
2. Add the mushrooms, cabbage, and celery with the soy sauce, saké, and sugar. Stir-fry for another minute. Drain the excess liquid. Transfer the contents to a bowl.
3. Add the bean sprouts to the mixture, and cool in the refrigerator.
4. To assemble each egg roll, shape approximately 2 ounces of filling into a 4-inch by 1-inch cylinder. Place the filling diagonally across the center of a wrapper. Lift the lower triangular flap over the filling. Tuck the point under on the far side, leaving the upper point of the wrapper exposed. Bring each of the two small end flaps, one at a time, up to the top of the enclosed filling, and press the points down firmly. Brush the upper and exposed triangle of the dough with some of the lightly beaten egg. Roll the wrapper into a neat package—the beaten egg will seal the edges and keep the wrapper intact.
5. Place the filled egg rolls on a plate, and cover with a dry kitchen towel.
6. Bring the wok temperature to 350° F, and deep-fry the egg rolls for 3 to 4 minutes. Place them on absorbent paper.
7. Serve with mustard sauce on the side.

Pork Rib Stew with Green Chili Sauce and Pico de Gallo

YIELD: 10 SERVINGS. SERVING SIZE: 8 OUNCES (BONE-IN).

COOKING TECHNIQUES:

Stew, Sauté

Stew:

1. Sear, sauté, sweat, or blanch the main food product.
2. Deglaze the pan, if desired.
3. Cover the food product with simmering liquid.
4. Remove the bouquet garni.

Sauté:

1. Heat the sauté pan to the appropriate temperature.
2. Evenly brown the food product.
3. For a sauce, pour off any excess oil, reheat, and deglaze.

GLOSSARY:

Fork-tender: without resistance
Al dente: to the bite
Chinois: cone-shaped strainer
Nappé: coated

HACCP:

Hold at 140° F.

HAZARDOUS FOODS:

Pork ribs
Farmer cheese

NUTRITION:

Calories: 497
Fat: 35.1 g
Protein: 29.9 g

INGREDIENTS:

5 pounds (10 count)	Double-cut pork ribs
2 quarts	Water
1 pound	Onions, peeled and cut into 20 wedges
8 cloves	Garlic, peeled and mashed into a purée
2 teaspoons	Salt, or to taste
1 head	Green cabbage, washed, cored, and cut into thin wedges
3 each	Poblano peppers, washed
1 tablespoon	Vegetable oil
3 each	Corn tortillas
	Freshly ground black pepper, to taste
4 ounces	Fresh cilantro leaves, chopped
4 ounces	Farmer cheese, crumbled
	Pico de gallo (see page 671)

METHOD OF PREPARATION:

1. In a large stewing pot, combine pork, water, onions, garlic, and salt. Bring the liquid to a boil; then lower heat to a simmer, cover, and cook for 1 hour.
2. Add the cabbage, re-cover, and continue to simmer another 30 minutes, or until the ribs are **fork-tender** and the cabbage is **al dente.**
3. Meanwhile, brush the peppers with oil, and grill or char over an open flame for about 10 minutes, turning frequently. Allow the peppers to cool enough to handle; then peel and remove the stems and seeds.
4. In a medium sauté pan, dry-roast the tortillas on both sides. Cool them enough to handle; then tear them into small pieces, and set aside.
5. Remove the ribs and cabbage from the pot, and place in a hotel pan. Hold loosely covered at 140° F or higher.
6. Place the stewing pan back over high heat, and reduce liquid to half, or to about 1½ pints. Strain off 8 ounces of the liquid, and reserve.

ENTRÉES

7. In a food processor, purée the remaining mixture along with the poblano peppers and tortilla pieces, adding the reserved liquid as needed to achieve the proper consistency. Season, to taste, with additional salt and pepper.
8. Strain the sauce through a **chinois** into a clean saucepan. Place over medium heat, return to a boil, and then pour the sauce over the ribs and cabbage; continue to hold at 140° F or higher, or serve immediately.
9. For service, arrange one rib with two wedges of cabbage on a preheated dinner plate. **Nappé** with some of the sauce, and garnish with cilantro, cheese, and pico de gallo on the side.

Roast Loin of Pork

YIELD: 50 SERVINGS. SERVING SIZE: SIX OUNCES.

INGREDIENTS:

37 pounds	Pork loin, deboned and trimmed
1¼ heads	Garlic, peeled
	Salt and freshly ground white pepper, to taste
1 teaspoon	Ginger, ground
5 ounces	All-purpose flour
2 quarts	Brown veal stock, heated to a boil (see page 7)
3 quarts	Veal demi-glace, heated to a boil (see page 26)

Mirepoix:

2½ pounds	Onions, peeled
1¼ pounds	Carrots, washed and peeled
6 stalks	Celery, washed and trimmed

METHOD OF PREPARATION:

1. Debone and **truss** the pork loin. Chop the bones into small pieces, and place them in a roasting pan.
2. Place the mirepoix and garlic in the roasting pan.
3. Mix together the seasonings, and rub them on the meat. Place the meat in the roasting pan on top of the bones and mirepoix. Roast in a 375° F oven for 20 minutes.
4. Reduce the heat to 300° F. Continue to roast until the meat reaches an internal temperature of 155° F, approximately 1 hour. Hold at 140° F.
5. Remove the meat from the pan. Place the pan on the stove, and brown the mirepoix and bones. Remove the excess fat, leaving about 2 ounces of fat in the pan.
6. Add flour to the mirepoix to form a roux, and cook on low heat.
7. Add the hot stock and hot demi-glace to the mirepoix. Heat to a boil. Simmer the gravy, scraping the pan to loosen any particles. **Dépouiller,** and adjust the seasonings. Strain through a chinois, and hold for service.
8. Slice the roasted loin, two slices per serving. Place on warm dinner plates and serve nappé with the sauce.

COOKING TECHNIQUE:
Roast

Roast:
1. Sear the food product, and brown evenly.
2. Elevate the food product in a roasting pan.
3. Determine doneness, and consider carryover cooking.
4. Let the food product rest before carving.

GLOSSARY:
Mirepoix: roughly chopped vegetables
Truss: to tie or secure
Dépouiller: to skim impurities/grease

HACCP:
Cook to internal temperature of 155° F.
Hold at 140° F.

HAZARDOUS FOOD:
Pork loin

NUTRITION:
Calories: 529
Fat: 22.3 g
Protein: 66.7 g

Roast Loin of Pork with Glazed Apple Dressing

YIELD: 50 SERVINGS. SERVING SIZE: FIVE OUNCES.

COOKING TECHNIQUE:
Roast

Roast:
1. Sear the food product, and brown evenly.
2. Elevate the food product in a roasting pan.
3. Determine doneness, and consider carryover cooking.
4. Let the food product rest before carving.

GLOSSARY:
Truss: to tie or secure
Brunoise: ⅛-inch dice

HACCP:
Refrigerate at 40° F or below.
Hold at 140° F.

HAZARDOUS FOOD:
Roast loin of pork

NUTRITION:
Calories: 625
Fat: 24.9 g
Protein: 45.9 g

INGREDIENTS:

15 pounds	Porkloin, boned, trimmed
5 ounces	Butter
20 each	Apples, washed, cored, peeled, quartered, and stored in acidulated water
15 ounces	Light brown sugar
1 teaspoon	Nutmeg
2 quarts	Pork or white chicken stock, heated to a boil (see page 12)
12 ounces	Butter
	Salt and ground white pepper, to taste
1½ tablespoons	Ginger, ground
8 cloves	Garlic, peeled and slivered

Dressing:

5 pounds	Bread crumbs
2 pounds	Onions, peeled and diced **brunoise**
8 stalks	Celery, washed, trimmed, and diced brunoise
4 ounces	Celery leaves, chopped
	Salt and ground black pepper, to taste
1½ tablespoons	Fresh sage leaves, washed and chopped
1½ tablespoons	Fresh thyme leaves, washed and chopped

METHOD OF PREPARATION:

1. Preheat the oven to 350° F.
2. To glaze the apples, heat the butter in a saucepan until the butter is foamy. Drain and add the apples first, then the sugar and nutmeg. Cook until the apples are tender and well glazed. Set aside to add to the dressing.
3. Combine the dressing ingredients in a large bowl, and toss with a wooden spoon. In a small saucepan, heat the stock, and then add the butter. Heat until the butter melts, and pour over the bread-crumb mixture. Mix until the stock is absorbed. Gently fold in the apples. Season, refrigerate, and store at 40° F or below.
4. Shape and **truss** the pork roast. Rub the outside of roast with the salt, pepper, ginger, and garlic. Place the roast in a roasting pan. Roast in the oven until an internal temperature of 155° F is achieved. Hold at 140° F.
5. Bake the stuffing until firm in a buttered roasting pan covered with parchment paper. Let it rest for 30 minutes. Portion the stuffing with the pork loin in a hotel pan.

Smoked Pork Tenderloin with Brown Meunière Sauce

YIELD: 10 SERVINGS. SERVING SIZE: SIX OUNCES.

INGREDIENTS:

4 pounds	Pork tenderloins, well trimmed
20 ounces	Brown meunière sauce, held at 140° F (see page 19)

Marinade:

6 ounces	Red wine
8 ounces	Liquid smoke
4 cloves	Garlic, peeled and finely minced
1 teaspoon	Thyme, dried, ground
2 teaspoons	Basil, dried, ground
	Salt and freshly ground black pepper, to taste

METHOD OF PREPARATION:

1. Combine the ingredients for the marinade, and mix well.
2. Place the tenderloins in the marinade, refrigerate, and allow to stand for 30 minutes. Turn the pork in the marinade two or three times during this period.
3. Preheat the oven to 400° F.
4. Transfer the tenderloins to a roasting pan, and add half of the marinade. Roast in the oven for approximately 20 minutes, or to an internal temperature of 155° F.
5. Remove from the oven, and hold at 140° F for 15 minutes before slicing.
6. For service, slice on a diagonal; overlap the slices on a preheated dinner plate, and **nappé** with sauce.

COOKING TECHNIQUE:

Roast

Roast:

1. Sear the food product, and brown evenly.
2. Elevate the food product in a roasting pan.
3. Determine doneness, and consider carryover cooking.
4. Let the food product rest before carving.

GLOSSARY:

Nappé: coated

HACCP:

Hold at 140° F.
Roast to an internal temperature of 155° F.

HAZARDOUS FOOD:

Pork

NUTRITION:

Calories: 649
Fat: 41.8 g
Protein: 51.7 g

CHEF NOTES:

1. If a smoker is available, liquid smoke can be eliminated, and tenderloins can be slow-cooked over wood chips.
2. As an alternative, loin of lamb can be used.

COOKING TECHNIQUE:

Braise

Braise:

1. Heat the braising pan to the proper temperature.
2. Sear and brown the food product to a golden color.
3. Degrease and deglaze.
4. Cook the food product in two thirds liquid until fork-tender.

GLOSSARY:

Julienne: matchstick strips
Sear: brown quickly
Fork-tender: without resistance

HACCP:

Hold at 140° F.

HAZARDOUS FOOD:

Pork tenderloin

NUTRITION:

Calories: 563
Fat: 21.5 g
Protein: 39.4 g

Spiced Pork Tenderloin

YIELD: 50 SERVINGS. SERVING SIZE: FOUR OUNCES.

INGREDIENTS:

1 pint	Vegetable oil
15 pounds	Pork tenderloins, trimmed and peeled
2 pounds	Medium onions, peeled and cut **julienne**
4 pounds	Cooking apples, washed, peeled, cored, and sliced
1 gallon	Apple juice
1½ ounces	Cinnamon
1 ounce	Nutmeg
3 pints	Honey
1 pint	Sherry
	Cornstarch, as needed
	Cold water or any flavoring liquid, as needed

METHOD OF PREPARATION:

1. Heat a braising pan, and add the vegetable oil. **Sear** the pork tenderloins until brown on all sides. Place in hotel pans.
2. Add the onion, and sauté in the same oil as the tenderloins. Add the apple slices, apple juice, cinnamon, nutmeg, honey, and sherry. Heat to a boil. Simmer 5 to 8 minutes, until the apples are tender and seasoned.
3. Divide the apple mixture on the tenderloins. Cover the hotel pans with foil, and place in a 350° F oven. Braise until the meat is **fork-tender.**
4. Remove the pork from the liquid, and hold at 140° F. Heat the cooking liquid to a boil, and thicken with the cornstarch-water mixture. Adjust the consistency and seasonings, as needed. Hold at 140° F.
5. Serve the pork, sliced on a bias, with the apple and spice sauce.

COOKING TECHNIQUE:

Bake

Bake:

1. Preheat the oven.
2. Place the food product on the appropriate rack.

GLOSSARY:

Sear: to brown quickly

HACCP:

Hold at 140° F.

HAZARDOUS FOOD:

Pork tenderloin

NUTRITION:

Calories: 472
Fat: 20.2 g
Protein: 42 g

Thai-Style Pork Tenderloin

YIELD: 50 SERVINGS. SERVING SIZE: FIVE OUNCES.

INGREDIENTS:

16 pounds	Pork tenderloin, trimmed and silverskin removed
1 pint	Oil
12 ounces	Honey
1 tablespoon	Coriander
2 quarts	Oriental peanut sauce (see page 39)

Marinade:

1 quart	Soy sauce
1 pint	Honey
8 ounces	Brown sugar
3 tablespoons	Coriander, ground
1 head	Garlic, peeled and minced
1 tablespoon	Cumin
½ teaspoon	Cayenne pepper
1 tablespoon	Ground black pepper
6 ounces	Frangelico (hazelnut liqueur)

METHOD OF PREPARATION:

1. Preheat the oven to 375° F.
2. Combine all of the ingredients for the marinade. Place the tenderloins in the marinade; cover and chill for 1 hour.
3. Remove the pork from the marinade, and pat dry with a towel.
4. Strain the marinade into a saucepan, and reduce by boiling to half. To the reduction, add 12 ounces of honey and 1 tablespoon of coriander. Cover and keep warm.
5. In a sauté pan, heat the oil, and **sear** the tenderloin on all sides; remove. Place the meat on a sheet pan, baste with the marinade, and place in the oven. Cook until the meat has reached an internal temperature of 165° F. Hold at 140° F.
6. Slice the pork tenderloins at an angle and place four to five slices on top of 1 ounce of peanut sauce on a preheated dinner plate. Drizzle the marinade reduction over the meat.

Baked Chicken in Foil

(Pollo en Camisa)

YIELD: 10 SERVINGS. SERVING SIZE: TWO PIECES.

INGREDIENTS:

4 ounces	Olive oil
	Salt, to taste
20 each	Chicken thighs and legs
2 each	Onions, peeled and thinly sliced
10 cloves	Garlic, peeled and mashed into a purée
5 each	Green chili peppers, washed, cored, and thinly sliced
10 ounces	Tomato **concassé**
6 ounces	Cilantro leaves, washed and coarsely chopped

METHOD OF PREPARATION:

1. Preheat the oven to 375° F.
2. Combine the oil and salt and dip the chicken into the mixture.
3. Place one thigh and one leg of oiled chicken on a 10-inch square of aluminum foil.
4. Divide all of the other ingredients over the chicken pieces, sprinkle with salt, and add the remaining oil.
5. Close the aluminum foil to avoid leaking and/or steam escaping. Place on a baking pan, and bake approximately 45 minutes, or until chicken is **fork-tender.**
6. Serve immediately on a preheated plate, or hold at 140° F or above.

COOKING TECHNIQUE:

Bake

Bake:

1. Preheat the oven.
2. Place the food product on the appropriate rack.

GLOSSARY:

Concassé: peeled, seeded, and roughly chopped
Fork-tender: without resistance

HACCP:

Hold at 140° F or above.

HAZARDOUS FOOD:

Chicken

NUTRITION:

Calories: 433
Fat: 30.4 g
Protein: 32.8 g

CHEF NOTE:

This dish is a specialty of Mexico. It is served in aluminum foil with white rice and fried plantains.

Baked Stuffed Chicken Leg with Sauce Suprême

YIELD: 50 SERVINGS. SERVING SIZE: SIX OUNCES.

COOKING TECHNIQUES:

Sauté, Bake

Sauté:

1. Heat the sauté pan to the appropriate temperature.
2. Evenly brown the food product.
3. For a sauce, pour off any excess oil, reheat, and deglaze.

Bake:

1. Preheat the oven.
2. Place the food product on the appropriate rack.

GLOSSARY:

Brunoise: ⅛-inch dice
Nappé: coated

HACCP:

Hold at 140° F.

HAZARDOUS FOODS:

Chicken legs
Ground pork
Ground veal
Milk
Egg yolks

NUTRITION:

Calories: 346
Fat: 21.1 g
Protein: 20.6 g

INGREDIENTS:

50 each	Chicken legs
1½ pounds	Ground pork
1 pound	Ground veal shoulder
1 pound	Bread, fresh, cut into ½-inch pieces
1½ pints	Milk
12 ounces	Onions, peeled and diced **brunoise**
3 stalks	Celery, washed, trimmed, and diced brunoise
4 each	Egg yolks
8 ounces	Dry bread crumbs
6 ounces	Butter, clarified
3 ounces	Freshly chopped parsley, excess moisture removed
½ ounce	Vegetable oil
	Salt, ground white pepper, and sage, to taste
3 quarts	Sauce suprême (see page 68)

METHOD OF PREPARATION:

1. Preheat the oven to 350° F.
2. Wash the chicken legs in cold running water, dry well, coat with oil, line up on a sheet pan, and refrigerate.
3. Sauté the onions and celery in the clarified butter until the onions are translucent. Cool under refrigeration to an internal temperature of 40° F.
4. In a mixing bowl, place the pork, veal, bread softened in milk, onions, celery, and sage. Add the egg yolks, chopped parsley, and bread crumbs. Mix throroughly until well blended. Season, to taste, with salt and ground white pepper.
5. Place the stuffing in a hotel pan, and bake in the oven until the internal temperature reaches 165° F. Hold at 140° F.
6. Season the chicken legs, and place in the oven for 1 hour, or until the chicken reaches an internal temperature of 165° F. Hold at 140° F.
7. Serve the chicken legs on scooped out stuffing, and **nappé** with sauce suprême.

COOKING TECHNIQUES:

Broil/Grill

Grill/Broil:

1. Clean and heat the grill/broiler.
2. To prevent sticking, brush the food product with oil.

HAZARDOUS FOODS:

Chicken breast
Coconut milk

NUTRITION:

Calories: 212
Fat: 14.6 g
Protein: 16.7 g

Broiled Chicken with Peanut Sauce

(Sate Ajam)

YIELD: 10 SERVINGS. SERVING SIZE: ONE-HALF CHICKEN BREAST.

INGREDIENTS:

5 each	Chicken breasts, whole, boneless
10 each	Bamboo skewers
	Oriental peanut sauce (see page 39), heated to a boil, as needed
	Salt, to taste
	Fresh cilantro sprigs (optional)

Marinade:

2 ounces	White vinegar
2 cloves	Garlic, peeled and finely minced
½ ounce	Soy sauce
1 ounce	Mirin
2 ounces	Sesame seed oil

METHOD OF PREPARATION:

1. Combine the ingredients of the marinade, and marinate the chicken for 1 hour.
2. Split the chicken breasts in half, and thread on skewers.
3. Broil or grill the skewered chicken breasts, and baste with the peanut sauce.
4. To serve, place one-half chicken breast on a preheated dinner plate, surround with peanut sauce, and garnish with fresh cilantro.

CHEF NOTE:

This is an Indonesian dish.

Buffalo Wings

YIELD: 50 SERVINGS. SERVING SIZE: TWO WINGS.

COOKING TECHNIQUE:

Deep-fry

Deep-Fry:

1. Heat the frying liquid to the proper temperature.
2. Submerge the food product completely.
3. Fry the product until it is cooked throughout.

GLOSSARY:

Bâton: (¼ × ¼ × 3 inches)

HAZARDOUS FOOD:

Chicken wings

NUTRITION:

Calories: 472
Fat: 39.4 g
Protein: 26.1 g

INGREDIENTS:

12 pounds	Chicken wings, second and third joints, cut
1½ quarts	Hot sauce
1 pound	Melted butter
2 pounds	Celery, washed, trimmed, and cut **bâton**
6 ounces	Kale
1½ quarts	Bleu cheese dressing

METHOD OF PREPARATION:

1. Wash and dry the wings.
2. Fry the wings in a deep-fat fryer at 350° F. When thoroughly cooked, remove and drain.
3. In a bowl, coat the wings in hot sauce and melted butter.
4. Place the coated wings on a plate lined with kale. Serve bleu cheese dressing and celery sticks on the side.

Chicken à la Kiev

YIELD: 50 SERVINGS. SERVING SIZE: EIGHT OUNCES.

COOKING TECHNIQUES:
Deep-fry, Bake

Deep-Fry:
1. Heat the frying liquid to the proper temperature.
2. Submerge the food product completely.
3. Fry the product until it is cooked throughout.

Bake:
1. Preheat the oven.
2. Place the food product on the appropriate rack.

HACCP:
Hold at 140° F.

HAZARDOUS FOODS:
Chicken breasts
Eggs
Milk

NUTRITION:
Calories: 492
Fat: 34.4 g
Protein: 30.6 g

INGREDIENTS:

50 each	Chicken breasts, 4-ounce whole, boneless
4 pounds	Butter
1 head	Garlic, peeled and minced
3 ounces	Fresh chives, washed and sliced
	Salt and ground white pepper, to taste
	Bread flour, as needed
	Bread crumbs, as needed
8 each	Eggs
2 pints	Milk
8 each	Lemons, sliced

METHOD OF PREPARATION:

1. Place the butter in a robo-coupe, and mix at low speed until the butter turns smooth (but not too soft).
2. Add the minced garlic and chives. Season with the salt and pepper, and chill slightly.
3. Flatten the boneless chicken breasts, skin-side down, using a meat mallet.
4. In the center of each breast, place 1 to 2 teaspoons of butter; roll up and fold in the ends. Place in the freezer to firm the shape.
5. Dredge each breast through the flour, egg wash, and bread crumbs. Pat off excess crumbs.
6. Preheat a deep-fat fryer to 350° F. Brown the chicken lightly. Remove, drain on absorbent paper, and transfer to a sheet pan.
7. Preheat the oven to 350° F, and finish baking the chicken in the oven for 15 to 20 minutes.
8. Using a paring knife, cut a 1-inch slit in the top of each breast. Hold at 140° F.
9. Garnish each breast with a lemon slice, and serve.

Chicken à la King

YIELD: 50 SERVINGS. SERVING SIZE: SIX OUNCES.

INGREDIENTS:

50 (6-ounce)	Chicken breasts, skinned and boned
	White chicken stock, as needed (see page 12), heated to a boil
10 ounces	Butter, clarified
1¼ pounds	Green bell peppers, washed, cored, and diced
1¼ pounds	Mushrooms, cleaned and sliced
2½ gallons	Chicken velouté (see page 69), heated to a boil
	Salt and ground white pepper, as needed
1¼ pints	Heavy cream
15 each	Egg yolks
	Sherry, as needed

METHOD OF PREPARATION:

1. Poach the chicken breasts to fork-tender. Remove, cool, and dice into 1-inch pieces. Reserve the chicken stock for the sauce.
2. Sauté the peppers and mushrooms in the clarified butter, and reserve with the chicken.
3. Heat the chicken velouté to a boil. Reduce to a simmer. Let simmer for 20 minutes, until the proper flavor and consistency are achieved.
4. Strain through a **chinois** into a suitable container. Adjust the seasonings, and mix with the chicken, peppers, and mushrooms.
5. **Temper** the **liaison** of the heavy cream and egg yolks and add to the velouté. Add the sherry. Hold at 140° F. Serve with rice.

COOKING TECHNIQUES:

Sauté, Boil, Simmer

Sauté:

1. Heat the sauté pan to the appropriate temperature.
2. Evenly brown the food product.
3. For a sauce, pour off any excess oil, reheat, and deglaze.

Boil: (at sea level)

1. Bring the cooking liquid to a rapid boil.
2. Stir the contents, and cook the food product throughout.
3. Serve hot.

Simmer and Poach:

1. Heat the cooking liquid to the proper temperature.
2. Submerge the food product completely.
3. Keep the cooked product moist and warm.

GLOSSARY:

Chinois: fine cone-shaped strainer
Temper: to equalize two extreme temperatures
Liaison: binding agent

HACCP:

Hold at 140° F.

HAZARDOUS FOODS:

Chicken breasts
Heavy cream
Egg yolks

NUTRITION:

Calories: 413
Fat: 28 g
Protein: 30 g

CHEF NOTE:

Chicken breasts can be replaced by pulled, frozen chicken meat.

Chicken Bayou, Creole Style

YIELD: 50 SERVINGS. SERVING SIZE: EIGHT OUNCES.

INGREDIENTS:

25 each	Chicken breast, boneless, split in two
Stuffing:	
50 ounces	Shrimp, medium, peeled, deveined, and chopped
5 each	Onions, peeled and diced **brunoise**
10 stalks	Celery, washed, trimmed, and diced brunoise
5 each	Red bell pepper, seeded and diced brunoise
10 teaspoons	Garlic, peeled, washed, and diced brunoise
2 bunches	Scallions, washed and diced brunoise
10 tablespoons	Pernod
1¼ quarts	Béchamel sauce (see page 16), heated to a boil
	Salt and ground black pepper, to taste
1¼ quarts	Seasoned flour
	Oil, as needed
1 gallon	Creole sauce (see page 23)
	Scallion flowers and strawberry fans, for garnish, as needed

METHOD OF PREPARATION:

1. Pound the chicken breast lightly with a mallet. Refrigerate.
2. For the stuffing, combine all of the ingredients except the flour and creole sauce. Mix well.
3. Place the stuffing on each breast; roll and **truss.**
4. **Dredge** the breasts in seasoned flour, and brown on all sides uniformly in hot oil.
5. Add the chicken breasts to the creole sauce in a roasting pan.
6. Cook the chicken for 10 to 12 minutes until it reaches an internal temperature of 165° F. Hold at 140° F. Adjust the seasonings, as needed.
7. Slice chicken breast into 3 slices. Pour sauce on warm dinner plate and arrange slices on the sauce. Garnish with the scallions and strawberry.

COOKING TECHNIQUE:
Sauté

Sauté:
1. Heat the sauté pan to the appropriate temperature.
2. Evenly brown the food product.
3. For a sauce, pour off any excess oil, reheat, and deglaze.

GLOSSARY:
Brunoise: ⅛-inch dice
Truss: to tie or secure
Dredge: to coat with flour

HACCP:
Hold at 140° F.

HAZARDOUS FOODS:
Chicken breast
Shrimp

NUTRITION:
Calories: 319
Fat: 14.6 g
Protein: 34 g

Chicken Braised in Red Wine

(Coq au Vin)

YIELD: 10 SERVINGS. SERVING SIZE: ONE-FOURTH CHICKEN.

COOKING TECHNIQUES:

Sauté, Braise

Sauté:

1. Heat the sauté pan to the appropriate temperature.
2. Evenly brown the food product.
3. For a sauce, pour off any excess oil, reheat, and deglaze.

Braise:

1. Heat the braising pan to the proper temperature.
2. Sear and brown the food product to a golden color.
3. Degrease and deglaze.
4. Cook the food product in two thirds liquid until fork-tender.

GLOSSARY:

Dredge: to coat with flour
Mirepoix: roughly chopped vegetables
Blanch: to par cook
Sear: to brown quickly
Flambé: to flame
Fork-tender: without resistance
Chinois: cone-shaped strainer
Dépouiller: to skim impurities/grease
Nappé: coated

HAZARDOUS FOOD:

Chicken

NUTRITION:

Calories: 674
Fat: 46.5 g
Protein: 50.3 g

INGREDIENTS:

10 each	Oil, as needed
	Chicken thighs and legs
	Seasoned flour, as needed for **dredging**
Mirepoix:	
5 ounces	Onions, peeled and chopped
5 ounces	Carrots, washed, peeled, and chopped
5 ounces	Celery, washed, trimmed, and chopped
3 ounces	Tomatoes, washed, cored, **blanched,** peeled, and seeded
3 cloves	Garlic, peeled and finely minced
1 ounce	Cognac
1 pint	Red wine
1 quart	Chicken demi-glace, heated to a boil (see page 26)
2 tablespoons	Clarified butter
6 ounces	Small mushrooms, cleaned, trimmed, and quartered
3 ounces	Pearl onions, washed, peeled, and blanched
	Salt and ground white pepper, to taste

METHOD OF PREPARATION:

1. Preheat the oven to 350° F.
2. In a braising pan, heat the oil. Dredge the chicken in seasoned flour, shaking off the excess. **Sear** in the hot oil until browned. Remove the chicken from the pan, and reserve.
3. Sauté the mirepoix and garlic in the same pan. **Flambé** with the cognac, and deglaze with the wine.
4. Return the chicken into the braising pan, and add the demi-glace.

CHEF NOTE:

The sauce can be thickened with beurre manié, which is added after the sauce is strained.

5. Cover and braise in the oven until the chicken is **fork-tender** and the juices are clear.
6. When the chicken is done, transfer pieces to a hotel pan, and hold covered at 140° F. Strain the sauce through a **chinois** into a saucepan.
7. Heat the sauce to a boil. Reduce to a simmer, and **dépouiller,** as needed. Adjust the consistency and seasoning, as desired.
8. In a sauté pan, heat the butter, and sauté the mushrooms and onions until golden brown. Reserve.
9. Add the mushrooms and onions to the sauce. Return to a boil, and pour over the chicken. Hold at 140° F or above.
10. To serve, place 1 thigh and 1 leg of chicken on a preheated plate, and **nappé** with sauce.

Chicken Breast in Crust

(Poitrine de Volaille Farcie en Croûte)

YIELD: 10 SERVINGS. SERVING SIZE: EIGHT OUNCES.

COOKING TECHNIQUES:

Sauté, Bake

Sauté:

1. Heat the sauté pan to the appropriate temperature.
2. Evenly brown the food product.
3. For a sauce, pour off any excess oil, reheat, and deglaze.

Bake:

1. Preheat the oven.
2. Place the food product on the appropriate rack.

GLOSSARY:

Render: to melt fat

HAZARDOUS FOODS:

Chicken breast
Bacon
Ground pork
Eggs
Velouté

NUTRITION:

Calories: 607
Fat: 26.7 g
Protein: 46.2 g

INGREDIENTS:

5 each	Chicken breasts, boneless and skinless
	Salt and ground white pepper, to taste
½ teaspoon	Ginger, ground
10 ounces	Bacon, chopped and **rendered**
3 ounces	Shallots, peeled and diced
3 cloves	Garlic, peeled and minced
10 ounces	Ground pork
20 ounces	Frozen spinach, thawed and drained
2 pounds	Sour cream pastry dough (see page 841)
3 each	Eggs
1 quart	Chicken velouté sauce (see page 68)

METHOD OF PREPARATION:

1. Split each chicken breast in half and butterfly. Pound the breast, using a mallet, to ¼-inch thickness.
2. Lay out the pounded chicken breasts by overlapping each by approximately 25% on the next. Season with salt, pepper, and ginger. Cover and refrigerate until needed.
3. Render the bacon; add the shallots and sauté until translucent; add garlic, ground pork, spinach. Sauté for 10 minutes. Season with the salt, pepper, ginger, and set aside to cool.
4. Roll out the dough to ¼-inch thickness.
5. Remove the chicken breasts from the refrigerator and cover chicken with the spinach mixture. Roll the chicken breasts into a roulade.
6. Place the chicken roulade on the dough. Roll and close ends of the dough by moistening with eggwash.
7. Decorate the roulade with dough strips, brush with eggwash, and bake in a 375° F oven until done. (When pierced, the stuffing will release clear moisture.)
8. To serve, allow 2 slices per serving; ladle some sauce on a preheated dinner plate, and place the chicken on top.

Chicken Breast Stuffed with Salmon Mousse

(Suprême de Volaille à la Wolseley)

YIELD: 10 SERVINGS. SERVING SIZE: SIX OUNCES.

INGREDIENTS:

6 ounces	Flour
10 each	Chicken breasts
1 quart	Sauce suprême, heated and held at 140° F (see page 67)

Stuffing:

10 ounces	Salmon
5 ounces	Panada
2 each	Egg yolks
2 ounces	Heavy cream
¼ teaspoon	Ginger
	Salt and ground white pepper, to taste
4 ounces	Butter, clarified

METHOD OF PREPARATION:

1. Make a mousse with the salmon and panada in a robo-coupe. Gradually add the egg yolks, heavy cream, ginger, salt, and pepper.
2. Butterfly the breast on its thick side with a fillet knife.
3. Place a tablespoon of salmon mousse on each breast, and fold back to its original shape.
4. **Dredge** in flour, and sauté in butter. Turn once, and bake in a 375° F oven for 20 to 25 minutes.
5. Place on a dinner plate, and **nappé** with sauce suprême.

COOKING TECHNIQUE:

Simmer

Simmer and Poach:

1. Heat the cooking liquid to the proper temperature.
2. Submerge the food product completely.
3. Keep the cooked product moist and warm.

GLOSSARY:

Dredge: to coat with flour
Nappé: coated

HAZARDOUS FOODS:

Egg yolks
Heavy cream

NUTRITION:

Calories: 411
Fat: 19.6 g
Protein: 32.4 g

Chicken Curry

YIELD: 50 SERVINGS. SERVING SIZE: EIGHT OUNCES.

COOKING TECHNIQUE:
Bake

Bake:
1. Preheat the oven.
2. Place the food product on the appropriate rack.

GLOSSARY:
Dredge: to coat with flour
Dépouiller: to skim impurities/grease

HACCP:
Hold at 140° F.

HAZARDOUS FOOD:
Whole chicken

NUTRITION:
Calories: 468
Fat: 21 g
Protein: 58.9 g

INGREDIENTS:

50 each	Chicken thighs and legs
10 ounces	Oil
	Seasoned flour (with salt and curry powder), as needed
	Curry sauce (see page 25) as needed, heated to a boil
	Apple rings, toasted coconut, for garnish

METHOD OF PREPARATION:

1. Preheat the oven to 350° F.
2. Heat the oil in a large sauté pan.
3. **Dredge** the chicken pieces in the seasoned flour, and shake off any excess. Brown the chicken on both sides in the hot oil. Remove, and drain off the excess oil. Place in hotel pans.
4. Pour the curry sauce over the browned chicken pieces. Cover the hotel pans with parchment paper and foil. Place in the oven for approximately 45 minutes, or until the chicken is fork-tender. Hold at 140° F.
5. **Dépouiller** the sauce prior to service.
6. Garnish with sliced, baked apple rings and toasted coconut.

Chicken Paprikash

YIELD: 10 SERVINGS. SERVING SIZE: ONE-HALF BREAST.

INGREDIENTS:

5 each	Boneless, skinless chicken breasts, split in half lengthwise
	Seasoned flour, as needed
4 ounces	Clarified butter
1 pound	Onions, peeled and diced **brunoise**
2 each	Green bell peppers, washed, seeded, and diced brunoise
1 pound	Ripe tomatoes, washed, cored, blanched, peeled, seeded, and roughly chopped
1½ ounces	Hungarian sweet paprika
	Hungarian hot paprika, to taste
12 ounces	White chicken stock, heated to a boil (see page 12)
	Salt, to taste
1 pound	Sour cream

METHOD OF PREPARATION:

1. Preheat the oven to 350° F.
2. In a sauté pan, heat the butter. **Dredge** chicken breasts in flour, and sauté until lightly colored; then transfer the chicken to a hotel pan.
3. Add the onions to sauté pan, and sauté until they are translucent. Add the peppers and tomatoes, and sauté for another 10 minutes.
4. Sprinkle the vegetables with the paprika, continue to sauté 2 to 3 minutes, then add the stock. Heat the liquid to a boil, and pour the mixture over the chicken breasts.
5. Cover tightly and place in the oven to braise until the chicken is **fork-tender,** or about 30 minutes.
6. Transfer the chicken to another pan, and hold covered at 140° F or above.
7. Strain the sauce through a **chinois** into a saucepan. Return to the heat, and reduce to a proper consistency. Temper the sour cream with some of the sauce, and whisk until smooth. Add the cream to the sauce, and heat to a minimum of 165° F or hold at 140° F.

COOKING TECHNIQUES:

Sauté, Braise

Sauté:

1. Heat the sauté pan to the appropriate temperature.
2. Evenly brown the food product.
3. For a sauce, pour off any excess oil, reheat, and deglaze.

Braise:

1. Heat the braising pan to the proper temperature.
2. Sear and brown the food product to a golden color.
3. Degrease and deglaze.
4. Cook the food product until fork-tender.

GLOSSARY:

Brunoise: ⅛-inch dice
Dredge: to coat with flour
Fork-tender: without resistance
Chinois: cone-shaped strainer

HACCP:

Hold covered at 140° F or above.
Heat to a minimum of 165° F.

HAZARDOUS FOODS:

Boneless, skinless chicken breasts
Sour cream

NUTRITION:

Calories: 379
Fat: 29.6 g
Protein: 17.9 g

CHEF NOTE:

This dish can be made "peasant" style, using bone-in quartered chicken pieces. The vegetables would be cut as a mirepoix, and the sauce would not be strained.
The recommended starch for this dish is spätzel.

Chicken Stew

YIELD: 10 SERVINGS. SERVING SIZE: ONE-FOURTH CHICKEN.

COOKING TECHNIQUE:

Braise

Braise:

1. Heat the braising pan to the proper temperature.
2. Sear and brown the food product to a golden color.
3. Degrease and deglaze.
4. Cook the food product in two thirds liquid until fork-tender.

GLOSSARY:

Brunoise: ⅛-inch diced
Sear: to brown quickly
Fork-tender: without resistance
Nappé: coated

HACCP:

Hold at 140° F

HAZARDOUS FOODS:

Chicken thighs and legs

NUTRITION:

Calories: 339
Fat: 19.5 g
Protein: 30.3 g

INGREDIENTS:

10 each	Chicken thighs
10 each	Chicken legs
	Oil, as needed
8 ounces	Onions, peeled and diced **brunoise**
2 pounds	Tomatoes, washed, cored, blanched, peeled, seeded, and cut into wedges
4 cloves	Garlic, peeled and mashed into a purée
6 ounces	Tomato purée
5 ounces	Pitted black olives, sliced
3 ounces	Fresh oregano leaves, washed and roughly chopped
½ teaspoon	Dried marjoram

Marinade:

2 ounces	Lemon juice
¼ teaspoon	Cloves, ground
¼ teaspoon	Cinnamon, ground
1 ounce	Salt
1 teaspoon	Paprika
½ teaspoon	Freshly ground black pepper

METHOD OF PREPARATION:

1. Preheat the oven to 350° F.
2. In a large bowl, combine the ingredients for the marinade, and mix well. Add the chicken, cover, and marinate at 40° F or below for 1 hour. Remove the chicken, and pat it dry.
3. In a braising pan, add the oil, and coat the bottom of the pan; **sear** the chicken pieces on all sides until lightly browned. Transfer the chicken to a hotel pan.

CHEF NOTE:

This is one of the many chicken dishes from Greece. Usually, the meat is cooked to such tenderness that it falls off of the bones.

4. In the same braising pan, add the onions, and sauté until they are translucent. Add the tomatoes and garlic, and sauté for 3 minutes; then add the balance of the ingredients, including the reserved marinade. Season as needed.
5. Return the chicken to the pan, and heat the liquid to a boil. Cover and transfer to the oven. Braise in the oven until **fork-tender,** or about 45 minutes.
6. Transfer the chicken to another hotel pan, and hold at 140° F.
7. Place the braising pan over high heat, and reduce the liquid to the desired flavor. Adjust the seasoning as necessary, and then pour the sauce over the chicken. Serve immediately, or hold at 140° F or above.
8. To serve, place two pieces of chicken on a preheated dinner plate, and **nappé** with sauce.

Chicken with Dumplings

YIELD: 10 SERVINGS. SERVING SIZE: ONE-FOURTH CHICKEN.

INGREDIENTS:

3 (2½- to 3-pound)	Chickens, rinsed in cold water and patted dry
1 gallon	White chicken stock, heated to a boil (see page 12)
2 each	Bay leaves
	Salt and freshly ground black pepper, to taste
4 ounces	Roux blonde
2 cups	Milk, heated to a boil

Mirepoix:

8 ounces	Carrots, washed and peeled
1 pound	Onions, peeled
1 pound	Celery, washed and trimmed
8 ounces	Leeks (white part only), washed and trimmed

Dumplings:

2 cups	All-purpose flour, or as needed
1 teaspoon	Salt
2 teaspoons	Baking powder
2 tablespoons	Butter
4 tablespoons	Vegetable oil
4 ounces	Cream

METHOD OF PREPARATION:

1. In a stockpot, combine the chicken and the stock. Heat to a boil, and after about 5 boiling minutes, **depouiller.** Add the mirepoix, bay leaves, and seasoning, to taste. Cover and cook until **fork-tender,** about 1 hour. Remove the chicken, and hold at 140° F or above. Strain the stock through a **chinois.**
2. To prepare the dumplings, combine the flour, salt, and baking powder. Blend the butter and oil into the flour, and add the cream to make a dough stiff enough to roll out. Roll the dough out to ¼-inch thickness. Cut the dough into 2-inch × 2-inch squares.

COOKING TECHNIQUES:
Stew, Simmer

Stew:
1. Sear, sauté, sweat, or blanch the main food product.
2. Deglaze the pan, if desired.
3. Cover the food product with simmering liquid.
4. Remove the bouquet garni.

Simmer and Poach:
1. Heat the cooking liquid to the proper temperature.
2. Submerge the food product completely.
3. Keep the cooked product moist and warm.

GLOSSARY:
Mirepoix: roughly chopped vegetables
Dépouiller: to skim impurities/grease
Fork-tender: without resistance
Chinois: cone-shaped strainer

HACCP:
Hold at 140° F or above.

HAZARDOUS FOODS:
Chicken
Cream
Milk

NUTRITION:
Calories: 447
Fat: 26 g
Protein: 18.2 g

CHEF NOTE:
Reserve extra stock for other uses.

3. Heat 2 quarts of the strained stock to a boil; reduce the heat to a simmer, and drop in the dumplings. When they rise to the top, cover and simmer for 10 minutes, or until they are cooked inside. Remove, and hold at 140° F or above.
4. Heat the roux, and add the hot milk. Heat to a boil, and then reduce the heat, and simmer for about 10 minutes, or until the starch is fully cooked and mixture is smooth. Strain the sauce through a chinois, and hold at 140° F.
5. Cut the chickens into quarters, and partially debone. Keep warm in seasoned stock. Hold covered at 140° F.
6. To serve, place a portion of chicken on a preheated dinner plate, arrange the dumplings around the chicken, and nappé with the sauce.

Cider-Marinated Duck Breast with Spicy Mango Chutney

YIELD: 10 SERVINGS. SERVING SIZE: HALF BREAST.

INGREDIENTS:

5 each	Boneless duck breasts, skin on
	Salt and freshly ground black pepper, to taste
1 pint	Apple cider
1 clove	Garlic, peeled and mashed to a purée
2 ounces	Granulated sugar
1 ounce	Cider vinegar
½ teaspoon	Chili peppers, dried, crushed, bruised
¼ teaspoon	Thyme, dried

Spicy Mango Chutney:

2 ounces	Clarified butter

Macédoine:

4 ounces	Onions, peeled
6 ounces	Granny Smith apples, washed, peeled, and cored
8 ounces	Mango, washed and peeled

METHOD OF PREPARATION:

1. Preheat the grill.
2. Season the duck breasts with salt and pepper, and place them in a noncorrosive container. Add the cider, refrigerate at 40° F or lower, and allow to marinate for at least 1 hour.
3. In a small saucepan, heat the butter; add the onions, and sauté until translucent.

COOKING TECHNIQUES:
Grill, Sauté, Simmer

Grill/Broil:
1. Clean and heat the grill/broiler.
2. To prevent sticking, brush the food product with oil.

Sauté:
1. Heat the sauté pan to the appropriate temperature.
2. Evenly brown the food product.
3. For a sauce, pour off any excess oil, reheat, and deglaze.

Simmer and Poach:
1. Heat the cooking liquid to the proper temperature.
2. Submerge the food product completely.
3. Keep the cooked product moist and warm.

GLOSSARY:
Macédoine: ¼-inch dice
Bain-marie: hot-water bath

HACCP:
Refrigerate at 40° F or lower.
Hold at 140° F or higher.

HAZARDOUS FOOD:
Duck

NUTRITION:
Calories: 225
Fat: 9.37 g
Protein: 16.9 g

CHEF NOTES:
1. Duck breasts should not be held in a **bain-marie** for more than 10 minutes, or they will become overcooked and dry.
2. Chutney holds refrigerated at 40° F, or lower, for at least 10 days. Return to room temperature for service.

4. Add the remaining ingredients, and sauté until wilted; then reduce the heat, and simmer, stirring frequently, until the fruits are very soft and the mixture is thick. Taste, and adjust the seasoning as needed. Hold at 140° F or higher.
5. Remove the duck breasts from the marinade, and pat dry. Grill each duck breast, skin-side down first, until skin is crisp and browned; then turn, and grill the flesh side for 2 minutes, or to an internal temperature of 165° F. Remove, and place skin-side up in a hotel pan; hold at 140° F for 10 minutes before serving.
6. For service, thinly slice the duck breast on a diagonal, and fan out the meat on a preheated dinner plate. Place a 2-ounce scoop of chutney in the center of the slices, and serve immediately.

Creole Chicken Stew

YIELD: 10 SERVINGS. SERVING SIZE: TWO PIECES.

COOKING TECHNIQUE:
Sauté

Sauté:
1. Heat the sauté pan to the appropriate temperature.
2. Evenly brown the food product.
3. For a sauce, pour off any excess oil, reheat, and deglaze.

GLOSSARY:
Fork-tender: without resistance
Nappé: coated

HACCP:
Hold at 140° F or above.

HAZARDOUS FOOD:
Chicken

NUTRITION:
Calories: 523
Fat: 32.7 g
Protein: 30 g

INGREDIENTS:

20 each	Chicken legs or thighs
	Flour, seasoned, as needed
4 ounces	Vegetable oil
3 ounces	Green bell pepper, washed, seeded, and diced
3 ounces	Red bell pepper, washed, seeded, and diced
2 (3-ounce)	Fresh tomatoes, washed, cored, and cut into wedges
6 ounces	Onions, peeled and diced
8 ounces	Fresh okra, washed and cut into ½-inch rings
1 teaspoon	Spanish paprika
1½ teaspoons	Chili peppers, crushed
¼ teaspoon	Caraway seeds, bruised
5 ounces	Black olives, pitted and cut in half
1½ quarts	Chicken demi-glace, heated to a boil (see page 26)
	Salt, to taste
4 ounces	Fresh parsley leaves, washed, excess moisture removed, and roughly chopped

METHOD OF PREPARATION:

1. Preheat the oven to 350° F.
2. Heat oil in a braising pan. Dredge the chicken in the seasoned flour, and brown in hot oil.
3. Place the chicken pieces in a 4-inch deep hotel pan.
4. In a sauté pan, heat the oil, and sauté all of the vegetables together until softened; then add the seasonings. Remove from the heat, and cool slightly.
5. Evenly cover the chicken pieces with the cooked vegetables. Add the olives and demi-glace.
6. Cover the pan tightly with aluminum foil, and place in the oven for approximately 1 hour, or until **fork-tender.** When done, adjust the seasoning with salt, and sprinkle chopped parsley evenly over the surface of the pan.
7. Serve immediately; arrange two pieces of the chicken on a preheated dinner plate, and **nappé** with some of the sauce. Hold at 140° F or above.

CHEF NOTE:
This dish can be served with different levels of spiciness, but it always must be served with rice as a starch.

Duckling à l'Orange

(Caneton Bigarade)

YIELD: 10 SERVINGS. SERVING SIZE: ONE-FOURTH DUCK.

INGREDIENTS:

3 each	Ducks, giblets removed
2 ounces	Grand Marnier
½ teaspoon	Ginger, ground
3 sprigs each	Thyme, marjoram, rosemary
1 pint	Orange juice
	Salt and freshly ground black pepper, to taste

Mirepoix:

8 ounces	Onions, peeled
8 ounces	Carrots, washed and peeled
8 ounces	Celery, washed

Sauce:

2 ounces	Grand Marnier
8 ounces	Orange juice
2 ounces	Orange marmalade
2 ounces	Apricot jam
1 pint	Duck or chicken demi-glace, seasoned
1-ounce zest	Orange
5 each	Oranges, segmented

METHOD OF PREPARATION:

1. Preheat oven to 425°.
2. Season ducks inside and outside. Place the herbs in the cavities, along with some orange peelings, and **truss.**
3. Rub Grand Marnier on the skin of the ducks, and set on the mirepoix in the roasting pan.
4. Place to roast in 425° F oven for 20 minutes, baste, and reduce temperature to 350° F. Roast the ducks until, when pierced in the joints, juice is released.
5. When the ducks are done, set to cool, and then disjoint and partially debone.
6. Place the bones and the interior parts from the ducks into the roasting pan.

COOKING TECHNIQUE:

Roast

Roast:

1. Sear the food product, and brown evenly.
2. Elevate the food product in a roasting pan.
3. Determine doneness, and consider carryover cooking.
4. Let the food product rest before carving.

GLOSSARY:

Mirepoix: roughly chopped vegetables
Truss: to tie or secure

HACCP:

Hold at 140° F or above.

HAZARDOUS FOOD:

Duck giblets

NUTRITION:

Calories: 517
Fat: 14.1 g
Protein: 32.4 g

7. Place the pan on the stove, and roast for a few minutes; add the orange juice and demi-glace. Simmer for 15 minutes, and then strain. Taste, season, and degrease.
8. Add the remaining ingredients to the strained liquid, heat to a boil, and hold at 140° F or above.
9. Place the orange segments in a sauté pan, add a little sauce, and heat.
10. To serve, reheat the quartered ducks to 140° F. Place the ducks on preheated dinner plates, nappé with the sauce, and ladle orange segments on top.

Grilled Chicken Teriyaki

YIELD: 50 SERVINGS. SERVING SIZE: FOUR OUNCES.

INGREDIENTS:

25 each	Whole boneless chicken breasts, washed, skinned, and split in half
6 pints	Teriyaki sauce (see page 75)
	Sesame oil, as needed

METHOD OF PREPARATION:

1. Ensure that the cooking surface of the grill or broiler is clean; then preheat.
2. Wash the portioned chicken breasts in cold running water, and pat dry. Lightly oil each chicken breast with the sesame oil.
3. Grill or broil the chicken breasts to order, as needed, brushing with the teriyaki sauce while broiling.
4. Cook until the chicken breasts are golden brown and firm to the touch. Cook to an internal temperature of 165° F. Serve one piece per serving (½ breast).

COOKING TECHNIQUES:

Grill or Broil

Grill/Broil:

1. Clean and heat the grill/broiler.
2. To prevent sticking, brush the food product with oil.

HACCP:

Cook to an internal temperature of 165° F.

HAZARDOUS FOOD:

Chicken breasts

NUTRITION:

Calories: 200
Fat: 9.5 g
Protein: 17.8 g

CHEF NOTES:

1. If this recipe is made using a broiler instead of a grill, the dish would be called Broiled Chicken Teriyaki.
2. Chicken tenders may be removed to be used in another recipe.

Herbed Baked Chicken

YIELD: 50 SERVINGS. SERVING SIZE: ¼ CHICKEN.

INGREDIENTS:

1 pint	Salad oil
1 tablespoon	Salt
1 tablespoon	Paprika
1 tablespoon	Thyme, dry
1 tablespoon	Marjoram, dry
1 tablespoon	Oregano, dry
1 tablespoon	Basil, dry
16 pounds	Chicken, thighs and legs

METHOD OF PREPARATION:

1. Preheat the oven to 350° F.
2. Mix all of the spices and herbs in oil, and let stand for half an hour.
3. Place the chicken pieces on a sheet pan.
4. Brush with herbed oil.
5. Bake in the oven until **fork-tender** and the internal temperature reaches 165° F.
6. Transfer baked chicken eighths or parts into 4-inch hotel pans. Hold at 165° F.

COOKING TECHNIQUE:

Bake

Bake:

1. Preheat the oven.
2. Place the food product on the appropriate rack.

GLOSSARY:

Fork-tender: without resistance

HACCP:

Hold at 165° F.

HAZARDOUS FOOD:

Chicken

NUTRITION:

Calories: 437
Fat: 31.3 g
Protein: 36.6 g

CHEF NOTE:

Verify that the 16 pounds of chicken have at least 50 pieces of chicken. Chicken may be ordered by the unit e.g., 50 each thighs, 50 each legs.

COOKING TECHNIQUES:

Stir-Fry, Simmer

HACCP:

Hold at 140° F or above.

HAZARDOUS FOOD:

Chicken breasts

NUTRITION:

Calories: 349
Fat: 15.1 g
Protein: 17.6 g

Hot and Spicy Chicken

(Ayam Panggang Gahru)

YIELD: 10 SERVINGS. SERVING SIZE: ONE-HALF CHICKEN.

INGREDIENTS:

5 each	Boneless chicken breasts, skin removed, split in half, and cut into ½-inch cubes
1 teaspoon	Ginger root, peeled and minced
1 ounce	Mirin
2 ounces	Soy sauce
3 cloves	Garlic, peeled and mashed into a purée
	Salt, to taste
2 ounces	Soybean oil
8 ounces	White chicken stock, heated to a boil (see page 12)
1 teaspoon	Cornstarch
2 teaspoons	Water
	Oriental noodles or rice, as needed
10 each	Scallions, for scallion fans

Sauce:

3 ounces	Scallions, washed, trimmed, and thinly sliced
1 teaspoon	Chili pepper flakes, crushed, dried
1 tablespoon	Anise, ground
8 ounces	Coconut milk
1 ounce	White wine vinegar
1 ounce	Sugar
1 ounce	Sesame seed oil
	Salt, to taste

METHOD OF PREPARATION:

1. In a large bowl, combine the chicken with the ginger root, mirin, soy sauce, and garlic, and season, to taste. Allow to marinate for 30 minutes.
2. In a hot wok or skillet, stir-fry the chicken in soybean oil for 3 to 5 minutes. Remove the chicken, and reserve. Add the chicken stock and the remaining sauce ingredients to the marinade and simmer for about 10 minutes.

CHEF NOTE:

Hot and spicy foods are common in the Southern Pacific region.

3. Dissolve the cornstarch in cold water. Thicken the sauce with this mixture.
4. Serve with Oriental noodles or rice, and garnish with scallion fans.
5. Serve immediately on a preheated dinner plate, or hold at 140° F or above.

Maple-Smoked Turkey

YIELD: 50 SERVINGS. SERVING SIZE: FIVE OUNCES.

INGREDIENTS:

2 each	Turkey breast, bone in
	Trimix (salt, white pepper, and ginger), as needed
2 quarts	Chicken demi-glace, heated to a boil (see page 26)
	Liquid smoke, as needed
3 ounces	Cornstarch

Mirepoix:

1 pound	Onions, peeled and sliced
1 pound	Carrots, washed, peeled, and sliced
8 ounces	Celery, washed and chopped
4 cloves	Garlic, left whole

METHOD OF PREPARATION:

1. Preheat the oven to 375° F.
2. Bone out the breast, rub the seasoning into the flesh, and **truss** with butcher's twine.
3. Arrange the mirepoix in the roasting plaque. Place the tied turkey breast on the mirepoix, and place the roast in the oven with the bones placed around the meat.
4. After 20 minutes of roasting, reduce the oven temperature to 325° F. Baste the breasts with the hot demi-glace, and brush with liquid smoke. Continue roasting and basting until done. When piercing the meat, clear moisture will appear (or cook until the internal temperature reaches 165° F.)
5. Let the meat rest for 30 minutes before slicing.

COOKING TECHNIQUE:

Roast

Roast:

1. Sear the food product, and brown evenly.
2. Elevate the food product in a roasting pan.
3. Determine doneness, and consider carryover cooking.
4. Let the food product rest before carving.

GLOSSARY:

Mirepoix: roughly chopped vegetables
Truss: to tie or secure

HAZARDOUS FOOD:

Turkey breast

NUTRITION:

Calories: 238
Fat: 6.43 g
Protein: 37.7 g

CHEF NOTE:

Using a mallet, chop breast bone to small pieces.

Marinated Grilled Breast of Chicken in a Roasted Garlic Tart and Roasted Red Pepper Coulis

YIELD: 10 SERVINGS. SERVING SIZE: SIX OUNCES (CHICKEN).

COOKING TECHNIQUES:

Grill/Roast, Steam

Grill/Broil:

1. Clean and heat the grill/broiler.
2. To prevent sticking, brush the food product with oil.

Roast:

1. Sear the food product, and brown evenly.
2. Elevate the food product in a roasting pan.
3. Determine doneness, and consider carryover cooking.
4. Let the food product rest before carving.

Steam: (Traditional)

1. Place a rack over a pot of water.
2. Prevent steam vapors from escaping.
3. Shock or cook the food product throughout.

HACCP:

Chill to 40° F.
Hold at 140° F.

HAZARDOUS FOODS:

Chicken
Cream
Egg yolks

NUTRITION:

Calories: 851
Fat: 62.1 g
Protein: 43.4 g

CHEF NOTES:

1. The garlic tart can be prepared in one 10-inch tart pan and cut into wedges for service.
2. It is important to steam the tart at a lower temperature to prevent the custard from separating.

INGREDIENTS:

10 (6-ounce) pieces	Chicken breasts, boneless, skinless, split in half lengthwise
	Roasted red pepper coulis
	Pico de gallo (see page 671)
10 each	Fresh cilantro sprigs, washed and drained

Marinade:

4 ounces	Olive oil
1 ounce	Freshly squeezed lime juice
2 each	Serrano peppers, washed, stemmed, seeded, and finely minced
	Salt and freshly ground black pepper, to taste

Ancho Chili Purée:

2 each	Ancho chilies, dried
4 ounces	Hot water
½ ounce	Olive oil
	Salt, to taste

Garlic Tart Dough:

12 ounces	Masa harina
2 ounces	Yellow cornmeal
2 teaspoons	Salt
2 teaspoons	Cumin, ground

¼ teaspoon	Cayenne pepper or Tabasco sauce
3 ounces	Vegetable shortening
2½ ounces	Ancho chili purée (recipe above)
	All-purpose flour, as needed

Roasted Garlic Custard:

½ ounce	Olive oil
6 cloves	Garlic, peeled and mashed into a purée
1½ pints	Heavy cream
4 ounces	Pasteurized egg yolks
1 teaspoon	Salt, or to taste
	Freshly ground white pepper, to taste

METHOD OF PREPARATION:

1. Combine the ingredients for the marinade, and mix well. Add the chicken, and marinate, refrigerated, at 40° F or lower for at least 1 hour.
2. To prepare the ancho chili purée, soak the ancho chilies in water for at least 1 hour or until they have softened. Drain off the water, reserving the liquid.
3. Remove the stems, seeds, and ribs from the chilies. In a food processor, combine the chilies, olive oil, and salt, and begin to purée. Add the reserved liquid, as necessary, to produce a thick purée. Remove from processor, place in a suitable storage container, and refrigerate at 40° F or lower, until needed.
4. To prepare the garlic tart dough, in a bowl of an electric mixer, combine the dry ingredients, including the spices and seasonings. Start to mix, and quickly blend in the shortening. Add the chili purée to taste, and blend just to incorporate, but do not overmix the dough. Transfer the dough to a lightly floured work surface, and shape it quickly into a log.
5. Cut the log into 10 equal portions. Press each portion into an individual 3-inch tart pan, smoothing to a thickness of about ¼ inch. Trim off the top edges of the tarts. Place the tart shells on a 2-inch perforated hotel pan, and chill to 40° F or below while preparing the custard.
6. Preheat the oven to 325° F.
7. In a small sauté pan, heat the olive oil, and add the garlic. Toss the garlic in the sauté pan; then cover with aluminum foil, and place in the oven to roast until golden brown and soft, or for about 15 to 20 minutes. Cool slightly; then transfer the oil and garlic to a food processor, and purée until smooth.
8. In a medium-sized saucepan, heat the cream to a boil; reduce the heat, and simmer for 10 minutes, reducing the cream by one fourth. Cool slightly; then add the cream to the garlic purée in the blender.
9. Lightly whisk the egg yolks; then slowly incorporate them into the garlic-cream mixture. Season to taste with salt and pepper.

10. Fill the tart shells three-fourths full with custard. Individually, totally wrap each tart with plastic wrap, and place back into the perforated pan. Place the pan over a 4-inch hotel pan filled to a depth of 2 inches with water; cover tightly with aluminum foil and place in the oven to steam until the custard is set, or for about 20 minutes. Unwrap the tarts, and hold at 140° F.
11. Remove the chicken breasts from the marinade, and pat off the excess oil. Grill the chicken breasts, turning once, until just firm and cooked through. Hold at 140° F.
12. For service, slice the chicken breasts, to order, on a diagonal, and fan out on a preheated dinner plate. Place one tart at the top of the plate, and garnish with red pepper coulis, pico de gallo, and cilantro sprigs.

Roast Chicken with Paprika

YIELD: 50 SERVINGS. SERVING SIZE: ONE-FOURTH CHICKEN.

COOKING TECHNIQUES:
Roast, Simmer

Roast:
1. Sear the food product, and brown evenly.
2. Elevate the food product in a roasting pan.
3. Determine doneness, and consider carryover cooking.
4. Let the food product rest before carving.

Simmer and Poach:
1. Heat the cooking liquid to the proper temperature.
2. Submerge the food product completely.
3. Keep the cooked product moist and warm.

GLOSSARY:
Truss: tie or secure
Mirepoix: roughly chopped vegetables

HAZARDOUS FOOD:
Chicken

NUTRITION:
Calories: 488
Fat: 27.3 g
Protein: 51.4 g

CHEF NOTE:
Natural reduction should be sufficient for the sauce, but cornstarch and water may be added, if necessary.

INGREDIENTS:

13 each	Chickens, 2½ pound, whole, washed, cavity seasoned, and **trussed**
	Seasoned, to taste

Mirepoix:

3 pounds	Onions, peeled and roughly chopped
1½ pounds	Celery, washed, trimmed, and roughly chopped
1½ pounds	Carrots, washed, peeled, and roughly chopped
10 ounces	Paprika
6 ounces	Clarified butter
2½ quarts	White chicken stock, heated (see page 12)
5 ounces	Cornstarch
5 ounces	Water, cold
	Seasoned, to taste

METHOD OF PREPARATION:

1. Preheat the oven to 350° F.
2. Place the chicken on top of the mirepoix in a roasting pan, and place in the oven.
3. Combine the paprika and butter, and baste the chicken at 15-minute intervals.
4. Roast the chicken until tender. The juices should be clear when the chicken is fully cooked.
5. Remove the chicken from the pan. Drain the juices from the cavity onto the mirepoix, and keep warm. Place the roasting pan on the stove, and caramelize the mirepoix in the pan.
6. Add 1 quart of hot chicken stock to deglaze the pan. Heat the stock to a boil. Reduce to a simmer. Skim the fat, and season. Strain through a fine mesh chinois into a saucepan.
7. Simmer, and skim the fat, as needed. Simmer until the proper flavor and consistency are achieved.
8. Portion the chicken into quarters, and serve with the sauce.

Roast Duck with Green Peppercorns

(Canard Rôti au Poivre Vert)

YIELD: 10 SERVINGS. SERVING SIZE: ONE-FOURTH DUCK.

COOKING TECHNIQUE:

Roast

Roast:

1. Sear the food product, and brown evenly.
2. Elevate the food product in a roasting pan.
3. Determine doneness, and consider carryover cooking.
4. Let the food product rest before carving.

GLOSSARY:

Mirepoix: roughly chopped
Truss: to tie or secure
Deglaze: to add liquid to a hot pan
Reduction: evaporation of liquid by boiling

HACCP:

Cook to an internal temperature of 165° F.

HAZARDOUS FOOD:

Whole ducks

NUTRITION:

Calories: 306
Fat: 19.3 g
Protein: 22.1 g

INGREDIENTS:

3–4 pounds	Whole ducks, giblets removed
	Salt and freshly ground black pepper, to taste
	Ginger, to taste
Zest of 1	Orange
4 ounces	White wine
4 ounces	Red wine
1½ pints	Duck or chicken stock (see page 12)
1 pint	Demi-glace (see page 26)
2 tablespoons	Green peppercorns
4 ounces	Butter, raw and cut into small pieces

Mirepoix:

4 ounces	Carrots, washed and peeled
4 ounces	Celery, washed
4 ounces	Onions, peeled

METHOD OF PREPARATION:

1. Trim the wing tips. Season the cavities with salt, white pepper, ginger, and orange zest. **Truss** the ducks.
2. Place the ducks on a wire rack in a roasting pan, along with the neck and gizzards. Roast the ducks in a preheated 400° F oven for 30 minutes.
3. Remove the ducks from the pan, and pour off the excess fat. Place the mirepoix under the wire rack, replace the ducks in the pan, and roast at 350° F for up to 1 hour, or until the juices run clear and the internal temperature in the thigh area is 165° F. Then remove the ducks and hold warm.

4. Place the pan on the stove. Heat until the mirepoix is golden brown, and pour off any excess fat. **Deglaze** with the wines; **reduce** by half, add the chicken or duck stock and demi-glacé, heat to a boil, dépouiller, and simmer until the proper flavor is achieved. Strain through a fine chinois mousseline into a saucepan. In the saucepan, heat to a boil; then reduce to a simmer, and add the green peppercorns. Gradually flake in the raw butter, and shake the pan. Adjust the seasonings, and keep warm for service.
5. Remove the string from the ducks. Pour the excess moisture from the cavities. Portion the ducks into quarters, and partially debone. Place a portion on a preheated plate, nappé with peppercorn sauce, and serve.

Roast Duckling with Peaches

(Caneton Rôti aux Pêches)

YIELD: 10 SERVINGS. SERVING SIZE: ONE-HALF DUCK.

INGREDIENTS:

3 each	Ducklings
½ teaspoon	Ginger, ground
	Salt and ground white pepper, to taste
3 each	Oranges, washed and cut in half
1 quart	Brown duck or chicken stock, heated to a boil (see page 7)
10 each	Peach halves, sliced into a fan
	Peach sauce (see page 41)

Mirepoix:

12 ounces	Onions, peeled
12 ounces	Carrots, washed and peeled
12 ounces	Celery, washed and trimmed

METHOD OF PREPARATION:

1. Preheat the oven to 425° F.
2. Remove the first two joints of wings and the giblets from the cavities of the birds. Rinse the ducks in cold water, and pat dry.
3. Season inside and out with ginger, salt, and pepper. Place two orange halves into each cavity, and **truss.**
4. Place the mirepoix in a roasting pan, and arrange the birds on top.
5. Roast in the oven for 25 to 30 minutes; then reduce the temperature to 375° F. Continue to roast, basting with stock, approximately 1½ hours more.
6. When **fork-tender,** remove from the pan, and hold at 140° F.
7. Degrease the pan, place on the stove, and brown mirepoix. Add the remaining stock, and heat to a boil. When the desired flavor is achieved, strain through a **chinois.** Pour over the ducks to hold moisture.
8. To serve, split the ducklings into quarters, and debone partially. Arrange two pieces on a preheated dinner plate. **Nappé** with peach sauce, and decorate with sliced peaches.

COOKING TECHNIQUES:
Roast, Sauté

Roast:
1. Sear the food product, and brown evenly.
2. Elevate the food product in a roasting pan.
3. Determine doneness, and consider carryover cooking.
4. Let the food product rest before carving.

Sauté:
1. Heat the sauté pan to the appropriate temperature.
2. Evenly brown the food product.
3. For a sauce, pour off any excess oil, reheat, and deglaze.

GLOSSARY:
Mirepoix: roughly chopped vegetables
Truss: to tie or secure
Fork-tender: without resistance
Chinois: cone-shaped strainer
Nappé: coated

HACCP:
Hold at 140° F.

HAZARDOUS FOOD:
Ducklings

NUTRITION:
Calories: 264
Fat: 10.3 g
Protein: 23.8 g

Roasted Chicken

YIELD: 50 SERVINGS. SERVING SIZE: ONE-FOURTH CHICKEN.

INGREDIENTS:

15 each	Chickens, whole
1 pound	All-purpose flour
2½ gallons	Brown chicken stock, heated to a boil (see page 7)
20 ounces	Curry spice

Mirepoix:

4 pounds	Onions, peeled and chopped
3 pounds	Carrots, washed and peeled
3 pounds	Celery stalks, washed and peeled

METHOD OF PREPARATION:

1. Preheat the oven to 350° F.
2. Season the chicken inside and out; then **truss.**
3. Place the mirepoix in a roasting pan, and arrange the chickens on top.
4. Roast until, when pierced at the thigh, the juices are clear, approximately 1 hour.
5. Lift the chickens from the roasting pan with a roasting fork, draining the juices from the cavities back into the pan.
6. Allow the chickens to cool slightly; then cut them into quarters, removing the backbones, and partially debone. Hold at 140° F.
7. Brown the mirepoix (with the bones) on the stove.
8. Add the flour, and brown. Add the chicken stock, and heat to a boil. Cook until the desired flavor is achieved.
9. Strain the liquid through a **chinois,** remove the fat, and adjust the seasonings. Hold at 140° F.

COOKING TECHNIQUE:

Roast

Roast:

1. Sear the food product, and brown evenly.
2. Elevate the food product in a roasting pan.
3. Determine doneness, and consider carryover cooking.
4. Let the food product rest before carving.

GLOSSARY:

Mirepoix: roughly chopped vegetables
Truss: to tie or secure
Chinois: cone-shaped strainer

HACCP:

Hold at 140° F.

HAZARDOUS FOOD:

Chicken

NUTRITION:

Calories: 549
Fat: 43 g
Protein: 44 g

Roasted Chicken with Tarragon Sauce

YIELD: 10 SERVINGS. SERVING SIZE: ONE-QUARTER CHICKEN.

INGREDIENTS:

3 (2½ pounds each)	Chickens
	Salt and ground white pepper, to taste
½ teaspoon	Ginger, ground
1½ quarts	Brown chicken stock, heated to a boil (see page 7)
2 ounces	Cornstarch, diluted in 2 ounces of cold water
3 ounces	Fresh tarragon leaves, washed and chopped

Mirepoix:

8 ounces	Onion, peeled
6 ounces	Carrots, washed and peeled
5 ounces	Celery stalks, washed and trimmed

METHOD OF PREPARATION:

1. Preheat the oven to 350° F.
2. Remove the giblets from the cavity of the chickens; then wash the chickens in cold water and pat dry.
3. Combine the seasonings and sprinkle on the chickens inside and out; then **truss** each.
4. Place the mirepoix in a roasting pan, and arrange the chickens on top. Roast in the oven for approximately 1 hour, or until done and juices run clear.
5. Remove the chickens from the roasting pan, draining any accumulated juices from the cavities back into pan. Hold at 140° F for at least 15 minutes before portioning.
6. Place the roasting pan over the heat on top of the stove; **deglaze** with boiling stock, and cook until the desired taste is achieved. Strain through a **chinois** into another saucepot.
7. Heat the strained sauce to a boil, and thicken it with the cornstarch mixture. Simmer for 5 to 6 minutes; then add the tarragon, and adjust the seasoning. Transfer to a suitable container, and hold at 140° F.
8. Portion the chickens, removing their back bones and breast bones. Keep bones for future use.
9. To serve, arrange a chicken quarter on a preheated dinner plate, and **nappé** with sauce.

COOKING TECHNIQUE:
Roast

Roast:
1. Sear the food product, and brown evenly.
2. Elevate the food product in a roasting pan.
3. Determine doneness, and consider carryover cooking.
4. Let the food product rest before carving.

GLOSSARY:
Truss: to tie or secure
Deglaze: to add liquid to hot pan
Chinois: cone-shaped strainer
Nappé: coated

HACCP:
Hold at 140° F.

HAZARDOUS FOOD:
Chicken

NUTRITION:
Calories: 788
Fat: 51.8 g
Protein: 64.8 g

CHEF NOTES:
1. Adding the tarragon at the last moment will retain its fresh flavor and color.
2. Be careful not to reduce the liquid too much (less than 1 quart).

Roasted Turkey with Giblet Gravy and Cranberry Relish

YIELD: 10 SERVINGS. SERVING SIZE: FIVE OUNCES.

INGREDIENTS:

1(12- to 14-pound)	Turkey, wing tips and giblets removed
	Salt and ground white pepper, to taste
½ teaspoon	Ginger, ground
3 tablespoons	All-purpose flour
2 quarts	White chicken or turkey stock, heated to a boil (see page 12)
	Cranberry relish (see page 643)

Mirepoix:

8 ounces	Onions, peeled
8 ounces	Celery stalks, washed and trimmed
6 ounces	Carrots, washed and peeled

METHOD OF PREPARATION:

1. Preheat the oven to 325° F.
2. Remove the giblets from the turkey cavity, and rinse, inside and out, under cold water; wipe dry.
3. Season the cavity with salt, pepper, and ginger. Truss the legs by tying or secure together with tail skin.
4. Place the mirepoix in bottom of a roasting pan, and position the turkey on top, breast-side up.
5. Add the giblets to the roasting pan, sprinkle lightly with additional salt and pepper, and roast for 3 to 3½ hours, or until an internal temperature of 165° F is reached and a clear juice flows when a leg joint is pierced.
6. Remove the turkey from the pan, and hold in a warm area for 20 to 30 minutes before carving.
7. Remove the giblets; cut the liver and gizzard into a ¼-inch dice. Remove the meat from the neck bones and cut into small pieces; then combine with the liver and gizzard, and set aside.
8. Place the roasting pan over medium heat, and caramelize the mirepoix. Sprinkle flour over the vegetables, and stir until the fat is absorbed, cooking for 3 minutes.

COOKING TECHNIQUES:

Roast, Simmer

Roast:

1. Sear the food product, and brown evenly.
2. Elevate the food product in a roasting pan.
3. Determine doneness, and consider carryover cooking.
4. Let the food product rest before carving.

Simmer and Poach:

1. Heat the cooking liquid to the proper temperature.
2. Submerge the food product completely.
3. Keep the cooked product moist and warm.

GLOSSARY:

Mirepoix: roughly chopped vegetables
Dépouiller: to skim impurities/grease
Chinois: cone-shaped strainer
Nappé: coated

HACCP:

Heat to an internal temperature of 165° F.
Hold at 140° F.

HAZARDOUS FOOD:

Turkey

NUTRITION:

Calories: 612
Fat: 5.7 g
Protein: 122 g

CHEF NOTES:

1. If it is desired that the turkey be stuffed with dressing, do not do so until the bird is ready to go into the oven.
2. Cooking time for a stuffed bird will need to be increased.
3. The internal temperature must be 165° F and meat and gravy holding temperature at least 140° F.

9. Gradually add half of the stock to the roasting pan, stirring constantly. Bring to a boil; reduce heat, and simmer for 15 to 20 minutes. **Dépouiller** the surface, as needed.
10. Strain the liquid through a **chinois,** and discard the vegetables. Add the reserved giblets, and adjust the seasoning as needed. Hold at 140° F or higher, until service.
11. Slice to order. Remove the breast meat from the turkey. Slice on a diagonal, across the grain. Remove the legs and thighs, debone (reserving the bones for stock), and slice the dark meat against the grain.
12. To serve, place one slice of dark and one slice of white meat on each preheated dinner plate, and **nappé** with the gravy.

COOKING TECHNIQUE:

Sauté

Sauté:

1. Heat the sauté pan to the appropriate temperature.
2. Evenly brown the food product.
3. For a sauce, pour off any excess oil, reheat, and deglaze.

HACCP:

Hold warm at 140° F or above.

HAZARDOUS FOOD:

Chicken breasts

NUTRITION:

Calories: 462
Fat: 34.8 g
Protein: 18 g

Sautéed Chicken Breasts with Artichokes and Béarnaise Sauce

(Supreme de Volaille Henry IV)

YIELD: 10 SERVINGS. SERVING SIZE: FIVE OUNCES.

INGREDIENTS:

5 each	Chicken breasts, split in half lengthwise
	Seasoned flour, as needed
6 ounces	Clarified butter
4 ounces	Vegetable oil
	Salt and ground white pepper, to taste
8 ounces	Demi-glace, heated to a boil (see page 26)
10 each	Artichoke bottoms
10 ounces	Béarnaise sauce, at 140° F (see page 15)
10 sprigs	Fresh tarragon

METHOD OF PREPARATION:

1. Coat the chicken with flour, and shake off any excess.
2. In a large skillet, heat the butter and oil, and sauté the chicken for 8 to 10 minutes, or until browned and cooked. Season, to taste.
3. Transfer the chicken to a hotel pan, and hold covered at 140° F or above. Place the demi-glace and artichoke bottoms in the skillet, and heat for 3 minutes; then hold at 140° F.
4. To serve, place the chicken on preheated dinner plates. Top with artichoke filled with béarnaise sauce. Garnish with tarragon leaves.

Southern-Style Fried Chicken

YIELD: 50 SERVINGS. SERVING SIZE: ONE-FOURTH CHICKEN.

INGREDIENTS:

50 each	Chicken thighs and legs
	Seasoned flour, as needed
	Oil, as needed

METHOD OF PREPARATION:

1. Heat the oil in a braising pan.
2. **Dredge** the chicken in flour, and shake off any excess.
3. Place the chicken into the hot oil, and fry evenly.
4. Remove chicken from oil, and place on a sheet pan with wirerack.
5. Put the sheet pan into a 350° F oven. Bake until the chicken is fork-tender and juices run clear.

COOKING TECHNIQUE:

Shallow-Fry

Shallow-Fry:

1. Heat the cooking medium to the proper temperature.
2. Cook the food product throughout.
3. Season, and serve hot.

GLOSSARY:

Dredge: to coat with flour

HACCP:

Hold at 140° F.

HAZARDOUS FOOD:

Chicken

NUTRITION:

Calories: 545
Fat: 39.7 g
Protein: 40.9 g

CHEF NOTES:

1. This fried chicken is almost the same as Maryland fried chicken, but the garnish for the Maryland fried chicken is small corn fritters or corn oysters with broiled bacon.
2. In many places in the South, this dish is served with a cream gravy that includes strained particles from the frying oil.

Stuffed Duckling with Sweet and Sour Sauce

YIELD: 10 SERVINGS. SERVING SIZE: SIX OUNCES.

INGREDIENTS:

3 four-pound	Ducklings, deboned, held at 40° F or below
Stuffing:	
8 ounces	Ground pork
8 ounces	Ground veal
4 ounces	Zucchini, washed and diced **brunoise**
1 pound	Frozen spinach, thawed, excess moisture removed, and chopped
4 ounces	Ham, brunoise
4 ounces	Genoa salami, brunoise
2 ounces	Pistachio nuts, shelled and peeled
2 cloves	Garlic, peeled and finely minced
4 ounces	Egg whites
	Salt and ground white pepper, to taste
	Sweet and sour sauce no. 3, heated to a boil, held at 140° F (see page 73)

METHOD OF PREPARATION:

1. Preheat the oven to 350° F. Combine all of the stuffing ingredients, seasoning generously, and stuff the ducks.
2. **Truss** to secure the cavity, and roll the ducks tightly in aluminum foil, forming an oblong tube.
3. Place them on a baking pan, and roast in the oven for 45 minutes; then remove and allow to cool 15 minutes.
4. Remove the foil, and discard the fat.
5. Return the duck (unwrapped) to roast in the oven for an additional 45 minutes. Remove from the oven, and hold at 140° F for at least 15 minutes before slicing.
6. To serve, slice the duck as a **galantine,** arrange overlapping slices on a preheated dinner plate, and **nappé** with sauce.

COOKING TECHNIQUE:

Roast

Roast:

1. Sear the food product, and brown evenly.
2. Elevate the food product in a roasting pan.
3. Determine doneness, and consider carryover cooking.
4. Let the food product rest before carving.

GLOSSARY:

Brunoise: ⅛-inch dice
Truss: to tie or secure
Galantine: boned and stuffed
Nappé: coated

HACCP:

Hold at 40° F or below.
Hold warm at 140° F.

HAZARDOUS FOODS:

Ducklings
Ground Pork
Ground veal
Egg whites

NUTRITION:

Calories: 687
Fat: 35.5 g
Protein: 55.4 g

CHEF NOTE:

Duck should be sliced to order rather than pre-sliced and held in a bain-marie.

Teriyaki Grilled Chicken with Pineapple

YIELD: 50 SERVINGS. SERVING SIZE: EIGHT OUNCES.

INGREDIENTS:

25 each	Chicken breasts, boneless, whole, split in half
6 pints	Teriyaki sauce
50 each	Pineapple slices
	Sesame oil, as needed

METHOD OF PREPARATION:

1. Ensure that the cooking surface of the grill is clean, then preheat it.
2. Lightly oil each chicken breast with the sesame oil.
3. Grill or broil the chicken breasts to order as needed, brushing with the teriyaki sauce while broiling.
4. Cook until the chicken breasts are golden brown and firm to the touch, and reach 165° F.
5. Serve one piece per serving, and garnish each with grilled pineapple slice.

COOKING TECHNIQUES:

Grill or Broil

Grill/Broil:

1. Clean and heat the grill/broiler.
2. To prevent sticking, brush the food product with oil.

HACCP:

Hold at 140° F.

HAZARDOUS FOOD:

Chicken breast

NUTRITION:

Calories: 182
Fat: 8.35 g
Protein: 17.8 g

CHEF NOTE:

The same recipe can be made by using a broiler instead of a grill, and the dish would then be called Teriyaki Broiled Chicken.

COOKING TECHNIQUE:
Bake

Bake:
1. Preheat the oven.
2. Place the food product on the appropriate rack.

GLOSSARY:
Nappé: coated

HACCP:
Hold at 140° F.

HAZARDOUS FOODS:
Sole fillets

NUTRITION:
Calories: 329
Fat: 10.6 g
Protein: 29 g

Baked Fish, Turkish Style

YIELD: 10 SERVINGS. SERVING SIZE: FIVE OUNCES.

INGREDIENTS:

10 five-ounce	Sole fillets
2 teaspoons	Ground marjoram
10 each	Scallions, washed, trimmed, and cut into 2-inch pieces
20 each	Grape leaves, rinsed of brine
	Vegetable oil, as needed
	Salt and freshly ground black pepper, to taste
16 ounces	White wine
3 ounces	Freshly squeezed lemon juice
8 ounces	Golden raisins, soaked overnight in water, then drained

METHOD OF PREPARATION:

1. Preheat the oven to 350° F.
2. Sprinkle each fillet with marjoram. Place a piece of scallion on the tail end, and roll the fillet toward its head end.
3. Spread the leaves open. Place a rolled fillet on each leaf, and enclose it in the leaf, tucking in the ends of the leaf as you roll. Place each roll in an oiled roasting pan, season with salt and pepper, pour the white wine and lemon juice over the top, and sprinkle the raisins over all.
4. Place in the oven, uncovered, and bake for 18 to 25 minutes, or to an internal temperature of 140° F. Serve immediately, or hold at 140° F.
5. To serve, place one wrapped fish on a preheated dinner plate, and **nappé** with raisin sauce.

CHEF NOTES:
1. Other fish, such as cod, haddock, whitefish, perch, and red snapper, also can be used.
2. Grape leaves are salty. Check level of saltiness before adding additional salt.

Baked Flounder Newburg

YIELD: 50 SERVINGS. SERVING SIZE: FIVE OUNCES.

INGREDIENTS:

10 pounds	Flounder fillets, cut into 3-ounce portions
5 ounces	Lemon juice
10 ounces	White wine, dry
	Salt and ground white pepper, to taste
3 quarts	Newburg sauce (see page 61)

METHOD OF PREPARATION:

1. Preheat the oven to 325° F.
2. Place the portioned fish on a sheet pan. Sprinkle with the lemon juice, white wine, salt, and pepper. Let marinate for 1 hour.
3. Thirty minutes before service, bake the fish in the oven for 20 minutes, or until done. Transfer the fish to 2-inch hotel pans, and hold at 165° F.
4. Serve on preheated dinner plates with Newburg sauce.

COOKING TECHNIQUE:

Bake

Bake:

1. Preheat the oven.
2. Place the food product on the appropriate rack.

HACCP:

Hold at 165° F.

HAZARDOUS FOOD:

Flounder fillets

NUTRITION:

Calories: 341
Fat: 20.7 g
Protein: 22.2 g

CHEF NOTE:

1. The flounder can be replaced with any other white-fleshed fish.
2. Flounder is a flaky-fleshed fish; handle with care after baking to avoid crumbling.

Baked Salmon with a Mustard-Dill Crust

YIELD: 10 SERVINGS. SERVING SIZE: SIX OUNCES.

INGREDIENTS:

2 ounces	Butter
3 ounces	Fresh dill, washed, excess moisture removed, and chopped
3 cloves	Garlic, peeled and finely minced
2 ounces	Dijon mustard
8 each	Eggs
	Salt and ground white pepper, to taste
20 ounces	Bread crumbs, or as needed
	All-purpose flour, as needed
10 six-ounce	Salmon fillets, skinned and bones removed

METHOD OF PREPARATION:

1. Preheat the oven to 375° F.
2. In a mixing bowl, combine all of the ingredients except the flour and salmon, and work the mixture into a dough-like mixture.
3. Dust a table with flour, and roll out the mixture into a ¼-inch-thick leaf.
4. Cut the dough into portions the shape of the salmon fillets.
5. Lay out the salmon fillets, flesh-side up, on an oiled baking pan, place the dough pieces on top of each fillet, and bake for 10 minutes, or until the fish is firm and cooked. Serve immediately, or hold at 140° F or higher.
6. To serve, carefully transfer the fillet to a preheated dinner plate.

COOKING TECHNIQUE:

Bake

Bake:

1. Preheat the oven.
2. Place the food product on the appropriate rack.

HAACP:

Hold at 140° F or higher.

HAZARDOUS FOODS:

Eggs
Salmon fillets

NUTRITION:

Calories: 619
Fat: 23.2 g
Protein: 47.7 g

Baked Scrod with Lemon Butter

YIELD: 50 SERVINGS. SERVING SIZE: FOUR TO FIVE OUNCES.

INGREDIENTS:

16 pounds	Scrod fillets, bones removed and cut into 5-ounce portions
	Fish stock, as needed (see page 10), heated to a boil
	Dry white wine, as needed
	Lemon juice, as needed
2½ pounds	Butter, clarified
2½ pounds	Bread crumbs
	Salt and ground white pepper, to taste
6 ounces	Freshly chopped parsley, excess moisture removed
9 each	Lemons, fresh, cut into wedges

METHOD OF PREPARATION:

1. Preheat the oven to 375° F.
2. In a large sauté pan, melt 1 pound of butter. Add the bread crumbs, and brown lightly. Season with salt and ground white pepper.
3. Lay the fish fillets out on a hotel pan. Add the white wine, lemon juice, and fish stock to the pan to moisten. Sprinkle the bread crumbs on the fish. Bake the fish in the oven for 15 minutes, or until firm to the touch and reaches an internal temperature of 170° F.
4. Combine the parsley, lemon juice, and 1½ pounds butter. Pour over the fish when serving. Garnish the fish with lemon wedges.

COOKING TECHNIQUE:

Bake

Bake:

1. Preheat the oven.
2. Place the food product on the appropriate rack.

HACCP:

Hold at 140° F.

HAZARDOUS FOOD:

Scrod fillets

NUTRITION:

Calories: 326
Fat: 11.6 g
Protein: 35.4 g

CHEF NOTE:

In quantity food production, the fish can be placed on sheet pans with parchment paper. When finished, the cooked fish can be transferred to hotel pans.

Bouillabaisse Marseillaise

YIELD: 10 SERVINGS. SERVING SIZE: ONE PINT.

INGREDIENTS:

20 ounces	Cod, skinned and boneless, cut into 10 cubes
20 ounces	Catfish, cut into 10 cubes
20 each	Scallops
20 each	Mussels
20 each	Little neck clams
2 pounds	Chicken lobster, tail cut into three pieces and two claws
½ teaspoon	Saffron
4 ounces	Olive oil
1 teaspoon	Salt
½ teaspoon	Freshly ground black pepper
3 quarts	Concentrated fish stock (see page 10)
20 each	Garlic-flavored croutons

METHOD OF PREPARATION:

1. Place all of the seafood in the cooking vessel; season and mix. Place in the cooler, and marinate for 1 hour.
2. Pour the cold, flavored, seasoned fish stock to cover the seafood, and poach until the shells open. Taste, and adjust the seasonings.
3. Serve in a bowl with a small plate on which to place the shells, and garlic toast.

COOKING TECHNIQUE:

Simmer

Simmer and Poach:

1. Heat the cooking liquid to the proper temperature.
2. Submerge the food product completely.
3. Keep the cooked product moist and warm.

HAZARDOUS FOODS:

Cod
Catfish
Scallops
Mussels
Little neck clams

NUTRITION:

Calories: 349
Fat: 19.8 g
Protein: 33 g

CHEF NOTES:

1. Bouillabaisse is a traditional fisherman's dish from the city of Marseilles in southern France. In some port villages, it is prepared with the head and bones in the soup.
2. If no bone in fish is used, the ingredients can be portioned and cooked to order.

Braised Stuffed Filet of Sole

(Filet de Sole Farci Dugléré)

YIELD: 10 SERVINGS. SERVING SIZE: FIVE OUNCES.

INGREDIENTS:

3½ pounds	Dover Sole fillets
	Salt and ground white pepper, to taste
1 pint	White wine
8 ounces	Fish stock, heated to a boil (see page 10)
2 ounces	Shallots, peeled and diced **brunoise**
20 ounces	Fish velouté (see page 68)
1 pound	Tomatoes, washed, cored, **blanched,** peeled, seeded, and diced
2–3 ounces	Butter, as needed
2 ounces	Freshly chopped parsley, excess moisture removed

METHOD OF PREPARATION:

1. Preheat the oven to 350° F.
2. Lay out fillets on a sheet pan. Season to taste. Roll into **paupiettes.**
3. Place the paupiettes into a 2-inch hotel pan, and add the wine and fish stock. Sprinkle shallots over the fish.
4. Cover tightly with aluminum foil, and braise in the oven until firm, about 10 to 12 minutes.
5. Remove from the oven, and strain 12 ounces of liquid through a **chinois;** reserve at 140° F or above. Re-cover the pan, and hold the sole at 140° F or above.
6. In a skillet, combine the velouté and the reserved cooking liquid. Heat to a simmer. Add the tomatoes, and reduce until a velvety texture is achieved. Season to taste, and hold at 140° F or above.
7. To serve, place a paupiette on a preheated dinner plate, and **nappé** with sauce. Garnish with parsley.

COOKING TECHNIQUE:

Braise

Braise:

1. Heat the braising pan to the proper temperature.
2. Sear and brown the food product to a golden color.
3. Degrease and deglaze.
4. Cook the food product in two thirds liquid until fork-tender.

GLOSSARY:

Brunoise: ⅛-inch dice
Blanch: to par cook
Paupiette: rolled meat or fish
Chinois: cone-shaped strainer
Nappé: coated

HACCP:

Hold at 140° F or above.

HAZARDOUS FOOD:

Sole

NUTRITION:

Calories: 220
Fat: 7.82 g
Protein: 28 g

CHEF NOTES:

1. The fillets can be spread with fish stuffing or salpicon before rolling. Sautéed mushrooms can be added to the sauce.
2. Flounder, gray sole, and lemon sole can be substituted for the sole.
3. This dish was named for a famous eighteenth-century French chef, Dugléré.

Broiled Catfish

YIELD: 50 SERVINGS. SERVING SIZE: FIVE OUNCES.

INGREDIENTS:

1 quart	Salad oil
16 pounds	Catfish fillets, any bones removed, cut into 5-ounce portions, if necessary
	Cajun spices, to taste
	Worcestershire sauce, as needed
	Salt and freshly ground white pepper, to season
13 each	Lemons, cut into wedges
2 ounces	Fresh parsley, washed and excess moisture removed

METHOD OF PREPARATION:

1. Rub the catfish fillets with oil, and season with the Cajun spice, Worcestershire sauce, and salt and pepper.
2. Preheat the broiler. Broil the fillets until golden brown and firm to the touch. Hold at 140° F.
3. Place the fish on warm plates, and serve with lemon wedges and parsley sprigs.

COOKING TECHNIQUE:
Broil

Grill/Broil:
1. Clean and heat the grill/broiler.
2. To prevent sticking, brush the food product with oil.

HACCP:
Cook to an internal temperature of 140° F.

HAZARDOUS FOOD:
Catfish

NUTRITION:
Calories: 269
Fat: 17.8 g
Protein: 24 g

CHEF NOTES:
1. Blackening is a method of rubbing the product with dry spice and placing it in a dry, very hot skillet to sear the item.
2. Do not prepare in advance as it will dry out.

Fried Fillets of Cod with Tartar Sauce

YIELD: 50 SERVINGS. SERVING SIZE: FIVE OUNCES.

INGREDIENTS:

16 pounds	Cod fillets, rinsed, bones removed, and cut into 5-ounce portions
10 each	Lemons, washed, cut into wedges
6 ounces	Freshly chopped parsley, excess moisture removed
2 quarts	Tartar sauce (see page 74)

Beer Batter:

15 each	Whole eggs, separated
4 pounds	All-purpose flour
5 12-ounce cans	Beer
1 teaspoon	Salt
1 teaspoon	Ground white pepper
	Seasoned flour, as needed

METHOD OF PREPARATION:

1. Preheat the oil in a deep-fat fryer to 350° F.
2. Combine the flour, beer, salt, pepper, and egg yolks. Blend until smooth.
3. Whip the egg whites to a soft peak in a stainless steel bowl.
4. Fold the egg whites into the batter.
6. **Dredge** the fillets in the seasoned flour, dip them in the batter, and place them into the deep-fat fryer, frying until golden brown and thoroughly cooked.
6. Remove, and drain on absorbent paper.
7. Serve on a preheated dinner plate garnished with lemon wedges, parsley, and tartar sauce.

COOKING TECHNIQUE:

Deep-Fry

Deep-Fry:

1. Heat the frying liquid to the proper temperature.
2. Submerge the food product completely.
3. Fry the product until it is cooked throughout.

GLOSSARY:

Dredge: to coat with flour
Bain-marie: hot-water bath

HACCP:

Hold at 140° F.

HAZARDOUS FOODS:

Cod fillets
Eggs

NUTRITION:

Calories: 787
Fat: 57.8 g
Protein: 32.8 g

CHEF NOTE:

Fried battered fish does not hold well in a **bain-marie.** If it is necessary to hold the item, place it in a 140° F oven with an open vent.

Broiled Whitefish with Browned Butter and Capers

COOKING TECHNIQUE:
Broil

GLOSSARY:
Nappé: coated

HACCP:
Hold at 140° F.

HAZARDOUS FOOD:
Whitefish

NUTRITION:
Calories: 384
Fat: 27.8 g
Protein: 31 g

YIELD: 10 SERVINGS. SERVING SIZE: SIX OUNCE FILLET.

INGREDIENTS:

4 pounds	Whitefish fillets, cut into 6-ounce portions
	salt and ground white peper, to taste
1 pound	Butter
5 ounces	Capers, rinsed of brine
10 each	Lemon crowns

METHOD OF PREPARATION:

1. Season the fish on both sides with salt and pepper.
2. Melt 4 ounces butter, and brush the fish on both sides; then broil to order until just firm.
3. Heat the remaining butter until golden brown (beurre noisettes), then add the capers. Hold at 140° F.
4. To serve, place a fillet of whitefish on a preheated dinner plate, and **nappé** with butter. Garnish with lemon crowns.

CHEF NOTES:

1. Whitefish is a staple of the mid-west. It has a high fat content and, therefore, is also great for smoking.
2. For the same method of preparation, flounder, sole, scrod fillets, or catfish fillets can be used.

Flounder Meunière

YIELD: 50 OUNCES. SERVING SIZE: FIVE OUNCES.

COOKING TECHNIQUE:
Sauté

GLOSSARY:
Chimois: cone-shaped strainer

HAZARDOUS FOOD:
Flounder

NUTRITION:
Calories: 258
Fat: 16.4 g
Protein: 30 g

INGREDIENTS:

16 pounds	Flounder, cut into 5-ounce fillets
	Seasoned flour, as needed
3 pounds	Lightly salted butter
8 ounces	Lemon juice
3 ounces	parsley, washed, chopped, and excess moisture removed
8 each	Fresh lemons
	Salt and ground white pepper, as needed

METHOD OF PREPARATION:

1. Heat half of the butter in a tilting skillet.
2. Dredge the fillets in seasoned flour, and shake off any excess flour.
3. Sauté the fish on both sides, turning only once to avoid breaking or crumbling.
4. Arrange the fish in a 2-inch hotel pan in an overlaying method, and keep warm.
5. Add the second half of the butter into the tilting skillet. Brown, and add lemon juice. Strain, using a fine **chinois.**
6. Add the parsley to the lemon butter; taste the sauce, and season, as needed.
7. Zest the lemons, slice them into round slices, and serve on the fish.

CHEF NOTE:
If the recipe calls for *belle meunière*, small deepfried croutons are added on top of the fish. In Provençe, fried eggplant is added.

Garlic Mussels

YIELD: 50 SERVINGS. SERVING SIZE: TEN EACH.

COOKING TECHNIQUE:
Simmer

Simmer and Poach:
1. Heat the cooking liquid to the proper temperature.
2. Submerge the food product completely.
3. Keep the cooked product moist and warm.

GLOSSARY:
Brunoise: ⅛-inch diced

HAZARDOUS FOOD:
Mussels

NUTRITION:
Calories: 169
Fat: 6 g
Protein: 11.5 g

INGREDIENTS:

1 quart	Dry white wine
1 pound	Onions, peeled and diced **brunoise**
4 heads	Garlic, peeled and chopped
10 pounds	Fresh mussels, scrubbed, debearded, and washed
2 pounds	Chorizo sausage, sliced (optional)
2 quarts	Fish stock, concentrated and seasoned (see page 10)
	Salt and freshly ground black pepper, to taste
	Garlic toast, as needed

METHOD OF PREPARATION:

1. Place the wine, onions, and garlic in a tilting skillet to simmer.
2. Add the mussels and chorizo sausage (optional); mix well, cover, and let steam until the mussels open.
3. Add the seasoned fish stock, and continue to simmer for 6 to 8 minutes, stirring frequently.
4. Check all of the mussels to see if they are open. Serve on a preheated soup plate with some of the cooking liquid.

CHEF NOTES:
1. Garlic toast matches very well with this dish.
2. Discard all mussels that do not open.

Grilled Red Snapper with Floridian Fruit Vinaigrette

COOKING TECHNIQUE:
Grill

Grill/Broil:
1. Clean and heat the grill/broiler.
2. To prevent sticking, brush the food product with oil.

GLOSSARY:
Nappé: coated

HACCP:
Place in a cooler at 40° F or lower.

HAZARDOUS FOOD:
Red snapper

NUTRITION:
Calories: 186
Fat: 8.56 g
Protein: 25.2 g

YIELD: 10 SERVINGS. SERVING SIZE: FIVE OUNCES.

INGREDIENTS:

4 pounds	Red snapper, cut into 5-ounce fillets (skin on)
	Salt and ground white pepper, to taste
3 ounces	Freshly squeezed lemon juice
3 ounces	Vegetable oil or clarified butter
	Floridian fruit vinaigrette (see page 225)

METHOD OF PREPARATION:

1. Place the snapper fillets in a hotel pan. Sprinkle the fillets on both sides with salt and pepper, and drizzle lemon juice over each. Place in cooler at 40° F or lower, and let marinate for 30 minutes, or until service.
2. For service, brush the fillets with oil or clarified butter, and grill to order, starting skin-side down. Brush additionally with oil or butter during the grilling period. Turn the fish only once during grilling. The total grilling time will be 5 to 7 minutes.
3. To serve, place the grilled fillet on a preheated dinner plate, and **nappé** with Floridian vinaigrette.

Grilled Salmon with Lemon-Ginger Marinade

YIELD: 10 SERVINGS. SERVING SIZE: SIX OUNCES.

INGREDIENTS:

10 (6-ounce)	Fillet of salmon, skin on, scaled, rinsed in cold water, and patted dry or drip-dried
	Salt and ground white pepper, to taste
	Vegetable oil, as needed
2 ounces	Freshly chopped parsley
2 each	Lemons, washed and cut into wedges

Lemon-Ginger Marinade:

8 ounces	Freshly squeezed lemon juice, strained
3 cloves	Garlic, peeled and diced **brunoise**
2 ounce	Fresh ginger root, peeled and diced brunoise
1 ounce	Fresh thyme, washed, leaves removed, and chopped
2 ounces	Scallions, washed, trimmed, and cut in ¼-inch rings

METHOD OF PREPARATION:

1. Combine the ingredients for the marinade, and mix well.
2. Season the salmon fillets with salt and pepper, and lay out in a hotel pan. Pour the marinade over the fish, and refrigerate at 40° F or lower for 30 minutes. Turn once in marinade.
3. Preheat the grill or broiler.
4. Remove the fish from the marinade, and pat each dry. Brush the fillets with oil.
5. Grill the fillets to order, skin-side down first, until the skin is crisp; then brush the flesh with marinade, and turn. Grill on the flesh side until the fish is just firm and opaque, or to an internal temperature of 140° F.
6. Strain the remaining juice from the marinade through a **chinois** into a small saucepot, and bring to a boil. Reduce to the desired strength, and hold at 140° F.
7. For service, place a fillet on a preheated dinner plate, spoon the juice from the marinade over the fillet, and garnish with parsley and a lemon wedge. Serve immediately.

COOKING TECHNIQUE:
Grill

Grill/Broil:
1. Clean and heat the grill/broiler.
2. To prevent sticking, brush the food product with oil.

GLOSSARY:
Brunoise: ⅛-inch dice
Chinois: cone-shaped strainer

HACCP:
Refrigerate at 40° F or lower.
Cook to an internal temperature of 140° F.
Hold at 140° F.

HAZARDOUS FOOD:
Salmon

NUTRITION:
Calories: 313
Fat: 16.7 g
Protein: 34.4 g

CHEF NOTE:
This preparation also works very well with salmon trout, or brook trout, served whole.

COOKING TECHNIQUES:

Steam, Deep-Fry

Steam: (Traditional)

1. Place a rack over a pot of water.
2. Prevent steam vapors from escaping.
3. Shock or cook the food product throughout.

Deep-Fry:

1. Heat the frying liquid to the proper temperature.
2. Submerge the food product completely.
3. Fry the product until it is cooked throughout.

GLOSSARY:

Bain-marie: hot-water bath

HAZARDOUS FOODS:

Shrimp
Egg whites

NUTRITION:

Calories: 634
Fat: 44.1 g
Protein: 11.9 g

CHEF NOTES:

1. *Ajishio* is a mixture of salt, white pepper, and monosodium glutamate, which is dry-cooked very lightly in a sauté pan.
2. This dish will lose texture if held in a **bain-marie.**

Japanese-Style Deep-Fried Shrimp Coated with Rice

(Domyoji Age)

YIELD: 10 SERVINGS. SERVING SIZE: FOUR SHRIMP.

INGREDIENTS:

2½ pounds	Japanese rice, precooked and dried (Domyoji)
40 each (16/20)	Shrimp, peeled and deveined
	Seasoned flour, as needed
7 each	Egg whites, lightly beaten
	Peanut oil, as needed
2 each (8-ounce)	Green bell peppers, washed, seeded, and sliced into ½-inch-wide strips
	Ajishio, as needed (see chef note 1)
	Sweet and sour sauce no. 2, as needed, heated to a boil (see page 72)
2 each	Lemons, washed, sliced, and seeds removed

METHOD OF PREPARATION:

1. Cook the rice the day before; spread it on a sheet pan, and refrigerate overnight.
2. Using food handler gloves, separate the grains by gently rubbing with the palms of your hands; then chop the rice coarsely.
3. Make three cuts, ½-inch apart, across the inner curve of the shrimp. (This prevents the shrimp from curling.)
4. Coat the shrimp in flour, and shake off any excess.
5. Dip the shrimp into the egg whites and then into the coarse rice. (Be sure that the rice adheres firmly.)
6. In a wok or skillet, heat the oil over moderate heat to 350° F. Deep-fry the shrimp until golden brown. Drain well on absorbent paper; reserve and keep warm at 140° F or above.
7. Add the green peppers to the wok or skillet and fry until tender. Remove, and drain well.
8. To serve, arrange four shrimp on green pepper strips on a preheated plate, sprinkle with ajishio seasoning and garnish with a lemon slice.
9. Offer sweet and sour sauce separately.

Lobster Thermidor

(Homard Thermidor)

YIELD: 10 SERVINGS. SERVING SIZE: ONE-HALF LOBSTER.

INGREDIENTS:

5 each (1½ pound)	Lobsters
2 quarts	Court-bouillon
5 ounces	Roux blanc
4 ounces	Heavy cream
½ ounce	Mustard powder, hot
	Salt if needed

METHOD OF PREPARATION:

1. Boil lobster in flavored court-bouillon; do not overcook.
2. Chill and split lobster lengthwise, removing sac and tail meat.
3. Make a cream sauce from the strained court-bouillon, flavor with powdered mustard, add heavy cream, and pour a little into the whole length.
4. Slice lobster tail, place into shell, cover with sauce, and brown in a salamander.
5. Cut shell of claw lengthwise to expose flesh, and repeat same process as with the tail.

COOKING TECHNIQUES:

Boil, Bake

Boil: (at sea level)

1. Bring the cooking liquid to a rapid boil.
2. Stir the contents, and cook the food product throughout.
3. Serve hot.

Bake:

1. Preheat the oven.
2. Place the food product on the appropriate rack.

HACCP:

Hold warm at 140° F.

HAZARDOUS FOOD:

Lobster

NUTRITION:

Calories: 352
Fat: 22.8 g
Protein: 28.2 g

CHEF NOTES:

1. Lobster meat can be sautéed instead of boiled. In this case use fish stock fortified with fumet for the cream sauce.
2. Adding cheese during browning is at the discretion of the chef. The recipe does not call for it.

Moroccan Baked Fish

YIELD: 10 SERVINGS. SERVING SIZE: TEN OUNCES.

COOKING TECHNIQUE:
Bake

Bake:
1. Preheat the oven.
2. Place the food product on the appropriate rack.

GLOSSARY:
Blanch: to par cook

HACCP:
Hold at 140° F or above.

HAZARDOUS FOOD:
Scrod fillets

NUTRITION:
Calories: 263
Fat: 7.4 g
Protein: 28.9 g

INGREDIENTS:

20 ounces	Potatoes, washed, peeled, and cut into ¼-inch slices
20 ounces	Carrots, washed, peeled, sliced diagonally into ¼-inch slices, and **blanched**
12 cloves	Garlic, peeled and thinly sliced
50 six-ounce	Scrod fillets
2 ounces	Olive oil
½ teaspoon	Hot paprika
1 teaspoon	Sweet paprika
	Salt, to taste
20 ounces	Water
3 ounces	Fresh parsley, washed, excess moisture removed, and chopped

METHOD OF PREPARATION:

1. Preheat the oven to 350° F.
2. Layer the potatoes on the bottom of a 2-inch hotel pan.
3. Add the sliced carrots over the potatoes.
4. Place the fish fillets on the carrots.
5. Sprinkle the garlic over the fillets.
6. In a bowl, combine the remaining ingredients, except the parsley, and whisk together. Pour the mixture evenly over the fish.
7. Add water, place the pan in the oven, and bake until all of the water evaporates.
8. Serve immediately on a preheated dinner plate garnished with parsley, or hold at 140° F or above.

CHEF NOTE:
In Morocco, this dish is made with whole fish and is served as a meal. It usually is prepared very spicy. In some areas, halved lemons are baked with the fish, and in other areas, mainly near the Spanish Morocco border, olives are cooked with the fish.

Oriental Shrimp Curry

(Jhinga Kari)

YIELD: 10 SERVINGS. SERVING SIZE: FOUR SHRIMP.

COOKING TECHNIQUE:

Stir-Fry
Simmer

Stir-Fry:

1. Heat the oil in a wok until hot but not smoking.
2. Keep the food in constant motion; use the entire cooking surface.

GLOSSARY:

Julienne: matchstick strips

HACCP:

Hold at 140° F or above.

HAZARDOUS FOOD:

Shrimp

NUTRITION:

Calories: 309
Fat: 27.4 g
Protein: 9.03 g

INGREDIENTS:

40 each	Shrimp (16/20 count)
	Salt, to taste
2 tablespoons	White vinegar
8 ounces	Grated coconut
1 each	Coconut milk (15-ounce can)
4 ounces	Coriander seeds, ground
3 ounces	Vegetable oil
2 tablespoons	Ginger root, peeled and finely chopped
4 ounces	Onions, peeled and cut **julienned**
2 tablespoons	Garlic, peeled and finely chopped
1 teaspoon	Turmeric
½ teaspoon	Cumin
	Cayenne pepper, to taste
	Freshly ground black pepper, to taste

METHOD OF PREPARATION:

1. Peel the shrimp, leaving the tail section on. Devein and wash.
2. Combine the salt and white vinegar, add the shrimp, and marinate for 1 hour. Strain the shrimp, and reserve the marinade.
3. In a food processor, purée the coconut, coconut milk, and coriander seeds. Strain through a cheesecloth, and reserve.
4. In a wok, over high heat, add the oil; sauté the ginger root and onions. Then add the garlic, and stir-fry for about 2 minutes. Add the turmeric, cumin, cayenne pepper, and black pepper.
5. Add the shrimp marinade; heat to a boil, and simmer for 2 minutes.
6. Add the shrimp and coconut milk mixture, and simmer until the shrimp are thoroughly cooked.
7. To serve, ladle the soup into a preheated cup, and serve immediately, or hold the soup at 140° F or above.

CHEF NOTES:

1. This dish is typical of the cuisine of India.
2. If the soup will be held for a period of time, withhold the shrimp and add just before service.

Pasta, Shrimp, and Scallops in Chive-Ginger Sauce

(Pâtes aux Fruits de Mer)

YIELD: 10 SERVINGS. SERVING SIZE: TEN OUNCES.

COOKING TECHNIQUES:

Simmer, Boil, Sauté

Simmer and Poach:

1. Heat the cooking liquid to the proper temperature.
2. Submerge the food product completely.
3. Keep the cooked product moist and warm.

Boil: (at sea level)

1. Bring the cooking liquid to a rapid boil.
2. Stir the contents, and cook the food product throughout.
3. Serve hot.

Sauté:

1. Heat the sauté pan to the appropriate temperature.
2. Evenly brown the food product.
3. For a sauce, pour off any excess oil, reheat, and deglaze.

GLOSSARY:

Al dente: to the bite
Reduction: evaporation of liquid by boiling

HAZARDOUS FOODS:

Shrimp
Bay scallops
Heavy cream

NUTRITION:

Calories: 432
Fat: 31.6 g
Protein: 18.7 g

INGREDIENTS:

1 pound	Fettuccine
	Olive oil, as needed
2 ounces	Clarified butter, divided in half
4 ounces	Almonds, sliced
20 each	Raw shrimp, washed, shelled, deveined, and split in half
40 each	Bay scallops
24 ounces	Heavy cream
4 cloves	Garlic, peeled and finely chopped
1 teaspoon	Fresh ginger
2 ounces	Fresh chives, washed and thinly sliced
2 ounces	Lemon juice
Zest of 1	Lemon
	Salt and ground white pepper, to taste
5 ounces	Parmesan cheese, grated

METHOD OF PREPARATION:

1. Cook the fettuccine in boiling, salted water until **al dente.** Rinse, drain, and toss in a small amount of olive oil; reserve.
2. Heat half of the butter in a sauté pan, and sauté the almonds until golden brown; reserve. Discard the butter if brown; add fresh butter, and sauté the shrimp, then the scallops, over medium-high heat until barely done.
3. Mix the garlic and ginger with the cream. **Reduce** over medium-high heat until the sauce thickens and lightly coats the back of a spoon. Stir in the chives, lemon juice, lemon zest, salt, and pepper.
4. Add the seafood and heat through; add the almonds, and toss with hot, cooked pasta. Top with grated cheese, and serve.

Poached Flounder with Caper Sauce

YIELD: 50 SERVINGS. SERVING SIZE: FIVE OUNCES.

INGREDIENTS:

16 pounds	Flounder fillets, cut into 5-ounce portions
4 quarts	Fish stock (see page 10), heated to a boil
5 ounces	Lemon juice
2 pints	Dry white wine
	Salt, to taste
3 quarts	Caper sauce

METHOD OF PREPARATION:

1. Place the fish stock, wine, and lemon juice in a braising pan. Heat to a simmer.
2. Poach the fish approximately 7 minutes, or until it is firm to the touch.
3. Hold at 140° F.
4. For service, place the fish on a preheated dinner plate. **Nappé** with the sauce.

COOKING TECHNIQUES:

Poach or Simmer

Simmer and Poach:

1. Heat the cooking liquid to the proper temperature.
2. Submerge the food product completely.
3. Keep the cooked product moist and warm.

GLOSSARY:

Nappé: coated

HACCP:

Hold at 140° F.

HAZARDOUS FOOD:

Flounder fillets

NUTRITION:

Calories: 225
Fat: 6.43 g
Protein: 31.2 g

CHEF NOTES:

1. Sole, turbot, and whiting can be used as substitutes for the flounder.
2. In quantity food production, the seafood may be steamed.

Polynesian Sweet and Sour Fish

(L'a Momona a Paakai)

YIELD: 10 SERVINGS. SERVING SIZE: SIX OUNCES.

INGREDIENTS:

2½ pounds	Mahi-mahi, cut into ½-inch cubes
	Seasoned flour, as needed
	Vegetable oil, as needed
20 ounces	Sweet and sour sauce no. 2 (Tim Sourn Wu) (see page 72)

Marinade:

1 ounce	Ginger root, peeled and diced **brunoise**
4 ounces	Mirin (available in Oriental markets)
1 ounce	Soy sauce

Batter:

4 ounces	Flour
4 ounces	Water
2 ounces	Soy sauce
1 teaspoon	Sugar
1 each	Egg
¼ teaspoon	Baking soda
½ teaspoon	Garlic powder

METHOD OF PREPARATION:

1. Combine the ingredients for the marinade, and marinate the fish for 30 minutes. Drain and **dredge** in seasoned flour.
2. Mix together the batter ingredients.
3. In a wok, heat the oil as needed. Coat the fish with the batter, and deep-fry until golden brown.
4. In a separate wok, heat the sauce to a boil. Add the fish, and heat throughout.
5. Serve immediately on a preheated dinner plate, or hold at 140° F or above.

COOKING TECHNIQUE:

Deep-Fry

Deep-Fry:

1. Heat the frying liquid to the proper temperature.
2. Submerge the food product completely.
3. Fry the product until it is cooked throughout.

GLOSSARY:

Brunoise: ⅛-inch dice
Dredge: coat with flour

HACCP:

Hold at 140° F or above.

HAZARDOUS FOODS:

Mahi-mahi
Egg

NUTRITION:

Calories: 727
Fat: 45.3 g
Protein: 31.4 g

CHEF NOTE:

This dish is also typical of cuisine of the Philippines.

Roasted Tuna with Fresh Garden Vegetables

YIELD: 10 SERVINGS. SERVING SIZE: SIX OUNCES.

INGREDIENTS:

10 (6-ounce)	Tuna steaks
	Olive oil, as needed
3 medium	Onions, peeled and thinly sliced
3 each	Italian peppers (cubanelle), washed, seeded, and cut into ½-inch strips
3 each	Red bell peppers, washed, seeded, and cut into ¼-inch strips
3 medium	Firm, ripe tomatoes, washed, cored, blanched, peeled, seeded, and cut into ½-inch wedges
½ teaspoon	Marjoram, dried
1 teaspoon	Basil, fried
	Salt and ground white pepper, to taste
2 cloves	Garlic, peeled and mashed into a purée
2½ quarts	Vegetable stock, heated to a boil (see page 11)
2 tablespoons	Cornstarch
2 tablespoons	White wine

METHOD OF PREPARATION:

1. Preheat the oven to 350° F.
2. Heat a large sauté pan, coat the bottom with a minimum of olive oil, and sear the tuna steaks quickly on both sides. Transfer to a hotel pan.
3. Add the onions to the same pan, and sauté for 3 minutes; then add the peppers, and sauté for an additional 3 minutes. Transfer the vegetables to a hotel pan, spreading them over the tuna steaks.
4. Add more oil to the sauté pan, and when hot, add the tomatoes. Add the herbs and seasonings, and sauté for 2 to 3 minutes; then transfer to a hotel pan, spreading the mixture over the tuna.
5. Mix the garlic into the stock, and pour it over the fish.
6. Place the pan in the oven, and roast until the fish is firm, approximately 10 to 15 minutes, or until the internal temperature reaches 140° F.
7. Carefully transfer the tuna steaks to another hotel pan, cover it with parchment paper, and hold at 140° F.

COOKING TECHNIQUES:

Roast, Sauté

Roast:

1. Sear the food product, and brown evenly.
2. Elevate the food product in a roasting pan.
3. Determine doneness, and consider carryover cooking.
4. Let the food product rest before carving.

Sauté:

1. Heat the sauté pan to the appropriate temperature.
2. Evenly brown the food product.
3. For a sauce, pour off any excess oil, reheat, and deglaze.

GLOSSARY:

Nappé: coated

HACCP:

Heat to 140° F.
Hold at 140° F.

HAZARDOUS FOOD:

Tuna

NUTRITION:

Calories: 343
Fat: 6.06 g
Protein: 57.2 g

CHEF NOTES:

1. It is important that both the pan and the oil are hot before adding the fish steaks, so that they are seared very quickly.
2. Careful handling of the tuna is necessary to prevent overcooking or drying the texture of the fish.
3. Alternative fish, such as catfish, fillet of carp, cod, or haddock fillets, can be used.

8. Dilute the cornstarch in the wine. Place the hotel pan with the juices and vegetables over the heat, and heat to a boil. Stir in the cornstarch mixture, as needed, until the sauce is thickened to a syrupy consistency. Adjust the seasonings, and pour the mixture over the fish. Serve immediately for the best texture, or hold at 140° F.
9. Serve the tuna steak on a preheated dinner plate, and **nappé** with sauce and vegetables.

Roast Tuna Niçoise

YIELD: 10 SERVINGS. SERVING SIZE: SIX OUNCES.

COOKING TECHNIQUE:
Bake

Bake:
1. Preheat the oven.
2. Place the food product on the appropriate rack.

HAZARDOUS FOOD:
Tuna steaks

NUTRITION:
Calories: 355
Fat: 23.5 g
Protein: 26.5 g

INGREDIENTS:

10 (6-ounces each)	Tuna steaks (approximately 1-inch thick)
8 ounces	Olive oil
	Salt and ground white pepper, to taste
1 teaspoon	Thyme, ground
1 teaspoon	Bay leaves, ground
4 each	Tomatoes, washed, peeled, and sliced
2 each	Onions (medium), peeled and sliced
1 pint	White wine
3 ounces	Black olives
1 head	Garlic, peeled and chopped fine

METHOD OF PREPARATION:

1. Rub oil on both sides of each tuna steak. Season with the salt, pepper, thyme, and bay leaves.
2. Place the steaks in roasting pan.
3. Arrange two slices of tomatoes on each steak, and place a few rings of onion on top of the tomatoes.
4. Pour ¼ inch of wine in the roasting pan. (Adjust the amount according to the pan size.) Sprinkle the black olives and garlic on top.
5. Bake in a 375° F oven for 20 to 25 minutes, or until the fish is done. (Do not burn the tomatoes. If necessary, cover them with foil during the cooking process.)
6. Remove the steaks from the pan with a spatula; discard the cooking liquid.
7. Serve on a warm plate.

CHEF NOTE:
If tuna is not available, use swordfish, bluefish, or cod, adjusting the cooking time accordingly.

Salmon Steak with Sorrel Sauce

(Darné de Saumon à l'Oseille)

YIELD: 10 SERVINGS. SERVING SIZE: SIX OUNCES.

COOKING TECHNIQUE:
Sauté

Sauté:
1. Heat the sauté pan to the appropriate temperature.
2. Evenly brown the food product.
3. For a sauce, pour off any excess oil, reheat, and deglaze.

GLOSSARY:
Fumet: concentrated fish stock
Blanch: parcook
Temper: to equalize two extreme temperatures
Nappé: coated

HAZARDOUS FOODS:
Salmon
Heavy cream

NUTRITION:
Calories: 475
Fat: 29.9 g
Protein: 44.3 g

INGREDIENTS:

1½ pounds	Fresh sorrel, washed and stemmed
	Salted water, as needed
1 pint	White wine
12 ounces	Fish stock (fish **fumet**), heated to a boil (see page 10)
3 ounces	Shallots, peeled and chopped fine
12 ounces	Heavy cream
	Salt and ground white pepper, to taste
3 ounces	Oil
2 ounces	Butter
10 each (6-ounce)	Salmon steaks

METHOD OF PREPARATION:

1. **Blanch** the sorrel in boiling, salted water for 20 seconds. Drain, remove excess moisture, coarsely chop, and reserve.
2. In a sauté pan, combine the wine, fish fumet, and shallots. Slowly heat to a boil; then reduce the heat until the liquid becomes syrupy.
3. **Temper** the cream, and add to the reduction. Add the sorrel, and continue to reduce slowly until reaching a proper consistency. Adjust the seasoning with salt and pepper. Hold at 140° F or above.
4. In a large skillet, heat the oil and butter. Season the salmon steak with salt and pepper. Place the salmon in the skillet carefully, and cook for 2 to 3 minutes on each side. Do not overcook fish.
5. Remove the fish from the skillet. Place on clean paper towels to absorb the fat, and hold at 140° F.
6. To serve, place a salmon steak on a preheated dinner plate, and **nappé** with sorrel sauce.

CHEF NOTE:
Decorate the plates with additional fresh sorrel leaves or puff pastry fleurons.

Sautéed Dover Sole Meunière

(Sole de Douvres Sauté Meunière)

YIELD: 10 SERVINGS. SERVING SIZE: EIGHT OUNCES.

COOKING TECHNIQUE:
Sauté

Sauté:
1. Heat the sauté pan to the appropriate temperature.
2. Evenly brown the food product.
3. For a sauce, pour off any excess oil, reheat, and deglaze.

GLOSSARY:
Dredge: coat with flour
Nappé: coated

HAZARDOUS FOOD:
Dover sole

NUTRITION:
Calories: 354
Fat: 21.8 g
Protein: 37 g

INGREDIENTS:

5 each (15–20 oz.)	Dover sole
2 ounces	Fresh lemon juice
	Salt and ground white pepper, to taste
3 ounces	Butter
3 ounces	Oil
	All-purpose flour, as needed
5 ounces	Butter
3 ounces	Fresh lemon juice
3 ounces	Freshly chopped parsley, excess moisture removed
3 each	Lemons, sliced

METHOD OF PREPARATION:

1. Scale the Dover sole (white face only) with a chef knife from tail to head. Peel the black skin from the Dover sole, starting from the tail.
2. Cut the side fins with scissors, and remove the egg sack and clean the fish under running water.
3. In a large sauté pan, heat the butter and oil. **Dredge** the Dover sole in flour, and shake off any excess.
4. Sauté the sole on the white skin until golden brown; turn and bake in the oven for 7 minutes. Hold at 140° F or above.
5. Just before service, in a clean sauté pan, heat the butter until it foams; add the lemon juice and parsley.
6. To serve, place the sole on a preheated dinner plate. **Nappé** with beurre meunière, and garnish with sliced lemons.

CHEF NOTES:
1. A similar dish, called Belle Meunière, is prepared with the addition of fried eggplant.
2. *Meunière* means "in the style of the miller's wife."

Scallops with Mushroom Sauce and Duchess Potatoes

(Coquille St. Jacques)

YIELD: 10 SERVINGS. SERVING SIZE: ONE SHELL.

COOKING TECHNIQUES:
Simmer/Poach, Broil

Simmer and Poach:
1. Heat the cooking liquid to the proper temperature.
2. Submerge the food product completely.
3. Keep the cooked product moist and warm.

Grill/Broil:
1. Clean and heat the grill/broiler.
2. To prevent sticking, brush the food product with oil.

GLOSSARY:
Brunoise: ⅛-inch dice
Reduction: evaporation of liquid by boiling
Temper: equalize two extreme temperatures

HAZARDOUS FOODS:
Scallops
Milk
Egg yolks
Heavy cream

NUTRITION:
Calories: 361
Fat: 19.9 g
Protein: 24.2 g

INGREDIENTS:	
1 pint	White wine
3 each	Bay leaves
1 ounce	Shallots, peeled and diced **brunoise**
2 pounds	Scallops, rinsed in cold water and dried
1 pound	Mushrooms, washed, trimmed, and sliced
	Fish fumet, as needed
	Salt and ground white pepper, to taste
2 ounces	Butter
2 ounces	All-purpose flour
10 ounces	Fish fumet, heated to a boil
4 each	Egg yolks
8 ounces	Heavy cream
5 ounces	Swiss cheese, grated
20 ounces	Duchesse potatoes (see page 479)

CHEF NOTE:
Traditionally this is served in a scallop shell. The potato is piped to create a high rim to keep the sauce from overflowing.

METHOD OF PREPARATION:

1. In a saucepan, combine the wine, bay leaves, and shallots. Bring to a boil. Reduce the heat, and simmer for 4 minutes.
2. Add the scallops and mushrooms; cover with the fish fumet. Season, to taste.
3. Cover and poach for 5 minutes. Remove the scallops and mushrooms with a slotted spoon, and reserve. Discard the bay leaves. **Reduce** the cooking liquid to about 2 ounces.
4. In a saucepan, melt the butter. Add the flour, stirring constantly, to make a white roux. Add the fish fumet to the roux; stir the roux with a whisk. Slowly add the reduced liquid. Heat to a boil. Reduce the heat, and simmer for 5 minutes.
5. In a mixing bowl, whisk the egg yolks with the cream. **Temper** the mixture, and add to the sauce. Season with salt and pepper. Remove from the heat.
6. Pipe the Duchess potatoes around the edge of the casserole. Hold at 40° F, or until needed for service. Mix the scallops and mushrooms with two thirds of the sauce. Add the remaining sauce on top of each casserole, and sprinkle with grated cheese.
7. To serve, brown the casseroles under a salamander or broiler until they are heated throughout. Serve immediately on a preheated plate.

Shrimp and Vegetable Fritter

(Bah-Wan)

YIELD: 10 SERVINGS. SERVING SIZE: FIVE EACH.

COOKING TECHNIQUE:

Deep-Fry

Deep-Fry:

1. Heat the frying liquid to the proper temperature.
2. Submerge the food product completely.
3. Fry the product until it is cooked throughout.

GLOSSARY:

Julienne: matchstick strips
Brunoise: ⅛-inch dice

HACCP:

Hold at 140° F or above.

HAZARDOUS FOODS:

Eggs
Shrimp

NUTRITION:

Calories: 356
Fat: 6.02 g
Protein: 32.7 g

INGREDIENTS:

2½ pounds	Shrimp (26/30 count), peeled, deveined, and split lengthwise
8 ounces	Carrots, washed, peeled and cut **julienne**
3 ounces	Celery, washed and diced **brunoise**
4 cloves	Garlic, peeled and finely minced
2 teaspoons	Coriander, ground
¼ teaspoon	Cumin, ground
¼ teaspoon	White pepper, ground
	Salt, to taste
	Oil for frying, heated to 350° F as needed

Batter:

1 pound	All-purpose flour
6 each	Eggs
12 ounces	Water, or as needed to make a smooth batter

Vinegar and Garlic Dip:

4 ounces	White vinegar
4 cloves	Garlic, peeled and finely minced
1 tablespoon	Soy sauce

METHOD OF PREPARATION:

1. Place the flour in a bowl, beat the eggs, and add to the flour. Add water, as needed, to the mixture, and work it into a smooth batter.
2. Add the shrimp, carrots, celery, garlic, coriander, cumin, white pepper, and salt, to taste. Mix well, and let it rest for 10 minutes.
3. Scoop 1-ounce sized balls of mixture into the hot oil, and fry until golden brown. Dry on absorbent paper. Hold at 140° F or above.
4. Combine the ingredients for the dip, and mix well. Hold at room temperature.
5. To serve, arrange five fritters on a preheated plate, and offer dip in a small dish on the side.

CHEF NOTES:

1. This is an Indonesian dish.
2. When vegetables are used, cut very thin. The shrimp needs only 3 to 4 minutes to cook.

Shrimp Créole

(Camarones à la Créole)

YIELD: 10 SERVINGS. SERVING SIZE: FIVE OUNCES.

INGREDIENTS:

4 ounces	Lime juice
1 tablespoon	Salt
4 pounds	Shrimp (16 to 20 count)
	Cold water, as needed
6 ounces	Shallots, peeled and diced **brunoise**
6 ounces	Scallions, peeled and thinly sliced
12 cloves	Garlic, peeled and finely minced
3 each	Chili peppers, washed, seeded, and diced brunoise
1 pound	Tomato **concassé**
3 ounces	Fresh parsley, washed, excess moisture removed, and chopped
½ teaspoon	Thyme, ground
3 each	Bay leaves
	Salt, to taste
6 ounces	Olive oil

METHOD OF PREPARATION:

1. Preheat the oven to 375° F.
2. Combine the lime juice and salt, add the shrimp, and cover with water; marinate and refrigerate at 40° F or below for 1 hour.
3. Mix all of the remaining ingredients, and allow to stand for 1 hour.
4. Drain the shrimp, and place in a baking pan. Pour the olive oil mixture over the shrimp.
5. Cover the pan tightly with aluminum foil, and braise in the oven, approximately 8 to 10 minutes, or until the shrimp are firm.
6. Serve immediately on a preheated plate, and **nappé** with the vegetable mixture. Remove bay leaves before serving.

COOKING TECHNIQUE:

Braise

Braise:

1. Heat the braising pan to the proper temperature.
2. Sear and brown the food product to a golden color.
3. Degrease and deglaze.
4. Cook the food product in two thirds liquid until fork-tender.

GLOSSARY:

Brunoise: ⅛-inch dice
Concassé: peeled, seeded, and roughly chopped
Nappé: coated

HACCP:

Refrigerate at 40° F or below.

HAZARDOUS FOOD:

Shrimp

NUTRITION:

Calories: 319
Fat: 18 g
Protein: 29.6 g

CHEF NOTES:

1. This dish of Martinique is normally presented in a baking dish, and each guest helps him- or herself.
2. It can be prepared with other shellfish or any kind of white-fleshed fish.

COOKING TECHNIQUE:
Bake

Bake:
1. Preheat the oven.
2. Place the food product on the appropriate rack.

GLOSSARY:
Nappé: coated

HACCP:
Hold at 40° F or below.
Hold at 140° F or above.

HAZARDOUS FOODS:
Sole
Shrimp

NUTRITION:
Calories: 528
Fat: 36.7 g
Protein: 36.5 g

Sole with Shrimp and Artichokes in Vermouth Sauce

YIELD: 10 SERVINGS. SERVING SIZE: FIVE OUNCES.

INGREDIENTS:

2 ounces	Olive oil, or as needed
10 four-ounce	Sole fillets
20 each	Shrimp, peeled and deveined
1½ pounds	Canned artichoke hearts, drained and quartered
	Salt and freshly ground black pepper, to taste
3 ounces	Freshly squeezed lemon juice
	Vermouth sauce (see page 77)

METHOD OF PREPARATION:

1. Preheat the oven to 350° F.
2. Arrange the fish fillets in oiled baking dish.
3. Split the shrimp lengthwise, and arrange on the fish fillets, alternating with artichoke heart pieces.
4. Season, to taste, and drizzle the lemon juice and remaining olive oil on each.
5. Place in the oven, and bake for 7 minutes, or until firm. Remove when done; drain, and reserve the liquid. Hold the fish at 140° F or above.
6. To serve, transfer the fish to a preheated dinner plate, and **nappé** with the sauce.

Spanish Rice Casserole

(Paella)

YIELD: 10 SERVINGS. SERVING SIZE: TEN OUNCES.

INGREDIENTS:

4 ounces	Bacon, diced **brunoise** and **rendered**
2 ounces	Olive oil
10 each	Chicken thighs
5 ounces	Fresh peapods, washed and trimmed
8 ounces	Tomato **concassé**
½ teaspoon	Spanish paprika
1 pound	Long-grained rice, soaked in cool water for 30 minutes, then drained
1 quart	White chicken stock, heated to a boil (see page 12)
	Salt and ground white pepper, to taste
½ teaspoon	Saffron
3 four-ounce	Shell-on lobster tails, each cut into four pieces
10 each	Snails, cleaned
5 ounces	Squid, cleaned and sliced
10 each	Small clams, scrubbed, opened, and upper shell removed

METHOD OF PREPARATION:

1. In a large braising pan, combine the bacon fat and oil, heated to almost smoking, and brown the chicken thighs on all sides. Remove the chicken, and hold at 140° F.
2. **Blanch** the peapods in the oil, and then add the tomato concassé.
3. Add the paprika, and sauté until the fat turns red.
4. Add the drained rice, and sauté for 5 minutes, stirring frequently.
5. Return the chicken pieces to the pan, and add the boiling stock. Season, to taste, and add the saffron.
6. Cover and simmer for 10 minutes. Remove the cover, and place all of the seafood on top. Re-cover and continue to simmer until the chicken and seafood are tender and the liquid is absorbed by the rice.
7. Serve immediately, or hold at 140° F or above.
8. Dish onto preheated dinner plates, being sure that each portion has rice, chicken, lobster, snails, squid, and clams.
9. Place casserole in the center of the table so guests may serve themselves.

COOKING TECHNIQUES:

Sauté, Simmer

Sauté:

1. Heat the sauté pan to the appropriate temperature.
2. Evenly brown the food product.
3. For a sauce, pour off any excess oil, reheat, and deglaze.

Simmer and Poach:

1. Heat the cooking liquid to the proper temperature.
2. Submerge the food product completely.
3. Keep the cooked product moist and warm.

GLOSSARY:

Brunoise: ⅛-inch dice
Render: melt fat
Concassé: peeled, seeded, roughly chopped
Blanch: par cook

HACCP:

Hold at 140° F.

HAZARDOUS FOODS:

Chicken thighs
Lobster
Snails
Squid
Clams

NUTRITION:

Calories: 560
Fat: 28 g
Protein: 31.9 g

CHEF NOTES:

1. This is a national dish of Spain.
2. This dish is served in a special paella casserole. In restaurants it is prepared to order.

Spicy Grilled Catfish Steaks with Black Bean Salsa

YIELD: 10 SERVINGS. SERVING SIZE: EIGHT OUNCES (BONE IN).

INGREDIENTS:

10 (8-ounce)	Catfish steaks, cut with bone in
	Salt and ground white pepper, to taste
2 ounces	Vegetable oil or clarified butter
1½ ounces	Cajun seasoning, to taste
	Black bean salsa, as needed (see page 636)

METHOD OF PREPARATION:

1. Season the catfish steaks lightly with salt and pepper; then brush with oil or butter.
2. Rub each steak generously with Cajun seasoning. Lay out on a parchment-lined baking tray, and refrigerate at 40° F or lower, until needed for service.
3. Preheat the grill or a sauté pan lightly coated with oil.
4. Grill the catfish steaks, turning once, until just firm to the touch. Be careful, because Cajun seasoning tends to flame up with high heat. Transfer the grilled fish to a hotel pan, and hold at 140° F.
5. For service, place a catfish steak on a preheated dinner plate, and spoon salsa around the edge.

COOKING TECHNIQUE:

Grill

Grill/Broil:

1. Clean and heat the grill/broiler.
2. To prevent sticking, brush the food product with oil.

GLOSSARY:

Bain-marie: hot water bath
À la minute: cooked to order

HACCP:

Chill to 40° F.
Heat to an internal temperature of 140° F.
Hold at 140° F.

HAZARDOUS FOOD:

Catfish steaks

NUTRITION:

Calories: 310
Fat: 12.1 g
Protein: 39.3 g

CHEF NOTES:

1. Fish should be cooked until just firm and not dried out, or to an internal temperature of 140° F.
2. Fish is not an item that holds well in a **bain-marie;** it is best cooked and served **à la minute.**

Stuffed Clams

YIELD: 50 SERVINGS. SERVING SIZE: TWO QUAHOGS.

INGREDIENTS:

100 each	Hard-shell cherrystone clams, opened, liquid reserved, removed from the shell, and chopped
5 ounces	Oil
10 ounces	Butter
7 ounces	Garlic, peeled and diced **brunoise**
4 ounces	Paprika
5 each	Onions, medium, peeled and diced brunoise
10 ounces	Lemon juice
4 pounds	Bread crumbs
10 tablespoons	Freshly chopped parsley, excess moisture removed
	Salt, ground white pepper, and paprika, to taste
5 teaspoons	Oregano
2½ pounds	Parmesan cheese, grated
50 each	Lemon wedges
	Parsley, as needed for garnish

METHOD OF PREPARATION:

1. Preheat the oven to 425° F to 450° F.
2. In a skillet, heat the oil and butter. Sauté the onions and garlic on medium heat for 2 to 3 minutes, stirring constantly. Add the clam juice, reduce by half, and cool to an internal temperature of 40° F.
3. In a mixing bowl, place the chilled onion and garlic mixture. Add the lemon juice, bread crumbs, parsley, paprika, oregano, and Parmesan cheese. Mix well but gently, using a flat paddle. Add the chopped clams, and mix.
4. Take the shells, and stuff them with the clam mixture. Sprinkle some paprika over each.
5. Bake in the oven until the tops are brown. Serve with a lemon wedge and parsley.

COOKING TECHNIQUES:

Sauté, Bake

Sauté:

1. Heat the sauté pan to the appropriate temperature.
2. Evenly brown the food product.
3. For a sauce, pour off any excess oil, reheat, and deglaze.

Bake:

1. Preheat the oven.
2. Place the food product on the appropriate rack.

GLOSSARY:

Brunoise: ⅛-inch dice

HACCP:

Chill to 40° F.

HAZARDOUS FOOD:

Hard-shell cherrystone clams

NUTRITION:

Calories: 241
Fat: 15.5 g
Protein: 19.8 g

CHEF NOTES:

1. The bigger the clams, the tougher they will be, and also the more filling there will be. Clams can be baked on top of rock salt, which will steady them. (Do not use rock salt for seasoning, because it is unpurified.)
2. Any sausage (e.g., chorizo or andouille) may be added.

Terrine of Sole and Salmon with Lobster and Sorrel Sauce

YIELD: 10 SERVINGS (2 TERRINES). SERVING SIZE: SIX OUNCES.

COOKING TECHNIQUES:
Bake, Steam

Bake:
1. Preheat the oven.
2. Place the food product on the appropriate rack.

Steam: (Traditional)
1. Place a rack over a pot of water.
2. Prevent steam vapors from escaping.
3. Shock or cook the food product throughout.

HACCP:
Chill to 40° F or below.
Hold at 140° F or higher.

HAZARDOUS FOODS:
Sole
Salmon
Lobster
Eggs
Heavy cream

NUTRITION:
Calories: 498
Fat: 35.4 g
Protein: 39.6 g

INGREDIENTS:

3 ounces	Clarified butter, melted
2 pounds	Sole fillets, skinless, boneless
	Salt and ground white pepper, to taste
1½ pounds	Salmon fillets, all bones removed and flesh cut into 2-inch pieces
2 each	Eggs
12 ounces	Heavy cream, divided: 8 ounces, 4 ounces
8 ounces	Lobster meat, diced
2 ounces	Fresh dill leaves, washed and finely snipped
1 ounce	Kirschwasser
	Sorrel sauce (recipe follows)

METHOD OF PREPARATION:

1. Preheat the oven to 350° F.
2. Butter two terrines (9-inch by 5-inch by 3-inch), and then line each with plastic wrap. Butter the plastic wrap.
3. Season the sole fillets with salt and pepper, and line the terrines with fish, cutting and piecing as necessary.
4. In a food processor, purée the salmon; add the eggs, and season with salt and pepper, to taste. Scrape down the sides of the work bowl, and with the motor running, slowly add 8 ounces of the cream. Purée until smooth, stopping to scrape down the bowl one more time.
5. Transfer the salmon mixture to a bowl; cover and chill to 40° F or below for at least 30 minutes.
6. Whip the remaining cream until stiff.
7. Fold the whipped cream, lobster, dill, and Kirschwasser into the salmon mixture, and mix well.
8. Divide the mixture between the two terrines. Tap the terrines on a counter to eliminate air bubbles.
9. Place a buttered sheet of parchment paper over the top of each terrine. Cover the terrines tightly with foil.

10. Place the terrines in a shallow hotel pan, and add hot water to halfway up the sides of the terrines. Bake in the oven for 35 to 45 minutes, or until firm and a skewer inserted into the center of the terrines comes out clean. Remove from the oven and from the water bath, allowing the terrines to stand for 10 minutes before unmolding. Hold covered at 140° F or higher.
11. To serve, slice the terrines, and arrange two slices of sole on a preheated dinner plate. Serve with sorrel sauce.

Sorrel Sauce

INGREDIENTS:

1 pound	Sour cream
8 ounces	Sorrel leaves, washed and stemmed
2 ounces	Fresh dill leaves, washed and finely snipped
1 ounce	Freshly squeezed lemon juice
1 tablespoon	Dijon mustard
	Salt and freshly ground black pepper, to taste

METHOD FOR PREPARATION:

1. In a food processor, combine the ingredients, and blend until smooth.
2. Hold at 40° F or lower, until needed for service.

Tilapia Baked in a Salt Crust

YIELD: 10 SERVINGS. SERVING SIZE: EIGHT TO TEN OUNCES.

INGREDIENTS:

10 (1-pound)	Tilapia, cleaned, with head on, rinsed and dried
	Freshly ground black pepper, to taste
10 pounds	Coarse kosher salt
6 ounces	Spanish olive oil
5 each	Lemons, cut in crowns or wedges

METHOD OF PREPARATION:

1. Preheat the oven to 400° F.
2. Season the cavities of the fish with pepper.
3. Place 8 ounces of salt in the bottom of a 2-inch deep individual baking dish. Press a fish into the salt; then cover completely with additional salt.
4. Spray the top of the salt with water to help form a crust. Pat down the surface to follow the shape of the fish. Repeat the procedure for the remaining fish.
5. Bake for 20 to 25 minutes, depending on the exact size of the fish. Remove from the oven, and allow to stand for 5 minutes.
6. Carefully break the salt crust away from the fish, and lift the fish out to a sheet pan. Discard the salt. Hold the fish warm at 140° F.
7. To serve, fillet the fish, and arrange on a preheated dinner plate. Drizzle with olive oil, season with additional freshly ground pepper, and garnish with lemon.

COOKING TECHNIQUE:

Bake

Bake:

1. Preheat the oven.
2. Place the food product on the appropriate rack.

HACCP:

Hold the fish warm at 140° F.

HAZARDOUS FOOD:

Tilapia

NUTRITION:

Calories: 536
Fat: 24.9 g
Protein: 72.2 g

CHEF NOTES:

1. In Spain, this dish is commonly made with a fish called dorade.
2. Baking in a salt crust results in a very moist product that does not retain a salty flavor.

Trout Almandine

(Truite aux Amandes)

YIELD: 10 SERVINGS. SERVING SIZE: ONE FISH.

INGREDIENTS:

10 each	Fresh trout, cleaned (bone in)
	Salt and ground white pepper, to taste
	Seasoned flour, as needed
5 ounces	Clarified butter
10 ounces	Sliced almonds, toasted in butter
5 ounces	Raw butter
4 ounces	White wine
4 ounces	Fresh lemon juice
3 ounces	Fresh parsley, washed, chopped fine, and excess moisture removed
2 each	Lemons, washed, ends cut off, and cut into 20 wedges

METHOD OF PREPARATION:

1. Season the trout with salt and pepper, and **dredge** in flour. Shake off any excess flour.
2. In a sauté pan, heat the clairified butter, and sauté the trout until golden brown, turning only once.
3. Remove the trout from the pan, and hold at 140° F or above.
4. Sauté the sliced almonds in the butter; then remove, and drain on absorbant paper.
5. **Deglaze** the pan with wine and lemon juice, and reducc by half.
6. To serve, place a fish on a preheated plate, and top with almonds and butter sauce. Garnish with parsley and lemon wedges.

COOKING TECHNIQUE:

Sauté

Sauté:

1. Heat the sauté pan to the appropriate temperature.
2. Evenly brown the food product.
3. For a sauce, pour off any excess oil, reheat, and deglaze.

GLOSSARY:

Dredge: coat with flour
Deglaze: add liquid to hot pan

HACCP:

Hold at 140° F or above.

HAZARDOUS FOOD:

Trout

NUTRITION:

Calories: 629
Fat: 36.4 g
Protein: 48.5 g

CHEF NOTE:

If trout will be deboned and skinned tableside, reserve the almonds for the server to sprinkle on after filleting.

Tuna with Tomato and Anchovy Sauce

YIELD: 10 SERVINGS. SERVING SIZE: FIVE OUNCES.

INGREDIENTS:

10 (5-ounce)	Tuna steaks
	Olive oil, as needed
Pinch	Salt
¼ teaspoon	Freshly ground black pepper
8 ounces	Onions, peeled and diced **brunoise**
4 cloves	Garlic, peeled and finely minced
1 pound	Tomato **concassé**
1 quart	Tomato sauce, heated to a boil (see page 76)
2 ounces	Anchovy paste or minced fillets, drained of oil and puréed
10 ounces	Fresh bread crumbs
3 ounces	Fresh parsley, washed, excess moisture removed, and chopped
2 each	Lemons, washed, ends removed, sliced into wedges, and seeds removed

METHOD OF PREPARATION:

1. Preheat the oven to 375° F.
2. Brush the tuna steaks with oil, sprinkle with salt, and refrigerate at 40° F or below until needed.
3. In a sauté pan, heat 2 ounces of the oil, and sauté the onions until translucent. Add the garlic and tomato concassé, and continue to sauté for 5 minutes.
4. Add the tomato sauce and anchovy paste or puréed anchovy fillets. Heat the mixture to a boil; reduce the heat, and simmer for 10 to 15 minutes.
5. Taste, and season as needed. Remove from the heat, transfer to a suitable container, and hold at 140° F or above.
6. Dredge the marinated tuna steaks in fresh bread crumbs.
7. In a sauté pan, heat the remaining olive oil, and quickly **sear** the tuna; then remove, and drain on absorbent paper.
8. Divide the tomato sauce into individual, oiled baking dishes. Place the tuna steaks on top, and bake to order until firm, or to an internal temperature of 165° F. The amount of time for baking will depend on the thickness of the tuna. Hold unbaked portions at 40° F or below.
9. To serve, transfer the tuna to a preheated dinner plate, nappé with the sauce, and garnish with parsley and lemon wedges. Serve immediately.

COOKING TECHNIQUES:
Sauté, Bake

Sauté:
1. Heat the sauté pan to the appropriate temperature.
2. Evenly brown the food product.
3. For a sauce, pour off any excess oil, reheat, and deglaze.

Bake:
1. Preheat the oven.
2. Place the food product on the appropriate rack.

GLOSSARY:
Brunoise: ⅛-inch dice
Concassé: peeled, seeded, roughly chopped
Sear: brown quickly
Bain-marie: hot-water bath

HACCP:
Refrigerate at 40° F or below.
Hold at 140° F or above.
Bake to an internal temperature of 165° F.

HAZARDOUS FOOD:
Tuna steaks

NUTRITION:
Calories: 444
Fat: 19.7 g
Protein: 34.6 g

CHEF NOTE:
Fish can be prepared in one hotel pan, but it does not hold well in a **bain-marie.** If service will be quick, hold at 140° F.

Braised Veal Roast

(Fricandeau de Veau Braisé)

YIELD: 10 SERVINGS. SERVING SIZE: FIVE OUNCES.

INGREDIENTS:

8 ounces	Salt pork, cut into ¼-inch ribbons
5 pounds	Veal, bottom round, trimmed
	Salt and ground white pepper, to taste
½ teaspoon	Ginger, ground
2 ounces	Oil
2 ounces	Butter
3 cloves	Garlic, peeled and sliced
2 each	Ripened tomatoes, washed, cored, and sliced
1 teaspoon	Paprika
2 quarts	Brown veal stock, heated to boil (see page 7)
1½ ounces	Cornstarch, dissolved in 3 ounces of cold water

Mirepoix:

4 ounces	Onions, peeled
4 ounces	Carrots, washed and peeled
4 ounces	Celery, washed

METHOD OF PREPARATION:

1. Preheat the oven to 325° F.
2. Lay out ribbons of salt pork on a plastic-lined tray, and freeze until firm.
3. Using a larding needle, insert frozen salt pork into the veal.
4. Rub larded meat with salt, pepper, and ginger.
5. In a braising pan, heat the oil and butter. **Sear** the meat on all sides until browned. Remove, and reserve.
6. Sauté the mirepoix in the same fat in the braising pan. Reserve until the onions become translucent. Add garlic and tomatoes, and continue to sauté for 5 minutes.
7. Add the paprika, and sauté for 1 minute. Add the veal stock.
8. Return the meat to the pan, and cover the braiser. Place in the oven to braise until the meat is **fork-tender,** turning every 20 minutes, approximately 1½ hours.
9. Remove the meat, and hold at 140° F.
10. Strain the sauce through a **chinois** into a saucepan. **Reduce** the sauce until the desired taste is achieved.

COOKING TECHNIQUES:
Braise, Sauté

Braise:
1. Heat the braising pan to the proper temperature.
2. Sear and brown the food product to a golden color.
3. Degrease and deglaze.
4. Cook the food product in two thirds liquid until fork-tender.

Sauté:
1. Heat the sauté pan to the appropriate temperature.
2. Evenly brown the food product.
3. For a sauce, pour off any excess oil, reheat, and deglaze.

GLOSSARY:
Mirepoix: roughly chopped vegetables
Sear: to brown quickly
Reduction: evaporation of liquid by boiling
Fork-tender: without resistance
Chinois: cone-shaped strainer

HAZARDOUS FOOD:
Veal, bottom round

NUTRITION:
Calories: 362
Fat: 24.8 g
Protein: 25.5 g

11. Taste, and adjust the seasoning.
12. Whisk the cornstarch mixture into the sauce, and continue to simmer for 10 minutes. Hold at 140° F.
13. To serve, slice the meat against the grain. Ladle the sauce onto a preheated dinner plate, and arrange slices, overlapping, on the sauce.

Braised Veal Shanks

(Osso Bucco)

YIELD: 10 SERVINGS. SERVING SIZE: TWELVE OUNCES (BONE IN).

INGREDIENTS:

10 (2-inch thick)	Veal shanks
	Seasoned flour, as needed for **dredging**
	Oil, as needed
6 cloves	Garlic, peeled and mashed into a purée
1 tablespoon	Sage, crushed, dried
3 pounds	Canned tomatoes, drained, seeded, and cut into a ½-inch dice
4 ounces	Tomato purée
2 each	Bay leaves
	Salt and freshly ground black pepper, to taste
8 ounces	Almond Marsala
1 quart	White veal stock, heated to a boil (see page 12)
1 pint	Veal demi-glace, heated to a boil (see page 26)
	Gremolada, recipe follows

Macédoine:

8 ounces	Onions, peeled
8 ounces	Carrots, washed and peeled
4 ounces	Celery, washed and trimmed

METHOD OF PREPARATION:

1. Preheat the oven to 350° F.
2. Dredge the shanks in flour, shaking off any excess. In a braising pan, heat enough oil to cover the bottom of the pan, and **sear** the shanks on all sides; then transfer the shanks to a hotel pan.
3. In the same braising pan, sauté the macédoine of vegetables; then add the garlic, sage, tomatoes, tomato purée, and bay leaves, and season, to taste.
4. **Deglaze** the pan with Marsala; then add the stock and demi-glace. Heat the liquid to a boil; reduce the heat, and simmer for 15 minutes.
5. Pour the sauce over the veal shanks. Cover the hotel pan with aluminum foil, and place in the oven to braise until the meat is fork-tender, or for approximately 1½ hours.
6. Remove the veal shanks, and hold covered at 140° F or higher.
7. Strain the sauce through a chinois into a saucepot, pressing down on the vegetables with a ladle to extract all of the flavor. Heat the sauce to a boil;

COOKING TECHNIQUES:

Braise, Sauté

Braise:

1. Heat the braising pan to the proper temperature.
2. Sear and brown the food product to a golden color.
3. Degrease and deglaze.
4. Cook the food product in two thirds liquid until fork-tender.

Sauté:

1. Heat the sauté pan to the appropriate temperature.
2. Evenly brown the food product.
3. For a sauce, pour off any excess oil, reheat, and deglaze.

GLOSSARY:

Dredge: to coat with flour
Macédoine: ¼-inch dice
Sear: to brown quickly
Deglaze: to add liquid to hot pan

HACCP:

Hold covered at 140° F or higher.

HAZARDOUS FOOD:

Veal shanks

NUTRITION:

Calories: 496
Fat: 11.7 g
Protein: 80.2 g

CHEF NOTES:

1. If almond Marsala is not available, substitute regular Marsala.
2. For a heavier-bodied sauce, purée the sauce; then strain through a chinois.

adjust the seasoning as needed, and then reduce the heat, and simmer until the proper consistency is achieved. Dépouiller any fat from the surface. Pour the sauce back over the veal shanks, and hold at 140° F (minimum).

8. To serve, place one shank on a preheated dinner plate, nappé with the sauce, and sprinkle with some of the gremolada.

Gremolada Garnish

INGREDIENTS:

3 ounces	Zest of lemon, minced
6 to 8 cloves	Garlic, peeled and finely minced
3 ounces	Fresh parsley, washed, excess moisture removed, and chopped

METHOD OF PREPARATION:

Combine the ingredients, and reserve as garnish.

Roast Loin of Veal

(Côte de Veau Prince Orloff)

YIELD: 8 SERVINGS. SERVING SIZE: FIVE OUNCES.

INGREDIENTS:

10 to 12 pounds	Veal loin, double, fat trimmed, rib bone, flank, and tenderloins removed
	Salt and ground white pepper, to taste
	Vegetable oil, as needed
1 quart	Sauce soubise (see page 66)
8 ounces	Gruyére cheese, grated

Mirepoix:

6 ounces	Onions, peeled and diced **parmentier**
4 ounces	Carrots, washed, trimmed, and diced parmentier
8 ounces	White wine
1 quart	Veal stock (see page 7)
3 ounces	Arrowroot
	Bread crumbs, as needed

METHOD OF PREPARATION:

1. Trim, season, and tie the veal loin.
2. In a roasting pan, heat the oil until smoking, and **sear** the veal on all sides.
3. Place the veal on top of the mirepoix and trimmings in a roasting pan. Place in a preheated 375° F oven. Cook until medium-rare and an internal temperature of 140° F is reached.
4. Remove, and let cool. Remove the veal loins, slice, spread with the sauce soubise, and reassemble into the veal loin. Cover the veal loin with more sauce soubise.
5. Sprinkle veal with the cheese and bread crumbs. Return it to the oven, and bake until golden brown. Then remove the veal from the pan, and keep it warm.
6. Place the roasting pan on the stove, and caramelize the mirepoix. Skim any excess fat, add the wine, and deglaze.
7. Add the veal stock, and reduce. Season with salt and white pepper.
8. Thicken with arrowroot, strain, and hold at 140° F.
9. Ladle 1 ounce of sauce onto preheated dinner plates, and top with the sliced veal.

COOKING TECHNIQUE:

Roast

Roast:

1. Sear the food product, and brown evenly.
2. Elevate the food product in a roasting pan.
3. Determine doneness, and consider carryover cooking.
4. Let the food product rest before carving.

GLOSSARY:

Mirepoix: roughly chopped vegetables
Parmentier: ½-inch dice
Sear: to brown quickly

HAZARDOUS FOOD:

Veal loin

NUTRITION:

Calories: 1172
Fat: 63 g
Protein: 107 g

CHEF NOTE:

The FDA recommends that veal be cooked to an internal temperature of 140° F.

Rolled Veal Scallops

(Paupiettes de Veau)

YIELD: 10 SERVINGS. SERVING SIZE: SIX OUNCES.

INGREDIENTS:

10 (5-ounce)	Veal cutlets, pounded thin
6 ounces	Onion, peeled and diced **brunoise**
8 ounces	Carrots, washed, peeled, and diced brunoise
8 ounces	Celery stalks, washed, trimmed, and diced brunoise
3 cloves	Garlic, peeled and sliced
1 pound	Canned, crushed tomatoes
2 ounces	Vegetable oil
1 pint	Dry white wine
1 pint	Veal demi-glace, heated to a boil (see page 26)
	Salt and freshly ground black pepper, to taste

Farce:

8 ounces	Ground veal
8 ounces	Ground pork
8 ounces	Rice, cooked
2 each	Eggs
½ teaspoon	Oregano, dried
½ teaspoon	Basil, dried
¼ teaspoon	Ginger, ground
	Salt and ground white pepper, to taste

METHOD OF PREPARATION:

1. Combine the farce ingredients, and mix well.
2. Lay out the pounded cutlets, and spread farce mixture on each.
3. Roll the cutlets, and secure with butcher twine or toothpicks.
4. In a sauté pan, heat the oil, and **sear** the veal scallops on all sides; then transfer the rolls to a hotel pan.
5. Sauté the vegetables in the same oil used to sear the veal. Add the wine and veal demi-glace.
6. Season the sauce, and cook 10 minutes.
7. Pour the sauce over the veal; seal the pan with aluminum foil, and simmer over low heat until **fork-tender,** approximately 20 minutes. Hold at 140° F.
8. To serve, remove twine or toothpicks. Strain the sauce. Place one "scallop" on a preheated dinner plate, and **nappé** with the sauce.

COOKING TECHNIQUES:
Sauté, Simmer

Sauté:
1. Heat the sauté pan to the appropriate temperature.
2. Evenly brown the food product.
3. For a sauce, pour off any excess oil, reheat, and deglaze.

Simmer and Poach:
1. Heat the cooking liquid to the proper temperature.
2. Submerge the food product completely.
3. Keep the cooked product moist and warm.

GLOSSARY:
Brunoise: ⅛-inch dice
Sear: to brown quickly
Fork-tender: without resistance
Nappé: coated

HACCP:
Hold at 140° F.

HAZARDOUS FOODS:
Ground veal
Ground pork
Eggs

NUTRITION:
Calories: 517
Fat: 27 g
Protein: 39 g

Stuffed Breast of Veal

YIELD: 50 SERVINGS. SERVING SIZE: EIGHT OUNCES.

INGREDIENTS:

5 each	Veal breasts weighing 7 to 9 pounds each, breastbones removed, trimmed, and pockets cut for stuffing
	Flour, as needed
	Olive oil, as needed
13 ounces	Carrots, washed and chopped
1¼ pounds	Onions, chopped
4 stalks	Celery, washed, trimmed and chopped
½ bunch	Freshly chopped parsley, stems and excess moisture removed
5 each	Bay leaves
3 tablespoons	Thyme, dried
1¼ pints	White wine
2 gallons	Veal demi-glace, heated to a boil (see page 26)
	Salt and ground white pepper, to taste

Stuffing (veal mixture):

1¼ pounds	White rice, cooked
5 ounces	Butter
2 pounds	Onions, peeled and diced **brunoise**
7 ounces	White chicken stock (see page 12)
5 pounds	Veal, ground
	Salt and ground white pepper, to taste
5 pounds	Spinach, washed and trimmed
1¼ pounds	Carrots, washed, peeled, and diced brunoise
1¼ pounds	Mushrooms, cleaned and thinly sliced
5 ounces	Butter

METHOD OF PREPARATION:

1. Preheat the oven to 450° F.
2. Sauté onions in butter until translucent. Add all the vegetables and sauté for 10 minutes. Remove from heat and place in a large bowl. Add all other ingredients. Season and mix well.

COOKING TECHNIQUE:
Braise

Braise:
1. Heat the braising pan to the proper temperature.
2. Sear and brown the food product to a golden color.
3. Degrease and deglaze.
4. Cook the food product in two thirds liquid until fork-tender.

GLOSSARY:
Brunoise: ⅛-inch dice
Truss: to tie or secure
Dépouiller: to skim impurities
Fork-tender: without resistance

HACCP:
Hold at 140° F.

HAZARDOUS FOODS:
Veal breasts
Ground veal

NUTRITION:
Calories: 808
Fat: 32 g
Protein: 108 g

3. Stuff the breasts with the veal mixture, and **truss.** Rub the breasts with the oil, and place in a braising pan. Brown in the oven, turning once. Then remove the veal, and spread the vegetables and herbs in the pan.
4. Place the veal on top of the vegetables, and add the wine and stock to cover meat by two thirds. Cover the pan, and braise in a 375° F oven for about 3 hours, or until **fork-tender.** Cook uncovered during the last ½ hour, basting frequently with the pan liquid. Remove the veal, and allow it to rest. Hold at 140° F.
5. Strain the braising liquid into a saucepan, and **dépouiller.** Adjust the seasonings and consistency to make the sauce.
6. For each serving, place one slice of veal on a preheated plate with some sauce. Garnish with additional parsley.

Veal Chops Pojarski

YIELD: 50 SERVINGS. SERVING SIZE: FIVE OUNCES.

COOKING TECHNIQUE:

Shallow-Fry

Shallow-Fry:

1. Heat the cooking medium to the proper temperature.
2. Cook the food product throughout.
3. Season, and serve hot.

GLOSSARY:

Brunoise: ⅛-inch dice

HACCP:

Hold at 140° F.

HAZARDOUS FOODS:

Ground veal leg
Whole eggs
Milk
Heavy cream

NUTRITION:

Calories: 537
Fat: 23.3 g
Protein: 36 g

INGREDIENTS:

12 pounds	Veal leg, ground
3 pounds	Bread crumbs
5 each	Whole eggs
¼ teaspoon	Nutmeg, ground
	Salt and ground white pepper, to taste
	Heavy cream, as needed
3 each	Onions, peeled, diced **brunoise,** and sautéed in clarified butter
	Oil, as needed
1 pint	Butter, clarified
6 each	Lemons, sliced

Breading:

	Seasoned flour, as needed
25 each	Whole eggs, slightly beaten
16 ounces	Milk, combined with eggs
	Bread crumbs, as needed

METHOD OF PREPARATION:

1. Mix the bread crumbs, cream, onions, nutmeg, and eggs to form a paste. Add to the ground veal meat. Season with salt and pepper and incorporate.
2. Form the mixture into 5-ounce chop-shaped portions. Refrigerate for 1 hour.
3. Dredge the chops in the flour, and dip in the egg wash and then into the bread crumbs.
4. Shallow-fry the chops in the oil until golden brown and an internal temperature of 160° F is achieved. Serve immediately on a preheated dinner plate, or hold at 140° F.
5. Serve garnished with lemon slices.

Veal Cordon Bleu

(Escalope de Veau Cordon Bleu)

YIELD: 10 SERVINGS. SERVING SIZE: FIVE OUNCES.

COOKING TECHNIQUES:
Sauté, Bake

Sauté:
1. Heat the sauté pan to the appropriate temperature.
2. Evenly brown the food product.
3. For a sauce, pour off any excess oil, reheat, and deglaze.

Bake:
1. Preheat the oven.
2. Place the food product on the appropriate rack.

GLOSSARY:
Dredge: to coat with flour

HAZARDOUS FOODS:
Veal, top round cutlets
Eggs
Milk

NUTRITION:
Calories: 762
Fat: 42.6 g
Protein: 38.9 g

INGREDIENTS:

10 (5-ounce)	Veal, top round cutlets
10 (1-ounce)	Ham slices
10 (½-ounce)	Swiss cheese slices
	Seasoned flour, as needed for **dredging**
3 each	Eggs
1 pound	Fresh bread crumbs, made from the white of the bread only and seasoned with salt and white pepper
4 ounces	Oil
4 ounces	Butter
5 ounces	Veal demi-glace (see page 26)

METHOD OF PREPARATION:

1. Pound the veal cutlets into thin, even slices. Place a ham slice on top of each veal cutlet and a cheese slice on top of each ham slice. Then fold the veal over, and seal the edges.
2. Bread the veal by dipping it in the flour, egg wash, and bread crumbs.
3. Heat the oil and butter in a sauté pan. Sauté the veal to order, until golden brown on all sides. Finish by baking in a 350° F oven for 5 to 7 minutes.
4. To serve, place a veal cutlet on warm dinner plate, surrounded with a cordon of demi-glace.

CHEF NOTE:
Veal cordon bleu may be served with a variety of sauces, including veal demi-glacé (see page 26).

Veal Cutlets with Cream and Mushroom Sauce

(Escalope de Veau à la Crème)

YIELD: 10 SERVINGS. SERVING SIZE: FIVE OUNCES.

INGREDIENTS:

4 pounds	Veal cutlets, pounded with a mallet
	Seasoned flour, as needed for **dredging**
5 ounces	Butter
2 ounces	Oil
12 ounces	Mushrooms, cleaned
3 ounces	Cognac
2 ounces	Meat glaze (see page 36)
1½ pints	Heavy cream
	Salt and ground white pepper, to taste

METHOD OF PREPARATION:

1. Coat the veal cutlets with flour, and shake off any excess. In a large skillet, heat the butter and oil. Sauté the veal on each side for 2 minutes. Remove the veal, and keep warm.
2. Add the mushrooms, and sauté for 4 minutes. Add the cognac, and **flambé.**
3. Add the meat glaze and cream to the mushrooms. Stir, heat, and **reduce** the sauce until fairly thick but still liquid; season with salt and pepper.
4. Arrange the veal on a preheated dinner plate, and spoon the mushrooms and sauce over it.

COOKING TECHNIQUE:
Sauté

Sauté:
1. Heat the sauté pan to the appropriate temperature.
2. Evenly brown the food product.
3. For a sauce, pour off any excess oil, reheat, and deglaze.

GLOSSARY:
Dredge: to coat with flour
Flambé: to flame
Reduction: evaporation of liquid by boiling

HAZARDOUS FOODS:
Veal cutlets
Heavy cream

NUTRITION:
Calories: 475
Fat: 33.2 g
Protein: 28.4 g

CHEF NOTE:
Morels, chanterelles, cépes, or any wild mushrooms can be used.

COOKING TECHNIQUE:

Sauté

Sauté:

1. Heat the sauté pan to the appropriate temperature.
2. Evenly brown the food product.
3. For a sauce, pour off any excess oil, reheat, and deglaze.

GLOSSARY:

Brunoise: ⅛-inch dice
Dredge: to coat with flour

HACCP:

Hold at 140° F or above.

HAZARDOUS FOODS:

Veal
Heavy cream

NUTRITION:

Calories: 644
Fat: 43.7 g
Protein: 38.6 g

Veal Cutlets with Mushrooms

(Escalope de Veau Chimay)

YIELD: 10 SERVINGS. SERVING SIZE: FIVE OUNCES.

INGREDIENTS:

4 pounds	Veal cutlets, trimmed and silverskin removed
	Seasoned bread flour, as needed
	Salt and ground white pepper, to taste
3 ounces	Vegetable oil
3 ounces	Clarified butter
4 ounces	Shallots, peeled and diced **brunoise**
12 ounces	Mushrooms, washed

Glacage:

8 ounces	Hollandaise sauce (see page 30)

METHOD OF PREPARATION:

1. Slice the veal into 2½-ounce cutlets, and pound each lightly with a meat mallet.
2. **Dredge** the cutlets in seasoned flour, and shake off any excess.
3. In a sauté pan, heat the oil, and sauté the veal, turning once, until lightly browned. Remove, place in a hotel pan, and hold at 140° F.
4. Add the butter to the sauté pan, and sauté the shallots.
5. Purée the mushrooms in a robot coupe. Add to the shallots. Simmer until almost dry; season. Remove, and let cool.
6. Lay out half of the sautéed cutlets on a half sheet pan. Spread the mushroom mixture evenly over the cutlets and place cutlets on mushrooms.
7. Cover the layered cutlets with the Hollandaise sauce, and glaze under a salamander or broiler. Serve immediately on a preheated dinner plate.

Veal Paupiettes with Roasted Garlic and Sun-Dried Tomato Purée

YIELD: 10 SERVINGS. SERVING SIZE: EIGHT OUNCES.

INGREDIENTS:

20 (3-ounce)	Veal top round cutlets, pounded lightly to tenderize
4 ounces	Sour dough bread, crusts removed and torn into pieces
2 ounces	Milk
1 pound	Veal, finely ground
2 each	Eggs, cracked and lightly beaten
1 ounce	Italian (flat-leafed) parsley, washed, finely chopped
	Salt and ground white pepper, to taste
4 ounces	Hot mustard
6 ounces	Carrots, washed, peeled, cut **julienne,** and steamed or blanched
6 ounces	Brine-cured pickles, cut julienne
	Seasoned flour, as needed
4 ounces	Clarified butter
1 pint	Red wine
1 pint	Veal demi-glace, heated to a boil (see page 26)
	Roasted garlic and sun-dried tomato purée (see page 48)

METHOD OF PREPARATION:

1. Preheat the oven to 350° F.
2. Lay out the veal slices on a plastic-lined tray, and hold refrigerated at 40° F or lower.
3. Soak the bread in the milk to soften; then squeeze out any excess liquid.
4. Place the ground veal in a mixing bowl; add the soaked bread, eggs, parsley, and seasonings, to taste. Mix well, and keep chilled at 40° F or lower until needed.
5. Brush the veal cutlets with the mustard, and place about 2 ounces of the stuffing on each cutlet. Place strips of carrot and pickle over the filling, and roll. Tie each **paupiette** with butcher twine, or secure with toothpicks.

COOKING TECHNIQUES:

Sauté, Braise, Roast

Sauté:

1. Heat the sauté pan to the appropriate temperature.
2. Evenly brown the food product.
3. For a sauce, pour off any excess oil, reheat, and deglaze.

Braise:

1. Heat the braising pan to the proper temperature.
2. Sear and brown the food product to a golden color.
3. Degrease and deglaze.
4. Cook the food product in two thirds liquid until fork-tender.

Roast:

1. Sear the food product, and brown evenly.
2. Elevate the food product in a roasting pan.
3. Determine doneness, and consider carryover cooking.
4. Let the food product rest before carving.

GLOSSARY:

Paupiette: rolled meat or fish
Reduction: evaporation of liquid by boiling
Julienne: matchstick strips
Deglaze: to add liquid to hot pan

HACCP:

Hold refrigerated at 40° F or higher.
Hold at 140° F or higher

HAZARDOUS FOODS:

Veal
Milk
Eggs

NUTRITION:

Calories: 748
Fat: 46.4 g
Protein: 56.2 g

6. Dip each roll into seasoned flour, and shake off any excess.
7. In a sautoir, heat the butter; add the veal rolls, and sauté until browned on all sides. As the rolls are browned, transfer them to a hotel pan.
8. **Deglaze** the pan with red wine, then pour over the veal. Add the demi-glacé, cover the pan tightly, and place in the oven. Braise until the veal is fork-tender, or about 1 hour.
9. Once tender, transfer the paupiettes to another hotel pan, cover loosely, and hold at 140° F or higher.
10. Strain the sauce through a chinois into a small saucepan. Place on medium heat, and **reduce** to the desired consistency and flavor. Taste, and adjust the seasonings. Pour the sauce back over the veal, and hold at 140° F.
11. For service, remove twine or toothpicks, arrange two paupiettes on a preheated plate, nappé with the sauce, and garnish the plate with roasted garlic and sun-dried tomato aioli, piped through a star tube or squeezed through a squirt bottle.

Veal with Prunes

YIELD: 10 SERVINGS. SERVING SIZE: FIVE OUNCES.

INGREDIENTS:

10 five-ounce	Top round veal cutlets, pounded with a mallet to tenderize and flatten
	Seasoned flour, as needed
2 ounces	Vegetable oil
1 pound	Onions, peeled and diced **brunoise**
12 ounces	Pitted prunes, split in half lengthwise
1 quart	Veal demi-glace, heated to boil (see page 26)
	Salt and white pepper to taste
½ teaspoon	Ginger, ground

METHOD OF PREPARATION:

1. Preheat the oven to 350° F.
2. Dredge the veal cutlets in flour, shaking off any excess.
3. Heat the oil in a sauté pan, and **sear** the meat on both sides. Remove, and arrange in a shallow braising pan.
4. In the same fat in which the meat was seared, sauté the onions until they become translucent. Add the prunes, and sauté for an additional 5 minutes.
5. Add the demi-glace to the onions and prunes, heat to a boil, season, and then pour over the veal. Cover the pan with foil, and braise in the oven for 20 minutes, or until the veal is **fork-tender.**
6. To serve, arrange a veal cutlet on a preheated dinner plate, surround it with prunes, and nappé with the sauce.

COOKING TECHNIQUES:

Sauté, Braise

Sauté:

1. Heat the sauté pan to the appropriate temperature.
2. Evenly brown the food product.
3. For a sauce, pour off any excess oil, reheat, and deglaze.

Braise:

1. Heat the braising pan to the proper temperature.
2. Sear and brown the food product to a golden color.
3. Degrease and deglaze.
4. Cook the food product in two thirds liquid until fork-tender.

GLOSSARY:

Brunoise: ⅛-inch dice
Sear: to brown quickly
Fork-tender: without resistance

HAZARDOUS FOOD:

Top round veal cutlets

NUTRITION:

Calories: 587
Fat: 25.1 g
Protein: 38.1 g

CHEF NOTE:

Turkish dishes utilize a lot of dried fruits. Dates could be substituted for the prunes.

Veal with Sherry

YIELD: 10 SERVINGS. SERVING SIZE: FIVE OUNCES.

COOKING TECHNIQUES:

Sauté, Braise

Sauté:

1. Heat the sauté pan to the appropriate temperature.
2. Evenly brown the food product.
3. For a sauce, pour off any excess oil, reheat, and deglaze.

Braise:

1. Heat the braising pan to the proper temperature.
2. Sear and brown the food product to a golden color.
3. Degrease and deglaze.
4. Cook the food product in two thirds liquid until fork-tender.

GLOSSARY:

Brunoise: ⅛-inch dice
Blanch: to par cook
Dredge: to coat with flour
Fork-tender: without resistance

HACCP:

Hold at 140° F.

HAZARDOUS FOOD:

Top round veal cutlets

NUTRITION:

Calories: 406
Fat: 25.5 g
Protein: 30.3 g

INGREDIENTS:

10 five-ounce	Top round veal cutlets, pounded thin
	Seasoned flour, as needed
3 ounces	Olive oil
1½ ounces	Shallots, peeled and diced **brunoise**
3 ounces	Lean smoked ham, diced brunoise
2 ounces	Green peppers, washed, seeded, and diced brunoise
4 ounces	Tomatoes, washed, cored, **blanched,** peeled, seeded, and roughly chopped
4 ounces	Mushrooms, cleaned, stemmed, and sliced
4 ounces	Pitted black olives, rinsed and drained
8 cloves	Garlic, peeled and finely minced
4 ounces	Dry sherry
1 quart	Veal demi-glace, heated to a boil (see page 26)
	Salt and ground white pepper, to taste

METHOD OF PREPARATION:

1. Preheat the oven to 325° F.
2. **Dredge** the veal in the flour, and shake off any excess. In a sauté pan, heat the olive oil, and sauté the veal cutlets until golden brown, turning once. Transfer the veal into a hotel pan.
3. In the same sauté pan, sauté the shallots, ham, green peppers, tomatoes, mushrooms, olives, and garlic. Add the sherry and demi-glace, and heat to a boil; season to taste.
4. Pour the sauce over the browned veal, cover the pan tightly with aluminum foil, and place in the oven for 15 to 20 minutes, or until the veal is **fork-tender.**
5. Serve immediately, or hold at 140° F.
6. For service, place a veal cutlet on a preheated dinner plate, and nappé with the sauce.

CHEF NOTE:

This dish features two of Spain's most popular products, olives and sherry from the Jerez region, as well as one of the favorite meat products, which is veal.

Wiener Schnitzel

YIELD: 10 SERVINGS. SERVING SIZE: FIVE OUNCES.

COOKING TECHNIQUE:

Shallow-Fry

Shallow-Fry:

1. Heat the cooking medium to the proper temperature.
2. Cook the food product throughout.
3. Season, and serve hot.

HACCP:

Refrigerate to 40° F or lower.
Hold at 140° F or above.

HAZARDOUS FOODS:

Veal cutlets
Milk
Eggs

NUTRITION:

Calories: 823
Fat: 60.6 g
Protein: 43.4 g

INGREDIENTS:

10 (5-ounce)	Veal cutlets, trimmed of excess fat
	Salt and ground white pepper, to taste
4 ounces	Milk
	Flour, seasoned, as needed
4 each	Large eggs, cracked and lightly beaten
1 pound	White bread crumbs
8 ounces	Clarified butter
6 ounces	Vegetable oil
3 each	Lemons
10 each	Anchovies, drained of oil
30 each	Capers, rinsed of brine
2 ounces	Fresh parsley, washed, excess moisture removed, and chopped

METHOD OF PREPARATION:

1. Pound each veal cutlet lightly with a meat mallet to tenderize, and season each to taste.
2. Bread the cutlets as follows: Dip each first in milk; then lightly coat with flour, shaking off any excess. Dip in the egg wash, then coat with the crumbs. Lay the breaded cutlets out on parchment-lined trays, and refrigerate at 40° F or lower until needed.
3. Heat the sauté pan, and when hot, add the butter and oil. Fry the veal until golden brown and crisp on both sides, turning once. Remove, and drain on absorbent paper. Hold at 140° F or above.
4. Slice the lemons, and arrange a rolled anchovy on each slice. Place three capers in the center of each anchovy.
5. To serve, place a cutlet on a preheated dinner plate, and set a garnished lemon slice in the center of each.

CHEF NOTE:

Wiener schnitzel also can be prepared with pork and is often served with cottage-fried potatoes (see page 692). For best result, bread cutlets just before frying.

Baked Macaroni with Mornay Sauce

YIELD: 50 SERVINGS. SERVING SIZE: FOUR OUNCES.

INGREDIENTS:

6 pounds	Elbow macaroni
3 quarts	Mornay sauce, heated to 165° F (see page 37)
7½ gallons	Water, heated to a boil

METHOD OF PREPARATION:

1. Cook the macaroni **al dente** in boiling, salted water. Rinse, and drain well.
2. Combine the mornay sauce with the macaroni. Bake in a 350° F oven for 20 to 30 minutes, or until the macaroni is heated through.
3. Hold at 140° F.

COOKING TECHNIQUES:

Boil, Bake

Boil: (at sea level)

1. Bring the cooking liquid to a rapid boil.
2. Stir the contents, and cook the food product throughout.
3. Serve hot.

Bake:

1. Preheat the oven.
2. Place the food product on the appropriate rack.

GLOSSARY:

Al dente: to the bite

HACCP:

Heat to 165° F.
Hold at 140° F.

NUTRITION:

Calories: 338
Fat: 12 g
Protein: 12.4 g

CHEF NOTE:

Baked macaroni may be topped with bread crumbs, melted butter, and Parmesan cheese to produce a crisp topping.

Baked Ravioli with Tomato Sauce

YIELD: 50 SERVINGS. SERVING SIZE: FOUR OUNCES.

INGREDIENTS:

10 ounces	Vegetable oil
1 pound	Onions, peeled and diced very fine
1 pound	Carrots, washed, peeled, and diced very fine
1 pound	Celery, washed, and diced very fine
1 pound	Green peppers, washed, deseeded, and diced fine
1 head	Garlic, peeled and minced
1½ gallon	Tomato sauce
	Salt, black pepper to taste
10 pounds	Cheese ravioli
1 pound	Butter, melted
1 pound	Parmesan cheese, grated

METHOD OF PREPARATION:

1. Sauté all the vegetables in oil until two thirds tender.
2. Add tomato sauce, boil, and simmer 30 minutes.
3. Cook ravioli in boiling salted water. When done, drain.
4. Mix drained ravioli with butter.
5. Pour sauce on bottom of hotel pan. Fill two thirds with buttered raviolis and cover with sauce.
6. Sprinkle cheese on top and bake until the cheese melts.

COOKING TECHNIQUES:

Sauté, Boil, Simmer, Bake

Sauté:

1. Heat the sauté pan to the appropriate temperature.
2. Evenly brown the food product.
3. For a sauce, pour off any excess oil, reheat, and deglaze.

Boil: (at sea level)

1. Bring the cooking liquid to a rapid boil.
2. Stir the contents, and cook the food product throughout.
3. Serve hot.

Simmer and Poach:

1. Heat the cooking liquid to the proper temperature.
2. Submerge the food product completely.
3. Keep the cooked product moist and warm.

Bake:

1. Preheat the oven.
2. Place the food product on the appropriate rack.

HACCP:

Cook to an internal temperature of 155° F
Hold at 140° F

NUTRITION:

Calories: 439
Fat: 28.4 g
Protein: 27.3 g

Broccoli Quiche

YIELD: 50 SERVINGS. SERVING SIZE: ONE-SIXTH PIE.

COOKING TECHNIQUE:

Bake

Bake:

1. Preheat the oven.
2. Place the food product on the appropriate rack.

HAZARDOUS FOODS:

Eggs
Milk
Half and half

NUTRITION:

Calories: 240
Fat: 16.2 g
Protein: 8.61 g

INGREDIENTS:

4 pounds	Broccoli, frozen or fresh
8 ounces	Leeks (white part only), peeled and chopped
4 ounces	Butter
20 each	Eggs
20 ounces	Milk
20 ounces	Half and half
20 ounces	Swiss cheese, grated
	Salt and ground white pepper, to taste
1 teaspoon	Nutmeg, ground
9 each	Pie shells, baked

METHOD OF PREPARATION:

1. Steam the broccoli; drain, shock, and roughly chop.
2. Sauté the leeks in the butter. When they are translucent, add the broccoli. Sauté and season.
3. Combine the eggs, milk, and half and half, and mix into a smooth liquid. Season, to taste.
4. Divide the broccoli evenly into the pie shells.
5. Pour the egg mixture to fill the pie shells, leaving ¼ inch of crust free.
6. Sprinkle the cheese evenly on top, and bake in a preheated 325° F oven until the liquid is solidified.
7. Let rest 10 to 15 minutes before slicing.

Crêpes with Vegetables

YIELD: 50 SERVINGS. SERVING SIZE: ONE CRÊPE.

INGREDIENTS:

1 pound	Butter, melted
24 ounces	Swiss cheese, grated

Crêpe Batter:

24 ounces	All-purpose flour
8 each	Whole eggs
1½ quarts	Milk
1 teaspoon	Salt
2 ounces	Salad oil

Stuffing:

1 pound	Leeks, white part only, washed, and sliced
1 pound	Carrots, washed, peeled, and diced **brunoise**
1 pound	Celery, washed, trimmed, and diced brunoise
2 pounds	Mushrooms, cleaned and sliced
1 pound	Corn kernels, frozen
	Salt, ground white pepper, and nutmeg, as needed
1 quart	Béchamel sauce, heavy (see page 16)
8 each	Egg yolks

METHOD OF PREPARATION:

1. Place the flour and salt into a mixing bowl.
2. Mix the eggs, milk, and oil together in a separate bowl. Add the mixture to the flour, and stir until a smooth batter is formed.
3. Let the batter rest for 30 minutes.
4. Using 9-inch crêpe pans, fry thin crêpes, and set aside.
5. Preheat the oven to 325° F.
6. Place the vegetables into a perforated hotel pan, and steam until tender.
7. Transfer the steamed vegetables into a bowl. Add the béchamel sauce, egg yolks, and seasonings. Mix well.
8. Lay out the crêpes on a clean, dry table. Pipe the stuffing down the middle by using a pastry bag.
9. Roll the crêpes, and place on a buttered sheet pan. Brush the butter onto the crêpes, and sprinkle with cheese.
10. Bake in the oven until the cheese melts. Serve immediately.

COOKING TECHNIQUES:

Shallow-Fry, Steam, Bake

Shallow-Fry:

1. Heat the cooking medium to the proper temperature.
2. Cook the food product throughout.
3. Season, and serve hot.

Steam: (Traditional)

1. Place a rack over a pot of water.
2. Prevent steam vapors from escaping.
3. Shock or cook the food product throughout.

Bake:

1. Preheat the oven.
2. Place the food product on the appropriate rack.

GLOSSARY:

Brunoise: ⅛-inch dice

HAZARDOUS FOODS:

Whole eggs
Milk
Egg yolks

NUTRITION:

Calories: 256
Fat: 16.6 g
Protein: 8.92 g

CHEF NOTES:

1. Half of the milk used in the crêpe batter can be replaced with club soda.
2. The vegetables in the stuffing can be sautéed instead of steamed.
3. Frozen, prepared crêpe skins may be purchased.

Eggplant Roulades

YIELD: 50 SERVINGS. SERVING SIZE: TWO ROLLS.

COOKING TECHNIQUE:
Bake

Bake:
1. Preheat the oven.
2. Place the food product on the appropriate rack.

GLOSSARY:
Brunoise: ⅛-inch dice
Al dente: to the bite

HAZARDOUS FOOD:
Egg yolks

NUTRITION:
Calories: 271
Fat: 16.6 g
Protein: 15.5 g

INGREDIENTS:

7 pounds	Eggplants, washed and sliced in length ⅕-inch thick
8 ounces	Salad oil
8 ounces	Onions, peeled and diced **brunoise**
12 ounces	Celery, washed and diced brunoise
12 ounces	Red bell peppers, washed, cored, and diced brunoise
2 pounds	Rice, cooked **al dente**
2 quarts	Béchamel sauce, heavy (see page 16)
10 each	Egg yolks
	Salt and ground black pepper, as needed
3 quarts	Tomato sauce (see page 76)
5 pounds	Mozzarella cheese, shredded

METHOD OF PREPARATION:

1. Preheat the oven to 375° F.
2. Lay out the eggplant slices, and sprinkle on the salt to marinate for 30 minutes.
3. Heat the oil in a skillet, add the onions, and sauté until translucent.
4. Add the peppers and celery, and sauté for 10 minutes.
5. Add the cooked rice, béchamel sauce, and egg yolks. Season, and mix well.
6. Soak up the moisture from the eggplants with paper towels.
7. Pipe the stuffing onto the eggplant slices, and roll into a roulade.
8. Place the roulades in 2-inch hotel pans, and bake, until the eggplants turn light brown.
9. Pour tomato sauce to cover two thirds of the roulades, and continue to bake for 15 to 20 minutes.
10. The roulades may be topped with mozzarella cheese. Melt the cheese under the salamander at service time.

CHEF NOTE:
Stuffing for the roulade can be varied according to the chef's discretion.

Fresh Vegetable Cannelloni

YIELD: 50 SERVINGS. SERVING SIZE: SIX OUNCES.

INGREDIENTS:

4 pounds	Pasta sheets, frozen and precooked
4 ounces	Vegetable oil
4 ounces	Margarine
1½ pounds	Onions, peeled and diced **brunoise**
3 teaspoons	Garlic, peeled and minced
4 pounds	Broccoli, washed and cut into small flowerettes
2 pounds	Mushrooms, washed and quartered
2 pounds	Zucchini, washed and diced
1 pound	Carrots, washed, peeled and diced
1½ pints	Vegetable stock, heated to a boil (see page 11)
1¼ pints	Fresh basil, washed and finely chopped
4½ ounces	Scallions, washed, peeled, and chopped
1 pound	Frozen peas
8 ounces	Butter, melted
1½ pints	Low-fat yogurt
1 pint	Buttermilk
1½ pounds	Parmesan cheese, grated
	Salt and freshly ground black pepper, as needed
1 gallon	Tomato sauce, heated and seasoned (see page 76)

METHOD OF PREPARATION:

1. Preheat the oven to 350° F.
2. Heat the oil in a large sauté pan.
3. Sauté the onions lightly until translucent.
4. Add the garlic, broccoli, mushrooms, zucchini, and carrots. Sauté for 3 minutes.
5. Add the vegetable stock and basil. Blend well. Heat to a boil.
6. Add the scallions and peas, and simmer until the peas are tender. Toss to blend. Season.
7. Layout pasta squares on a clean surface. Place one ounce of stuffing on pasta; roll, and place in buttered hotel pan.

COOKING TECHNIQUES:

Boil, Sauté

Boil: (at sea level)

1. Bring the cooking liquid to a rapid boil.
2. Stir the contents, and cook the food product throughout.
3. Serve hot.

Sauté:

1. Heat the sauté pan to the appropriate temperature.
2. Evenly brown the food product.
3. For a sauce, pour off any excess oil, reheat, and deglaze.

GLOSSARY:

Brunoise: ⅛-inch dice

HACCP:

Heat to 140° F.

HAZARDOUS FOODS:

Yogurt
Buttermilk

NUTRITION:

Calories: 199
Fat: 9.18 g
Protein: 11.2 g

8. Mix yogurt and buttermilk; brush the rolls with the mixture. Sprinkle with grated cheese and bake to an internal temperature of 140° F.
9. Serve with tomato sauce.

Maultaschen of Wild Mushrooms with Herb Sauce

(Wild Mushroom Ravioli)

YIELD: 10 SERVINGS. SERVING SIZE: SIX OUNCES.

INGREDIENTS:

1 recipe	Basic pasta II (see page 469)
	Freshly ground black pepper, to taste
	Boiling salted water, to cook

Ravioli Filling:

4 ounces	Butter
2 ounces	Shallots, peeled and diced **brunoise**
2½ pounds	Fresh wild mushrooms (combination of at least three types), washed, trimmed, and diced brunoise
	Salt and freshly ground black pepper, to taste

Herb Sauce:

2 ounces	Fresh chervil leaves, washed
2 ounces	Fresh (curly-leafed) parsley, washed and stemmed
2 ounces	Fresh Italian (flat-leafed) parsley, washed and stemmed
3 ounces	Red wine vinegar
8 ounces	Chilled butter, cut into small pieces

METHOD OF PREPARATION:

1. In a sauté pan, melt the butter; add the shallots, and sauté until they become translucent.
2. Add the mushrooms, increase the heat, and sauté, shaking the pan frequently, until the liquid evaporates and the mushrooms are lightly browned. Season, to taste, and spread the mixture out on a sheet pan to cool.
3. Roll the pasta dough into sheets, and cut into 2-inch squares.
4. Mound 1 tablespoon of mushroom mixture in the center of half of the squares. Moisten the exposed pasta edge with a pastry brush dipped in water; then cover with another square of dough and press down around the filling.

COOKING TECHNIQUES:
Sauté, Simmer

Sauté:
1. Heat the sauté pan to the appropriate temperature.
2. Evenly brown the food product.
3. For a sauce, pour off any excess oil, reheat, and deglaze.

Simmer and Poach:
1. Heat the cooking liquid to the proper temperature.
2. Submerge the food product completely.
3. Keep the cooked product moist and warm.

GLOSSARY:
Brunoise: ⅛-inch dice
Emulsion: incorporation of two liquids
Bain-marie: hot-water bath
Chinoise: cone-shaped strainer

HACCP:
Hold at 140° F.

NUTRITION:
Calories: 620
Fat: 35.1 g
Protein: 14.8 g

CHEF NOTES:
1. Additional mushrooms, split in half or quartered, depending on the size, can be sautéed for garnish on top of each ravioli.
2. Because of the amount of butter, this sauce cannot be held in a **bain-marie** for more than 10 to 15 minutes.
3. It is important to finish the sauce just before service in order to maintain the vivid green color.

5. Cut the excess dough with a cutter to form a uniform-shaped ravioli. Lay out the prepared ravioli on a parchment-lined tray that has been lightly dusted with flour. Keep covered with a towel to prevent drying of the dough.
6. Heat a small pot of water to a boil; add 1 teaspoon of salt and the herbs. Return to a boil, and cook for 5 minutes. Remove from the heat, strain the liquid through a **chinois,** discard all but 2 ounces of the water, and reserve the herbs.
7. Transfer the herbs to a food processor, and purée, adding the reserved water to make a smooth paste.
8. Just before service, place the purée back on the heat, add the vinegar, and heat to a boil. Gradually whisk in butter, piece by piece, until a smooth **emulsion** is achieved.
9. Season with pepper, to taste, and hold minimally at 140° F.
10. In a large pot, heat water to a boil. Add the salt, and cook the ravioli until tender, about 2 to 3 minutes. Remove with a slotted spoon, and drain on towels.
11. To serve, ladle the sauce over the bottom of a preheated dinner plate, and arrange four to five ravioli on top.

Mixed Vegetables

(Pisto)

YIELD: 10 SERVINGS. SERVING SIZE: SIX OUNCES.

COOKING TECHNIQUE:
Sauté

Sauté:
1. Heat the sauté pan to the appropriate temperature.
2. Evenly brown the food product.
3. For a sauce, pour off any excess oil, reheat, and deglaze.

GLOSSARY:
Brunoise: ⅛-inch dice
Bâtons: stick-like cut
Al dente: to the bite

HACCP:
Hold at 140° F or above.

NUTRITION:
Calories: 159
Fat: 11.9 g
Protein: 2.96 g

INGREDIENTS:

4 ounces	Spanish olive oil
6 ounces	Onions, peeled and diced **brunoise**
3 each	Red bell peppers, washed, seeded, and thinly sliced
3 each	Green bell peppers, washed, seeded, and thinly sliced
3 pounds	Whole canned tomatoes, drained, seeded, and cut in half (liquid reserved for other use)
2 pounds	Zucchini, washed and cut in **bâtons**
	Spanish paprika, to taste
	Salt and cayenne pepper, to taste

METHOD OF PREPARATION:

1. In a large sauté pan, heat the oil; add the onions, and sauté until translucent.
2. Add the peppers, and continue to sauté for 5 minutes; then add the tomatoes and zucchini.
3. Season, to taste, and sauté for another 10 minutes, or until the vegetables are **al dente.**
4. Serve immediately, or hold at 140° F or above.

Oven-Roasted Tomato and Grilled Eggplant Terrine

YIELD: 10 SERVINGS. SERVING SIZE: SIX OUNCES.

INGREDIENTS:

3 ounces	Olive oil, or as needed
5 pounds	Ripe Roma tomatoes, washed, cored, and cut lengthwise into thirds
	Salt and freshly ground black pepper, to taste
1 tablespoon	Basil, dried
5 each	Eggplants, washed, trimmed, and cut lengthwise into ¼-inch slices
10 ounces	Red bell pepper coulis
10 ounces	Yellow bell pepper coulis
10 sprigs	Fresh basil, washed

METHOD OF PREPARATION:

1. Preheat the oven to 200° F.
2. Brush a baking pan with olive oil, and lay out the tomatoes, skin-side down when possible. Generously season the tomatoes with salt, pepper, and basil, and drizzle additional oil over each.
3. Place the tomatoes in the oven, and roast until dried but still soft, or about 1½ hours. Remove the tomatoes, and hold at room temperature, covered with parchment paper.
4. Meanwhile, brush the eggplant slices with olive oil, season with salt and pepper, and grill until tender.
5. Line individual 3- to 4-ounce ramekins with plastic wrap, allowing additional wrap to overhang the sides.
6. Cut and line the molds with eggplant slices, allowing enough eggplant extending over the sides to cover tops of the ramekins when filled.
7. Layer the tomatoes and eggplant to fill the ramekins, pressing down to make sure they are tightly packed.
8. Fold the extended eggplant over the top, and wrap tightly with plastic wrap. Refrigerate at 40° F or lower until service.
9. For service, remove the plastic wrap from the top, and unmold a ramekin onto the center of a serving plate.
10. Alternate red and yellow pepper coulis around the ramekin, and garnish with a basil sprig.

COOKING TECHNIQUES:

Grill or Roast

Grill/Broil:

1. Clean and heat the grill/broiler.
2. To prevent sticking, brush the food product with oil.

Roast:

1. Sear the food product, and brown evenly.
2. Elevate the food product in a roasting pan.
3. Determine doneness, and consider carryover cooking.
4. Let the food product rest before carving.

HACCP:

Refrigerate at 40° F or lower.
Hold at 140° F.

NUTRITION:

Calories: 214
Fat: 9.98 g
Protein: 4.72 g

CHEF NOTE:

This dish is meant to be served chilled but could be heated in a water bath in a 350° F oven. Heat to a minimum of 165° F and hold at 140° F.

Potato Turnovers

YIELD: 50 SERVINGS. SERVING SIZE: ONE EACH.

INGREDIENTS:

6 pounds	All-purpose potatoes, washed, peeled, and cut into quarters
1 pound	Onions, peeled and diced **brunoise**
12 ounces	Margarine
12 each	Whole eggs
6 pounds	Puff pastry, frozen
	Salt and ground white pepper, to taste
¼ teaspoon	Nutmeg, ground

METHOD OF PREPARATION:

1. Preheat the oven to 375° F.
2. Boil the potatoes in salted water. When tender, drain.
3. Sauté the onions in margarine until translucent.
4. Purée the potatoes, onions, six eggs, salt, pepper, and nutmeg in a mixing bowl.
5. Use the remaining six eggs to make an egg wash.
6. Cut the puff pastry into squares, and pipe approximately 2 ounces of potato mixture on each square. Brush the edges of the puff pastry, and fold over to make a triangular shape.
7. Place the turnovers on a wet sheet pan, and brush them with egg wash. Bake for approximately 15 to 20 minutes.

COOKING TECHNIQUE:

Bake

Bake:

1. Preheat the oven.
2. Place the food product on the appropriate rack.

GLOSSARY:

Brunoise: ⅛-inch dice

HAZARDOUS FOOD:

Whole eggs

NUTRITION:

Calories: 415
Fat: 27.5 g
Protein: 6.71 g

CHEF NOTE:

If no sauce or gravy is served with the turnovers, sprinkle some sesame seeds onto the egg wash.

Spaghetti Alio-Olio

YIELD: 50 SERVINGS. SERVING SIZE: SIX OUNCES.

COOKING TECHNIQUES:
Boil, Sauté

Boil: (at sea level)
1. Bring the cooking liquid to a rapid boil.
2. Stir the contents, and cook the food product throughout.
3. Serve hot.

Sauté:
1. Heat the sauté pan to the appropriate temperature.
2. Evenly brown the food product.
3. For a sauce, pour off any excess oil, reheat, and deglaze.

GLOSSARY:
Al dente: to the bite

NUTRITION:
Calories: 395
Fat: 11.5 g
Protein: 13.6 g

INGREDIENTS:

8 pounds	Spaghetti
	Salt, as needed in 4 gallons of water
2 heads	Garlic, peeled and minced
1 pound	Black olives, pitted and sliced
12 ounces	Olive oil
	Salt and ground black pepper, as needed
1 pound	Parmesan cheese, shredded

METHOD OF PREPARATION:

1. Place the spaghetti in boiling, salted water, and cook until **al dente.** Strain, and do not rinse.
2. Sauté the garlic in the oil (do not burn). Toss the spaghetti into the garlic oil, add the olives, and season.
3. Transfer the spaghetti into hotel pans, and sprinkle with parmesan cheese.

CHEF NOTE:
Chopped anchovies can be mixed into the spaghetti as desired.

Spinach Lasagna

YIELD: 2 HOTEL PANS. SERVING SIZE: FIVE OUNCES.

INGREDIENTS:

1 pound	Onions, peeled and diced **brunoise**
1 head	Garlic, peeled and minced
8 ounces	Vegetable oil
5 pounds	Spinach, cleaned, washed, and drained
	Salt and black pepper, to taste
18 each	Frozen lasagna sheets
5 pounds	Ricotta cheese
3 quarts	Basil-flavored tomato sauce
1 pound	Romano cheese, shredded
1 pound	Mozzarella cheese, shredded

METHOD OF PREPARATION:

1. Preheat the oven to 325° F.
2. Sauté the onions and garlic in oil until the onions are translucent. Add the spinach, and steam under a lid until the spinach wilts. Season, to taste.
3. In a 2-inch hotel pan, lay three lasagna sheets.
4. Evenly spread a thin layer of ricotta cheese on the lasagna sheets.
5. Lay out half of the spinach to cover the ricotta.
6. Place three sheets to cover the spinach, and pour on tomato sauce.
7. Repeat the steps again in the same manner.
8. On top of the upper layer, sprinkle the Romano cheese. Bake in the oven approximately 40 to 45 minutes, or until done.
9. Top with mozzarella and Romano cheese; bake until cheese melts, and serve.

COOKING TECHNIQUE:

Bake

Bake:

1. Preheat the oven.
2. Place the food product on the appropriate rack.

GLOSSARY:

Brunoise: ⅛-inch dice

HAZARDOUS FOODS:

Ricotta cheese
Mozzarella cheese

NUTRITION:

Calories: 274
Fat: 15.5 g
Protein: 16.7 g

CHEF NOTE:

Let the lasagna rest for 15 minutes before cutting into portions.

Spinach Turnovers

YIELD: 50 SERVINGS. SERVING SIZE: ONE EACH.

COOKING TECHNIQUE:
Bake

Bake:
1. Preheat the oven.
2. Place the food product on the appropriate rack.

GLOSSARY:
Brunoise: ⅛-inch dice

HAZARDOUS FOOD:
Eggs

NUTRITION:
Calories: 389
Fat: 27.9 g
Protein: 7.69 g

INGREDIENTS:

6 pounds	Spinach, frozen
1 pound	Onions, peeled and diced **brunoise**
12 ounces	Margarine
12 each	Eggs
	Salt and pepper, to taste
	Nutmeg, to taste
6 pounds	Puff pastry, frozen

METHOD OF PREPARATION:

1. Preheat the oven to 375° F.
2. Steam the frozen spinach for 6 minutes. Shock, and drain. Hold below 40° F.
3. Sauté the onions in the margarine until they are translucent.
4. Place the spinach, onions, six eggs, salt, pepper, and nutmeg into a mixing bowl. Purée to a smooth texture.
5. Cut the puff pastry into squares. Pipe approximately 2 ounces of the mixture on each square. Brush the edges of the pastry with water, and fold over to create a triangle shape.
6. Place the turnovers on a sheet pan lined with parchment paper. Brush each turnover with eggwash. Bake for approximately 15 to 20 minutes.

CHEF NOTE:
If no sauce or gravy is served with the turnover, sprinkle sesame seeds over the turnover after it is egg-washed.

Tomato and Pepper Tart

YIELD: 10 SERVINGS. SERVING SIZE: TWO WEDGES.

INGREDIENTS:

24 ounces	Bread or pizza dough (see page 901)
4 ounces	Olive oil
6 ounces	Onions, peeled and diced **brunoise**
6 cloves	Garlic, peeled and minced
1½ pounds	Canned crushed tomatoes, drained (liquid reserved for other use)
¼ teaspoon	Marjoram, dried
¼ teaspoon	Oregano, dried
¼ teaspoon	Thyme, dried
¼ teaspoon	Freshly ground black pepper
2 each	Red bell peppers, washed and roasted
2 each	Green bell peppers, washed and roasted
2 each	Yellow bell peppers, washed and roasted
3 ounces	Pitted black olives

METHOD OF PREPARATION:

1. Preheat the oven to 475° F.
2. Divide the dough into four equal parts, and roll each into a round shape to fit a 9-inch pan.
3. Lightly oil the bottoms of each pan, and press the dough into each, creating a rim around the edges. Brush the crust with the oil, and place to **proof.**
4. In a sauté pan, heat half of the oil, and sauté the onions with the garlic. When the onions become translucent, add the tomatoes, and continue to sauté for about 10 minutes.
5. Add the herbs and seasonings, and remove from the heat. Cool slightly; then divide the mixture evenly over the dough within the rim.
6. Peel the roasted peppers; core and cut **julienne.**
7. Arrange the julienned peppers over the tomato mixture, and add the olives in an eye-appealing manner. Sprinkle some of the remaining oil over the top of each, and brush the rims with oil.
8. Bake in the oven until crisp and the dough is cooked throughout.

COOKING TECHNIQUES:

Sauté, Bake

Sauté:

1. Heat the sauté pan to the appropriate temperature.
2. Evenly brown the food product.
3. For a sauce, pour off any excess oil, reheat, and deglaze.

Bake:

1. Preheat the oven.
2. Place the food product on the appropriate rack.

GLOSSARY:

Brunoise: ⅛-inch dice
Proof: to allow to rise
Julienne: matchstick strips

NUTRITION:

Calories: 360
Fat: 16.3 g
Protein: 8.45 g

CHEF NOTE:

For additional flavor, cheese can be added to the tarts.

Vegetable Baklava

YIELD: 50 SERVINGS. SERVING SIZE: FIVE OUNCES.

COOKING TECHNIQUES:

Sauté, Bake

Sauté:

1. Heat the sauté pan to the appropriate temperature.
2. Evenly brown the food product.
3. For a sauce, pour off any excess oil, reheat, and deglaze.

Bake:

1. Preheat the oven.
2. Place the food product on the appropriate rack.

HAZARDOUS FOOD:

Whole eggs

NUTRITION:

Calories: 144
Fat: 11 g
Protein: 3.05 g

INGREDIENTS:

3 pounds	Onions, peeled and roughly chopped
3 pounds	Carrots, washed, peeled, and roughly chopped
3 pounds	Celery (white part only), washed and roughly chopped
3 pounds	Broccoli, cleaned, washed, and roughly chopped
2 heads	Garlic, peeled
2 boxes	Filo dough, frozen
8 each	Whole eggs, slightly beaten
1 pint	Salad oil or melted clarified butter
3 ounces	Sesame seeds (optional)
	Salt, ground black pepper, and ground ginger, as needed

METHOD OF PREPARATION:

1. Preheat the oven to 350° F.
2. Place all of the vegetables in a robo-coupe, and chop into a barley-sized mixture.
3. Sauté the chopped vegetables in 8 ounces of oil until almost tender; season and cool.
4. Lay out one layer of filo dough on an oiled sheet pan, and brush oil on the dough with pastry brush. Repeat this three times to have three layers of dough sheets.
5. Spread the sautéed vegetables evenly over the dough sheets, and layer out three sheets of dough as before on top of the vegetable mixture. Do not brush or add oil on the last layer of dough.
6. Brush the eggs on the top layer of filo dough, and sprinkle sesame seeds on it.
7. Bake in the oven until done.

CHEF NOTE:

1. Using a serrated knife, cut the dough into portions before baking.
2. Add some liquid to vegetables when sautéing to create steam.

Vegetable Calzone

YIELD: 50 PORTIONS. SERVING SIZE: SIX OUNCES.

INGREDIENTS:

5 pounds	Whole wheat crust (see page 442)
	Flour, as needed
Filling:	
	Blanched vegetables, as per vegetable medley (see page 615)
6 cloves	Garlic, peeled and mashed into a purée
6¼ pounds	Parmesan cheese, grated
1 gallon	Tomato sauce (see page 76)
	Salt and freshly ground black pepper, to taste

METHOD OF PREPARATION:

1. Preheat the oven to 375° F.
2. Place the dough on sheet pan lined with parchment paper until needed.
3. In a large bowl, combine the vegetables, garlic, cheese, seasoning, and 1 pint of tomato sauce, and mix well. Hold the remaining tomato sauce at 140° F or above.
4. Place approximately 2 ounces of the filling off-center on each round of dough. Fold the dough over the filling, and firmly press the edges of the dough together. Transfer the calzones to the parchment-lined sheet pan.
5. Bake in oven until the crust is browned and cooked, or approximately 20 to 30 minutes. Hold at 140° F.
6. For service, split a calzone in half on a diagonal, arrange the pieces on a preheated plate, and surround with additional sauce.

COOKING TECHNIQUES:

Bake, Steam or Boil

Bake:

1. Preheat the oven.
2. Place the food product on the appropriate rack.

Steam: (Traditional)

1. Place a rack over a pot of water.
2. Prevent steam vapors from escaping.
3. Shock or cook the food product throughout.

Boil: (at sea level)

1. Bring the cooking liquid to a rapid boil.
2. Stir the contents, and cook the food product throughout.
3. Serve hot.

HACCP:

Refrigerate at 40° F or below.
Hold at 140° F or above.

NUTRITION:

Calories: 457
Fat: 19 g
Protein: 31 g

Whole Wheat Crust

COOKING TECHNIQUE:
Not applicable

YIELD: 2 POUNDS. SERVING SIZE: NOT APPLICABLE.

INGREDIENTS:

2 ounces	Compressed yeast
8 ounces	Water, lukewarm
1 ounce	Olive oil
½ pound	Whole wheat flour
1 pound	All-purpose flour
½ teaspoon	Salt

METHOD OF PREPARATION:

1. In a mixing bowl, combine the yeast, water, and oil, and mix well.
2. In a clean mixing bowl, combine the flour and the salt. Mix; then make a well in the center, add the yeast mixture, and blend into a dough. Knead for 4 to 5 minutes. Place the dough in an oiled bowl.
3. Cover with a clean, damp towel, and allow the dough to rise until doubled in size.
4. Roll to the desired size or number.

Vegetable Couscous

YIELD: 50 SERVINGS. SERVING SIZE: TWELVE OUNCES.

INGREDIENTS:

2 pounds	Yellow squash, peeled, and diced **brunoise**
2 pounds	Turnips, peeled, and diced brunoise
2 pounds	Carrots, washed, peeled, and diced brunoise
2 pounds	Parsnips, washed, peeled, and diced brunoise
1 pound	Onions, peeled and diced brunoise
3 pounds	Green peas
5 pounds	Couscous
5 quarts	Water
8 ounces	Butter
	Salt, ground white pepper, and yellow saffron, to taste

METHOD OF PREPARATION:

1. Steam all of the vegetables to **al dente,** and keep warm.
2. Heat the water to a boil. Add the salt and pepper.
3. Pour the couscous into the boiling water. Mix well. Remove from the stove, and let stand for 5 minutes with the lid on.
4. Add the butter, and mix with a braising fork.
5. Place the couscous into a large bowl. Add the vegetables and yellow saffron.
6. Mix together. Taste, and adjust the seasoning. Transfer into 2-inch hotel pans, and keep warm at 165° F.

COOKING TECHNIQUES:

Boil, Sauté, Steam

Boil: (at sea level)

1. Bring the cooking liquid to a rapid boil.
2. Stir the contents, and cook the food product throughout.
3. Serve hot.

Sauté:

1. Heat the sauté pan to the appropriate temperature.
2. Evenly brown the food product.
3. For a sauce, pour off any excess oil, reheat, and deglaze.

Steam: (Traditional)

1. Place a rack over a pot of water.
2. Prevent steam vapors from escaping.
3. Shock or cook the food product throughout.

GLOSSARY:

Brunoise: ⅛-inch dice
Al dente: to the bite

HACCP:

Hold at 165° F.

NUTRITION:

Calories: 144
Fat: 4.63 g
Protein: 4.03 g

Vegetable Fritters

YIELD: 50 SERVINGS. SERVING SIZE: FOUR OUNCES.

COOKING TECHNIQUE:

Deep-Fry

Deep-Fry:

1. Heat the frying liquid to the proper temperature.
2. Submerge the food product completely.
3. Fry the product until it is cooked throughout.

GLOSSARY:

Julienne: matchstick strips
Dredge: to coat with flour

HAZARDOUS FOOD:

Egg whites

NUTRITION:

Calories: 354
Fat: 14.2 g
Protein: 8.82 g

INGREDIENTS:

40 each	Egg whites
5 pounds	Bread flour, sifted
7½ pints	Beer
5 each	Onions, peeled, and cut **julienne**
15 each	Carrots, washed, peeled, and cut julienne
1 bunch	Celery, washed, trimmed, and cut julienne
5 each	Zucchini, washed, and cut julienne
5 each	Yellow squash, washed, and cut julienne
½ head	Garlic, peeled and minced
	Ground coriander, cumin, salt, and ground white pepper, to taste
	Frying oil, as needed
	Vinegar and garlic dip, as needed

METHOD OF PREPARATION:

1. Mix the beer and flour in a mixing bowl. Season, to taste.
2. Whip the egg whites to a soft peak in a clean stainless steel bowl. Fold the egg whites into the batter.
3. Blend all of the vegetables together. **Dredge,** and pour the batter over them. Blend well.
4. Heat the oil, and fry the fritters. Drain on absorbent paper. Serve with vinegar and garlic dip.

Vinegar and Garlic Dip

INGREDIENTS:

20 ounces	White vinegar
20 cloves	Garlic, peeled and crushed
5 tablespoons	Soy sauce

METHOD OF PREPARATION:

1. Blend all of the ingredients.
2. Serve with vegetable fritters.

Vegetable Quiche

YIELD: 50 SERVINGS. SERVING SIZE: ONE-SIXTH PIE.

INGREDIENTS:

8 ounces	Butter
8 ounces	Leeks, peeled and sliced into rings
1 pound	Carrots, washed, peeled, and diced **brunoise**
1 pound	Green peppers, washed, cored, and diced brunoise
2 pounds	Green peas, frozen
1 (#10 can)	Pimientos, drained
20 each	Whole eggs
20 ounces	Milk
20 ounces	Half-and-half
20 ounces	Swiss cheese, grated
	Salt and ground white pepper, to taste
9 each	Pie shells, baked

METHOD OF PREPARATION:

1. Preheat the oven to 325° F.
2. Sauté vegetables in the butter. When **al dente,** add the pimientoes, and season.
3. Combine the eggs, milk, and half and half. Mix into a smooth liquid. Season, to taste.
4. Divide the sautéed vegetables into the pie shells.
5. Pour the egg mixture to fill the pie shells, leaving ¼ inch of crust free.
6. Sprinkle the cheese evenly on top, and bake until the liquid solidifies.

COOKING TECHNIQUE:
Bake

Bake:
1. Preheat the oven.
2. Place the food product on the appropriate rack.

GLOSSARY:
Julienne: matchstick strips
Brunoise: ⅛-inch dice
Al dente: to the bite

HAZARDOUS FOODS:
Eggs
Milk
Half-and-half

NUTRITION:
Calories: 279
Fat: 18.2 g
Protein: 9.24 g

CHEF NOTE:
Drain the cooked vegetables well prior to portioning into the baked pie shells.

Vegetarian Burrito

YIELD: 50 SERVINGS. SERVING SIZE: SIX OUNCES.

COOKING TECHNIQUES:

Sauté, Simmer, Bake

Sauté:

1. Heat the sauté pan to the appropriate temperature.
2. Evenly brown the food product.
3. For a sauce, pour off any excess oil, reheat, and deglaze.

Simmer and Poach:

1. Heat the cooking liquid to the proper temperature.
2. Submerge the food product completely.
3. Keep the cooked product moist and warm.

Bake:

1. Preheat the oven.
2. Place the food product on the appropriate rack.

GLOSSARY:

Chiffonade: ribbons of leafy greens
Allumette: thin matchstick strips

HACCP:

Hold at 140° F.

HAZARDOUS FOOD:

Cheese

NUTRITION:

Calories: 365
Fat: 13 g
Protein: 14 g

INGREDIENTS:

Refried Beans:

2½ pounds	Pinto beans, sorted, washed, and soaked overnight in cold water
3 tablespoons	Cumin
2 tablespoons	Chili powder
1 (#2½ can)	Green chilies, chopped
	Salt, to taste

Burritos:

	Oil, as needed
3 pounds	Yellow squash, washed and diced
3 pounds	Zucchini, washed, and diced
2 pounds	Onions, peeled and diced
1 gallon	Red chili salsa
50 each	Flour tortillas, 10-inch
1 pound	Monterey Jack cheese, shredded
1 pound	Cheddar cheese, shredded
1 pound	Black olives, sliced
4 heads	Iceberg lettuce, cut **chiffonade**
6 each	Red bell peppers, washed, blistered, peeled, seeded, and cut **allumette**
6 each	Green bell peppers, washed, blistered, peeled, seeded, and cut allumette
	Salsa and guacamole, as needed
	Salt and ground black pepper, to taste

METHOD OF PREPARATION:

1. Preheat the oven to 350° F.
2. Drain the water from the beans, and replace with fresh water. Heat the beans and water to a boil. Reduce to a simmer.
3. Add the cumin, chili powder, and green chilies. Add salt when the beans are tender.
4. Simmer gently, stirring occasionally, until the beans are very tender. Drain the liquid, and cool to 40° F or lower. Reserve the liquid from the beans, because it may be needed later to adjust the consistency.

5. Heat the oil in a sauté pan. Sauté the vegetables. Season with the salt, pepper and red chili salsa. Chill to an internal temperature of 40° F or lower.
6. To assemble, on each layer of the flour tortillas, place 2 ounces of the bean and vegetable mixture; then sprinkle with 2 tablespoons of cheese and 1 tablespoon of olives. Roll the tortillas, and place them folded-side down on a lightly oiled hotel pan. Brush oil over them.
7. Bake in the oven until heated thoroughly, approximately 5 minutes. Take care not to brown the tortillas. Hold at 140° F.
8. Garnish with shredded lettuce, red peppers, green peppers, salsa, and guacamole.

Vegetarian Chili

YIELD: 10 SERVINGS. SERVING SIZE: EIGHT OUNCES.

COOKING TECHNIQUES:

Sauté, Simmer

Sauté:

1. Heat the sauté pan to the appropriate temperature.
2. Evenly brown the food product.
3. For a sauce, pour off any excess oil, reheat, and deglaze.

Simmer and Poach:

1. Heat the cooking liquid to the proper temperature.
2. Submerge the food product completely.
3. Keep the cooked product moist and warm.

GLOSSARY:

Brunoise: ⅛-inch dice
Marmite: stockpot

HACCP:

Hold to 140° F.

NUTRITION:

Calories: 215
Fat: 13 g
Protein: 5.18 g

INGREDIENTS:

4 ounces	Olive oil
12 ounces	Onions, peeled and diced **brunoise**
6 ounces (2 each)	Red bell peppers, washed, seeded, and diced
4 ounces (2 each)	Pablano peppers, washed, seeded, and diced
2 each	Jalapeño peppers, washed, seeded, and diced
6 ounces	Celery stalks, washed, trimmed, and diced
6 cloves	Garlic, peeled and finely minced
3 pounds	Canned tomatoes, drained, seeded, and coarsely chopped
4 ounces	Tomato paste
2 ounces	Hot chili powder
½ ounce	Cumin, ground
½ ounce	Dried oregano
	Salt and freshly ground black pepper, to taste
2 quarts	Vegetable stock, heated to a boil (see page 11)
1 pound	Zucchini, washed, seeded, and cut into 1-inch pieces
3 ounces	Fresh cilantro leaves, washed and roughly chopped

METHOD OF PREPARATION:

1. In a **marmite,** heat half of the olive oil. Add the onions, and sauté until translucent.
2. Add the peppers, celery, and garlic, and continue to sauté, stirring frequently, for 10 to 15 minutes, or until the vegetables soften.
3. Add the tomatoes and strained juice, tomato paste, and seasonings. Sauté another 10 minutes.
4. Add the stock, and bring to a boil; reduce the heat, and simmer for 30 minutes. Remove, and hold at 140° F or above.
5. In sauté pan, heat the remaining olive oil, add the zucchini, and sauté until lightly browned. Season with additional salt and pepper. Remove, and hold on the side.
6. For service, add the zucchini as needed, in order to hold the color, and ladle into preheated bowls. Garnish with cilantro.

Vegetarian Lasagna

YIELD: 10 SERVINGS. SERVING SIZE: EIGHT OUNCES.

INGREDIENTS:

12 each	Fresh ripe Roma tomatoes, washed, cored, and cut in half lengthwise
2 ounces	Olive oil, plus additional for brushing vegetables
½ tablespoon	Oregano, dried
	Salt and freshly ground black pepper, to taste
1 medium-sized	Eggplant, washed and stemmed
4 each	Red bell peppers, washed
4 each	Yellow bell peppers, washed
1 pound	Shiitake mushrooms, washed and trimmed
1 pound	Zucchini, washed and trimmed
1 pound	Fresh spinach leaves, washed and **blanched**
2½ quarts	Tomato sauce, heated to a boil (see page 76)
5 sheets	Frozen lasagna (12 inches by 12 inches), thawed
12 ounces	Mozzarella cheese, grated
4 ounces	Parmesan cheese, grated

METHOD OF PREPARATION:

1. Preheat the oven to 350° F.
2. Place the tomatoes, cut-side up, on an oiled sheet pan, and brush with olive oil. Sprinkle with oregano, salt, and pepper. Roast until slightly dried, or about 30 to 40 minutes. This can be done 1 day ahead.
3. Cut the eggplant and zucchini lengthwise into ¼-inch slices. Brush the slices with olive oil, and grill on both sides until marked and slightly softened.
4. Brush the peppers with oil, and grill until charred. Peel, de-seed, and cut into quarters.
5. Grill the mushrooms until cooked.
6. Lightly oil the bottom of a half hotel pan, and coat with some of the sauce.
7. Layer the lasagna as follows: pasta, mozzarella cheese, eggplant, sauce, and pasta. Continue to layer, alternating the vegetables for color.
8. Finish with a small amount of sauce coating the top sheet of the pasta, and sprinkle on the Parmesan cheese. Hold the remaining amount of sauce at 140° F.
9. Bake until bubbling and cooked throughout, about 45 minutes. The internal temperature should be a minimum of 165° F.
10. Remove from the oven, and allow to stand for 15 minutes before cutting into portions.
11. Serve on preheated dinner plates, surrounded with additional sauce, or hold at 140° F.

COOKING TECHNIQUES:

Grill, Bake, Roast

Grill/Broil:

1. Clean and heat the grill/broiler.
2. To prevent sticking, brush the food product with oil.

Bake:

1. Preheat the oven.
2. Place the food product on the appropriate rack.

Roast:

1. Sear the food product, and brown evenly.
2. Elevate the food product in a roasting pan.
3. Determine doneness, and consider carryover cooking.
4. Let the food product rest before carving.

GLOSSARY:

Blanch: to parboil

HACCP:

Heat to a minimum temperature of 165° F.
Hold at 140° F.

HAZARDOUS FOOD:

Mozzarella cheese

NUTRITION:

Calories: 636
Fat: 17.1 g
Protein: 30.6 g

CHEF NOTES:

1. Roasting the tomatoes will intensify the tomato flavor, as well as eliminate excess moisture.
2. Dried pasta can be substituted, but it must be precooked before layering.
3. Vegetables can be pan-fried or roasted, but the intensity of the flavors will be less.

Vegetarian Pizza

Three-Cheese with Sautéed Onions and Olives

YIELD: 10 SERVINGS. SERVING SIZE: FOUR OUNCES (EIGHT-INCH PIZZAS).

COOKING TECHNIQUES:
Sauté, Bake

Sauté:
1. Heat the sauté pan to the appropriate temperature.
2. Evenly brown the food product.
3. For a sauce, pour off any excess oil, reheat, and deglaze.

Bake:
1. Preheat the oven.
2. Place the food product on the appropriate rack.

GLOSSARY:
Blanch: to par cook
Al dente: to the bite

HACCP:
Hold at 140° F.

NUTRITION:
Calories: 352
Fat: 15.1 g
Protein: 13.7 g

INGREDIENTS:

2 ounces	Olive oil
6 ounces	Onion, peeled and thinly sliced
	Salt and freshly ground black pepper, to taste
8 ounces	Broccoli rabe, washed, trimmed, **blanched** in salted water, and chilled in a blast of cold water
2 cloves	Garlic, peeled and finely chopped
¼ teaspoon	Hot pepper flakes, crushed
1½ pounds	Pizza dough, cut into 10 portions (see page 452)
	Flour, as needed
5 ounces	Tomato sauce, heated to a boil (see page 76)
8 ounces	Mozzarella cheese, grated
8 ounces	Niçoise olives, drained and pitted
2 each	Lemons, cut into five wedges each

METHOD OF PREPARATION:

1. Preheat the oven to 425° F.
2. In a small sauté pan, heat the olive oil; add the onion, and sauté until golden brown. Season with salt and pepper.
3. Add the broccoli and garlic, and cook until the broccoli is **al dente.** Season with salt and pepper, to taste.
4. Roll out the pizza dough to 8-inch circles, and place on a lightly floured baking tray.
5. Brush each circle of pastry lightly with oil, and thinly coat with tomato sauce.
6. Divide the broccoli over the top, season with pepper, and add the cheese and olives.
7. Bake in the oven until the crust is cooked and crisp.
8. For service, cut the pizza into fourths, and place on a preheated plate. Garnish with a lemon wedge. Hold at 140° F for no longer than 15 minutes.

Vegetarian Pizza

Grilled Tomato, Mozzarella, and Mizuna Lettuce

YIELD: 10 SERVINGS. SERVING SIZE: FOUR OUNCES (EIGHT-INCH PIZZAS).

INGREDIENTS:

1½ pounds	Pizza dough (see page 452)
	All-purpose flour, as needed
2 ounces	Olive oil
12 each	Ripe Roma tomatoes, washed, cored, and sliced lengthwise into ¼-inch slices
	Salt and freshly ground black pepper, to taste
5 ounces	Tomato sauce, heated to a boil (see page 76)
16 ounces	Fresh mozzarella cheese, cut into 30 even slices
1 pound	Mizuna, washed and torn into pieces

METHOD OF PREPARATION:

1. Preheat the oven to 450° F and preheat the grill.
2. Divide the dough into 10 portions; roll each portion on a lightly floured table into an 8-inch circle. Place the rolled dough on baking sheets. Hold chilled at 40° F or below.
3. Brush the olive oil on the tomatoes, season with salt and pepper, and grill over medium-high heat until charred on each side. Remove, and cool.
4. Brush the dough with the remaining olive oil; then brush with the tomato sauce. Sprinkle the cheese over sauce, and arrange the tomatoes over the top with the mozzarella.
5. Place a small bunch of mizuna in the center of each pizza, and drizzle with additional oil.
6. Bake to order for 10 to 15 minutes, or until the crust is cooked. Cut the pizza into quarters, and place on a preheated plate. Hold at 140° F for no longer than 15 minutes.

COOKING TECHNIQUES:
Sauté, Bake, Grill

Sauté:
1. Heat the sauté pan to the appropriate temperature.
2. Evenly brown the food product.
3. For a sauce, pour off any excess oil, reheat, and deglaze.

Bake:
1. Preheat the oven.
2. Place the food product on the appropriate rack.

Grill/Broil:
1. Clean and heat the grill/broiler.
2. To prevent sticking, brush the food product with oil.

HACCP:
Hold at 140° F.

NUTRITION:
Calories: 400
Fat: 16.7 g
Protein: 20.2 g

Pizza Dough

COOKING TECHNIQUE:
Not applicable

YIELD: 2 POUNDS. SERVING SIZE: APPROXIMATELY TWELVE SERVINGS.

INGREDIENTS:

2 ounces	Compressed yeast
6 ounces	Water, lukewarm
2 ounces	Milk
1 ounce	Olive oil
1½ pounds	All-purpose flour
1 teaspoon	Salt

METHOD OF PREPARATION:

1. In a mixing bowl, combine the yeast, water, and milk. Mix well. Add the oil.
2. In a clean mixing bowl, combine the flour and the salt. Mix; then make a well in the center, add the yeast mixture, and blend into a dough. Knead for 4 to 5 minutes. Place the dough in an oiled bowl.
3. Cover with a clean, damp towel, and allow the dough to rise until doubled in size.
4. Roll to the desired size or number.

Wild Mushroom Strudel with Rosemary-Orange Mayonnaise

YIELD: 10 SERVINGS. SERVING SIZE: SIX OUNCES.

INGREDIENTS:

3 pounds	Wild mushrooms (an assortment of at least three varieties)
	Olive oil, as needed
8 ounces	Clarified butter, melted
4 ounces	Shallots, peeled and minced
12 ounces	Fresh spinach, blanched in 2 ounces of boiling salted water, excess moisture removed
8 ounces	Béchamel sauce (see page 16)
4 ounces	Dry sherry
6 ounces	Seasoned bread crumbs, lightly toasted
	Salt and freshly ground black pepper, to taste
3 ounces	Freshly chopped parsley, excess moisture removed
10 sheets	Phyllo dough
	Rosemary-orange mayonnaise (see page 50)

METHOD OF PREPARATION:

1. Wash the mushrooms, and trim the stems. In a sauté pan, heat butter; add mushrooms and sauté. Cool slightly, chop coarsely, and place in a mixing bowl.
2. In a sauté pan, heat 2 ounces of the butter, and sauté the shallots until they are translucent. Transfer to the bowl with the mushrooms.
3. Chop the spinach, and add to the mushrooms.
4. Add the béchamel sauce, sherry, half of the bread crumbs, salt, pepper, and parsley, and mix well.
5. Preheat the oven to 400° F.
6. Place a sheet of phyllo dough on a clean towel. Brush lightly with some of the remaining butter, and lightly dust with bread crumbs. Repeat, using five sheets.
7. Evenly spread half of mushroom mixture along one long side of the dough;

COOKING TECHNIQUES:
Sauté, Bake

Sauté:
1. Heat the sauté pan to the appropriate temperature.
2. Evenly brown the food product.
3. For a sauce, pour off any excess oil, reheat, and deglaze.

Bake:
1. Preheat the oven.
2. Place the food product on the appropriate rack.

HACCP:
Hold at 140° F.

NUTRITION:
Calories: 742
Fat: 61.9 g
Protein: 10.3 g

CHEF NOTES:
1. Shiitake, portabello, morels, or lobster mushrooms all grill very well. If using oyster or a more fragile style of mushroom, it should be sautéed in butter or oil.
2. Mushrooms should be cooked enough to extract the excess moisture.

then roll with the aid of the towel, and turn out onto a half sheet pan lined with parchment paper. Brush the top of the dough with butter.

8. Repeat steps 6 and 7, forming another strudel.
9. Bake until the rolls are brown and flaky, or about 20 minutes. Hold at 140° F.
10. To serve, trim the ends; then slice. Place two slices on a preheated dinner plate, and serve with mayonnaise.

Hungarian Onion and Pepper Stew

(Lecsó)

YIELD: 10 SERVINGS. SERVING SIZE: THREE OUNCES.

COOKING TECHNIQUES:

Sauté, Stew

Sauté:

1. Heat the sauté pan to the appropriate temperature.
2. Evenly brown the food product.
3. For a sauce, pour off any excess oil, reheat, and deglaze.

Stew:

1. Sear, sauté, sweat, or blanch the main food product.
2. Deglaze the pan, if desired.
3. Cover the food product with simmering liquid.
4. Remove the bouquet garni.

GLOSSARY:

Brunoise: ⅛-inch dice
Julienne: matchstick strips
Blanch: to par cook

NUTRITION:

Calories: 219
Fat: 16.5 g
Protein: 8.66 g

INGREDIENTS:

4 ounces	Bacon, diced **brunoise**
1 ounce	Vegetable oil
1 pound	Onions, peeled and cut **julienne**
2 pounds	Italian (Cubanella) peppers, washed, seeded, and sliced into rings
1 teaspoon	Hungarian paprika, sweet
½ teaspoon	Hungarian paprika, hot
4 cloves	Garlic, peeled, and finely minced
1 pound	Tomatoes, washed, cored, **blanched,** peeled, seeded, and cut into wedges
	Salt, to taste
10 ounces	Kielbasa (smoked sausage), optional

METHOD OF PREPARATION:

1. In a sauté pan, combine the bacon and oil, and render the bacon fat. When the bacon is brown and crisp, add the onions, and sauté until they are translucent.
2. Add the pepper rings, and continue to sauté until the peppers are wilted.
3. Add the paprika and garlic, and sauté for 1 minute. Add the tomatoes, season, cover, and stew over low heat until the vegetables are tender but not overcooked.
4. If kielbasa is added, it should be sliced and added just before removing from the heat.

CHEF NOTE:

This preparation is a base for a variety of dishes. Boiled rice or beaten eggs can be added to make a light luncheon treat.

Moroccan Stew

(Couscous)

YIELD: 10 SERVINGS. SERVING SIZE: TEN OUNCES.

INGREDIENTS:

3 quarts	White beef stock, heated to a boil (see page 12)
20 ounces	Beef chuck, trimmed and cut into 2-ounce portions
10 ounces	Dried chick peas, soaked overnight in cold water, then drained
10 pieces	Chicken legs
2 each	Turnips, washed, peeled, sliced, and cut into ¼-inch slices
2 each	Carrots, washed, peeled, and cut into 1-inch pieces
10 ounces	Squash, peeled, and cut into 1-inch pieces
1 head	Green cabbage, trimmed, washed, and cut into wedges
4 ounces	Celeriac, washed, peeled, and cut into ¼-inch slices
	Salt and ground white pepper, to taste
¼ teaspoon	Saffron
1 pound	Couscous semolina (quick cooking), steamed according to directions
	Harissa, as needed

METHOD OF PREPARATION:

1. In a large pot, combine the beef stock, beef, and chick peas, and heat the liquid to a boil. **Dépouiller** as needed; reduce the heat, and simmer until the meats are three fourths tender; then add the chicken legs and vegetables.
2. Continue to simmer until all ingredients are **fork-tender.**
3. Season to taste, add the saffron, and hold at 140° F or above.
4. Hold the steamed semolina (couscous) at 140° F or above.
5. For service, separate the meats, vegetables, and liquid; place the couscous in a preheated bowl. Add a portion of the chicken, beef, vegetables, and chickpeas, and ladle the soup over the entire dish. Serve, offering harissa on the side.

COOKING TECHNIQUES:
Boil, Steam

Boil: (at sea level)
1. Bring the cooking liquid to a rapid boil.
2. Stir the contents, and cook the food product throughout.
3. Serve hot.

Steam: (Traditional)
1. Place a rack over a pot of water.
2. Prevent steam vapors from escaping.
3. Shock or cook the food product throughout.

GLOSSARY:
Dépouiller: to skim impurities/grease
Fork-tender: without resistance

HACCP:
Hold at 140° F.

HAZARDOUS FOODS:
Beef chuck
Chicken legs

NUTRITION:
Calories: 476
Fat: 18.6 g
Protein: 34.3 g

CHEF NOTES:
1. Harissa can be found in jars in Middle Eastern markets.
2. Although the term *couscous* is often used to define steamed semolina, the authentic Moroccan dish is prepared with meat and vegetables, and is a complete meal.
3. In family-style cooking, all the vegetables and meat are cooked whole and served as is.

Philippino Stir-Fried Rice Noodles

(Pansit Grisado)

YIELD: 10 SERVINGS. SERVING SIZE: EIGHT OUNCES.

INGREDIENTS:	
4 ounces	Soybean oil
2 cloves	Garlic, peeled and finely minced
1 four-ounce	Medium-sized onion, peeled and cut **julienned**
1 pound	Pork loin, boneless, cut in long, narrow strips
1 pound	Chicken breast, boneless, cut in long, narrow strips
4 ounces	Soy sauce
1 tablespoon	Chicken base, diluted with warm water
½ ounce	Oyster sauce
12 ounces	Carrots, washed, peeled, and cut julienne
12 ounces	Celery stalks, washed, trimmed, and cut julienne
12 ounces	Green cabbage, washed and shredded
8 ounces	Chinese sausage, cut in narrow strips
2 pounds	Rice noodles, soaked in cold water for 20 minutes
	Salt and ground white pepper, to taste
1 tablespoon	Monosodium glutamate, optional
3 each	Scallions, washed, trimmed, and coarsely chopped
2 each	Lemons, washed, tips cut off, sliced into wedges

METHOD OF PREPARATION:

1. In a wok or skillet, heat the oil. Add the garlic and onions, and stir-fry until the onions are translucent.
2. Add the pork and chicken, and **sear.**
3. Add the soy sauce, chicken base, oyster sauce, carrots, celery, cabbage, and sausage to the wok or skillet. Reduce the heat, and simmer for 5 minutes, or until the vegetables are **al dente** and the wok or skillet is almost dry.
4. Drain the rice noodles, stir into the wok or skillet, and cook for 10 minutes with the meat and vegetables.
5. Season, to taste, and serve immediately on a preheated plate, garnished with scallions and a lemon wedge. Hold at 140° F or above.

COOKING TECHNIQUES:

Stir-Fry, Simmer

Stir-Fry:

1. Heat the oil in a wok until hot but not smoking.
2. Keep the food in constant motion; use the entire cooking surface.

Simmer and Poach:

1. Heat the cooking liquid to the proper temperature.
2. Submerge the food product completely.
3. Keep the cooked product moist and warm.

GLOSSARY:

Julienne: matchstick strips
Sear: to brown quickly
Al dente: to the bite

HACCP:

Hold at 140° F or above.

HAZARDOUS FOODS:

Boneless pork loin
Boneless chicken breast
Chinese sausage

NUTRITION:

Calories: 688
Fat: 36.9 g
Protein: 33.8 g

CHEF NOTES:

1. This dish is a meal in itself. It has protein, vegetables, and starch.
2. This dish is popular in the Philippines.

Pork with Clams

(Cataplana)

YIELD: 10 SERVINGS. SERVING SIZE: SIX OUNCES.

INGREDIENTS:

4 pounds	Pork loin, boneless, trimmed of fat, and cut into 1-ounce cubes
6 ounces	Spanish olive oil
3 tablespoons	Spanish paprika
1 pound	Onions, peeled and thinly sliced
4 pounds	Cherrystone clams, scrubbed

Marinade:

24 ounces	Dry white wine
6 cloves	Garlic, peeled and minced
1 tablespoon	Piri piri (see page 672)
	Salt and freshly ground black pepper, to taste
4 each	Bay leaves

METHOD OF PREPARATION:

1. In a mixing bowl, place the pork cubes. Combine all of the ingredients for the marinade, and pour the mixture over the pork. Marinate refrigerated at 40° F or below for at least 2 hours. Drain, and reserve the marinade; then pat the meat dry.
2. In a saucepan, heat 4 ounces of the olive oil until it is almost smoking; add the pork, and brown, turning the pieces frequently.
3. Add the paprika and the marinade, and heat to a boil; reduce the heat, and simmer until almost all of the liquid evaporates, approximately 45 minutes. Remove the bay leaves.
4. In a separate saucepan, heat the remaining olive oil, and sauté the onions until they are translucent. Add the clams, season with salt and pepper, and cook over high heat, covered, for 5 minutes, or until the clams open. Add the clam mixture (with their juice) to the pork, and cook until done. Hold at 140° F.
5. To serve, place portion on a preheated dinner plate.

COOKING TECHNIQUES:

Sauté, Simmer

Sauté:

1. Heat the sauté pan to the appropriate temperature.
2. Evenly brown the food product.
3. For a sauce, pour off any excess oil, reheat, and deglaze.

Simmer and Poach:

1. Heat the cooking liquid to the proper temperature.
2. Submerge the food product completely.
3. Keep the cooked product moist and warm.

HACCP:

Heat to a minimum internal temperature of 165° F.
Hold at 140° F.

HAZARDOUS FOODS:

Pork
Clams

NUTRITION:

Calories: 733
Fat: 34 g
Protein: 78 g

CHEF NOTE:

Preparation should begin early, so the pork can be marinated according to the recipe.

Stewed Beef and Pork

(Apritada Carne at Baboy)

YIELD: 10 SERVINGS. SERVING SIZE: EIGHT OUNCES.

COOKING TECHNIQUE:

Stir-Fry, Simmer

Stir-Fry:

1. Heat the oil in a wok until hot but not smoking.
2. Keep the food in constant motion; use the entire cooking surface.

Simmer and Poach:

1. Heat the cooking liquid to the proper temperature.
2. Submerge the food product completely.
3. Keep the cooked product moist and warm.

GLOSSARY:

Brunoise: ⅛-inch dice
Sear: to brown quickly

HACCP:

Hold at 140° F or above.

HAZARDOUS FOODS:

Ground round beef
Pork butt

NUTRITION:

Calories: 534
Fat: 25.9 g
Protein: 54.2 g

INGREDIENTS:

2 ounces	Peanut oil
2½ pounds	Bottom round beef, cut into ½-inch cubes
6 cloves	Garlic, peeled and finely minced
12 ounces	Onions, peeled and diced **brunoise**
1¼ quarts	Tomato sauce, heated to a boil (see page 76)
1 tablespoon	Ham base, diluted in warm water
1 tablespoon	Beef base, diluted in warm water
1 pound	Pork butt, cut into ½-inch cubes
3 each	Potatoes, washed, peeled, washed again, and cut into 1-inch by 2-inch fingers
3 each	Green peppers, washed, peeled, and cut into 1-inch squares
10 ounces	Green pitted olives, sliced
	Salt and freshly ground black pepper, to taste

METHOD OF PREPARATION:

1. Heat the oil in a wok or skillet; **sear** and brown the beef, and set aside.
2. In the same oil, sauté the garlic and onions; add tomato sauce, and let the mixture simmer for 5 minutes.
3. Add the diluted meat bases to the sauce, and simmer for 1 minute. Return the browned beef to the sauce, and simmer for 30 minutes.
4. Add the pork cubes, and continue to simmer for 20 minutes more.
5. Add the potatoes, and simmer 20 minutes more.
6. Add the peppers and olives, simmer for 2 minutes more, and season, to taste.
7. Serve immediately, or hold at 140° F or above.

CHEF NOTE:

This is an Indonesian dish.

Stuffed Cabbage

YIELD: 50 SERVINGS. SERVING SIZE: TWO ROLLS.

INGREDIENTS:

6 each	Large cabbages, cored, outer leaves removed
1½ gallons	Tomato sauce, heated to a boil (see page 76)

Stuffing:

5 ounces	Oil
30 ounces	Onions, peeled, and diced **brunoise**
3 pounds	Beef, ground
3 pounds	Pork, ground
13 cloves	Garlic, peeled and mashed
2 pounds	Cooked rice
12 ounces	Tomato purée
7 each	Eggs
1 bunch	Freshly chopped parsley, excess moisture removed
1 teaspoon	Ground black pepper
	Salt, to taste

METHOD OF PREPARATION:

1. Preheat the oven to 350° F.
2. Heat the oil in a sauté pan, and sauté the onions until they are transparent. Cool to below 40° F, and reserve.
3. Place the ground meats in a mixing bowl, and add the sautéed onions. Add the garlic, cooked rice, tomato purée, eggs, and seasonings. Blend the ingredients well.
4. **Blanch** the cabbage in salted, boiling water. Remove the leaves, shock them in cold water, and drain well. Trim the heavy vein near the core end.
5. Place 2 ounces of stuffing onto each leaf. Roll, and tuck both ends.
6. Ladle the tomato sauce in a hotel pan, enough to coat the bottom, and place the stuffed cabbages on top. Lightly coat them with tomato sauce.
7. Cover the pan with parchment paper and aluminum foil. Bake in the oven until the cabbage is tender, about two hours. Hold at 140° F.
8. Serve with boiled potatoes.

COOKING TECHNIQUES:

Sauté, Bake

Sauté:

1. Heat the sauté pan to the appropriate temperature.
2. Evenly brown the food product.
3. For a sauce, pour off any excess oil, reheat, and deglaze.

Bake:

1. Preheat the oven.
2. Place the food product on the appropriate rack.

GLOSSARY:

Brunoise: ⅛-inch dice
Blanch: to par cook

HACCP:

Cool to below 40° F.
Hold at 140° F.

HAZARDOUS FOODS:

Ground beef
Ground pork
Eggs

NUTRITION:

Calories: 267
Fat: 13.2 g
Protein: 17.4 g

CHEF NOTE:

As an alternative method of cooking cabbage, freeze the cored cabbage head for a few days. Then, place it under cool running water, and proceed to the trimming step.

Veal and Lobster Rouladen with Parsley Sauce

YIELD: 10 SERVINGS. SERVING SIZE: ONE EACH.

INGREDIENTS:

10 three-ounce	Veal cutlets, pounded thin
	Salt and ground white pepper, to taste
3 ounces	Butter
3 ounces	Vegetable oil
1 each	Large eggplant, washed, trimmed, and cut into ⅙-inch slices
10 ounces	Ham, cut into 10 thin slices
1 pound	Spinach leaves, washed, steamed, and well drained
5 eight-ounce	Lobster tails, split lengthwise
	Seasoned flour, as needed
5 each	Eggs, cracked and lightly beaten
	Bread crumbs, as needed
	Parsley sauce (see page 40)

METHOD OF PREPARATION:

1. Preheat the oven to 350° F.
2. On a sheet pan, lay out the veal cutlets, and lightly season with salt and pepper. Place in the refrigerator at 40° F or lower, until needed.
3. In a sauté pan, heat 1 ounce each of butter and oil, and fry the eggplant slices until golden on each side. Remove, and set aside to cool.
4. Place one slice of eggplant on each veal cutlet, and top with ham.
5. Wrap the lobster tail pieces in spinach leaves, and place on the ham; season again, and roll into a **paupiette.** Secure each with a toothpick, or tie with butcher twine.
6. **Dredge** the paupiettes in flour, and dip in eggwash; then coat with bread crumbs.
7. Heat the remaining butter and oil; then brown the paupiettes on all sides. Transfer the paupiettes into a 2-inch deep hotel pan. Place in the oven to finish cooking, or for approximately 20 minutes. Serve immediately with parsley sauce, or hold at 140° F. To serve, arrange one paupiette on a preheated dinner plate, and **nappé** with sauce.

COOKING TECHNIQUES:

Shallow-Fry, Sauté, Bake

Shallow-Fry:

1. Heat the cooking medium to the proper temperature.
2. Cook the food product throughout.
3. Season, and serve hot.

Sauté:

1. Heat the sauté pan to the appropriate temperature.
2. Evenly brown the food product.
3. For a sauce, pour off any excess oil, reheat, and deglaze.

Bake:

1. Preheat the oven.
2. Place the food product on the appropriate rack.

GLOSSARY:

Paupiette: rolled meat or fish
Dredge: to coat with flour
Nappé: coated

HACCP:

Preheat the oven to 350° F.
Refrigerate to 40° F or lower.
Hold at 140° F.

HAZARDOUS FOODS:

Veal
Ham
Lobster tails
Eggs

NUTRITION:

Calories: 548
Fat: 33.9 g
Protein: 52.7 g

Veal Sweetbreads with Grapes

(Ris de Veau aux Raisins)

YIELD: 10 SERVINGS. SERVING SIZE: ONE-HALF SWEETBREAD.

INGREDIENTS:

5 each	Veal sweetbreads
	Court boullion, as needed, heated to a boil (see page 9)
6 ounces	Butter
2 ounces	Vegetable oil
1 each	Onions, peeled and diced **brunoise**
	Seasoned flour, as needed
10 ounces	White veal stock, heated to a boil (see page 12)
	Salt and ground white pepper, to taste
8 ounces	White grape juice
4 ounces	Madeira wine
8 ounces	Seedless grapes, washed, peeled, and cut in half lengthwise

METHOD OF PREPARATION:

1. Preheat the oven to 375° F.
2. In a saucepan, combine the sweetbreads and court-boullion. Heat to a gentle boil, and cook for 15 minutes. Drain, and rinse in cold water; then remove the membranes.
3. In a braising pan, heat the butter and oil; add the onions, and sauté until they are translucent.
4. Cut the sweetbreads in half lengthwise, and coat with flour, shaking off any excess.
5. Add the sweetbreads to the onions, and brown on each side for 3 minutes. Moisten with the veal stock, and season to taste. Cover the braising pan tightly, and place in the oven for 10 minutes. Turn the sweetbreads from time to time.
6. Remove from the oven, transfer the sweetbreads to a tray, cover, and hold at 140° F.
7. Place the braising pan on an open flame. Add the grape juice and Madeira wine. Cook and reduce over high heat. If too thin, add a little beurre manié.
8. To serve, place a portion of sweetbread on a preheated dinner plate. **Nappé** with sauce, and garnish with grapes.

COOKING TECHNIQUES:

Braise, Sauté, Boil

Braise:

1. Heat the braising pan to the proper temperature.
2. Sear and brown the food product to a golden color.
3. Degrease and deglaze.
4. Cook the food product in two thirds liquid until fork-tender.

Sauté:

1. Heat the sauté pan to the appropriate temperature.
2. Evenly brown the food product.
3. For a sauce, pour off any excess oil, reheat, and deglaze.

Boil: (at sea level)

1. Bring the cooking liquid to a rapid boil.
2. Stir the contents, and cook the food product throughout.
3. Serve hot.

GLOSSARY:

Brunoise: ⅛-inch dice
Nappé: coated

HACCP:

Hold warm at 140° F.

HAZARDOUS FOOD:

Veal sweetbreads

NUTRITION:

Calories: 360
Fat: 30.3 g
Protein: 17 g

6

Starches

Starches

Good cookery is the food of clear conscience.
Des Essarts
French gastronome and actor

As we approach the year 2000, we face the greatest challenge that the culinary world has ever experienced. As nutritional trends are constantly gaining more and more popularity, we, as chefs, not only have to restructure our attitudes, but also constantly update ourselves on education as it relates to nutrition. If we begin to approach cooking as a science, understanding all components, we can then develop a product that the public demands: good-tasting food for the health-minded. As chefs, we have a great responsibility to set the example by providing exciting, nutritional dishes.

Johnny Rivers
Distinguished Visiting Chef
January 17–19, 1988
Executive Chef
Resorts Research and Development
Walt Disney World Company

The HACCP Process

The following recipe illustrates how starches flow through the HACCP process.

Baked Stuffed Potatoes

Receiving

- Potatoes, Idaho—no wet spots; smooth and clean. Potatoes should be delivered at 45° F to 50° F (7.2° C to 10° C).
- Salad oil—packaging intact
- Ham—lean portions, pink in color. Ham should be delivered at 40° F (4.4° C) or lower.
- Butter—should have a sweet flavor, firm texture, and uniform color. Butter should be free of specks and mold, and the packaging should be intact. Butter should be delivered at 40° F (4.4° C) or lower.
- Egg, whole—Shells should not be cracked or dirty. Containers should be intact. Whole shell eggs should be delivered at 40° F (4.4° C) or lower.
- Herbs (parsley)—no visible signs of discoloration
- Sour cream—packaging intact. Sour cream should be delivered at 40° F (4.4° C) or lower.
- Cheese (Parmesan and cheddar)—Check to see that each type has its characteristics, flavor, and texture, as well as uniform color. Cheese should be delivered at 40° F (4.4° C) or lower.
- Spices (salt and pepper)—packaging intact

Storage

- Potatoes, Idaho—Store in dry storage at 50° F (10° C) with a relative humidity of 50% to 60%.
- Salad oil—Store in dry storage at 50° F (10° C) with a relative humidity of 50% to 60%.
- Ham—Store under refrigeration, with a product temperature not to exceed 40° F (4.4° C). Store below already cooked foods.
- Butter—Store under refrigeration at 40° F (4.4° C) or lower and above and away from raw, potentially hazardous foods.
- Eggs, whole—Store under refrigeration at 40° F (4.4° C) or lower. Store above and away from raw, potentially hazardous foods.
- Herbs (parsley)—Store under refrigeration at 40° F (4.4° C) or lower.
- Sour cream—Store under refrigeration at 40° F (4.4° C) or lower. Store above and away from raw, potentially hazardous foods.
- Cheese (Parmesan and cheddar)—Store under refrigeration, with a product temperature not to exceed 40° F (4.4° C). Store above and away from raw, potentially hazardous foods.

- Spices (salt and pepper)—Store in dry storage at 50° F (10° C) with a relative humidity of 50% to 60%. Keep dry.

Preparation and Cooking

- Wash and dry the parsley.
- Chop the parsley on a clean and sanitized cutting board with clean and sanitized utensils.
- Pierce the potatoes with a fork before baking.
- Rub the potatoes with oil. Place them on a baking sheet. Bake for 1 hour in a 350° F oven until fork-tender. Remove the potatoes; let them cool for 15 minutes.
- Cut the baked potatoes in half lengthwise. Remove the pulp.
- Pass the pulp through a food mill or ricer; then place it in a mixing bowl.
- Add the remaining ingredients, except the cheese and butter. Mix well.
- Place the potato shells in a baking pan.
- Using a pastry bag with a star tube, pipe the potato mixture into the potato skins.
- Sprinkle cheese on the potatoes.
- Drizzle with butter, and bake in a 350° F oven until heated through and golden brown. Cook to an internal temperature of 165° F (73.9° C).

Holding and Service

- Hold at 140° F (60° C) or higher.
- Serve immediately.

Nutritional Notes

Grains and pastas are excellent sources of complex carbohydrates. They are naturally very low in fat and are replacing meat as the mainstay of the meal in contemporary menus. In the past, starches had mistakenly been regarded as fattening; actually, ounce for ounce, starch contains less that half the calories of fat. In addition to starch, whole-grain pastas, burgul, buckwheat, brown rice, quinoa, amaranth, and wild rice also provide fiber, vitamins, minerals, and terrific flavor. Enriched, refined grain products, such as white pasta and white rice, also provide some vitamins and iron.

- *Cook starches in unsalted water or stock. Cook grains in the least amount of liquid possible. Do not rinse either before or after cooking.*
- *Minimize the amount of added fat.*
- *Use mono- and polyunsaturated oils when fat is necessary.*
- *Use low-fat sauces and flavorings.*

Baked Potatoes

YIELD: 50 SERVINGS. SERVING SIZE: ONE POTATO.

INGREDIENTS:

50 each	Baking potatoes, washed, scrubbed, and dried
8 ounces	Oil
1 quart	Sour cream
1 ounce	Chives, dried

METHOD OF PREPARATION:

1. Preheat the oven to 375° F.
2. Rub oil on the washed and scrubbed potatoes. Place the potatoes on a sheet pan. Pierce the potatoes with a braising fork. Bake in the oven for 1 hour, or until **fork-tender.**
3. Combine the sour cream and chives. Cut the potatoes lengthwise, half deep, and press both ends of the potato together to expose the flesh. Serve with sour cream and chives on the side.
4. Hold at 140° F.

COOKING TECHNIQUE:

Bake

Bake:

1. Preheat the oven.
2. Place the food product on the appropriate rack.

GLOSSARY:

Fork-tender: without resistance

HACCP:

Hold at 140° F.

HAZARDOUS FOOD:

Sour cream

NUTRITION:

Calories: 190
Fat: 8.26 g
Protein: 3 g

CHEF NOTE:

Different types of potatoes can be used, such as large red bliss potatoes.

Basic Pasta Recipe I

COOKING TECHNIQUE:
Not applicable

HAZARDOUS FOOD:
Eggs

NUTRITION:
Calories: 326
Fat: 2.12 g
Protein: 11.3 g

YIELD: 2 POUNDS. SERVING SIZE: FOUR OUNCES, COOKED.

INGREDIENTS:

12 ounces	Semolina flour
12 ounces	Durum flour
1 tablespoon	Salt
3 ounces	Water, or as needed
2 each	Eggs

METHOD OF PREPARATION:

1. Mix the dry ingredients in the hopper of a pasta machine.
2. Slowly add the water and eggs.
3. Mix for 10 minutes; the dough should be crumbly.
4. Extrude pasta as desired.

Basic Pasta Recipe II

YIELD: 2 POUNDS. SERVING SIZE: FOUR OUNCES, COOKED.

COOKING TECHNIQUE:

Not Applicable

Hazardous Food:

Egg yolks

NUTRITION:

Calories: 385
Fat: 8.23 g
Protein: 10.9 g

INGREDIENTS:

1½ pounds	All-purpose flour
6 each	Egg yolks
1 ounce	Olive oil
1 tablespoon	Salt

METHOD OF PREPARATION:

1. Place the flour in a mixing bowl, make a well in the center of the flour, and add the remaining ingredients.
2. Mix from the center until all ingredients are incorporated.
3. Knead into a smooth ball. Wrap in plastic wrap and let the dough rest until needed.
4. Prepare the pasta according to the recipe.

Basmati Rice Pilaf

YIELD: 50 SERVINGS. SERVING SIZE: FOUR OUNCES.

INGREDIENTS:

1 pound	Onions, peeled, and diced
3 each	Bay leaves
	Salad oil, as needed
3 pounds	Basmati rice, long grain
3 quarts	White chicken stock (see page 12)
5 tablespoons	Freshly chopped parsley, excess moisture removed
	Salt and ground white pepper, to taste

METHOD OF PREPARATION:

1. In a medium saucepan, heat the salad oil, and sauté the onion. Add the bay leaf and the Basmati rice until coated with oil.
2. Add the chicken stock, stirring constantly. Heat to a boil. Add the parsley, pepper, and salt.
3. Cover, reduce the heat, and simmer until the liquid has been absorbed and the rice is tender, about 20 minutes.
4. You also may cover the pan and finish in the oven at 350° F for 18 to 20 minutes.
5. Hold at 140° F.
6. Remove bay leaves before service.

COOKING TECHNIQUES:

Sauté, Boil, Simmer

Sauté:

1. Heat the sauté pan to the appropriate temperature.
2. Evenly brown the food product.
3. For a sauce, pour off any excess oil, reheat, and deglaze.

Boil: (at sea level)

1. Bring the cooking liquid to a rapid boil.
2. Stir the contents, and cook the food product throughout.
3. Serve hot.

Simmer and Poach:

1. Heat the cooking liquid to the proper temperature.
2. Submerge the food product completely.
3. Keep the cooked product moist and warm.

HACCP:

Hold at 140° F.

NUTRITION:

Calories: 42.5
Fat: 2.31 g
Protein: 2.23 g

CHEF NOTE:

Basmati rice is an aromatic rice from northern India. When cooking, this rice gives off a pleasant aroma, much like that of popcorn.

Boiled New Potatoes

YIELD: 50 SERVINGS. SERVING SIZE: SIX OUNCES.

INGREDIENTS:

20 pounds	New potatoes, washed and scored with a channel knife
	Water, as needed
	Seasoned, to taste
2 pounds	Clarified butter, salted
12 ounces	Freshly chopped parsley, excess moisture removed

METHOD OF PREPARATION:

1. In a stockpot, place the potatoes in cold, salted water. Heat to a boil. Cook until the potatoes are **fork-tender.**
2. Drain and air dry. Put the potatoes into a mixing bowl.
3. Melt the butter in a saucepan, and drizzle over the potatoes.
4. Adjust the seasoning. Add chopped parsley, and mix well. Place in a hotel pan, and keep warm for service.

COOKING TECHNIQUE:

Boil

Boil: (at sea level)

1. Bring the cooking liquid to a rapid boil.
2. Stir the contents, and cook the food product throughout.
3. Serve hot.

GLOSSARY:

Fork-tender: without resistance

NUTRITION:

Calories: 249
Fat: 14.9 g
Protein: 4.53 g

CHEF NOTES:

1. For better flavor, add the chopped parsley to the butter, sauté half a minute, and toss the potatoes in the butter.
2. For boiled new potatoes with dill, prepare same as above, but replace parsley with fresh dill.

Boulangére Potatoes

(Pommes de Terre Boulangére)

YIELD: 10 SERVINGS. SERVING SIZE: FOUR OUNCES.

INGREDIENTS:

3 pounds	Potatoes, washed, peeled and cut into ¼-inch slices
	Cold salted water, as needed
3 ounces	Butter
12 ounces	Onions, peeled and cut **julienne**
	Salt and ground white pepper, to taste
1 pint	Brown veal stock, seasoned and heated to a boil (see page 7)
3 ounces	Freshly chopped parsley, excess moisture removed

METHOD OF PREPARATION:

1. Preheat the oven to 375° F.
2. In a saucepan, place the potatoes, and cover with water. Boil, and cook for approximately 8 to 10 minutes; then drain.
3. In a sauté pan, melt the butter, add the onions, season, to taste, and sauté until smothered.
4. Place the potatoes in a hotel pan, season with salt and pepper, and top with the smothered onions.
5. Pour the veal stock over, and bake until the potatoes are **fork-tender.** Hold at 140° F.
6. Sprinkle with chopped parsley before serving.

COOKING TECHNIQUES:

Boil, Sauté, Bake

Boil: (at sea level)

1. Bring the cooking liquid to a rapid boil.
2. Stir the contents, and cook the food product throughout.
3. Serve hot.

Sauté:

1. Heat the sauté pan to the appropriate temperature.
2. Evenly brown the food product.
3. For a sauce, pour off any excess oil, reheat, and deglaze.

Bake:

1. Preheat the oven.
2. Place the food product on the appropriate rack.

GLOSSARY:

Julienne: matchstick strips
Fork-tender: without resistance

HACCP:

Hold at 140° F.

NUTRITION:

Calories: 173
Fat: 7.22 g
Protein: 3.23 g

CHEF NOTE:

Lamb stock is often used with these potatoes to accompany a lamb dish. Chicken stock also may be substituted to complement the items to be served.

Buttered Egg Noodles

(Pâtes au Beurre)

YIELD: 10 SERVINGS. SERVING SIZE: FOUR OUNCES.

INGREDIENTS:

1 gallon	Water
1 tablespoon	Salt
30 ounces	Egg noodles (medium)
4 ounces	Butter

METHOD OF PREPARATION:

1. Heat the salted water to a boil; add the egg noodles, and cook **al dente.**
2. Remove the noodles from the heat; strain, rinse with cold water, and drain well.
3. Using a wire basket, reheat the noodles in boiling salted water before service. Shake off all water, and place in a hotel pan.
4. Melt the butter, and mix with the noodles. Season as needed.

COOKING TECHNIQUES:

Boil

Boil: (at sea level)

1. Bring the cooking liquid to a rapid boil.
2. Stir the contents, and cook the food product throughout.
3. Serve hot.

GLOSSARY:

Al dente: to the bite

NUTRITION:

Calories: 119
Fat: 9.61 g
Protein: 1.45 g

Château Potatoes

(Pommes de Terre Château)

YIELD: 10 SERVINGS. SERVING SIZE: FOUR OUNCES.

INGREDIENTS:

10 each	Idaho potatoes
	Cold water, to cover
	Salt, as needed
2 slices	Lemon
6 ounces	Clarified butter
3 ounces	Freshly chopped parsley, excess moisture removed

METHOD OF PREPARATION:

1. Preheat the oven to 400° F.
2. **Tourné** the potatoes to uniform, oval shapes. Place in a saucepan with water to cover.
3. Add the salt and lemon slices to the water, and heat to a boil. When the water is foaming, remove from the heat. Rinse the potatoes in cold water, drain, and dry.
4. In a roasting pan, heat the butter, and sauté the potatoes until golden brown. Transfer the potatoes to the oven, and bake until the potatoes are tender. Remove from the oven, and hold at 140° F or above.
5. To serve, garnish the potatoes with parsley.

COOKING TECHNIQUE:

Bake

Bake:

1. Preheat the oven.
2. Place the food product on the appropriate rack.

GLOSSARY:

Tournéed: trimmed to a large olive shape

HACCP:

Hold at 140° F.

NUTRITION:

Calories: 238
Fat: 13.9 g
Protein: 2.46 g

CHEF NOTE:

For quantity cooking, no sautéing is required. Place the potatoes on buttered sheet pans, season, and roast in a 425° F oven until brown; then lower the temperature to 350° F. Continue to bake until the potatoes are tender.

Corn Fritters

YIELD: 50 SERVINGS. SERVING SIZE: TWO FRITTERS.

INGREDIENTS:

	Oil, as needed
15 each	Eggs
1¼ quart	Milk
1¼ ounces	Granulated sugar
1¼ teaspoons	Salt
2½ pounds	Bread flour, sifted
2½ ounces	Baking powder
3¾ pounds	Frozen kernel corn, thawed and drained

METHOD OF PREPARATION:

1. Separate the eggs.
2. Add the yolks to the corn. Blend well.
3. Whip the whites to a stiff peak.
4. Add the milk to the corn mixture. Mix the flour and baking powder together in another bowl, and add to the corn mixture and incorporate.
5. Fold in the whipped egg whites. Blend well.
6. To prepare the fritters, place the batter by the spoonful in the preheated oil. Fry until golden brown and cooked thoroughly. Remove, and drain on absorbent paper.
7. Serve with warm maple syrup and sprinkled powdered sugar, if desired.

COOKING TECHNIQUE:

Deep-Fry

Deep-Fry:

1. Heat the frying liquid to the proper temperature.
2. Submerge the food product completely.
3. Fry the product until it is cooked throughout.

HAZARDOUS FOODS:

Eggs
Milk

NUTRITION:

Calories: 523
Fat: 38 g
Protein: 8.1 g

CHEF NOTES:

1. Pasteurized egg yolks and egg whites may be used.
2. Fry only the quantity needed for immediate service.

Croquette, Potato

(Pommes de Terre Croquettes)

YIELD: 10 SERVINGS. SERVING SIZE: ONE CROQUETTE.

INGREDIENTS:

2 pounds	Potatoes, washed, peeled, and cut into fourths
	Salted cold water, as needed
4 each	Egg yolks
½ teaspoon	Nutmeg, ground
	Salt, to taste
	Seasoned flour, as needed
5 each	Eggs, cracked and beaten
	Seasoned bread crumbs, as needed

METHOD OF PREPARATION:

1. Place the potatoes in salted water to cover, and heat to a boil. Cook until tender; then drain and dry on the stove.
1. Purée the potatoes; then add the egg yolks and nutmeg. Season, to taste.
2. Shape the potatoes into cylinders 1-inch by 3-inches long.
3. Roll the cylinders into the flour, dip in the eggs, and coat with bread crumbs.
5. Preheat the fryer, and deep-fry the croquettes **á la minute** until crisp and golden brown. Serve immediately.

COOKING TECHNIQUES:

Boil, Deep-Fry

Boil: (at sea level)

1. Bring the cooking liquid to a rapid boil.
2. Stir the contents, and cook the food product throughout.
3. Serve hot.

Deep-Fry:

1. Heat the frying liquid to the proper temperature.
2. Submerge the food product completely.
3. Fry the product until it is cooked throughout.

GLOSSARY:

À la minute: cooked to order
Bain-marie: hot-water bath

HAZARDOUS FOOD:

Eggs

NUTRITION:

Calories: 413
Fat: 26.8 g
Protein: 7.97 g

CHEF NOTE:

Do not hold croquettes in a **bain-marie.** Fried items should never be covered.

Cumin Rice

YIELD: 10 SERVINGS. SERVING SIZE: FOUR OUNCES.

INGREDIENTS:

2 ounces	Bacon fat
6 ounces	Onions, peeled and diced **brunoise**
6 ounces	Green pepper, washed, seeded, and diced brunoise
3 cloves	Garlic
1 teaspoon	Cumin
1 pound	Long-grained rice
1 quart	White chicken stock, heated to a boil (see page 12)
	Salt and ground white pepper, to taste

METHOD OF PREPARATION:

1. In a saucepot, heat the bacon fat; add the onion, and sauté until translucent. Add the peppers and garlic, and continue to sauté for 2 more minutes.
2. Add the cumin, and mix well; then add the rice, and stir to coat the grains.
3. Add the stock, stir, taste, and adjust the seasoning as needed.
4. Cover, and steam on low heat or in an oven until the liquid is absorbed and the rice is tender, or approximately 20 minutes.
5. For service, mix the rice with a braising fork, and serve immediately, or hold at 140° F or higher.

COOKING TECHNIQUES:
Sauté, Steam

Sauté:
1. Heat the sauté pan to the appropriate temperature.
2. Evenly brown the food product.
3. For a sauce, pour off any excess oil, reheat, and deglaze.

Steam: (Traditional)
1. Place a rack over a pot of water.
2. Prevent steam vapors from escaping.
3. Shock or cook the food product throughout.

GLOSSARY:
Brunoise: ⅛-inch dice

HACCP:
Hold at 140° F.

NUTRITION:
Calories: 245
Fat: 6.55 g
Protein: 5.66 g

CHEF NOTE:
Always use a braising fork to stir the rice after cooking.

Delmonico Potatoes

COOKING TECHNIQUE:
Bake

Bake:
1. Preheat the oven.
2. Place the food product on the appropriate rack.

GLOSSARY:
Al dente: to the bite
Fork-tender: without resistance

HACCP:
Heat to 165° F.
Hold at 140° F.

NUTRITION:
Calories: 256
Fat: 6.68 g
Protein: 4.88 g

YIELD: 50 SERVINGS. SERVING SIZE: FOUR OUNCES.

INGREDIENTS:

20 pounds	Potatoes, washed, peeled, eyes removed, and diced
1 gallon	Béchamel sauce, heated to 165° F and seasoned (see page 16)
1 (#303 can)	Pimientos, rinsed and diced
12 ounces	Bread crumbs

METHOD OF PREPARETION:

1. In a stockpot, place the potatoes in cold, salted water. Heat to a boil. Cook **al dente,** and drain. Place on a sheet pan to air dry. When dry, place in a mixing bowl.
2. Combine the potatoes with the seasoned béchamel sauce.
3. Add the diced pimientos to the potatoes, and mix well. Place the mixture in a buttered hotel pan.
4. Sprinkle the bread crumbs on the potatoes.
5. Bake in a 350° F oven until lightly browned and the potatoes are **fork-tender.** Hold at 140° F.

Duchess Potatoes

(Pommes de Terre Duchesse)

YIELD: 10 SERVINGS. SERVING SIZE: FOUR OUNCES.

COOKING TECHNIQUES:
Boil, Bake

Boil: (at sea level)
1. Bring the cooking liquid to a rapid boil.
2. Stir the contents, and cook the food product throughout.
3. Serve hot.

Bake:
1. Preheat the oven.
2. Place the food product on the appropriate rack.

HACCP:
Hold at 140° F

HAZARDOUS FOOD:
Egg yolks

NUTRITION:
Calories: 182
Fat: 6.83 g
Protein: 3.52 g

INGREDIENTS:

3 pounds	Potatoes, washed, peeled, and cut into ½-inch pieces
4 each	Egg yolks
	Salt and ground white pepper, to taste
	Nutmeg, to taste
4 ounces	Melted butter

METHOD OF PREPARATION:

1. In a large saucepan, cover the potatoes in cold, salted water. Heat to a boil; reduce the heat, and simmer until tender. Drain and dry on the stove.
2. Pass the potatoes through a food mill or ricer.
3. Place the potatoes and egg yolks in a mixing bowl, season, to taste, and blend well.
4. Preheat the oven to 350° F.
5. Using a pastry bag with a star tube, pipe 2 ounces of the potatoe mixture out onto a parchment-papered sheet pan.
6. Place the potatoes in the oven; after 10 minutes, remove, and brush the potatoes with melted butter. Finish baking until the potatoes are golden brown and heated throughout, and hold at 140° F.

CHEF NOTE:
If the potatoes are brushed with egg yolks and browned they become *pommes mont'd'or*.

English-Style Potatoes

(Pommes de Terre Anglaise)

YIELD: 10 SERVINGS. SERVING SIZE: FOUR OUNCES.

INGREDIENTS:

10 each	All-purpose or chefs' potatoes
	Cold salted water, as needed
1 slice	Lemon
3 ounces	Freshly chopped parsley, excess moisture removed

METHOD OF PREPARATION:

1. Tourné the potatoes, and leave whole.
2. In a saucepan, cover the potatoes with water, and add the lemon.
3. Heat the water to a boil. Reduce the heat, and simmer until the potatoes are **fork-tender.**
4. To serve, sprinkle with parsley. Hold at 140° F.

COOKING TECHNIQUE:
Boil

Boil: (at sea level)
1. Bring the cooking liquid to a rapid boil.
2. Stir the contents, and cook the food product throughout.
3. Serve hot.

GLOSSARY:
Fork-tender: without resistance

NUTRITION:
Calories: 118
Fat: .18 g
Protein: 2.49 g

CHEF NOTE:
If available, a steamer can be used to cook the potatoes.

COOKING TECHNIQUE:
Boil, Sauté

GLOSSARY:
Brunoise: finely diced
Julienne: matchstick strips
Render: to melt fat

HACCP:
Hold at 140° F or above.

Nutrition:
Calories: 346
Fat: 21.4 g
Protein: 23.6 g

Fava Beans, Ribatejo Style

YIELD: 10 SERVINGS. SERVING SIZE: FOUR OUNCES.

INGREDIENTS:

8 ounces	Bacon, cut into ½-inch pieces
8 ounces	Onions, peeled and diced **brunoise**
6 cloves	Garlics, peeled and mashed into a purée
2 quarts	Boiling salted water
2 pounds	Fava beans, fresh or frozen
8 ounces	Proscuitto, cut **julienned**
8 ounces	Cured chorizo (Spanish sausage), diced
2 ounces	Fresh cilantro leaves, chopped
	Salt and ground white pepper, to taste

METHOD OF PREPARATION:

1. In a sauté pan, **render** the bacon, and add the onion and garlic. Sauté the onions until they turn a golden brown, then set aside.
2. Add the beans to the water, and cook until they are tender, which will require about 10 to 15 minutes. Drain the beans, and combine with the bacon mixture.
3. Add the proscitto and sausage and sauté until heated throughout.
4. Add the cilantro, and season, to taste. Serve immediately, or hold at 140° F or above.

CHEF NOTES:

1. Smoked pork is used to prepare Spanish chorizo; fresh pork is used in Mexican chorizo.
2. If fava beans are not available, butter (lima) beans can be substituted.

Flour Dumplings

(Egg Spätzel)

YIELD: 10 SERVINGS. SERVING SIZE: FOUR OUNCES.

INGREDIENTS:

1 pound	All-purpose flour
¼ teaspoon	Nutmeg, ground
1 teaspoon	Salt
2 each	Eggs
1 ounce	Vegetable oil
8 ounces	Milk
	Boiling salted water, as needed
3 ounces	Butter

METHOD OF PREPARATION:

1. In the bowl of a food processor, combine the flour, nutmeg, and salt.
2. Beat together the eggs, oil, and milk in a mixing bowl.
3. With the machine running, gradually add the egg mixture, and process to make a smooth, elastic batter. Let the batter rest for about 30 minutes before cooking.
4. Press the batter through a large-holed colander or spätzel maker into the boiling water. Cook uncovered for 5 to 8 minutes, or until tender. Cooking time will vary, depending on the size of the spätzel.
5. Use a **skimmer** or slotted spoon, and transfer spätzel to a bowl with cold water, to stop the cooking process.
6. Just before service, drain the spätzel. In a large sauté pan heat the butter, and sauté spätzel until thoroughly heated. Serve immediately, or hold at 140° F or higher.

COOKING TECHNIQUE:
Boil

Boil: (at sea level)
1. Bring the cooking liquid to a rapid boil.
2. Stir the contents, and cook the food product throughout.
3. Serve hot.

GLOSSARY:
Skimmer: long-handled, round, perforated tool

HACCP:
Hold at 140° F or higher.

HAZARDOUS FOODS:
Eggs
Milk

NUTRITION:
Calories: 280
Fat: 11.9 g
Protein: 6.81 g

CHEF NOTE:
Spätzel is usually served with stews such as Hungarian goulash or chicken ragout with paprika.

Fondant Potatoes

(Pomme de Terre Fondante)

YIELD: 10 SERVINGS. SERVING SIZE: FOUR OUNCES.

COOKING TECHNIQUE:
Bake

Bake:
1. Preheat the oven.
2. Place the food product on the appropriate rack.

GLOSSARY:
Fork-tender: without resistance

NUTRITION:
Calories: 183
Fat: 9.44 g
Protein: 2.52 g

INGREDIENTS:

4 pounds	Russet potatoes, uniform, medium potatoes selected, washed, peeled, and re-washed
4 ounces	Butter
	White stock, seasoned, as needed (see page 12)
	Salt and white pepper, to taste

METHOD OF PREPARATION:

1. Cut the whole potatoes uniformly. Place in a small roasting pan.
2. Melt the butter, and brush the potatoes with it.
3. Pour the seasoned white stock into the pan with the potatoes to cover half the height of the potatoes. Add the salt and white pepper.
4. Bake the potatoes in a 420° F oven; brush the potatoes with butter periodically.
5. The potatoes will be ready when the stock is absorbed and the upper part off the potatoes are golden brown and **fork-tender.**

French Fried Potatoes

(Pommes Frites)

YIELD: 50 SERVINGS. SERVING SIZE: FOUR OUNCES.

COOKING TECHNIQUE:
Deep-Fry

Deep-Fry:
1. Heat the frying liquid to the proper temperature.
2. Submerge the food product completely.
3. Fry the product until it is cooked throughout.

NUTRITION:
Calories: 295
Fat: 17.6 g
Protein: 2.89 g

INGREDIENTS:

22 pounds	Potatoes, washed, peeled, cut into strips 2 inches long and ¼-inch thick
	Oil, as needed
	Salt, to taste

METHOD OF PREPARATION:

1. Keep the potatoes in cold water for 1 hour to remove some of the starch. Drain and pat dry with absorbent paper.
2. Half-fill the frying basket with potatoes. Dip the potatoes in the oil in a deep-fat fryer (340° F) and blanch until almost done. Drain, remove, and spread on a sheet pan lined with absorbent paper. Hold until needed.

To order:

1. Cook only the quantity of potatoes needed for immediate service. Half-fill the frying baskets with potatoes.
2. Dip the basket in the oil in a deep-fat fryer set at 375° F.
3. Fry to a golden brown: The potatoes should be crisp outside but soft inside.
4. Drain on absorbent paper.
5. Lightly salt the potatoes.

CHEF NOTE:
Allow 1 minute to pass before dipping subsequent basket into the fryer to avoid a drop in fat temperature.

Fried Yam Patties

(Frituras de Ñame)

YIELD: 10 SERVINGS. SERVING SIZE: THREE OUNCES.

INGREDIENTS:

3 pounds	Fresh yams, washed, peeled, and grated
3 ounces	Onions, peeled and diced **brunoise**
3 ounces	Fresh parsley, washed, excess moisture removed, and chopped
8 each	Egg yolks
	Salt and freshly ground black pepper, to taste
	Oil for frying, as needed

METHOD OF PREPARATION:

1. Place all ingredients except the oil in a bowl; mix well. Taste and adjust seasoning. Test fry 1 pattie.
2. In a sauté pan, heat the oil; scoop 1½ ounces of yam mixture into the oil, press to flatten, and fry on both sides until lightly browned and the yams are tender.
3. Transfer the patties to a hotel pan layered with absorbent paper, and hold at 140° F or above.

COOKING TECHNIQUE:

Shallow-Fry

Shallow-Fry:

1. Heat the cooking medium to the proper temperature.
2. Cook the food product throughout.
3. Season, and serve hot.

GLOSSARY:

Brunoise: ⅛-inch dice

HACCP:

Hold at 140° F or above.

HAZARDOUS FOOD:

Egg yolks

NUTRITION:

Calories: 221
Fat: 7.32 g
Protein: 4.84 g

CHEF NOTE:

This side accompaniment is a specialty in Jamaica.

Greek Pasta

YIELD: 50 SERVINGS. SERVING SIZE: SIX OUNCES.

COOKING TECHNIQUES:
Boil, Sauté

Boil: (at sea level)
1. Bring the cooking liquid to a rapid boil.
2. Stir the contents, and cook the food product throughout.
3. Serve hot.

Sauté:
1. Heat the sauté pan to the appropriate temperature.
2. Evenly brown the food product.
3. For a sauce, pour off any excess oil, reheat, and deglaze.

GLOSSARY:
Concassé: peeled, seeded, roughly chopped

HAZARDOUS FOOD
Feta cheese

NUTRITION:
Calories: 705
Fat: 24.5 g
Protein: 24 g

INGREDIENTS:

12 pounds	Fettuccine
	Virgin olive oil, as needed
3 heads	Garlic, peeled and chopped
3 pounds	California olives, ripe, washed and sliced
7 pounds	Tomatoes, washed, cored and **concassé**
4 tablespoons	Oregano, dried, or to taste
	Salt and ground white pepper, to taste
3 pounds	Feta cheese, crumbled
50 each	Parsley sprigs, wash and dry

METHOD OF PREPARATION:

1. Cook the fettuccine in boiling, salted water, with no oil, until **al dente.** Drain. Rinse under cold water, oil slightly, and set aside for service.
2. Heat the oil over medium heat; add the garlic, and sauté briefly. Add the olives, tomatoes, oregano, salt, and pepper. Toss lighly. Use very little salt, because feta cheese is very salty.
3. Add the drained pasta and feta to the pan. Return the pan to the stove, and toss the pasta and feta well over medium heat, until heated throughout. Serve, using a slotted spoon to avoid having too much oil on the reheated dinner plate. Garnish with a sprig of parsley.

Grilled Sliced Potatoes with Sesame Oil

YIELD: 10 SERVINGS. SERVING SIZE: FOUR OUNCES.

INGREDIENTS:

3 pounds	Idaho potatoes, washed and baked or steamed **al dente**
3 ounces	Sesame oil, or as needed
2 ounces	Freshly chopped parsley, excess moisture removed
	Salt and freshly ground black pepper, to taste

METHOD OF PREPARATION:

1. Preheat the grill or broiler.
2. Cut the potatoes on a diagonal into ½-inch slices. Score the slices on one side, and brush with oil.
3. Grill the potatoes until golden and tender, turning once.
4. Transfer to a steam table pan, and hold at 140° F or higher.
5. Serve, sprinkled with parsley.

COOKING TECHNIQUES:

Grill, Bake or Steam

Grill/Broil:

1. Clean and heat the grill/broiler.
2. To prevent sticking, brush the food product with oil.

Bake:

1. Preheat the oven.
2. Place the food product on the appropriate rack.

Steam: (Traditional)

1. Place a rack over a pot of water.
2. Prevent steam vapors from escaping.
3. Shock or cook the food product throughout.

GLOSSARY:

Al denté: to the bite

HACCP:

Hold at 140° F or higher.

NUTRITION:

Calories: 170
Fat: 8.65 g
Protein: 2.14 g

Grits Pilaf

YIELD: 10 SERVINGS. SERVING SIZE: FOUR OUNCES.

INGREDIENTS:

1 quart	Boiling water
½ teaspoon	Salt
6 ounces	Grits
2 ounces	Butter
5 ounces	Bacon, diced
10 ounces	Onions, peeled and diced **brunoise**

METHOD OF PREPARATION:

1. In a suitable **marmite,** place the water and salt, and heat to a boil.
2. Pour the grits into the boiling water, stirring to avoid lumps, and cook approximately 20 minutes, depending on the grain size.
3. Add the butter to cover the top of the grits. The fat will prevent a skin from forming on the top of grits. Hold at 140° F.
4. In a sauté pan, fry the bacon, **rendering** the fat. When browned, remove, and reserve the bacon pieces.
5. Sauté the onions in the bacon fat until they are golden in color.
6. Add the reserved bacon to the onions, and mix well. Hold at 140° F.
7. For service, keep a 4-ounce open ring mold in clean warm water and a teaspoon in the onions. Fill the ring mold with the grits, release on a preheated dinner plate, and top with a spoonful of the onions and bacon.

COOKING TECHNIQUES:

Boil, Shallow-Fry, Sauté

Boil: (at sea level)

1. Bring the cooking liquid to a rapid boil.
2. Stir the contents, and cook the food product throughout.
3. Serve hot.

Shallow-Fry:

1. Heat the cooking medium to the proper temperature.
2. Cook the food product throughout.
3. Season, and serve hot.

Sauté:

1. Heat the sauté pan to the appropriate temperature.
2. Evenly brown the food product.
3. For a sauce, pour off any excess oil, reheat, and deglaze.

GLOSSARY:

Brunoise: ⅛-inch dice
Marmite: stockpot
Render: to melt fat

HACCP:

Hold at 140° F.

NUTRITION:

Calories: 141
Fat: 7.15 g
Protein: 3.3 g

CHEF NOTE:

Paprika can be added at the discretion of the chef.

Indonesian Fried Rice

YIELD: 10 SERVINGS. SERVING SIZE: SIX OUNCES.

INGREDIENTS:

8 ounces	Peanut oil
3 cloves	Garlic, peeled and minced
3 tablespoons	Ketchup
3 tablespoons	Soy sauce
	Salt, to taste
10 ounces	Chicken, beef, or pork, cooked and diced
2 pounds	Rice, cooked and cooled

Macédoine:

12 ounces	Onions, peeled
3 each	Green bell peppers, washed and seeded
1 pound	Shrimp, washed, peeled, and deveined

METHOD OF PREPARATION:

1. In a wok, over moderate heat, add the peanut oil.
2. When the oil is hot, stir-fry the garlic, onions, and green peppers for 3 minutes.
3. Add the shrimp, and stir-fry for 1 minute. Add the ketchup, soy sauce, and salt. Then add the chicken, beef, or pork, and stir-fry until heated throughout.
4. Add the cooked rice, and stir-fry for 5 minutes, until the rice is heated throughout.
5. Taste, and adjust seasonings.
6. Serve immediately, on a preheated dinner plate, or hold at 140° F.

COOKING TECHNIQUE:
Stir-Fry

Stir-Fry:
1. Heat the oil in a wok until hot but not smoking.
2. Keep the food in constant motion; use the entire cooking surface.

GLOSSARY:
Macédoine: ¼-inch dice

HACCP:
Hold at 140° F.

HAZARDOUS FOODS:
Shrimp
Chicken
Beef
Pork

NUTRITION:
Calories: 434
Fat: 24 g
Protein: 21.4 g

Jalapeño Rice

YIELD: 10 SERVINGS. SERVING SIZE: FOUR OUNCES.

COOKING TECHNIQUES:
Sauté, Steam

Sauté:
1. Heat the sauté pan to the appropriate temperature.
2. Evenly brown the food product.
3. For a sauce, pour off any excess oil, reheat, and deglaze.

Steam: (Traditional)
1. Place a rack over a pot of water.
2. Prevent steam vapors from escaping.
3. Shock or cook the food product throughout.

GLOSSARY:
Render: to melt fat
Marmite: stockpot

HACCP:
Hold at 140° F.

NUTRITION:
Calories: 338
Fat: 9.02 g
Protein: 5.68 g

INGREDIENTS:

3 ounces	**Rendered** bacon fat
6 ounces	Onions, peeled and finely diced
2 each	Jalapeño peppers, washed, cored, and thinly sliced
1 each	Red chili pepper
1 teaspoon	Spanish paprika
24 ounces	Long-grained rice
1½ quarts	White beef stock, heated to a boil (see page 12)
	Salt, to taste

METHOD OF PREPARATION:

1. In an appropriate-sized **marmite,** heat the bacon fat, add the onions and peppers, and sauté until the onions are translucent. Add the paprika, and sauté for 1 more minute.
2. Add the rice, and stir until it becomes golden brown.
3. Pour in the stock, cover, and steam on low heat until the rice is tender, which will require about 20 minutes.
4. Toss together with a braising fork, and season, to taste. Serve immediately, or hold at 140° F.

Korean Rice and Bean Sprouts

(Kun-na-mul-bob)

YIELD: 10 SERVINGS. SERVING SIZE: FOUR OUNCES.

INGREDIENTS:

1 ounce	Peanut oil
2 ounces	Toasted sesame seeds, crushed
5 each	Scallions, washed, trimmed, and diced **brunoise**
3 cloves	Garlic, peeled and finely minced
10 ounces	Fresh bean sprouts
2 pounds	Oriental rice, cooked
1 ounce	Soy sauce
	Salt and ground white pepper, to taste

METHOD OF PREPARATION:

1. In a wok, heat the oil, and sauté the sesame seeds, scallions, and garlic.
2. Add the bean sprouts, and heat thoroughly.
3. Add the rice, soy sauce, and salt. Toss together, and heat thoroughly.
4. Serve immediately, or hold at 140° F or above.

COOKING TECHNIQUE:

Stir-Fry

Stir-Fry:

1. Heat the oil in a wok until hot but not smoking.
2. Keep the food in constant motion; use the entire cooking surface.

GLOSSARY:

Brunoise: ⅛-inch dice

HACCP:

Hold at 140° F or above.

NUTRITION:

Calories: 188
Fat: 5.98 g
Protein: 4.61 g

Layered Potatoes

YIELD: 10 SERVINGS. SERVING SIZE: FOUR OUNCES.

INGREDIENTS:

2 pounds	All-purpose potatoes, washed and preboiled until two-thirds cooked
6 each	Eggs, hard-boiled and peeled
	Clarified butter, as needed
6 ounces	Ham, diced **brunoise**
	Salt and ground white pepper, to taste
8 ounces	Heavy cream
2 each	Eggs

METHOD OF PREPARATION:

1. Preheat the oven to 350° F.
2. Peel the potatoes, and slice ¼-inch thick.
3. Slice the hard-boiled eggs.
4. In a buttered baking pan layer the potatoes and eggs, distributing the diced ham between each layer, and season each layer with salt and white pepper.
5. Place the pan in the oven, and bake for 15 minutes.
6. Combine the cream and eggs, and mix well. Pour the mixture over the potatoes, and shake the pan to let the cream penetrate.
7. Return to the oven, and bake until the cream mixture is set.
8. Remove, and hold for 15 minutes at 140° F, then cut into square portions. Serve immediately, or continue to hold at 140° F.

COOKING TECHNIQUES:

Boil, Bake

Boil: (at sea level)

1. Bring the cooking liquid to a rapid boil.
2. Stir the contents, and cook the food product throughout.
3. Serve hot.

Bake:

1. Preheat the oven.
2. Place the food product on the appropriate rack.

GLOSSARY:

Brunoise: ⅛-inch dice

HACCP:

Hold at 140° F.

HAZARDOUS FOODS:

Heavy cream
Eggs

NUTRITION:

Calories: 285
Fat: 18.9 g
Protein: 9.78 g

Lemon Orzo

YIELD: 10 SERVINGS. SERVING SIZE: FOUR OUNCES.

INGREDIENTS:

2 quarts	White chicken stock, heated to a boil (see page 12)
1½ pounds	Orzo pasta, dried
2 tablespoons	Peanut oil
3 ounces	Fresh chives, washed and snipped
1 tablespoon	Grated zest of lemon
1 tablespoon	Fresh lemon juice
	Salt and ground white pepper, to taste

METHOD OF PREPARATION:

1. Place the stock in a saucepan, and heat to a boil.
2. Add the orzo, reduce the heat, and simmer until cooked **al dente,** approximately 7 to 8 minutes. Drain, and rinse under cold water.
3. In a saucepan, heat the oil, and add the cooked orzo and the remaining ingredients. Toss to combine; then sauté to a minimum temperature of 165° F. Season, to taste, and serve immediately, or hold at 140° F or above.

COOKING TECHNIQUES:

Boil, Simmer

Boil: (at sea level)

1. Bring the cooking liquid to a rapid boil.
2. Stir the contents, and cook the food product throughout.
3. Serve hot.

Simmer and Poach:

1. Heat the cooking liquid to the proper temperature.
2. Submerge the food product completely.
3. Keep the cooked product moist and warm.

GLOSSARY:

Al dente: to the bite

HACCP:

Heat to a minimum temperature of 165° F.
Hold at 140° F.

NUTRITION:

Calories: 273
Fat: 4.22 g
Protein: 5.67 g

Lentils with Prosciutto

YIELD: 10 SERVINGS. SERVING SIZE: FOUR OUNCES.

INGREDIENTS:

24 ounces	Dried lentils, cleaned, washed, and drained
	Salted water, as needed
3 ounces	Olive oil
5 ounces	Leeks (white part only), trimmed, split lengthwise, and thinly sliced
5 ounces	Prosciutto, diced **brunoise**
¼ teaspoon	Rosemary, ground
	Salt and freshly ground black pepper, to taste

METHOD OF PREPARATION:

1. In a saucepan, place the lentils, add water to cover, and heat to a boil. Cook until the lentils are tender; then drain, and reserve them.
2. In a sauté pan, heat the oil; add the leeks, and sauté until they are translucent. Add the proscuitto, and sauté until browned.
3. Add the rosemary, and sauté for 1 more minute.
4. Add the lentils, mix, and heat throughout. Season, to taste, and serve immediately, or hold at 140° F or above.

COOKING TECHNIQUES:

Sauté, Simmer

Sauté:

1. Heat the sauté pan to the appropriate temperature.
2. Evenly brown the food product.
3. For a sauce, pour off any excess oil, reheat, and deglaze.

Simmer and Poach:

1. Heat the cooking liquid to the proper temperature.
2. Submerge the food product completely.
3. Keep the cooked product moist and warm.

GLOSSARY:

Brunoise: ⅛-inch dice

HACCP:

Hold at 140° F.

NUTRITION:

Calories: 329
Fat: 9.51 g
Protein: 22.1 g

CHEF NOTE:

This is a perfect accompaniment to lamb, instead of serving a starch.

Linguini with Tomatoes, Olives, Goat Cheese, and Olive Oil

YIELD: 10 SERVINGS. SERVING SIZE: EIGHT OUNCES.

INGREDIENTS:	
6 ounces	Olive oil
8 cloves	Garlic
3 pounds	Fresh ripe tomatoes, washed, cored, blanched, peeled, seeded, and diced **brunoise**
	Salt and freshly ground black pepper, to taste
1 pound	Pitted Niçoise olives
4 pounds	Linguini, cooked **al dente,** drained and shocked in cold water
2 ounces	Fresh basil leaves, washed and cut **julienne**
1 pound	Fresh goat cheese, crumbled

METHOD OF PREPARATION:

1. In a large sauté pan, heat the olive oil and garlic together, and sauté until the garlic begins to color. Add the tomatoes, and sauté until the tomatoes soften and most of the liquid evaporates.
2. Add the olives and linguine, and toss together until thoroughly heated. Add the basil and cheese, and toss for 2 more minutes. Season with salt and pepper.
3. Serve immediately on preheated plates, or hold at 140° F., or higher.

COOKING TECHNIQUES:

Boil, Sauté

Boil: (at sea level)

1. Bring the cooking liquid to a rapid boil.
2. Stir the contents, and cook the food product throughout.
3. Serve hot.

Sauté:

1. Heat the sauté pan to the appropriate temperature.
2. Evenly brown the food product.
3. For a sauce, pour off any excess oil, reheat, and deglaze.

GLOSSARY:

Brunoise: ⅛-inch dice
Al dente: to the bite
Julienne: matchstick strips

HACCP:

Hold at 140° F or higher.

NUTRITION:

Calories: 492
Fat: 23.6 g
Protein: 10.5 g

CHEF NOTE:

Add basil and cheese only to quantity served.

Lyonnaise-Style Potatoes

(Pommes de Terre Lyonnaise)

YIELD: 10 SERVINGS. SERVING SIZE: FOUR OUNCES.

INGREDIENTS:

4 pounds	Potatoes, washed and peeled
	Salted cold water, as needed
8 ounces	Butter
1 pound	Onions, peeled and cut **julienne**
	Salt and ground white pepper, to taste
3 ounces	Freshly chopped parsley, excess moisture removed

METHOD OF PREPARATION:

1. In a saucepan, place the potatoes, and add water to cover. Heat to a boil, and cook until the potatoes are half done. Drain, and slice crosswise ¼-inch thick.
2. In a sauté pan, melt the butter; add the onions, and sauté until they become brown. Remove and reserve the onions.
3. Add the potatoes in only one layer. Sauté until browned on each side, turning only once. Transfer to a hotel pan. Season, to taste, and top with the onions. Hold at 140° F or higher.
4. To serve, place a portion on a preheated plate. Garnish with chopped parsley.

COOKING TECHNIQUES:

Sauté

Sauté:

1. Heat the sauté pan to the appropriate temperature.
2. Evenly brown the food product.
3. For a sauce, pour off any excess oil, reheat, and deglaze.

GLOSSARY:

Julienne: matchstick strips

HACCP:

Hold at 140° F or higher.

NUTRITION:

Calories: 337
Fat: 18.7 g
Protein: 3.92 g

CHEF NOTES:

1. In large-quantity cooking, this dish is finished by baking.
2. The word *Lyonnaise* means "in the style of Lyon," and denotes onions.

Mashed Potatoes

(Pommes de Terre Purée)

YIELD: 10 SERVINGS. SERVING SIZE: FOUR OUNCES.

INGREDIENTS:

4 pounds	Potatoes, chef, washed, peeled, and quartered
	Water with salt, as needed
	Seasoned, to taste
4 ounces	Butter
	Milk, as needed, heated to 165° F.

METHOD OF PREPARATION:

1. Place the potatoes in a saucepan, cover with cold salted water, and boil.
2. Drain the potatoes when done; place in a mixing bowl, and mash on low speed until lump-free.
3. Add the butter. Season with salt and white pepper. Add the hot milk, if needed, to adjust the consistency. Keep in a **bain-marie** at 140° F.
4. Use a pastry bag fixed with a star tube, and pipe onto preheated plates, creating a pyramid.

COOKING TECHNIQUE:

Boil

Boil: (at sea level)

1. Bring the cooking liquid to a rapid boil.
2. Stir the contents, and cook the food product throughout.
3. Serve hot.

GLOSSARY:

Bain-marie: hot-water bath

NUTRITION:

Calories: 224
Fat: 7.7 g
Protein: 3.4 g

CHEF NOTES:

1. For additional flavor, add a pinch of nutmeg.
2. This recipe can be enhanced additionally with roasted garlic and/or minced fresh scallions.
3. The amount of milk used will depend on the dryness of the potatoes.
4. Sour cream can be used instead of milk.

Moroccan Potato Cake

YIELD: 10 SERVINGS. SERVING SIZE: FOUR OUNCES.

INGREDIENTS:

	Salted cold water, as needed
2 pounds	Potatoes, washed, peeled, and cut into fourths
5 each	Eggs
	Salt and ground white pepper, to taste
2 ounces	Vegetable oil

METHOD OF PREPARATION:

1. Preheat the oven to 425° F.
2. Boil the potatoes in salted water until tender. Drain the potatoes well; then purée.
3. Break the eggs into a bowl, and beat lightly; then add them to the potatoes. Season, to taste, and blend well without overmixing.
4. In a baking pan, heat the oil to the smoking point. Pour the potato mixture into the oil, and spread out evenly.
5. Bake in the oven until firm and golden brown.
6. Hold at 140° F or higher for 20 minutes before cutting into portions. Serve immediately, or continue to hold at 140° F.

COOKING TECHNIQUES:

Boil, Bake

Boil: (at sea level)

1. Bring the cooking liquid to a rapid boil.
2. Stir the contents, and cook the food product throughout.
3. Serve hot.

Bake:

1. Preheat the oven.
2. Place the food product on the appropriate rack.

HACCP:

Hold at 140° F.

HAZARDOUS FOOD:

Eggs

NUTRITION:

Calories: 163
Fat: 8.04 g
Protein: 4.68 g

CHEF NOTES:

1. The pan size should be such that the potato mixture will be 2 inches thick when placed in the pan.
2. This dish reflects the city of Rabat. Each city has his own way to make it. For example, in Fezz, sautéed liver is added; in Casablanca, green peas.

Moselle Braised Potatoes

YIELD: 10 SERVINGS. SERVING SIZE: FOUR OUNCES.

INGREDIENTS:

5 (8-ounce)	Maine or eastern potatoes of equal size, washed and peeled (held in cold water)
	Salted water, to cover
8 ounces	Bacon, diced **brunoise**
2 ounces	Butter
8 ounces	Onions, peeled and diced brunoise
3 ounces	Liverwurst
2 each	Egg yolks, lightly beaten
	Salt and freshly ground black pepper, to taste
3 ounces	Fresh parsley, washed, excess moisture removed, and chopped
1 tablespoon	Vegetable oil
16 ounces	Moselle wine

METHOD OF PREPARATION:

1. Preheat the oven to 325° F.
2. Place the potatoes in a large pot, and add salted water to cover. Place over high heat, cover, and boil until the potatoes are three-fourths cooked; then cool slightly, and cut each in half.
3. Using a melon baller, scoop out the centers, leaving a ¼-inch-thick shell. Dice the centers, and reserve.
4. Place the bacon in a sauté pan, and **render.** Use a slotted spoon to remove the bacon. Reserve the bacon pieces in a large bowl.
5. Add the butter to the bacon fat. Sauté the onions until browned and cooked. Add to the bacon. In the same sauté pan, brown the diced potato; then combine with the bacon and onions.
6. Add the liverwurst, egg yolks (seasoning to taste) and parsley; then mix well.
7. Lightly oil a 2-inch hotel pan. Divide the filling mixture among the potato shells, and place in a roasting pan.
8. Pour the wine over the potatoes, cover the pan tightly with aluminum foil, and place in the oven. Bake until fully tender and the liquid is reduced by half.
9. For service, spoon the remaining reduced wine over the potatoes, and serve immediately, or hold at 140° F or higher.

COOKING TECHNIQUES:

Braise, Sauté, Boil

Braise:

1. Heat the braising pan to the proper temperature.
2. Sear and brown the food product to a golden color.
3. Degrease and deglaze.
4. Cook the food product in two-thirds liquid until fork-tender.

Sauté:

1. Heat the sauté pan to the appropriate temperature.
2. Evenly brown the food product.
3. For a sauce, pour off any excess oil, reheat, and deglaze.

Boil: (at sea level)

1. Bring the cooking liquid to a rapid boil.
2. Stir the contents, and cook the food product throughout.
3. Serve hot.

GLOSSARY:

Brunoise: ⅛-inch dice
Render: to melt fat

HACCP:

Hold at 140° F or higher.

HAZARDOUS FOOD:

Egg yolks

NUTRITION:

Calories: 364
Fat: 20.8 g
Protein: 11.2 g

CHEF NOTES:

1. If Moselle wine is not available, substitute a dry white wine.
2. Maine or eastern potatoes are recommended for this dish because they will not collapse during the baking period.

Oriental Fried Rice

YIELD: 50 SERVINGS. SERVING SIZE: EIGHT OUNCES.

INGREDIENTS:

4 pounds	Converted rice
½ pint	Vegetable oil
2½ pints	Green bell peppers, washed, seeded, and diced
1½ pints	Green onions, peeled and sliced
2 tablespoons	Granulated garlic
1½ pounds	Lean ham, diced
2½ pints	Mushrooms, washed and sliced
6 each	Eggs, beaten
½ pint	Low-sodium soy sauce
3 teaspoons	Pepper
2½ pounds	Shrimp (titi), cooked

METHOD OF PREPARATION:

1. Cook the rice in two parts of boiling water to one part of converted rice. Dry and cool the rice by spreading on a sheet pan and refrigerating.
2. Preheat a tilting skillet to 425° F. Add the vegetable oil.
3. Add the peppers, onions, garlic, ham, and mushrooms in recipe order, and stir-fry quickly. Add the cooked rice.
4. Combine the eggs, soy sauce, and peppers. Pour over the rice mixture, and stir-fry until the eggs are cooked. Add the shrimp (do not overcook), and serve.

COOKING TECHNIQUE:

Stir-Fry

Stir-Fry:

1. Heat the oil in a wok until hot but not smoking.
2. Keep the food in constant motion; use the entire cooking surface.

HACCP:

Cook to 140° F.

HAZARDOUS FOODS:

Shrimp
Ham
Eggs

NUTRITION:

Calories: 243
Fat: 7.4 g
Protein: 12 g

Oriental Steamed Rice (Chinese Style)

YIELD: 10 SERVINGS. SERVING SIZE: FOUR OUNCES.

INGREDIENTS:

1 pound	Oriental rice
1 quart	Water

METHOD OF PREPARATION:

Place the rice in a rice cooker. Add the water and slowly cook, with a hermetically closed lid, until all of the water is absorbed and the rice is tender.

COOKING TECHNIQUE:

Steam

Steam: (Traditional)

1. Place a rack over a pot of water.
2. Prevent steam vapors from escaping.
3. Shock or cook the food product throughout.

NUTRITION:

Calories: 166
Fat: .299 g
Protein: 3.24 g

CHEF NOTE:

The Chinese do not add salt or any other seasonings to this rice.

Polenta No. 1

YIELD: 10 SERVINGS. SERVING SIZE: FOUR OUNCES.

INGREDIENTS:

1½ quarts	White chicken stock, heated to a boil (see page 12)
	Salt and ground white pepper, to taste
4 ounces	Butter
12 ounces	Cornmeal
4 ounces	Parmesan cheese, freshly grated

METHOD OF PREPARATION:

1. In a saucepan, combine the stock, salt, pepper, and butter, and heat to a boil.
2. Slowly pour the cornmeal into the boiling stock, stirring constantly. Reduce the heat, and simmer 20 minutes, or until the liquid is absorbed.
3. Spread the cooked polenta onto a plastic- or parchment-lined tray, and cool to 40° F or lower.
4. Cut the polenta into triangle portions.
5. Preheat the oven to 350° F.
6. For service, arrange the polenta on parchment-lined trays, sprinkle the wedges with Parmesan, place in the oven, and bake until the cheese browns and the polenta is hot. Hold at 140° F, or serve immediately.

COOKING TECHNIQUES:

Simmer, Bake

Simmer and Poach:

1. Heat the cooking liquid to the proper temperature.
2. Submerge the food product completely.
3. Keep the cooked product moist and warm.

Bake:

1. Preheat the oven.
2. Place the food product on the appropriate rack.

HACCP:

Cool to 40° F.
Hold at 140° F.

NUTRITION:

Calories: 281
Fat: 14 g
Protein: 10.7 g

CHEF NOTES:

1. Polenta pieces also can be browned under a salamander or sautéed in butter.
2. Polenta also can be served as a hot hors d'oeuvre with tomato sauce.
3. This dish is also called *Gnocchi al Romana.*

Polenta No. 2

YIELD: 10 SERVINGS. SERVING SIZE: FOUR OUNCES.

INGREDIENTS:

2 quarts	Water
1½ teaspoons	Salt
1 pound	Cornmeal, medium-ground

METHOD OF PREPARATION:

1. In a medium-sized saucepan, heat the water to a boil; add the salt, and gradually add the cornmeal, stirring continuously with a wooden spoon.
2. When blended without the lumps, lower the heat, and simmer until thickened, approximately 30 minutes. When done, the polenta will pull away from the side of the pot.
3. Pour the polenta into an oiled pan, and spread to a ½-inch thickness.
4. Allow to rest a few minutes; then cut into portions. Hold at 140° F or higher.

COOKING TECHNIQUE:

Boil

Boil: (at sea level)

1. Bring the cooking liquid to a rapid boil.
2. Stir the contents, and cook the food product throughout.
3. Serve hot.

HACCP:

Hold at 140° F or above.

NUTRITION:

Calories: 176
Fat: 2.98 g
Protein: 3.69 g

Polish Noodles and Cabbage

YIELD: 10 SERVINGS. SERVING SIZE: FOUR OUNCES.

INGREDIENTS:

6 ounces	Clarified butter
2 ounces	Onions, peeled and diced **brunoise**
1 small head	White cabbage, washed, trimmed, cored, and shredded
2 teaspoons	Caraway seeds
	Salt and freshly ground black pepper, to taste
1 pound	Egg noodles, cooked **al dente** (see page 473)
8 ounces	Sour cream

METHOD OF PREPARATION:

1. In a saucepan, heat the butter; add the onions, and sauté until they are translucent.
2. Add the cabbage, and sauté for 5 minutes, or until wilted but still crisp. Add the caraway seeds, and season, to taste.
3. Add the noodles, and sauté until heated throughout.
4. Stir the sour cream until smooth; then blend into the noodles and cabbage. Continue to cook for 5 minutes, stirring often.
5. Serve immediately, or hold at 140° F or above.

COOKING TECHNIQUES:
Sauté, Boil

Sauté:
1. Heat the sauté pan to the appropriate temperature.
2. Evenly brown the food product.
3. For a sauce, pour off any excess oil, reheat, and deglaze.

Boil: (at sea level)
1. Bring the cooking liquid to a rapid boil.
2. Stir the contents, and cook the food product throughout.
3. Serve hot.

GLOSSARY:
Brunoise: ⅛-inch dice
Al dente: to the bite

HACCP:
Hold at 140° F or above.

HAZARDOUS FOOD:
Sour cream

NUTRITION:
Calories: 284
Fat: 22.7 g
Protein: 4.41 g

Polynesian Fried Rice

YIELD: 10 SERVINGS. SERVING SIZE: EIGHT OUNCES.

INGREDIENTS:

4 ounces	Bacon, diced
5 each	Eggs, lightly beaten
4 ounces	Peanut oil
8 ounces	Bean sprouts
2 pounds	Rice, cooked
6 ounces	Scallions, washed and chopped
	Salt and white pepper, to taste
4 ounces	Lean pork, diced
4 ounces	Chicken breast, diced
4 ounces	Shrimp, sliced
4 ounces	Ham, diced
4 ounces	Mushrooms, sliced

METHOD OF PREPARATION:

1. In a wok, **render** the bacon.
2. Pour the beaten eggs on the rendered bacon, and make a large, thin omelet. Remove from the wok, cool slightly, cut **julienne,** and set aside.
3. Clean the wok, add the oil, and heat. Stir-fry the diced ingredients and bean sprouts until the meats and seafood are cooked.
4. Add the rice and scallions, and continue to stir-fry until the rice is thoroughly heated. Season, to taste, and mix into the shredded omelet.
5. Serve immediately on a preheated dinner plate, or hold at 140° F or above.

COOKING TECHNIQUE:

Stir-Fry

Stir-Fry:

1. Heat the oil in a wok until hot but not smoking.
2. Keep the food in constant motion; use the entire cooking surface.

GLOSSARY:

Render: to melt fat
Julienne: matchstick strips

HACCP:

Hold at 140° F or above.

HAZARDOUS FOODS:

Eggs
Lean pork
Chicken breast
Shrimp

NUTRITION:

Calories: 397
Fat: 22.1 g
Protein: 20.4 g

Potato Cake

YIELD: 10 SERVINGS. SERVING SIZE: FOUR OUNCES.

COOKING TECHNIQUES:

Boil, Bake

Boil: (at sea level)

1. Bring the cooking liquid to a rapid boil.
2. Stir the contents, and cook the food product throughout.
3. Serve hot.

Bake:

1. Preheat the oven.
2. Place the food product on the appropriate rack.

HAZARDOUS FOOD:

Heavy cream

NUTRITION:

Calories: 350
Fat: 21.3 g
Protein: 11.7 g

INGREDIENTS:

3 pounds	Potatoes, washed
	Salted water, to cover
6 each	Eggs, cracked and lightly beaten
8 ounces	Heavy cream
6 ounces	Swiss cheese, grated
	Salt and ground white pepper, to taste
½ teaspoon	Nutmeg, ground
2 ounces	Butter

METHOD OF PREPARATION:

1. Preheat the oven to 350° F.
2. In a saucepan, combine the potatoes and water, and heat the water to a boil. Cook until the potatoes are tender; then drain and peel them.
3. In a 4-quart bowl, combine four of the eggs, the cream, and the cheese, and mix well.
4. Mash the potatoes, and mix them with the egg mixture. Season, to taste, with salt, pepper, and nutmeg.
5. Butter a baking pan or individual soufflé cups. Fill with the potato mixture, and smooth the surface.
6. Brush the top of the potato cake (or cakes) mixture with the remaining eggs, and bake in the oven for approximately 30 minutes. Let set for 10 minutes before cutting the cake into wedges, or if soufflé cups are used, serve the potatoes in the cups.

Potato Corn Cakes

YIELD: 10 SERVINGS. SERVING SIZE: FOUR OUNCES.

INGREDIENTS:

3 pounds	Purée of potatoes (see page 498)
3 each	Fresh ears of corn, shucked and silks removed
6 ounces	Clarified butter
2 each	Egg yolks, lightly beaten
2 ounces	All-purpose flour
2 ounces	Fresh chives, washed and snipped
	Salt and freshly ground black pepper, to taste

METHOD OF PREPARATION:

1. Preheat the oven to 475° F.
2. Prepare the potatoes as per the recipe, but do not add too much liquid. The purée should be stiff.
3. Brush the corn with some of the butter, and roast in the oven until tender and lightly browned, or grill, turning frequently, until tender and lightly browned.
4. Let the corn cool slightly, and cut the kernels from the cob.
5. Combine the potato purée, corn kernels, egg yolks, flour, and chives; season, to taste. Mix until well blended.
6. In a large sauté pan, heat the remaining butter. Scoop portions of potato mixture into the hot butter, and flatten to a "cake" that is ½-inch thick. Cook until browned and crisp; turn, and brown on the opposite side.
7. Remove from the sauté pan, drain on paper towels, and serve immediately, or hold at 140° F.

COOKING TECHNIQUES:
Boil, Roast, Grill, Shallow-Fry

Boil: (at sea level)
1. Bring the cooking liquid to a rapid boil.
2. Stir the contents, and cook the food product throughout.
3. Serve hot.

Roast:
1. Sear the food product, and brown evenly.
2. Elevate the food product in a roasting pan.
3. Determine doneness, and consider carryover cooking.
4. Let the food product rest before carving.

Grill/Broil:
1. Clean and heat the grill/broiler.
2. To prevent sticking, brush the food product with oil.

Shallow-Fry:
1. Heat the cooking medium to the proper temperature.
2. Cook the food product throughout.
3. Season, and serve hot.

HACCP:
Hold at 140° F.

HAZARDOUS FOOD:
Egg yolks

NUTRITION:
Calories: 250
Fat: 9.35 g
Protein: 5.53 g

CHEF NOTES:
Potato cakes can be made from shredded or grated potatoes.

Potato Croquettes

YIELD: 10 SERVINGS. SERVING SIZE: ONE PIECE.

INGREDIENTS:

2 pounds	All-purpose potatoes, washed and peeled
	Water, to cover
	Salt, to taste
1 ounce	Pasteurized egg yolks
¼ teaspoon	Nutmeg, ground
4 ounces	Seasoned flour
4 each	Eggs, cracked and lightly beaten
10 ounces	Bread crumbs
	Fat, as needed for frying, heated to 350° F.

METHOD OF PREPARATION:

1. In a saucepan, place the potatoes, and cover with salted water. Heat to a boil, and cook until tender.
2. Strain off the water, and return the pan to the stove to allow the potatoes to dry.
3. Purée the potatoes in a ricer or mixer using a flat paddle. Add the egg yolks, salt, and nutmeg, and mix, but do not whip.
4. Shape the potatoes into cylinders 1-inch thick by 3-inches long.
5. Dip the cylinders in the flour, then the egg wash, and coat with the bread crumbs.
6. Lay the croquettes out on a parchment-lined tray, and refrigerate at 40° F or lower, until needed for service.
7. Deep-fry to order, and serve.

COOKING TECHNIQUES:

Boil, Deep-Fry

Boil: (at sea level)

1. Bring the cooking liquid to a rapid boil.
2. Stir the contents, and cook the food product throughout.
3. Serve hot.

Deep-Fry:

1. Heat the frying liquid to the proper temperature.
2. Submerge the food product completely.
3. Fry the product until it is cooked throughout.

GLOSSARY:

À la minute: cooked to order
Bain-marie: hot-water bath

HACCP:

Refrigerate at 40° F or lower.

HAZARDOUS FOODS:

Egg yolks
Eggs

NUTRITION:

Calories: 372
Fat: 15.7 g
Protein: 9.31 g

CHEF NOTE:

Do not hold breaded, deep-fried items in a **bain-marie;** always prepare **à la minute.**

Potato Moussaka

YIELD: 10 SERVINGS. SERVING SIZE: SIX OUNCES.

INGREDIENTS:

2 pounds	Potatoes, washed, peeled, and sliced ¼-inch thick
	Cold salted water, as needed
4 ounces	Olive oil
1 pound	Ground lamb
6 ounces	Onions, peeled and diced **brunoise**
8 cloves	Garlic, peeled and finely minced
6 ounces	Fresh tomatoes, washed, cored, **blanched,** peeled, seeded, and chopped
	Salt and freshly ground black pepper, to taste
2 ounces	Fresh parsley, washed, excess moisture removed, and chopped
½ teaspoon	Marjoram, dried
1½ pints	Thick béchamel sauce, heated to 140° F (see page 16)
6 ounces	Feta cheese, crumbled
20 ounces	Tomato sauce, heated to a boil, then held at 140° F (see page 76)

METHOD OF PREPARATION:

1. Preheat the oven to 350° F.
2. In a saucepan, place the potatoes, and cover with water. Heat to a boil, and cook until half done. Drain, place on a sheet pan, and allow to air-dry and cool.
3. In a sauté pan, heat the oil; add one layer of potatoes. Brown on both sides; then remove, and drain on absorbent paper. Oil the bottom of a hotel pan, and cover with a layer of potatoes. Repeat, browning the remaining potatoes, and reserve.
4. In the same pan, drain the excess fat; add the meats, and brown for 5 minutes. Add the onions, garlic, and tomatoes, and season the mixture. Cook until the onions are translucent. Drain any excess fat.
5. Add the heated béchamel sauce, and incorporate.
6. Pour the mixture over the potatoes. Add a second layer of potatoes. Sprinkle the feta cheese on top, and bake for 20 minutes, or until cooked throughout and the internal temperature is a minimum of 165° F.
7. Remove from the oven, and allow to rest for 15 minutes; then cut into portions.
8. Serve immediately, or hold at 140° F.
9. To serve, ladle the sauce on a preheated plate, and place a square of moussaka on the top.

COOKING TECHNIQUE:

Sauté, Bake

Sauté:

1. Heat the sauté pan to the appropriate temperature.
2. Evenly brown the food product.
3. For a sauce, pour off any excess oil, reheat, and deglaze.

Bake:

1. Preheat the oven.
2. Place the food product on the appropriate rack.

GLOSSARY:

Brunoise: ⅛-inch dice
Blanch: to parcook

HACCP:

Heat béchamel and tomato sauces to 140° F.
Cook to an internal temperature of 165° F.
Hold at 140° F.

HAZARDOUS FOODS:

Ground beef
Ground pork

NUTRITION:

Calories: 377
Fat: 25.3 g
Protein: 14.4 g

CHEF NOTES:

1. If desired, feta cheese can be sprinkled over the moussaka before the final baking.
2. This is a typical Greek preparation.

Potato Nests

(Nids de Pommes de Terre)

YIELD: 10 SERVINGS. SERVING SIZE: FOUR OUNCES.

INGREDIENTS:

	Oil, as needed
3 pounds	Chefs' potatoes, washed, peeled, and re-washed
	Salt, to taste

METHOD OF PREPARATION:

1. Cut the peeled potatoes into fine strips, or use mandoline for uniformity.
2. Line a wire basket with the potatoes, and press with the upper part of the basket, pressing hard on the handle. Secure with the basket ring.
3. Dip the basket into the oil in a 340° F deep-fat fryer, and fry the potatoes to a golden brown.
4. When done, remove from the basket, and place potatoes on absorbent paper. Sprinkle with salt.

COOKING TECHNIQUE:

Deep-Fry

Deep-Fry:

1. Heat the frying liquid to the proper temperature.
2. Submerge the food product completely.
3. Fry the product until it is cooked throughout.

NUTRITION:

Calories: 217
Fat: 11.5 g
Protein: 2.39 g

Potato Pancakes

(Galettes de Pommes de Terre)

YIELD: 10 SERVINGS. SERVING SIZE: TWO EACH.

INGREDIENTS:

3 pounds	Potatoes, washed and peeled
6 ounces	Onions, peeled
6 ounces	All-purpose flour
6 each	Eggs
¼ teaspoon	Nutmeg, ground
	Salt and freshly ground black pepper, to taste
	Oil, as needed to fry

METHOD OF PREPARATION:

1. Finely grate the potatoes and onions.
2. Add all of other ingredients, except the oil. Blend well. In a large sauté pan, heat the oil.
3. Using a 2-ounce scoop, make a mound of the mixture in the frying pan, and flatten to a ⅓-inch pancake.
4. Fry on both sides to a golden brown. Transfer to absorbent paper to drain. Serve immediately.

COOKING TECHNIQUE:

Shallow-Fry

Shallow-Fry:

1. Heat the cooking medium to the proper temperature.
2. Cook the food product throughout.
3. Season, and serve hot.

HAZARDOUS FOOD:

Eggs

NUTRITION:

Calories: 321
Fat: 13.9 g
Protein: 7.6 g

CHEF NOTE:

Potato Galettes will lose crispness if held in a bain-marie.

Potato Puffs

(Pommes de Terre Dauphine)

YIELD: 10 SERVINGS. SERVING SIZE: SIX EACH.

INGREDIENTS:

24 ounces	Duchesse potatoes (see page 479)
12 ounces	Pâte a Choux (see pages 837 and 838)
	Vegetable oil, add as needed

METHOD OF PREPARATION:

1. Combine the Duchesse potatoes and pâte a Choux.
2. Place the mixture in a pastry bag with a medium-sized star tube.
3. Pipe small rosettes on an oiled parchment paper–lined sheet pan.
4. For service, heat the oil to 350° F.
5. Fry the potatoes, à la minute, serving immediately.

COOKING TECHNIQUE:
Shallow-Fry

Shallow-Fry:
1. Heat the cooking medium to the proper temperature.
2. Cook the food product throughout.
3. Season, and serve hot.

NUTRITION:
Calories: 338
Fat: 29.7 g
Protein: 3.57 g

CHEF NOTE:
The parchment paper should be the size of the frying pan, and turned into the pan upside down.

Potatoes, Berrichonne Style

(Pommes de Terre Berrichonne)

YIELD: 10 SERVINGS. SERVING SIZE: FOUR OUNCES.

INGREDIENTS:

6 ounces	Bacon, diced small
6 ounces	Onions peeled and diced **brunoise**
	Salt and ground white pepper, to taste
4 pounds	Red bliss potatoes (small), washed
	White chicken stock, as needed (see page 12)

Bouquet Garni:

5 each	Black peppercorns
2 each	Bay leaves
1½ ounces	Fresh parsley stems
½ teaspoon	Thyme leaves

METHOD OF PREPARATION:

1. **Render** the diced bacon in a braising pan, and brown.
2. Add the diced onions, and brown to a golden color. Season.
3. Peel a ½-inch band around the potatoes, leaving the rest of the skin intact. Add the potatoes to the braising pan, toss, and season.
4. Add the white stock to cover half of the potatoes. Add the **bouquet garni,** and cook until the potatoes are **fork-tender.**

COOKING TECHNIQUES:
Sauté, Simmer

Sauté:
1. Heat the sauté pan to the appropriate temperature.
2. Evenly brown the food product.
3. For a sauce, pour off any excess oil, reheat, and deglaze.

Simmer and Poach:
1. Heat the cooking liquid to the proper temperature.
2. Submerge the food product completely.
3. Keep the cooked product moist and warm.

GLOSSARY:
Brunoise: ⅛-inch dice
Bouquet garni: bouquet of herbs and spices
Render: to melt the fat
Fork-tender: without resistance

HAZARDOUS FOOD:
Bacon

NUTRITION:
Calories: 197
Fat: 6.1 g
Protein: 8.2 g

Potatoes Colombine

(Pommes de Terre Colombine)

YIELD: 10 SERVINGS. SERVING SIZE: FIVE OUNCES.

COOKING TECHNIQUE:
Sauté

Sauté:
1. Heat the sauté pan to the appropriate temperature.
2. Evenly brown the food product.
3. For a sauce, pour off any excess oil, reheat, and deglaze.

GLOSSARY:
Julienne: matchstick strips

NUTRITION:
Calories: 346
Fat: 21.4 g
Protein: 3.48 g

INGREDIENTS:

8 ounces	Butter, melted
4 ounces	Oil
4 pounds	Potatoes, washed, peeled, and sliced ¼-inch thick
3 each	Pimientos, cored, washed, and cut **julienne**
½ bunch	Fresh parsley, washed, excess moisture removed, and chopped
	Salt and pepper, to taste

METHOD OF PREPARATION:

1. In a heavy skillet, melt the butter. When the butter is hot, add the potatoes, and cook for 2 minutes.
2. Shake the skillet, and season with salt and pepper; cook for 2 more minutes.
3. Add the pimientos. Stir slowly, and cook for 4 more minutes.
4. Lower the heat, cover the skillet with a lid, and cook for 15 to 18 minutes, shaking the skillet from time to time. Pierce the potatoes with a paring knife to check doneness, and sprinkle parsley over them when they are done.

CHEF NOTE:
Add oil to prevent the butter from burning.

Rice Cakes

YIELD: 10 SERVINGS. SERVING SIZE: FOUR OUNCES.

INGREDIENTS:

8 ounces	Wild rice, washed
1 quart	Boiling salted water
8 ounces	Long-grained white rice
4 ounces	Carrots, peeled, washed, grated, and **blanched**
6 ounces	Pistachio nuts, unblanched, peeled, and chopped
6 ounces	Whole-wheat flour
3 each	Eggs, cracked and lightly beaten
	Salt and freshly ground black pepper, to taste
12 ounces	Clarified butter
2 ounces	Fresh parsley, washed, excess moisture removed, and chopped

METHOD OF PREPARATION:

1. In a small, heavy saucepan, place the wild rice, and add 1½ pints of the boiling water. Cover, and cook over low heat until tender, or about 30 to 40 minutes. Add more water, if necessary, but the liquid should be absorbed at the end of the cooking time.
2. In a separate saucepot, cook the white rice in 12 ounces of water over low heat until tender, about 20 minutes, and the liquid is absorbed.
3. Combine both rices in a large mixing bowl, and cool to room temperature.
4. Add all of the remaining ingredients to the rice except the butter, and mix well.
5. In a large sauté pan, melt the butter. Use a 4-ounce scoop or cup, and tightly pack in the rice mixture. Invert the mixture into the sauté pan, and flatten the cake to a ½-inch thickness. Fry, turning once, until golden brown and crisp on both sides. Remove, and drain on towels. Serve immediately, or hold at 140° F or higher.

COOKING TECHNIQUES:

Boil, Shallow-Fry

Boil: (at sea level)

1. Bring the cooking liquid to a rapid boil.
2. Stir the contents, and cook the food product throughout.
3. Serve hot.

Shallow-Fry:

1. Heat the cooking medium to the proper temperature.
2. Cook the food product throughout.
3. Season, and serve hot.

GLOSSARY:

Blanch: to parcook

HACCP:

Hold at 140° F or higher.

HAZARDOUS FOOD:

Eggs

NUTRITION:

Calories: 629
Fat: 41.2 g
Protein: 13.3 g

CHEF NOTES:

1. The amount of boiling water needed will depend on the size and heaviness of the cooking utensils.
2. For added flavor, the rice can be cooked in stock instead of water.

Rice Croquette

(Croquette de Riz)

YIELD: 10 SERVINGS. SERVING SIZE: TWO EACH.

INGREDIENTS:

10 ounces	Rice, cooked **al dente,** washed in cold water, and drained
10 ounces	Béchamel sauce (thick) (see page 16)
6 each	Egg yolks
½ teaspoon	Nutmeg, ground
	Salt and ground white pepper, as needed
	All-purpose flour, as needed
	Egg wash, as needed
	Bread crumbs, as needed

METHOD OF PREPARATION:

1. Mix the first three ingredients; season, to taste.
2. Place flat on a sheet pan, and let cool.
3. Shape rice mixture into cylinders 1 × 3 inches.
4. **Dredge** in the flour, egg wash, and bread crumbs.
5. Deep-fry to order.

COOKING TECHNIQUE:

Deep-Fry

Deep-Fry:

1. Heat the frying liquid to the proper temperature.
2. Submerge the food product completely.
3. Fry the product until it is cooked throughout.

GLOSSARY:

Al dente: to the bite
Dredge: to coat with flour

HAZARDOUS FOOD:

Egg yolks and wash

NUTRITION:

Calories: 459
Fat: 30.3 g
Protein: 9.06 g

CHEF NOTE:

Do not hold in covered bain-marie.

Rice Pilaf

YIELD: 10 SERVINGS. SERVING SIZE: FOUR OUNCES.

INGREDIENTS:

2 ounces	Clarified butter
8 ounces	Onion, peeled and diced **brunoise**
1 pound	Long-grained rice
1 quart	White chicken stock, heated to a boil (see page 12)
	Salt and ground white pepper, to taste

METHOD OF PREPARATION:

1. Preheat the oven to 350° F.
2. Place the butter in a braising pan and heat.
3. Add the onions, and sauté until translucent.
4. Add the rice, and stir to coat all of the grains with butter.
5. Add the chicken stock and seasonings, and heat the liquid to a boil.
6. Cover the pot with a lid, and place in the oven. Steam for 17 to 20 minutes, or until all of the stock is absorbed and the rice is tender.
7. When done, mix with a fork, and adjust the seasoning, if necessary. Serve immediately, or hold at 140° F.

COOKING TECHNIQUES:

Sauté, Simmer, Steam

Sauté:

1. Heat the sauté pan to the appropriate temperature.
2. Evenly brown the food product.
3. For a sauce, pour off any excess oil, reheat, and deglaze.

Simmer and Poach:

1. Heat the cooking liquid to the proper temperature.
2. Submerge the food product completely.
3. Keep the cooked product moist and warm.

Steam: (Traditional)

1. Place a rack over a pot of water.
2. Prevent steam vapors from escaping.
3. Shock or cook the food product throughout.

GLOSSARY:

Brunoise: ⅛-inch dice

HACCP:

Hold at 140° F.

NUTRITION:

Calories: 232
Fat: 6.04 g
Protein: 3.81 g

CHEF NOTES:

1. For saffron rice, add ½ teaspoon of saffron threads to the stock.
2. For buttered saffron rice, toss with an additional 2 ounces of heated clarified butter just before serving.
3. For alternatives, see the recipe for rice pilaf, Greek style, (page 518), and cumin rice (page 477).

Rice Pilaf, Greek Style

YIELD: 10 SERVINGS. SERVING SIZE: FOUR OUNCES.

INGREDIENTS:

2 ounces	Olive oil
4 ounces	Onions, peeled and diced **brunoise**
4 ounces	Chicken livers, trimmed and diced brunoise
1 ounce	Pine nuts
1 each	Fresh tomato, washed, cored, blanched, peeled, seeded, and diced brunoise
1 ounce	Currants, dried
½ teaspoon	Fresh sage leaves, finely chopped
12 ounces	Long-grained rice
1 quart	White chicken stock, heated to a boil (see page 12)

METHOD OF PREPARATION:

1. In a saucepan, combine the oil and onion, and sauté until the onion is translucent.
2. Add the chicken livers, and continue to sauté briefly.
3. Add the pine nuts, tomato, currants, and seasonings, and sauté for an additional 5 minutes; then stir in the rice.
4. Add the chicken stock, stir, and heat the liquid to a boil. Cover tightly, reduce the heat, and simmer until the liquid is absorbed and the rice is tender.
5. Remove from the heat when done, and stir with a braising fork.
6. Serve immediately, or hold at 140° F or above.

COOKING TECHNIQUES:

Sauté, Simmer

Sauté:

1. Heat the sauté pan to the appropriate temperature.
2. Evenly brown the food product.
3. For a sauce, pour off any excess oil, reheat, and deglaze.

Simmer and Poach:

1. Heat the cooking liquid to the proper temperature.
2. Submerge the food product completely.
3. Keep the cooked product moist and warm.

GLOSSARY:

Brunoise: ⅛-inch dice

HACCP:

Hold at 140° F or above.

HAZARDOUS FOOD:

Chicken livers

NUTRITION:

Calories: 213
Fat: 7.68 g
Protein: 4.61 g

COOKING TECHNIQUE:

Steam

Steam: (Traditional)

1. Place a rack over a pot of water.
2. Prevent steam vapors from escaping.
3. Shock or cook the food product throughout.

GLOSSARY:

Brunoise: ⅛-inch dice

NUTRITION:

Calories: 228
Fat: 5.53 g
Protein: 5.72 g

Riz au Champignons

(Rice with Mushrooms)

YIELD: 10 SERVINGS. SERVING SIZE: FOUR OUNCES.

INGREDIENTS:

1 ounce	Butter
1 pound	Rice, long-grain
1 quart	White chicken stock, heated to a boil (see page 12)
1 ounce	Butter
4 ounces	Shallots, peeled and diced **brunoise**
10 ounces	Mushrooms, cleaned and sliced

METHOD OF PREPARATION:

1. Heat the butter in a suitable pot.
2. Sauté the shallots in the butter. When the shallots are translucent, add the sliced mushrooms, and sauté for 3 minutes.
3. Add the rice, and heat.
4. Add the boiling, seasoned white chicken stock.
5. Heat to a boil. Cover, and cook in 375° F oven 18 minutes or until all the liquid is absorbed and rice is tender. Mix with braising fork.

Risotto alla Valdostana

YIELD: 10 SERVINGS.

INGREDIENTS:

4 ounces	Butter
4 ounces	Shallots, minced
1¼ pounds	Arborio rice
40 ounces	Chicken stock (see page 12)
6 ounces	Parmesan cheese, grated
4 ounces	Fontina cheese, grated
4 ounces	Heavy cream
2 ounces	Italian parsley, washed, excess moisture removed, and minced
	Salt and pepper, to taste

METHOD OF PREPARATION:

1. Heat 2 ounces of the butter in a large sauté pan that is set over medium heat.
2. Add the shallots, and sauté until translucent.
3. Add the rice, and stir until well coated.
4. Add 2 cups of the chicken stock, and simmer for 10 minutes, stirring occasionally.
5. Add the remaining stock, and simmer the rice until all of the liquid is absorbed and the rice is tender and creamy, stirring periodically.
6. Add the cheeses, and stir well. Add the butter, cream and parsley, and adjust the seasonings.
7. Serve at once.

COOKING TECHNIQUES:

Sauté, Simmer

Sauté:

1. Heat the sauté pan to the appropriate temperature.
2. Evenly brown the food product.
3. For a sauce, pour off any excess oil, reheat, and deglaze.

Simmer and Poach:

1. Heat the cooking liquid to the proper temperature.
2. Submerge the food product completely.
3. Keep the cooked product moist and warm.

HAZARDOUS FOOD:

Heavy cream

NUTRITION:

Calories: 479
Fat: 21.8 g
Protein: 14.3 g

Risotto Milanese

YIELD: 10 SERVINGS. **SERVING SIZE:** FOUR OUNCES.

INGREDIENTS:

2 ounces	Olive oil
8 ounces	Onion, peeled and diced **brunoise**
20 ounces	Arborio rice
2 quarts	White chicken stock, or as needed, heated to boil (see page 12)
5 ounces	Butter
3 ounces	Parmesan cheese, grated
	Salt and ground white pepper, to taste

METHOD OF PREPARATION:

1. In a sauté pan, heat the oil; add the onions, and sauté until translucent.
2. Add the rice, and stir to coat with oil.
3. Add 1 quart of stock, stir, and cook on low heat, uncovered, until the stock is absorbed.
4. Add additional stock, and continue to cook, stirring occasionally.
5. Continue to add stock, stirring frequently, until the rice is soft but not mushy.
6. Remove from the heat, and stir 8 ounces of stock into the rice in a rapid movement.
7. Add the butter and cheese, and incorporate. Taste, and add the seasoning.

COOKING TECHNIQUES:

Sauté, Simmer

Sauté:

1. Heat the sauté pan to the appropriate temperature.
2. Evenly brown the food product.
3. For a sauce, pour off any excess oil, reheat, and deglaze.

Simmer and Poach:

1. Heat the cooking liquid to the proper temperature.
2. Submerge the food product completely.
3. Keep the cooked product moist and warm.

GLOSSARY:

Brunoise: ⅛-inch dice

NUTRITION:

Calories: 302
Fat: 20.9 g
Protein: 9.15 g

CHEF NOTE:

This rice can be made with many different additions or variations. The most popular is saffron.

Roasted New Potatoes with Garlic and Rosemary

YIELD: 10 SERVINGS. SERVING SIZE: THREE POTATOES.

INGREDIENTS:

30 small	Red bliss potatoes, washed
3 ounces	Olive oil
4 cloves	Garlic, peeled and minced
1 ounce	Fresh rosemary leaves, washed and chopped
	Salt and freshly ground black pepper, to taste

METHOD OF PREPARATION:

1. Preheat the oven to 350° F.
2. Shape the potatoes as desired, and hold them in cold water.
3. In a mixing bowl, combine all of the remaining ingredients. Drain the potatoes well, and then toss in the oil mixture until they are well coated.
4. Place the potatoes in a shallow roasting pan, and place in the oven. Roast about 45 minutes or until the potatoes are **fork-tender.**

COOKING TECHNIQUE:

Roast

Roast:

1. Sear the food product, and brown evenly.
2. Elevate the food product in a roasting pan.
3. Determine doneness, and consider carryover cooking.
4. Let the food product rest before carving.

GLOSSARY:

Fork-tender: without resistance
Tournéed: trimmed to a large olive shape

NUTRITION:

Calories: 132
Fat: 8.62 g
Protein: 2.05 g

CHEF NOTE:

Potatoes can be cut into various shapes, including mushrooms shaped **tournéed,** or simply scored with a zester.

Roasted Potatoes

COOKING TECHNIQUE:
Bake

Bake:
1. Preheat the oven.
2. Place the food product on the appropriate rack.

GLOSSARY:
Fork-tender: without resistance

NUTRITION:
Calories: 219
Fat: 7.17 g
Protein: 3.29 g

YIELD: 10 SERVINGS. SERVING SIZE: FOUR OUNCES.

INGREDIENTS:

4 pounds	Russet potatoes, washed, peeled, and cut in half
	Salted water, to cover
3 ounces	Butter
1 tablespoon	Spanish paprika
	Salt, to taste

METHOD OF PREPARATION:

1. Preheat the oven to 350° F.
2. In a stockpot, place the potatoes in cold, salted water to cover, and heat to a boil. Cook the potatoes for approximately 10 minutes, then drain, and place in a large mixing bowl. Do not shock the potatoes in cold water.
3. In a sauté pan, melt the butter, and add the paprika and salt. When the paprika is dissolved in the butter, pour the mixture over the potatoes, and gently toss together.
4. Spread the potatoes out in a shallow hotel pan, and place in the oven.
5. Bake until **fork-tender.**

Spaghetti Putanesca

YIELD: 10 SERVINGS. SERVING SIZE: FIVE OUNCES.

INGREDIENTS:

4 ounces	Olive oil
12 cloves	Garlic, peeled and coarsely chopped
2 ounces	Anchovy fillets, oil drained and chopped
2 pounds	Tomato **concassé**
2 ounces	Tomato purée
5 ounces	Black pitted olives, cut into ⅓-inch slices
½ tablespoon	Freshly ground black pepper
	Salt, to taste
1½ pounds	Spaghetti, cooked in boiling, salted water until **al dente**

METHOD OF PREPARATION:

1. In a sauté pan, heat the oil, and sauté garlic for 2 to 3 minutes. Add the anchovies, and sauté for 1 minute more.
2. Add the remaining ingredients, except the spaghetti, and simmer until the tomatoes are reduced to a sauce, approximately 20 minutes.
3. Add the spaghetti to the sauce, and toss together; taste, and adjust the seasoning, if necessary. Serve immediately, or hold at 140° F or above.

COOKING TECHNIQUES:
Sauté, Boil, Simmer

Sauté:
1. Heat the sauté pan to the appropriate temperature.
2. Evenly brown the food product.
3. For a sauce, pour off any excess oil, reheat, and deglaze.

Boil: (at sea level)
1. Bring the cooking liquid to a rapid boil.
2. Stir the contents, and cook the food product throughout.
3. Serve hot.

Simmer and Poach:
1. Heat the cooking liquid to the proper temperature.
2. Submerge the food product completely.
3. Keep the cooked product moist and warm.

GLOSSARY:
Concassé: peeled, seeded, roughly chopped
Al dente: to the bite

HACCP:
Hold at 140° F or above.

NUTRITION:
Calories: 244
Fat 14.4 g
Protein 6.07 g

CHEF NOTE:
This dish is served as a starch without grated cheese.

Spicy Paprika Potatoes

YIELD: 10 SERVINGS. SERVING SIZE: FOUR OUNCES.

INGREDIENTS:

2 ounces	Spanish olive oil
6 cloves	Garlic, peeled and minced
2 ounces	Hot Spanish paprika
2½ pounds	New red potatoes, washed, peeled, and steamed until tender
	Salt and ground white pepper, to taste

METHOD OF PREPARATION:

1. In a sauté pan, combine the oil and garlic, and heat together. Sauté the garlic over low heat until golden brown and crunchy.
2. Add the paprika to the garlic; then add the potatoes.
3. Shaking the pan constantly, heat the potatoes to an internal temperature of 165° F, coating with the paprika oil. Season, to taste, and serve immediately, or hold at 140° F or above.

COOKING TECHNIQUES:

Steam, Sauté

Steam: (Traditional)

1. Place a rack over a pot of water.
2. Prevent steam vapors from escaping.
3. Shock or cook the food product throughout.

Sauté:

1. Heat the sauté pan to the appropriate temperature.
2. Evenly brown the food product.
3. For a sauce, pour off any excess oil, reheat, and deglaze.

HACCP:

Heat the potatoes to an internal temperature of 165° F.
Hold at 140° F or above.

NUTRITION:

Calories: 142
Fat: 6.53 g
Protein: 3.57 g

CHEF NOTE:

Add the garlic to the oil before heating the oil, to allow the bits to slow-cook and become "toasted" without becoming bitter.

Spiral Pasta with Broccoli

YIELD: 10 SERVINGS. SERVING SIZE: FIVE OUNCES.

INGREDIENTS:

4 ounces	Olive oil
12 cloves	Garlic, peeled and coarsely minced
1 pound	Broccoli florettes, washed and **blanched**
1½ pounds	Spiral pasta, cooked **al dente** in boiling, salted water; then rinsed and drained
	Salt and freshly ground black pepper, to taste
5 ounces	Freshly grated Romano cheese

METHOD OF PREPARATION:

1. In a sauté pan, heat the oil, and sauté the garlic. Add the broccoli, and toss together.
2. Add the pasta to the garlic and broccoli, season, to taste, and heat throughout.
3. Serve immediately on a preheated dinner plate, or hold at 140° F or above, and garnish with Romano cheese.

COOKING TECHNIQUES:

Sauté, Boil

Sauté:

1. Heat the sauté pan to the appropriate temperature.
2. Evenly brown the food product.
3. For a sauce, pour off any excess oil, reheat, and deglaze.

Boil: (at sea level)

1. Bring the cooking liquid to a rapid boil.
2. Stir the contents, and cook the food product throughout.
3. Serve hot.

GLOSSARY:

Blanch: to parcook
Al dente: to the bite
Bain-marie: hot-water bath

HACCP:

Hold at 140° F.

NUTRITION:

Calories: 269
Fat: 15.8 g
Protein: 9.35 g

CHEF NOTE:

Hold a minimal time in a **bain-marie,** because dish will lose color and texture.

Steamed Semolina with Vegetables

YIELD: 10 SERVINGS. SERVING SIZE: FOUR OUNCES.

INGREDIENTS:

8 ounces	Boiling water
8 ounces	White chicken stock, heated to a boil (see page 12)
2 teaspoons	Curry powder, or to taste
	Salt and ground white pepper, to taste
8 ounces	Couscous semolina (quick-cooking)
6 ounces	Fennel bulbs, washed, trimmed, and diced
1 each	Red bell pepper, washed, seeds removed, and diced
1 each	Yellow bell pepper, washed, seeds removed, and diced

METHOD OF PREPARATION:

1. Add the fennel to the water, and poach for about 5 minutes, or until **al dente.** Add the peppers, and allow them to cook for another 2 minutes.
2. Add the stock and seasonings, and return to a boil.
3. Stir in the semolina, cover, and remove from the heat.
4. Allow to stand for about 10 minutes, or until the liquid is absorbed.
5. Mix with a fork before serving. Hold at 140° F.

COOKING TECHNIQUE:
Steam

Steam: (Traditional)
1. Place a rack over a pot of water.
2. Prevent steam vapors from escaping.
3. Shock or cook the food product throughout.

GLOSSARY:
Al dente: to the bite
Cocotte: small oven-proof dish

HACCP:
Hold at 140° F.

NUTRITION:
Calories: 93.6
Fat: .372 g
Protein: 3.33 g

CHEF NOTES:
1. Semolina can be molded into buttered **cocottes** for service, and then unmolded onto preheated dinner plates.
2. The feathery tops of the fennel can be used as sprigs for garnish.

Stuffed Pasta

YIELD: 10 SERVINGS: SERVING SIZE: EIGHT OUNCES.

INGREDIENTS:

20 each	Frozen 6-inch square pasta sheets
3 ounces	Butter, melted
1 quart	Tomato sauce, heated to a boil (see page 76)
5 ounces	Parmesan cheese, grated

Filling:

30 ounces	Ricotta cheese
6 ounces	Romano cheese
2 tablespoons	Fresh parsley, washed, excess moisture removed, and chopped
½ teaspoon	Nutmeg, ground
½ teaspoon	Basil, ground
6 each	Egg yolks, or 3 ounces of pasteurized egg yolks
	Salt, to taste

METHOD OF PREPARATION:

1. Precook the pasta to **al dente,** cool in cold water, and drain.
2. Combine the filling ingredients, and mix until smooth.
3. Using a tubeless pastry bag, pipe the filling along one edge of the pasta sheets, and roll into a tube shape.
4. Butter the individual baking dishes, and place two pasta rolls in each dish. Refrigerate, covered, at 40° F or below until needed for service.
5. Preheat the oven to 425° F.
6. Ladle tomato sauce around the rolled pasta shells, and sprinkle Parmesan cheese over the pasta. Bake until heated throughout and the cheese is browned, about 15 minutes.
7. Serve in a baking dish, or transfer to a preheated dinner plate. Hold at 140° F.

COOKING TECHNIQUES:

Boil, Bake

Boil: (at sea level)

1. Bring the cooking liquid to a rapid boil.
2. Stir the contents, and cook the food product throughout.
3. Serve hot.

Bake:

1. Preheat the oven.
2. Place the food product on the appropriate rack.

GLOSSARY:

Al dente: to the bite

HACCP:

Refrigerate at 40° F or below.
Hold at 140° F.

HAZARDOUS FOODS:

Ricotta cheese
Egg yolks

NUTRITION:

Calories: 527
Fat: 26.4 g
Protein: 29.2 g

Sweet Rice Cake

(Kue Kochi Dawn)

YIELD: 10 SERVINGS. SERVING SIZE: FOUR OUNCES.

INGREDIENTS:

4 ounces	Coconut, grated
2 ounces	Brown sugar
2 ounces	Water
¼ teaspoon	Cinnamon
1 pound	Glutinous rice flour
1 pint	Coconut milk
	Salt, to taste
	Banana leaves, as needed
	Peanut oil, as needed

METHOD OF PREPARATION:

1. In a saucepan, combine the coconut, brown sugar, water, and cinnammon, and simmer, stirring continuously, until the moisture is absorbed and the coconut has a moist consistency. Set aside to cool.
2. Combine the rice flour, coconut milk, and salt in a bowl, and mix to form dough.
3. Select a piece of banana leaf big enogh to wrap a 3-ounce portion of the dough.
4. Oil the leaf, and place one portion of the dough in the center of the leaf. Add a tablespoon of the coconut mixture, and wrap the dough around it.
5. Fold the leaf around the dough, completely encasing it, and steam for 15 minutes.
6. Serve immediately on a preheated plate, or hold at 140° F or above.

COOKING TECHNIQUES:

Simmer, Steam

Simmer and Poach:

1. Heat the cooking liquid to the proper temperature.
2. Submerge the food product completely.
3. Keep the cooked product moist and warm.

Steam: (Traditional)

1. Place a rack over a pot of water.
2. Prevent steam vapors from escaping.
3. Shock or cook the food product throughout.

HACCP:

Hold at 140° F or above.

NUTRITION:

Calories: 316
Fat: 14.1 g
Protein: 4 g

CHEF NOTE:

This is an Indonesian side dish.

Texas Fries

YIELD: 10 SERVINGS. SERVING SIZE: FOUR OUNCES.

INGREDIENTS:

	Oil for frying, heated to 350° F, as needed
3 pounds	Idaho or baking potatoes, washed, cut into lengthwise wedges, steamed until **al dente,** and cooled
1 tablespoon	Salt
1 tablespoon	Spanish paprika
1 teaspoon	Caraway seeds, ground

METHOD OF PREPARATION:

1. Deep-fry the potatoes **á la minute** to golden brown.
2. Sprinkle the mixture of seasonings on the potatoes, and serve immediately.

COOKING TECHNIQUE:

Deep-Fry

Deep-Fry:

1. Heat the frying liquid to the proper temperature.
2. Submerge the food product completely.
3. Fry the product until it is cooked throughout.

GLOSSARY:

Al dente: to the bite
À la minute: cooked to order

NUTRITION:

Calories: 240
Fat: 13.9 g
Protein: 2.48 g

Timbales of Grains and Sprouts

YIELD: 10 SERVINGS. SERVING SIZE: FOUR OUNCES.

INGREDIENTS:

3 ounces	Wild rice
3 ounces	White rice, long-grained
3 ounces	Barley
3 ounces	Quinoa
2 quarts	White veal or chicken stock, heated to a boil (see page 12)
	Salt and freshly ground black pepper, to taste
3 ounces	Freshly chopped parsley, excess moisture removed
6 ounces	Fresh pea sprouts

METHOD OF PREPARATION:

1. In separate pots, cook the grains in the stock until tender. Drain if necessary; then cool.
2. Combine the grains, season, to taste, and add the parsley. Mix well; then pack into timbales.
3. Preheat the oven to 375° F.
4. For service, reheat in the oven to an internal temperature of 165° F. Hold at 140° F.
5. Serve unmolded on preheated plates, garnished with fresh pea sprouts.

COOKING TECHNIQUES:

Boil, Steam

Boil: (at sea level)

1. Bring the cooking liquid to a rapid boil.
2. Stir the contents, and cook the food product throughout.
3. Serve hot.

Steam: (Traditional)

1. Place a rack over a pot of water.
2. Prevent steam vapors from escaping.
3. Shock or cook the food product throughout.

HACCP:

Heat to an internal temperature of 165° F.
Hold at 140° F.

NUTRITION:

Calories: 168
Fat: 1.17 g
Protein: 6.47 g

CHEF NOTES:

1. Alternate grains can be substituted, but it is important to cook each separately, so that the proper textures can be maintained.
2. Alternative green sprouts or small sprigs of fresh herbs can be used.

Wild Rice Pilaf

YIELD: 10 SERVINGS. SERVING SIZE: FOUR OUNCES.

INGREDIENTS:

8 ounces	Wild rice, parboiled
1½ ounce	Olive oil
1 each	Medium onion, peeled and diced
1 each	Bay leaf
16 ounces	Converted rice, long-grain
4 ounces	Dry white wine
1 quart	White chicken stock, heated to a boil (see page 12)
2 tablespoons	Chopped parsley
	Salt and freshly ground black pepper, to taste

METHOD OF PREPARATION:

1. In a medium saucepan, combine the oil and onion, and sauté until the onion is translucent. Add the bay leaf and rice, and coat the grains with oil.
2. Add the white wine, white chicken stock, and seasonings. Stir to combine. Heat the liquid to a boil.
3. Reduce the heat to a simmer, cover, and cook until the rice is tender, or about 40 minutes. Strain off any remaining liquid.
4. Hold at 140° F.

COOKING TECHNIQUES:
Sauté, Simmer

Sauté:
1. Heat the sauté pan to the appropriate temperature.
2. Evenly brown the food product.
3. For a sauce, pour off any excess oil, reheat, and deglaze.

Simmer and Poach:
1. Heat the cooking liquid to the proper temperature.
2. Submerge the food product completely.
3. Keep the cooked product moist and warm.

HACCP:
Hold at 140° F.

NUTRITION:
Calories: 157
Fat: 5.76 g
Protein: 2.96 g

CHEF NOTES:
1. Wild rice is actually an aquatic grass that is native to North America. Wash wild rice before cooking. While cooking, make sure the liquid does not evaporate completely.
2. Add 1 cup of wild rice that has been washed and parboiled for 20 minutes to the white rice in step 1. Do not adjust the liquid amounts.

7

Vegetables

7

Vegetables

Away with all this slicing, this dicing, this grating, this peeling of truffles . . . Eat it like the vegetable it is, hot and served in munificent quantities.

Colette
Prisons et Paradis

I have been actively involved with Johnson & Wales University since its beginning as a culinary school. Through the years, I have watched with great pride as Johnson & Wales has developed into one of the finest culinary schools in the world. The quality of education at Johnson & Wales not only covers the basics of the culinary field, but also stresses the professionalism and realities of being a chef. The variety of programs, such as the Distinguished Visiting Chef program, offers students expertise from world-renowned chefs and an opportunity to work with some of the finest chefs first-hand.

Johnson & Wales has continually strived to produce professional chefs for the hospitality industry. I have had the pleasure to work with many Johnson & Wales graduates and consider them well qualified.

We only get out of life or school what we put into it, but *Johnson & Wales University, College of Culinary Arts, offers all of the basics, plus a variety of special programs and opportunities from which one can learn and excel.*

Stanley James Nicas, CEC, AAC
Distinguished Visiting Chef
November 21–24, 1982
Executive Chef/Owner
Castle Restaurant
Leicester, MA

The HACCP Process

The following recipe illustrates how vegetables flow through the HACCP process.

Broccoli Soufflé

Receiving

- Vegetables (broccoli)—packaging intact, no cross-contamination from other foods on the truck; no signs of insect or rodent activity
- Béchamel sauce (see page 16)
- Eggs, whole—shells not cracked or dirty. Container should be intact. Whole shell eggs should be delivered at 40° F (4.4° C) or lower.
- Spices (salt, pepper, and nutmeg)—packaging intact
- Butter should have a sweet flavor, firm texture, and uniform color. Butter should be free of specks and mold, and packaging should be intact. Butter should be delivered at 40° F (4.4° C) or lower.

Storage

- Vegetables (broccoli)—Store under refrigeration, with a product temperature not to exceed 40° F (4.4° C). Store above and away from raw, potentially hazardous foods.
- Béchamel sauce—Store under refrigeration, with a product temperature not to exceed 40° F (4.4° C). Store above and away from raw, potentially hazardous foods.
- Eggs, whole—Store under refrigeration at 40° F (4.4° C) or lower. Store above and away from raw, potentially hazardous foods.
- Spices (salt, pepper, and nutmeg)—Store in dry storage at 50° F (10° C) with a relative humidity of 50% to 60%. Keep dry.
- Butter—Store under refrigeration at 40° F (4.4° C) or lower. Store above and away from raw, potentially hazardous foods.

Preparation and Cooking

- Separate broccoli florets and stems. Wash the broccoli.
- Chop the broccoli on a clean and sanitized cutting board with clean and sanitized utensils.

- Place the broccoli in boiling, salted water. Cook until fork-tender. Drain, and cool. Purée in a food processor.
- Incorporate the béchamel sauce with the puréed broccoli.
- Mix the egg yolks, salt, white pepper, and nutmeg with the broccoli.
- Whip the egg whites until stiff, and fold into the broccoli mixture.
- Melt the butter, and brush a soufflé dish.
- Spoon the mixture into the soufflé dish until it is two-thirds full.
- Place in a bain-marie, and bake in a 400° oven for approximately 20 minutes. Cook to an internal temperature of 145° F (62.8° C).

Holding and Service

- Serve immediately.

NUTRITIONAL NOTES

Vegetables add color, texture, vitamins, minerals, and fiber to a meal. They can be used to introduce infinite variety to the menu. Vegetables are cholesterol-free and, with the exception of avocados, are low in fat. For healthful vegetable cooking:

- *Use low-fat cooking methods such as steaming, broiling, baking, and grilling.*
- *Cook in as little water as possible for the shortest amount of time to retain vitamins and minerals as well as texture. Prolonged exposure to heat destroys many vitamins. If there is any cooking liquid, use it. It contains vitamins and minerals from the vegetables.*
- *Cook and serve vegetables as whole as possible. Leave as much skin on as possible to minimize nutrient loss.*

Baked Eggplant

YIELD: 10 SERVINGS. SERVING SIZE: FOUR OUNCES.

INGREDIENTS:

	Oil, as needed, heated to 350° F.
10 ounces	Onions, peeled and diced **brunoise**
5 each	Long narrow eggplants, washed, trimmed, pulp removed, split in half lengthwise and diced
10 ounces	Tomato **concassé**
8 cloves	Garlic, peeled and finely minced
1 ounce	Granulated sugar
1 ounce	Freshly squeezed lemon juice
	Salt and freshly ground black pepper, to taste
3 ounces	Fresh parsley, washed, excess moisture removed, and chopped

METHOD OF PREPARATION:

1. Preheat the oven to 350° F.
2. In a sauté pan, heat the oil, and sauté the onion and diced eggplant until the onion is translucent.
3. Add the tomato concassé and garlic, and continue to sauté until the tomatoes are cooked. Add the sugar and lemon juice, and season, to taste. Cook until the sugar is melted.
4. Place the eggplant in a baking pan. Cover with the sautéed mixture, and bake 15 minutes.
5. Serve immediately, and sprinkle with chopped parsley, or hold at 140° F.

COOKING TECHNIQUES:
Shallow-Fry, Sauté, Bake

Shallow-Fry:
1. Heat the cooking medium to the proper temperature.
2. Cook the food product throughout.
3. Season, and serve hot.

Sauté:
1. Heat the sauté pan to the appropriate temperature.
2. Evenly brown the food product.
3. For a sauce, pour off any excess oil, reheat, and deglaze.

Bake:
1. Preheat the oven.
2. Place the food product on the appropriate rack.

GLOSSARY:
Brunoise: ⅛-inch dice
Concassé: peeled, seeded, roughly chopped

HACCP:
Hold at 140° F.

NUTRITION:
Calories: 206
Fat: 11.6 g
Protein: 3.24 g

Baked Plum Tomatoes

YIELD: 10 SERVINGS. SERVING SIZE: FOUR OUNCES.

INGREDIENTS:

20 each	Plum tomatoes, washed, cored
	Salt, to taste
	Salted water, to cover
2 heads	Celeriac, washed, peeled, and diced **brunoise**
2 ounces	Egg whites
8 ounces	Frozen green peas, **blanched** and drained
8 ounces	Heavy cream
4 ounces	Butter
1 pound	Onions, peeled and diced brunoise
½ teaspoon	Paprika
6 ounces	Vegetable stock, heated to a boil (see page 11)

METHOD OF PREPARATION:

1. Preheat the oven to 350° F.
2. Season the inside of the tomatoes with salt and turn upside down on a wire rack to drain.
3. Cook the celeriac in lightly salted water, to cover, until tender; then drain, and purée.
4. Lightly whip the egg whites and add the puréed celeriac and green peas.
5. Add half of the heavy cream, and salt, to taste; mix, and fill the tomatoes.
6. In a sauté pan, melt the butter, and sauté the onions until they become translucent. Add the paprika, and sauté for 1 more minute.
7. Add the stock, and heat the liquid to a boil. **Temper** the remaining heavy cream, and add it to the onions.
8. Transfer the onions into a half hotel pan, and spread the mixture evenly. Arrange the stuffed tomatoes on the onions, place in the oven, and bake until the stuffing is heated thoroughly, or to an internal temperature of 165° F. Serve immediately as a side dish with some of the onions, or hold at 140° F.

COOKING TECHNIQUES:
Bake, Sauté

Bake:
1. Preheat the oven.
2. Place the food product on the appropriate rack.

Sauté:
1. Heat the sauté pan to the appropriate temperature.
2. Evenly brown the food product.
3. For a sauce, pour off any excess oil, reheat, and deglaze.

GLOSSARY:
Brunoise: ⅛-inch dice
Blanch: to parcook
Temper: to equalize two extreme temperatures

HACCP:
Heat thoroughly or to an internal temperature of 165°.
Hold at 140° F.

HAZARDOUS FOODS:
Egg whites
Heavy cream

NUTRITION:
Calories: 243
Fat: 18.9 g
Protein: 4.27 g

CHEF NOTE:
At the discretion of the chef, the onions can be puréed, thinned with seasoned liquid, and served as a sauce over the tomatoes.

COOKING TECHNIQUE:

Bake

Bake:

1. Preheat the oven.
2. Place the food product on the appropriate rack.

GLOSSARY:

Brunoise: ⅛-inch dice

NUTRITION:

Calories: 143
Fat: 9.81 g
Protein: 2.21 g

Baked Stuffed Tomatoes

(Tomato Soubise)

YIELD: 10 SERVINGS. SERVING SIZE: ONE TOMATO.

INGREDIENTS:

10 each	Tomatoes, washed, cored, and hollowed out
	Salt, as needed
4 ounces	Butter
1 pound	Onions, peeled and diced **brunoise**
3 cloves	Garlic, washed and mashed into to a purée
2 ounces	Freshly chopped parsley, washed, excess moisture removed
8 ounces	Chicken demi-glacé, heated to a boil (see page 26)

METHOD OF PREPARATION:

1. Preheat the oven to 375° F.
2. Season the inside of the tomatoes with salt, and turn upside down on a tray to drain.
3. In a sauté pan, melt the butter, add onions, and sauté until they are translucent. Add the garlic, and continue to sauté for 2 to 3 minutes. Add the parsley, and season, to taste.
4. Stuff the tomatoes with the onion mixture.
5. Top each tomato with 1½ tablespoons of the demi-glacé.
6. Place the tomatoes in a braising pan. Bake for 10 minutes, or until heated throughout. Serve immediately, or hold minimally at 140° F.

CHEF NOTE:

Do not overbake the tomatoes or hold longer than 15 minutes in a bain-marie, because they will collapse for service.

Beer-Battered Fried Okra with Cajun Tartar Sauce

YIELD: 10 SERVINGS. SERVING SIZE: FOUR OUNCES.

INGREDIENTS:

	Vegetable oil, as needed, for frying, heated to 350° F
3 pounds	Okra, washed, trimmed, and dried

Beer Batter:

2 pounds	All-purpose flour
2 teaspoons	Salt
1 teaspoon	Garlic powder
	White pepper, to taste
8 each	Eggs, cracked and separated
24 ounces	Beer

METHOD OF PREPARATION:

1. In a large bowl, combine the flour and seasonings, and mix well.
2. Lightly beat the egg yolks, and add to the flour, along with the beer and oil, and mix to a smooth batter.
3. Whip the egg whites to a soft peak, and then fold whites into the batter.
4. For service, dip the okra in the batter, and deep-fry until golden brown and crisp. Drain on absorbent paper, and serve immediately, or hold at 140° F or in a 200° F oven with the door open. Offer tartar sauce on the side.

COOKING TECHNIQUE:

Deep-Fry

Deep-Fry:

1. Heat the frying liquid to the proper temperature.
2. Submerge the food product completely.
3. Fry the product until it is cooked throughout.

HACCP:

Hold at 140° F.

HAZARDOUS FOOD:

Eggs

NUTRITION:

Calories: 662
Fat: 27 g
Protein: 17.6 g

CHEF NOTE:

This batter can be used on other vegetables, including zucchini, mushrooms, sweet potatoes, cauliflower, and winter spinach.

Bouquet of Vegetables

(Bouquetiere de Légumes)

YIELD: 10 SERVINGS. SERVING SIZE: EIGHT OUNCES.

INGREDIENTS:

1 pound	Carrots, washed, peeled, and cut into uniform shapes
1 pound	Green beans, washed and trimmed
1 head	Cauliflower, core removed, washed, and cut into flowerets (reserve stems for alternate use)
1 pound	Asparagus, washed, trimmed, and stalks peeled
8 ounces	Butter
	Salt and ground white pepper, to taste
30 each	Cherry tomatoes
5 ounces	Hollandaise sauce, heated to 140° F (see page 30)
5 ounces	Mimosa butter (recipe follows)

METHOD OF PREPARATION:

1. In separate saucepans, cook the carrots, green beans, cauliflower, and asparagus in boiling, salted water until **al dente.** Shock in ice water, and drain.
2. At the time of service, in sauté pans, heat the butter, and separately reheat or sauté all of the vegetables, seasoning, to taste.
3. To serve, arrange the vegetables on preheated dinner plates. **Nappé** the asparagus with Hollandaise sauce and the cauliflower with mimosa butter.

Mimosa Butter

INGREDIENTS:

4 ounces	Butter, melted
1 each	Egg, hard-boiled, peeled, and grated
1 tablespoon	Freshly chopped parsley, excess moisture removed
	Salt and ground white pepper, to taste

METHOD OF PREPARATION:

1. Melt the butter in a sauté pan.
2. Add the egg and parsley, and season, to taste.
3. Toss together, remove, and serve.

COOKING TECHNIQUES:

Boil, Sauté

Boil: (at sea level)

1. Bring the cooking liquid to a rapid boil.
2. Stir the contents, and cook the food product throughout.
3. Serve hot.

Sauté:

1. Heat the sauté pan to the appropriate temperature.
2. Evenly brown the food product.
3. For a sauce, pour off any excess oil, reheat, and deglaze.

GLOSSARY:

Al dente: to the bite
Nappé: coated

NUTRITION:

Calories: 86
Fat: 1.8 g
Protein: 4 g

Braised Cabbage Rolls

YIELD: 10 SERVINGS. SERVING SIZE: ONE BALL.

INGREDIENTS:

1 head	White cabbage, cored and blanched
	Salt and freshly ground black pepper, to taste
2 ounces	Vegetable oil or butter
12 ounces	Canned crushed tomatoes, drained
3 cloves	Garlic, peeled and minced
1 pint	Demi-glacé, heated to a boil (see page 26)

Mirepoix:

8 ounces	Onions, peeled
8 ounces	Carrots, washed and peeled
8 ounces	Celery stalks, washed and trimmed

METHOD OF PREPARATION:

1. Preheat the oven to 350° F.
2. Separate the blanched cabbage leaves, and remove the heavy vein near the core end.
3. Season the leaves with salt and pepper. Roll and tuck both ends.
4. In a sauté pan, heat the oil or butter, and add the mirepoix. Sauté until the onions become translucent, then add the tomatoes and garlic.
5. Add the demi-glacé, and simmer until the vegetables are softened. Season, to taste.
6. Place the cabbage rolls in a half hotel pan, and ladle the sauce over the top of each roll.
7. Cover tightly, and place pan in the oven. Braise for 20 minutes, or until cabbages are tender. Serve immediately, or hold at 140° F or above.

COOKING TECHNIQUES:

Sauté, Braise

Sauté:

1. Heat the sauté pan to the appropriate temperature.
2. Evenly brown the food product.
3. For a sauce, pour off any excess oil, reheat, and deglaze.

Braise:

1. Heat the braising pan to the proper temperature.
2. Sear and brown the food product to a golden color.
3. Degrease and deglaze.
4. Cook the food product in two-thirds liquid until fork-tender.

GLOSSARY:

Mirepoix: roughly chopped vegetables

HACCP:

Hold at 140° F or above.

NUTRITION:

Calories: 287
Fat: 19.4 g
Protein: 5.72 g

Braised Fennel

YIELD: 10 SERVINGS. SERVING SIZE: FOUR OUNCES.

COOKING TECHNIQUE:

Braise

Braise:

1. Heat the braising pan to the proper temperature.
2. Sear and brown the food product to a golden color.
3. Degrease and deglaze.
4. Cook the food product in two-thirds liquid until fork-tender.

GLOSSARY:

Brunoise: ⅛-inch dice
Nappé: coated

NUTRITION:

Calories: 187
Fat: 9.18 g
Protein: 3.93 g

INGREDIENTS:

3 ounces	Olive oil
8 ounces	Onions, peeled and diced **brunoise**
4 cloves	Garlic, peeled and finely minced
1½ pounds	Canned crushed tomatoes
8 ounces	Dry white wine
2 ounces	Basil leaves, washed and chopped
	Salt and freshly ground black pepper, to taste
10 each	Fresh fennel bulbs, washed, trimmed, and quartered
10 sprigs	Fresh basil, washed

METHOD OF PREPARATION:

1. Preheat the oven to 325° F.
2. In a sauté pan, heat the oil, and sauté the onions until they are translucent. Add the garlic and tomatoes, and sauté for another 5 minutes. Add the wine and basil, and season, to taste.
3. Arrange the fennel in a hotel pan, ladle the sauce over each quarter, and cover tightly.
4. Place in the oven, and braise until tender, approximately 30 minutes.
5. To serve, place the fennel on a preheated plate; **nappé** with the sauce, and garnish with basil sprigs.

Braised Green Beans with Pears

YIELD: 10 SERVINGS. SERVING SIZE: FOUR OUNCES.

INGREDIENTS:

1 quart	Water, boiling, lightly salted
8 ounces	Lean bacon, diced
2½ pounds	Fresh green beans, washed and ends trimmed
2 pounds	Small, firm, ripe pears, peeled, halved, cored, cut into ¼-inch slices (held in **acidulated** cold water)
1 tablespoon	Fresh thyme leaves, chopped
	Freshly ground black pepper, to taste
2 ounces	Italian (flat-leafed) parsley, washed, stemmed, and finely chopped

METHOD OF PREPARATION:

1. Preheat the oven to 300° F.
2. In a braising pan, heat the water to a boil, and add the bacon. Blanch the bacon for 2 to 3 minutes, remove with a **skimmer,** and chill in an ice water bath.
3. Return the water to a boil; add the beans, and remove from the heat.
4. Drain the pears, and place in a layer on top of the beans. Add the bacon on top of the pears, and sprinkle in herbs and seasonings.
5. Cover the pan tightly, and place in the oven to braise for at least 1 hour.
6. Remove from the oven, and if too much liquid remains, place over high heat and **reduce.**
7. Toss all of the ingredients together with a fork, and serve immediately, or hold at 140° F or higher.

COOKING TECHNIQUES:
Braise, Boil

Braise:
1. Heat the braising pan to the proper temperature.
2. Sear and brown the food product to a golden color.
3. Degrease and deglaze.
4. Cook the food product in two-thirds liquid until fork-tender.

Boil: (at sea level)
1. Bring the cooking liquid to a rapid boil.
2. Stir the contents, and cook the food product throughout.
3. Serve hot.

GLOSSARY:
Acidulated: with acid added
Skimmer: long-handled, round, perforated tool
Reduction: evaporation of liquid by boiling

HACCP:
Hold at 140° F or higher.

NUTRITION:
Calories: 224
Fat: 11.8 g
Protein: 9.6 g

CHEF NOTES:
1. This German-style preparation is somewhat sweet and sour with the pears and bacon.
2. Long, slow cooking still produces a final product with lots of texture and flavor.

Braised Red Cabbage

(Choux Rouge Braisé)

YIELD: 10 SERVINGS. SERVING SIZE: FOUR OUNCES.

INGREDIENTS:

4 ounces	Duck fat
4 ounces	Onions, peeled and diced **brunoise**
4 pounds	Red cabbage, shredded
1 ounce	Red wine vinegar
3 ounces	Sugar
	Salt, to taste
10 ounces	Sour apples, washed, peeled, and grated

METHOD OF PREPARATION:

1. Heat the duck fat in a braising pan. Add the onions to the duck fat, and sauté.
2. Mix the shredded cabbage with red wine vinegar and sugar.
3. Add the red cabbage to the sautéed onions.
4. Add the apples to the cabbage. Cover, and braise until tender.

COOKING TECHNIQUE:

Braise

Braise:

1. Heat the braising pan to the proper temperature.
2. Sear and brown the food product to a golden color.
3. Degrease and deglaze.
4. Cook the food product in two-thirds liquid until fork-tender.

GLOSSARY:

Brunoise: ⅛-inch dice

NUTRITION:

Calories: 17
Fat: 9.1 g
Protein: 2.83 g

CHEF NOTES:

1. Red wine (dry or sweet) or applesauce may be added at the beginning of braising.
2. To keep the original color, acid (i.e., lemon or vinegar) should be added at the beginning of braising.
3. Hold the grated apples in lemon juice until needed, to prevent discoloration of the apples.

Braised Romaine

(Romaine Braisée)

YIELD: 10 SERVINGS. SERVING SIZE: FOUR OUNCES.

INGREDIENTS:

3 heads	Romaine lettuce, washed, cored, and separated
	Salt, as needed
6 ounces	Onions, peeled and diced **brunoise**
6 ounces	Carrots, washed, peeled, and diced brunoise
3 ounces	Clarified butter
4 ounces	Baked ham, diced brunoise
5 ounces	Demi-glacé, seasoned and heated to a boil (see page 26)

METHOD OF PREPARATION:

1. In a stockpot, **blanch** the romaine in boiling, salted water. Drain, and shock in an ice bath. Remove when cold.
2. In a sauté pan, heat the butter; sauté the onions and carrots until the onions are translucent.
3. Add the ham, and sauté for another 3 minutes, and reserve.
4. Stuff the romaine lettuce, and roll.
5. Place the romaine in a hotel pan. Add the demi-glacé, and braise in a 350° F oven until the romaine is tender.

COOKING TECHNIQUES:

Sauté, Braise

Sauté:

1. Heat the sauté pan to the appropriate temperature.
2. Evenly brown the food product.
3. For a sauce, pour off any excess oil, reheat, and deglaze.

Braise:

1. Heat the braising pan to the proper temperature.
2. Sear and brown the food product to a golden color.
3. Degrease and deglaze.
4. Cook the food product in two-thirds liquid until fork-tender.

GLOSSARY:

Brunoise: ⅛-inch dice
Blanch: to parcook

HAZARDOUS FOOD:

Baked ham

NUTRITION:

Calories: 112
Fat: 6.58 g
Protein: 4.76 g

Broccoli au Gratin

YIELD: 50 PORTIONS. SERVING SIZE: FIVE OUNCES.

INGREDIENTS:

20 pounds	Broccoli, washed, trimmed, stems peeled
2½ gallons	Mornay sauce, heated to 165° F (see page 37)
20 ounces	Cheddar cheese, grated
20 ounces	Butter, melted
20 ounces	Bread crumbs, sifted

METHOD OF PREPARATION:

1. Preheat the oven to 350° F.
2. Cut the broccoli into flowers, and cut the stems on a **bias.** Steam to **al dente,** about 6 minutes. Shock in an ice bath, and hold.
3. Place the broccoli florets in a hotel pan, and mix with Mornay sauce.
4. Place in a 350° F oven until heated through. Remove, and sprinkle with cheddar cheese, melted butter, and bread crumbs.
5. Place under a salamander to brown.

COOKING TECHNIQUES:

Boil, Bake

Boil: (at sea level)

1. Bring the cooking liquid to a rapid boil.
2. Stir the contents, and cook the food product throughout.
3. Serve hot.

Bake:

1. Preheat the oven.
2. Place the food product on the appropriate rack.

GLOSSARY:

Bias: at an angle
Al dente: to the bite

HAZARDOUS FOODS:

Cheese
Mornay sauce

NUTRITION:

Calories: 539
Fat: 42.1 g
Protein: 22.2 g

Broccoli Hollandaise

YIELD: 10 SERVINGS. SERVING SIZE: FOUR OUNCES.

COOKING TECHNIQUE:

Boil

Boil: (at sea level)

1. Bring the cooking liquid to a rapid boil.
2. Stir the contents, and cook the food product throughout.
3. Serve hot.

GLOSSARY:

Nappé: coated
Al dente: to the bite

NUTRITION:

Calories: 103
Fat: 4.97 g
Protein: 6.48 g

INGREDIENTS:

4 pounds	Broccoli, trimmed to 3-inch heads
	Cold salted water, as needed
20 ounces	Hollandaise sauce, heated to 140° F (see page 30)

METHOD OF PREPARATION:

1. Place the broccoli in boiling, salted water. Cook until **al dente.**
2. Drain well, and hold at 140° F.
3. To serve, place a portion of broccoli on a preheated plate, and **nappé** with the sauce.

CHEF NOTES:

1. If the broccoli is held for a long period in a bain-marie, its color and texture will be lost.
2. As an alternate, shock the broccoli in an ice water bath, drain, and reserve. For service, reheat it in boiling water.

Broccoli with Lemon Butter

COOKING TECHNIQUES:
Boil, Sauté

Boil: (at sea level)
1. Bring the cooking liquid to a rapid boil.
2. Stir the contents, and cook the food product throughout.
3. Serve hot.

Sauté:
1. Heat the sauté pan to the appropriate temperature.
2. Evenly brown the food product.
3. For a sauce, pour off any excess oil, reheat, and deglaze.

GLOSSARY:
Al dente: to the bite
À la minute: cooked to order

NUTRITION:
Calories: 104
Fat: 7.84 g
Protein: 4.17 g

YIELD: 50 SERVINGS. SERVING SIZE: FOUR OUNCES.

INGREDIENTS:

15 pounds	Broccoli, washed, trimmed, stems peeled
1 pound	Butter
2 each	Lemon, peeled, membranes removed, and chopped
	Salt and ground white pepper, to taste
	Worcestershire sauce, to taste
½ ounce	Lemon juice

METHOD OF PREPARATION:

1. In a stockpot, cook the broccoli in boiling, salted water until **al dente.** Drain. Shock in an ice bath. Remove, and separate into florets.
2. Sauté the broccoli florets **à la minute** in butter.
3. Add the chopped lemon, lemon juice, and Worcestershire sauce to the butter in the pan, and season. Pour this sauce over the broccoli florets when serving.

Broccoli Polonaise

YIELD: 10 SERVINGS. SERVING SIZE: FOUR OUNCES.

COOKING TECHNIQUES:
Boil, Sauté

Boil: (at sea level)
1. Bring the cooking liquid to a rapid boil.
2. Stir the contents, and cook the food product throughout.
3. Serve hot.

Sauté:
1. Heat the sauté pan to the appropriate temperature.
2. Evenly brown the food product.
3. For a sauce, pour off any excess oil, reheat, and deglaze.

GLOSSARY:
Al dente: to the bite

HAZARDOUS FOOD:
Eggs

NUTRITION:
Calories: 223
Fat: 12.9 g
Protein: 7.64 g

INGREDIENTS:

5 pounds	Broccoli, washed, trimmed, and stems peeled
	Salted water, to cover

Garniture Polonaise:

5 ounces	Butter
8 ounces	Bread crumbs
5 each	Eggs, hard-boiled, peeled, and grated
3 ounces	Fresh parsley, washed, excess moisture removed and chopped
	Salt and ground white pepper, to taste
3 ounces	Clarified butter

METHOD OF PREPARATION:

1. In a stockpot, cook the broccoli in boiling, salted water until **al dente;** then drain, and shock in an ice bath. When cold, drain and remove.
3. In a sauté pan, melt 5 ounces of the butter. Add the bread crumbs, and sauté until golden brown, tossing frequently; then remove and cool.
4. Add the eggs and parsley to the bread crumbs, and season, to taste.
5. For service, heat the broccoli in the clarified butter, and garnish with the polonaise.

CHEF NOTE:
To prepare hard-boiled eggs, place the eggs in a pan with cold water; heat to a boil, and simmer for 10 minutes. Drain; then cool under cold running water.

Broiled Tomatoes

YIELD: 10 SERVINGS. SERVING SIZE: ½ TOMATO.

INGREDIENTS:

	Salt and ground white pepper, to taste
1 teaspoon	Garlic powder
5 each	Beefsteak tomatoes, washed, cored, and split in half
5 ounces	Butter, cut into ½-ounce pieces

METHOD OF PREPARATION:

1. Preheat the broiler.
2. Mix together the salt, pepper, and garlic powder, and season the tomato halves on the cut side.
3. Place one piece of butter on each tomato, and broil to order.

COOKING TECHNIQUE:
Broil

Grill/Broil:
1. Clean and heat the grill/broiler.
2. To prevent sticking, brush the food product with oil.

HACCP:
Hold at 140° F.

NUTRITION:
Calories: 115
Fat: 11.7 g
Protein: .681 g

CHEF NOTES:
1. Do not broil the tomatoes too early, or they will collapse under the holding temperature of 140° F.
2. In large-quantity preparation, heat the oven to 400° F, and bake the tomatoes 10 minutes or less, depending on their size.

Brussel Sprouts with Bacon

YIELD: 50 SERVINGS. SERVING SIZE: FOUR OUNCES.

INGREDIENTS:

12½ pounds	Brussel sprouts, outer leaves removed, trimmed, bottoms scored, and washed
2 pounds	Smoked bacon, diced **brunoise**
1 pound	Onion, peeled and diced brunoise
	Salt and ground black pepper, to taste

METHOD OF PREPARATION:

1. Place the brussel sprouts in boiling, salted water. Cook for about 5 minutes, until tender. Drain, and shock in ice water. Remove when cold.
2. In a sauté pan, **render** the bacon until lightly browned. Add the onions, and sauté until they are translucent.
3. Add the cooked brussel sprouts, and sauté until they are well heated.
4. Season with salt and ground black pepper. Serve hot.

COOKING TECHNIQUES:
Boil, Sauté

Boil: (at sea level)
1. Bring the cooking liquid to a rapid boil.
2. Stir the contents, and cook the food product throughout.
3. Serve hot.

Sauté:
1. Heat the sauté pan to the appropriate temperature.
2. Evenly brown the food product.
3. For a sauce, pour off any excess oil, reheat, and deglaze.

GLOSSARY:
Brunoise: ⅛-inch dice
Render: to melt fat

HAZARDOUS FOOD:
Smoked bacon

NUTRITION:
Calories: 152
Fat: 9.52 g
Protein: 8.55 g

CHEF NOTE:
The bacon is optional.

Buttered Broccoli

YIELD: 50 SERVINGS. SERVING SIZE: FIVE OUNCES.

INGREDIENTS:

15 pounds	Broccoli, washed and trimmed, stems peeled
1 pound	Butter
	Salt and ground white pepper, to taste
1 teaspoon	Garlic powder

METHOD OF PREPARATION:

1. Place the broccoli in boiling, salted water or a steamer. Cook until **al dente.** Drain, and then shock in an ice bath. When cold, remove and separate into flowerets. Refrigerate.
2. Heat the butter in a sauté pan. Add the broccoli, and sauté. Season, and serve.

COOKING TECHNIQUES:

Boil, Sauté

Boil: (at sea level)

1. Bring the cooking liquid to a rapid boil.
2. Stir the contents, and cook the food product throughout.
3. Serve hot.

Sauté:

1. Heat the sauté pan to the appropriate temperature.
2. Evenly brown the food product.
3. For a sauce, pour off any excess oil, reheat, and deglaze.

GLOSSARY:

Al dente: to the bite

NUTRITION:

Calories: 103
Fat: 7.83 g
Protein: 4.2 g

Buttered Fresh Corn on the Cob

YIELD: 50 SERVINGS. SERVING SIZE: ONE EAR OF CORN.

INGREDIENTS

25 ears	Fresh corn, silks and husks removed
	Salted water, to cover and boil
	Sugar, as needed
	Butter, as needed, melted
	Salt and ground white pepper, to taste

METHOD OF PREPARATION:

1. In a large stockpot, heat the salted water to a boil. Add the sugar and butter.
2. Add the corn to the pot, and bring back to a boil. Cook for 6 to 12 minutes until the kernels are tender.
3. Remove from the heat, and hold for service in the cooking liquid. Season, to taste. Serve with melted butter.

COOKING TECHNIQUE:
Boil

Boil: (at sea level)
1. Bring the cooking liquid to a rapid boil.
2. Stir the contents, and cook the food product throughout.
3. Serve hot.

NUTRITION:
Calories: 68.5
Fat: 4.08 g
Protein: 1.09 g

CHEF NOTES:
1. Young, fresh corn should cook in about 6 minutes. When cooking older corn, add sugar to the water for sweetness.
2. Corn on the cob also is available individually quick frozen (IQF).

Buttered Spinach

(Epinard en Brache au Beurre)

YIELD: 16 SERVINGS. SERVING SIZE: FOUR OUNCES.

COOKING TECHNIQUE:
Sauté

Sauté:
1. Heat the sauté pan to the appropriate temperature.
2. Evenly brown the food product.
3. For a sauce, pour off any excess oil, reheat, and deglaze.

NUTRITION:
Calories: 184
Fat: 17 g
Protein: 4.3 g

INGREDIENTS:

5 pounds	Spinach, fresh, washed, stems removed
12 ounces	Butter
	Salt and pepper, to taste

METHOD OF PREPARATION:

1. In a saucepan, place the spinach in a small amount of boiling, salted water. Cook only until wilted. Shock in ice water. Drain well and reserve.
2. Sauté the spinach in butter in a large skillet, and season with salt and pepper.

Calabrian Tomatoes

YIELD: 10 SERVINGS. SERVING SIZE: ONE-HALF TOMATO.

COOKING TECHNIQUES:

Bake, Broil

Bake:

1. Preheat the oven.
2. Place the food product on the appropriate rack.

Grill/Broil:

1. Clean and heat the grill/broiler.
2. To prevent sticking, brush the food product with oil.

NUTRITION:

Calories: 290
Fat: 23.9 g
Protein: 4.35 g

INGREDIENTS:

5 medium-sized	Ripe tomatoes, washed, cored, and split in half lengthwise
	Croutons
1 clove	Garlic, peeled and mashed into a purée
2 ounces	Fresh Italian (flat-leafed) parsley, washed, excess moisture removed, and chopped
1 teaspoon	Thyme leaves, dried
	Salt and freshly ground black pepper, to taste
3 ounces	Olive oil

METHOD OF PREPARATION:

1. Preheat the oven to 350° F.
2. Using a Parisienne scoop, scoop out the seeds and some of the center pulp from the tomato halves. Discard the seeds, and dice the pulp.
3. In a food processor, combine the croutons, garlic, parsley, thyme, salt and pepper, to taste, and olive oil. Process until the croutons become crumbs and the seasonings are well blended. Fold in the tomato pulp.
4. Place the tomato halves on a baking sheet, and divide crumb mixture over the tomatoes, filling the cavities.
5. Place the tomatoes in the oven, and bake until cooked throughout but still firm. Hold at room temperature.
6. For service, place under a broiler for 30 seconds to 1 minute, or until heated throughout and crumbs golden brown.

CHEF NOTE:

It is important not to overbake the tomatoes, because they will become too soft and collapse.

Caramelized Squash or Pumpkin

YIELD: 10 SERVINGS. SERVING SIZE: FOUR OUNCES.

INGREDIENTS:

3 pounds	Winter squash or pumpkin, peeled, seeded, and cut into a 1-inch dice
6 ounces	Brown sugar
3 ounces	Butter
½ teaspoon	Cinnamon

METHOD OF PREPARATION:

1. **Blanch** the squash or pumpkin in boiling, salted water, or steam until **al dente.** Drain, and hold at 140° F or above.
2. In a sauté pan, melt the sugar with the butter, and cook until a light caramel.
3. Fold the squash into the caramel, sprinkle with cinnamon, and serve immediately on a preheated plate, or hold at 140° F or above.

COOKING TECHNIQUE:

Sauté

Sauté:

1. Heat the sauté pan to the appropriate temperature.
2. Evenly brown the food product.
3. For a sauce, pour off any excess oil, reheat, and deglaze.

GLOSSARY:

Blanch: to parcook
Al dente: to the bite

HACCP:

Hold warm at 140° F or above.

NUTRITION:

Calories: 152
Fat: 6.99 g
Protein: 1.05 g

Carrots and Cabbage

YIELD: 50 SERVINGS. SERVING SIZE: FOUR OUNCES.

INGREDIENTS:

4 pounds	Cabbage, washed, cored, and cut **chiffonade**
8 pounds	Carrots, washed, peeled, and cut on a bias
	Salt and ground white pepper, to taste
2 gallons	Water or vegetable stock (see page 11)
20 ounces	Clarified butter
12 ounces	Freshly chopped parsley, excess moisture removed

METHOD OF PREPARATION:

1. In a stockpot, place the carrots in cold, salted water or vegetable stock. Heat to a boil, and cook until **al dente.** Drain, shock in an ice bath, drain again, and refrigerate.
2. In a sauté pan, heat the butter. Sauté the cabbage until tender. Add the carrots.
3. Season the mixture with salt, pepper, and parsley.
4. Repeat steps 2 and 3 as necessary.

COOKING TECHNIQUES:

Boil, Sauté

Boil: (at sea level)

1. Bring the cooking liquid to a rapid boil.
2. Stir the contents, and cook the food product throughout.
3. Serve hot.

Sauté:

1. Heat the sauté pan to the appropriate temperature.
2. Evenly brown the food product.
3. For a sauce, pour off any excess oil, reheat, and deglaze.

GLOSSARY:

Chiffonade: ribbons of leafy greens
Al dente: to the bite

CHEF NOTE:

This recipe may replace the vegetable in the corned beef recipe.

Carrot Soufflé

(Soufflé aux Carottes)

YIELD: 10 SERVINGS. SERVING SIZE: FOUR OUNCES.

INGREDIENTS:

12 ounces	Potatoes, washed and peeled
	Cold salted water, as needed
1 pound	Carrots, washed and peeled
1 tablespoon	Sugar, or to taste
	Fresh lemon juice, to taste
5 each	Eggs, cracked and separated
	Salt, to taste
¼ teaspoon	Nutmeg, ground
	Butter, as needed

METHOD OF PREPARATION:

1. Preheat the oven to 400° F.
2. In a saucepan, place the potatoes in water, and heat to a boil. Cook until **fork-tender.**
3. In a separate saucepan, place the carrots, sugar, and lemon juice, and add water to cover. Heat the water to a boil, and cook until the carrots are tender.
4. Drain each vegetable well, and pass them through a food mill until puréed.
5. Add the yolks to the puréed vegetables, and mix well. Season with salt and nutmeg.
6. Whip the egg whites to a peak; then fold into the purée.
7. Butter a soufflé dish, fill two thirds with purée, and place in a **bain-marie.** Bake in the oven for 8 to 10 minutes, or until puffed and lightly browned.
8. Serve immediately on preheated plates.

COOKING TECHNIQUES:

Boil, Bake

Boil: (at sea level)

1. Bring the cooking liquid to a rapid boil.
2. Stir the contents, and cook the food product throughout.
3. Serve hot.

Bake:

1. Preheat the oven.
2. Place the food product on the appropriate rack.

GLOSSARY:

Fork-tender: without resistance
Bain-marie: hot-water bath

HAZARDOUS FOOD:

Eggs

NUTRITION:

Calories: 103
Fat: 4.64 g
Protein: 3.86 g

CHEF NOTE:

Soufflés can be prepared in individual soufflé dishes, but reduce the cooking time.

Carrots with Raisins

YIELD: 10 SERVINGS. SERVING SIZE: FOUR OUNCES.

INGREDIENTS:

2½ pounds	Carrots, washed, peeled, and grated
3 ounces	Butter, melted
2 ounces	Granulated sugar
3 ounces	Spanish raisins
8 ounces	Grape juice

METHOD OF PREPARATION:

1. In a sauté pan, combine the carrots, butter, and sugar. Sauté until the sugar melts.
2. Add all of the other ingredients, and mix well; cover, and simmer until the carrots are tender. Hold at 140° F or above.

COOKING TECHNIQUES:

Sauté, Simmer

Sauté:

1. Heat the sauté pan to the appropriate temperature.
2. Evenly brown the food product.
3. For a sauce, pour off any excess oil, reheat, and deglaze.

Simmer and Poach:

1. Heat the cooking liquid to the proper temperature.
2. Submerge the food product completely.
3. Keep the cooked product moist and warm.

HACCP:

Hold at 140° F or above.

NUTRITION:

Calories: 182
Fat: 7.21 g
Protein: 1.9 g

CHEF NOTE:

This dish also can be made with the zest and juice of Valencia oranges.

Carrots with Tarragon

YIELD: 50 SERVINGS. SERVING SIZE: FOUR OUNCES.

INGREDIENTS:

12 pounds	Carrots, washed, peeled, and cut on a bias
	Salt, to taste
1 pound	Sugar
2 ounces	Lemon juice
1 pound	Butter
2 tablespoons	Fresh tarragon, washed, dried, and chopped
2 ounces	Fresh parsley, washed, dried, and chopped

METHOD OF PREPARATION:

1. In a saucepan, place the carrots in cold water. Heat to a boil. Cook **al dente.** Drain, and shock in an ice bath. Remove when cold, and refrigerate until service.
2. In a sauté pan, heat the butter, and sauté the carrots.
3. Season the carrots with salt, sugar, lemon juice, and tarragon.
4. Sprinkle parsley on the carrots when serving.

COOKING TECHNIQUES:

Boil, Sauté

Boil: (at sea level)

1. Bring the cooking liquid to a rapid boil.
2. Stir the contents, and cook the food product throughout.
3. Serve hot.

Sauté:

1. Heat the sauté pan to the appropriate temperature.
2. Evenly brown the food product.
3. For a sauce, pour off any excess oil, reheat, and deglaze.

GLOSSARY:

Al dente: to the bite

NUTRITION:

Calories: 148
Fat: 7.58 g
Protein: 1.27 g

Cauliflower Mimosa

YIELD: 50 SERVINGS. SERVING SIZE: FOUR OUNCES.

INGREDIENTS:

8 heads	Cauliflower, washed, soaked in salt water for 1 hour
	Water and oil, as needed
2 pounds	Butter, clarified

METHOD OF PREPARATION:

1. In a stockpot, place the cauliflower in boiling, unsalted water with lemon. Cook until **al dente.** Drain, shock in an ice bath, and remove when cold. Cut out the hard core, and separate the cauliflower into florets.
2. Heat up the clarified butter, and sauté the cauliflower until hot.
3. Garnish with Mimosa butter.

Mimosa Butter

INGREDIENTS:

1 pint	Butter, melted
5 each	Whole eggs, hard-boiled, peeled, and grated
5 tablespoons	Freshly chopped parsley, excess moisture removed
	Salt and ground white pepper, to taste

METHOD OF PREPARATION:

1. Melt the butter in a sauté pan.
2. Stir the eggs into the butter.
3. Add the parsley, and season, to taste.

COOKING TECHNIQUES:

Boil, Sauté

Boil: (at sea level)

1. Bring the cooking liquid to a rapid boil.
2. Stir the contents, and cook the food product throughout.
3. Serve hot.

Sauté:

1. Heat the sauté pan to the appropriate temperature.
2. Evenly brown the food product.
3. For a sauce, pour off any excess oil, reheat, and deglaze.

GLOSSARY:

Al dente: to the bite

HAZARDOUS FOOD:

Whole eggs

NUTRITION:

Calories: 134
Fat: 15 g
Protein: .306 g

Cauliflower Polonaise

YIELD: 10 SERVINGS. SERVING SIZE: FOUR OUNCES.

INGREDIENTS:

1 large or 2 medium heads	Cauliflower
	Water with lemon, to cover
3 ounces	Clarified butter

Garniture Polonaise:

4 ounces	Butter
8 ounces	Bread crumbs
4 each	Eggs, hard-boiled, peeled, and grated
3 ounces	Fresh parsley, washed, excess moisture removed, and chopped
	Salt and ground white pepper, to taste

METHOD OF PREPARATION:

1. Wash and soak the cauliflower in salted water for 1 hour; then drain and rinse.
2. In a stockpot, place the cauliflower, and cover with boiling water with lemon. Cook until the heads are **al dente;** then drain, and shock in an ice bath. When cold, drain and remove, and separate into florets.
3. In a sauté pan, melt 4 ounces of the butter. Add the bread crumbs, and sauté until golden brown, tossing frequently; then remove and cool.
4. Add the eggs and parsley to the bread crumbs, and season, to taste.
5. For service, sauté the cauliflower in the clarified butter, and garnish with the polonaise. Serve immediately, or hold at 140° F or above.

COOKING TECHNIQUES:

Boil, Sauté

Boil: (at sea level)

1. Bring the cooking liquid to a rapid boil.
2. Stir the contents, and cook the food product throughout.
3. Serve hot.

Sauté:

1. Heat the sauté pan to the appropriate temperature.
2. Evenly brown the food product.
3. For a sauce, pour off any excess oil, reheat, and deglaze.

GLOSSARY:

Al dente: to the bite

HAZARDOUS FOOD:

Eggs

NUTRITION:

Calories: 295
Fat: 20.9 g
Protein: 7.01 g

CHEF NOTE:

Soaking the cauliflower helps to remove any imbedded insects.

Cauliflower with Mornay Sauce

(Chou-Fleur Mornay)

YIELD: 10 SERVINGS. SERVING SIZE: FOUR OUNCES.

INGREDIENTS:

2 heads	Cauliflower
	Salted water, to cover
	Water, to cover
1 each	Lemon, sliced
1 quart	Mornay sauce (see page 37)

METHOD OF PREPARATION:

1. Clean the cauliflower. Soak in salted water for 1 hour.
2. Boil the cauliflower in unsalted water with the lemon slices until tender. Drain, and separate into florets.
3. **Nappé** cauliflower florets with the Mornay sauce. Brown under a broiler or salamander.

COOKING TECHNIQUE:

Boil

Boil: (at sea level)

1. Bring the cooking liquid to a rapid boil.
2. Stir the contents, and cook the food product throughout.
3. Serve hot.

GLOSSARY:

Nappé: coated

NUTRITION:

Calories: 150
Fat: 11 g
Protein: 6.88 g

Cauliflower with Nuts

YIELD: 10 SERVINGS. SERVING SIZE: FOUR OUNCES.

INGREDIENTS:

2 heads	Cauliflower
	Salted water, as needed to soak cauliflower
4 ounces	Olive oil
2 each	Medium-sized onions, peeled and diced **brunoise**
3 ounces	Pistachio nuts, shelled, peeled, and coarsely chopped
1 pint	White chicken stock, seasoned, heated to a boil (see page 12)
	Salt and ground white pepper, to taste
3 ounces	Fresh parsley, washed, excess moisture removed, and chopped
5 ounces	Feta cheese, crumbled

METHOD OF PREPARATION:

1. In a saucepan, place the cauliflower, and cover with water. Heat to a boil, and cook **al dente.**
2. Drain, and shock in an ice water bath; then separate into florets.
3. In a sauté pan, heat the oil, and sauté the onions until they become translucent. Add chopped nuts, and sauté for 2 minutes more.
4. Add the cauliflower and chicken stock. Heat to a boil, and simmer until the cauliflower is tender; season, to taste.
5. Combine the parsley and cheese.
6. Serve immediately, or hold at 140° F or above, and garnish with the parsley and cheese mixture.

COOKING TECHNIQUES:

Sauté, Simmer

Sauté:

1. Heat the sauté pan to the appropriate temperature.
2. Evenly brown the food product.
3. For a sauce, pour off any excess oil, reheat, and deglaze.

Simmer and Poach:

1. Heat the cooking liquid to the proper temperature.
2. Submerge the food product completely.
3. Keep the cooked product moist and warm.

GLOSSARY:

Brunoise: ⅛-inch dice
Al dente: to the bite

HACCP:

Hold at 140° F or above.

NUTRITION:

Calories: 207
Fat: 18.7 g
Protein: 4.78 g

Cherry Tomatoes with Cilantro

YIELD: 10 SERVINGS. SERVING SIZE: FOUR TOMATOES.

INGREDIENTS:

40 each	Cherry tomatoes, washed, well drained, and cut in half
4 ounces	Butter
6 cloves	Garlic, peeled and minced
	Salt, to taste
3 ounces	Fresh cilantro leaves, washed and chopped

METHOD OF PREPARATION:

1. In a sauté pan, heat the butter as needed; then add the tomatoes, garlic, and salt.
2. Toss together, and sauté until the tomatoes are thoroughly heated.
3. Sprinkle the tomatoes with chopped cilantro, and serve immediately.

COOKING TECHNIQUE:

Sauté

Sauté:

1. Heat the sauté pan to the appropriate temperature.
2. Evenly brown the food product.
3. For a sauce, pour off any excess oil, reheat, and deglaze.

GLOSSARY:

À la minute: cooked to order
Bain-marie: hot-water bath

NUTRITION:

Calories: 99.9
Fat: 9.48 g
Protein: .993 g

CHEF NOTE:

Tomatoes should be cooked **à la minute.** Do not hold sautéed tomatoes in a **bain-marie,** because they will become soft and fall apart.

Eggplant Maite

YIELD: 10 SERVINGS. SERVING SIZE: FOUR OUNCES.

INGREDIENTS:

4 each	Small eggplants, washed and trimmed
1½ pounds	Chorizo (Spanish sausage), thinly sliced
6 ounces	Flour
2 ounces	Water
1 tablespoon	White vinegar
6 each	Eggs, cracked and lightly beaten
	Salt, to taste
	Olive oil, as needed (1-inch deep in cooking vessel), heated to 350° F

METHOD OF PREPARATION:

1. Slice the eggplant into ¼-inch-thick rounds. Sprinkle with salt and let stand at room temperature for 1 hour. Pat the slices dry.
2. Sandwich the slices of chorizo between two slices of eggplant.
3. Prepare a batter with the flour, water, vinegar, eggs, and salt.
4. Coat the eggplant sandwiches with the batter, and fry until golden brown.
5. Drain, and serve immediately, or hold minimally at 140° F.

COOKING TECHNIQUE:

Shallow-Fry

Shallow-Fry:

1. Heat the cooking medium to the proper temperature.
2. Cook the food product throughout.
3. Season, and serve hot.

HACCP:

Heat to 350° F.
Hold at 140° F.

HAZARDOUS FOOD:

Eggs

NUTRITION:

Calories: 257
Fat: 35.4 g
Protein: 23.7 g

CHEF NOTES:

1. Spanish chorizo is prepared using smoked pork, as opposed to Mexican style, which uses fresh pork.
2. Soaking the eggplant in salted water removes its bitterness.

Eggplant Scapece

YIELD: 10 SERVINGS. SERVING SIZE: FOUR OUNCES.

COOKING TECHNIQUE:

Shallow-Fry

Shallow-Fry:

1. Heat the cooking medium to the proper temperature.
2. Cook the food product throughout.
3. Season, and serve hot.

GLOSSARY:

Chiffonade: ribbons of leafy greens
Nappé: coated

HACCP:

Chill to 40° F or lower.

HAZARDOUS FOOD:

Anchovies

NUTRITION:

Calories: 159
Fat: 9.38 g
Protein: 3 g

INGREDIENTS:

	Oil, as needed
4 medium-sized	Eggplants, trimmed and cut into ¼-inch slices
2 ounces	Fresh basil leaves, cut **chiffonade**

Anchovy Dressing:

8 fillets	Anchovies, drained of oil
2 cloves	Garlic, peeled and mashed into a purée
1 ounce	Fresh parsley, washed, excess moisture removed, and chopped
2 tablespoons	Granulated sugar
1 teaspoon	Hot pepper paste, such as Thai chili sauce
	Salt and freshly ground black pepper, to taste
1 cup	Red wine vinegar

METHOD OF PREPARATION:

1. In a saucepan, add ½-inch of oil, and heat to 350° F.
2. Fry the eggplant slices. Drain the slices on towels.
3. Purée together the anchovies, garlic, parsley, and hot pepper paste, and season, to taste. Add the sugar and vinegar, and blend well to form a dressing.
4. Arrange the slices of eggplant, overlapping, in a circle on individual salad plates. **Nappé** with anchovy dressing, and chill to 40° F or lower for a minimum of 1½ hours. Return to room temperature for service.
5. Garnish with fresh basil before serving.

CHEF NOTES:

1. A combination of zucchini and eggplant can be used as an alternative.
2. Finely diced red onion can be added to the dressing.
3. If hot pepper paste is not available, substitute a few drops of hot sauce mixed with 1 teaspoon of tomato paste.

COOKING TECHNIQUE:
Braise

Braise:
1. Heat the braising pan to the proper temperature.
2. Sear and brown the food product to a golden color.
3. Degrease and deglaze.
4. Cook the food product in two-thirds liquid until fork-tender.

GLOSSARY:
Brunoise: ⅛-inch dice
Blanch: to parcook

HACCP:
Hold at 140° F or above.

NUTRITION:
Calories: 127
Fat: 5.56 g
Protein: 3.78 g

Farmer's-Style Celery

(Celeri à la Fermière)

YIELD: 10 SERVINGS. SERVING SIZE: FOUR OUNCES.

INGREDIENTS:

3 heads	Celery, washed
2 ounces	Butter
4 ounces	Onions, peeled and diced **brunoise**
4 ounces	Carrots, peeled and diced brunoise
4 ounces	Ham, diced brunoise
1 quart	Chicken demi-glacé, heated to a boil (see page 26)
	Salt and ground white pepper, to taste

METHOD OF PREPARATION:

1. Preheat the oven to 350° F.
2. Trim the celery, and remove the leaves. Cut into 3-inch lengths and **blanch.** Drain, cool, and place in a hotel pan.
3. In a sauté pan, melt the butter, add the onions, and sauté until translucent. Add the carrots, sauté 2 more minutes.
4. Add the ham and demi-glacé to the sautéed onions and carrots. Heat the liquid to a boil. Season, to taste, and pour over the celery.
5. Braise in the oven for 25 to 30 minutes, or until the celery is tender. Serve immediately, or hold at 140° F or above.

Farmer's-Style Peas

(Petits Pois Fermière)

YIELD: 10 SERVINGS. SERVING SIZE: FOUR OUNCES.

INGREDIENTS:

4 ounces	Butter
6 ounces	Onions, peeled and diced **brunoise**
6 ounces	Mushrooms, washed, stemmed, and sliced
6 ounces	Ham, diced brunoise
6 ounces	Boston lettuce, washed, trimmed, and cut **chiffonade**
16 ounces	Heavy cream
2 pounds	Frozen green peas, thawed and steamed
	Salt and freshly ground black pepper, to taste

METHOD OF PREPARATION:

1. In a sauté pan, melt the butter, and sauté onions until translucent.
2. Add the mushrooms, ham, and lettuce. Continue to sauté until the lettuce is wilted.
3. Temper with the heavy cream, add to the mixture, and reduce to a proper consistency.
4. Add the peas, season, to taste, and serve immediately, or hold at 140° F or above.

COOKING TECHNIQUE:

Sauté

Sauté:

1. Heat the sauté pan to the appropriate temperature.
2. Evenly brown the food product.
3. For a sauce, pour off any excess oil, reheat, and deglaze.

GLOSSARY:

Brunoise: ⅛-inch dice
Chiffonade: ribbons of leafy greens

HACCP:

Hold at 140° F or above.

HAZARDOUS FOOD:

Heavy cream

NUTRITION:

Calories: 190
Fat: 10.5 g
Protein: 9.16 g

CHEF NOTE:

This recipe can be prepared with pearl onions, carrots, and a chiffonade of lettuce, but under the same name, *fermière.*

Fava Bean Sauté

YIELD: 10 SERVINGS. SERVING SIZE: FOUR OUNCES.

INGREDIENTS:

1½ pounds	Dried fava beans, sorted, rinsed, soaked in cold water overnight, and then drained
	Salt and freshly ground black pepper, to taste
3 ounces	Olive oil
2 ounces	Fresh coriander leaves, washed and chopped

METHOD OF PREPARATION:

1. In a saucepan, place the beans, and add cold water to cover. Heat the liquid to a boil.
2. Once boiling, **dépouiller;** then reduce the heat to a simmer.
3. Cook the beans until tender, or about 45 minutes. When tender, drain, and hold covered at 140° F. Just before service, heat the olive oil in a sauté pan, add the beans, and toss together; then season and add the coriander. Toss well, and serve immediately.

COOKING TECHNIQUES:

Boil, Sauté

Boil: (at sea level)

1. Bring the cooking liquid to a rapid boil.
2. Stir the contents, and cook the food product throughout.
3. Serve hot.

Sauté:

1. Heat the sauté pan to the appropriate temperature.
2. Evenly brown the food product.
3. For a sauce, pour off any excess oil, reheat, and deglaze.

GLOSSARY:

Dépouiller: to skim impurities/grease

HACCP:

Hold at 140° F.

NUTRITION:

Calories: 138
Fat: 8.35 g
Protein: 5.17 g

CHEF NOTE:

If fresh coriander is not available, use ground coriander, and garnish with chopped fresh parsley.

Fava Beans with Garlic

YIELD: 10 SERVINGS. SERVING SIZE: FOUR OUNCES.

INGREDIENTS:

2 quarts	Salted water
2 pounds	Frozen fava beans, thawed
3 ounces	Olive oil
5 cloves	Garlic, peeled and finely minced
12 ounces	Canned pimientos, drained and diced **brunoise**
1 ounce	Fresh marjoram leaves, washed and minced
	Salt and freshly ground black pepper, to taste

METHOD OF PREPARATION:

1. In a saucepan, heat the water to a boil; add the beans, and cook until tender; then drain.
2. In a sauté pan, heat the oil; add the garlic, and sauté for 2 to 3 minutes. Add the pimientos, and sauté for 1 minute more; then add the beans.
3. Add the marjoram, season, to taste, and heat thoroughly.
4. Serve immediately, or hold at 140° F or above.

COOKING TECHNIQUES:

Boil, Sauté

Boil: (at sea level)

1. Bring the cooking liquid to a rapid boil.
2. Stir the contents, and cook the food product throughout.
3. Serve hot.

Sauté:

1. Heat the sauté pan to the appropriate temperature.
2. Evenly brown the food product.
3. For a sauce, pour off any excess oil, reheat, and deglaze.

GLOSSARY:

Brunoise: ⅛-inch dice
Brain-marie: hot-water bath

HACCP:

Hold at 140° F or above.

NUTRITION:

Calories: 395
Fat: 10 g
Protein: 24.2 g

CHEF NOTE:

Because this is a "green" vegetable, color and texture will be lost if held in a **bain-marie.**

Fried Cauliflower

YIELD: 50 SERVINGS. SERVING SIZE: FOUR OUNCES.

INGREDIENTS:

8 heads	Cauliflower, washed, soaked in salt water for 1 hour
	Water, to cover
	Oil, as needed

Batter:

2½ pints	Milk
10 each	Eggs
2 pounds	Flour
	Salt, to taste
	Seasoned flour for dredging, as needed

METHOD OF PREPARATION:

1. In a stockpot, place the cauliflower in boiling, unsalted water with lemon. Cook until **al dente.** Drain, shock in an ice bath, and remove when cold. Cut out the hard core, and separate the cauliflower into florets.
2. For the batter, place the milk in a mixing bowl, and add the eggs and flour. Mix well, and season.
3. Coat the cauliflower in the flour, shaking off any excess.
4. Dip the cauliflower in the batter, and deep-fry in the oil at 350° F **à la minute.** Place the fried cauliflower on absorbent paper to soak up the fat, and serve.

COOKING TECHNIQUES:
Boil, Deep-Fry

Boil: (at sea level)
1. Bring the cooking liquid to a rapid boil.
2. Stir the contents, and cook the food product throughout.
3. Serve hot.

Deep-Fry:
1. Heat the frying liquid to the proper temperature.
2. Submerge the food product completely.
3. Fry the product until it is cooked throughout.

GLOSSARY:
Al dente: to the bite
À la minute: cooked to order

HAZARDOUS FOODS:
Milk
Eggs

NUTRITION:
Calories: 206
Fat: 17.6 g
Protein: 8.46 g

Fried Eggplant Fingers

YIELD: 50 SERVINGS. SERVING SIZE: SIX OUNCES.

COOKING TECHNIQUE:
Deep-Fry

Deep-Fry:
1. Heat the frying liquid to the proper temperature.
2. Submerge the food product completely.
3. Fry the product until it is cooked throughout.

GLOSSARY:
Frite: ¼ × ¼ × 2½-inch cut

HAZARDOUS FOOD:
Egg wash

NUTRITION:
Calories: 300
Fat: 16.4 g
Protein: 6.4 g

INGREDIENTS:

21 pounds	Eggplants, washed, peeled, cut into **frite**
	Salt, as needed
	Seasoned flour, as needed
	Egg wash, as needed
	Bread crumbs, as needed

METHOD OF PREPARATION:

1. To remove bitterness, salt the eggplant and let stand for ½ hour. Rinse; pat dry.
2. Coat the eggplant through the standard breading procedure.
3. Deep-fry, to order, at 350° F, and serve immediately.

Glazed Acorn Squash

YIELD: 50 SERVINGS. SERVING SIZE: FOUR OUNCES.

INGREDIENTS:

25 each	Medium-sized acorn squash, washed, seeds removed, and cut in half lengthwise
2 pounds	Clarified butter, melted
1½ pounds	Brown sugar
	Salt, to taste
2½ teaspoons	Ground ginger
1 pint	Honey

METHOD OF PREPARATION:

1. Preheat the oven to 350° F.
2. Remove a thin slice from the rounded side of each squash half. Brush a sheet pan with some of the butter, and place the squash on the pan, skin-side down.
3. Combine the brown sugar, salt, ginger, and honey, and divide the mixture among the squash cavities.
4. Drizzle the remaining butter over the squash, and place in the oven. Bake until **fork-tender,** or approximately 45 minutes. Halfway through the cooking time, spoon the glaze over the edges of the squash.

COOKING TECHNIQUE:

Bake

Bake:

1. Preheat the oven.
2. Place the food product on the appropriate rack.

GLOSSARY:

Fork-tender: without resistance
Blanch: to parcook

HACCP:

Hold above 140° F.

NUTRITION:

Calories: 226
Fat: 11.4 g
Protein: 1.03 g

CHEF NOTES:

1. Squash can be **blanched** in a streamer and then glazed and baked to finish.
2. Butternut squash or any winter squash can be substituted in this recipe. The method of cutting would be altered, depending on size of the squash.
3. Hold above 140° F.

Glazed Carrots

YIELD: 50 SERVINGS. SERVING SIZE: FOUR OUNCES.

COOKING TECHNIQUES:
Boil, Sauté

Boil: (at sea level)
1. Bring the cooking liquid to a rapid boil.
2. Stir the contents, and cook the food product throughout.
3. Serve hot.

Sauté:
1. Heat the sauté pan to the appropriate temperature.
2. Evenly brown the food product.
3. For a sauce, pour off any excess oil, reheat, and deglaze.

GLOSSARY:
Fork-tender: without resistance

NUTRITION:
Calories: 114
Fat: 4.31 g
Protein: 1.25 g

INGREDIENTS:

13 pounds	Carrots, washed, peeled, and cut into uniform shapes
	Water, as needed
	Salt, to taste
6 ounces	Lemon juice
8 ounces	Clarified butter
12 ounces	Sugar

METHOD OF PREPARATION:

1. Place the carrots in a stockpot with cold water, to cover. Add the salt and lemon juice. Bring to a boil, and cook until **fork-tender.** Drain, rinse under cold water, and hold in a refrigerator.
2. Heat the butter in a skillet. Add the carrots and sugar. Sauté until the carrots become shiny.

CHEF NOTE:
A mixture of a small amount of cornstarch and cold water can be added for shine.

Glazed Turnips

(Navets Glacés)

YIELD: 10 SERVINGS. SERVING SIZE: FOUR OUNCES.

COOKING TECHNIQUE:

Simmer

Simmer and Poach:

1. Heat the cooking liquid to the proper temperature.
2. Submerge the food product completely.
3. Keep the cooked product moist and warm.

GLOSSARY:

Blanch: to parcook
Parmentier: ½-inch dice

NUTRITION:

Calories: 214
Fat: 20.1 g
Protein: .964 g

INGREDIENTS:

1½ gallons	Salted water, heated to a boil
3 pounds	Purple turnips, peeled and dice **parmentier**
5 ounces	Butter
3 ounces	Oil
20 ounces	White beef stock (see page 12)
2 ounces	Sugar
	Salt and ground white pepper, to taste
2 tablespoons	Fresh parsley, washed, excess moisture removed, and chopped

METHOD OF PREPARATION:

1. **Blanch** the turnips in boiling water. Drain, and dry well.
2. In a heavy skillet, heat 3 ounces of butter and 3 ounces of oil. Add the turnips, and sauté for 4 to 5 minutes until lightly browned. Add the beef stock to barely cover the turnips.
3. Add the remaining butter and sugar. Cover, and heat to a boil. Simmer for 25 to 30 minutes, or until the turnips are tender. Season with salt and pepper.
4. If the beef stock is not syrupy, remove the lid and boil the stock rapidly. Toss the turnips until completely coated with a glaze. Sprinkle with parsley, and serve.

Glazed Vegetables with Sultana Raisins

YIELD: 10 SERVINGS. SERVING SIZE: FOUR OUNCES.

INGREDIENTS:

4 ounces	Golden sultana raisins
4 ounces	Sweet white wine
30 each	Miniature carrots, washed and peeled
30 each	Miniature turnips, washed and peeled
½ teaspoon	Thyme, dried
1 each	Bay leaf
1 ounce	Lemon juice
2 ounces	Sugar
	Salted water, to cover
20 each	Pearl onions, peeled and blanched; then soaked in hot water
1 tablespoon	Green peppercorns, drained
	Salt and freshly ground black pepper, to taste

METHOD OF PREPARATION:

1. In a small saucepan, combine the raisins and wine, and allow to stand in a warm (140° F) place for 30 minutes.
2. In a stainless steel **russe,** combine the carrots, turnips, thyme, bay leaf, lemon juice, and 1 ounce of sugar, and add just enough water to cover the vegetables.
3. Heat the water to a boil; cover with a lid, reduce the heat, and simmer for 15 minutes. Then remove the lid, and continue to simmer until all of the liquid evaporates.
4. Remove the bay leaf, and add the butter; drain the pearl onions, and add them; then add the green peppercorns and raisins.
5. Add the remaining sugar, and sauté until the sugar melts and glazes the vegetables.
6. Taste, and add the seasoning, to taste.

COOKING TECHNIQUES:

Simmer, Sauté

Simmer and Poach:

1. Heat the cooking liquid to the proper temperature.
2. Submerge the food product completely.
3. Keep the cooked product moist and warm.

Sauté:

1. Heat the sauté pan to the appropriate temperature.
2. Evenly brown the food product.
3. For a sauce, pour off any excess oil, reheat, and deglaze.

GLOSSARY:

Russe: long-handled saucepan

HACCP:

Hold at 140° F.

NUTRITION:

Calories: 97
Fat: .312 g
Protein: 1.36 g

Green Beans in Garlic Sauce

YIELD: 10 SERVINGS. SERVING SIZE: FOUR OUNCES.

INGREDIENTS:

3 ounces	Butter, melted
8 cloves	Garlic, peeled and minced
1 pound	Canned crushed tomatoes
3 pounds	Fresh green beans, washed, ends trimmed, and cut in half
1 pint	White chicken stock, heated to a boil (see page 12)
	Salt and freshly ground black pepper, to taste

METHOD OF PREPARATION:

1. In a saucepan, place the green beans in boiling, salted water. Cook until **al dente.** Drain and shock in an ice bath. When cold, remove and drain.
2. In a sauté pan, heat the butter, and sauté the garlic. Add the crushed tomatoes, and sauté for 5 minutes.
3. Add the green beans and chicken stock to the tomatoes.
4. Simmer until al dente. Season, to taste, and serve, or hold at 140° F.

COOKING TECHNIQUES:

Sauté, Simmer

Sauté:

1. Heat the sauté pan to the appropriate temperature.
2. Evenly brown the food product.
3. For a sauce, pour off any excess oil, reheat, and deglaze.

Simmer and Poach:

1. Heat the cooking liquid to the proper temperature.
2. Submerge the food product completely.
3. Keep the cooked product moist and warm.

GLOSSARY:

Al dente: to the bite
Bain-marie: hot-water bath

HACCP:

Hold at 140° F.

NUTRITION:

Calories: 132
Fat: 7.93 g
Protein: 3.73 g

CHEF NOTE:

Green vegetables lose their color when held in a **bain-marie.**

Green Beans with Almonds

(Haricots Verts aux Amandes)

YIELD: 10 SERVINGS. SERVING SIZE: FIVE OUNCES.

COOKING TECHNIQUE:

Sauté

Sauté:

1. Heat the sauté pan to the appropriate temperature.
2. Evenly brown the food product.
3. For a sauce, pour off any excess oil, reheat, and deglaze.

GLOSSARY:

À la minute: cooked to order
Al dente: to the bite
Bain-marie: hot-water bath

NUTRITION:

Calories: 156
Fat: 11.8 g
Protein: 3.76 g

INGREDIENTS:

3 pounds	Fresh green beans, washed and both ends trimmed
	Boiling, salted water, as needed
4 ounces	Clarified butter
4 ounces	Sliced almonds, toasted
	Salt and ground white pepper, to taste

METHOD OF PREPARATION:

1. In a saucepan, cook the beans in the salted water until **al dente.** Drain, and shock in an ice bath. Drain, and reserve until needed for service.
2. In a sauté pan, heat 1 tablespoon of butter, and sauté **à la minute.** Add the almonds, and season, to taste.
3. Serve immediately.

CHEF NOTE:

Green vegetables do not keep their color or texture when held in a **bain-marie.**

COOKING TECHNIQUES:

Boil, Sauté

Boil: (at sea level)

1. Bring the cooking liquid to a rapid boil.
2. Stir the contents, and cook the food product throughout.
3. Serve hot.

Sauté:

1. Heat the sauté pan to the appropriate temperature.
2. Evenly brown the food product.
3. For a sauce, pour off any excess oil, reheat, and deglaze.

GLOSSARY:

Al dente: to the bite

NUTRITION:

Calories: 140
Fat: 11.7 g
Protein: 2.11 g

Green Beans with Garlic

(Haricots Verts à l'Ail)

YIELD: 10 SERVINGS. SERVING SIZE: FOUR OUNCES.

INGREDIENTS:

3 pounds	Green beans, washed and both ends trimmed
6 ounces	Butter
6 cloves	Garlic, peeled and minced
	Salt and ground white pepper, to taste

METHOD OF PREPARATION:

1. In a saucepan, place the green beans in boiling, salted water.
2. Cook until **al dente.**
3. Drain, and shock in an ice bath. Remove when cold, and drain.
4. Melt the butter in a sauté pan, and sauté the beans.
5. Add the garlic to the beans, and heat thoroughly. Adjust the seasonings, and serve.

Green Beans with Pimientos

YIELD: 50 SERVINGS. SERVING SIZE:FOUR OUNCES.

COOKING TECHNIQUE:
Sauté

Sauté:
1. Heat the sauté pan to the appropriate temperature.
2. Evenly brown the food product.
3. For a sauce, pour off any excess oil, reheat, and deglaze.

GLOSSARY:
Al dente: to the bite
Bain-marie: hot-water bath

NUTRITION:
Calories: 114
Fat: 8.27 g
Protein: 4.24 g

INGREDIENTS:

1 pound	Bacon, chopped
1 pound	Onions, peeled and finely diced
1 pound	Canned pimientos, drained
8 ounces	Margarine
10 pounds	Green beans, frozen
	Salt and freshly ground black pepper, to taste

METHOD OF PREPARATIONS:

1. Render and sauté the bacon to light brown.
2. Add the diced onions, and sauté until they are translucent.
3. Add the drained pimientos, heat, and set aside.
4. Wash the frozen green beans under running cold water, and drain.
5. Heat the margarine in a tilting skillet, add the drained green beans, and sauté until cooked **al dente.**
6. Add the sautéed pimientos, and incorporate.
7. Season with salt and pepper, to taste.

CHEF NOTES:
1. For additional flavor, fresh minced garlic can be added, or use a tri-mix of salt, pepper, and garlic powder.
2. If the beans are prepared and kept in a **bain-marie** for an extended time, they will become wilted and lose color.
3. Bacon may be removed from this recipe.

Green Beans with Tomatoes

YIELD: 50 SERVINGS. SERVING SIZE: FOUR OUNCES.

INGREDIENTS:

15 pounds	Green beans, washed, ends cut off
1 pound	Bacon, finely chopped
1 pound	Onions, peeled and diced **brunoise**
5 cloves	Garlic, peeled and finely chopped
3 (#2½ cans)	Peeled tomatoes, strained and chopped
	Salt and ground white pepper, to taste

METHOD OF PREPARATION:

1. Cook the green beans in boiling, salted water until **al dente.** Remove from the heat, drain, and shock in an ice bath. Drain when cold, and refrigerate until service.
2. In a sauté pan, render the bacon, and add the onions. Add the garlic and tomatoes to the onions, and sauté. Season, to taste. Simmer for 15 minutes.
3. Add the green beans, and simmer for an additional 5 minutes. Serve when thoroughly heated.

COOKING TECHNIQUES:

Boil, Sauté

Boil: (at sea level)

1. Bring the cooking liquid to a rapid boil.
2. Stir the contents, and cook the food product throughout.
3. Serve hot.

Sauté:

1. Heat the sauté pan to the appropriate temperature.
2. Evenly brown the food product.
3. For a sauce, pour off any excess oil, reheat, and deglaze.

GLOSSARY:

Brunoise: ⅛-inch dice
Al dente: to the bite

NUTRITION:

Calories: 99
Fat: 4.72 g
Protein: 5.35 g

Grilled Fresh Asparagus

YIELD: 10 SERVINGS. SERVING SIZE: FOUR OUNCES.

INGREDIENTS:

3 pounds	Asparagus, washed, trimmed, and stems peeled
	Salt, to taste
5 ounces	Butter, melted

METHOD OF PREPARATION:

1. Tie the asparagus into 4-ounce bundles.
2. Place the asparagus bundles in boiling, salted water, and blanch until **al dente.** Shock in ice water, and drain.
3. Just before service, brush the bundles with melted butter, and grill or broil, turning frequently. Serve immediately, or hold briefly at 140° F.

COOKING TECHNIQUE:

Grill

Grill/Broil:

1. Clean and heat the grill/broiler.
2. To prevent sticking, brush the food product with oil.

GLOSSARY:

Al dente: to the bite

HACCP:

Hold at 140° F.

NUTRITION:

Calories: 133
Fat: 11.8 g
Protein: 3.23 g

CHEF NOTES:

1. Blanched strips of fresh leek leaves can be used to tie the bundles.
2. As an alternative to grilling or broiling, place the bundles in individual baking dishes, brush with butter, and place in a preheated 450° F oven to finish cooking.
3. Other vegetables can be substituted for the asparagus.

Grilled Fresh Vegetables

YIELD: 10 SERVINGS. SERVING SIZE: FOUR OUNCES.

INGREDIENTS:

6 ounces	Carrots, washed, peeled, and cut on a diagonal into 1-inch pieces
6 ounces	Onions, peeled and cut into 1-inch cubes
6 ounces	Zucchini, washed, trimmed, and cut on a diagonal into 1-inch pieces
6 ounces	Yellow squash, washed, trimmed, and cut on a diagonal into 1-inch pieces
1 pound	Mushrooms, washed and stemmed
1 pound	Cherry tomatoes, washed and stemmed
	Olive oil, as needed

METHOD OF PREPARATION:

1. Preheat the grill or broiler.
2. **Blanch** or steam the carrots and onions until **al dente.**
3. Arrange the vegetables on skewers, alternating colors for eye appeal.
4. Brush the vegetables with olive oil, and grill, turning frequently, until they are tender but remain crisp. Serve immediately, or hold minimally at 140° F.

COOKING TECHNIQUES:

Grill, Steam

Grill/Broil:

1. Clean and heat the grill/broiler.
2. To prevent sticking, brush the food product with oil.

Steam: (Traditional)

1. Place a rack over a pot of water.
2. Prevent steam vapors from escaping.
3. Shock or cook the food product throughout.

GLOSSARY:

Blanch: to parcook
Al dente: to the bite

HACCP:

Hold minimally at 140° F.

NUTRITION:

Calories: 90.5
Fat: 6.13 g
Protein: 2.12 g

CHEF NOTES:

1. Alternative available vegetables can be substituted.
2. To be tender, root vegetables need to be blanched or steamed before grilling.

Grilled Swiss Chard

YIELD: 10 SERVINGS. SERVING SIZE: ONE-FOURTH HEAD.

COOKING TECHNIQUE:

Grill

Grill/Broil:

1. Clean and heat the grill/broiler.
2. To prevent sticking, brush the food product with oil.

GLOSSARY:

Blanch: to parcook
À la minute: cooked to order

NUTRITION:

Calories: 79.1
Fat: 8.13 g
Protein: .673 g

INGREDIENTS:

3 heads	Swiss chard, washed, trimmed, and cut into quarters
	Boiling, salted water
	Salt, to taste
3 ounces	Olive oil
1 ounce	Red wine vinegar

METHOD OF PREPARATION:

1. **Blanch** the Swiss chard in boiling salted water; then shock in an ice water bath.
2. Remove, and drain each portion on absorbent paper.
3. Combine the olive oil and vinegar.
4. **À la minute,** brush or drizzle the olive oil mixture over portions of Swiss chard, and grill until marked and hot. Serve immediately.

CHEF NOTE:

If "baby" chards are available, use five heads, and cut each in half.

Grilled Vegetable Kebabs

YIELD: 50 SERVINGS. SERVING SIZE: ONE SKEWER.

COOKING TECHNIQUE:
Broil or Grill

Grill/Broil:
1. Clean and heat the grill/broiler.
2. To prevent sticking, brush the food product with oil.

NUTRITION:
Calories: 77.4
Fat: 4.86 g
Protein: 1.56 g

INGREDIENTS:

2 pounds	Mushrooms
2 pounds	Green bell peppers
2 pounds	Red bell peppers
2 pints	Cherry tomatoes
3 pounds	Onions (Vidalia/Bermuda or Imperial Sweet)
2 pounds	Eggplant
2 pounds	Zucchini
1 quart	Olive oil
8 ounces	Lemon juice
	Seasoned, to taste

METHOD OF PREPARATION:

1. Wash all of the vegetables. Clean the mushrooms, and remove the stems. Seed the peppers, and cut them into large diced chunks. Peel the onions, and cut them into large diced chunks. Remove the stems from the cherry tomatoes. Slice the zucchini into ½-inch slices. Cut the eggplant into large diced chunks.
2. Thread the vegetables onto 6- to 8-inch-long metal skewers, alternating colors for attractive presentation.
3. In a bowl, combine the lemon juice, salt, and pepper. Whip in the olive oil in a fine stream to create a marinade.
4. Brush the marinade over the skewered vegetables, and grill 4 inches from the hot coals, turning and basting frequently. Cook for about 12 to 15 minutes.

CHEF NOTE:
If wooden skewers are used, soak them in water for 30 minutes before using.

Honey Lime Carrots

YIELD: 50 SERVINGS. SERVING SIZE: FOUR OUNCES.

INGREDIENTS:

15 pounds	Carrots, washed, peeled, and sliced round
10 ounces	Brown sugar
4 each	Limes, washed and split in half
1 tablespoon	Salt
12 ounces	Honey
4 ounces	Lime juice
8 ounces	Butter
3 each	Limes, washed and zested
6 ounces	Freshly chopped parsley, excess moisture removed

METHOD OF PREPARATION:

1. Add the sugar, lime, and salt. Heat to a boil. Skim off the foam with a **skimmer,** and reduce the heat to a simmer. Cook the carrots until tender.
2. Heat the honey in a small saucepan. Add the lime juice, butter, and lime zest.
3. Drain the carrots, and place in a hotel pan. Add the honey mixture, and incorporate evenly, paying attention not to mash the carrots.
4. Taste, and adjust the flavor.

COOKING TECHNIQUES:

Boil, Simmer

Boil: (at sea level)

1. Bring the cooking liquid to a rapid boil.
2. Stir the contents, and cook the food product throughout.
3. Serve hot.

Simmer and Poach:

1. Heat the cooking liquid to the proper temperature.
2. Submerge the food product completely.
3. Keep the cooked product moist and warm.

GLOSSARY:

Skimmer: long-handled, round, perforated tool

NUTRITION:

Calories: 150
Fat: 3.96 g
Protein: 1.72 g

CHEF NOTE:

This dish's flavor should be sweet.

Korean Marinated Celery Cabbage

(Kim Chee)

YIELD: 10 SERVINGS. SERVING SIZE: THREE OUNCES.

COOKING TECHNIQUE:
Not applicable

GLOSSARY:
Julienne: matchstick strips

HACCP:
Refrigerate at 40° F or below.

NUTRITION:
Calories: 70.6
Fat: .349 g
Protein: 2.86 g

INGREDIENTS:

1 head (1 ½ pounds)	Celery cabbage, washed, trimmed, and cut into 1-inch cubes
3 tablespoons	Salt
4 each	Scallions, washed, trimmed, and cut **julienne**
10 cloves	Garlic, peeled and finely minced
2 each (4 ounces)	Red hot peppers, washed, seeded, and cut julienne
2 each (4 ounces)	Green hot chili peppers, washed, seeded, and cut julienne
1 teaspoon	Ginger root, finely minced
6 ounces	Red hot bean paste (available in Oriental markets)
	Water, as needed

METHOD OF PREPARATION:

1. In a mixing bowl, combine the celery cabbage and salt; mix well, and let stand 30 for minutes.
2. Rinse the cabbage under cold water, and remove the excess moisture. Drain thoroughly.
3. In an appropriate container, combine the cabbage, scallions, garlic, red and green chilies, ginger root, and bean paste, adding enough water to completely cover the mixture.
4. Cover, and marinate at room temperature for 7 to 10 days. Taste for acidity; then cover again, and refrigerate at 40° F or below until needed for service.

CHEF NOTE:
This is a Korean national dish that is best served at room temperature.

Leek Tart

YIELD: 10 SERVINGS. SERVING SIZE: FOUR OUNCES.

COOKING TECHNIQUES:
Sauté, Bake

Sauté:
1. Heat the sauté pan to the appropriate temperature.
2. Evenly brown the food product.
3. For a sauce, pour off any excess oil, reheat, and deglaze.

Bake:
1. Preheat the oven.
2. Place the food product on the appropriate rack.

HACCP:
Refrigerate at 40° F or lower.
Hold at 140° F or above.

HAZARDOUS FOODS:
Eggs
Milk
Half-and-half

NUTRITION:
Calories: 535
Fat: 43 g
Protein: 15.4 g

INGREDIENTS:

10 sheets	Phyllo dough, thawed
10 ounces	Butter, melted

Filling:

1½ ounces	Butter
5 each (2 pounds)	Leeks (white part only), trimmed, washed, and thinly sliced
10 ounces	Emmentaler cheese, grated
4 each	Eggs, cracked and lightly beaten
12 ounces	Milk
12 ounces	Half-and-half
1½ ounces	Kirschwasser
1 teaspoon	Salt
¼ teaspoon	Nutmeg
½ teaspoon	Ground white pepper

METHOD OF PREPARATION:

1. Preheat the oven to 325° F.
2. In a sauté pan, melt the butter, add the leeks, and sauté until golden brown; then cool to room temperature.
3. Combine the leeks and cheese, and refrigerate at 40° F or lower. Combine the eggs, milk, half-and-half, Kirschwasser, and seasonings, and refrigerate at 40° F or lower.
4. Brush the bottom and sides of a baking pan with butter.
5. Place a sheet of phyllo dough in the pan, allowing an overhang on opposite sides, and brush with butter.
6. Place a second sheet of phyllo dough in the pan, allowing the overhang to go in the opposite direction, and brush again with butter.
7. Repeat steps 5 and 6 with the remaining sheets of phyllo dough.
8. Place the leek and cheese mixture into the pan, and pour in the egg mixture.
9. Place in the oven, and bake uncovered for 50 to 60 minutes, or until set.
10. Cool the tart in the pan for 20 minutes before portioning. Serve immediately, or hold at 140° F or above.

CHEF NOTE:
To test for doneness, insert a toothpick into the center of the tart, and then remove it. If the tart is done, the toothpick will come out clean.

Leek Timbales

YIELD: 10 SERVINGS. SERVING SIZE: ONE RAMEKIN.

INGREDIENTS:

16 ounces	Chicken stock, heated to a boil (see page 12)
2 pounds	Leeks, washed, outside leaves removed, and diced **brunoise**
8 each	Eggs, cracked and lightly beaten
1½ pints	Half-and-half
8 ounces	Gruyère cheese, grated
¼ teaspoon	Freshly grated nutmeg
	Salt and ground white pepper, to taste
	Clarified butter, as needed, melted

METHOD OF PREPARATION:

1. Preheat the oven to 350° F.
2. In a medium saucepan, bring the stock to a simmer, add the leeks. Cover and simmer until tender, or about 20 minutes. Drain, reserving the stock for other use. Cool the leeks thoroughly.
3. Place the eggs in mixing bowl, add the half-and-half, cheese, and seasonings. Mix well; then add the leeks.
4. Butter the timbales or ramekins and fill two-thirds full with the leek mixture. Place the containers in a shallow hotel pan, and fill the pan with water to half of the depth of the molds.
5. Place the pan in the oven, and cook for about 20 minutes, or until the custard is set. Remove, and hold at 140° F or higher, or serve immediately.

COOKING TECHNIQUES:

Simmer, Steam

Simmer and Poach:

1. Heat the cooking liquid to the proper temperature.
2. Submerge the food product completely.
3. Keep the cooked product moist and warm.

Steam: (Traditional)

1. Place a rack over a pot of water.
2. Prevent steam vapors from escaping.
3. Shock or cook the food product throughout.

GLOSSARY:

Brunoise: ⅛-inch dice

HACCP:

Hold at 140° F or higher.

HAZARDOUS FOODS:

Eggs
Half-and-half

NUTRITION:

Calories: 342
Fat: 24.1 g
Protein: 15.3 g

Mushrooms with Garlic and Parsley

YIELD: 10 SERVINGS. SERVING SIZE: FOUR OUNCES.

COOKING TECHNIQUE:
Sauté

Sauté:
1. Heat the sauté pan to the appropriate temperature.
2. Evenly brown the food product.
3. For a sauce, pour off any excess oil, reheat, and deglaze.

HACCP:
Hold at 140° F.

NUTRITION:
Calories: 156
Fat: 14.3 g
Protein: 2.91 g

INGREDIENTS:

6 ounces	Butter
2½ pounds	Mushrooms, washed and stems trimmed
6 cloves	Garlic, peeled and minced
3 ounces	Fresh parsley, washed, excess moisture removed, and chopped
	Salt and freshly ground black pepper, to taste

METHOD OF PREPARATION:

1. In a sauté pan, melt the butter, add the mushrooms, and sauté, tossing frequently until cooked but still firm.
2. Add the garlic and parsley, continuing to sauté for 5 more minutes.
3. Season, to taste, and serve immediately, or hold at 140° F.

CHEF NOTE:
When sautéing the mushrooms, it is important to toss frequently, and the heat should be high enough to allow all the liquid to evaporate.

Provencal Tomatoes

(Tomatoes Provençale)

YIELD: 10 SERVINGS. SERVING SIZE: ONE EACH.

INGREDIENTS:

10 each	Medium-sized tomatoes, washed, cored, and cut in half
2 ounces	Butter, melted
5 ounces	Seasoned bread crumbs
5 cloves	Garlic, peeled and finely minced
3 ounces	Freshly chopped parsley, excess moisture removed

METHOD OF PREPARATION:

1. Preheat the oven to 350° F.
2. Squeeze the seeds out of the tomato halves.
3. In a sauté pan, heat the butter, and sauté the bread crumbs until golden brown. Add the garlic as the crumbs begin to brown. Remove from the heat, and cool slightly. Add the parsley to the crumb mixture. Fill the tomatoes with the mixture.
4. Just before service, bake or broil the tomatoes until thoroughly heated throughout and browned. Serve immediately on a preheated plate, or hold minimally at 140° F.

COOKING TECHNIQUES:

Bake or Broil

Bake:

1. Preheat the oven.
2. Place the food product on the appropriate rack.

Grill/Broil:

1. Clean and heat the grill/broiler.
2. To prevent sticking, brush the food product with oil.

NUTRITION:

Calories: 124
Fat: 5.45 g
Protein: 3.46 g

CHEF NOTE:

Do not overcook the tomatoes or finish baking too much in advance, or they will collapse when handling.

Pumpkin with Sautéed Onions

YIELD: 50 SERVINGS. SERVING SIZE: FOUR OUNCES.

INGREDIENTS:

20 pounds	Pumpkin or seasoned squash, washed, peeled, seeded, and cut **parmentier**
20 ounces	Bacon, diced (optional)
5 tablespoons	Butter
3 pounds	Onions, peeled and chopped
7 tablespoons	Thyme, dried
1½ tablespoons	Pepper, black, crushed
5 cloves	Garlic, peeled and minced
5 tablespoons	Parsley, washed and chopped (no stems)
	Seasoned, to taste

METHOD OF PREPARATION:

1. In a saucepan, cover the pumpkin with lightly salted water, and cook until **al dente** (about 10 minutes). Drain the water, and keep the pumpkin in the saucepan.
2. In a sauté pan that has been heated over medium-high heat, render the bacon until brown, and remove. Pour off any excess fat; add the butter to the sauté pan, and heat over medium-high heat.
3. Add the onions, garlic, thyme, salt, and pepper. Sauté until the onions are translucent. Add this mixture to the pumpkin. Cover, and cook for another 5 minutes.
4. Add cooked bacon and serve.

COOKING TECHNIQUES:
Boil, Sauté

Boil: (at sea level)
1. Bring the cooking liquid to a rapid boil.
2. Stir the contents, and cook the food product throughout.
3. Serve hot.

Sauté:
1. Heat the sauté pan to the appropriate temperature.
2. Evenly brown the food product.
3. For a sauce, pour off any excess oil, reheat, and deglaze.

GLOSSARY:
Parmentier: ½-inch dice
Al dente: to the bite

HAZARDOUS FOOD:
Bacon

NUTRITION:
Calories: 46.7
Fat: 1.97 g
Protein: 1.8 g

CHEF NOTES:
1. Pumpkin is a good, hardy fall vegetable and deserves to be considered as much more than just a pie ingredient.
2. Pumpkin needs powerful herbs and spices that can stand up to its flavor (e.g., garlic, onions, fresh herbs, etc.).

Ratatouille

YIELD: 10 SERVINGS. SERVING SIZE: FIVE OUNCES.

COOKING TECHNIQUES:

Sauté, Stew

Sauté:

1. Heat the sauté pan to the appropriate temperature.
2. Evenly brown the food product.
3. For a sauce, pour off any excess oil, reheat, and deglaze.

Stew:

1. Sear, sauté, sweat, or blanch the main food product.
2. Deglaze the pan, if desired.
3. Cover the food product with simmering liquid.
4. Remove the bouquet garni.

GLOSSARY:

Brunoise: ⅛-inch dice
Blanch: to parcook

NUTRITION:

Calories: 155
Fat: 11.8 g
Protein: 2.18 g

INGREDIENTS:

4 ounces	Olive oil
4 ounces	Onions, peeled and diced **brunoise**
1 pound	Green bell peppers, washed, seeded, and sliced into rings
1 pound	Tomatoes, washed, cored, and cut into wedges
1 pound	Eggplant, washed and diced
1 pound	Zucchini, washed and diced
1 pound	Green beans, washed, **blanched,** and cooled
	Tomato purée
	Salt and freshly ground black pepper, to taste
¼ head	Garlic, peeled and mashed into a purée

METHOD OF PREPARATION:

1. Sauté all vegetables for 5 minutes, and place in a rondeau.
2. Season, cover, and bake in a 350° F oven until the vegetables are tender.

Ratatouille-Stuffed Artichoke Bottom

YIELD: 10 SERVINGS. SERVING SIZE: FOUR OUNCES (2 ARTICHOKE BOTTOMS).

COOKING TECHNIQUE:
Sauté

Sauté:
1. Heat the sauté pan to the appropriate temperature.
2. Evenly brown the food product.
3. For a sauce, pour off any excess oil, reheat, and deglaze.

GLOSSARY:
Macédoine: ¼-inch dice
Blanch: to parcook

HACCP:
Hold at 140°F or higher.

NUTRITION:
Calories: 312
Fat: 18.9 g
Protein: 9.62 g

CHEF NOTES:
1. If using frozen artichoke bottoms, steam or sauté them until tender. If using canned product, rinse them thoroughly in cold water; then drain, and proceed with recipe.
2. Ratatouille is best when prepared and allowed to stand at least 30 minutes to allow its flavors to mingle.
3. If desired, a small amount of vinegar can be added to enhance flavors.

INGREDIENTS:

4 ounces	Olive oil, as needed
½ ounce	Fresh basil leaves, washed and finely chopped
1 teaspoon	Thyme leaves, dried
	Salt and freshly ground black pepper, to taste
20 each	Frozen or canned artichoke bottoms (see chef notes)
	Additional oil or clarified butter, as needed

Macédoine:

1 each, small	Eggplant, washed and trimmed
4 ounces	Onions, peeled
3 each	Red bell peppers, washed, trimmed, and seeded
2 each, small	Zucchinis, washed and trimmed
1 each, small	Yellow squash, washed and trimmed
6 cloves	Garlic, peeled and finely minced
6 ounces	Fresh tomato, washed, cored, **blanched,** peeled, and seeded

METHOD OF PREPARATION:

1. In a suitably sized sautoir, heat half of the olive oil. When the oil is near the smoking point, add the eggplant, and toss. Sauté until the eggplant is browned and tender; then use a skimmer to transfer the eggplant to a bowl or half hotel pan.
2. Add the onion to the pan, and sauté until translucent; then add the peppers, zucchini, squash, and garlic. Add more oil as needed, and sauté until the vegetables begin to brown. Transfer the vegetable mixture to the container with eggplant.
3. Add the tomatoes, herbs, and seasonings to the pan, and sauté until the tomatoes are softened and the liquid has evaporated. Add to the vegetable mixture, mix well, taste, and adjust the seasonings, as needed. Hold at 140° F or higher.
4. For service, brush the artichoke bottoms with oil or butter, season lightly, and warm in a sauté pan or in a preheated oven. Spoon the ratatouille into the shells, and serve immediately.

Roasted Beets

YIELD: 10 SERVINGS. SERVING SIZE: FOUR OUNCES.

INGREDIENTS:

4 pounds	Fresh beets, washed and trimmed, leaving 1-inch of stems and root intact
4 ounces	Clarified butter
	Salt and freshly ground black pepper, to taste

METHOD OF PREPARATION:

1. Preheat the oven to 350° F.
2. Wrap the beets individually in aluminum foil, and place in a baking pan. Roast in the oven until fork-tender, which will take about 1½ hours.
3. Remove from the oven, open the foil, and allow the beets to cool enough to handle. Trim the ends, and peel; then cut or slice as desired.
4. For service, heat a saucepan, and melt the butter. Add the beets, and sauté until heated thoroughly. Season, to taste, and serve, or hold at 140° F.

COOKING TECHNIQUES:
Roast, Sauté

Roast:
1. Sear the food product, and brown evenly.
2. Elevate the food product in a roasting pan.
3. Determine doneness, and consider carryover cooking.
4. Let the food product rest before carving.

Sauté:
1. Heat the sauté pan to the appropriate temperature.
2. Evenly brown the food product.
3. For a sauce, pour off any excess oil, reheat, and deglaze.

HACCP:
Hold at 140° F.

NUTRITION:
Calories: 178
Fat: 11.6 g
Protein: 2.98 g

CHEF NOTES:
1. Leaving 1-inch of stem and root attached during the roasting period prevents bleeding of color.
2. Depending on the size of the beets, they can be left whole, cut into wedges, sliced, or julienned.
3. Herbs of choice can be added for additional flavor.

Sautéed Butternut Squash

YIELD: 50 SERVINGS. SERVING SIZE: FOUR OUNCES.

COOKING TECHNIQUES:
Boil, Sauté

Boil: (at sea level)
1. Bring the cooking liquid to a rapid boil.
2. Stir the contents, and cook the food product throughout.
3. Serve hot.

Sauté:
1. Heat the sauté pan to the appropriate temperature.
2. Evenly brown the food product.
3. For a sauce, pour off any excess oil, reheat, and deglaze.

GLOSSARY:
Bâtonet: stick-like cut
Al dente: to the bite

NUTRITION:
Calories: 199
Fat: 18.1 g
Protein: 1.32 g

INGREDIENTS:

12 pounds	Butternut squash, washed, peeled, and cut **bâtonet**
	Salted water to cover, heated to a boil
2 pints	Clarified butter
	Salt and ground white pepper, to taste
	Nutmeg, to taste

METHOD OF PREPARATION:

1. In a large stockpot, heat the water to a boil.
2. Add the squash. Heat to a boil, and cook just until **al dente.**
3. Drain at once, and shock in an ice bath. When chilled, drain well, and reserve.
4. For service, heat the butter in a sauté pan. Add the squash, and sauté until hot. Season, to taste, and serve.

Sautéed Celery

YIELD: 10 SERVINGS. SERVING SIZE: FOUR OUNCES.

INGREDIENTS:

2 pounds (3 bunches)	Celery, trimmed and washed
3 ounces	Olive oil
8 cloves	Garlic, peeled and finely minced
	Salt and freshly ground black pepper, to taste

METHOD OF PREPARATION:

1. Cut the celery on a diagonal into 1½-inch pieces. **Blanch** until **al dente.**
2. In a sauté pan, heat the oil, add the garlic, and lightly sauté. Add the celery, and sauté until tender.
3. Season, to taste, and serve immediately, or hold at 140° F.

COOKING TECHNIQUES:

Sauté

Sauté:

1. Heat the sauté pan to the appropriate temperature.
2. Evenly brown the food product.
3. For a sauce, pour off any excess oil, reheat, and deglaze.

GLOSSARY:

Blanch: to parcook
Al dente: to the bite
Julienne: matchstick strips
Bain-marie: hot-water bath

HACCP:

Hold at 140° F.

NUTRITION:

Calories: 93.5
Fat: 8.65 g
Protein: .848 g

CHEF NOTES:

1. For added flavor the following can be added: Genoa salami, **julienned** and sautéed; prosciutto, julienned and sautéed; precooked pork roast, julienned; and/or freshly grated Parmesan cheese.
2. Celery will lose color if held in a **bain-marie** for more than 20 minutes.

Sautéed Peas with Mushrooms

YIELD: 50 SERVINGS. SERVING SIZE: FOUR OUNCES.

COOKING TECHNIQUE:
Sauté

Sauté:
1. Heat the sauté pan to the appropriate temperature.
2. Evenly brown the food product.
3. For a sauce, pour off any excess oil, reheat, and deglaze.

GLOSSARY:
Brunoise: ⅛-inch dice
Al dente: to the bite

HACCP:
Hold at 165° F.

NUTRITION:
Calories: 130
Fat: 5.89 g
Protein: 5.42 g

INGREDIENTS:

12 ounces	Butter
1 pound	Onions, peeled, and diced **brunoise**
3 pounds	Mushrooms, cleaned, and sliced
10 pounds	Peas, washed and drained
	Salt and ground black pepper, to taste

METHOD OF PREPARATION:

1. Sauté the onions in a tilting skillet with the butter.
2. Add the sliced mushrooms, and sauté for 2 minutes.
3. Add the green peas, and sauté until **al dente.**
4. Season with the salt and pepper.
5. Transfer the vegetables into 2-inch hotel pans.
6. Hold at 165° F.

Sautéed Peas with Onions

(Petit Pois au Oignons)

YIELD: 50 SERVINGS. SERVING SIZE: FOUR TO FIVE OUNCES.

COOKING TECHNIQUES:
Sauté

Sauté:
1. Heat the sauté pan to the appropriate temperature.
2. Evenly brown the food product.
3. For a sauce, pour off any excess oil, reheat, and deglaze.

GLOSSARY:
Blanch: to parcook
Al dente: to the bite

NUTRITION:
Calories: 132
Fat: 5 g
Protein: 6 g

INGREDIENTS:

1 pound	Fresh pearl onions, small, peeled and washed
8–12 ounces	Clarified butter
12½ pounds	Frozen peas, defrosted and drained
8 ounces	Pimientos, diced brunoise
	Salt and ground white pepper, to taste

METHOD OF PREPARATION:

1. Peel and wash the onions, and **blanch** until **al dente.** Shock, drain, and reserve.
2. In a sauté pan, heat the butter, and sauté the onions and peas until heated. Add the pimientos, and toss to blend.
3. Season, to taste, and serve immediately.

Sautéed Peas with Pimiento

YIELD: 50 SERVINGS. SERVING SIZE: FOUR–FIVE OUNCES.

COOKING TECHNIQUE:
Sauté

Sauté:
1. Heat the sauté pan to the appropriate temperature.
2. Evenly brown the food product.
3. For a sauce, pour off any excess oil, reheat, and deglaze.

GLOSSARY:
Macédoine: ¼-inch dice

NUTRITION:
Calories: 129
Fat: 5 g
Protein: 6 g

INGREDIENTS:

8 to 12 ounces	Clarified butter
12½ pounds	Frozen peas, defrosted and drained
12 ounces	Pimientos, drained and cut **macédoine**

METHOD OF PREPARATION:

1. In a sauté pan, heat the butter, and sauté the peas and pimientos until well heated.
2. Season, to taste, and serve immediately.

Sautéed Red and Yellow Peppers

YIELD: 10 SERVINGS. SERVING SIZE: FOUR OUNCES.

INGREDIENTS:

2 ounces	Vegetable oil
1 pound	Onions, peeled and cut **julienne**
1½ pounds	Red bell peppers, washed, seeded, and cut julienne
1½ pounds	Yellow bell peppers, washed, seeded, and cut julienne
	Salt and freshly ground black pepper, to taste
3 ounces	Fresh parsley, washed, excess moisture removed, and chopped

METHOD OF PREPARATION:

1. In a sauté pan, heat the oil, and sauté the onions for 3 minutes, or until translucent.
2. Add the peppers; season, to taste, and sauté until the peppers are cooked but still remain **al dente.**
3. Serve immediately, garnished with parsley.

COOKING TECHNIQUE:

Sauté

Sauté:

1. Heat the sauté pan to the appropriate temperature.
2. Evenly brown the food product.
3. For a sauce, pour off any excess oil, reheat, and deglaze.

GLOSSARY:

Julienne: matchstick strips
Al dente: to the bite
Bain-marie: hot-water bath

NUTRITION:

Calories: 105
Fat: 5.85 g
Protein: 1.99 g

CHEF NOTES:

1. At the discretion of the chef, chilies or jalapeño peppers can be added as additional flavor.
2. Do not prepare this dish early, because holding it in a **bain-marie** will change the color and texture of the peppers.

Semolina-Stuffed Onions

YIELD: 10 SERVINGS. SERVING SIZE: FOUR OUNCES.

INGREDIENTS:

10 medium-sized	Onions, peeled, one third of top removed and reserved for alternate use
	Salt and freshly ground black pepper, to taste
10 ounces	Veal demi-glacé, heated to a boil (see page 26)
10 ounces	Clarified butter, melted
1 recipe	Semolina with vegetables (see page 527)
2 ounces	Freshly chopped parsley, excess moisture removed

METHOD OF PREPARATION:

1. Preheat the oven to 375° F.
2. Use a paring knife, and remove the center of the onions; reserve for alternative use.
3. Cut 10-inch squares of aluminum foil, and place one onion in the center of each. Season each with salt and pepper, and ladle demi-glacé over the top of each.
4. Drizzle each with butter, and wrap tightly with the foil.
5. Place the onions in a baking pan, and put in the oven. Roast for about 1 hour, or until tender.
6. Remove from the oven, open the foil, and transfer the onions to a hotel pan. Pour the accumulated juices from each package over the semolina; then fill the onions with the semolina mixture. Hold at 140° F or higher, until service.
7. Sprinkle with parsley and serve.

COOKING TECHNIQUES:

Roast, Steam

Roast:

1. Sear the food product, and brown evenly.
2. Elevate the food product in a roasting pan.
3. Determine doneness, and consider carryover cooking.
4. Let the food product rest before carving.

Steam: (Traditional)

1. Place a rack over a pot of water.
2. Prevent steam vapors from escaping.
3. Shock or cook the food product throughout.

H.A.C.C.P.:

Hold at 140° F or higher.

NUTRITION:

Calories: 404
Fat: 26.5 g
Protein: 6.41 g

CHEF NOTE:

As an alternative, onions can be peeled, their tops removed, steamed, and separated into cups to be filled with semolina. In this case, two onions will be enough product.

Snow Peas with Ginger

YIELD: 10 SERVINGS. SERVING SIZE: FOUR OUNCES.

INGREDIENTS:

3 pounds	Snow peas, washed and ends removed
	Salt, as needed
4 ounces	Peanut oil
4 ounces	Ginger root, peeled and grated
	Salt and ground white pepper, to taste

METHOD OF PREPARATION:

1. **Blanch** the snow peas in boiling, salted water; then shock in an ice water bath.
2. In a wok, heat the oil, add the ginger, and sauté for 30 seconds. Then add the snow peas, and toss until thoroughly heated.
3. Season, to taste, and serve immediately on a preheated plate. Hold minimally at 140° F or above.

COOKING TECHNIQUES:

Boil, Sauté

Boil: (at sea level)

1. Bring the cooking liquid to a rapid boil.
2. Stir the contents, and cook the food product throughout.
3. Serve hot.

Sauté:

1. Heat the sauté pan to the appropriate temperature.
2. Evenly brown the food product.
3. For a sauce, pour off any excess oil, reheat, and deglaze.

GLOSSARY:

Blanch: to parcook
Bain-marie: hot-water bath

HACCP:

Hold minimally at 140° F or above.

NUTRITION:

Calories: 166
Fat: 11.7 g
Protein: 4.68 g

CHEF NOTE:

Prepare the snow peas just minutes before service, because they will wilt quickly and lose color in a **bain-marie.**

Spanish-Style Peas with Ham

YIELD: 10 SERVINGS. SERVING SIZE: FOUR OUNCES.

COOKING TECHNIQUE:

Sauté

Sauté:

1. Heat the sauté pan to the appropriate temperature.
2. Evenly brown the food product.
3. For a sauce, pour off any excess oil, reheat, and deglaze.

GLOSSARY:

Brunoise: ⅛-inch dice
Julienne: matchstick strips
Blanch: to parcook

HACCP:

Hold at 140° F or above.

NUTRITION:

Calories: 189
Fat: 9.87 g
Protein: 9.58 g

INGREDIENTS:

3 ounces	Spanish olive oil
8 ounces	Onion, peeled and diced **brunoise**
4 ounces	Carrot, peeled and grated
8 ounces	Ham, cut **julienne**
2 pounds	Frozen peas or (shelled) fresh peas, **blanched**
	Salt and ground white pepper, to taste

METHOD OF PREPARATION:

1. In a sauté pan, heat the oil, and sauté the onion and carrot. When the onion begins to brown, add the ham.
2. Sauté the above ingredients for 10 minutes; then add the peas. Season, to taste, and continue to cook until the peas are hot.
3. Serve immediately, or hold at 140° F or above.

Spinach Custard

(Dariole d'Épinards)

YIELD: 10 SERVINGS. SERVING SIZE: ONE EACH.

INGREDIENTS:

1 pound	Spinach, stemmed, washed, and drained
6 each	Eggs
10 ounces	Milk
10 ounces	Heavy cream
½ teaspoon	Nutmeg
¼ teaspoon	Freshly ground white pepper
2 ounces	Butter, melted

METHOD OF PREPARATION:

1. Preheat the oven to 300° F.
2. Steam the cleaned spinach for 1 minute. Plunge in ice water, and drain.
3. Combine the eggs, milk, cream, and seasonings in a bowl, and whisk together.
4. Melt half of the butter in a suitable sauté pan; add the spinach, and cook for 5 minutes. Drain the excess moisture. Cool slightly; then add the spinach to the custard, and mix well.
5. Using a brush, butter the dariole molds, and fill with the spinach custard to within ⅓-inch of the top.
6. Using a 2-inch hotel pan, place a sheet of parchment paper or a kitchen towel in the bottom of the pan, and place the darioles on it.
7. Fill the pan with water to half the depth of the molds, and bake for 20 to 30 minutes, or until the custard is firm.

COOKING TECHNIQUE:

Bake

Bake:

1. Preheat the oven.
2. Place the food product on the appropriate rack.

HAZARDOUS FOODS:

Eggs
Milk
Heavy cream

NUTRITION:

Calories: 177
Fat: 15.5 g
Protein: 6.3 g

CHEF NOTE:

If using parchment paper as a liner, cut incisions into it to avoid steam from developing under the liner, which will elevate the liner and cause the darioles to tip.

Spinach Soufflé

YIELD: 10 SERVINGS. SERVING SIZE: ONE COCOTTE.

INGREDIENTS:

1½ pounds	Spinach, washed and stemmed
4 ounces	Clarified butter, plus additional butter to butter the **cocottes**
4 ounces	All-purpose flour
20 ounces	Milk, heated to the scalding stage
	Salt, to taste
½ teaspoon	Nutmeg, ground
8 each	Eggs, cracked and separated

METHOD OF PREPARATION:

1. Preheat the oven to 300° F.
2. Steam the spinach until tender; then drain, and press out any excess moisture. Allow to cool; then finely chop or purée in a food processor.
3. In a saucepan, melt the butter. When hot, add the flour, and prepare a roux blanc.
4. Temper the roux with the scalding milk, and cook for 10 minutes. Season, to taste, with salt, and add the nutmeg. Allow to cool to 140° F.
5. Incorporate the egg yolks into the sauce; then add the spinach, and mix well. Taste, and adjust the seasoning, as needed.
6. Whip the egg whites until they form peaks; then fold the egg whites into the spinach mixture.
7. Brush the inside of the cocottes with butter, and fill two-thirds full with the spinach mixture.
8. Line a 2-inch-deep hotel pan with a towel, and place the prepared cocottes in the pan. Add water to a depth of half the cocottes, and place in the oven.
9. Bake for approximately 12 minutes. Test for doneness by inserting a toothpick into the center of the soufflé. Serve immediately.

COOKING TECHNIQUES:

Boil, Steam, Bake

Boil: (at sea level)

1. Bring the cooking liquid to a rapid boil.
2. Stir the contents, and cook the food product throughout.
3. Serve hot.

Steam: (Traditional)

1. Place a rack over a pot of water.
2. Prevent steam vapors from escaping.
3. Shock or cook the food product throughout.

Bake:

1. Preheat the oven.
2. Place the food product on the appropriate rack.

GLOSSARY:

Cocotte: small oven-proof dish

HACCP:

Allow to cool to 140° F.

HAZARDOUS FOODS:

Milk
Eggs

NUTRITION:

Calories: 233
Fat: 15.5 g
Protein: 10.1 g

CHEF NOTE:

For best results—to prevent the soufflés from collapsing—do not remove them at once from the oven; instead, open the oven door, and keep it open for 1 minute; then remove the soufflés.

Stuffed Zucchini

YIELD: 50 SERVINGS. SERVING SIZE: ONE EACH.

INGREDIENTS:

50 each	Zucchini, small
12 ounces	Carrots, washed, peeled, and diced **brunoise**
12 ounces	Celery, washed, trimmed, and diced brunoise
12 ounces	Zucchini flesh, diced brunoise
6 cloves	Garlic, peeled and minced
24 ounces	Rice, uncooked
8 each	Whole eggs
	Salt and ground black pepper, to taste
3 quarts	Tomato sauce (see page 76)

METHOD OF PREPARATION:

1. Preheat the oven to 300° F.
2. Wash the zucchini. Cut off both ends, and hollow out, using an apple corer.
3. Place all of the vegetables in a robot coupe, and chop to the size of a boiled grain of rice.
4. Mix the chopped vegetables with the rice. Add the seasoning and the quantity of liquid needed to steam the rice, and mix again.
5. Cover the cooking vessel with foil, and steam the rice until **al dente.**
6. Let the rice cool to room temperature, and then incorporate the eggs. Taste, and adjust the seasonings. Stuff the zucchini.
7. Arrange the stuffed zucchini in a hotel pan. Pour the tomato sauce over it to cover three fourths of the zucchini. Bake until the zucchini is done.
8. Serve hot.

COOKING TECHNIQUES:

Steam, Bake

Steam: (Traditional)

1. Place a rack over a pot of water.
2. Prevent steam vapors from escaping.
3. Shock or cook the food product throughout.

Bake:

1. Preheat the oven.
2. Place the food product on the appropriate rack.

GLOSSARY:

Brunoise: ⅛-inch dice
Al dente: to the bite

HAZARDOUS FOOD:

Whole eggs

NUTRITION:

Calories: 100
Fat: 1.1 g
Protein: 4.32 g

CHEF NOTES:

1. This dish can be served as a hot appetizer or cut into 2-inch lengths as a buffet item.
2. For a non-vegetarian dish, replace the vegetables with ground beef or other kinds of ground meat.
3. A variation of this dish can be made using ricotta cheese and fresh herbs. In this case, the rice would be omitted.

Summer Squash with Herbs

YIELD: 50 SERVINGS. SERVING SIZE: FOUR OUNCES.

INGREDIENTS:

15 pounds	Squash in season, washed, peeled, and cut **parmentier**
1 pound	Butter or margarine, melted
3 ounces	Freshly chopped marjoram, excess moisture removed
3 ounces	Freshly chopped basil, excess moisture removed
	Salt and ground black pepper, to taste

METHOD OF PREPARATION:

1. Cook the diced squash in salted water until **al dente.** Drain.
2. Melt the butter or margarine in a tilting skillet. Add the squash, and sauté.
3. Sprinkle with the herbs and spices, to taste.
4. Transfer into a 2-inch hotel pan, and hold at 165° F.

COOKING TECHNIQUE:

Sauté

Sauté:

1. Heat the sauté pan to the appropriate temperature.
2. Evenly brown the food product.
3. For a sauce, pour off any excess oil, reheat, and deglaze.

GLOSSARY:

Al dente: to the bite
Parmentier: ½-inch dice

HACCP:

Hold at 165° F.

NUTRITION:

Calories: 92.8
Fat: 7.65 g
Protein: 1.75 g

CHEF NOTE:

The squash may be steamed in perforated hotel pans.

Sweet and Sour Savoy Cabbage

YIELD: 10 SERVINGS. SERVING SIZE: FOUR OUNCES.

INGREDIENTS:

2 ounces	Butter
2 ounces	Vegetable oil
1 pound	Onions, peeled and diced **brunoise**
4 pounds	Savoy cabbage, washed, cored, and cut **chiffonade**
1½ pounds	Carrots, washed, peeled, and grated
3 tablespoons	Sugar
	Salt and freshly ground black pepper, to taste
12 ounces	Cider vinegar

METHOD OF PREPARATION:

1. In a large sauté pan, heat the butter and oil; add the onions, and sauté until lightly browned.
2. Add the cabbage, and mix well; then add the carrots.
3. Reduce the heat to low, cover, and steam the cabbage until tender.
4. Season, to taste, with sugar, salt, and pepper; then add the vinegar, and toss gently but mix well.
5. Raise the heat to moderate, and simmer uncovered, stirring until the juices are reduced by half.
6. Serve immediately, or hold at 140° F or higher.

COOKING TECHNIQUES:

Sauté, Steam, Simmer

Sauté:

1. Heat the sauté pan to the appropriate temperature.
2. Evenly brown the food product.
3. For a sauce, pour off any excess oil, reheat, and deglaze.

Steam: (Traditional)

1. Place a rack over a pot of water.
2. Prevent steam vapors from escaping.
3. Shock or cook the food product throughout.

Simmer and Poach:

1. Heat the cooking liquid to the proper temperature.
2. Submerge the food product completely.
3. Keep the cooked product moist and warm.

GLOSSARY:

Brunoise: ⅛-inch dice
Chiffonade: ribbons of leafy greens

HACCP:

Hold at 140° F or higher.

NUTRITION:

Calories: 204
Fat: 10.4 g
Protein: 4.94 g

CHEF NOTE:

Green or red cabbage can be substituted in this recipe.

Tomato Chutney

(Tamatar Chatni)

YIELD: 1 QUART. SERVING SIZE: AS NEEDED.

COOKING TECHNIQUE:
Boil

Boil: (at sea level)

1. Bring the cooking liquid to a rapid boil.
2. Stir the contents, and cook the food product throughout.
3. Serve hot.

NUTRITION:
Calories: 61.6
Fat: 2.16
Protein: .634 g

INGREDIENTS:

1 pound	Tomatoes, washed, cored, and coarsely chopped
8 ounces	Palm vinegar
4 ounces	Onions, peeled and finely chopped
1 inch	Cinnamon stick
	Salt, to taste
4 ounces	Brown sugar
1 tablespoon	Molasses
½ ounce	Ginger root, peeled and finely minced
2 cloves	Garlic, peeled and finely minced
8 each	Whole cloves
½ teaspoon	Fresh chili pepper, seeded and minced
1 tablespoon	Coriander seeds, ground
1 ounce	Vegetable oil
1 tablespoon	Black mustard seeds

METHOD OF PREPARATION:

1. In a saucepan over medium heat, combine the tomatoes, vinegar, onions, cinnamon stick and salt, and heat to a boil, stirring constantly.
2. Add the brown sugar, molasses, ginger root, garlic, cloves, chili pepper, and coriander seeds, and cook for 5 minutes.
3. In a sauté pan, heat the vegetable oil, and sauté the mustard seeds. Add the seeds to the chutney.
4. Cool, and serve as a relish.

CHEF NOTE:
This is a common side dish of India.

Tomatoes Clamart

YIELD: 50 SERVINGS. SERVING SIZE: ONE TOMATO.

INGREDIENTS:

50	Tomatoes
	Salt and ground white pepper, to taste
20 ounces	Butter
2 each	Large onion, peeled and finely diced
15 cloves	Garlic, peeled and mashed
7½ pounds	Green peas

METHOD OF PREPARATION:

1. Preheat the oven to 350° F.
2. Wash the tomatoes; cut off the tops and hollow, using a parisienne scoop. Season the inside of the tomatoes with salt and pepper. Turn the tomatoes upside down to drain the juice from the flesh.
3. Melt the butter, and sauté the onions and garlic until the onions are translucent. Add the green peas, and season with salt and pepper.
4. Place the tomatoes on a half sheet pan, and fill with peas.
5. Bake **à la minute** for approximately 10 minutes, or until heated throughout.
6. Serve immediately.

COOKING TECHNIQUES:
Sauté, Bake

Sauté:
1. Heat the sauté pan to the appropriate temperature.
2. Evenly brown the food product.
3. For a sauce, pour off any excess oil, reheat, and deglaze.

Bake:
1. Preheat the oven.
2. Place the food product on the appropriate rack.

GLOSSARY:
À la minute: cooked to order

NUTRITION:
Calories: 166
Fat: 9.88 g
Protein: 4.97 g

CHEF NOTES:
1. Do not overbake the tomatoes, or they will collapse and fall apart.
2. Do not use overripe tomatoes.

Tomatoes Stuffed with Mushrooms

(Tomates aux Champignons)

YIELD: 10 SERVINGS. SERVING SIZE: ONE TOMATO.

INGREDIENTS:

2 ounces	Butter
2 ounces	Shallots, peeled and diced **brunoise**
10 ounces	Mushrooms, washed and trimmed
8 ounces	Mornay sauce, heated (see page 37)
	Salt and ground white pepper, to taste
10 each	Tomatoes, washed, cored, and tops cut off

METHOD OF PREPARATION:

1. Preheat the oven to 375° F.
2. Remove a small scoop of tomato pulp from the cut side.
3. In a sauté pan, melt the butter, add the shallots, and **sweat.**
4. Robot coupe the mushrooms to a watery stage.
5. Add the mushrooms to the shallots, and sauté until all of the liquid evaporates.
6. Season the mixture, to taste.
7. Fill the tomatoes with the mushroom mixture. Place on a sheet pan, and bake for 5 minutes, or until heated throughout.
8. Nappé with Mornay sauce, sprinkle with grated Parmesan, and gratiner under a salamander or broiler.
9. Serve immediately, or hold at 140° F.

COOKING TECHNIQUES:

Bake, Sauté

Bake:

1. Preheat the oven.
2. Place the food product on the appropriate rack.

Sauté:

1. Heat the sauté pan to the appropriate temperature.
2. Evenly brown the food product.
3. For a sauce, pour off any excess oil, reheat, and deglaze.

GLOSSARY:

Brunoise: ⅛-inch dice
Sweat: to sauté under a cover
Bain-marie: hot-water bath

HACCP:

Hold at 140° F.

NUTRITION:

Calories: 95.7
Fat: 7.03 g
Protein: 1.95 g

CHEF NOTE:

Do not overbake the tomatoes or hold too long in a **bain-marie,** or they will collapse for service.

Vegetable Medley

YIELD: 50 SERVINGS. SERVING SIZE: FOUR OUNCES.

INGREDIENTS:

Macédoine:

4 pounds	Carrots, washed and peeled
2 pounds	Turnips, washed and peeled
2 pounds	Fresh green beans, washed, ends trimmed, and cut in ½-inch lengths
2 pounds	Frozen kernel corn
1½ pounds	Clarified butter
	Salt and ground white pepper, to taste

METHOD OF PREPARATION:

1. **Blanch** or steam all of the vegetables until **al dente.** Shock, and hold until ready to sauté.
2. In a large sauté pan, heat the butter. Add all of the vegetables, and sauté until thoroughly heated. Do not overcook the vegetables.
3. Season, to taste. Serve immediately, or hold at 140° F.

COOKING TECHNIQUES:

Boil, Sauté

Boil: (at sea level)

1. Bring the cooking liquid to a rapid boil.
2. Stir the contents, and cook the food product throughout.
3. Serve hot.

Sauté:

1. Heat the sauté pan to the appropriate temperature.
2. Evenly brown the food product.
3. For a sauce, pour off any excess oil, reheat, and deglaze.

GLOSSARY:

Macédoine: ¼-inch dice
Blanch: to parcook
Al dente: to the bite

HACCP:

Hold at 140° F.

NUTRITION:

Calories: 160
Fat: 13.7 g
Protein: 1.5 g

CHEF NOTE:

Additional herbs and/or spices can be added, depending on usage.

Zucchini Fritters

YIELD: 50 SERVINGS. SERVING SIZE: THREE EACH.

COOKING TECHNIQUE:
Deep-fry

Deep-Fry:
1. Heat the frying liquid to the proper temperature.
2. Submerge the food product completely.
3. Fry the product until it is cooked throughout.

HAZARDOUS FOODS:
Eggs
Milk
Peanut oil

NUTRITION:
Calories: 319
Fat: 27.1 g
Protein: 3.67 g

INGREDIENTS:

25 (7 pounds)	Zucchini
1 tablespoon	Salt
4 cups	Flour
2 tablespoons	Baking powder
7 each	Whole eggs, beaten
13 ounces	Milk
8 ounces	Onion, peeled and diced
1 gallon	Peanut oil, for frying
	or
6 gallons	Vegetable oil in fryolator

METHOD OF PREPARATION:

1. Wash, trim, and grate the zucchini. Do not peel. Sprinkle with salt.
2. Drain in a colander for 30 minutes.
3. Squeeze the zucchini, a handful at a time, to remove as much moisture as possible.
4. In a separate bowl, mix the flour and baking powder.
5. In another bowl, combine the eggs, milk, and onions.
6. Add the flour mixture to the egg mixture, and whip until smooth.
7. Add the grated zucchini. Stir until incorporated.
8. Drop by spoonfuls into oil heated to 350° F until golden brown on all sides.
9. Drain on absorbent towels, and serve.

CHEF NOTES:
1. A garnish of lemon wedges may be used.
2. To lighten the product, add eight beaten egg whites to the fritter mixture.

Zucchini Sauté

YIELD: 10 SERVINGS. SERVING SIZE: FOUR OUNCES.

COOKING TECHNIQUES:

Sauté, Simmer

Sauté:

1. Heat the sauté pan to the appropriate temperature.
2. Evenly brown the food product.
3. For a sauce, pour off any excess oil, reheat, and deglaze.

Simmer and Poach:

1. Heat the cooking liquid to the proper temperature.
2. Submerge the food product completely.
3. Keep the cooked product moist and warm.

GLOSSARY:

Brunoise: ⅛-inch dice

NUTRITION:

Calories: 78.9
Fat: 5.7 g
Protein: 2.13 g

INGREDIENTS:

2 ounces	Olive oil
8 ounces	Onions, peeled and diced **brunoise**
3 pounds	Zucchini, washed and cut into ½-inch slices
1 teaspoon	Turmeric (Moroccan paprika)
3 ounces	Fresh parsley, washed, excess moisture removed, and chopped
	Salt and ground white pepper, to taste
½ teaspoon	Garlic powder

METHOD OF PREPARATION:

1. In a sauté pan, heat the olive oil, and add the onions. Sauté the onions until they are translucent. Add the zucchini and the remaining ingredients, and continue to sauté for 2 to 3 more minutes.
2. Cover the pan, and let the zucchini simmer over low heat until tender, or for about 10 to 15 minutes.

CHEF NOTE:

The Moroccans like their vegetables very tender, almost overcooked.

8

Garde Manger

Garde Manger

There is nothing much better in the Western world than a fine, unctuous, truffled pâté.

M. F. K. Fisher
An Alphabet for Gourmets

The future of American gastronomy is in your hands. American chefs have, over recent years, demonstrated to the culinary world that American cuisine has developed to the stage at which it is uniquely identifiable to the United States. Your education at one of the world's utmost culinary schools, Johnson & Wales, will equip you to go forward and further develop this native cuisine. You have made the most important step by choosing to complete your culinary degree at Johnson & Wales. This education will enable you, with the help of the faculty, to be all that you can be as a chef. Remember: There is much more to learn. Seek out new information, and be prepared to watch and listen always. I wish you well.

Noel Cullen, CMC
Distinguished Visiting Chef
October 18–21, 1986
Associate Professor
Boston University

The HACCP Process

The following recipe illustrates how garde manger items flow through the HACCP process.

Smoked Duck Pâté

Receiving

- Duck—Check for firm flesh. There should be no signs of odor or stickiness. Duck should be delivered at 40° F (4.4° C) or lower.
- Vegetables (garlic)—packaging intact; no cross-contamination from other foods on the truck; no signs of insect or rodent activity
- Cream cheese—packaging intact. Cream cheese should be delivered at 40° F (4.4° C) or lower.
- Wine, Madeira—packaging intact
- Mustard, Dijon—packaging intact
- Herbs and spices (white pepper, nutmeg, and thyme)—packaging intact

Storage

- Duck—Store under refrigeration, with a product temperature not to exceed 40° F (4.4° C). Store below already cooked foods.
- Vegetables (garlic)—Store under refrigeration, with a product temperature not to exceed 40° F (4.4° C). Store above and away from raw, potentially hazardous foods.
- Cream cheese—Store under refrigeration, with a product temperature not to exceed 40° F (4.4° C). Store above and away from raw, potentially hazardous foods.
- Wine, Madeira—Store in dry storage at 50° F (10° C) with a relative humidity of 50% to 60%.
- Mustard, Dijon—Store in dry storage at 50° F (10° C) with a relative humidity of 50% to 60%.
- Herbs and spices (white pepper, nutmeg, and thyme)—Store in dry storage at 50° F (10° C) with a relative humidity of 50% to 60%. Keep dry.

Preparation and Cooking

- Smoke the duck to an internal temperature of 155° F (73.9° C).
- Debone, and remove the skin from the duck. Put the duck into a hermetically sealed bag, and immerse into an ice bath. Cool the duck to an internal temperature of 40° F (4.4° C) or less within 6 hours.
- In a food processor, coarsely grind the duck meat.

- Add the garlic and cream cheese, blending slowly.
- Add the remaining ingredients, blending on the "pulse" setting. Refrigerate.
- Use on canapés, pipe onto plastic wrap and make a roll, or place in a mold of choice.

Holding and Service

- Serve immediately at 40° F (4.4° C).

Nutritional Notes

Garde manger can be an avenue of creativity and artistic expression for the chef and a visual delight for the diner. An elaborate buffet table or artistic plate presentation can (and often does) make a meal memorable and the chef renowned. Garde manger not only provides the flair and artistry that separates the experienced professional from the amateur, but also encompasses the preservation of food through chilling, smoking, or curing and the production of forcemeats.

A food is not nutritious until it is eaten. Beautifully presented food entices the diner to taste it and sets a positive tone, heightening expectations of a pleasurable flavor experience. On the one hand, this makes it easier to offer healthy substitutions and smaller portion sizes. On the other hand, beautiful presentation increases the challenge to make the food delicious so as not to disappoint the diner. Garde manger is the perfect solution to this quandary: It can add variety and nutritional quality to the meal. Rich colors and a blend of textures and shapes delight the eye as well as the palate, and the vegetables and fruits used to enhance visual presentation also add vitamins, minerals, and fiber with few calories. (The more colorful fruits and vegetables are generally higher in vitamins.)

Aspic is often used as a binding agent, for flavor, to retain moisture, or for decoration. All forms of aspic are low in fat and calories and are, therefore, an excellent means of adding volume to the meal without adding fat and calories. Aspic made from the connective tissue of animals and fish is a source of protein. (Note, however, that unflavored gelatin is a refined product made from animal connective tissue and is a source of poor-quality protein, and aspic made from agar is a source of soluble fiber, not protein.)

Many sauces can be made using less fat, or can be prepared to eliminate fat. Chaud-froid sauces can be made using demi-glacé or reduced vegetable pulp and gelatin or agar. Arrowroot, cornstarch, instant flour, or other starches can be used in place of a roux. Emulsified sauces require an emulsifier, usually provided in the form of lecithin from egg yolks, which are high in cholesterol, and oil. Because it is difficult to produce good-quality, low-fat or fat-free forms of emulsified sauces in the kitchen, it is recommended that commercial forms of low-fat or fat-free mayonnaise or emulsified dressing be used when these are called for in a recipe. Evaporated skim milk thickened with a bit of cornstarch; pursed low-fat, mild, soft cheese; or drained nonfat yogurt can be used in place of cream or sour cream.

Many forcemeats can be made lower in fat by reducing or eliminating the fat called for in the recipe, using lean cuts of meat or fish, and using low-fat alternatives for cream and sour cream. Remember that the removal of fat will affect flavor and moisture, and so this must be taken into account when making substitutions and cooking.

The dough used for pâtés is high in fat, and it is difficult to achieve a similar result using a low-fat or oil version. The filling for the pâté, however, can be made lower in fat by reducing the amount of fat called for in the recipe and substituting lean cuts of meat and lower-fat ingredients. Legumes (particularly soy) and other vegetables can be used as extenders or in place of meat as a filling.

Mousses, galantines, and quenelles can easily be made low in fat by substituting low-fat ingredients, as mentioned. An additional binding agent may be required in some cases, however, to replace the fat.

Marinades made with little or no fat are an excellent way of adding flavor to low-fat dishes. Note, though, that the salt content of marinades may be of concern to some people.

Smoked meat and fish can be low in fat if lean cuts of meat are used. Fatty varieties of fish will be higher in calories, but the kind of fat contained in fish is considered to be healthy. Again, the salt used in curing the product may be of concern to some people.

A healthy diet is one that is well balanced—no one food or product is unhealthy by itself. Some items simply cannot be made successfully in a low-fat form; thus these items should be left as they are. High-fat garde manger items used as part of a presentation can be balanced by low-fat offerings. The primary objective is to offer choices that can provide a balanced, varied, interesting, and enjoyable meal.

All-Purpose Cure

YIELD: 2½ POUNDS. SERVING SIZE: ONE OUNCE.

COOKING TECHNIQUE:

Not applicable

NUTRITION:

Calories: 54.9
Fat: 0 g
Protein: 0 g

INGREDIENTS:

1¼ pounds	Kosher salt
1¼ to 1½ pounds	Granulated sugar

METHOD OF PREPARATION:

Combine the ingredients, and keep them in an air-tight container until ready for use.

CHEF NOTES:

1. Light brown sugar can be used instead of granulated sugar.
2. Any seasoning, herb, or combinations of seasonings can be used to add flavor.

Antipasto

COOKING TECHNIQUE:
Not applicable

NUTRITION:
Calories: 782
Fat: 68.3 g
Protein: 23 g

YIELD: 50 SERVINGS. SERVING SIZE: EIGHT OUNCES.

INGREDIENTS:

2 pounds	Salami, thinly sliced and cut into thin strips
2 pounds	Pepperoni, thinly sliced
2 pounds	Capocollo, thinly sliced
2 pounds	Canned tuna, flaked
1 pound	Cucumbers, peeled and thinly sliced
2 heads	Broccoli, cut into flowerettes
3 pounds	Mushrooms, washed and cut in half
2 pounds	Mozzarella cheese, thinly sliced and cut into thin strips
1 pound	Black olives
1 pound	Green olives
1 pound	Green peppers, washed, seeded, and cut into thin strips
1 pound	Red peppers, washed, seeded, and cut into thin strips
8 pounds	Tomatoes, washed, cored, and cut into wedges
1 pound	Red onions, thinly sliced
	Salad greens, as needed

Dressing:

2½ quarts	Olive oil
1½ quarts	Red wine vinegar
4 ounces	Granulated sugar
2 ounces	Fresh basil, washed and roughly chopped
2 ounces	Fresh oregano, washed and roughly chopped
½ teaspoon	Marjoram
3 cloves	Garlic, peeled and mashed into a paste
	Salt and black pepper, to taste
2 ounces	Fresh parsley, washed, excess moisture removed, and chopped

METHOD OF PREPARATION:

1. Decoratively arrange the meats, vegetables, and cheese on a plate lined with greens.
2. Mix the dressing ingredients together, and serve with the salad.

Apple Chutney

YIELD: ABOUT 1 QUART. SERVING SIZE: TWO OUNCES.

COOKING TECHNIQUE:

Not applicable

NUTRITION:

Calories: 84
Fat: 0 g
Protein: 0 g

INGREDIENTS:

1 pound	Cooking apples, unpeeled, cored, and sliced
1 each	Spanish onion, sliced
8 ounces	Raisins
18¼ ounces	Cider vinegar
16 ounces	Honey
2⅔ tablespoons	Fresh ginger, minced
2 tablespoons	Lemon juice
1 large clove	Garlic, minced
2 teaspoons	Sweet paprika
¾ teaspoon	Salt
½ teaspoon	Cinnamon
¼ teaspoon	Mace
⅛ teaspoon	Cloves
⅛ teaspoon	Cayenne pepper
½ pound	Cooking apples, unpeeled, cored, and sliced

METHOD OF PREPARATION:

Combine all of the ingredients except the last ½ pound of apples, and heat to a boil. Simmer for 40 to 45 minutes. Add the remaining apples, and simmer for another 15 minutes, or until quite thick, and then chill.

CHEF NOTE:

You can substitute quince for the apples.

COOKING TECHNIQUE:

Simmer

Simmer and Poach:

1. Heat the cooking liquid to the proper temperature.
2. Submerge the food product completely.
3. Keep the cooked product moist and warm.

GLOSSARY:

Reduction: evaporation of liquid by boiling

HAZARDOUS FOODS:

Veal bones
Pigs feet
Neck meat of beef
Egg whites

NUTRITION:

Calories: 11.1
Fat: 0 g
Protein: 1.27 g

Aspic

(Classical or Traditional Method)

YIELD: 2 GALLONS. SERVING SIZE: ONE OUNCE.

INGREDIENTS:

Stage 1:

10 pounds	Veal bones, washed
10 pounds	Pig feet
4 gallons	White veal stock, heated to a boil (see page 12)
1 pound	Onions, peeled and roughly chopped
½ pound	Carrots, washed, peeled, and roughly chopped
½ stalk	Celery, washed and roughly chopped
3 cloves	Garlic, peeled and mashed
3 each	Bay leaves
10 each	Peppercorns
3 each	Cloves

Stage 2:

3 pounds	Neck meat of beef, ground
12 each	Egg whites
4 ounces	Tomato purée
	Salt, to taste
1 pint	White wine

METHOD OF PREPARATION:

1. Heat the veal bones to a boil. Remove the scum as it forms on the top.
2. Combine all of the ingredients listed under stage 1. Simmer overnight. Strain and cool.
3. Mix all of the ingredients listed under stage 2. Pour into the cold strained liquid from stage 1, and slowly heat to a boil. Simmer for 2 hours.
4. Pour 2 ounces of clarified liquid on a precooled 7-inch plate. Place in a refrigerator to cool.
5. Check for firmness. If not firm, return the liquid to the stove. **Reduce** by simmering.
6. From time to time, cool 2 ounces of liquid. Retest the firmness by repeating steps 4 and 5.

Aspic

(Modern Method)

YIELD: 1 GALLON. SERVING SIZE: ONE OUNCE.

COOKING TECHNIQUE:
Simmer

Simmer and Poach:
1. Heat the cooking liquid to the proper temperature.
2. Submerge the food product completely.
3. Keep the cooked product moist and warm.

NUTRITION:
Calories: 6.18
Fat: 0 g
Protein: 1.35 g

INGREDIENTS:

1 gallon	Consommé or clarified liquid
4 to 8 ounces	Unflavored gelatin, depending on intended use of the aspic

METHOD OF PREPARATION:

1. Heat 3 quarts of consommé to a boil.
2. In the remaining quart, dissolve the gelatin.
3. Add the gelatin mixture to the boiling consommé.
4. Simmer until the liquid becomes clear again.

Aspic

(Quick or Convenient Method)

YIELD: 1 GALLON. SERVING SIZE: ONE OUNCE.

INGREDIENTS:

1 gallon	Water
8 ounces	Aspic powder
2 ounces	Unflavored gelatin

METHOD OF PREPARATION:

1. Heat 3 quarts of water to a boil.
2. In the remaining 1 quart of cold water, dissolve the unflavored gelatin and aspic powder.
3. Add the gelatin mixture to the boiling water. Return to a boil.

COOKING TECHNIQUE:

Boil

Boil: (at sea level)

1. Bring the cooking liquid to a rapid boil.
2. Stir the contents, and cook the food product throughout.
3. Serve hot.

NUTRITION:

Calories: 4.76
Fat: 0 g
Protein: 1.21 g

CHEF NOTE:

The liquid must be heated to a boil to ensure clarity. It does not have to be simmered or reduced.

Aspic Color Sheets

YIELD: 1½ CUPS. SERVING SIZE: NOT APPLICABLE.

COOKING TECHNIQUE:
Not applicable

Glossary:
Bain-marie: hot-water bath

NUTRITION:
Calories: 6.18
Fat: 0 g
Protein: 1.35 g

INGREDIENTS:

8 ounces	Coloring elements (organic)
1½ ounces	Unflavored gelatin
6 ounces	Aspic, heated

METHOD OF PREPARATION:

Method 1:

1. Cook the food product.
2. Place all of the ingredients into a stainless steel bowl, and place over **bain-marie** until the gelatin crystals are dissolved, approximately 1 to 3 minutes.
3. Place the mixture into blender, and purée. As a result of overmixing, the texture and color will change.
4. Place the puréed product on a tray lined with plastic wrap, and spread evenly over the tray by tipping. The thickness should not exceed ⅛ inch.
5. Refrigerate until firm, and roll for easier storage.

Method 2:

1. Combine all of the ingredients in a blender and purée.
2. Heat in bain-marie until the gelatin crystals are dissolved.
3. Further blending may be necessary to remove lumps.
4. Sieve the liquid, if necessary, and pour onto plastic-wrapped sheet pans.
5. A solidified sheet must be no more than ⅛-inch thick. Leftover scraps can be melted down and poured out again.

CHEF NOTES:

1. The sheet pans should be flat and straight.
2. Aspic color sheets are used when a vegetable cannot be used directly on another food product.

Basic Mousseline—Raw Farce

COOKING TECHNIQUE:
Not applicable

HACCP:
Cook to an internal temperature of 165° F.

HAZARDOUS FOODS:
Lean fillet of fish
Shellfish
Veal
Poultry
Egg whites
Heavy cream

NUTRITION:
Calories: 213
Fat: 15.8 g
Protein: 15.3 g

YIELD: 10 SERVINGS. SERVING SIZE: FOUR OUNCES.

INGREDIENTS:

1½ pounds	Lean fillets of fish, shellfish, veal, or poultry, cut into 1-inch pieces
1 to 1½ teaspoons	Salt
½ teaspoon	Nutmeg
1 pinch	Cayenne
¼ teaspoon	or Tabasco sauce
2 to 3 each	Egg whites (less gelatinous fish and poultry require three egg whites)
1¼ pint	Heavy cream

METHOD OF PREPARATION:

1. Combine the fish or meat and seasonings in the bowl of a food processor fitted with the metal blade. Process by frequently turning the machine on and off and scraping the mixture from the sides of the bowl until a smooth consistency is achieve.
2. With the food processor running, add the egg whites, and slowly add the heavy cream through the feed tube. Process until the mixture is well blended and fluffy. Test the seasonings, and correct, if necessary.
3. Mousseline keeps refrigerated for about 1 day or for up to 3 months if frozen. Mold, and cook in a water bath to an internal temperature of 165° F.

CHEF NOTE:

1. For better result, freeze or semi-freeze the protein ingredients before puréeing.
2. Place bowl on ice when mixing ingredients together.

Basic Pâté Spice I

COOKING TECHNIQUE:
Not applicable

YIELD: APPROXIMATELY 6 OUNCES. SERVING SIZE: AS NEEDED TO SEASON PÂTÉ.

INGREDIENTS:

⅔ ounce	Thyme
⅔ ounce	Bay leaf
⅓ ounce	Marjoram
⅓ ounce	Rosemary
1¼ ounces	Nutmeg
1¼ ounces	Cloves
⅔ ounce	Cayenne pepper
⅔ ounce	White pepper

METHOD OF PREPARATION:

Place all of the dry ingredients in the bowl of an electric coffee grinder; cover, and pulverize to a fine powder.

Basic Pâté Spice II

YIELD: APPROXIMATELY 6½ OUNCES. SERVING SIZE: AS NEEDED TO SEASON PÂTÉ.

COOKING TECHNIQUE:
Not applicable

INGREDIENTS:

1 ounce	White peppercorns
1 ounce	Black peppercorns
1 ounce	Mild (sweet) paprika
1 ounce	Hot paprika
½ ounce	Marjoram
½ ounce	Thyme
½ ounce	Basil
½ ounce	Nutmeg
½ ounce	Mace
1 ounce	Bay leaves
1 ounce	Cloves
½ ounce	Ginger

METHOD OF PREPARATION:

Place all of the dry ingredients in the bowl of an electric coffee grinder; cover, and pulverize to a fine powder.

Basic Pâté Spice III

COOKING TECHNIQUE:
Not applicable

YIELD: APPROXIMATELY 4 1/2 OUNCES.

SERVING SIZE: AS NEEDED TO SEASON PÂTÉ.

INGREDIENTS:

⅔ ounce	Whole black peppercorns
⅔ ounce	Whole white peppercorns
½ ounce	Sweet paprika
⅓ ounce	Allspice
⅓ ounce	Nutmeg
⅓ ounce	Cloves
⅓ ounce	Thyme
⅓ ounce	Ginger
¼ ounce	Savory
⅓ ounce	Mace
⅓ ounce	Hot paprika
⅛ ounce	Bay leaves

METHOD OF PREPARATION:

Place all of the dry ingredients in the bowl of an electric coffee grinder; cover, and pulverize to a fine powder.

Beef and Pork Sausage

YIELD: 3 POUNDS. SERVING SIZE: THREE OUNCES.

INGREDIENTS:

3 yards	Pork casings, washed

Sausage Stuffing:

2 pounds	Boneless beef chuck or beef scraps, cut into strips or cubes
1 pound	Boneless pork butt marbled with fat, cut into strips or cubes
3 each	Crushed ice cubes (preferably made from brown stock, see page 7)
1 to 2 tablespoons	Sage, crumbled
1 to 1½ tablespoons	Coriander, ground
½ tablespoon	Nutmeg
2 teaspoons	Kosher salt
1½ teaspoons	Pepper melange
1½ teaspoons	Mustard, ground
1½ teaspoons	Fennel greens, chopped

METHOD OF PREPARATION:

1. Combine all of the stuffing ingredients, and pass them through a coarse grinding disk. Cook a small amount in ¼ teaspoon of oil to test the seasonings, and adjust if necessary.
2. Stuff the casings, and form 3-inch links. Refrigerate the sausage, and uncover for 5 to 6 hours before cooking, to develop its flavor.

COOKING TECHNIQUES:

Shallow-Fry, Broil

Shallow-Fry:

1. Heat the cooking medium to the proper temperature.
2. Cook the food product throughout.
3. Season, and serve hot.

Grill/Broil:

1. Clean and heat the grill/broiler.
2. To prevent sticking, brush the food product with oil.

HAZARDOUS FOODS:

Beef chuck or beef scraps
Pork butt

NUTRITION:

Calories: 268
Fat: 14.2 g
Protein: 32.7 g

CHEF NOTE:

For best results, mix all ingredients together and refrigerate over night.

Black Bean Salsa

COOKING TECHNIQUE:
Not applicable

YIELD: 10 SERVINGS. SERVING SIZE: NOT APPLICABLE.

INGREDIENTS:

6 ounces	Cooked black beans, drained
1 each	Mango, peeled and coarsely chopped
6 ounces	Red onion, peeled and diced brunoise
1 each	Jalapeño chili, washed, seeded, and finely minced
2 ounces	Fresh cilantro leaves, coarsely chopped
2 ounces	Freshly squeezed lime juice, strained
	Salt and freshly ground black pepper, to taste

METHOD OF PREPARATION:

1. In a noncorrosive bowl, combine all of the ingredients for the salsa, and mix well.
2. Cover, and refrigerate at 40° F or lower for at least 1 hour, to allow the flavors to mingle.
3. Bring to room temperature for service.

Brine for Duck

YIELD: 5 QUARTS. SERVING SIZE: NOT APPLICABLE.

COOKING TECHNIQUE:
Boil

Boil: (at sea level)
1. Bring the cooking liquid to a rapid boil.
2. Stir the contents, and cook the food product throughout.
3. Serve hot.

NUTRITION:
Calories: 10.8
Fat: 0 g
Protein: 0 g

INGREDIENTS:

1 gallon	Water
½ pound	Kosher salt
8 ounces	Light port wine
1 pint	Honey
2 tablespoons	Whole cloves
2 tablespoons	Orange zest
1 tablespoon	Thyme
2 tablespoons	Fresh ground pepper melange

METHOD OF PREPARATION:

Combine all of the ingredients, and heat to a boil. Cool before using.

Brine for Meat, Poultry, or Fish

YIELD: 2 GALLONS. SERVING SIZE: ONE OUNCE.

INGREDIENTS:

2 gallons	Water
1½ pounds	Kosher salt
2 each	Bay leaves
1 tablespoon	Thyme
1 tablespoon	Peppercorns
1 tablespoon	Juniper berries
1 teaspoon	Ginger
1 teaspoon	Dry mustard powder
12 ounces	Sugar

METHOD OF PREPARATION:

Combine all of the ingredients, and boil for 3 minutes. Cool and strain before use.

COOKING TECHNIQUE:

Boil

Boil: (at sea level)

1. Bring the cooking liquid to a rapid boil.
2. Stir the contents, and cook the food product throughout.
3. Serve hot.

NUTRITION:

Calories: 5.61
Fat: 0 g
Protein: 0 g

Brine for Pork Loins or Butts

YIELD: 3½ GALLONS. SERVING SIZE: ONE OUNCE.

INGREDIENTS:

3 gallons	Water
2¼ pounds	Kosher salt
1 pound	Sugar

METHOD OF PREPARATION:

Heat the ingredients to a short boil. Cool before use.

COOKING TECHNIQUE:
Boil

Boil: (at sea level)
1. Bring the cooking liquid to a rapid boil.
2. Stir the contents, and cook the food product throughout.
3. Serve hot.

NUTRITION:
Calories: 4.66
Fat: 0 g
Protein: 0 g

CHEF NOTES:
1. This is a basic brine. Additional flavorings, such as ginger, ground mustard, or cardamom, can be added to intensify the flavor.
2. The sugar can be replaced with light brown sugar or honey.

Chanterelle Terrine

YIELD: 2½ POUNDS. SERVING SIZE: TWO OUNCES.

COOKING TECHNIQUE:

Bake

Bake:

1. Preheat the oven.
2. Place the food product on the appropriate rack.

GLOSSARY:

Bain-marie: hot-water bath
Brunoise: ⅛-inch dice

HACCP:

Cook to 165° F.

HAZARDOUS FOODS:

Egg white
Boneless lean veal
Heavy cream

NUTRITION:

Calories: 7.69
Fat: 4.89 g
Protein: 3.93 g

INGREDIENTS:

4 ounces	Shallots, diced **brunoise**
4 tablespoons	Olive oil
1 each	Egg white
3 slices	White bread, crusts removed
8 ounces	Lean, boneless veal
6 ounces	Heavy cream
1½ pound	Fresh chanterelles, washed
1 teaspoon	Caraway seeds, chopped
2 tablespoons	Fresh parsley, washed, excess moisture removed, and chopped
1 tablespoon	Fresh chervil, washed, excess moisture removed, and chopped
½ teaspoon	Cardamom
	Salt and pepper, to taste

METHOD OF PREPARATION:

1. Sauté the shallots in olive oil, and cool.
2. Add the egg white to the bread, and soften for panada.
3. Grind the veal in a robot coupe, add the shallots and panada, and incorporate completely.
4. Slowly add the heavy cream until a velvety consistency is achieved, and chill.
5. Fold the fresh chanterelles into the veal farce, and add the remaining seasonings. Cook a small amount to check the seasoning. Adjust, if necessary.
6. Grease a 1-quart terrine mold. Add the forcemeat, and bake in a **bain-marie** until an internal temperature of 165° F is reached.
7. Cool, and store in the mold until ready for use.

CHEF NOTE:

1. The temperature of the bain-marie should not exceed 176° F.
2. Also can be checked for doneness by inserting a toothpick and making sure liquids run clear.

Chaud-Froid Sauce

YIELD: 104 OUNCES. SERVING SIZE: TWENTY OUNCES.

INGREDIENTS:

7 ounces	Clarified butter
7 ounces	Flour
1¼ quarts	White veal stock, heated to a boil (see page 12)
1 quart	Milk, scalded
6 ounces	Unflavored gelatin
12 ounces	Cold water
8 ounces	White wine
8 ounces	Heavy cream
	Salt, to taste
	Tabasco sauce, to taste

METHOD OF PREPARATION:

1. Make a roux from the clarified butter and flour. Cook for 4 minutes. Do not brown.
2. Add the veal stock.
3. Add the scalded milk, and blend until smooth. Heat to a boil, reduce the heat, and simmer for 15 minutes.
4. Dilute the gelatin with the water. Add it to the mixture, and blend well. Simmer for 5 minutes.
5. Blend in the white wine and heavy cream.
6. Season. Pass through a fine strainer or **chinois.**

COOKING TECHNIQUE:

Boil

Boil: (at sea level)

1. Bring the cooking liquid to a rapid boil.
2. Stir the contents, and cook the food product throughout.
3. Serve hot.

GLOSSARY:

Chinois: cone-shaped strainer
Temper: to equalize two extreme temperatures

HAZARDOUS FOODS:

Milk
Heavy cream

NUTRITION:

Calories: 85.7
Fat: 5.63 g
Protein: 4.29 g

CHEF NOTES:

1. If a whiter product is desired, replace the 1¼ quarts of veal stock with 1¼ quarts of milk.
2. **Temper** the chaud-froid over an ice bath to achieve the proper temperature. When tempering, stir slowly to prevent air bubbles.
3. Utensils should be properly sanitized before using to prevent contaminating the chaud-froid with food particles, fat, and so on.
4. Place the food product on a wire rack or coating pan. Clean the drip pan.
5. Use only moderate amounts of chaud-froid to avoid contamination of large amounts.
6. Strain the chaud-froid before coating food products.
7. Always pour the chaud-froid in a continuous movement, starting with the part nearest you. It should have a smooth surface without lumps or air bubbles.

Country-Style Pâté

YIELD: 1 PÂTÉ, ABOUT 2¾ POUNDS. SERVING SIZE: TWO OUNCES.

COOKING TECHNIQUE:

Bake

Bake:

1. Preheat the oven.
2. Place the food product on the appropriate rack.

HAZARDOUS FOODS:

Pork liver
Milk
Ground pork
Ground veal
Boneless lean pork
Fresh fatback

NUTRITION:

Calories: 179
Fat: 13.6 g
Protein: 12 g

INGREDIENTS:

½ pound	Pork liver, cut into ⅝-inch cubes
6 ounces	Milk
½ pound	Ground pork
½ pound	Ground veal
2 each	Eggs
½ pound	Fresh fatback, cut into ⅝-inch cubes
2 tablespoons	Brandy, heated to evaporate the alcohol, and cooled
3 cloves	Garlic, minced
Zest of 1	Orange, grated
1½ tablespoons	Pâté seasoning II
¾ pound	Fatty bacon strips to line mold
3 each	Juniper sprig
2 each	Bay leaves
1 each	Fresh thyme sprig
1 each	Fresh sage sprig

METHOD OF PREPARATION:

1. Soak the diced pork liver in the milk for 4 hours, or overnight. Combine the ground pork and veal with the eggs. Mix in the diced pork and fatback, brandy, garlic, orange zest, and pâté seasoning. Drain and discard the milk from the liver; then add the liver to the farce.
2. Line the bottom and sides of a 6-cup rectangular mold with the bacon strips, and fill the mold with farce. Crisscross any remaining bacon strips over the farce. Place the juniper berries, bay leaves, and herb sprigs on top. Cover the mold with a well-greased piece of parchment paper cut to fit the top.
3. Bake the pâté in a preheated 425° F oven for about 1 hour, or bake it at 350° F to 375° F in a water bath for 1 to 1 ½ hours.
4. For service, garnish the pâté with a dab of coarse, spicy mustard, cornichons, and pickled pearl onions.

CHEF NOTES:

1. This country-style pâté, known as *pâté de campagne*, actually is a terrine and is baked as such. Its coarse texture is reminiscent of the flavorful, no-fuss loaves made by French farmers.
2. For a different variation, eliminate the ground veal, and use ¼ pound diced pork liver and ¼ pound ground pork. Another variation is to line the terrine with caul fat or fatback instead of bacon strips.

Cranberry Relish

YIELD: 10 SERVINGS. SERVING SIZE: TWO OUNCES.

INGREDIENTS:

1 each	Pear, washed, peeled, cored, and quartered
1 each	Apple, prepared same as above
1 each	Orange, zest removed and reserved, then **segmented**
1 each	Lemon, peel removed and segmented
¾ pound	Fresh cranberries, washed and stemmed
3 ounces	Granulated sugar, or to taste
1 ounce	Brandy
1 ounce	Cranberry liqueur

METHOD OF PREPARATION:

1. Combine all of the fruits in a food processor, and grind to a fine consistency.
2. Transfer the mixture to a nonreactive container, such as an earthenware crock. Add the sugar, and mix well.
3. Add the brandy and liqueur, mix well, and cover. Marinate for at least 24 hours, holding at a temperature of 40° F or below.

COOKING TECHNIQUE:

Not applicable

GLOSSARY:

Segmented: membranes removed

NUTRITION:

Calories: 88.3
Fat: .217 g
Protein: .411 g

CHEF NOTES:

1. Preparation should be started 1 day in advance of service to allow flavors to develop.
2. Fresh cranberries can be replaced with the same amount of frozen berries.

Croute Dough No. 1

YIELD: 1 DOUGH. SERVING SIZE: TWO OUNCES.

COOKING TECHNIQUE:
Not applicable

NUTRITION:
Calories: 208
Fat: 10 g
Protein: 4 g

INGREDIENTS:

12 ounces	Butter
35 ounces	Flour
3 each	Eggs
10 to 14 ounces	Water, lukewarm
¾ ounce	Salt

METHOD OF PREPARATION:

Mix or flake the butter into the flour. Add and mix the remaining ingredients. Allow to rest for 2 hours.

CHEF NOTES:

1. If a stronger dough is needed, 2 ounces of oil can be added to the recipe.
2. For a more flavorful product, add 2 tablespoons of chopped fresh herbs.

Croute Dough No. 2

YIELD: 2 DOUGHS. SERVING SIZE: ONE OUNCE.

COOKING TECHNIQUE:
Not applicable

NUTRITION:
Calories: 100
Fat: 4 g
Protein: 2 g

INGREDIENTS:

8 ounces	Butter
2 pounds	Flour
1 each	Egg
4 to 6 ounces	Water
1 teaspoon	Salt

METHOD OF PREPARATION:

Mix or flake the butter into the flour. Add and mix the remaining ingredients. Allow to rest for 2 hours.

CHEF NOTES:

1. If a stronger dough is needed, 2 ounces of oil can be added to the recipe.
2. For a more flavorful product, add 2 tablespoons of chopped fresh herbs.

Curried Cauliflower and Broccoli Terrine

YIELD: 1 THREE-POUND TERRINE. SERVING SIZE: THREE OUNCES.

INGREDIENTS:

2 pounds	Cauliflower, cleaned, washed, and hard core removed
4 each	Egg whites
2 teaspoons	Salt
¼ teaspoon	Ground white pepper
1 cup	Heavy cream, whipped
1 to 1 ½ bunches	Broccoli florets, steamed and cooled
2½ cups melted	Vegetable aspic

METHOD OF PREPARATION:

1. Steam the cauliflower until very tender; then drain and cool it in a colander. Purée the cauliflower, egg whites, curry, salt, and pepper in a food processor until smooth. Fold in the whipped cream a little at a time.
2. Grease a 6-cup mold (8½ × 4½ × 2½ inches), and spoon in half of the cauliflower mousseline. Arrange a cluster of broccoli florets down the center of the mousseline. Cover them with the remaining mousseline. Seal the mold with greased aluminum foil.
3. Bake in a water bath in a preheated 350° F oven for 1 hour. Chill the terrine. Unmold and decorate the cold terrine with a layer of the remaining broccoli florets. Glaze with cooled, melted aspic.
4. For service, serve the terrine with Dal, a lentil dish of India.

COOKING TECHNIQUES:
Steam, Bake

Steam: (Traditional)
1. Place a rack over a pot of water.
2. Prevent steam vapors from escaping.
3. Shock or cook the food product throughout.

Bake:
1. Preheat the oven.
2. Place the food product on the appropriate rack.

HACCP:
Bake in a preheated 350° F oven.

HAZARDOUS FOODS:
Egg whites
Heavy cream

NUTRITION:
Calories: 60
Fat: 4 g
Protein: 2.8 g

CHEF NOTES:
1. If you prepare the terrine a day or two in advance, leave it in its mold, and once chilled, seal it in plastic wrap. Unmold and decorate the terrine the night before or on the day you serve it.
2. Follow this basic recipe, and substitute another choice of vegetables for the mousseline and garniture for different variations.

Dried-Fruit Compote

YIELD: ABOUT 1 QUART. SERVING SIZE: ONE OUNCE.

INGREDIENTS:

1 pound	Mixed dried fruits, such as appricots, prunes, figs, peaches, pears, apples, and raisins
Zest of 1	Lemon, cut **julienne**
1 tablespoon	Lemon juice
1 each	Cinnamon stick
2 each	Whole cloves
16 ounces	Water

METHOD OF PREPARATION:

Combine all of the ingredients, and soak overnight. Heat the compote to a boil, and simmer for about 20 minutes, or until the fruit becomes tender. Add more water if necessary. Serve warm or cold.

COOKING TECHNIQUES:

Boil, Simmer

Boil: (at sea level)

1. Bring the cooking liquid to a rapid boil.
2. Stir the contents, and cook the food product throughout.
3. Serve hot.

Simmer and Poach:

1. Heat the cooking liquid to the proper temperature.
2. Submerge the food product completely.
3. Keep the cooked product moist and warm.

GLOSSARY:

Julienne: matchstick strips

NUTRITION:

Calories: 69
Fat: 0 g
Protein: .5 g

CHEF NOTE:

This compote goes well with duck.

Dry Cure for Fish

YIELD: 3½ POUNDS. SERVING SIZE: ONE OUNCE.

COOKING TECHNIQUE:

Not applicable

NUTRITION:

Calories: 54.5
Fat: 0 g
Protein: 0 g

INGREDIENTS:

1½ pounds	Kosher salt
1½ to 2 pounds	Light brown sugar
3 tablespoons	Lemon zest
1 teaspoon	Cayenne pepper
1 teaspoon	Fresh ground pepper melange
3 tablespoons	Dill, chopped

METHOD OF PREPARATION:

Combine all of the ingredients, and keep in an air-tight container until ready for use.

CHEF NOTE:

If the cure is too salty, increase the amount of all ingredients except salt.

Duck Terrine with Veal and Green Peppercorns

COOKING TECHNIQUE:
Not applicable

GLOSSARY:
Galantine: deboned and stuffed

HAZARDOUS FOODS:
Duck
Veal eye of round
Sausage meat

NUTRITION:
Calories: 420
Fat: 33 g
Protein: 25 g

CHEF NOTE:
Because duck contains a lot of fat, there is no need to line the mold with fatback in this recipe.

YIELD: 1 TERRINE, ABOUT 3¼ POUNDS. SERVING SIZE: FOUR OUNCES.

INGREDIENTS:

1 (5-pound)	Duck, boned as for a galantine (see page 651) (dice the liver and add to the farce)
¾ pound	Veal eye of round, sliced into fillets
1 small	Onion, minced
2 cloves	Garlic, pressed
1 tablespoon	Orange, grated zest
2 tablespoons	Cognac
1¼ pounds	Sausage meat or basic veal and pork farce
1 to 2 tablespoons	Pâté seasoning I
1 teaspoon	Dijon-style mustard
1 tablespoon	Green peppercorns, coarsely chopped and drained
½ teaspoon	Cracked black pepper
¼ teaspoon	Thyme
⅛ teaspoon	Cinnamon
Pinch	Nutmeg
Pinch	Cloves
1 each	Bay leaf
1 each	Fresh rosemary sprig
1 each	Fresh thyme sprig
3 each	Juniper berries

Marinade:

2 tablespoons	Vegetable oil
2 ounces	Strong red wine (Cabernet)
1 teaspoon	Rosemary
6 each	Juniper berries, crushed
1 each	Bay leaf
	Grated black pepper

METHOD OF PREPARATION:

1. Combine the ingredients for the marinade, and pour over the boned duck and the veal fillets. Cover and marinate; refrigerate for 6 hours to 3 days, occasionally turning the meat.
2. Heat the onion, garlic, and orange zest in the cognac to soften the ingredients and evaporate the alcohol. Mix into the sausage meat, along with the pâté seasoning, mustard, green peppercorns, cracked black pepper, thyme, cinnamon, nutmeg, and cloves. Test the seasonings and correct if necessary, and chill.
3. Drain off and discard the marinade. Line a 6-cup triangular mold (7 ½ × 3¾ × 3½ inches) with the boned duck, skin-side out. Press half of the farce into the bottom of the mold, forming a slight groove down the center of the farce. Place the marinated veal fillets in the groove. Add the remaining farce. Fold over the edges of the duck and sew them closed, or secure them together with toothpicks. Place the bay leaf, rosemary and thyme sprigs, and juniper berries on top. Seal the mold with aluminum foil.

Galantine of Capon

YIELD: 10 SERVINGS. SERVING SIZE: THREE OUNCES.

COOKING TECHNIQUE:

Simmer

Simmer and Poach:

1. Heat the cooking liquid to the proper temperature.
2. Submerge the food product completely.
3. Keep the cooked product moist and warm.

GLOSSARY:

Brunoise: finely diced
Galantine: boned and stuffed

HACCP:

Heat to an internal temperature of 165 ° F.

HAZARDOUS FOODS:

Whole capon
Chicken meat
Egg whites
Heavy cream

NUTRITION:

Calories: 163
Fat: 9.5 g
Protein: 17 g

INGREDIENTS:

1 seven-pound	Capon, whole
3½ to 4 pounds	Chicken meat
1 ounce	Onion (medium), diced **brunoise**
8 each	Shallots, finely chopped
2 ounces	Clarified butter
8 each	Egg whites
12 ounces	Heavy cream
2 tablespoons	Poultry seasonings
	Salt and fresh ground black pepper, to taste
4 ounces	Olives, black, or truffles, diced
4 ounces	Pistachio nuts, shelled (whole, blanched, peeled)
1 teaspoon	Fresh thyme
1 teaspoon	Fresh sage
2 ounces	Sherry
6 each	Ham bars, ¼ × ¼ × 6 inches long
3½ gallons	White chicken stock, heated to a boil (see page 12)

METHOD OF PREPARATION:

1. Debone and skin the capon, reserving the skin. Remove the meat, trimming the fat and tendons. Pass through the grinder twice, and chill.
2. In a sauté pan, heat the butter, and saute the onions until translucent and shallots. Cool; then add the chicken meat, and run the mixture through a robot coupe.
3. Slowly add the egg whites, heavy cream, and poultry seasoning, and season, to taste. Remove from robot coupe.
4. Fold in the olives, pistachios, thyme, sage, and sherry. Adjust the seasonings accordingly. Place the chicken skin outside, facing down on a cheesecloth or plastic wrap. Spread one fourth of the forcemeat on the skin evenly.
5. Insert the ham bars into the forcemeat. Cover the bars with the forcemeat. Place the rest of the bars into the forcemeat evenly. Covering with the balance of the forcemeat mixture, roll up and tie.
6. Add the galantine to the simmering stock, and simmer for approximately 1½ hours, or until an internal temperature of 165° F is reached. Then cool the galantine in the stock.

Game Pâté Spice

COOKING TECHNIQUE:
Not applicable

YIELD: APPROXIMATELY 9 OUNCES. SERVING SIZE: AS NEEDED TO SEASON PÂTÉ.

INGREDIENTS:

½ ounce	Whole white peppercorns
½ ounce	Whole black peppercorns
¼ ounce	Mild (sweet) paprika
¼ ounce	Hot paprika
¼ ounce	Fresh marjoram
½ ounce	Fresh basil
½ ounce	Nutmeg
½ ounce	Mace
10 each	Bay leaves
½ ounce	Cloves
½ ounce	Lovage
½ ounce	Ginger, ground
½ ounce	Fresh thyme
1 ounce	Juniper berries
2 ounces	Dried mousseron (pickled mushrooms)

METHOD OF PREPARATION:

Place all of the dry ingredients in the bowl of an electric coffee grinder; cover, and pulverize to a fine powder.

Garlic Sausage

COOKING TECHNIQUE:
Not applicable

HAZARDOUS FOOD:
Boneless lean pork
Pork fat

NUTRITION:
Calories: 455
Fat: 34 g
Protein: 31 g

YIELD: 3¾ POUNDS. SERVING SIZE: FOUR OUNCES.

INGREDIENTS:	
3 pounds	Boneless lean pork, cut into cubes
¾ pound	Pork fat, cut into cubes
4 ounces	Ice cubes
8 cloves	Garlic, minced
2 ounces	Brandy, chilled
1½ teaspoon	Salt
¾ teaspoon	Fresh ground black pepper
⅛ teaspoon	Nutmeg
⅛ teaspoon	Ginger, ground
⅛ teaspoon	Cloves, ground
4 yards	Pork casing (medium), soaked in water

METHOD OF PREPARATION:

Combine all of the ingredients except the casings and pass through a medium grinding disk. Place the mixture in the refrigerator, covered for 3 to 4 hours. Using a sausage press, stuff the casings, and form 4-inch links. Refrigerate until ready to cook.

Gravlaks with Mustard Sauce

COOKING TECHNIQUE:
Not applicable

HAZARDOUS FOOD:
Norwegian salmon fillet

NUTRITION:
Calories: 563
Fat: 49.4 g
Protein: 18.4 g

YIELD: 10 SERVINGS. SERVING SIZE: FIVE OUNCES.

INGREDIENTS:

2 pounds	Norwegian salmon fillet with ground black pepper on skin
3 tablespoons	Kosher salt
5 tablespoons	Sugar
2 tablespoons	Aquavit, if available; if not, use Schnaps
2 tablespoons	Caraway seeds
8 ounces	Dill, fresh and chopped

Mustard Sauce:

1 pint	Olive oil
2 tablespoons	Dijon mustard
2 ounces	Grain mustard
4 tablespoons	Sugar
2 tablespoons	White wine vinegar
1 each	Lemon
4 tablespoons	Dill, chopped
	Salt and pepper, to season

METHOD OF PREPARATION:

1. Marinate the salmon with the pepper, dill, salt, sugar, and aquavit for 2 to 3 days in the refrigerator. Turn the fillets twice each day.
2. Slice the gravlaks very thin.
3. Mix the mustard, sugar, vinegar, lemon juice, dill, salt, and pepper. Add olive oil as if making a mayonnaise. If the sauce gets too thick, add lukewarm water. The sauce should be made 1 day ahead to get the best flavor.

Grilled Gravlaks with Dill Sauce

YIELD: 10 SERVINGS. SERVING SIZE: SIX OUNCES.

INGREDIENTS:

2 pounds	Salmon
3 tablespoons	Salt
5 tablespoons	Sugar
3 tablespoons	Pepper, black, freshly ground
3 ounces	Dill, fresh and chopped

Dill Sauce:

3 each	Shallots
1 ounce	Butter
2 ounces	Dijon-style mustard
10 ounces	Fish stock or broth, heated to a boil (see page 10)
10 ounces	Cream
3 teaspoons	Lemon juice
3 tablespoons	Dill, fresh and chopped
1 teaspoon	Pepper melange

METHOD OF PREPARATION:

1. Marinate the salmon fillet with pepper, dill, salt, and sugar for 12 to 16 hours. Keep refrigerated, and turn the fillet once during the process.
2. Remove the spices, cut the gravlaks in serving portions, and grill medium.
3. Start the sauce by cooking the onions and butter in the mustard, add the fish stock and cream, and reduce to half of the original amount.
4. Season with salt and pepper, and add some drops of lemon juice to give the sauce a touch of acidity. Add the dill at the last minute. Cool to room temperature.
5. Serve the grilled gravlaks with dill sauce, fresh vegetables or salad, and boiled potatoes.

COOKING TECHNIQUE:

Grill

Grill/Broil:

1. Clean and heat the grill/broiler.
2. To prevent sticking, brush the food product with oil.

HAZARDOUS FOODS:

Salmon
Cream

NUTRITION:

Calories: 261
Fat: 15.9 g
Protein: 19.8 g

CHEF NOTE:

Pepper melange is a mixture of black, green, and pink peppercorns.

Guacamole

YIELD: 50 SERVINGS. SERVING SIZE: ONE OUNCE.

COOKING TECHNIQUE:
Not applicable

GLOSSARY:
Concassé: peeled, seeded, roughly chopped
Brunoise: finely diced

HAZARDOUS FOOD:
Sour cream

NUTRITION:
Calories: 81
Fat: 7.2 g
Protein: 1.14 g

INGREDIENTS:

10 each	Avocado, ripened
1 pound	Tomato, monder and **concassé**
1 pound	Onion, peeled and diced **brunoise**
2 each	Jalapeño peppers, washed, seeded, and diced brunoise
8 ounces	Sour cream
2 ounces	Lemon juice
	Salt and ground black pepper, to taste

METHOD OF PREPARATION:

1. Cut the avocado in half. Remove the seed, and remove the flesh from the skin. If holding the flesh, addition of lemon or lime juice will prevent discoloration.
2. Purée the flesh in a robot coupe until smooth.
3. Blend the puréed mixture with the remaining ingredients.
4. Tightly cover with plastic wrap to prevent discoloration, and refrigerate until service.

Mango Relish

YIELD: 1 QUART. SERVING SIZE: THREE OUNCES.

COOKING TECHNIQUE:
Not applicable

HACCP:
Refrigerate at 40° F or below.

NUTRITION:
Calories: 94
Fat: 5.27 g
Protein: .674 g

INGREDIENTS:

5 each	Large mangoes, washed
5 each	Hot chili peppers, seeded and finely minced
5 cloves	Garlic, peeled and finely minced
	Salt, to taste
3 ounces	Olive oil

METHOD OF PREPARATION:

1. Peel the mangoes, and cut the flesh from the seed.
2. Chop the flesh to a pulp, add the remaining ingredients, and mix well.
3. Hold in a covered stainless steel bowl and refrigerate at 40° F or below.

No-Cook Apple Chutney

COOKING TECHNIQUE:
Not applicable

GLOSSARY:
Brunoise: finely diced

NUTRITION:
Calories: 56
Fat: 0 g
Protein: .5 g

YIELD: ABOUT 44 OUNCES. SERVING SIZE: TWO OUNCES.

INGREDIENTS:

1 pound	Cooking apples, cored and grated with peel
2 tablespoons	Lemon juice mixed with 2 tablespoons of water
½ pound	White onions, grated
1¼ ounces	Golden raisins
1 each	Green bell pepper, diced **brunoise**
2 tablespoons	Pimiento, diced brunoise
2 ounces	Pitted dates, diced brunoise
3 tablespoons	Crystallized ginger, finely chopped
2 tablespoons	Cider vinegar
1 teaspoon	Honey
1 teaspoon	Salt
¼ teaspoon	Cardamom

METHOD OF PREPARATION:

Toss the grated apples in the lemon water. Drain, and discard the liquid. Combine the apples with the remaining ingredients, and mix well.

COOKING TECHNIQUE:
Bake

Bake:
1. Preheat the oven.
2. Place the food product on the appropriate rack.

GLOSSARY:
Julienne: matchstick strips

HACCP:
Bake in a 350° F oven.

NUTRITION:
Calories: 22
Fat: 0 g
Protein: 0 g

Orange Poached Plums

YIELD: ABOUT 16 OUNCES. SERVING SIZE: ONE OUNCE.

INGREDIENTS:

Juice of 1	Orange
½ each	Orange, peeled and diced
Zest of ¼ each	Orange, cut **julienned**
½ pound	Victoria plums, halved and pitted
2 tablespoons	Honey
1 each	Cinnamon stick
2 each	Cloves

METHOD OF PREPARATION:

Combine all of the ingredients in a casserole dish. Bake in a 350° F oven for about 15 minutes, or until the plums become tender. Serve warm or chilled.

Pâté en Croute

YIELD: 1 MOLD. SERVING SIZE: THREE OUNCES.

COOKING TECHNIQUE:
Bake

Bake:
1. Preheat the oven.
2. Place the food product on the appropriate rack.

GLOSSARY:
Brunoise: ⅛-inch dice

NUTRITION:
Calories: 219
Fat: 13.5 g
Protein: 15 g

INGREDIENTS:

4 ounces	Clarified butter
6 ounces	Onions, diced **brunoise**
3 ounces	Carrots, diced brunoise
8 each	Shallots, diced brunoise
1 teaspoon	Poultry seasoning
10 ounces	Port wine
2 each	Bay leaves
12 ounces	Ham bars, ¼ × ¼ × 6 inches long
1 pound	Veal, trimmed, ground two times through a ⅛-inch plate, and chilled
1 pound	Ground pork
6 to 8 ounces	Heavy cream
4 each	Egg whites
2 teaspoons	Parsley, chopped
½ pound	Salt pork, diced brunoise
4 ounces	Pistachio nuts, whole, blanched, and peeled

METHOD OF PREPARATION:

1. Preheat the oven to 400° F.
2. In a sauté pan, heat the butter, and sauté the vegetables until the onions are translucent, and then cool. Transfer to a suitable container, and add the seasonings and 7 ounces of the wine, and mix well. Add the ham bars, and marinate overnight. Remove the bars and bay leaves, and drain, reserving the vegetables and 3 ounces of the wine.
3. Pass the veal through a food processor until it becomes a smooth paste; add the pork and vegetables; then add the heavy cream and egg whites slowly with the reserved wine, and remove from the food processor. Fold in the parsley, salt pork, and pistachio nuts, and adjust the seasonings.
4. Roll out the previously prepared dough to ⅛-inch thick. Grease a pâté mold, and dust with flour. Cut out one piece to cover the bottom and overlap on both sides. Cut two pieces larger than the ends, allowing for overlap, and cut one piece for the lid, but do not allow it to overlap.
5. Place the large piece in the mold first. The overlap should hang over the sides. Place the ends in next. Egg-wash the areas that meet so they glue together. Place 1-inch-thick forcemeat in the bottom of the dough lining. Place four bars of ham on the forcement. Repeat with another layer of forcemeat and ham bars. Put the remaining forcemeat in the mold, filling

to the top. Fold and overlap the dough up over the top of the forcemeat; egg wash, and place the dough lid on top.

6. Cut two holes, and line with chimneys made with foil, allowing for the steam to escape. Decorate the top cover with dough that was cut into different shapes. Use an egg wash to hold the dough to dough.
7. Bake in the oven for approximately 15 minutes, until the top is brown. Cover loosely with foil, and reduce the oven to 350° F. Continue baking for approximately 1 hour. Check for doneness by piercing the chimney with a needle. Liquid should run clear.
8. When done, chill. Fill with tempered aspic, and chill. When the aspic is set, unmold the pâté. Cut into slices ½-inch thick, and aspic all the slices. Leave ⅓ of the pâté en croute unsliced for decoration on the platter.

Pâté of Salmon

YIELD: 16 SERVINGS. SERVING SIZE: THREE OUNCES.

INGREDIENTS:

2 ounces	Milk
3 ounces	Flour
3 each	Egg yolks
1½ ounces	Butter
2 pounds	Salmon fillet, diced
1 each	Egg white
2 ounces	Brandy
1 ounce	Fresh dill, minced

METHOD OF PREPARATION:

1. Preheat the oven to 350° F.
2. In a sauce pan, bring the milk and butter to a boil. Add the flour all at once and cook by stirring with a wooden spoon. It is done when the dough shapes into a ball and separates from the side of the pan. Remove, add the yolks, and incorporate.
3. In a food processor, place the diced salmon, panada, egg white, and brandy, and emulsify until smooth.
4. Stir in the dill. Line a terrine pan with plastic wrap. Fill with the salmon mixture. Cover the pan with aluminum foil. Set the pan in a water bath, and bake for 1½ hours, or until an internal temperature of 150° F degrees is reached.
5. Cool, placing a pan on top of the pâté to press down and firm the pâté.
6. Let rest for 3 hours; then unmold, decorate, and slice.

COOKING TECHNIQUES:

Boil, Bake

Boil: (at sea level)

1. Bring the cooking liquid to a rapid boil.
2. Stir the contents, and cook the food product throughout.
3. Serve hot.

Bake:

1. Preheat the oven.
2. Place the food product on the appropriate rack.

HACCP:

Preheat the oven to 350° F.
Cook to an internal temperature of 150° F.

HAZARDOUS FOODS:

Milk
Egg yolks
Salmon fillet
Egg white

NUTRITION:

Calories: 142
Fat: 6.92 g
Protein: 12.7 g

CHEF NOTE:

Diced shrimp or scallops can be added to the fish mixture to add extra garnish and color appeal.

Pâté Seasoning I

YIELD: 4 ½ OUNCES

INGREDIENTS:

⅔ ounce	Whole black peppercorns
⅔ ounce	Whole white peppercorns
½ ounce	Sweet paprika
⅓ ounce	Allspice
⅓ ounce	Basil
⅓ ounce	Cloves
⅓ ounce	Ginger
⅓ ounce	Thyme
¼ ounce	Hot paprika
¼ ounce	Oregano
⅛ ounce	Bay leaves

METHOD OF PREPARATION:

Place the herbs and spices in a blender or coffee grinder, and blend well. Place the spice blends in a sifter to remove any large bits, and cover in an air-tight container.

CHEF NOTE:

This is a good, basic seasoning for any charcuterie.

Pâté Seasoning II

YIELD: ABOUT 6 OUNCES

INGREDIENTS:	
1 to 1¼ ounces	Dried cepes or shiitake mushrooms
¾ ounce	Whole white peppercorns
¾ ounce	Basil
¾ ounce	Nutmeg
¾ ounce	Thyme
½ each	Vanilla bean, scraped
⅓ ounce	Allspice
⅓ ounce	Coriander seeds
⅓ ounce	Mace
¼ ounce	Cinnamon
¼ ounce	Cloves
⅛ ounce	Bay leaves
⅛ ounce	Cardamom

METHOD OF PREPARATION:

Place the herbs and spices in a blender or coffee grinder, and blend well. Place the spice blends in a sifter to remove any large bits, and cover in an air-tight container.

CHEF NOTE:

For a light and fragrant blend, use this recipe to season white meats such as veal, chicken, and sweetbreads.

Pâté Seasoning III

YIELD: ABOUT 5½ OUNCES

INGREDIENTS:

1½ ounces	Whole green peppercorns, dried
⅔ ounce	Sweet paprika
½ ounce	Summer savory
½ ounce	Thyme
⅓ ounce	Allspice
⅓ ounce	Basil
⅓ ounce	Coriander seeds
⅓ ounce	Mace
¼ ounce	Cloves
¼ ounce	Ginger
¼ ounce	Rosemary
⅛ ounce	Bay leaves

METHOD OF PREPARATION:

Place the herbs and spices in a blender or coffee grinder, and blend well. Place the spice blends in a sifter to remove any large bits, and cover in an air-tight container

CHEF NOTES:

1. Stored properly, these spices will keep for up to 3 months.
2. Spicy and aromatic, this mixture accentuates the flavor of a coarse pork pâté, terrine, or sausage.

Pâté Spice for Coarse Pork Pâté, Terrine, or Sausage

COOKING TECHNIQUE:
Not applicable

YIELD: APPROXIMATELY 5½ OUNCES. SERVING SIZE: AS NEEDED TO SEASON PÂTÉ.

INGREDIENTS:

1½ ounces	Dried peppercorn melange
⅔ ounce	Sweet paprika
½ ounce	Savory
⅓ ounce	Allspice
½ ounce	Thyme
⅓ ounce	Dried orange zest
⅓ ounce	Coriander seeds
⅓ ounce	Mace
¼ ounce	Cloves
⅛ ounce	Ginger
¼ ounce	Rosemary
⅛ ounce	Bay leaves

METHOD OF PREPARATION:

Place all of the dry ingredients in the bowl of an electric coffee grinder; cover, and pulverize to a fine powder.

COOKING TECHNIQUE:
Not applicable

Pâté Spice for Delicately Flavored Meats

YIELD: APPROXIMATELY 5.5 OUNCES. SERVING SIZE: AS NEEDED TO SEASON PÂTÉ.

INGREDIENTS:

¾ ounce	Whole white peppercorns
½ ounce	Coriander seeds
1 ounce	Thyme
1 leaf	Basil
¼ ounce	Cloves
¾ ounce	Nutmeg
½ ounce	Bay leaves
¼ ounce	Allspice
¼ ounce	Mace
1 ounce	Dried mushrooms

METHOD OF PREPARATION:

Place all of the dry ingredients in the bowl of an electric coffee grinder; cover, and pulverize to a fine powder.

COOKING TECHNIQUE:
Not applicable

Pâté Spice for Highly Flavored, Country-Style Pâtés

YIELD: APPROXIMATELY 4 OUNCES. SERVING SIZE: AS NEEDED TO SEASON PÂTÉ.

INGREDIENTS:

1½ ounces	Dried green peppercorns
½ ounce	Allspice
½ ounce	Mace
½ ounce	Mild (sweet) paprika
6 each	Bay leaves
⅛ ounce	Coriander seeds
⅛ ounce	Fresh thyme
⅛ ounce	Fresh rosemary
½ ounce	Fresh basil
½ ounce	Fresh marjoram
1 teaspoon	Cloves

METHOD OF PREPARATION:

Place all of the ingredients in the bowl of an electric coffee grinder; cover, and pulverize to a fine powder.

COOKING TECHNIQUE:
Not applicable

Pâté Spice for White Meats

(Veal, Chicken, Sweetbreads)

YIELD: APPROXIMATELY 4 OUNCES. SERVING SIZE: AS NEEDED TO SEASON PÂTÉ.

INGREDIENTS:

¾ ounce	Whole white peppercorns
⅓ ounce	Allspice
⅓ ounce	Coriander seeds
⅓ ounce	Mace
⅓ ounce	Juniper berries
¾ ounce	Nutmeg
¾ ounce	Thyme
¼ ounce	Cinnamon
¼ ounce	Cloves
⅛ ounce	Bay leaves
⅛ ounce	Cardamom

METHOD OF PREPARATION:

Place all of the dry ingredients in the bowl of an electric coffee grinder; cover, and pulverize to a fine powder.

Pickled Salmon

YIELD: 15 SERVINGS. SERVING SIZE: FIVE OUNCES.

COOKING TECHNIQUE:

Boil

Boil: (at sea level)

1. Bring the cooking liquid to a rapid boil.
2. Stir the contents, and cook the food product throughout.
3. Serve hot.

HAZARDOUS FOOD:

Salmon

NUTRITION:

Calories: 300
Fat: 3.87 g
Protein: 12.1 g

INGREDIENTS:

2 pounds	Salmon with skin and bones
3 tablespoons	Salt
1 quart	Water
24 ounces	Apple cider vinegar
26 ounces	Sugar
1 tablespoon	Black pepper
1 tablespoon	Mustard seeds
10 each	Whole cloves
5 each	Laurel leaves
1 each	Onion, medium, sliced
2 ounces	Parsley stems
1 ounce	Dill stems
1 each	Orange, juiced and zest

METHOD OF PREPARATION:

1. Cut the salmon into small cubes (2×2 cm) or 1-inch-thick slices. Add the salt, and leave for 1 hour. Rinse in cold water.
2. Heat the water to a boil with the rest of the ingredients for 10 minutes. Strain, and add the salmon fillet.
3. The salmon fillets are ready to serve after 5 to 10 minutes, depending on their size, but they will be best left in the marinade overnight.
4. Serve the pickled salmon with a cold horseradish sauce (made with sour cream) and assorted greens in a vinaigrette.

Pico de Gallo

(Salsa)

COOKING TECHNIQUE:
Not applicable

GLOSSARY:
Blanch: to par cook
Brunoise: ⅛-inch dice

HACCP:
Chill to 40° F.

NUTRITION:
Calories: 66.3
Fat: 5.59 g
Protein: .694 g

YIELD: 20 OUNCES. SERVING SIZE: TWO OUNCES.

INGREDIENTS:

12 ounces	Fresh ripe tomatoes, washed, cored, **blanched,** peeled, seeded, and diced **brunoise**
6 ounces	Onions, peeled and diced brunoise
4 to 6 cloves	Garlic, peeled and finely minced
2 each	Jalapeño peppers, washed, seeded, and finely minced
2 ounces	Freshly squeezed lemon juice
2 ounces	Olive oil
1 ounce	Fresh cilantro leaves, washed, dried, and roughly chopped
	Salt, to taste

METHOD OF PREPARATION:

1. In a noncorrosive bowl, combine all of the ingredients, and mix well.
2. Refrigerate at 40° F or below for at least 30 minutes before using to allow flavors to blend. Hold at 40° F or below, until service.

CHEF NOTE:
Serrano peppers can be substituted for jalapeñno.

Piri Piri

COOKING TECHNIQUE:

Not applicable

YIELD: 20 OUNCES.

INGREDIENTS:

2 ounces	Hot red peppers, dried
2 ounces	Dry sherry
16 ounces	Olive oil

METHOD OF PREPARATION:

Combine all of the ingredients, and cover. The longer this stands, the better it is. It can be kept up to 1 month. Add more oil as the quantity is reduced.

CHEF NOTE:

Preparation should begin early so that the pork can be marinated according to the recipe.

Pork and Chicken Boudin with Fine Herbs

YIELD: 3½ POUNDS. SERVING SIZE: FIVE OUNCES.

INGREDIENTS:

¼ to ½ pound	Ground pork fat
12 ounces	Onions, diced
8 ounces	Milk
8 ounces	Bread crumbs
1½ pounds	Ground, boneless, lean pork from shoulder or loin
2 each	Eggs, whole
3 each	Egg whites
1 tablespoon	All-purpose pâté seasoning
1 to 2 teaspoons	Kosher salt
8 ounces	Heavy cream
2 tablespoons	Fresh parsley, washed, excess moisture removed, and chopped
2 teaspoons	Dill seeds
1 pound	Boneless, skinless chicken breasts, diced
	Water, half milk/half water, or white stock (see page 12), for poaching
2 teaspoons	Chervil, chopped
3½ yards	Medium pork casings, rinsed

METHOD OF PREPARATION:

1. Melt half of the ground fat in a skillet over medium-low heat, and add the onions. Cover, and slowly cook for 15 to 20 minutes. Cool.
2. Meanwhile, make a panada by heating the milk to a boil. Add the bread crumbs. Stir constantly over medium heat until the mixture forms a thick paste and almost holds its own shape. Cool completely before using.
3. Mix together the remaining pork fat, sautéed onions, ground pork, panada, eggs, egg whites, pâté spice, and salt. Process the mixture in a robot coupe until it develops a velvety, thick consistency. Chill for about 15 minutes.
4. Return the mixture to the robot coupe. With the machine running, slowly

COOKING TECHNIQUES:

Sauté, Boil, Poach

Sauté:

1. Heat the sauté pan to the appropriate temperature.
2. Evenly brown the food product.
3. For a sauce, pour off any excess oil, reheat, and deglaze.

Boil: (at sea level)

1. Bring the cooking liquid to a rapid boil.
2. Stir the contents, and cook the food product throughout.
3. Serve hot.

Simmer and Poach:

1. Heat the cooking liquid to the proper temperature.
2. Submerge the food product completely.
3. Keep the cooked product moist and warm.

GLOSSARY:

Macedoine: ¼-inch dice

HACCP:

Cook to 165° F.

HAZARDOUS FOODS:

Ground pork fat
Lean pork
Eggs and egg whites
Heavy cream
Chicken

NUTRITION:

Calories: 285
Fat: 16 g
Protein: 20 g

pour in the heavy cream in a steady stream. Cook a small amount to test the seasonings, and adjust, if necessary. Fold in the cubed chicken, and chill.

5. Loosely stuff the mixture into casings to form it into cylinders wrapped in greased plastic wrap. Tie the ends with string.
6. Poach in the liquid until an internal temperature of 165° F is reached. Serve immediately.

Pork Sausage with Sage

YIELD: 3 POUNDS. SERVING SIZE: FIVE OUNCES.

INGREDIENTS:

3 yards	1-inch pork casings, washed

Sausage Stuffing:

2½ pounds	Boneless lean pork, cut into strips or cubes
½ pound	Fresh pork fat, cut into strips or cubes
3 each	Crushed ice cubes (preferably made from white stock, see page 12)
4 to 5 teaspoons	Sage, crumbled
1½ tablespoons	Black pepper
1 to 2 teaspoons	Kosher salt
⅛ teaspoon	Cayenne
2 tablespoons	Orange zest

METHOD OF PREPARATION:

1. Combine all of the ingredients for the stuffing, and pass them through a fine grinding disk twice. Cook a small amount to test the seasonings, and adjust, if necessary.
2. Stuff the casings, and form 3-inch links. Refrigerate the sausage for 3 to 6 hours before cooking to develop its flavor.

COOKING TECHNIQUES:
Shallow-Fry

Bake:
1. Preheat the oven.
2. Place the food product on the appropriate rack.

Grill/Broil:
1. Clean and heat the grill/broiler.
2. To prevent sticking, brush the food product with oil.

Shallow-Fry:
1. Heat the cooking medium to the proper temperature.
2. Cook the food product throughout.
3. Season, and serve hot.

HAZARDOUS FOODS:
Pork and pork fat

NUTRITION:
Calories: 433
Fat: 31.9 g
Protein: 33.9 g

CHEF NOTE:
Securely wrapped, this sausage keeps for 2 to 3 days refrigerated or about 1 month if frozen.

Sole Terrine with Vegetables and Shrimp

YIELD: 4 POUNDS. SERVING SIZE: TWO OUNCES.

COOKING TECHNIQUE:
Bake

Bake:
1. Preheat the oven.
2. Place the food product on the appropriate rack.

GLOSSARY:
Blanch: to parcook
Bain-marie: hot-water bath

HACCP:
Preheat the oven to 350° F.

HAZARDOUS FOODS:
Ground sole
Eggs
Heavy cream
Shrimp
Sole fillets

NUTRITION:
Calories: 71.8
Fat: 4.76 g
Protein: 6.13 g

INGREDIENTS:

1 ten-ounce	Frozen spinach, thawed, squeezed dry, and chopped
2 tablespoons	Fresh coriander, chopped
4 ounces	Tomato concassé
2 tablespoons	Truffle peelings
8 ounces	Small shrimp, cleaned and quickly **blanched**
½ pound	Sole fillets, lightly pounded
8 ounces	Fish aspic, to glaze

Farce:

1½ pounds	Finely ground sole
4 each	Eggs
¾ teaspoon	White pepper
	Salt, to taste
1¼ pints	Heavy cream, semi-whipped

METHOD OF PREPARATION:

1. Preheat the oven to 350° F.
2. Combine the ground sole, eggs, white pepper, and salt. Purée the mixture in a food processor until smooth, and chill.
3. Fold in the whipped cream, test the seasonings, and correct if necessary.
4. Purée the spinach with the coriander, and reserve. Divide the farce into three parts. Mix the puréed spinach into one third, the tomato concassé into another third, and the truffle peelings into the last third.
5. Bake in a **bain-marie** in the oven for 1 to 1¼ hours.
6. Cool, and store in the mold until ready to use.

CHEF NOTES:
1. Another firm, non-oily, white fish can be substituted for the ground sole.
2. Another fish combination can be used instead of small shrimp and sole fillets to separate the layers of mousseline, such as mussels and smoked eel.

Spiced Creole Sausage

YIELD: 3 POUNDS. SERVING SIZE: THREE OUNCES.

INGREDIENTS:

3 yards	Lamb or pork casings, washed

Sausage Stuffing:

2 pounds	Boneless lean pork or boneless lean pork butt, diced **brunoise**
1 pound	Fresh pork fat
1 each	Large Spanish onion, minced
2 to 3 cloves	Garlic, minced
1 tablespoon	Pepper melange
1 teaspoon	Coarse or kosher salt
1 teaspoon	Hot paprika
1 teaspoon	Sweet paprika
1 teaspoon	Red pepper flakes, dried
½ teaspoon	Cayenne
3 sprigs	Fresh parsley, minced
2 each	Fresh sage leaves, minced
1 sprig	Fresh thyme, minced
1 pinch	Allspice
1 pinch	Mace

METHOD OF PREPARATION:

1. Combine all of the ingredients for the stuffing, and mix well. Cook a small amount of the forcemeat to check the seasoning.
2. Stuff the casings, and form any length of link desired.

COOKING TECHNIQUES:

Shallow-Fry

Shallow-Fry:

1. Heat the cooking medium to the proper temperature.
2. Cook the food product throughout.
3. Season, and serve hot.

GLOSSARY:

Brunoise: finely diced

HAZARDOUS FOODS:

Pork or pork butt
Pork fat

NUTRITION:

Calories: 460
Fat: 46.9 g
Protein: 6.81 g

CHEF NOTE:

Securely wrapped, this sausage keeps for 2 to 3 days refrigerated or about 1 month if frozen.

Spicy Kielbasa

YIELD: 5 POUNDS. SERVING SIZE: SIX OUNCES.

COOKING TECHNIQUES:
Shallow-Fry

Shallow-Fry:
1. Heat the cooking medium to the proper temperature.
2. Cook the food product throughout.
3. Season, and serve hot.

HAZARDOUS FOOD:
Pork butt

NUTRITION:
Calories: 684
Fat: 47.8 g
Protein: 57.7 g

INGREDIENTS:

5 yards	2-inch-wide pork casings, washed

Sausage Stuffing:

5 pounds	Boneless pork butt, cut into strips
5 each	Crushed ice cubes (preferably made from brown stock; see page 7)
5 cloves	Garlic, pressed
1 to 2 tablespoons	Sweet paprika
1 tablespoon	Coarse or kosher salt
1 tablespoon	Coarsely ground black pepper
2 teaspoons	Marjoram
1 teaspoon	Summer savory
1 teaspoon	Allspice
1 to 2 tablespoons	Hot paprika

METHOD OF PREPARATION:

1. Combine the ingredients for the stuffing, and pass them through a coarse grinding disk. Cook a small amount to test the seasonings, and adjust, if necessary.
2. Stuff the casings, and form 18- to 24-inch links. Tie the two ends of each together to form a ring of sausage. Refrigerate the sausage overnight before cooking, to develop its flavor.

CHEF NOTE:
Securely wrapped, kielbasa keeps for 2 to 3 days refrigerated or about 1 month, if frozen.

Terrine de la Maison

YIELD: 2½ POUNDS. SERVING SIZE: TWO OUNCES.

INGREDIENTS:

2 pounds	Boneless pork, fat and sinews trimmed
1 each	Onion, diced **brunoise**
½ ounce	Chopped parsley
1 tablespoon	Chervil
1 teaspoon	Pâté spice
4 ounces	White bread panada
1 ounce	White wine
4 each	Egg whites
4 ounces	Heavy cream
	Salt and white pepper, to taste
8 ounces	Salt pork, diced brunoise
6 ounces	Salt pork, thinly sliced

METHOD OF PREPARATION:

1. Preheat the oven to 400° F.
2. Pass the boneless pork through a meat grinder twice, and chill.
3. Sauté the onions, and chill. Add to the ground pork, and place in food processor. Run until the meat is a smooth paste.
4. Add the parsley, chervil, and pâté spice to the ground pork.
5. To the ground pork mixture, slowly add the white bread panada, white wine, egg whites, heavy cream, salt, and pepper. When mixed, remove from processor, and adjust the seasonings.
6. Fold in the diced salt pork, and cool the mixture in a terrine mold overnight, or until ready for use.
7. Place the terrine in a **bain-marie,** and bake in a 400° F oven for 1½ hours, or until an internal temperature of 165° F is reached. Cool, and store the terrine in the mold until use.

COOKING TECHNIQUES:
Sauté, Bake

Sauté:
1. Heat the sauté pan to the appropriate temperature.
2. Evenly brown the food product.
3. For a sauce, pour off any excess oil, reheat, and deglaze.

Bake:
1. Preheat the oven.
2. Place the food product on the appropriate rack.

GLOSSARY:
Brunoise: finely diced
Bain-marie: hot-water bath

HACCP:
Preheat the oven to 400° F.
Cook to an internal temperature of 165° F.

HAZARDOUS FOODS:
Boneless pork
Egg whites
Heavy cream
Salt pork

NUTRITION:
Calories: 172
Fat: 13.4 g
Protein: 10.4 g

Venison Sausage

YIELD: 5 POUNDS. SERVING SIZE: FIVE OUNCES.

COOKING TECHNIQUES:

Boil, Bake, Shallow-Fry

Boil: (at sea level)

1. Bring the cooking liquid to a rapid boil.
2. Stir the contents, and cook the food product throughout.
3. Serve hot.

Bake:

1. Preheat the oven.
2. Place the food product on the appropriate rack.

Shallow-Fry:

1. Heat the cooking medium to the proper temperature.
2. Cook the food product throughout.
3. Season, and serve hot.

GLOSSARY:

Brunoise: finely diced

HACCP:

Cook to an internal temperature of 165° F.

HAZARDOUS FOODS:

Boneless venison
Boneless pork butt
Pork fat

NUTRITION:

Calories: 550
Fat: 46.4 g
Protein: 30.3 g

INGREDIENTS:

Marinade:

1 cup	Red wine
½ cup	Cider vinegar
1 each	Onion, diced **brunoise**
3 cloves	Garlic, chopped
1 each	Carrot, diced brunoise
2 stalks	Celery, diced brunoise
1 each	Bay leaf
1 teaspoon	Marjoram
4 each	Juniper berries, cracked
6 each	Peppercorns, cracked
1 teaspoon	Kosher salt

Sausage Stuffing:

2½ pounds	Boneless venison, cut into strips or cubes
1½ pounds	Boneless pork butt, cut into strips or cubes
1 pound	Fresh pork fat, cut into strips or cubes
5 each	Crushed ice cubes (preferably made from game stock; see page 12)
2 to 3 cloves	Garlic, minced or pressed
	Salt, to taste
1½ teaspoons	Ground juniper berries
1 teaspoon	Pepper melange
1 teaspoon	Hot paprika
1 teaspoon	Thyme
1 teaspoon	Chervil
5 yards	Pork casings, rinsed

METHOD OF PREPARATION:

1. Combine the ingredients for the marinade. Bring to a boil, and cool. Pour the marinade over the pieces of venison. Marinate the meat in the refrigerator overnight.
2. Drain the venison, and discard the marinade.
3. Combine the stuffing ingredients, and pass them through a coarse grinding disk twice. Cook a small amount to test the seasoning, and adjust, if necessary.
4. Stuff the casings, and form 4-inch links. Refrigerate the sausage; uncover overnight before cooking, to develop its flavor.
5. Cook to an internal temperature of 165° F.

Winter Terrine

YIELD: 1 TERRINE, ABOUT 3 POUNDS (10 TO 12 SERVINGS). SERVING SIZE: FOUR OUNCES.

COOKING TECHNIQUE:
Bake

Bake:
1. Preheat the oven.
2. Place the food product on the appropriate rack.

HACCP:
Bake in a preheated 325° F oven.

HAZARDOUS FOODS:
Eggs
Heavy cream

NUTRITION:
Calories: 126
Fat: 7 g
Protein: 4.6 g

INGREDIENTS:

1 pound	Celeriac, cooked, peeled, and diced
¾ pound	Potatoes, peeled, cooked, and diced
6 each	Eggs
1 teaspoon	Salt
¼ teaspoon	White pepper
¼ teaspoon	Allspice
1½ tablespoons	Fresh lovage, chopped
1 cup	Heavy cream, whipped
8 large	Cabbage or grape leaves, blanched
6 to 8 each	Pumpkin or acorn squash, cooked and cut into wedges
2 stalks	Broccoli florets, crisply cooked
2 thin	Carrots, scraped and cooked
3 each	Parsnips or small turnips, cooked and cut into wedges

METHOD OF PREPARATION:

1. Combine the celeriac, potatoes, eggs, salt, pepper, and allspice. Purée the mixture in a food processor until smooth. Remove the purée to a bowl, and fold in the lovage and whipped cream. Test the seasonings, and correct if necessary. Chill.
2. Grease a 6½-cup mold (8½ × 4½ × 3 inches), and line it with cabbage leaves. Spread a quarter of the mousseline in the bottom of the lined mold. Arrange a layer of pumpkin wedges, and cover them with another quarter of the mousseline. Center a row of carrots, flanked on each side by a row of pumpkin wedges. Add a final layer of the mousseline. Cover with the remaining cabbage leaves. Seal the mold with greased aluminum foil.
3. Bake in a water bath in a preheated 325° F oven for 1 to 1½ hours, or until cooked. After baking, chill the terrine.
4. For service, serve with tomato chutney.

CHEF NOTES:
1. Substitute another vegetable, such as cauliflower, for the celeriac.
2. Use another assortment of vegetables as a garniture in the terrine.

9

Breakfast Foods

Breakfast Foods

The joys of the table belong equally to all ages, conditions, countries, and times. They mix with all other pleasures, and remain the last to console us for their loss.

Jean-Anthelme Brillat-Savarin
The Physiology of Taste

Opportunities in the culinary arts in America are endless. The entire world is looking at what is happening with food in America—and for good reason. Quality food products are now very accessible, and the art of cooking is one that reflects sincere dedication and professionalism.

Programs at Johnson & Wales University offer the opportunity to acquire the basic knowledge so important for growing and prospering in the industry. Along with learning the basics, Johnson & Wales University also offers a unique outlook to students that provides them with the degree of humbleness so very necessary when first beginning in the industry.

A good knowledge of the basics, combined with a winning attitude and a sincere desire to work hard, is sure to lead to success in the many avenues the industry provides. Pride and professionalism are two qualities that can ensure your success in this wonderful industry.

Carolyn Buster
Distinguished Visiting Chef
November 16–18, 1986
Chef and Co-owner
The Cottage Restaurant
Calumet City, IL

The HACCP Process

The following recipe illustrates how breakfast foods flow through the HACCP process.

Omelette Paysanne

Receiving

- Eggs, whole—shells should not be cracked or dirty. Container should be intact. Whole shell eggs should be delivered at 40° F (4.4° C) or lower.
- Spices (salt and pepper)—packaging intact
- Butter, clarified—should have a sweet flavor, firm texture, and uniform color. Butter should be free of specks and mold, and the packaging should be intact. Butter should be delivered at 40° F (4.4° C) or lower.
- Ham—lean portions, pink in color. Ham should be delivered at 40° F (4.4° C) or lower.
- Vegetables (onions, mushrooms, potatoes, and peppers)—packaging intact; no cross-contamination from other foods on the truck; no signs of insect or rodent activity
- Herbs (parsley)—no visible signs of discoloration

Storage

- Eggs, whole—Store under refrigeration at 40° F (4.4° C) or lower. Store above and away from raw, potentially hazardous foods.
- Spices (salt and pepper)—Store in dry storage at 50° F (10° C) with a relative humidity of 50% to 60%. Keep dry.
- Butter—Store under refrigeration at 40° F (4.4° C) or lower. Store above and away from raw, potentially hazardous foods.
- Ham—Store under refrigeration, with a product temperature not to exceed 40° F (4.4° C). Store below already cooked foods.
- Vegetables (onions, mushrooms, and peppers)—Store under refrigeration, with a product temperature not to exceed 40° F (4.4° C). Store above and away from raw, potentially hazardous foods.
- Potatoes—Store in a dry storage at 50° F (10° C) with a relative humidity of 50% to 60%.
- Herbs (parsley)—Store under refrigeration at 40° F (4.4° C) or lower.

Preparation and Storage

- Peel the outer layer of the onions, and wash.
- Wash the mushrooms and potatoes.

- Wash and seed the peppers.
- Wash and dry the parsley.
- Dice, slice, and chop the appropriate vegetables and parsley on a clean and sanitized cutting board with clean and sanitized utensils.
- Dice the ham on a clean and sanitized cutting board with clean and sanitized utensils.
- Cook the potatoes.
- Place 5 ounces of butter in a sauce pan, and sauté the onions; add the ham, mushrooms, cooked potatoes, and peppers, and sauté.
- Heat an omelette pan, place ½ ounce of butter in the pan. Add the eggs (three eggs per portion). Use a 6 ounce ladle; season.
- Shake the pan, and mix the eggs until they begin to firm. Turn them over. Place the filling in the center of the egg.
- Shape by folding the eggs to an oval shape; cook to an internal temperature of 145° F (62.8° C). Roll the omelette out onto a preheated dinner plate, and sprinkle with chopped parsley.

Holding and Service

- Serve immediately.

Nutritional Notes

Breakfast is the most important meal of the day. The old adage, "Eat breakfast like a king, lunch like a prince, and supper like a pauper," is no less true today than when most Americans lived and worked on farms. Breakfast provides the energy needed to get through the most active part of the day. This is also the meal at which most people get their daily allowance of vitamin C, riboflavin, and calcium. A balanced breakfast should be high in complex carbohydrates, low in fat, and contain some protein. Breakfast can help stave off the hunger that often leads to overeating by those who skip breakfast.

While breakfast can consist of any food from fish to pasta, most Americans prefer traditional breakfast items such as pancakes, bacon, eggs, and home fries. Many of these items are high in fat and can transform a healthy breaking of the fast into a heartbreaker.

The following are some suggestions for hearty and healthy breakfast menu choices:

- *Whole-grain, low-fat, hot or cold cereals. Granola cereals are generally high in fat.*
- *Skim or 1% fat milk in place of whole milk or cream for cereal or beverages.*
- *Pancakes, waffles, French toast, and crêpes using whole-grain flours; or add wheat germ, bran, or fruit for fiber, vitamins, and minerals. Replace eggs in the recipes with egg whites or egg substitutes, and minimize fat used in cooking. Serve with nonfat yogurt or cottage cheese, low-fat ricotta, fresh fruit, or fruit sauces. Remember, honey and maple syrup contain as much sugar as pancake syrup and table sugar. None of these contains fat, but all are considered empty calories (i.e., calories that provide little or no nutritional value)*
- *Omelettes can offer the highest-quality protein with little or no fat when made with egg whites or egg substitutes. Add flavorful ingredients or dried butter solids to replace the flavor of the missing yolks. Cook on a nonstick or well-seasoned pan with no added fat.*
- *Substitute lean Canadian bacon for regular bacon to reduce fat. Make extra-lean sausage in-house by using ground turkey or chicken blended with very lean pork.*
- *Most quick breads are high in fat. Try reducing the fat in the recipe as much as possible. Other ingredients, such as corn syrup or fruit, may be required to retain moisture.*
- *Offer plenty of fresh fruits or sulfate-free dried fruit as a source of complex carbohydrates, vitamins, and minerals.*
- *Bagels are an excellent low-fat source of complex carbohydrates. Whole-grain bagels provide fiber as well as flavor.*
- *Offer reduced-fat, soft margarines, and low-fat cheese spreads in place of butter and cream cheese.*
- *Lean dough can be used to make buns and sweet rolls to replace doughnuts or rich-dough buns.*
- *Bake hash-brown potatoes rather than frying them, or offer potato pancakes.*

Baked Ham Slices

YIELD: 50 SERVINGS. SERVING SIZE: ONE AND ONE-HALF OUNCES.

INGREDIENT:

5 pounds	Virginia ham, trimmed of fat

METHOD OF PREPARATION:

1. Preheat the broiler or oven to 350° F.
2. Slice the ham into 1½-ounce portions, and lay out on sheet pans.
3. Bake for 10 minutes, or until heated throughout. Remove, and transfer to hotel pans for service. Hold at 140° F or above.

COOKING TECHNIQUE:

Bake

Bake:

1. Preheat the oven.
2. Place the food product on the appropriate rack.

HACCP:

Hold at 140° F or above.

HAZARDOUS FOOD:

Ham

NUTRITION:

Calories: 82.6
Fat: 5.71 g
Protein: 7.34 g

CHEF NOTE:

Ham also can be grilled or cooked on a flat top or in a tilting skillet.

Broiled Bacon Slices

YIELD: 50 SERVINGS. SERVING SIZE: THREE SLICES.

INGREDIENT:

7 pounds (150 slices)	Sliced bacon, layed out on a sheet pan covered with parchment paper

METHOD OF PREPARATION:

1. Preheat the broiler and/or oven to 425° F.
2. Place the sheet pan under the broiler, and broil the bacon until golden brown and crispy, or place the sheet pan in the oven, and bake the bacon until golden brown and crispy.
3. Transfer the bacon slices to a perforated hotel pan with a solid hotel pan underliner, and hold at 140° F or above, but do not cover.

COOKING TECHNIQUES:

Broil or Bake

Grill/Broil:

1. Clean and heat the grill/broiler.
2. To prevent sticking, brush the food product with oil.

Bake:

1. Preheat the oven.
2. Place the food product on the appropriate rack.

HACCP:

Hold at 140° F or above.

NUTRITION:

Calories: 120
Fat: 10.3 g
Protein: 6.4 g

CHEF NOTE:

Rendered bacon fat can be reserved for cooking, if desired.

Buttermilk Waffles

YIELD: 50 SERVINGS. SERVING SIZE: TWO AND ONE-HALF OUNCES.

INGREDIENTS:

4 ounces	Pasteurized egg whites
1 pound	Whole-wheat flour
2 pound	All-purpose flour
3 ounces	Baking powder
1 tablespoon	Salt
2 quarts	Buttermilk
8 ounces	Safflower oil

METHOD OF PREPARATION:

1. Preheat a waffle iron.
2. Beat the egg whites until stiff but not dry; then set aside.
3. In a medium bowl, sift together the flours, baking powder, and salt.
4. Add the buttermilk; stir together, but do not overmix. Add the oil, and then fold in the egg whites.
5. Ladle 2½ ounces of batter onto a lightly oiled, hot waffle iron. Close, and bake until the steaming stops. Lift the waffle from the iron with a fork. Serve immediately, or hold at 140° F or above.
6. Repeat the procedure until all of the batter is used.

COOKING TECHNIQUE:
Bake

Bake:
1. Preheat the oven.
2. Place the food product on the appropriate rack.

HACCP:
Hold at 140° F or above.

HAZARDOUS FOODS:
Pasteurized egg whites
Buttermilk

NUTRITION:
Calories: 153
Fat: 5.05 g
Protein: 4.67 g

CHEF NOTE:
1. Keep the batter chilled to 40° F or below if not used immediately.
2. For best results, make waffles to order.

Canadian Bacon

YIELD: 50 SERVINGS. SERVING SIZE: ONE AND ONE-HALF OUNCES.

INGREDIENT:

5 pounds	Canadian bacon

METHOD OF PREPARATION:

1. Preheat the broiler or oven to 350° F.
2. Slice the bacon into 1½-ounce portions.
3. Lay out the slices on a sheet pan, and broil or bake to an internal temperature of 165° F. Serve immediately, or hold in a hotel pan at 140° F or above.

COOKING TECHNIQUES:

Broil or Bake

Grill/Broil:

1. Clean and heat the grill/broiler.
2. To prevent sticking, brush the food product with oil.

Bake:

1. Preheat the oven.
2. Place the food product on the appropriate rack.

HACCP:

Bake to 165° F.
Hold at 140° F or above.

HAZARDOUS FOOD:

Canadian bacon

NUTRITION:

Calories: 52.9
Fat: 2.41 g
Protein: 6.93 g

CHEF NOTE:

Bacon also can be cooked on a flat-top, grill, or in a tilting skillet.

Corned Beef Hash

YIELD: 50 SERVINGS. SERVING SIZE: FOUR OUNCES.

INGREDIENTS:

10 ounces	Clarified butter
2 pounds	Onions, peeled and diced **brunoise**
1½ pounds	Green bell peppers, washed, seeded, and diced brunoise
5 pounds	Cooked corned beef, processed through a food processor
2½ pounds	Potatoes, washed, peeled, blanched, and grated or diced brunoise
2½ pounds	Scallions, washed, trimmed, and diced brunoise
4 ounces	Fresh parsley, washed, excess moisture removed, and chopped
2 tablespoons	Thyme, dried
	Salt and freshly ground black pepper, to taste
	Vegetable oil, as needed

METHOD OF PREPARATION:

1. Preheat the oven to 350° F.
2. In a sauté pan, heat the butter; add the onions and peppers, and sauté until the onions are translucent. Remove, and place in a large bowl.
3. Add the beef, potatoes, scallions, parsley, and thyme, and season, to taste. Mix well.
4. In an oiled hotel pan, place the hash, and bake until the potatoes are tender and the mixture is lightly browned.
5. Serve immediately, or hold at 140° F or above.

COOKING TECHNIQUES:

Sauté, Bake

Sauté:

1. Heat the sauté pan to the appropriate temperature.
2. Evenly brown the food product.
3. For a sauce, pour off any excess oil, reheat, and deglaze.

Bake:

1. Preheat the oven.
2. Place the food product on the appropriate rack.

GLOSSARY:

Brunoise: ⅛-inch dice

HACCP:

Hold at 140° F or above.

HAZARDOUS FOOD:

Butter

NUTRITION:

Calories: 197
Fat: 14.1 g
Protein: 9.44 g

CHEF NOTES:

For roast beef hash, substitute roast beef for corned beef.

Cottage-Fried Potatoes

YIELD: 10 SERVINGS. SERVING SIZE: FOUR OUNCES.

INGREDIENTS:

3 pounds	Potatoes, washed and peeled (hold in cold water)
5 ounces	Vegetable oil
5 ounces	Butter, clarified, melted
8 ounces	Onions, peeled and diced **brunoise**
	Salt and ground white pepper, to taste
4 ounces	Fresh parsley, washed, excess moisture removed, and chopped

METHOD OF PREPARATION:

1. In a saucepan, place the potatoes, and cover with water. Bring to a boil, and cook until tender.
2. Drain the potatoes, and place on a sheet pan to air-dry and cool.
3. Cut the potatoes into ¼-inch-thick slices.
4. In a sauté pan, heat 4 ounces each of oil and butter. Arrange the potatoes in the pan in one layer, and brown both sides, turning only once.
5. Remove, and drain on absorbent paper; then transfer the potatoes to a hotel pan, and hold at 140° F.
6. In the same pan, heat the remaining oil and butter, and sauté the onions until lightly browned. Drain any excess fat.
7. Sprinkle the onions on the potatoes, and season, to taste.
8. Preheat the oven to 350° F.
9. For service, bake the potatoes for 10 to 15 minutes, or until hot and crisp. Serve immediately, or hold at 140° F. Garnish with parsley.

COOKING TECHNIQUES:

Boil, Sauté

Boil: (at sea level)

1. Bring the cooking liquid to a rapid boil.
2. Stir the contents, and cook the food product throughout.
3. Serve hot.

Sauté:

1. Heat the sauté pan to the appropriate temperature.
2. Evenly brown the food product.
3. For a sauce, pour off any excess oil, reheat, and deglaze.

GLOSSARY:

Brunoise: ⅛-inch dice

HACCP:

Hold at 140° F.

HAZARDOUS FOOD:

Butter

NUTRITION:

Calories: 383
Fat: 26.7 g
Protein: 3.39 g

Cream of Wheat

YIELD: 50 SERVINGS. SERVING SIZE: SIX OUNCES.

INGREDIENTS:

2 gallons	Water
1 pound	Butter, softened
1 tablespoon	Salt
3 pounds	Cream of wheat
2 quarts	Milk, heated to 165° F

METHOD OF PREPARATION:

1. In a large saucepan, combine the water, butter, and salt, and bring the mixture to a boil.
2. Slowly pour the cream of wheat into the boiling liquid.
3. Stir continuously, and simmer for 3 to 5 minutes, or until thickened and the cream of wheat is cooked.
4. Serve immediately in a preheated cereal bowl, with hot milk on the side, or hold at 140° F or above.

COOKING TECHNIQUE:
Simmer

Simmer and Poach:
1. Heat the cooking liquid to the proper temperature.
2. Submerge the food product completely.
3. Keep the cooked product moist and warm.

HACCP:
Hold at 140° F or above.

HAZARDOUS FOOD:
Milk

NUTRITION:
Calories: 185
Fat: 4.02 g
Protein: 4.63 g

CHEF NOTES:
1. The cereal can be topped with fresh fruit, brown sugar, or syrups.
2. For a richer flavor, replace part or all of the water with milk.

Egg, Bacon, and Cheese Croissant

YIELD: 50 SERVINGS. SERVING SIZE: SIX OUNCES.

INGREDIENTS:

50 each	Croissants, split in half and toasted
50 servings	Scrambled eggs (see page 715)
50 one-ounce slices	American or Swiss cheese
50 slices	Bacon, crisply fried, split in half and held at 140° F

METHOD OF PREPARATION:

1. Fill the croissants with 6 ounces of scrambled eggs, and top with one slice of cheese and two halves of bacon.
2. Serve immediately on a preheated plate, or hold at 140° F.

COOKING TECHNIQUE:

Shallow-Fry

Shallow-Fry:

1. Heat the cooking medium to the proper temperature.
2. Cook the food product throughout.
3. Season, and serve hot.

HACCP:

Hold at 140° F.

HAZARDOUS FOOD:

Eggs

NUTRITION:

Calories: 839
Fat: 64.6 g
Protein: 32.5 g

CHEF NOTES:

1. As an alternative, eggs, bacon, and cheese can be served on a muffin.
2. The croissant with cheese can be placed under a salamander until the cheese melts.
3. The bacon can be baked ahead of time, drained of fat, and held at 140° F.

Farina

YIELD: 50 SERVINGS. SERVING SIZE: SIX OUNCES.

INGREDIENTS:

3 quarts	Water
3 quarts	Milk
4 ounces	Butter
1 ounce	Salt
3 pounds	Farina
1 pound	Sugar
2 ounces	Cinnamon

METHOD OF PREPARATION:

1. In a saucepan, combine the water, milk, butter, and salt. Bring to a boil.
2. Slowly add the farina to the boiling liquid, stirring continuously. Cook for 5 minutes, or until thickened and the farina is tender. Hold at 140° F or above.
3. Mix together the sugar and cinnamon.
4. To serve, place a portion of farina in a preheated bowl, and sprinkle the top with the sugar mixture. Hold at 140° F or above.

COOKING TECHNIQUE:

Simmer

Simmer and Poach:

1. Heat the cooking liquid to the proper temperature.
2. Submerge the food product completely.
3. Keep the cooked product moist and warm.

HACCP:

Hold at 140° F or above.

HAZARDOUS FOOD:

Milk

NUTRITION:

Calories: 185
Fat: 4.02 g
Protein: 4.63 g

CHEF NOTE:

The farina can be topped with fresh fruit, brown sugar, or syrups.

French Toast

YIELD: 50 SERVINGS. SERVING SIZE: SIX STRIPS.

COOKING TECHNIQUE:
Shallow-Fry

Shallow-Fry:
1. Heat the cooking medium to the proper temperature.
2. Cook the food product throughout.
3. Season, and serve hot.

HACCP:
Hold at 140° F or above.

HAZARDOUS FOODS:
Milk
Pasteurized eggs

NUTRITION:
Calories: 500
Fat: 19.4 g
Protein: 9.63 g

CHEF NOTE:
This recipe originally called for 2-inch angle cut slices from a French baguette.

INGREDIENTS:

1 quart	Pasteurized eggs
2½ quarts	Milk
2 ounces	Cinnamon
1 tablespoon	Salt
4 ounces	Sugar
100 thick slices	Bread, crusts removed and cut into thirds
1½ pounds	Butter, clarified, melted
2 quarts	Maple syrup, heated and kept warm at 140° F

METHOD OF PREPARATION:

1. Preheat the griddle.
2. In a mixing bowl, whisk the eggs.
3. Add the milk, cinnamon, salt, and sugar, and mix well with a whisk. Hold at 40° F or below, or use immediately.
4. Dip the bread strips into the egg mixture, coating well.
5. Place on a well-buttered griddle or in a large skillet, and brown on one side. Turn over, and brown the other side.
6. Serve immediately, or hold at 140° F or above.
7. To serve, arrange six "fingers" on a plate, and offer warmed maple syrup.

Fried Eggs

YIELD: 1 SERVING. SERVING SIZE: TWO EGGS.

INGREDIENTS:

1 ounce	Butter, clarified
2 large	Eggs, broken into a cup

METHOD OF PREPARATION:

1. In a sauté pan, heat the butter.
2. When hot, but not browning, slip the eggs into the pan, and cook to order (e.g., sunny-side up, over easy, over medium, or as requested). Serve immediately on preheated plates.

COOKING TECHNIQUE:

Shallow-Fry

Shallow-Fry:

1. Heat the cooking medium to the proper temperature.
2. Cook the food product throughout.
3. Season, and serve hot.

HAZARDOUS FOOD:

Eggs

NUTRITION:

Calories: 373
Fat: 35.5 g
Protein: 12.6 g

CHEF NOTE:

Do not over cook or the eggs will get rubbery.

Home Fries

YIELD: 10 SERVINGS. SERVING SIZE: FOUR OUNCES.

INGREDIENTS:

3 pounds	Small potatoes, washed
	Cold water, to cover
1 teaspoon	Salt
1 ounce	Lemon juice
	Butter, clarified, as needed
	Salt and ground white pepper, to taste
3 ounces	Fresh parsley, washed, excess moisture removed, and chopped

METHOD OF PREPARATION:

1. In a saucepan, place the potatoes, and cover with cold water. Add the salt and lemon juice, and bring to a boil. Cook until **al dente.**
2. When done, drain and chill immediately.
3. Peel and slice the potatoes ¼-inch thick.
4. Coat the bottom of a sauté pan with clarified butter. Add one portion of potatoes, and sauté until golden brown.
5. Season, to taste, and sprinkle with some chopped parsley. Serve immediately.

COOKING TECHNIQUES:

Boil, Sauté

Boil: (at sea level)

1. Bring the cooking liquid to a rapid boil.
2. Stir the contents, and cook the food product throughout.
3. Serve hot.

Sauté:

1. Heat the sauté pan to the appropriate temperature.
2. Evenly brown the food product.
3. For a sauce, pour off any excess oil, reheat, and deglaze.

GLOSSARY:

Al dente: to the bite

NUTRITION:

Calories: 321
Fat: 25.7 g
Protein: 2.3 g

CHEF NOTE:

For garlic-fried potatoes, add 1 teaspoon of minced fresh garlic to the sauté pan with the potatoes.

Oatmeal

YIELD: 50 SERVINGS. SERVING SIZE: SIX OUNCES.

COOKING TECHNIQUE:

Simmer

Simmer and Poach:

1. Heat the cooking liquid to the proper temperature.
2. Submerge the food product completely.
3. Keep the cooked product moist and warm.

HACCP:

Hold at 140° F or above.

HAZARDOUS FOOD:

Milk

NUTRITION:

Calories: 162
Fat: 6.03 g
Protein: 6.68 g

INGREDIENTS:

4 quarts	Water
4 quarts	Milk
4 ounces	Butter
1 ounce	Salt
3 pounds	Oatmeal
	Additional milk, to taste, heated to 165° F and held at 140° F
	Sugar, optional

METHOD OF PREPARATION:

1. In a large saucepot, combine the water, milk, butter, and salt, and bring to a boil.
2. Add the oatmeal slowly, stirring continuously, and cook for 10 to 15 minutes, continuing to stir, or until the oatmeal is tender and the desired consistency is achieved.
3. Serve immediately in a preheated cereal bowl, offering additional milk or sugar on the side, or hold at 140° F or above.

CHEF NOTES:

1. The cereal also can be offered with fresh fruit, brown sugar, or syrups.
2. As an alternative, sauté peeled, cored, and diced apple in butter with sugar, and then add to the boiling liquid before adding the oatmeal. Season with cinnamon.

O'Brien Potatoes

COOKING TECHNIQUES:
Sauté, Deep-Fry

Sauté:
1. Heat the sauté pan to the appropriate temperature.
2. Evenly brown the food product.
3. For a sauce, pour off any excess oil, reheat, and deglaze.

Deep-Fry:
1. Heat the frying liquid to the proper temperature.
2. Submerge the food product completely.
3. Fry the product until it is cooked throughout.

GLOSSARY:
Brunoise: ⅛-inch dice
Render: to melt fat

HACCP:
Hold at 140° F or above.

NUTRITION:
Calories: 622
Fat: 49.8 g
Protein: 16.5 g

YIELD: 10 SERVINGS. SERVING SIZE: FOUR OUNCES.

INGREDIENTS:

1 pound	Bacon, cut into ½-inch dice
8 ounces	Onions, peeled and diced **brunoise**
8 ounces	Green bell peppers, washed, seeded, and cut into ¼-inch dice
4 ounces	Canned pimientos, rinsed in cold water and cut into ¼-inch dice
1 teaspoon	Paprika
	Oil, as needed
3 pounds	Potatoes, washed, peeled, and cut into ½-inch dice (held in cold water)
	Salt and ground white pepper, to taste

METHOD OF PREPARATION:

1. In a sauté pan, **render** the bacon until lightly browned.
2. Add the onions, and sauté until they are translucent. Add the peppers, and sauté until soft.
3. Add the pimientos and paprika, and blend; then remove from the heat, and hold at 140° F.
4. Heat the oil to 350° F.
5. Drain the potatoes thoroughly. Deep-fat fry the potatoes in oil until golden brown and tender.
6. Drain the potatoes of excess oil, and place them in a hotel pan. Add the bacon mixture, and mix together well. Season, to taste.
7. For service, hold at 140° F or above.

Omelette

YIELD: 10 SERVINGS. SERVING SIZE: SIX OUNCES.

INGREDIENTS:

30	Eggs, cracked into a bowl and whisked together
	Salt and ground white pepper, to taste
8 ounces	Milk
5 ounces	Clarified butter, melted
3 ounces	Fresh parsley, washed, excess moisture removed, and chopped

METHOD OF PREPARATION:

1. Season the eggs with salt and pepper. Add the milk, and **whisk** until the eggs are well combined.
2. Heat an omelette pan with ½ ounce of butter.
3. When hot, add a 6-ounce ladle of egg mixture.
4. Shake the pan, and mix the eggs until they begin to firm, lifting the edges to allow liquid to run underneath (see chef notes).
5. Roll in the pan, and turn out onto a preheated serving plate. Serve immediately, or hold minimally at 140° F.
6. Repeat the procedure until all of the eggs are cooked.
7. Garnish with chopped parsley.

COOKING TECHNIQUE:

Shallow-Fry

Shallow-Fry:

1. Heat the cooking medium to the proper temperature.
2. Cook the food product throughout.
3. Season, and serve hot.

GLOSSARY:

Whisk: to aerate with a whip

HACCP:

Hold cooked eggs at 140° F.
Hold uncooked mixture below 40° F.

HAZARDOUS FOODS:

Eggs
Milk

NUTRITION:

Calories: 336
Fat: 27.8 g
Protein: 18.8 g

CHEF NOTES:

1. When the eggs have set in the sauté pan, place the pan under a broiler for 10 to 15 seconds to finish cooking the eggs; then roll the omelette out of the pan and onto a preheated serving plate. This creates a fluffier presentation and ensures that the eggs are well done.
2. Hold the egg mixture below 40° F if not cooking immediately.

Omelette with Cheese

YIELD: 10 SERVINGS. SERVING SIZE: SIX OUNCES.

INGREDIENTS:

30 each	Eggs (see omelette, page 701)
1 pound	Cheese, **julienne**

METHOD OF PREPARATION:

1. Fry the omelettes according to the directions on page 701.
2. When the omelette is almost firm, turn it over.
3. Place the cheese in the center of the omelette, fold, and roll onto a preheated dinner plate. Serve immediately, or hold at 140° F.

COOKING TECHNIQUE:

Shallow-Fry

Shallow-Fry:

1. Heat the cooking medium to the proper temperature.
2. Cook the food product throughout.
3. Season, and serve hot.

GLOSSARY:

Julienne: matchstick strips

HACCP:

Hold at 140° F.

HAZARDOUS FOOD:

Eggs

NUTRITION:

Calories: 506
Fat: 41.9 g
Protein: 28.9 g

Omelette with Chives

YIELD: 10 SERVINGS. SERVING SIZE: SIX OUNCES.

INGREDIENTS:

3 ounces	Chives, washed and minced
30 each	Eggs (see omelette, page 701)

METHOD OF PREPARATION:

1. Combine the beaten eggs with the chives.
2. Fry the omelettes according to the directions on page 701.
3. Serve on a preheated dinner plate. Serve immediately, or hold at 140° F.

COOKING TECHNIQUE:

Shallow-Fry

Shallow-Fry:

1. Heat the cooking medium to the proper temperature.
2. Cook the food product throughout.
3. Season, and serve hot.

HACCP:

Hold at 140° F.

HAZARDOUS FOOD:

Eggs

NUTRITION:

Calories: 338
Fat: 27.8 g
Protein: 19.1 g

Omelette with Fine Herbs

YIELD: 10 SERVINGS. SERVING SIZE: SIX OUNCES.

COOKING TECHNIQUE:
Shallow-Fry

Shallow-Fry:
1. Heat the cooking medium to the proper temperature.
2. Cook the food product throughout.
3. Season, and serve hot.

HAZARDOUS FOOD:
Eggs

NUTRITION:
Calories: 337
Fat: 27.8 g
Protein: 18.9 g

INGREDIENTS:

30 each	Eggs (see omelette, page 701)
1 tablespoon	Fresh parsley, washed, excess moisture removed, and chopped
1 tablespoon	Scallions, minced
1 teaspoon	Tarragon, freshly chopped
1 tablespoon	Chives, freshly minced

METHOD OF PREPARATION:

1. Prepare the egg mixture for the omelettes, and add all of the herbs.
2. Fry the omelettes according to the recipe on page 701.

Omelette with Spinach

YIELD: 10 SERVINGS. **SERVING SIZE:** EIGHT OUNCES.

INGREDIENTS:

5 ounces	Butter, clarified
1 pound	Spinach leaves, thawed, drained, and roughly chopped
6 ounces	Onions, peeled and diced **brunoise**
⅓ teaspoon	Nutmeg, ground
	Salt and white pepper, to taste
1 tablespoon	Chives, freshly minced
30 each	Eggs (see omelette, page 701)
3 ounces	Fresh parsley, washed, excess moisture removed, and chopped

METHOD OF PREPARATION:

1. In a sauté pan, melt ½ ounce of butter, and bring it to the smoke point. Sauté the onions until they are translucent.
2. Add the spinach, and continue to sauté for 5 minutes.
3. Season with nutmeg; reserve.
4. Prepare the omelette according to omelette recipe on page 701. Add the chives; season with salt and white pepper.
5. When the eggs are almost cooked, add 1½ ounces of the spinach mixture.
6. Roll in the pan, and turn out onto a preheated serving plate. Serve immediately, or hold at 140° F or above.
7. Repeat the procedure until all of the eggs are cooked.
8. Garnish with chopped parsley.

COOKING TECHNIQUES:

Sauté, Shallow-Fry

Shallow-Fry:

1. Heat the cooking medium to the proper temperature.
2. Cook the food product throughout.
3. Season, and serve hot.

Sauté:

1. Heat the sauté pan to the appropriate temperature.
2. Evenly brown the food product.
3. For a sauce, pour off any excess oil, reheat, and deglaze.

GLOSSARY:

Brunoise: ⅛-inch dice

HACCP:

Hold at 140° F.

HAZARDOUS FOOD:

Eggs

NUTRITION:

Calories: 376
Fat: 30.5 g
Protein: 20.3 g

Omelette Paysanne

YIELD: 10 SERVINGS. SERVING SIZE: EIGHT OUNCES.

COOKING TECHNIQUE:

Shallow-Fry

Shallow-Fry:

1. Heat the cooking medium to the proper temperature.
2. Cook the food product throughout.
3. Season, and serve hot.

GLOSSARY:

Brunoise: ⅛-inch dice

HAZARDOUS FOOD:

Eggs

NUTRITION:

Calories: 415
Fat: 29.6 g
Protein: 26.4 g

INGREDIENTS:

10 ounces	Clarified butter
5 ounces	Onions, peeled and finely diced
10 ounces	Ham diced **brunoise**
5 ounces	Mushrooms, washed and sliced
10 ounces	Potatoes, washed, peeled, diced, and steamed
5 ounces	Red bell pepper, washed, seeded, and diced brunoise
30	Eggs, cracked and lightly beaten
	Salt and ground white pepper, to taste
3 ounces	Fresh parsley, washed, excess moisture removed, and chopped

METHOD OF PREPARATION:

1. Place 5 ounces of butter in a saucepan, and sauté the onion until it is translucent. Add the ham, mushrooms, potatoes, and peppers, and sauté for another 5 minutes.
2. Heat an omelette pan. Add ½ ounce of butter to the pan and 6 ounces of the eggs.
3. Shake the pan, lifting the edges until the eggs are firm.
4. Place some of the filling (2 to 3 ounces) in the center of the eggs.
5. Shape by folding the eggs to an oval shape, and roll out onto a preheated dinner plate. Garnish with chopped parsley. Serve immediately.

CHEF NOTE:

If the eggs are not used, keep them refrigerated at 40° F or below.

Omelette, Western Style

YIELD: 10 SERVINGS. SERVING SIZE: EIGHT OUNCES.

INGREDIENTS:

6 ounces	Butter, clarified
5 ounces	Onions, peeled and finely diced
10 ounces	Ham, diced **brunoise**
10 ounces	Green or red bell pepper, washed, seeded, and diced brunoise
3 ounces	Fresh parsley, washed, excess moisture removed, and chopped
16 ounces	Milk or water
30 each	Eggs, cracked in a bowl and whisked together (see omelette, page 701)
	Salt and white pepper, to taste

METHOD OF PREPARATION:

1. Heat 1 ounce of butter in a saucepan, and add the onions. Sauté until they are golden brown.
2. Add the peppers and ham to the onions, and sauté for another 5 minutes. Reserve the filling on the side.
3. Whisk the eggs and milk together until light and fluffy. Season with salt and white pepper, and whisk again.
4. Heat a saucepan, and add ½ ounce of butter. When near the smoke point, add a 6-ounce ladle of the egg mixture.
5. Shake the pan, and mix the eggs until they begin to firm, lifting the eggs to allow the liquid to run underneath. (See chef note.)
6. Add 2½ ounces of the filling to the eggs. Roll the omelette in the pan, and turn it out onto a heated serving plate. Serve immediately, or hold at 140° F.
7. Repeat until all of the egg mixture is cooked.
8. Garnish with chopped parsley.

COOKING TECHNIQUES:

Sauté, Shallow-Fry

Sauté:

1. Heat the sauté pan to the appropriate temperature.
2. Evenly brown the food product.
3. For a sauce, pour off any excess oil, reheat, and deglaze.

Shallow-Fry:

1. Heat the cooking medium to the proper temperature.
2. Cook the food product throughout.
3. Season, and serve hot.

GLOSSARY:

Brunoise: ⅛-inch dice

HAZARDOUS FOOD:

Eggs

NUTRITION:

Calories: 511
Fat: 42.3 g
Protein: 26 g

CHEF NOTE:

In some areas, potatoes are added to the filing.

Pancakes with Maple Syrup

YIELD: 50 SERVINGS. SERVING SIZE: FOUR EACH.

INGREDIENTS:

1 quart	Pasteurized eggs
3 quarts	Milk
2 tablespoons	Vanilla extract
6 pounds	All-purpose flour
8 ounces	Sugar
6 ounces	Baking powder
1 pound	Butter, melted
2 quarts	Maple syrup, heated and kept warm at 140° F.

METHOD OF PREPARATION:

1. Preheat the griddle.
2. In a mixing bowl, beat the eggs.
3. Add the milk and vanilla to the beaten eggs, and mix well. Set aside.
4. Mix all of the dry ingredients together. Add the egg mixture, and whisk to a smooth batter.
5. Stir the butter into the mixture.
6. Let the batter rest for 1 hour before using.
7. To cook, pour approximately 2 ounces of batter on a seasoned, lightly buttered griddle.
8. Cook until the bubbles appear on the top and the edges become dry.
9. Turn over, and bake the other side until done. Serve imediately, or hold at 140° F.
10. Hold the unused batter at 40° F or below if not used immediately.
11. Serve with warm syrup.
12. Repet the procedure until all of the batter is used.

COOKING TECHNIQUE:
Bake

Bake:
1. Preheat the oven.
2. Place the food product on the appropriate rack.

HACCP:
Hold at 140° F.

HAZARDOUS FOOD:
Milk
Pasteurized eggs

NUTRITION:
Calories: 478
Fat: 11.6 g
Protein: 9.63 g

CHEF NOTE:
For best results, make pancakes to order.

Poached Eggs

YIELD: 10 SERVINGS. SERVING SIZE: TWO EGGS.

COOKING TECHNIQUES:
Simmer, Poach

Simmer and Poach:
1. Heat the cooking liquid to the proper temperature.
2. Submerge the food product completely.
3. Keep the cooked product moist and warm.

HACCP:
Hold minimally in water at 140° F.

HAZARDOUS FOOD:
Eggs

NUTRITION:
Calories: 149
Fat: 10 g
Protein: 12.5 g

INGREDIENTS:

1 quart	Water
3 ounces	Vinegar
1 tablespoon	Salt
20 each	Eggs

METHOD OF PREPARATION:

1. In a poaching pan, combine the water, vinegar, and salt, and bring to a simmer.
2. Crack the eggs separately into a soup cup, and slide them into the simmering liquid. Poach for 3 minutes; then remove with a skimmer. Hold minimally in heated water at 140° F.

CHEF NOTES:
1. If the poached eggs are for use at a later time, transfer them from the poaching liquid into iced water to chill. Preheat them in a water bath of 140° F or above.
2. Poached eggs are frequently served with Mornay sauce or Hollandaise sauce (see pages 37 and 30).

Poached Eggs with Mornay Sauce

YIELD: 10 SERVINGS. SERVING SIZE: TWO EGGS.

INGREDIENTS:

10 portions	Poached eggs (see page 709)
20 slices	Warm toast, cut in half on a diagonal
20 ounces	Mornay sauce, heated to 165° F (see page 37)

METHOD OF PREPARATION:

1. Prepare the eggs according to the recipe.
2. To serve, arrange four half slices of toast on a preheated plate. Drain, and place two eggs on two half slices of toast, and **nappé** with the sauce. Serve immediately.
3. If necessary, hold the Mornay sauce at 140° F.

COOKING TECHNIQUES:

Simmer, Poach

Simmer and Poach:

1. Heat the cooking liquid to the proper temperature.
2. Submerge the food product completely.
3. Keep the cooked product moist and warm.

GLOSSARY:

Nappé: coated

HACCP:

Heat Mornay sauce to 165° F.

HAZARDOUS FOOD:

Eggs

NUTRITION:

Calories: 367
Fat: 18.5 g
Protein: 16.8 g

Poached Eggs with Spinach

YIELD: 10 SERVINGS. SERVING SIZE: TWO EGGS.

INGREDIENTS:

2 ounces	Butter, clarified
2 ounces	Onions, peeled and diced **brunoise**
2 pounds	Frozen spinach, thawed, drained, and chopped
2 cloves	Garlic, peeled, minced, and mashed into a purée
¼ teaspoon	Nutmeg, ground
	Salt and ground white pepper, to taste
10 servings	Poached eggs (see page 709)
	Mornay sauce, heated to 165° F (see page 37)

METHOD OF PREPARATION:

1. In a sauté pan, melt the butter. Add the onions, spinach, and garlic, and sauté until the onions become translucent.
2. Add the seasonings, remove from the heat, and hold at 140° F.
3. To serve, place a portion of the spinach on a preheated plate, and place two eggs on top. **Nappé** with sauce, and serve immediately.

COOKING TECHNIQUES:

Simmer, Poach, Sauté

Simmer and Poach:

1. Heat the cooking liquid to the proper temperature.
2. Submerge the food product completely.
3. Keep the cooked product moist and warm.

Sauté:

1. Heat the sauté pan to the appropriate temperature.
2. Evenly brown the food product.
3. For a sauce, pour off any excess oil, reheat, and deglaze.

GLOSSARY:

Brunoise: ⅛-inch dice
Nappé: coated

HACCP:

Hold at 140° F.

HAZARDOUS FOOD:

Eggs

NUTRITION:

Calories: 526
Fat: 37.3 g
Protein: 27.4 g

CHEF NOTES:

1. If the poached eggs are for use at a later time, transfer them from the poaching liquid into iced water to chill. Preheat them in a water bath of 140° F or above.
2. Poached eggs are frequently served with Mornay sauce or Hollandaise sauce.

Poached Eggs Touraine

YIELD: 10 SERVINGS. SERVING SIZE: TWO EGGS.

INGREDIENTS:

20 each	Poached eggs (see page 709)
20 each	Artichoke bottoms, steamed and held at 140° F
20 ounces	Hollandaise sauce, held at 140° F (see page 30)

METHOD OF PREPARATION:

1. Preheat a salamander.
2. Prepare the eggs according to the recipe.
3. For service, place two artichoke bottoms on a preheated dinner plate, and place one egg on each.
4. **Nappé** with the sauce, and glaze under the salamander. Serve immediately.

COOKING TECHNIQUE:

Poach

Simmer and Poach:

1. Heat the cooking liquid to the proper temperature.
2. Submerge the food product completely.
3. Keep the cooked product moist and warm.

GLOSSARY:

Nappé: coated

HACCP:

Hold at 140° F.

HAZARDOUS FOOD:

Eggs

NUTRITION:

Calories: 622
Fat: 59.1 g
Protein: 15.7 g

Rôsti

(Hashed, Browned Potatoes)

YIELD: 10 SERVINGS. SERVING SIZE: FOUR OUNCES.

COOKING TECHNIQUES:
Boil, Shallow-Fry

Boil: (at sea level)
1. Bring the cooking liquid to a rapid boil.
2. Stir the contents, and cook the food product throughout.
3. Serve hot.

Shallow-Fry:
1. Heat the cooking medium to the proper temperature.
2. Cook the food product throughout.
3. Season, and serve hot.

NUTRITION:
Calories: 292
Fat: 15.5 g
Protein: 3.46 g

INGREDIENTS:

	Salted water, to cover
3 pounds	Medium-sized potatoes
	Ground white pepper, to taste
6 ounces	Butter, clarified, as needed

METHOD OF PREPARATION:

1. In a large pot, bring the water to a boil; add the potatoes, and cook briskly for approximately 10 minutes, or until the point of a knife can be inserted about 1 inch into a potato before meeting any resistance.
2. Drain the potatoes, and when cool enough to handle, peel. Chill the potatoes to 40° F or lower for at least 1 hour.
3. Just before service, grate the potatoes into long strips on the tear-shaped side of a four-sided stand-up grater. Toss lightly with additional seasoning.
4. In a sauté pan (preferably one with a nonstick cooking surface), heat the butter over moderate heat until a drop of water flicked over the butter evaporates instantly.
5. Drop the potatoes into the pan, and spread out evenly. Fry, uncovered, for 8 to 10 minutes. When the underside of the potato cake is as brown as it can be without burning, turn it over. (If a pan with a nonstick surface is not being used, add more butter and oil before returning the potatoes to the pan.) Fry 6 to 8 minutes more, or until the bottom side of the cake is as evenly browned as the top and the edges are crisp.
6. Slide the potato cake onto a heated platter, and serve immediately.

CHEF NOTES:
1. If using a pan without a nonstick surface, be sure to add more butter before returning the potatoes to the pan.
2. The Swiss prepare this recipe with the addition of onions and bacon and call it *pommes de terre bearnaise.*

Sausage Links

YIELD: 50 SERVINGS. SERVING SIZE: THREE LINKS.

INGREDIENT:

150	Pork sausage links

METHOD OF PREPARATION:

1. Preheat the oven to 425° F.
2. Place the sausage in a perforated hotel pan, and place in a steamer for 5 minutes. Transfer to sheet pans, and cover with parchment paper.
3. Place in the oven, and brown until golden, about 10 minutes. The internal temperature must be 165° F or higher.
4. Hold in a perforated hotel pan with a solid hotel pan underliner at 140° F or above.

COOKING TECHNIQUES:

Bake, Steam

Bake:

1. Preheat the oven.
2. Place the food product on the appropriate rack.

Steam: (Traditional)

1. Place a rack over a pot of water.
2. Prevent steam vapors from escaping.
3. Shock or cook the food product throughout.

HACCP:

Heat to an internal temperature of 165° F or above.

HAZARDOUS FOOD:

Pork sausage links

NUTRITION:

Calories: 144
Fat: 12.2 g
Protein: 7.68 g

CHEF NOTE:

The sausage can be seasoned additionally with crumbled dry sage and freshly ground black pepper.

Scrambled Eggs

YIELD: 10 SERVINGS. SERVING SIZE: SIX OUNCES.

COOKING TECHNIQUE:

Shallow-Fry

Shallow-Fry:

1. Heat the cooking medium to the proper temperature.
2. Cook the food product throughout.
3. Season, and serve hot.

GLOSSARY:

Bain-marie: hot-water bath

HACCP:

Hold at 140° F or above.

HAZARDOUS FOODS:

Eggs
Milk

NUTRITION:

Calories: 493
Fat: 43 g
Protein: 21.2 g

INGREDIENTS:

30 each	Eggs
1½ pints	Milk
10 ounces	Butter, clarified

METHOD OF PREPARATION:

1. Break the eggs into a bowl, and whisk them lightly; then whisk in the milk.
2. In a sauté pan, heat 1 ounce of butter.
3. Add one 6-ounce ladle of the egg mixture, and cook, stirring gently, until the desired firmness is attained. Serve immediately, or hold at 140° F or above.
4. Repeat the procedure until all of the eggs are cooked.

CHEF NOTES:

1. If the eggs are to be held in a **bain-marie,** add 10% béchamel sauce (see page 16) to keep them moist.
2. When cooking eggs to order, determine, desired firmness.
3. Eggs will turn green where they touch the pan if kept for too long in a bain-marie.

Scrambled Eggs with Cheese

YIELD: 10 SERVINGS. SERVING SIZE: EIGHT OUNCES.

INGREDIENTS:

10 servings	Scrambled eggs (see page 715)
5 ounces	Grated cheese, of choice
	Paprika, for garnish

METHOD OF PREPARATION:

1. Prepare the egg mixture according to recipe, adding cheese.
2. Fry the eggs according to the recipe.
3. Serve immediately, garnished with paprika, or hold at 140° F.

COOKING TECHNIQUE:

Shallow-Fry

Shallow-Fry:

1. Heat the cooking medium to the proper temperature.
2. Cook the food product throughout.
3. Season, and serve hot.

HACCP:

Hold at 140° F.

HAZARDOUS FOOD:

Eggs

NUTRITION:

Calories: 546
Fat: 47.4 g
Protein: 24.4 g

Scrambled Eggs with Ham

YIELD: 10 SERVINGS. SERVING SIZE: EIGHT OUNCES.

INGREDIENTS:

3 ounces	Butter, clarified
8 ounces	Onions, peeled and diced **brunoise**
16 ounces	Ham, diced brunoise
10 servings	Scrambled eggs (see page 715)
10 sprigs	Fresh parsley, washed, excess moisture removed, and chopped

METHOD OF PREPARATION:

1. In a sauté pan, melt the butter, and add the onions. Sauté until the onions are translucent; then add the ham, and sauté for 5 more minutes. Reserve at 140° F.
2. Prepare the eggs according to recipe, adding 2½ ounces of the ham and onion mixture during the cooking process.
3. Serve as specified, garnished with parsley, or hold at 140° F or above.

COOKING TECHNIQUES:

Sauté, Shallow-Fry

Sauté:

1. Heat the sauté pan to the appropriate temperature.
2. Evenly brown the food product.
3. For a sauce, pour off any excess oil, reheat, and deglaze.

Shallow-Fry:

1. Heat the cooking medium to the proper temperature.
2. Cook the food product throughout.
3. Season, and serve hot.

GLOSSARY:

Brunoise: ⅛-inch dice

HACCP:

Hold at 140° F or above.

HAZARDOUS FOOD:

Eggs

NUTRITION:

Calories: 636
Fat: 53.2
Protein: 31.7

CHEF NOTE:

Regional ham styles can be utilized.

Scrambled Eggs with Mushrooms

YIELD: 10 SERVINGS. SERVING SIZE: EIGHT OUNCES.

INGREDIENTS:

3 ounces	Butter, clarified
5 ounces	Onions, peeled and diced **brunoise**
1 pound	Mushrooms, cleaned, drained, and sliced
	Salt and freshly ground pepper, to taste
10 servings	Scrambled eggs (see page 715)
3 ounces	Fresh parsley, washed, excess moisture removed, and chopped

METHOD OF PREPARATION:

1. In a sauté pan, melt the butter, and add the onions. Sauté until the onions are translucent; then add the mushrooms. Season, to taste, and continue to sauté until the mushrooms are dry. Hold at 140° F.
2. Prepare the eggs according to recipe, adding one portion of mushrooms with each.
3. Serve as specified, garnished with parsley, or hold at 140° F.

COOKING TECHNIQUES:

Sauté, Shallow-Fry

Sauté:

1. Heat the sauté pan to the appropriate temperature.
2. Evenly brown the food product.
3. For a sauce, pour off any excess oil, reheat, and deglaze.

Shallow-Fry:

1. Heat the cooking medium to the proper temperature.
2. Cook the food product throughout.
3. Season, and serve hot.

GLOSSARY:

Brunoise: ⅛-inch dice

HACCP:

Hold at 140° F.

HAZARDOUS FOOD:

Eggs

NUTRITION:

Calories: 468
Fat: 38.1
Protein: 22.6

Scrambled Eggs with Shrimp

YIELD: 10 SERVINGS. SERVING SIZE: EIGHT OUNCES.

INGREDIENTS:

2 ounces	Butter, clarified
20 ounces (40/ 50 count)	Shrimp, peeled and deveined
10 servings	Scrambled eggs (see page 715)

METHOD OF PREPARATION:

1. In a sauté pan, heat the butter, and sauté the shrimp for 3 minutes. Reserve, holding at 140° F.
2. Prepare the eggs according to recipe, adding one portion of shrimp just before the eggs are totally set.
3. Serve immediately on a preheated plate, or hold at 140° F.

COOKING TECHNIQUES:

Sauté, Shallow-Fry

Sauté:

1. Heat the sauté pan to the appropriate temperature.
2. Evenly brown the food product.
3. For a sauce, pour off any excess oil, reheat, and deglaze.

Shallow-Fry:

1. Heat the cooking medium to the proper temperature.
2. Cook the food product throughout.
3. Season, and serve hot.

HACCP:

Hold at 140° F.

HAZARDOUS FOODS:

Eggs
Shrimp

NUTRITION:

Calories: 444
Fat: 33.2 g
Protein: 29.8 g

Shirred Eggs Florentine

YIELD: 10 SERVINGS. SERVING SIZE: TWO EGGS.

INGREDIENTS:

16 ounces	Butter, clarified, melted
	Bread crumbs, as needed to coat the ramekin or **cocotte**
8 ounces	Onions, peeled and diced **brunoise**
2 pounds	Fresh spinach leaves, washed, **blanched,** drained, and chopped
2 cloves	Garlic, peeled and mashed into a purée
	Ground nutmeg, to taste
	Salt and ground white pepper, to taste
20 each	Eggs
8 ounces	Cream or milk

METHOD OF PREPARATION:

1. In a large sauté pan, melt 2 ounces of the clarified butter. Add the onions, and sauté until they are caramelized. Add the spinach and garlic, and sauté. Season with the salt, pepper, and nutmeg.
2. Brush the ramekin with clarified butter, and coat lightly with bread crumbs.
3. Fill each ramekin half full of the spinach mixture.
4. Break the eggs into a cup to check for quality; then put one egg each onto the spinach mixture.
5. Place the ramekins in a bain-marie, and bake in a 300° F oven for 10 minutes until the yolk is set.

COOKING TECHNIQUES:

Sauté, Bake

Sauté:

1. Heat the sauté pan to the appropriate temperature.
2. Evenly brown the food product.
3. For a sauce, pour off any excess oil, reheat, and deglaze.

Bake:

1. Preheat the oven.
2. Place the food product on the appropriate rack.

GLOSSARY:

Brunoise: ⅛-inch dice
Blanch: to parcook
Cocotte: small oven-proof dish

HACCP:

Hold minimally at 140° F.

HAZARDOUS FOODS:

Eggs
Cream/milk

NUTRITION:

Calories: 446
Fat: 30.8
Protein: 19.2

CHEF NOTE:

Do not overbake the eggs. There are many varieties of shirred eggs, but the eggs are always baked in a cocotte, or ramekin, and are always served over a bed of sautéed vegetables or hash.

Whole-Wheat Waffles

COOKING TECHNIQUE:
Bake

Bake:
1. Preheat the oven.
2. Place the food product on the appropriate rack.

HACCP:
Hold at 140° F or above.

HAZARDOUS FOODS:
Milk
Pasteurized eggs

NUTRITION:
Calories: 156
Fat: 4.04 g
Protein: 6.47 g

YIELD: 50 SERVINGS. **SERVING SIZE:** TWO AND ONE-HALF OUNCES.

INGREDIENTS:

2 pounds	Whole-wheat flour
1 pound	All-purpose flour
2½ ounces	Baking powder
1 tablespoon	Salt
1½ pounds	Pasteurized eggs
2 quarts	Milk
4 ounces	Vegetable oil
3 ounces	Honey

METHOD OF PREPARATION:

1. Preheat a waffle iron.
2. Sift together the flours, baking powder, and salt.
3. Beat the eggs with the milk, oil, and honey, and, blending well, stir into the dry ingredients.
4. Ladle 2½ ounces of the batter onto a lightly oiled, hot waffle iron.
5. Close the top, and bake until the steaming stops.
6. Lift the waffle from the iron with a fork. Serve immediately, or hold at 140° F or above.
7. Repeat the procedure until all of the batter is used.

CHEF NOTE:
Keep the batter chilled to 40° F or below if not used immediately.

10

Sandwiches

10

Sandwiches

When mighty roast beef was the Englishman's food, it ennobled our hearts and enriched our blood. Our soldiers were brave and our courtiers good. Oh! The roast beef of old England!

Richard Leveridge (1670–1758)
The Cook's Quotation Book
A Literary Feast

In America today, more people than ever before want good food—meals that are nutritionally balanced, well prepared, and light. As the topic of food and the study of cooking become more mainstream, we find a great number of our culinary inventions are the result of cross-cultural exchanges. American food has become the product of an amazing array of ethnic combinations.

As professionals, we live in a most exciting time. We are more open to new ideas and new challenges. Our willingness to explore and experience the food of many cultures gives freshness and new life to the many age-old classics.

I congratulate you, the new generation of food professionals, who have so much to work for and whose contributions will undoubtedly influence the culinary world.

Martin Yan
Distinguished Visiting Chef
May 6–8, 1990
President
Yan Can Cook

The HACCP Process

The following recipe illustrates how sandwiches flow through the HACCP process.

Texas-Style Chicken Sandwich

Receiving

- Rolls—packaging intact; no visible mold
- Chicken breasts, whole, boneless—check for firm flesh. There should be no sign of odor or stickiness. Chicken should be delivered at 40° F (4.4° C) or lower.
- Canadian bacon, frozen—temperature at or below 0° F (−17.8° C), with no signs of thawing; packaging intact
- Spices (salt and pepper)—packaging intact
- Cheese (Monterey Jack)—check for flavor, texture, and uniform color. Cheese should be delivered at 40° F (4.4° C) or lower.
- Barbecue sauce—packaging intact
- Lettuce (iceberg)—firm, semi-hard heads, free of rusty or burned tips; texture should be firm. Lettuce should be delivered at 40° F (4.4° C) or lower.
- Tomato slices—good color, firm, well-shaped, no blemishes, and smooth. Tomatoes should be delivered at 40° F (4.4° C) or lower.
- Pickles—packaging intact

Storage

- Rolls—Store in dry storage at 50° F (10° C) with a relative humidity of 50% to 60%.
- Chicken—Store under refrigeration, with a product temperature not to exceed 40° F (4.4° C). Store below already cooked foods.
- Spices (salt and pepper)—Store in dry storage at 50° F (10° C) with a relative humidity of 50% to 60%. Keep dry.
- Cheese (Monterey Jack)—Store under refrigeration, with a product temperature not to exceed 40° F (4.4° C). Store above and away from raw, potentially hazardous foods.
- Barbecue sauce—Store in dry storage at 50° F (10° C) with a relative humidity of 50% to 60%.
- Lettuce—Store under refrigeration, with a product temperature not to exceed 40° F (4.4° C). Store above and away from raw, potentially hazardous foods.
- Tomato slices—Store under refrigeration, with a product temperature not to exceed 40° F (4.4° C). Store

above and away from raw, potentially hazardous foods.

- Pickles—Store in dry storage at 50° F (10° C) with a relative humidity of 50% to 60%.

Thawing Canadian Bacon

- Thaw under refrigerated storage. The air temperature of refrigerated unit should be 38° F (3.30° C) or lower.

Preparation and Cooking

- Trim and wash the lettuce.
- Cut the tomatoes on a clean and sanitized cutting board with clean and sanitized utensils.
- Split the roll, and toast it on a grill. Remove. Season the chicken; then grill it evenly on both sides. Cook to an internal temperature of 165° F (73.9° C). Place it on the roll.
- Lightly grill the Canadian bacon. Place it on top of the chicken.
- Top with barbecue sauce and Monterey Jack cheese.
- Garnish the roll with lettuce, tomato slices, and a pickle spear.

Holding and Service

- Serve immediately.

Nutritional Notes

Many classic sandwiches are a nutritionist's nightmare. These sandwiches are piled high with ingredients that are high in total fat, saturated fat, cholesterol, and sodium. This need not be the case. Sandwiches made with moderate amounts of lean meats, poultry, seafood, low-fat and reduced-sodium cheese, whole-grain breads, fresh or cooked vegetables, and condiments low in fat and sodium can be delicious as well as nutritious.

- *Prepare burgers using ground turkey or a mixture of ground turkey and lean ground beef.*
- *Broil, bake, or roast meats for sandwiches rather than frying them.*
- *Avoid commercially prepared sandwich meats. Cook meats in-house.*
- *Offer low-fat or no-fat sandwich spreads such as low-fat or no-fat mayonnaise.*
- *Feature lean meats such as chicken breast on whole-grain breads to reduce fat and increase fiber.*
- *Prepare sandwich fillings such as tuna or seafood using low-fat mayonnaise, or substitute yogurt for the mayonnaise. A creative use of herbs and spices adds interest to these fillings.*
- *Offer open-faced sandwiches or a half sandwich with a fresh green salad combination to lower fat and calories while increasing vitamins and fiber. Be sure to follow nutrition guidelines for salads and salad dressings.*
- *Offer vegetarian sandwiches. Prepare vegetables with no fat by steaming—or better yet, use raw vegetables.*

American Grinder

COOKING TECHNIQUE:

Not applicable

Glossary:

Chiffonade: ribbons of leafy greens

HAZARDOUS FOODS:

Mayonnaise
Ham
Salami
Turkey
Provolone and American cheese

NUTRITION:

Calories: 937
Fat: 44.5 g
Protein: 47.6 g

YIELD: 50 SERVINGS. SERVING SIZE: ONE SANDWICH.

INGREDIENTS:

50 each	Submarine rolls, split
1½ pints	Mayonnaise
2 heads	Iceberg lettuce, cleaned and washed, cut **chiffonade**
7 pounds	Tomatoes, washed, cored, and sliced
6 pounds	Ham, sliced thin
3 pounds	Salami, sliced thin
6 pounds	Turkey, sliced thin
3 pounds	Provolone cheese, sliced
3 pounds	American cheese, sliced
50 each	Pickle spears

METHOD OF PREPARATION:

1. Split the submarine roll, spread with mayonnaise, and fill with shredded lettuce and tomato slices.
2. Fill with meats, alternating ham, salami, and turkey.
3. Top with sliced cheeses, cut in half, and serve with a pickle spear.

CHEF NOTES:

1. Serve with cross-cut, seasoned french fries and cole slaw.
2. In quantity food production, the mayonnaise should be served on the side.

Clam Roll

YIELD: 50 SERVINGS. SERVING SIZE: FOUR OUNCES.

INGREDIENTS:

12 pounds	Clams, whole frying, drained
	Seasoned flour, as needed
15 each	Whole eggs, for egg wash
1 quart	Milk, for egg wash
	Bread crumbs, as needed
2 quarts	Mayonnaise
50 each	Parker rolls, split in half
	Tartar sauce, (see page 74), as needed
	Lemon wedges, as needed
	Oil, as needed

METHOD OF PREPARATION:

1. **Dredge** the clams in seasoned flour, and shake off any excess.
2. Dip in the egg wash, and transfer into the bread crumbs.
3. Roll the clams in the bread crumbs, and cover well.
4. Fry the clams **à la minute** in the hot fryer until golden brown, and lay out on absorbant paper.
5. Transfer to a lined hotel pan, and keep the pan uncovered and hot.
6. Spread both halves of the rolls with mayonnaise, place fried clams on one half, top with the other half, and cut the sandwich in two.
7. Serve with tartar sauce and a lemon wedge.

COOKING TECHNIQUE:

Deep-Fry

Deep-Fry:

1. Heat the frying liquid to the proper temperature.
2. Submerge the food product completely.
3. Fry the product until it is cooked throughout.

GLOSSARY:

Dredge: to coat with flour.
À la minute: cooked to order.

HAZARDOUS FOODS:

Clams
Eggs
Milk
Mayonnaise
Tartar sauce

NUTRITION:

Calories: 818
Fat: 36.6 g
Protein: 30.7 g

CHEF NOTES:

1. The rolls can be toasted and topped with clams to make an open-faced sandwich. If served like this, add lemon wedges as a garnish.
2. The clams are also available via the Individual Quick Frozen (IQF) process.

French Dip Sandwich

YIELD: 50 SERVINGS. SERVING SIZE: ONE SANDWICH.

INGREDIENTS:

50 each	French bread, cut in 6-inch lengths
12½ pounds	Top round of beef, roasted, sliced thin
1½ quarts	Meat glaze (see page 36), heated to a boil

METHOD OF PREPARATION:

1. Toast the bread.
2. Heat the top round of beef in the meat glaze.
3. Place the bottom of a section of French bread on a preheated dinner plate and top with 4-ounces of beef. Cover with the top of the French bread, and serve with a side portion of meat glaze.

COOKING TECHNIQUE:

Simmer

Simmer and Poach:

1. Heat the cooking liquid to the proper temperature.
2. Submerge the food product completely.
3. Keep the cooked product moist and warm.

GLOSSARY:

Au jus: with natural juices

NUTRITION:

Calories: 304
Fat: 6.77 g
Protein: 39.2 g

CHEF NOTE:

See the recipe for roasted top round of beef (page 270).

Grilled California Sandwich

YIELD: 50 SERVINGS. SERVING SIZE: ONE SANDWICH.

INGREDIENTS:

15 pounds	Flap, cap, or flank steaks
10 loaves	Sour dough bread, sliced diagonally ½-inch thick
	Clarified butter, as needed
	Lettuce leaves, as needed
16 each	Tomatoes, washed, cored, and sliced
8 each	Onions, peeled and cut **julienne**
6 each	Avocado, peeled, pit removed, flesh sliced ⅛-inch thick, and held in lemon wash
	Oil, as needed

California Marinade:

3 pints	Red wine, dry
1 pint	Orange juice concentrate, thawed
2 ounces	Rosemary
2 ounces	Garlic, chopped
2 tablespoons	Dry mustard
1 tablespoon	Salt
1 tablespoon	Ground black pepper

METHOD OF PREPARATION:

1. Mix all of the ingredients of the marinade in a bowl. Pour the marinade over the beef. Cover, and refrigerate for at least 4 hours. Remove the beef from the marinade, and grill or broil to a minimum of 140° F. Slice ⅛-inch thick **à la minute.**
2. For each sandwich, brush both sides of the two slices of bread with oil. Grill or broil until golden.
3. Layer one slice of bread with lettuce, two slices of tomato, onion, three ounces of beef, and one slice of avocado. Top with the remaining bread slice. Serve warm with the two choices of sauce. Recipes follow.

COOKING TECHNIQUES:

Grill or Broil

Grill/Broil:

1. Clean and heat the grill/broiler.
2. To prevent sticking, brush the food product with oil.

GLOSSARY:

Julienne: matchstick strips
À la minute: to order

HAZARDOUS FOOD:

Beef top round

NUTRITION:

Calories: 395
Fat: 16.6 g
Protein: 32.2 g

Bistro Rémoulade

METHOD OF PREPARATION:

Mix 1¾ quarts of mayonnaise, 8 ounces of canned tomato paste, 4 tablespoons of lemon juice, 6 cloves of minced garlic, and 2 teaspoons of coarse ground black pepper. This recipe makes 2 quarts.

Golden Gate Grill Sauce

METHOD OF PREPARATION:

Mix 1¾ quarts of mayonnaise, 8 ounces of dry mustard, 6 tablespoons of snipped chives, and 2 ounces of finely grated fresh orange peel. This recipe makes 2 quarts.

Grilled Ham and Cheese Sandwich

YIELD: 50 SERVINGS. SERVING SIZE: ONE SANDWICH.

INGREDIENTS:

5 loaves	Sliced sandwich bread
4 pounds	Ham, boneless, cooked, cut into 1-ounce slices
3 pounds	American cheese, sliced

METHOD OF PREPARATION:

1. Lay out the bread slices on a clean dry table.
2. Place one slice of ham and cheese on each second slice of bread.
3. Cover with the second slice of bread, and cut on an angle to create two three angle sandwiches.
4. Lightly coat with butter or margarine.
5. Grill until golden brown on both sides.
6. Serve hot.

COOKING TECHNIQUE:

Grill

Grill/Broil:

1. Clean and heat the grill/broiler.
2. To prevent sticking, brush the food product with oil.

HAZARDOUS FOODS:

Ham
Cheese

NUTRITION:

Calories: 345
Fat: 14 g
Protein: 17.7 g

CHEF NOTE:

Do not prepare the sandwiches too far in advance, because they will become soggy.

Italian Grinder

COOKING TECHNIQUE:

Bake

Bake:

1. Preheat the oven.
2. Place the food product on the appropriate rack.

HAZARDOUS FOODS:

Ham
Salami
Bologna
Provolone

NUTRITION:

Calories: 688
Fat: 25 g
Protein: 32 g

YIELD: 50 SERVINGS. SERVING SIZE: ONE SANDWICH.

INGREDIENTS:

50 each	Submarine rolls, split lengthwise
2½ pounds	Ham, sliced thin
2½ pounds	Salami, sliced thin
2½ pounds	Bologna, sliced thin
2½ pounds	Provolone cheese, sliced thin
25 each	Tomatoes, washed, cored, and sliced
2½ pounds	Red onions, peeled, and sliced into thin rings
4 heads	Lettuce, washed and shredded

METHOD OF PREPARATION:

1. Preheat the oven to 375° F.
2. On an open-faced roll, place two slices of tomato and onion.
3. Add one slice each of the meats and cheese.
4. Place on sheet pans, and bake for 15 minutes.
5. Garnish each sandwich with shredded lettuce.

Monte Cristo Sandwich

YIELD: 50 SERVINGS. SERVING SIZE: FIVE OUNCES.

COOKING TECHNIQUE:

Grill

Grill/Broil:

1. Clean and heat the grill/broiler.
2. To prevent sticking, brush the food product with oil.

HAZARDOUS FOODS:

Whole eggs
Ham
Turkey
Swiss cheese

NUTRITION:

Calories: 600
Fat: 26.1 g
Protein: 43.7 g

INGREDIENTS:

50 two-ounce	Turkey breast, cooked, and sliced thin
50 two-ounce	Virginia ham, sliced
50 one-ounce	Swiss cheese, sliced
150 slices	White bread
25 each	Whole eggs, slightly beaten
1 pound	Butter or oil

METHOD OF PREPARATION:

1. Place a slice of bread on a sheet pan; place one slice of turkey on the bread, and then place a second slice of bread on the turkey.
2. Place one slice of ham and one slice of Swiss cheese on top of the second slice of bread. Cover with a third slice of bread.
3. Dip the sandwich in egg. Cook on both sides to a golden brown on a well-buttered griddle. Cut in half on the diagonal.

Philadelphia Steak Sandwich

YIELD: 50 SERVINGS. SERVING SIZE: ONE SANDWICH.

INGREDIENTS:

4 ounces	Vegetable oil
15 each	Medium onions, peeled and cut **julienne**
15 each	Green bell peppers, washed, seeded, and cut julienne
12½ pounds	Top round of beef, sliced ⅛- to ¼-inch thick
	Salt and ground black pepper, to taste
50 each	French-style rolls, split lengthwise
1½ pints	Swiss cheese, sliced and warm

METHOD OF PREPARATION:

1. Heat the oil in a large sauté pan over medium-high heat.
2. Add the onions and green peppers, and stir-fry for 4 minutes. Sauté the sliced top round of beef to an internal temperature of 140° F, and season, to taste.
3. Place the beef mixture on the bottom half of each roll. Top with cheese, and melt under a salamander or in an oven. Cover with the top of the roll, and serve.

COOKING TECHNIQUES:

Sauté, Stir-Fry

Sauté:

1. Heat the sauté pan to the appropriate temperature.
2. Evenly brown the food product.
3. For a sauce, pour off any excess oil, reheat, and deglaze.

Stir-Fry:

1. Heat the oil in a wok until hot but not smoking.
2. Keep the food in constant motion; use the entire cooking surface.

GLOSSARY:

Julienne: matchstick strips

HACCP:

Cook the beef to internal temperature of 140° F.

HAZARDOUS FOODS:

Top round of beef
Swiss cheese

NUTRITION:

Calories: 392
Fat: 12.8 g
Protein: 42.8 g

Texas-Style Chicken Sandwich

YIELD: 50 SERVINGS. SERVING SIZE: ONE SANDWICH.

COOKING TECHNIQUE:
Grill

Grill/Broil:
1. Clean and heat the grill/broiler.
2. To prevent sticking, brush the food product with oil.

HAZARDOUS FOODS:
Chicken breasts
Canadian bacon
Jack cheese

NUTRITION:
Calories: 904
Fat: 31.6 g
Protein: 64.2 g

INGREDIENTS:

50 each	Bulkie rolls
12½ pounds	4-ounce boneless chicken breasts
	Salt and ground white pepper, to taste
6 pounds	Canadian bacon
3 pounds	Monterey Jack cheese
6 pints	Barbecue sauce
50 leaves	Lettuce
150 each	Tomato slices
50 each	Pickle spears

METHOD OF PREPARATION:

1. Split the roll. Toast it on a grill, and remove. Season the chicken. Grill it evenly on both sides, and place it on a roll.
2. Lightly grill the Canadian bacon, and place it on top of the chicken.
3. Top with barbecue sauce and Monterey Jack cheese.
4. Garnish the roll with lettuce and tomato slices, and serve with a pickle spear.

CHEF NOTE:
Serve with seasoned french fries and cole slaw.

Tuna Melt

YIELD: 50 SERVINGS. SERVING SIZE: TWO SANDWICHES.

INGREDIENTS:

25 each	Large English muffins, halved
1½ pints	Mayonnaise
2 heads	Iceberg lettuce, washed and leaves separated
25 each	Tomatoes, washed, cored, and sliced
12½ pounds	Tuna salad
6 pounds	Provolone cheese, thinly sliced
2 heads	Lettuce, washed and leaves separated
150 each	Pineapple slices
150 each	Orange slices

METHOD OF PREPARATION:

1. Toast the English muffins halves, and spread with mayonnaise.
2. Place the lettuce, tomato, and tuna salad on the English muffin halves.
3. Put a slice of provolone cheese on the tuna and place sandwich in a salamander, until the cheese melts.
4. Garnish the plate with a leaf of lettuce. Place the pineapple slices on the lettuce. Shingle the orange slices between the pineapple slices.

COOKING TECHNIQUE:

Bake

Bake:

1. Preheat the oven.
2. Place the food product on the appropriate rack.

HAZARDOUS FOODS:

Mayonnaise
Tuna salad

NUTRITION:

Calories: 745
Fat: 35.8 g
Protein: 36.2 g

CHEF NOTE:

Serve with seasoned french fries and cole slaw.

11

Baked Goods and Pastries

Baked Goods and Pastries

The discovery of a new dish is more beneficial to humanity than the discovery of a new star.

Jean-Anthelme Brillat-Savarin
The Cook's Quotation Book
A Literary Feast

After every visit to Johnson & Wales, I can't help thinking how lucky are those who have had the opportunity to study at such a great university. Therefore, I would like to share the following few lines with the students of Johnson & Wales University.

As a young pastry chef in France, I felt a need to come up with a formula that would give me a professional edge. That formula is best summarized by the letters TOPS.

T *is for* think	*Use your brain, not your feet.*
O *is for* organization	*Organize your day to cut down on wasted time.*
P *is for* precision	*Every move you make should be productive.*
S *is for* speed	*Time yourself; speed is actually the key to success.*

At the close of every workday, I would ask myself if I had been on TOPS today.

Roland Mesnier
Distinguished Visiting Chef, October 22–24, 1989
Executive Pastry Chef, The White House
Washington, DC

The HACCP Process

The following recipe illustrates how desserts flow through the HACCP process.

Cheesecake, New York Style

Receiving

- Cream cheese—packaging intact. Cream cheese should be delivered at 40° F (4.4° C) or lower.
- Sugar, granulated—packaging intact.
- Butter, unsalted—should have a sweet flavor, firm texture, and uniform color. Butter should be free of specks and mold, and the packaging should be intact. Butter should be delivered at 40° F (4.4° C) or lower.
- Lemons—fine textured skin, heavy for size, and uniform color
- Vanilla extract—packaging intact
- Eggs, whole—shells not cracked or dirty. Container should be intact. Whole shell eggs should be delivered at 40° F (4.4° C) or lower.
- Heavy cream—should have a sweetish taste. Container should be intact. Heavy cream should be delivered at 40° F (4.4° C) or lower.
- Sour cream—packaging intact. Sour cream should be delivered at 40° F (4.4° C) or lower.
- Graham cracker crumbs—packaging intact

Storage

- Cream cheese—Store under refrigeration at 40° F (4.4° C) or lower. Store above and away from raw, potentially hazardous foods.
- Sugar, granulated—Store in dry storage at 50° F (10° C) with a relative humidity of 50% to 60%. Keep dry.
- Butter—Store under refrigeration at 40° F (4.4° C) or lower. Store above and away from raw, potentially hazardous foods.
- Lemons—Store under refrigeration at 40° F (4.4° C) or lower.
- Vanilla extract—Store in dry storage at 50° F (10° C) with a relative humidity of 50% to 60%.
- Eggs, whole—Store under refrigeration at 40° F (4.4° C) or lower. Store above and away from raw, potentially hazardous foods.
- Heavy cream—Store under refrigeration at 40° F (4.4° C) or lower. Store above and away from raw, potentially hazardous foods.
- Sour cream—Store under refrigeration at 40° F (4.4° C) or lower. Store above and away from raw, potentially hazardous foods.

- Graham cracker crumbs—Store in dry storage at 50° F (10° C) with a relative humidity of 50% to 60%. Keep dry.

Preparation and Cooking

- Cream the first five ingredients thoroughly, scraping the bowl as necessary.
- Add the eggs in three stages, scraping the bowl each time.
- When smooth, blend in the heavy cream and sour cream.
- Coat six 9-inch cake pans with butter and the graham cracker crumbs.
- Divide the cheesecake mixture equally among the six pans. The cake pans should be nearly full.
- Bake in a water bath at 325°, approximately 1 hour, 15 minutes. Cook to an internal temperature of 140° F (60° C).

Cooling

- Place the cheese cake in a chill blaster. Lower the temperature to 40° F (4.4° C) or lower within 4 hours.
- Cover loosely with plastic wrap when cooled.
- Label with the date, time, and name of the product.
- Store on the upper shelf of a refrigerated unit at 40° F (4.4° C) or lower.

Holding and Service

- Unmold by heating the pan slightly over an open fire or in oven and inverting the cheesecake on a plastic film–lined cardboard. Turn right-side up on another cardboard, gently peeling off the plastic film.
- Serve immediately.

Nutritional Notes

Desserts are the finale to the meal. The choice might be a subtle, light dessert to conclude a heavy, high-protein meal, or a more substantial dessert might be selected as the grand finale to a lighter meal; it is all a matter of balance.

In addition to ending the meal, desserts are often looked on as treats and rewards by the consumer and an expression of creativity by the chef. It is important to retain these qualities when adapting a dessert recipe to reduce total fat, saturated fat, cholesterol, and calories.

- *Whole eggs can be replaced with egg whites or egg substitute whenever possible.*
- *Meringues are an elegant, fat-free base for desserts.*
- *Angel food cakes also are fat-free.*
- *Solid carob contains saturated fat; replace it with carob powder.*
- *Replace saturated fats with mono- and polyunsaturated oils whenever possible. Be aware that this may change the texture of the final product.*
- *Filo leaves, sprayed with oil, can replace pastry in some recipes. For example, dessert shells and fruit turnovers can be made using filo leaves.*
- *Fat-free milk can be used in place of whole milk whenever possible. Evaporated skim milk can replace cream in some recipes. Mock sour cream or drained yogurt also can be used in place of sour cream.*
- *Heavy whipping cream can be replaced with a reduced-fat version, and less used. This is more appropriate than using nondairy toppings, which may be high in saturated fat.*
- *Whole-grain flours can be utilized when baking to increase flavor and add fiber.*
- *Reduced-fat or non-fat cheeses can be used.*
- *The amount of fat and sugar called for in recipes often can be reduced by one fourth to one half.*
- *The skins of fruits can be left on for added color and fiber.*
- *A number of fat-free dessert mixes are available. These can be used as the base for a more elaborate dessert.*
- *Low- or non-fat frozen yogurt or ice milk can replace ice cream.*

Almond Sponge Cake

Use with silpat and combed stencils

PASTRY TECHNIQUES:

Creaming, Whipping, Folding

Creaming:

1. Soften the fats on low speed.
2. Add the sugar(s) and cream; increase the speed slowly.
3. Add the eggs one at a time; scrape the bowl frequently.
4. Add the dry ingredients in stages.

Whipping:

1. Hold the whip at a 55-degree angle.
2. Create circles, using a circular motion.
3. The circular motion needs to be perpendicular to the bowl.

Folding:

Do steps 1, 2, and 3 in one continuous motion.

1. Run a bowl scraper under the mixture, across the bottom of the bowl.
2. Turn the bowl counterclockwise.
3. Bring the bottom mixture to the top.

HACCP:

Store at 40° F for 1 day. After 1 day, store at 0° F.

HAZARDOUS FOODS:

Eggs
Egg whites

CHEF NOTE:

This formula is a special sponge cake of low volume for use with combed stencils to fill in spaces.

YIELD:	1 pound, 3 ounces.	2 pounds, 6 ounces.
INGREDIENTS:		
Almond meal	4 ounces	8 ounces
Flour, cake	1 ounce	2 ounces
Sugar, confectionery	4 ounces	8 ounces
Eggs, whole	5 ounces	10 ounces
Butter, unsalted, melted	1 ounce	2 ounces
Egg whites, at room temperature	3½ ounces	7 ounces
Cream of tartar	Pinch	Pinch
Sugar, granulated	½ ounce	1 ounce

METHOD OF PREPARATION:

1. Gather the equipment and ingredients.
2. Sift together the almond meal, cake flour, and confectionery sugar.
3. Whisk the whole eggs lightly, and add to the dry ingredients.
4. Add the melted butter to the mixture, and combine well.
5. Whip the egg whites on high speed; add the cream of tartar when the whites foam up; add the sugar gradually as the whites increase in volume. Reduce the speed as the meringue thickens, to avoid overmixing.
6. Fold the whipped egg whites into the mixture in three stages.
7. Ready for use.

How to Use Almond Sponge Cake:

1. Spread the batter uniformly over the silpat mat.
2. Comb in the desired decoration.
3. Freeze until completely set.
4. Spread the sponge cake batter for the silpat on top evenly.
5. Bake at 400° F, but *do not overbake.*
6. Once cool, place the sheet on top, and invert the mat.
7. Remove the silpat mat.
8. Cut the cake to the desired shape and sizes.
9. Utilize, as directed, for siding and garnishes.

Angel Food Cake

PASTRY TECHNIQUE:
Whipping

Whipping:
1. Hold the whip at a 55-degree angle.
2. Create circles, using a circular motion.
3. The circular motion needs to be perpendicular to the bowl.

HACCP:
Store at 40° F for 1 day. After 1 day, store at 0° F.

HAZARDOUS FOOD:
Egg whites

YIELD:	4 pounds, 12¾ ounces. 5, 8-inch Cakes	9 pounds, 9½ ounces. 10, 8-inch Cakes
INGREDIENTS:		
Egg whites	2 pounds	4 pounds
Flour, cake	12 ounces	1 pound, 8 ounces
Sugar, confectionery	12 ounces	1 pound, 8 ounces
Cream of tartar	¼ ounce	½ ounce
Sugar, granulated	1 pound, 4 ounces	2 pounds, 8 ounces
Salt	⅛ ounce	¼ ounce
Extract, vanilla	¼ ounce	½ ounce
Extract, almond	⅛ ounce	¼ ounce

METHOD OF PREPARATION:

1. Gather the equipment and ingredients.
2. Bring the egg whites to room temperature; it may be necessary to warm them over a double boiler.
3. Sift together the cake flour and confectionery sugar.
4. Place the egg whites in a bowl, and whip on high speed; add the cream of tartar when the egg whites foam up; add the sugar gradually as the whites increase in volume. Reduce the speed as the meringue thickens, to avoid overmixing. After the meringue is mixed, add the salt, vanilla extract, and almond extract.
5. Whip to soft, moist peaks; *do not overmix*. Turn the mixer speed to medium during the process of whipping, to prevent overmixing.
6. Remove from the mixer, and fold in the sifted cake–confectionery sugar mixture.
7. Fill 8-inch angel food cake pans with 14 ounces of batter.
8. Bake at 375° F for 25 minutes, or until light golden brown on top and sides and set in center.
9. Remove from the oven, and invert to assist unmolding.

CHEF NOTE:
Chocolate angel food cake can be made by replacing 25% of the cake flour with cocoa powder.

Carrot Cake

PASTRY TECHNIQUE:
Combining

Combining:
Bringing together two or more components.
1. Prepare the components to be combined.
2. Add one to the other, using the appropriate mixing method (if needed).

HACCP:
Store at 40° F for 1 day. After 1 day, store at 0° F.

HAZARDOUS FOOD:
Eggs

YIELD:	11 pounds, 4 ounces. 5, 9-inch Cakes	22 pounds, 8 ounces. 10, 9-inch Cakes
INGREDIENTS:		
Flour, bread	2 pounds, 2 ounces	4 pounds, 4 ounces
Baking powder	1¼ ounces	2½ ounces
Baking soda	¾ ounce	1½ ounces
Cinnamon, ground	¾ ounce	1½ ounces
Salt	½ ounce	1 ounce
Nutmeg, ground	¼ ounce	½ ounce
Sugar, granulated	2 pounds, 7 ounces	4 pounds, 14 ounces
Oil, vegetable	2 pounds	4 pounds
Eggs, whole	1 pound, 8 ounces	3 pounds
Raisins, seedless	12 ounces	1 pound, 8 ounces
Carrots, grated	2 pounds, 4 ounces	4 pounds, 8 ounces

METHOD OF PREPARATION:

1. Gather the equipment and ingredients.
2. Grease the cake pans thoroughly.
3. Sift together the bread flour, baking powder, baking soda, cinnamon, salt, and nutmeg.
4. Place the granulated sugar and oil in a bowl, and combine using a paddle.
5. Add the eggs slowly in stages to the sugar-oil mixture, and combine.
6. Add the sifted ingredients into sugar-oil mixture; combine thoroughly.
7. Fold in the raisins and grated carrots.
8. Fill 9-inch by 3-inch cake pans with 2 pounds, 3 ounces of batter.
9. Bake at 350° F for 45 minutes, or until light golden brown.

Chocolate Applesauce Cake

PASTRY TECHNIQUE:

Combining

Combining:

Bringing together two or more components.

1. Prepare the components to be combined.
2. Add one to the other, using the appropriate mixing method (if needed).

HACCP:

Store at 40° F. for 1 day. After 1 day, store at 0° F.

HAZARDOUS FOODS:

Eggs
Buttermilk

YIELD:	8 pounds, 9½ ounces.	17 pounds, 3 ounces.
	6, 9-inch Cakes	12, 9-inch Cakes
	1 Sheet Cake	2 Sheet Cakes

INGREDIENTS:

Flour, cake, sifted	1 pound, 11 ounces	3 pounds, 6 ounces
Cocoa powder, sifted	1½ ounces	3 ounces
Baking soda, sifted	¾ ounce	1½ ounces
Baking powder, sifted	¾ ounce	1½ ounces
Salt	¾ ounce	1½ ounces
Cinnamon, ground	¾ ounce	1½ ounces
Sugar, brown	2 pounds, 4 ounces	4 pounds, 8 ounces
Oil, vegetable	1 pound, 5 ounces	2 pounds, 10 ounces
Eggs, whole	13 ounces	1 pound, 10 ounces
Applesauce	12 ounces	1 pound, 8 ounces
Buttermilk	1 pound, 8 ounces	3 pounds

METHOD OF PREPARATION:

1. Gather the equipment and ingredients.
2. Sift together the cake flour, cocoa powder, baking soda, baking powder, salt, and cinnamon.
3. Place all of the sifted ingredients and the brown sugar in bowl. Blend together using a paddle.
4. Slowly add the oil, to avoid lumps from forming, and mix into a paste.
5. Add the eggs in stages, scraping in between.
6. Add the applesauce slowly, scraping regularly.
7. Add the buttermilk, and mix until smooth.
8. Fill 9-inch cake pans with 1 pound, 7 ounces of batter, or fill 1 sheet pan with 8 pounds, 9½ ounces of batter.
9. Bake at 360° F until firm.

Chocolate Chiffon Genoise

YIELD:	4 pounds, 3 ounces. 3, 9-inch Cakes	12 pounds, 9½ ounces. 9, 9-inch Cakes

INGREDIENTS:		
Egg yolks	1 pound	3 pounds
Sugar, granulated	9⅓ ounces	1 pound, 12 ounces
Flour, cake	10⅔ ounces	2 pounds
Baking soda	Pinch	½ ounce
Baking powder	⅓ ounce	1 ounce
Salt	To taste	To taste
Butter, unsalted	4 ounces	12 ounces
Chocolate, dark, semisweet	9⅓ ounces	1 pound, 12 ounces
Egg whites	10⅔ ounces	2 pounds
Sugar, granulated	6⅔ ounces	1 pound, 4 ounces

METHOD OF PREPARATION:

1. Gather the equipment and ingredients.
2. Place the egg yolks and 1 pound, 12 ounces of the granulated sugar in a bowl, and place over a double boiler.
3. Stir continuously until it reaches 120° F.
4. Remove from the heat, and place on a mixer.
5. Whip on the highest speed to full volume.
6. Sift together the cake flour, baking soda, baking powder, and salt.
7. Melt the butter. Keep it on the cool side.
8. Melt the chocolate; do not heat it higher than 122° F.
9. Add the melted butter to the melted chocolate, and combine well.
10. Add the sifted dry ingredients into the whipped egg yolks.
11. Fold the combined butter-chocolate mixture into the egg yolk mixture.
12. In another bowl, whip the egg whites to medium peak; slowly add 1 pound, 4 ounces of granulated sugar to make a meringue.
13. Carefully fold the meringue into the batter.
14. Scale 1 pound, 6 ounces of batter into lightly greased, 9-inch cake pans.
15. Bake at 375° F until firm and dry in the center.

PASTRY TECHNIQUES:

Whipping, Melting, Folding

Whipping:

1. Hold the whip at a 55-degree angle.
2. Create circles, using a circular motion.
3. The circular motion needs to be perpendicular to the bowl.

Melting:

1. Prepare the food product to be melted.
2. Place the food product in an appropriate-sized pot over direct heat or over a double boiler.
3. Stir frequently or occasionally, depending on the delicacy of the product, until melted.

OR

1. Place the product on a sheet pan or in a bowl, and place in a low-temperature oven until melted.

Folding:

Do steps 1, 2, and 3 in one continuous motion.

1. Run a bowl scraper under the mixture, across the bottom of the bowl.
2. Turn the bowl counterclockwise.
3. Bring the bottom mixture to the top.

HACCP:

Store at 40° F for 1 day. After 1 day, store at 0° F.

HAZARDOUS FOODS:

Egg yolks
Egg whites

Chocolate Genoise

PASTRY TECHNIQUES:

Whipping, Melting, Folding

Whipping:

1. Hold the whip at a 55-degree angle.
2. Create circles, using a circular motion.
3. The circular motion needs to be perpendicular to the bowl.

Melting:

1. Prepare the food product to be melted.
2. Place the food product in an appropriate sized pot over direct heat or over a double boiler.
3. Stir frequently or occasionally, depending on the delicacy of the product, until melted.

OR

1. Place the product on a sheet pan or in a bowl, and place in a low-temperature oven until melted.

Folding:

Do steps 1, 2, and 3 in one continuous motion.

1. Run a bowl scraper under the mixture, across the bottom of the bowl.
2. Turn the bowl counterclockwise.
3. Bring the bottom mixture to the top.

HACCP:

Store at 40° F for 1 day. After 1 day, store at 0° F.

HAZARDOUS FOOD:

Eggs

CHEF NOTES:

1. The following methods test for doneness:
   (a) Check for color; a golden brown is good.
   (b) Check to see if the cake is pulling away from the sides of the pan.

YIELD:	3 pounds, 10⅞ ounces. 2, 9-inch by 3-inch Pans	7 pounds, 5¾ ounces. 4, 9-inch by 3-inch Pans
INGREDIENTS:		
Flour, cake	6 ounces	12 ounces
Cornstarch	6 ounces	12 ounces
Cocoa powder	4 ounces	8 ounces
Baking soda	1⁄16 ounce	⅛ ounce
Water	½ ounce	1 ounce
Butter, unsalted	4 ounces	8 ounces
Eggs, whole	1 pound, 8 ounces	3 pounds
Salt	⅛ ounce	¼ ounce
Extract, vanilla	¼ ounce	½ ounce
Sugar, granulated	14 ounces	1 pound, 12 ounces

METHOD OF PREPARATION:

1. Gather the equipment and ingredients.
2. Properly grease the 9-inch by 3-inch cake pans.
3. Sift together the cake flour, cornstarch, and cocoa powder; set aside.
4. Dissolve the baking soda in water, and set aside.
5. Melt the butter, and set aside. (Butter should be only lukewarm.)
6. Place the whole eggs, salt, vanilla extract, granulated sugar, and baking soda–water mixture in a bowl; immediately whisk over a warm water bath. (Never scale ingredients together in advance, or lumping will occur when the dry sugar comes in contact with the eggs.)
7. Heat the mixture to 110° F, constantly mixing. *Do not cook* the mixture.
8. Remove the bowl from the warm bath, and place on the mixer; whip on high speed to full volume. (Full volume can be determined by the "5-second track method." See chef note 3.)
9. When full volume is reached, remove the bowl from the mixer, and immediately fold the dry ingredients into the whipped egg mixture. Use your hand along with a plastic scraper to distribute the sifted dry ingredients evenly. This is a combination of the folding technique (turning the bowl counterclockwise with the left hand and scraping up from the bottom with the right hand in a clockwise rotation; finish out on top of the batter with your right hand and scraper palm up and shaking vigorously to distribute the flour and starch). Work quickly. Continue mixing until all of the ingredients are completely incorporated. (The longer it takes to combine the ingredients, the greater the chance that lumps will be trapped in the batter.)
10. Place a small amount of batter in another small bowl.

CHEF NOTES: continued

(c) Touch the center of cake, and it should spring back.

2. Sponge cake is a very delicate cake; do not slam the door or knock the pans together.
3. The following is the "5-second track method":
   Put ½ inch of your fingertip through the batter. If it holds a track for 5 seconds before closing, it is done. If it fails to hold a track, whip more. Check several times during mixing. Stay with the machine to monitor its progress. Keep track of mixing times and speeds. As the batter thickens, reduce the speed to medium, and check more frequently. *Do not overmix;* this will result in loss of volume or total collapse. Eggs should be cool at the end of this process.

11. Using your fingers, quickly whisk the melted butter into this batter until completely incorporated.

12. Fold this butter mixture, gently, back into the remaining batter.

13. Scale the batter at 1 pound, 11 ounces per 9-inch by 3-inch pan, approximately two thirds full.

14. Bake at 350° F for approximately 35 to 40 minutes. Test for doneness. (See chef note 1.)

15. When done, remove from the oven, and place on a cooling rack.

Chocolate Layer Cake

(High-Ratio)

PASTRY TECHNIQUES:

Combining

Combining:

Bringing together two or more components.

1. Prepare the components to be combined.
2. Add one to the other, using the appropriate mixing method (if needed).

HACCP:

Store at 40° F for 1 day. After 1 day, store at 0° F.

HAZARDOUS FOODS:

Egg whites
Eggs
Milk

YIELD:	4 pounds, 5 ounces.	12 pounds, 15 ounces.
	3, 9-inch Cakes	9, 9-inch Cakes

INGREDIENTS:

Ingredient	3, 9-inch Cakes	9, 9-inch Cakes
Flour, cake, sifted	13⅓ ounces	2 pounds, 8 ounces
Cocoa powder, sifted	2⅔ ounces	8 ounces
Baking soda, sifted	¼ ounce	¾ ounce
Baking powder, sifted	⅓ ounce	1 ounce
Salt	⅓ ounce	1 ounce
Sugar, granulated	1 pound, 4 ounces	3 pounds, 12 ounces
Shortening, high-ratio	6¾ ounces	1 pound, 4¼ ounces
Egg whites	5⅓ ounces	1 pound
Eggs, whole	2⅔ ounces	8 ounces
Milk, whole	1 pound, 1⅓ ounces	3 pounds, 4 ounces

METHOD OF PREPARATION:

1. Gather the equipment and ingredients.
2. Sift together the cake flour, cocoa powder, baking soda, baking powder, and salt.
3. Place the sifted dry ingredients, granulated sugar, shortening, egg whites, whole eggs, and one third of the milk into a bowl.
4. Mix together, using a paddle, on medium speed for 5 minutes.
5. Add the remaining milk in three parts; scrape the bowl between additions, and mix until smooth. Mix for 3 minutes.
6. Fill 9-inch cake pans with 1 pound, 7 ounces of batter.
7. Bake at 375° F for 25 to 30 minutes or until done.
8. Remove, and turn onto sugar-flour–dusted, paper-lined pans to cool.

Decorating Paste

YIELD:	14 ounces.	1 pound, 12 ounces.
INGREDIENTS:		
Sugar, confectionery, sifted	3½ ounces	7 ounces
Butter, unsalted, soft	3½ ounces	7 ounces
Egg whites	3½ ounces	7 ounces
Flour, pastry, sifted	3½ ounces	7 ounces
Coloring, food (optional)	As needed	As needed

METHOD OF PREPARATION:

1. Gather the equipment and ingredients.
2. Place the sugar, butter, and egg whites in a bowl; combine until smooth.
3. Add the pastry flour; blend until smooth and well incorporated.
4. *Optional:* Mix coloring, as desired.
5. Rest for 1 hour.

How to Use Decorating Paste:

1. Spread the paste uniformly over the silpat mat.
2. Comb in the desired decoration.
3. Freeze until completely set.
4. Spread the sponge cake batter for the silpat on top evenly.
5. Bake at 400° F, but *do not overbake.*
6. Once cool, place a sheet on top, and invert the mat.
7. Remove the silpat mat.
8. Cut the cake to the desired shape and sizes.
9. Utilize, as directed, for siding and garnishes.

PASTRY TECHNIQUES:

Combing, Spreading

Combing:

1. Prepare the item with the appropriate amount of icing.
2. Drag a clean comb across the surface.

Spreading:

1. Using an icing spatula or off-set spatula, smooth the icing or other spreading medium over the surface area.

HACCP:

Store at 40° F for 1 day. After 1 day, store at 0° F.

HAZARDOUS FOOD:

Egg whites

CHEF NOTES:

1. Prior to use, decorating paste should rest for about 1 hour to allow for absorption of the flour.
2. Consistency can be adjusted by adding either more flour to thicken or oil to thin.
3. Cocoa butter can be sprayed over the cake surface to prevent drying when used for siding and garnishes.
4. Decorating paste can be frozen until needed.

Decorating Paste, Chocolate

YIELD:	11¼ ounces.	1 pound, 6½ ounces.
INGREDIENTS:		
Sugar, confectionery, sifted	2¾ ounces	5½ ounces
Butter, soft	2¾ ounces	5½ ounces
Egg whites	2¾ ounces	5½ ounces
Flour, pastry, sifted	2 ounces	4 ounces
Cocoa powder, sifted	1 ounce	2 ounces
Almond sponge cake batter (see page 743)	As needed	As needed

METHOD OF PREPARATION:

1. Gather the equipment and ingredients.
2. Place the sugar, butter, and egg whites in a bowl; combine until smooth.
3. Sift together the pastry flour and cocoa powder.
4. Add the sifted mixture into the butter mixture, and blend until smooth and well incorporated.

How to Use Decorating Paste:

1. Spread the paste uniformly over the silpat mat.
2. Comb in the desired decoration.
3. Freeze until completely set.
4. Spread the almond sponge for the silpat on top evenly.
5. Bake at 400° F, but *do not overbake.*
6. Once cool, place the sheet on top, and invert the mat.
7. Remove the silpat mat.
8. Cut the cake to the desired shape and sizes.
9. Utilize, as directed, for siding and garnishes.

PASTRY TECHNIQUES:

Combing, Spreading

Combing:

1. Prepare the item with the appropriate amount of icing.
2. Drag a clean comb across the surface.

Spreading:

1. Using an icing spatula or off-set spatula, smooth the icing or other spreading medium over the surface area.

HACCP:

Store at 40° F for 1 day. After 1 day, store at 0° F.

HAZARDOUS FOOD:

Egg whites

CHEF NOTES:

1. Prior to use, decorating paste should rest for about 1 hour to allow for absorption of the flour.
2. Consistency can be adjusted by adding either more flour to thicken or oil to thin.
3. Cocoa butter can be sprayed over the cake surface to prevent drying when used for siding and garnishes.
4. Decorating paste can be frozen until needed.

Frangipane No. 1

YIELD:	4 pounds, 8½ ounces.	9 pounds, 1 ounce.
	½ Sheet Pan	1 Sheet Pan

INGREDIENTS:

Almond paste	2 pounds, 3 ounces	4 pounds, 6 ounces
Eggs, whole	1 pound, 1½ ounces	2 pounds, 3 ounces
Butter, unsalted, room temperature	1 pound, 1½ ounces	2 pounds, 3 ounces
Flour, cake	2½ ounces	5 ounces

METHOD OF PREPARATION:

1. Gather the equipment and ingredients.
2. Place the almond paste in a bowl with a paddle; soften by adding a small amount of eggs a little at a time until smooth.
3. In another bowl, soften the butter; scrape well.
4. Add the softened almond paste to the softened butter; scrape well.
5. Cream the butter and almond paste until light in color.
6. Fold in the sifted flour by hand.
7. Place 4 pounds, 8½ ounces of batter into a greased, paper-lined half sheet pan.
8. Bake at 325° F for 20 minutes, or until firm.

PASTRY TECHNIQUES:

Creaming, Combining, Folding

Creaming:

1. Soften the fats on low speed.
2. Add the sugar(s) and cream; increase the speed slowly.
3. Add the eggs one at a time; scrape the bowl frequently.
4. Add the dry ingredients in stages.

Combining:

Bringing together two or more components.

1. Prepare the components to be combined.
2. Add one to the other, using the appropriate mixing method (if needed).

Folding:

Do steps 1, 2, and 3 in one continuous motion.

1. Run a bowl scraper under the mixture, across the bottom of the bowl.
2. Turn the bowl counterclockwise.
3. Bring the bottom mixture to the top.

HACCP:

Store at 40° F, for 1 day. After 1 day, store at 0° F.

HAZARDOUS FOOD:

Eggs

Frangipane No. 2

YIELD:	7 pounds, 5 ounces.	14 pounds, 10 ounces.
INGREDIENTS:		
Butter, unsalted	2 pounds	4 pounds
Almond paste	3 pounds	6 pounds
Eggs, whole	2 pounds	4 pounds
Flour, cake, sifted	5 ounces	10 ounces

METHOD OF PREPARATION:

1. Gather the equipment and ingredients.
2. Place the butter in a bowl, and soften with a paddle; then remove.
3. Place the almond paste in another bowl, and mix on low speed, with paddle, to soften. Add one or two eggs to further soften the paste; scrape well.
4. When the almond paste and butter are the same consistency, add the butter into the almond paste, and incorporate well.
5. Mix on medium speed to aerate.
6. Add the remaining eggs in stages, and scrape well.
7. Fold in the sifted flour, by hand, and incorporate well.

Scale:

1. 7 pounds, 5 ounces for full, 1-inch sheet pan; bake at 325° F.
2. 2 pounds, 7 ounces for full, thin sheet pan; bake at 400° F.
3. 3 pounds, 10½ ounces for full, 1-inch half sheet pan; bake at 325° F.

PASTRY TECHNIQUES:

Creaming, Combining, Folding

Creaming:

1. Soften the fats on low speed.
2. Add the sugar(s) and cream; increase the speed slowly.
3. Add the eggs one at a time; scrape the bowl frequently.
4. Add the dry ingredients in stages.

Combining:

Bringing together two or more components.

1. Prepare the components to be combined.
2. Add one to the other, using the appropriate mixing method (if needed).

Folding:

Do steps 1, 2, and 3 in one continuous motion.

1. Run a bowl scraper under the mixture, across the bottom of the bowl.
2. Turn the bowl counterclockwise.
3. Bring the bottom mixture to the top.

HACCP:

Store at 40° F for 1 day. After 1 day, store at 0° F.

HAZARDOUS FOOD:

Eggs

German Chocolate Cake

PASTRY TECHNIQUES:

Creaming, Combining, Whipping, Folding

Creaming:

1. Soften the fats on low speed.
2. Add the sugar(s) and cream; increase the speed slowly.
3. Add the eggs one at a time; scrape the bowl frequently.
4. Add the dry ingredients in stages.

Combining:

Bringing together two or more components.

1. Prepare the components to be combined.
2. Add one to the other, using the appropriate mixing method (if needed).

Whipping:

1. Hold the whip at a 55-degree angle.
2. Create circles, using a circular motion.
3. The circular motion needs to be perpendicular to the bowl.

Folding:

Do steps 1, 2, and 3 in one continuous motion.

1. Run a bowl scraper under the mixture, across the bottom of the bowl.
2. Turn the bowl counterclockwise.
3. Bring the bottom mixture to the top.

HACCP:

Store at 40° F for 1 day. After 1 day, store at 0° F.

HAZARDOUS FOODS:

Eggs
Egg whites
Buttermilk

YIELD:	10 pounds. 7, 9-inch Cakes	20 pounds. 14, 9-inch Cakes.
INGREDIENTS:		
Cocoa powder, sifted	3 ounces	6 ounces
Extract, vanilla	1 ounce	2 ounces
Salt	¼ ounce	½ ounce
Water, cold	12 ounces	1 pound, 8 ounces
Sugar, granulated	3 pounds	6 pounds
Butter, unsalted, or margarine	12 ounces	1 pound, 8 ounces
Shortening, high-ratio	11 ounces	1 pound, 5 ounces
Baking soda, sifted	¾ ounce	1½ ounces
Eggs, whole	12 ounces	1 pound, 8 ounces
Egg whites	12 ounces	1 pound, 8 ounces
Flour, cake, sifted	1 pound, 8 ounces	3 pounds
Buttermilk	1 pound, 8 ounces	3 pounds
German chocolate cake icing (see page 799)	As needed	As needed

METHOD OF PREPARATION:

1. Gather the equipment and ingredients.
2. Place the sifted cocoa powder, vanilla extract, salt, cold water, and one eighth of the granulated sugar in mixing bowl; blend together, and set aside.
3. Place the remaining granulated sugar, butter, shortening, and baking soda in a bowl, and cream together, using a paddle, at medium speed for 5 minutes.
4. Slowly add the whole eggs; continue mixing, and scrape the bowl as needed; mix for an additional 5 minutes.
5. Place the egg whites in a bowl, and whip to a stiff peak.
6. Add the sifted cake flour, cocoa-water mixture, and buttermilk to the creaming mixture. Scrape the bowl and mix for 5 minutes.
7. Gently fold the beaten egg whites, by hand, into the creamed mixture; incorporate well.
8. Fill 9-inch cake pans with 1 pound, 7 ounces of batter.
9. Bake at 360° F until set in the center and firm in texture.
10. Turn out onto paper-lined sheet pans to cool, or cool in pans on wire racks.
11. Cover the cake with German chocolate cake icing.

Jelly Roll Sponge Cake

YIELD:	2 pounds, 1 ounce.	6 pounds, 3 ounces.
	1 Sheet Pan	3 Sheet Pans

INGREDIENTS:

Ingredient		
Flour, bread, sifted	4 ounces	12 ounces
Flour, cake, sifted	1⅓ ounces	4 ounces
Baking powder, sifted	Pinch	½ ounce
Salt	Pinch	¼ ounce
Simple syrup, warm (see page 966)	2⅔ ounces	8 ounces
Egg yolks	7 ounces	1 pound, 5 ounces
Honey	1⅔ ounces	5 ounces
Oil, vegetable	1⅔ ounces	5 ounces
Extract, vanilla	⅓ ounce	1 ounce
Extract, lemon	⅛ ounce	¼ ounce
Egg whites	10 ounces	1 pound, 14 ounces
Sugar, granulated	4⅔ ounces	14 ounces

METHOD OF PREPARATION:

1. Gather the equipment and ingredients.
2. Sift together the bread and cake flours, baking powder, and salt; set aside.
3. Prepare the simple syrup; after it is warm, set it aside.
4. Place the egg yolks, honey, oil, vanilla extract, and lemon extract in a bowl; whip to full volume.
5. When the egg yolk mixture just about reaches full volume, turn down the mixer to medium speed, and slowly add the warm simple syrup in a slow, steady stream.
6. In another bowl, place the egg whites, and whip to a medium peak, slowly adding the granulated sugar to make a meringue. (This procedure should be started when the egg yolks are about half whipped.)
7. When the egg yolks and egg whites are both done, remove the bowls from the mixers, and fold the meringue into the egg yolk mixture by hand. *Do not overmix.* (This process should be done very carefully.)
8. Fold in the dry ingredients carefully by hand.
9. Scale 2 pounds, 1 ounce of batter per pan into straight, parchment-lined sheet pans.

PASTRY TECHNIQUES:

Whipping, Folding, Spreading

Whipping:

1. Hold the whip at a 55-degree angle.
2. Create circles, using a circular motion.
3. The circular motion needs to be perpendicular to the bowl.

Folding:

Do steps 1, 2, and 3 in one continuous motion.

1. Run a bowl scraper under the mixture, across the bottom of the bowl.
2. Turn the bowl counterclockwise.
3. Bring the bottom mixture to the top.

Spreading:

1. Using an icing spatula or off-set spatula, smooth the icing or other spreading medium over the surface area.

HACCP:

Store at 40° F for 1 day. After 1 day, store at 0° F.

HAZARDOUS FOODS:

Egg yolks
Egg whites

CHEF NOTES:

Use the following methods to test for doneness:

1. Check for color; golden brown is good.
2. Check for firmness; the cake should spring back.

10. Spread evenly in the pan.
11. Bake at 420° F for approximately 10 to 15 minutes. Test for doneness. (See chef notes.)
12. Remove from the oven.
13. Roll up each jelly roll while still hot.
14. Cool in the roll position.
15. Unroll to fill and reroll.

Liquid Shortening Sponge Cake, Chocolate

PASTRY TECHNIQUE:
Blending

Blending:
1. Combine the dry ingredients on low speed.
2. Add the softened fat(s) and liquid(s).
3. Mix the ingredients on low speed.
4. Increase the speed gradually.

HACCP:
Store at 40° F for 1 day. After 1 day, store at 0° F.

HAZARDOUS FOODS:
Eggs
Milk

YIELD:	10 pounds, 13¾ ounces.	21 pounds, 11½ ounces.
	7, 9-inch Cakes	14, 9-inch Cakes

INGREDIENTS:

Eggs, whole	3 pounds, 5 ounces	6 pounds, 10 ounces
Shortening, liquid, high-ratio	1 pound, 4 ounces	2 pounds, 8 ounces
Milk, whole	1 pound, 4 ounces	2 pounds, 8 ounces
Extract, vanilla	1 ounce	2 ounces
Sugar, granulated	2 pounds, 12 ounces	5 pounds, 8 ounces
Flour, cake	1 pound, 10 ounces	3 pounds, 4 ounces
Baking powder	2 ounces	4 ounces
Salt	1 ounce	2 ounces
Baking soda	¾ ounce	1½ ounces
Cocoa powder	6 ounces	12 ounces

METHOD OF PREPARATION:

1. Gather the equipment and ingredients.
2. Grease the cake pans thoroughly.
3. Place all of the liquid ingredients in a mixing bowl.
4. Sift all of the dry ingredients together.
5. Place all of the sifted dry ingredients on top of the liquid ingredients in bowl.
6. Using a whip, whip on first speed until all of the ingredients are blended slightly, approximately 30 seconds.
7. Whip for 4 minutes on high speed; scrape the bowl well.
8. Whip for 3 minutes on medium speed; scrape the bowl well.
9. Fill 9-inch cake pans with 1 pound, 9 ounces of batter.
10. Bake at 350°F, until the cake is golden brown.
11. Cool in the pans for 5 to 10 minutes.
12. Turn out onto paper-lined sheet pans, and remove the pans.

Liquid Shortening Sponge Cake: Vanilla

PASTRY TECHNIQUE:

Blending

Blending:

1. Combine the dry ingredients on low speed.
2. Add the softened fat(s) and liquid(s).
3. Mix the ingredients on low speed.
4. Increase the speed gradually.

HACCP:

Store at 40° F for 1 day. After 1 day, store at 0° F.

HAZARDOUS FOODS:

Eggs
Milk

YIELD:	10 pounds, 6 ounces. 7, 9-inch Cakes	20 pounds, 12 ounces. 14, 9-inch Cakes
INGREDIENTS:		
Eggs, whole	3 pounds, 5 ounces	6 pounds, 10 ounces
Shortening, liquid, high-ratio	1 pound, 4 ounces	2 pounds, 8 ounces
Milk, whole	1 pound	2 pounds
Extract, vanilla	2 ounces	4 ounces
Sugar, granulated	2 pounds, 8 ounces	5 pounds
Flour, cake	2 pounds	4 pounds
Baking powder	2¼ ounces	4½ ounces
Salt	¾ ounce	1½ ounces

METHOD OF PREPARATION:

1. Gather the equipment and ingredients.
2. Grease the cake pans thoroughly.
3. Place all of the liquid ingredients in a mixing bowl.
4. Sift all of the dry ingredients together.
5. Place all of the sifted ingredients on top of the liquid ingredients in bowl.
6. Using a whip, whip on first speed until the ingredients are blended slightly, approximately 30 seconds.
7. Whip for 4 minutes on high speed; scrape the bowl well.
8. Whip for 3 minutes on medium speed; scrape the bowl well.
9. Fill 9-inch cake pans with 1 pound, 7 ounces of batter.
10. Bake at 350° F, until the cake is golden brown.
11. Cool in the pans for 5 to 10 minutes.
12. Turn out onto paper-lined sheet pans, and remove the pans.

Pound Cake

PASTRY TECHNIQUE:

Combining

Combining:

Bringing together two or more components.

1. Prepare the components to be combined.
2. Add one to the other, using the appropriate mixing method (if needed).

HACCP:

Store at 40° F for 1 day. After 1 day, store at 0° F.

HAZARDOUS FOOD:

Eggs

YIELD:	6 pounds, 6 ounces. 6 Loaves	12 pounds, 12 ounces. 12 Loaves

INGREDIENTS:

Flour, cake, sifted	1 pound, 9¾ ounces	3 pounds, 3½ ounces
Shortening, high-ratio	8 ounces	1 pound
Butter, unsalted, softened	8 ounces	1 pound
Sugar, granulated	1 pound, 12½ ounces	3 pounds, 9 ounces
Dry milk solids (DMS), sifted	2½ ounces	5 ounces
Salt	½ ounce	1 ounce
Baking powder, sifted	¼ ounce	½ ounce
Water	12½ ounces	1 pound, 9 ounces
Eggs, whole	1 pound	2 pounds

METHOD OF PREPARATION:

1. Gather the equipment and ingredients.
2. Place the cake flour, shortening, butter, granulated sugar, DMS, salt, and baking powder in a bowl.
3. Using a paddle, mix on low speed to blend all of the ingredients together until smooth.
4. Gradually add water to the mixture.
5. Add the whole eggs in stages; continue to mix until smooth, scraping the bowl as needed throughout. When all of the eggs have been added, continue mixing until the batter is smooth and creamy.
6. Fill each greased, aluminum loaf pan with 1 pound, 1 ounce of batter. Place the loaf pans on sheet pans for baking.
7. Bake at 375° F for 30 minutes, or until golden brown and the center splits. The center split must be dry before removing the cakes from the oven.

Sacher Cake: Classical

YIELD:	5 pounds.	10 pounds.
	2, 9-inch Cakes	4, 9-inch Cakes

INGREDIENTS:

Butter, unsalted, softened	1 pound	2 pounds
Extract, vanilla	To taste	To taste
Sugar, granulated	1 pound	2 pounds
Chocolate, dark, semi-sweet	1 pound	2 pounds
Egg yolks	6½ ounces	13 ounces
Egg whites	9½ ounces	1 pound, 3 ounces
Flour, cake	1 pound	2 pounds

METHOD OF PREPARATION:

1. Gather the equipment and ingredients.
2. Place the butter, vanilla extract, and sugar in a bowl; whip until light and fluffy.
3. Chop the chocolate, and melt using a double boiler.
4. Once the butter mixture has been whipped light and fluffy, add the melted chocolate.
5. Blend until smooth.
6. Gradually add the egg yolks into the chocolate-butter mixture, and combine until smooth.
7. In another bowl, whip the egg whites to a medium peak.
8. Sift the cake flour.
9. Alternately fold in the sifted flour and egg whites into the chocolate-butter mixture.
10. Scale 2 pounds, 8 ounces of batter into greased, paper-lined, 9-inch cake pans.
11. Bake at 325° F until firm and dry in the center.

PASTRY TECHNIQUES:

Melting, Whipping, Folding

Chopping:

1. Use a sharp knife.
2. Hold the food product properly.
3. Cut with a quick downward motion.

Melting:

1. Prepare the food product to be melted.
2. Place the food product an appropriate sized pot over direct heat or over a double boiler.
3. Stir frequently or occasionally, depending on the delicacy of the product, until melted.

OR

1. Place the product on a sheet pan or in a bowl, and place in a low oven until melted.

Whipping:

1. Hold the whip at a 55-degree angle.
2. Create circles, using a circular motion.
3. The circular motion needs to be perpendicular to the bowl.

Folding:

Do steps 1, 2, and 3 in one continuous motion.

1. Run a bowl scraper under the mixture, across the bottom of the bowl.
2. Turn the bowl counterclockwise.
3. Bring the bottom mixture to the top.

HACCP:

Store at 40° F for 1 day. After 1 day, store at 0° F.

HAZARDOUS FOODS:

Egg yolks
Egg whites

Sacher Cake: Modern

PASTRY TECHNIQUES:

Chopping, Melting, Whipping, Folding

Chopping:

1. Use a sharp knife.
2. Hold the food product properly.
3. Cut with a quick downward motion.

Melting:

1. Prepare the food product to be melted.
2. Place the food product an appropriate sized pot over direct heat or over a double boiler.
3. Stir frequently or occasionally, depending on the delicacy of the product, until melted.

OR

1. Place the product on a sheet pan or in a bowl, and place in a low oven until melted.

Whipping:

1. Hold the whip at a 55-degree angle.
2. Create circles, using a circular motion.
3. The circular motion needs to be perpendicular to the bowl.

Folding:

Do steps 1, 2, and 3 in one continuous motion.

1. Run a bowl scraper under the mixture, across the bottom of the bowl.
2. Turn the bowl counterclockwise.
3. Bring the bottom mixture to the top.

HACCP:

Store at 40° F for 1 day. After 1 day, store at 0° F.

HAZARDOUS FOODS:

Egg yolks
Egg whites

YIELD:	10 pounds, 10 ounces. 5, 9-inch Cakes	31 pounds, 14½ ounces. 15, 9-inch Cakes

INGREDIENTS:

Butter, unsalted, softened	2 pounds	6 pounds
Sugar, confectionery	12 ounces	2 pounds, 4 ounces
Chocolate, dark, semi-sweet	1 pound, 10½ ounces	5 pounds
Egg yolks	1 pound, 7½ ounces	4 pounds, 6½ ounces
Egg whites	2 pounds, 3 ounces	6 pounds, 9½ ounces
Sugar, granulated	12½ ounces	2 pounds, 5 ounces
Flour, bread	14 ounces	2 pounds, 10 ounces
Flour, cake	14 ounces	2 pounds, 10 ounces
Baking powder	½ ounce	1½ ounces

METHOD OF PREPARATION:

1. Gather the equipment and ingredients.
2. Place the butter and confectionery sugar in a bowl; whip until very light.
3. Chop the chocolate, and melt it using a double boiler.
4. Slowly add the melted chocolate to the whipped butter mixture.
5. Blend well.
6. Gradually add the egg yolks to the chocolate-butter mixture.
7. In another bowl, place the egg whites; whip to a medium stiff peak while slowly adding the granulated sugar to make a meringue.
8. Gently fold the meringue into the chocolate-butter mixture.
9. Sift the dry ingredients together.
10. Fold in the remaining sifted dry ingredients.
11. Scale 2 pounds, 2 ounces of batter into greased, parchment-lined, 9-inch cake pans.
12. Bake at 350° F, for approximately 40 minutes or until firm and dry in the center.

Vanilla Chiffon Genoise

PASTRY TECHNIQUES:
Whipping, Combining

Whipping:
1. Hold the whip at a 55-degree angle.
2. Create circles, using a circular motion.
3. The circular motion needs to be perpendicular to the bowl.

Combining:
Bringing together two or more components.
1. Prepare the components to be combined.
2. Add one to the other, using the appropriate mixing method (if needed).

HACCP:
Store at 40° F for 1 day. After 1 day, store at 0° F.

HAZARDOUS FOODS:
Egg yolks
Egg whites

YIELD:	10 pounds, 6 ounces. 7, 9-inch Cakes	20 pounds, 12 ounces. 14, 9-inch Cakes
INGREDIENTS:		
Egg yolks	2 pounds	4 pounds
Sugar, granulated	3 pounds	6 pounds
Oil, vegetable	12 ounces	1 pound, 8 ounces
Egg whites	2 pounds	4 pounds
Flour, cake, sifted	2 pounds, 4 ounces	4 pounds, 8 ounces
Baking powder	1 ounce	2 ounces
Water, room temperature	5 ounces	10 ounces
Extract, vanilla	To taste	To taste

METHOD OF PREPARATION:

1. Gather the equipment and ingredients.
2. Properly grease the cake pans.
3. Place the egg yolks and half of the granulated sugar in a bowl; whip to full volume.
4. Continue mixing on medium speed, and slowly incorporate the oil.
5. In another bowl, whip the egg whites to a medium peak; slowly add the remaining granulated sugar to make a meringue.
6. Sift together the cake flour and baking powder.
7. Combine the water and vanilla extract.
8. Alternately add the flour and water mixtures into the yolk mixture by hand.
9. Fold the meringue into the batter.
10. Scale 1 pound, 8 ounces batter into a greased paper-lined 9-inch cake pan.
11. Bake at 360° F until spongy in the center.

Vanilla Genoise

PASTRY TECHNIQUES:

Whipping, Melting, Folding

Whipping:

1. Hold the whip at a 55-degree angle.
2. Create circles, using a circular motion.
3. The circular motion needs to be perpendicular to the bowl.

Melting:

1. Prepare the food product to be melted.
2. Place the food product an appropriate sized pot over direct heat or over a double boiler.
3. Stir frequently or occasionally, depending on the delicacy of the product, until melted.

OR

1. Place the product on a sheet pan or in a bowl, and place in a low-temperature oven until melted.

Folding:

Do steps 1, 2, and 3 in one continuous motion.

1. Run a bowl scraper under the mixture, across the bottom of the bowl.
2. Turn the bowl counterclockwise.
3. Bring the bottom mixture to the top.

HACCP:

Store at 40° F for 1 day. After 1 day, store at 0° F.

HAZARDOUS FOOD:

Eggs

YIELD:	3 pounds, 10½ ounces. 22, 9-inch by 3-inch Pans	7 pounds, 5 ounces. 4, 9-inch by 3-inch Pans

INGREDIENTS:

Flour, cake	8 ounces	1 pound
Cornstarch	6 ounces	12 ounces
Butter, unsalted	6 ounces	12 ounces
Eggs, whole	1 pound, 8 ounces	3 pounds
Salt	⅛ ounce	¼ ounce
Extract, vanilla	¼ ounce	½ ounce
Extract, lemon	⅛ ounce	¼ ounce
Sugar, granulated	14 ounces	1 pound, 12 ounces

METHOD OF PREPARATION:

1. Gather the equipment and ingredients.
2. Properly grease the cake pans.
3. Sift the cake flour and cornstarch together; set aside.
4. Melt the butter, and set aside. (Butter should be only lukewarm.)
5. Place the whole eggs, salt, vanilla extract, lemon extract, and granulated sugar in a bowl; immediately whisk over a warm water bath. (Never scale the ingredients together in advance, or lumping will occur when the dry sugar comes into contact with the eggs.)
6. Heat the mixture to 110° F constantly mixing at all times. *Do not cook* the mixture.
7. Remove the bowl from the warm bath, and place on a mixer; whip on high speed to full volume. (Full volume can be determined by the "5 second track method." See chef note 3.)
8. When full volume is reached, remove the bowl from the mixer, and immediately fold the dry ingredients into the whipped egg mixture. Use your hand along with a plastic scraper to distribute the sifted dry ingredients evenly. This is a combination of the folding technique (turning the bowl counterclockwise with the left hand and scraping up from the bottom with the right hand in a clockwise rotation; finish out on top of the batter with your right hand and scraper palm up and shaking vigorously to distribute the flour and starch). Work quickly. Continue mixing until all of the ingredients are completely incorporated. (The longer it takes to combine ingredients, the greater the chance that lumps will be trapped in the batter.)
9. Place a small amount of batter in another small bowl.
10. Using your fingers, quickly whisk the melted butter into this batter until completely incorporated.

CHEF NOTES:

1. The following methods test for doneness:
   (a) Check for color; a golden brown is good.
   (b) Check to see if the cake is pulling away from the sides of the pan.
   (c) Touch the center of the cake; it should spring back.
2. Sponge cake is a very delicate cake; do not slam the door or knock the pans together.
3. The following is the "5-second track method":
   Put ½ inch of your fingertip through the batter. If it holds a track for 5 seconds before closing, it is done. If it fails to hold a track, whip more. Check several times during mixing. Stay with the machine to monitor its progress. Keep track of mixing times and speeds. As the batter thickens, reduce the speed to medium, and check more frequently. *Do not overmix;* this will result in loss of volume or total collapse. Eggs should be cool at the end of this process.

11. Fold this butter mixture, gently, back into the remaining batter.

12. Scale the batter at 1 pound, 11 ounces per 9-inch by 3-inch pan, approximately two-thirds full.

13. Bake at 350° F for approximately 35 to 40 minutes. Test for doneness. (See chef note 1.)

14. When done, remove from the oven, and place on a cooling rack.

Walnut Chiffon Genoise

PASTRY TECHNIQUES:

Whipping, Combining

Whipping:

1. Hold the whip at a 55-degree angle.
2. Create circles, using a circular motion.
3. The circular motion needs to be perpendicular to the bowl.

Combining:

Bringing together two or more components.

1. Prepare the components to be combined.
2. Add one to the other, using the appropriate mixing method (if needed).

HACCP:

Store at 40° F for 1 day. After 1 day, store at 0° F.

HAZARDOUS FOODS:

Egg yolks
Egg whites

YIELD:	5 pounds, 10 ounces.	11 pounds, 4 ounces.
	4, 9-inch Cakes.	8, 9-inch Cakes.

INGREDIENTS:

Egg yolks	1 pound	2 pounds
Sugar, granulated	1 pound, 8 ounces	3 pounds
Oil, vegetable	6 ounces	12 ounces
Egg whites	1 pound	2 pounds
Walnuts or hazelnuts, finely chopped	7 ounces	14 ounces
Flour, cake, sifted	1 pound, 2 ounces	2 pounds, 4 ounces
Baking powder	½ ounce	1 ounce
Water, room temperature	2½ ounces	5 ounces
Extract, vanilla	To taste	To taste

METHOD OF PREPARATION:

1. Gather the equipment and ingredients.
2. Properly grease the cake pans.
3. Place the egg yolks and half of the granulated sugar in a bowl; whip to full volume.
4. Continue mixing on medium speed, and slowly incorporate the oil.
5. In another bowl, whip the egg whites to a medium peak; slowly add the remaining granulated sugar to make a meringue.
6. Combine the nuts, sifted cake flour, and baking powder.
7. Combine the water and vanilla extract.
8. Alternately add the flour and water mixtures into the yolk mixture by hand.
9. Fold the meringue into the batter.
10. Fill each 9-inch greased cake pan with 1 pound, 6 ounces of batter.
11. Bake at 360° F until spongy in the center.

CHEF NOTE:

If hazelnut is used, the product will be *Hazelnut Chiffon Genoise*.

Yellow Layer Cake

(High-Ratio)

PASTRY TECHNIQUE:

Blending

Blending:

1. Combine the dry ingredients on low speed.
2. Add the softened fat(s) and liquid(s).
3. Mix the ingredients on low speed.
4. Increase the speed gradually.

HACCP:

Store at 40° F for 1 day. After 1 day, store at 0° F.

HAZARDOUS FOOD:

Eggs

YIELD:	5 pounds, 2½ ounces.	10 pounds, 5 ounces.
	2½, 9-inch Cakes.	5, 9-inch Cakes.
	1 Full Sheet Pan.	2 Full Sheet Pans.

INGREDIENTS:

Eggs, whole	1 pound	2 pounds
Water, cold	12 ounces	1 pound, 8 ounces
Flour, cake	1 pound, 5 ounces	2 pounds, 10 ounces
Salt	½ ounce	1 ounce
Dry milk solids (DMS)	2 ounces	4 ounces
Baking powder	1 ounce	2 ounces
Shortening, high-ratio	9 ounces	1 pound, 2 ounces
Sugar, granulated	1 pound, 5 ounces	2 pounds, 10 ounces

METHOD OF PREPARATION:

1. Gather the equipment and ingredients.
2. Combine together the whole eggs and two thirds of the water; whisk together, and set aside.
3. Sift together the cake flour, salt, DMS, and baking powder.
4. Place the sifted ingredients, shortening, sugar, and one third of the egg-water mixture into a mixing bowl.
5. Blend on low speed, with a paddle, for 4 minutes; scrape the bowl well.
6. Add one third more of the egg-water mixture, and mix for 4 minutes; scrape the bowl well.
7. Add the remaining third of the egg-water mixture, and mix for 4 minutes; scrape the bowl well.
8. Add the remaining one third water, and mix until smooth.
9. Bake at 375° F for 30 to 40 minutes, or until golden brown.

Scale:

1. 2 pounds, 1 ounce for 9-inch by 3-inch cake pans.
2. 5 pounds, 2½ ounces for full sheet pan.

Basic Buttercream

PASTRY TECHNIQUES:
Creaming, Whipping

Creaming:
1. Soften the fats on low speed.
2. Add the sugar(s) and cream; increase the speed slowly.
3. Add the eggs one at a time; scrape the bowl frequently.
4. Add the dry ingredients in stages.

Whipping:
1. Hold the whip at a 55-degree angle.
2. Create circles, using a circular motion.
3. The circular motion needs to be perpendicular to the bowl.

HACCP:
Store at 40° F.

HAZARDOUS FOOD:
Egg whites

YIELD:	4 pounds, 8 ounces.	18 pounds.

INGREDIENTS:

Butter, unsalted	12 ounces	3 pounds
Sugar, confectionery	2 pounds, 8 ounces	10 pounds
Shortening, high-ratio	12 ounces	3 pounds
Egg whites (pasteurized)	8 ounces	2 pounds
Extract, vanilla	To taste	To taste
Lemon juice	To taste	To taste

METHOD OF PREPARATION:

1. Gather the equipment and ingredients.
2. Place the butter in a bowl with half of the confectionery sugar; cream using a paddle to the consistency of the shortening.
3. Scrape bowl well.
4. Add the shortening and remaining confectionery sugar; paddle until light and airy.
5. Remove the paddle attachment, and replace with the whip attachment.
6. While whipping slowly add the egg whites and flavorings in stages; scrape the bowl between additions.
7. Combine ingredients well.
8. Continue whipping until light and airy.

Buttercream with Fondant

YIELD:	5 pounds.	10 pounds.
INGREDIENTS:		
Butter, unsalted, room temperature	1 pound	2 pounds
Shortening, high-ratio	1 pound	2 pounds
Fondant, prepared	3 pounds	6 pounds

METHOD OF PREPARATION:

1. Gather the equipment and ingredients.
2. Scale all of the ingredients.
3. Place the firmest ingredient in the bowl; paddle to the consistency of the next ingredient's consistency.
4. Scrape the bowl well.
5. Add the next ingredient, and paddle to the consistency of the last ingredient.
6. Scrape the bowl well.
7. Add the last ingredient; paddle until light and fluffy.
8. Remove the paddle attachment, and replace with the whip attachment.
9. Whip until light in consistency.

PASTRY TECHNIQUES:

Creaming, Whipping

Creaming:

1. Soften the fats on low speed.
2. Add the sugar(s) and cream; increase the speed slowly.
3. Add the eggs one at a time; scrape the bowl frequently.
4. Add the dry ingredients in stages.

Whipping:

1. Hold the whip at a 55-degree angle.
2. Create circles, using a circular motion.
3. The circular motion needs to be perpendicular to the bowl.

HACCP:

Store at 40° F.

CHEF NOTE:

Dry fondant can be substituted for prepared fondant by combining 2 pounds, 11 ounces of dry fondant with 5 ounces of warm water until it has a smooth consistency.

Buttercream with Meringue

PASTRY TECHNIQUES:

Cooking, Creaming, Whipping, Combining

Cooking:

Preparing food through the use of various sources of heating.

1. Choose the appropriate heat application: baking, boiling, simmering, etc.
2. Prepare the formula according to instructions.
3. Cook according to instructions.

Creaming:

1. Soften the fats on low speed.
2. Add the sugar(s) and cream; increase the speed slowly.
3. Add the eggs one at a time; scrape the bowl frequently.
4. Add the dry ingredients in stages.

Whipping:

1. Hold the whip at a 55-degree angle.
2. Create circles, using a circular motion.
3. The circular motion needs to be perpendicular to the bowl.

Combining:

Bringing together two or more components.

1. Prepare the components to be combined.
2. Add one to the other, using the appropriate mixing method (if needed).

HACCP:

Store at 40° F.

HAZARDOUS FOOD:

Egg whites

CHEF NOTE:

To rewhip cold buttercream: Melt one third of the amount. Place the remaining cold buttercream in a bowl. Add the melted buttercream, and, using a whip, combine at low speed. Whip at high speed until the cream becomes smooth again. If it is still lumpy, melt more buttercream, and add it to the bowl. Continue whipping the mixture.

YIELD: 17 pounds, 8½ ounces. 26 pounds, 4¾ ounces.

INGREDIENTS:

Water, hot	1 pound, 8 ounces	2 pounds, 4 ounces
Sugar, granulated	6 pounds	9 pounds
Egg whites (70° F)	3 pounds	4 pounds, 8 ounces
Butter, unsalted, room temperature	7 pounds	10 pounds, 8 ounces
Salt	½ ounce	¾ ounce
Extract, vanilla	To taste	To taste

METHOD OF PREPARATION:

1. Gather the equipment and ingredients.
2. Place the water and two thirds of the granulated sugar in a pot, and bring to a boil. As it begins to boil, skim the scum (impurities in foam on surface) with a ladle. Boil; do not stir. Wash down the sides of pot with a brush dipped in water to prevent crystals from forming.
3. Heat to 250° F.
4. Place the egg whites in a bowl, and whip to a stiff consistency; slowly add the remaining granulated sugar to form a meringue.
5. When the sugar syrup reaches 250° F, pour slowly into the meringue at medium speed.
6. Whip until cool.
7. In another bowl, cream the butter, salt, and vanilla extract until light and fluffy.
8. Add the creamed butter to the cooled meringue; whip on medium-to-high speed until all of the ingredients are well combined and the mixture is creamy.

Chocolate Buttercream

YIELD:	2 pounds, 8 ounces.	5 pounds.

INGREDIENTS:

Buttercream	2 pounds	4 pounds
Chocolate, dark, semi-sweet	8 ounces	1 pound
Extract, vanilla	To taste	To taste

METHOD OF PREPARATION:

1. Gather the equipment and ingredients.
2. Prepare the buttercream according to the directions (see page 768), and leave in mixing bowl with a whip.
3. Chop and melt the dark chocolate.
4. Turn the mixer on low speed, and slowly add the chocolate in a steady stream into the buttercream without hitting the whip or the sides of the bowl.
5. Incorporate the chocolate completely.
6. Add the vanilla extract.

Variations:

1. *Fudge Base:*
   a. Soften the fudge base. (Use 30% according to the weight of the buttercream.)
   b. Add to the buttercream.
2. *Cocoa Powder:*
   a. Place the cocoa powder in a bowl. (Use according to desired taste.)
   b. Using a whisk, add either rum or brandy; add enough to create a paste.
   c. Add to the buttercream.
3. *Chocolate:*
   a. If a rich chocolate taste is desired, increase the amount of the chocolate.
   b. Follow the procedure as listed above.

PASTRY TECHNIQUES:

Chopping, Melting, Whipping

Chopping:

1. Use a sharp knife.
2. Hold the food product properly.
3. Cut with a quick downward motion.

Melting:

1. Prepare the food product to be melted.
2. Place the food product in an appropriate sized pot over direct heat or over a double boiler.
3. Stir frequently or occasionally, depending on the delicacy of the product, until melted.

OR

1. Place the product on a sheet pan or in a bowl, and place in a low-temperature oven until melted.

Whipping:

1. Hold the whip at a 55-degree angle.
2. Create circles, using a circular motion.
3. The circular motion needs to be perpendicular to the bowl.

HACCP:

Store at 40° F.

Chocolate Fudge Icing

YIELD:	2 pounds, 4¼ ounces.	4 pounds, 8½ ounces.
	½ Sheet Pan.	1 Sheet Pan.

INGREDIENTS:

Corn syrup	1 ounce	2 ounces
Shortening, all-purpose	¾ ounce	1½ ounces
Water	4½ ounces	9 ounces
Fudge base	8 ounces	1 pound
Sugar, confectionery, sifted	1 pound, 6 ounces	2 pounds, 12 ounces

METHOD OF PREPARATION:

1. Gather the equipment and ingredients.
2. Place the corn syrup, shortening, and water in a pot, and bring to a boil; remove from the heat.
3. Place the fudge base and confectionery sugar in a bowl; paddle until well blended.
4. Add the boiling water mixture into the fudge base in stages.
5. Combine until well blended.

PASTRY TECHNIQUES:

Boiling, Combining

Boiling:

1. Bring the cooking liquid to a rapid boil.
2. Stir the contents.

Combining:

Bringing together two or more components.

1. Prepare the components to be combined.
2. Add one to the other, using the appropriate mixing method (if needed).

HACCP:

Store at 60° F to 65° F.

Chocolate Glaze/ Coating

YIELD:	1 pound, 14 ounces.	3 pounds, 12 ounces.
INGREDIENTS:		
Chocolate, dark, semi-sweet, finely chopped	1 pound, 4 ounces	2 pounds, 8 ounces
Shortening, all-purpose	10 ounces	1 pound, 4 ounces

METHOD OF PREPARATION:

1. Gather the equipment and ingredients.
2. Melt the chopped chocolate over a double boiler.
3. Melt the shortening in a pot.
4. Combine the chocolate and shortening; mix well.
5. If necessary, strain through a chinois mousseline or cheesecloth.
6. Store covered in a dry storage area.
7. To use, heat over a double boiler.

PASTRY TECHNIQUES:

Chopping, Melting, Combining

Chopping:

1. Use a sharp knife.
2. Hold the food product properly.
3. Cut with a quick downward motion.

Melting:

1. Prepare the food product to be melted.
2. Place the food product in an appropriate-sized pot over direct heat or over a double boiler.
3. Stir frequently or occasionally, depending on the delicacy of the product, until melted.

OR

1. Place the product on a sheet pan or in a bowl, and place in a low-temperature oven until melted.

Combining:

Bringing together two or more components.

1. Prepare the components to be combined.
2. Add one to the other, using the appropriate mixing method (if needed).

HACCP:

Store at 60° F to 65° F.

CHEF NOTE:

Do not, under any circumstances, *heat* the chocolate or chocolate coating above 122° F.

Cream Cheese Icing

PASTRY TECHNIQUE:

Creaming

Creaming:

1. Soften the fats on low speed.
2. Add the sugar(s) and cream; increase the speed slowly.
3. Add the eggs one at a time; scrape the bowl frequently.
4. Add the dry ingredients in stages.

HACCP:

Store at 40° F.

HAZARDOUS FOOD:

Cream cheese

YIELD:	1 pound, 11 ounces.	3 pounds, 6 ounces.

INGREDIENTS:

Butter, unsalted, room temperature	4 ounces	8 ounces
Cream cheese	8 ounces	1 pound
Shortening, high-ratio	1¾ ounces	3½ ounces
Sugar, confectionery, sifted	13 ounces	1 pound, 10 ounces
Lemon juice	¼ ounce	½ ounce

METHOD OF PREPARATION:

1. Gather the equipment and ingredients.
2. Place the butter in a bowl, and soften; remove.
3. Place the cream cheese in a bowl, and soften.
4. Add the shortening in stages to the softened cream cheese, and scrape the bowl well.
5. Add the softened butter in stages to the cream cheese and shortening mixture, and scrape the bowl well.
6. Add the sifted confectionery sugar and lemon juice; mix only until incorporated.

CHEF NOTE:

Excessive mixing will soften the mixture greatly.

French Cream Icing

YIELD:	7 pounds, 9¼ ounces.	15 pounds, 2½ ounces.

INGREDIENTS:

Ingredient		
Sugar, confectionery	3 pounds, 12 ounces	7 pounds, 8 ounces
Dry milk solids (DMS)	4½ ounces	9 ounces
Shortening, high-ratio	2 pounds, 8 ounces	5 pounds
Emulsion, lemon	To taste	To taste
Extract, vanilla	To taste	To taste
Water	1 pound	2 pounds
Color, egg shade (optional)	As needed	As needed

METHOD OF PREPARATION:

1. Gather the equipment and ingredients.
2. Sift together the confectionery sugar and DMS.
3. Place the sifted ingredients in a bowl, and add the shortening; using a paddle, combine well.
4. Cream the mixture until light.
5. Combine the lemon emulsion, vanilla extract, water, and egg color (optional) into one container.
6. Slowly add the liquids in stages to the creamed sugar mixture.
7. Incorporate well.
8. Allow to whip until light and fluffy.

PASTRY TECHNIQUES:

Creaming, Whipping

Creaming:

1. Soften the fats on low speed.
2. Add the sugar(s) and cream; increase the speed slowly.
3. Add the eggs one at a time; scrape the bowl frequently.
4. Add the dry ingredients in stages.

Whipping:

1. Hold the whip at a 55-degree angle.
2. Create circles, using a circular motion.
3. The circular motion needs to be perpendicular to the bowl.

HACCP:

Store at 40° F.

Royal Icing

PASTRY TECHNIQUE:
Mixing

Mixing:
1. Follow the proper mixing procedure: creaming, blending, whipping, or combination.

HACCP:
Store at 40° F.

HAZARDOUS FOOD:
Egg whites

YIELD:	4 pounds, 12¼ ounces.	9 pounds, 8½ ounces.
INGREDIENTS:		
Egg whites, room temperature	12 ounces	1 pound, 8 ounces
Sugar, confectionery	4 pounds	8 pounds
Cream of tartar	¼ ounce	½ ounce

METHOD OF PREPARATION:

1. Gather the equipment and ingredients.
2. Place the egg whites in a mixing bowl.
3. Sift all of the dry ingredients together twice or more until very fine.
4. Add half of the dry ingredients into the egg whites, and mix at low speed; use a paddle until well incorporated.
5. Add the remaining dry ingredients, and continue to mix on low speed; scrape the bowl often.
6. Mix for an additional 5 to 7 minutes on medium speed, or to a desired consistency.
7. Royal icing is ready for use, but if stored, it should be wrapped airtight and be refrigerated.

Black Forest Torte

PASTRY TECHNIQUES:

Filling, Icing

Filling:

1. Cut open the food product.
2. Carefully spread the filling using an icing spatula, or
3. Carefully pipe the filling using a pastry bag.

Icing:

1. Use a clean icing spatula.
2. Work quickly and neatly.

HACCP:

Store at 40° F.

HAZARDOUS FOOD:

Heavy cream

YIELD: 1, 9-inch Torte.

INGREDIENTS:

Cake layers, chocolate	3, 9-inch layers, ⅜-inch thick
Simple syrup, Kirschwasser (see page 967)	3 ounces
Cream, heavy, whipped	1 pound
Black forest cherry filling (see page 796)	4 ounces
Chocolate, dark, semi-sweet, shaved	1 ounce
Glazé red cherries	As needed

METHOD OF PREPARATION:

1. Gather the equipment and ingredients.
2. Moisten each cake layer with syrup during assembly.
3. Ice inside of the ring with whipped cream, about ¼-inch thick.
4. Place the ring over the cardboard circle (cut to fit the ring).
5. Cut the cake layers in half.
6. Using the seam of the ring as a guide, grasp half of the cake layer along the cut side, and place it into the ring. The cut should line up with the seam, and the outside edge should be up against the whipped cream.
7. Turn the ring 180 degrees, and repeat with the second half of the cake layer.
8. Pipe a target pattern in whipped cream on the cake layer. Pipe a dot in the center, a ¼-inch bead around the outside edge, and finally a ¼-inch circle between the dot and the perimeter.
9. In between these lines of whipped cream, spoon in the filling; place the halved cherries next to each other to completely fill in all of the space. You should now have alternating circles of whipped cream and cherries.
10. Give the ring a third turn, and place the second cake layer in it. Press to level.
11. Pipe a ¼-inch layer of whipped cream on top of the cake, in a spiral pattern, from the center to the outside.
12. Give the ring a third turn, and repeat the process with the final layer. Be sure to level the cake layer as you build.
13. There should be about an ⅛ inch of space left between the top of the cake and the top edge of the ring. Fill it with whipped cream, and level.
14. Refrigerate the torte for about 5 minutes, or until firm.
15. Unmold by setting the torte on a platform of smaller diameter. Heat the outside of the ring carefully with a torch or hot towel. Carefully slide the ring down.
16. Repair any defects.
17. Divide and mark 16 portions evenly.

18. Using whipped cream, pipe rosettes with a no. 3 star tube, ¼ inch from the edge of the torte, in the center of each portion.
19. Garnish the center of the torte and the outside bottom edge with shaved chocolate; reserve the largest pieces for the top and the crushed pieces for the bottom edge.
20. Garnish the top of the torte with glazé and cherries.

Ganache Torte

YIELD: 1, 9-inch Torte.

INGREDIENTS:

Vanilla chiffon genoise (see page 763)	3, 9-inch layers, ⅓-inch thick
Simple syrup, rum-flavored (see page 967)	3 ounces
Ganache, dark (see pages 955 and 956)	1 pound, 15 ounces
White chocolate, shaved	As needed
Marzipan (see page 965)	As needed
White chocolate, coating	As needed
Chocolate, dark, semi-sweet, piping	As needed

METHOD OF PREPARATION:

1. Gather the equipment and ingredients.
2. Moisten three vanilla chiffon genoise layers with flavored simple syrup, as needed, during assembly.
3. Fill each chiffon genoise layer with ganache; stack the layers.
4. Cover the cake with ganache.
5. Place the white chocolate shavings around the bottom edge of the torte.
6. Mark the cake into 16 portions.
7. Using a small, round tip, pipe a spiral of ganache. Start the spiral about 1½ inches away from the center until the outer edge is reached.
8. Set the torte aside.
9. Roll out the marzipan 1⁄12-inch thick.
10. Cut an 8 7⁄16-inch disk out of the marzipan.
11. Let the marzipan disk dry for 10 minutes before using.
12. Coat the marzipan disk with white chocolate.
13. Let it set, and then cut into 16 equal wedges.
14. Decorate each wedge of marzipan with dark piping chocolate, as instructed.
15. Place a marzipan wedge on top of each slice, slightly angled.

PASTRY TECHNIQUES:

Filling, Icing, Rolling

Filling:

1. Cut open the food product.
2. Carefully spread the filling using an icing spatula, or
3. Carefully pipe the filling using a pastry bag.

Icing:

1. Use a clean icing spatula.
2. Work quickly and neatly.

Rolling:

1. Prepare the rolling surface by dusting with the appropriate medium (flour, cornstarch, etc.).
2. Use the appropriate style pin (stick pin or ball bearing pin) to roll the dough to desired thickness; rotate the dough during rolling to prevent sticking.

HACCP:

Store at 40° F.

Linzer Torte

YIELD: 1, 8½-inch Torte.

PASTRY TECHNIQUES:

Rolling, Spreading

Rolling:

1. Prepare the rolling surface by dusting with the appropriate medium (flour, cornstarch, etc.).
2. Use the appropriate style pin (stick pin or ball bearing pin) to roll the dough to desired thickness; rotate the dough during rolling to prevent sticking.

Spreading:

1. Using an icing spatula or off-set spatula, smooth the icing or other spreading medium over the surface area.

HACCP:

Store at 40° F.

INGREDIENTS:

Linzer dough, bottom shell (see page 839)	1 pound, ½ ounce
Preserves, raspberry	3 ounces
Linzer dough, lattice top	1 pound, ¼ ounce
Egg wash (see page 964)	As needed

METHOD OF PREPARATION:

1. Gather the equipment and ingredients.
2. Roll out the dough for the bottom shell to ¼-inch thick.
3. Cut the dough to the shape of the pan, and place in the pan.
4. Spread 3 ounces of raspberry preserves on the dough.
5. Roll out the dough for the crust to ¼-inch thick; make a diamond-lattice top crust. Use the egg wash to glue the lattice top together.
6. Bake at 350° F until golden brown.
7. Cool in the pans.

Lutetia Torte

PASTRY TECHNIQUES:
Filling, Icing, Coating

Filling:
1. Cut open the food product.
2. Carefully spread the filling using an icing spatula, or
3. Carefully pipe the filling using a pastry bag.

Icing:
1. Use a clean icing spatula.
2. Work quickly and neatly.

Coating:
1. Use a coating screen, with a sheet pan underneath.
2. Ensure that the product is the correct temperature.
3. Coat the product; use an appropriately-sized utensil.

HACCP:
Store at 40° F.

YIELD: 1, 9-inch Torte.

INGREDIENTS:

Walnut chiffon genoise (see page 766)	3, 9-inch layers, ⅓-inch thick
Simple syrup, rum-flavored (see page 967)	2½ ounces
Preserves, raspberry	2½ ounces
Ganache (see pages 955 and 956)	1 pound, 9 ounces
Chantilly cream (see page 787)	4 ounces
Dark chocolate, coating	As needed
Walnuts, halves	16 each
Royal icing (see page 776)	As needed
Marzipan, yellow carnation (see page 965)	1 each
Marzipan, green leaves	3 each

METHOD OF PREPARATION:

1. Gather the equipment and ingredients.
2. Moisten three layers of walnut chiffon genoise cakes with flavored simple syrup during assembly.
3. Spread the raspberry preserves on the first layer of genoise.
4. Spread ganache on top of the raspberry preserves.
5. Place a second layer of genoise on top of the ganache.
6. Spread a layer of chantilly cream on top.
7. Top with a third layer of genoise.
8. Cover the cake with ganache.
9. Chill for about 20 minutes.
10. Coat the cake with a dark chocolate coating.
11. Mark into 16 portions.
12. Decorate with walnut halves and royal icing.
13. Place a marzipan carnation with leaves in the center.

Mocha Buttercream Torte No. 1

YIELD: 1, 9-inch Torte.

PASTRY TECHNIQUES:
Filling, Icing

Filling:
1. Cut open the food product.
2. Carefully spread the filling using an icing spatula, or
3. Carefully pipe the filling using a pastry bag.

Icing:
1. Use a clean icing spatula.
2. Work quickly and neatly.

HACCP:
Store at 40° F.

INGREDIENTS:

Cake layers, vanilla	3, 9-inch layers, ⅜-inch thick
Simple syrup, kahlua (see page 966)	3 ounces
Preserves, apricot	2½ ounces
Buttercream, mocha	1 pound, 8 ounces
Buttercream, vanilla	4 ounces
Mocha beans	16 each
Chocolate, decorettes	As needed

METHOD OF PREPARATION:

1. Gather the equipment and ingredients.
2. Moisten each layer of cake with syrup during assembly.
3. Spread the preserves evenly on the first layer of cake.
4. Spread mocha buttercream evenly on top.
5. Place a second layer of cake on top; center and level.
6. Spread mocha buttercream evenly on the second layer.
7. Place a third layer of cake on top; center and level.
8. Cover the entire torte with mocha buttercream; ice straight and level; leave no more than ⅛ inch of buttercream on top.
9. Divide and mark 16 portions evenly.
10. Using vanilla buttercream, pipe 16 rosettes with the no. 3 star tube, ¼ inch from the edge of the torte, in the center of each portion.
11. Place the mocha beans on the rosettes, with the dimple on the bean pointed toward the center of the torte.
12. Garnish the bottom outside edge with chocolate decorettes about ½ inch up the side of the torte.

Mocha Buttercream Torte No. 2

YIELD: 1, 9-inch Torte.

INGREDIENTS:

Vanilla chiffon genoise (see page 763)	3, 9-inch layers, ⅓-inch thick
Simple syrup, brandy-flavored (see page 967)	3 ounces
Preserves, apricot	2½ ounces
Buttercream, mocha	2 pounds
Mocha beans	16 each
Chocolate, dark, semi-sweet, piping	As needed
Almonds, light, blanched, sliced, and toasted	As needed

METHOD OF PREPARATION:

1. Gather the equipment and ingredients.
2. Moisten each layer of vanilla chiffon genoise with flavored simple syrup during assembly.
3. Spread the apricot preserves on the first layer of genoise, and top with mocha buttercream.
4. Top with the second layer of genoise.
5. Spread mocha buttercream on top.
6. Place the third layer of genoise.
7. Cover the torte with mocha buttercream; ice straight and level; leave no more than ⅛ inch of buttercream on top.
8. Mark into 16 portions.
9. Decorate with mocha buttercream lines and mocha beans.
10. Finish with piping chocolate over buttercream lines.
11. Place sliced almonds around the bottom edge and in the center of the torte.

PASTRY TECHNIQUES:

Filling, Icing

Filling:

1. Cut open the food product.
2. Carefully spread the filling using an icing spatula, or
3. Carefully pipe the filling using a pastry bag.

Icing:

1. Use a clean icing spatula.
2. Work quickly and neatly.

HACCP:

Store at 40° F.

Sacher Torte

YIELD: 1, 9-inch Torte.

PASTRY TECHNIQUES:

Filling, Icing, Coating

Filling:

1. Cut open the food product.
2. Carefully spread the filling using an icing spatula, or
3. Carefully pipe the filling using a pastry bag.

Icing:

1. Use a clean icing spatula.
2. Work quickly and neatly.

Coating:

1. Use a coating screen, with a sheet pan underneath.
2. Ensure that the product is the correct temperature.
3. Coat the product; use an appropriately-sized utensil.

HACCP:

Store at 40° F.

INGREDIENTS:

Sacher cake (see pages 761 and 762)	2, 9-inch layers, 1¼-inch thick
Simple syrup, rum-flavored (see page 967)	3 ounces
Preserves, apricot	3 ounces
Chocolate, coating	As needed
Chocolate, dark, semi-sweet, piping	As needed

METHOD OF PREPARATION:

1. Gather the equipment and ingredients.
2. Moisten both cake layers with flavored simple syrup during assembly.
3. Spread the first layer with the preserves.
4. Place the second layer on top.
5. Cover the torte completely with ganache.
6. Chill for about 20 minutes.
7. Coat the torte with chocolate coating.
8. Mark the torte into 16 portions.
9. Write the word *Sacher* on each slice with dark piping chocolate.

CHEF NOTE:

Instead of using chocolate ganache to cover the torte; you could cover the cake with a thin layer of apricot glaze and then coat cake with a chocolate coating.

Bavarian Cream No. 1

Using a Pâté à Bombe

PASTRY TECHNIQUES:

Blooming, Whipping, Boiling, Folding

Blooming:

Gelatin sheets or leaves:
1. Fan the sheets out.
2. Cover the sheets in liquid.
3. Sheets are bloomed when softened.

Granular gelatin:
1. Sprinkle the gelatin.
2. Gelatin is ready when it is cream of wheat consistency.

Whipping:

1. Hold the whip at a 55-degree angle.
2. Create circles, using a circular motion.
3. The circular motion needs to be perpendicular to the bowl.

Boiling:

1. Bring the cooking liquid to a rapid boil.
2. Stir the contents.

Folding:

Do steps 1, 2, and 3 in one continuous motion.
1. Run a bowl scraper under the mixture, across the bottom of the bowl.
2. Turn the bowl counterclockwise.
3. Bring the bottom mixture to the top.

HACCP:

Store at 40° F.

HAZARDOUS FOODS:

Milk
Heavy cream

CHEF NOTES:

1. Different flavorings can be added to basic bavarian cream. They can be added when folding in the pâté à bombe.
2. It is recommended to chill the molds in ice water before pouring the cream into them. They will then be easier to unmold.

YIELD: 2 pounds, 9 ounces. | 5 pounds, 1¾ ounces.

INGREDIENTS:		
Gelatin	¾ ounce	1½ ounces
Milk, whole	1 pound	2 pounds
Cream, heavy	1 pound	2 pounds
Salt	⅛ ounce	¼ ounce
Pâté à bombe	8 ounces	1 pound

METHOD OF PREPARATION:

1. Gather the equipment and ingredients.
2. Place the gelatin in a dry bowl, and bloom with one fourth of the milk.
3. Whip the cream to a soft-to-medium peak; set aside.
4. Place the remaining milk and salt in a pot; bring to a boil.
5. Remove from the heat, and pour over the bloomed gelatin.
6. Cool over an ice bath or in the refrigerator.
7. When it is cool, whisk in the pâté à bombe.
8. Fold in the whipped cream.
9. Pour into desired molds.
10. Refrigerate or freeze molds immediately.

Bavarian Cream No. 2

Using English Sauce

YIELD:	2 pounds, 9 ounces.	5 pounds, 3 ounces.

INGREDIENTS:		
Gelatin	½ ounce	1 ounce
Milk, whole	1 pound	2 pounds
Cream, heavy	1 pound	2 pounds
Sugar, granulated	4 ounces	8 ounces
Egg yolks	5 ounces	10 ounces
Extract, vanilla	To taste	To taste
Flavorings, as desired	To taste	To taste

METHOD OF PREPARATION:

1. Gather the equipment and ingredients.
2. Place the gelatin in a dry bowl, and bloom with one fourth of the milk.
3. Whip the cream to a soft-to-medium peak. Set aside.
4. Place the remaining milk and sugar in a pot; bring to a boil.
5. Temper the egg yolks, and add to the boiling milk. Heat, stirring constantly, until the liquid coats the back of a spoon; do not boil.
6. Remove from the heat, and strain.
7. Melt the bloomed gelatin over a double boiler.
8. Add the melted gelatin to the egg yolk mixture.
9. To quickly stop the cooking process, cool in an ice bath; stir often.
10. When cool, fold in the vanilla extract, flavorings, and whipped cream.
11. Fill the molds (see Chef Notes on p. 784).
12. Refrigerate or freeze the molds immediately.

PASTRY TECHNIQUES:

Blooming, Whipping, Boiling, Tempering, Folding

Blooming:

Gelatin sheets or leaves:
1. Fan the sheets out.
2. Cover the sheets in liquid.
3. Sheets are bloomed when softened.

Granular gelatin:
1. Sprinkle the gelatin.
2. Gelatin is ready when it is cream of wheat consistency.

Whipping:

1. Hold the whip at a 55-degree angle.
2. Create circles, using a circular motion.
3. The circular motion needs to be perpendicular to the bowl.

Boiling:

1. Bring the cooking liquid to a rapid boil.
2. Stir the contents.

Tempering:

1. Whisk the eggs vigorously while ladling hot liquid.

Folding:

Do steps 1, 2, and 3 in one continuous motion.
1. Run a bowl scraper under the mixture, across the bottom of the bowl.
2. Turn the bowl counterclockwise.
3. Bring the bottom mixture to the top.

HACCP:

Store at 40° F.

HAZARDOUS FOODS:

Milk
Egg yolks
Heavy cream

Chantilly Cream

PASTRY TECHNIQUE:

Whipping

Whipping:

1. Hold the whip at a 55-degree angle.
2. Create circles, using a circular motion.
3. The circular motion needs to be perpendicular to the bowl.

HACCP:

Store at 40° F.

HAZARDOUS FOOD:

Heavy cream

YIELD:	1 pound, 1 ounce.	2 pounds, 2 ounces.
INGREDIENTS:		
Cream, heavy	1 pound	2 pounds
Sugar, granulated	1 ounce	2 ounces
Abstract, vanilla	To taste	To taste

METHOD OF PREPARATION:

1. Gather the equipment and ingredients.
2. Place all of the ingredients in a cold bowl.
3. Whip at high speed to the desired thickness and consistency.
4. Use immediately, or refrigerate until needed.

Pastry Cream

YIELD:	3 pounds, 4½ ounces.	6 pounds, 9⅛ ounces.

INGREDIENTS:

Milk, whole	2 pounds	4 pounds
Sugar, granulated	8 ounces	1 pound
Salt	Pinch	Pinch
Cornstarch	3¾ ounces	7½ ounces
Flour, cake	½ ounce	1 ounce
Egg yolks	4 ounces	8 ounces
Eggs, whole	2 ounces	4 ounces
Extract, vanilla	To taste	To taste
Butter, unsalted	2 ounces	4 ounces

METHOD OF PREPARATION:

1. Gather the equipment and ingredients.
2. Place three fourths of the milk, half of the granulated sugar, and the salt in a pot; bring to a boil.
3. Place the cornstarch and flour in a dry bowl; slowly add the remaining fourth of milk slowly to form a smooth paste.
4. Add the egg yolks and whole eggs to the cornstarch mixture. Combine well.
5. Temper the cornstarch mixture, and add to the boiling milk.
6. Bring the mixture back to a second boil, constantly stirring; cook for 3 minutes.
7. Remove from the heat; stir in the vanilla extract and butter.
8. Whisk well.
9. Cool properly, and refrigerate.

PASTRY TECHNIQUES:

Boiling, Tempering, Thickening

Boiling:

1. Bring the cooking liquid to a rapid boil.
2. Stir the contents.

Tempering:

1. Whisk the eggs vigorously while ladling hot liquid.

Thickening:

1. Mix a small amount of sugar with the starches.
2. Create a slurry.
3. Whisk vigorously until thickened and translucent.

HACCP:

Store at 40° F.

HAZARDOUS FOODS:

Milk
Egg yolks
Eggs

CHEF NOTE:

Place buttered parchment paper on top of cream to avoid forming a hard skin on top.

Basic Fruit Mousse

PASTRY TECHNIQUES:

Blooming, Combining, Whipping, Folding

Blooming:

Gelatin sheets or leaves:
1. Fan the sheets out.
2. Cover the sheets in liquid.
3. Sheets are bloomed when softened.

Granular gelatin:
1. Sprinkle the gelatin.
2. Gelatin is ready when it is cream of wheat consistency.

Combining:

Bringing together two or more components.
1. Prepare the components to be combined.
2. Add one to the other, using the appropriate mixing method (if needed).

Whipping:
1. Hold the whip at a 55-degree angle.
2. Create circles, using a circular motion.
3. The circular motion needs to be perpendicular to the bowl.

Folding:

Do steps 1, 2, and 3 in one continuous motion.
1. Run a bowl scraper under the mixture, across the bottom of the bowl.
2. Turn the bowl counterclockwise.
3. Bring the bottom mixture to the top.

HACCP:

Store at 40° F.

HAZARDOUS FOOD:

Heavy cream

CHEF NOTES:
1. Some fruit purée, such as raspberry or strawberry, is sweeter than others and will require a reduction in the amount of sugar used.
2. Some fruits, such as pineapple and kiwi, contain acids, which interfere with the setting of the gelatin. They will require cooking to neutralize the enzymes.
3. Mousse can be made moldable by increasing the gelatin in the formula.

YIELD: 1 pound, 1¾ ounces. 3 pounds, 10½ ounces.

INGREDIENTS:		
Gelatin	¼ ounce	½ ounce
Water	1¼ ounces	2½ ounces
Purée, fruit	9½ ounces	1 pound, 3 ounces
Cream, heavy	1 pound	2 pounds
Sugar, confectionery	2 ounces (variable)	4 ounces (variable)

METHOD OF PREPARATION:

1. Gather the equipment and ingredients.
2. Place the gelatin in a dry bowl, and bloom with water.
3. Melt the bloomed gelatin over a double boiler.
4. Add the fruit purée to the gelatin.
5. Allow the mixture to cool, but do not let it become firm.
6. Whip the cream and sugar to a medium peak.
7. Fold gently together the whipped cream and purée mixture.
8. Allow the mixture to set slightly; then pipe into a ramekin or other appropriate serving dish.

Chocolate Mousse No. 1

PASTRY TECHNIQUES:

Whipping, Chopping, Melting, Folding

Whipping:

1. Hold the whip at a 55-degree angle.
2. Create circles, using a circular motion.
3. The circular motion needs to be perpendicular to the bowl.

Chopping:

1. Use a sharp knife.
2. Hold the food product properly.
3. Cut with a quick downward motion.

Melting:

1. Prepare the food product to be melted.
2. Place the food product in an appropriate-sized pot over direct heat or over a double boiler.
3. Stir frequently or occasionally, depending on the delicacy of the product, until melted.

OR

1. Place the product on a sheet pan or in a bowl, and place in a low-temperature oven until melted.

Folding:

Do steps 1, 2, and 3 in one continuous motion.

1. Run a bowl scraper under the mixture, across the bottom of the bowl.
2. Turn the bowl counterclockwise.
3. Bring the bottom mixture to the top.

HACCP:

Store at 40° F.

HAZARDOUS FOODS:

Heavy cream
Egg yolks
Egg whites

YIELD: 1 pound, 8 ounces. 3 pounds.

INGREDIENTS:		
Cream, heavy	8 ounces	1 pound
Chocolate, dark, semi-sweet	6 ounces	12 ounces
Egg yolks	2 ounces	4 ounces
Sugar, granulated	4½ ounces	9 ounces
Egg whites	2½ ounces	5 ounces
Flavor, brandy or rum	1 ounce	2 ounces

METHOD OF PREPARATION:

1. Gather the equipment and ingredients.
2. Whip the heavy cream to a firm peak, and hold in a refrigerator.
3. Melt the chocolate over a water bath, and hold at 98° F.
4. Heat the egg yolks with half of the sugar over a double boiler; whip constantly to prevent overheating. The egg yolks must reach 145° F.
5. In another bowl, whip the egg whites, and gradually add the remaining granulated sugar to make a meringue. Whip to a wet, medium peak.
6. Fold the chocolate into the whipped yolk mixture.
7. Add either the brandy or rum flavoring.
8. Fold the whipped cream into the chocolate mixture.
9. Add the whipped meringue into the chocolate mixture. *Do not overfold.*
10. Let it set in a refrigerator.
11. Pipe into glasses.

Chocolate Mousse No. 2

With Ganache, Pâté à Bombe, Italian Meringue

PASTRY TECHNIQUES:

Whipping, Combining

Whipping:

1. Hold the whip at a 55-degree angle.
2. Create circles, using a circular motion.
3. The circular motion needs to be perpendicular to the bowl.

Combining:

Bringing together two or more components.

1. Prepare the components to be combined.
2. Add one to the other, using the appropriate mixing method (if needed).

HACCP:

Store at 40° F.

HAZARDOUS FOODS:

Egg yolks
Egg whites
Heavy cream

YIELD:	2 pounds, 14½ ounces.	5 pounds, 13 ounces.

INGREDIENTS:		
Cream, heavy	1 pound, 6½ ounces	2 pounds, 13 ounces
Ganache, hard	8½ ounces	1 pound, 1 ounce
Rum	½ ounce	1 ounce
Pâté à Bombe:		
Sugar, granulated	6½ ounces	13 ounces
Water	2½ ounces	5 ounces
Egg yolks	2 ounces	4 ounces
Italian Meringue:		
Egg whites	3 ounces	6 ounces
Cream of tartar	Pinch	⅛ ounce
Sugar, granulated	1 ounce	2 ounces

METHOD OF PREPARATION:

1. Gather the equipment and ingredients.
2. Whip the cream to a soft peak.
3. Soften the ganache.
4. Add the pâté à bombe to the softened ganache.
5. Add the rum.
6. Fold in the Italian meringue.
7. Fold in the whipped cream.

Preparation of Pâté à Bombe:

1. Place the granulated sugar and water in a pot, and cook until it reaches 238° F.
2. Place the egg yolks in a bowl, and whip at high speed until light.
3. Decrease the mixer speed to medium; slowly pour 3 ounces of 238° F sugar syrup into the whipped yolks.
4. Whip until cool.

Preparation of Italian Meringue:

1. Return the remaining sugar syrup to the heat; continue to cook until a temperature of 250° F is reached.
2. Prepare a common meringue by warming the egg whites to 100° F. Whip at high speed, gradually adding the cream of tartar and granulated sugar.
3. When the sugar syrup reaches 250° F, add it to the meringue. Whip at medium speed until the meringue is cool.

PASTRY TECHNIQUES:

Blooming, Whipping, Scalding, Melting, Folding

Blooming:

Gelatin sheets or leaves:
1. Fan the sheets out.
2. Cover the sheets in liquid.
3. Sheets are bloomed when softened.

Granular gelatin:
1. Sprinkle the gelatin.
2. Gelatin is ready when it is cream of wheat consistency.

Whipping:
1. Hold the whip at a 55-degree angle.
2. Create circles, using a circular motion.
3. The circular motion needs to be perpendicular to the bowl.

Scalding:
1. Heat the liquid on high heat.
2. Do not boil the liquid.

Melting:
1. Prepare the food product to be melted.
2. Place the food product in an appropriate sized pot over direct heat or over a double boiler.
3. Stir frequently or occasionally, depending on the delicacy of the product, until melted.

OR

1. Place the product on a sheet pan or in a bowl, and place in a low-temperature oven until melted.

Folding:

Do steps 1, 2, and 3 in one continuous motion.
1. Run a bowl scraper under the mixture, across the bottom of the bowl.
2. Turn the bowl counterclockwise.
3. Bring the bottom mixture to the top.

HACCP:

Store at 40° F.

HAZARDOUS FOODS:

Egg yolks
Milk
Heavy cream

White Chocolate Mousse

YIELD:	15⅛ ounces.	1 pound, 14¼ ounces.
INGREDIENTS:		
Gelatin	⅛ ounce	¼ ounce
Egg yolks	2¼ ounces	4½ ounces
Sugar, granulated	1¼ ounces	2½ ounces
Milk, whole	2½ ounces	5 ounces
Chocolate, white, chopped	2¼ ounces	4½ ounces
Cream, heavy, whipped	6¾ ounces	13½ ounces
Extract, vanilla	To taste	To taste
Triple Sec or Grand Marnier	To taste	To taste

METHOD OF PREPARATION:

1. Gather the equipment and ingredients.
2. Bloom the gelatin, and set aside.
3. Place the egg yolks and granulated sugar in a bowl over a double boiler; heat to 140° F, whipping constantly.
4. Scald the milk.
5. Place the chopped chocolate in a bowl, and pour scalded milk over it. Stir until all of the chocolate has melted.
6. Pour the hot chocolate mixture over the bloomed gelatin; stir thoroughly to dissolve the gelatin. It may be necessary to place it over a double boiler if all of the gelatin is not dissolved.
7. Fold in the flavorings.
8. Cool the chocolate mixture until it begins to thicken slightly.
9. Fold in the whipped cream.
10. Let set in refrigerator.
11. Pipe into glasses.

Almond Filling

YIELD:	2 pounds, 13 ounces.	5 pounds, 10 ounces.

INGREDIENTS:

Almond paste	2 pounds	4 pounds
Butter, unsalted	8 ounces	1 pound
Egg whites	2 ounces	4 ounces
Flour, bread	3 ounces	6 ounces

METHOD OF PREPARATION:

1. Gather the equipment and ingredients.
2. Place the almond paste and butter in a bowl with a paddle; cream together.
3. Add the egg whites slowly, and scrape well.
4. Add the bread flour, and combine well.

PASTRY TECHNIQUES:

Creaming, Combining

Creaming:

1. Soften the fats on low speed.
2. Add the sugar(s) and cream; increase the speed slowly.
3. Add the eggs one at a time; scrape the bowl frequently.
4. Add the dry ingredients in stages.

Combining:

Bringing together two or more components.

1. Prepare the components to be combined.
2. Add one to the other, using the appropriate mixing method (if needed).

HACCP:

Store at 40° F.

HAZARDOUS FOOD:

Egg whites

Bakers' Cheese

(for Danish)

PASTRY TECHNIQUE:

Creaming

Creaming:

1. Soften the fats on low speed.
2. Add the sugar(s) and cream; increase the speed slowly.
3. Add the eggs one at a time; scrape the bowl frequently.

HACCP:

Store at 40° F.

HAZARDOUS FOODS:

Bakers' cheese
Eggs

YIELD:	4 pounds, 5/8 ounce.	8 pounds, 1¼ ounces.
INGREDIENTS:		
Sugar, granulated	10 ounces	1 pound, 4 ounces
Cheese, bakers'	2 pounds, 8 ounces	5 pounds
Salt	5/8 ounce	1¼ ounces
Eggs, whole	8 ounces	1 pound
Butter, unsalted, soft	5 ounces	10 ounces
Flour, bread	1½ ounces	3 ounces

METHOD OF PREPARATION:

1. Gather the equipment and ingredients.
2. Cream the granulated sugar and bakers' cheese with a paddle on the first speed in a 20-quart mixer until smooth.
3. Add the eggs slowly on low speed.
4. Add the butter, salt, and flour; mix until well combined.

Black Forest Cherry Filling

(Dark Sweet)

PASTRY TECHNIQUES:

Boiling, Thickening

Boiling:

1. Bring the cooking liquid to a rapid boil.
2. Stir the contents.

Thickening:

1. Mix a small amount of sugar with the starches.
2. Create a slurry.
3. Whisk vigorously until thickened and translucent.

HACCP:

Store at 40° F.

YIELD:	1 pound, 8½ ounces.	3 pounds, 1 ounce.
INGREDIENTS:		
Cherries, canned, dark, sweet, drained and halved	1 pound	2 pounds
Cherry syrup	8 ounces	1 pound
Cornstarch	½ ounce	1 ounce

METHOD OF PREPARATION:

1. Gather the equipment and ingredients.
2. Drain the cans of cherries, and reserve the liquid for the cherry syrup portion.
3. Place 13 ounces of syrup in a pot, and bring it to a boil.
4. Place the cornstarch in a dry bowl, and slowly add remaining syrup; whisk constantly to prevent lumps from forming.
5. Add the cornstarch mixture to the boiling syrup, and whisk constantly as it thickens. Continue to boil; whisk for 2 minutes to ensure the starch is cooked.
6. Add the halved cherries, and combine.
7. Cool.

CHEF NOTE:

If there is not enough syrup in the cherries, add water and sugar to make up the difference.

Cheese Filling

PASTRY TECHNIQUES:

Creaming, Combining

Creaming:

1. Soften the fats on low speed.
2. Add the sugar(s) and cream; increase the speed slowly.
3. Add the eggs one at a time; scrape the bowl frequently.

Combining:

Bringing together two or more components.

1. Prepare the components to be combined.
2. Add one to the other, using the appropriate mixing method (if needed).

HACCP:

Store at 40° F.

HAZARDOUS FOODS:

Cream cheese
Bakers' cheese
Egg yolks

YIELD:	2 pounds.	6 pounds.
INGREDIENTS:		
Cream cheese, room temperature	1 pound	3 pounds
Cheese, bakers', room temperature	8 ounces	1 pound, 8 ounces
Sugar, granulated	5¼ ounces	15¾ ounces
Egg yolks	3 ounces	9 ounces
Extract, vanilla	To taste	To taste
Orange zest	To taste	To taste
Lemon zest	To taste	To taste

METHOD OF PREPARATION:

1. Gather the equipment and ingredients.
2. Place the cream cheese, bakers' cheese, and granulated sugar in a bowl with a paddle.
3. Paddle until smooth; scrape often.
4. Add the yolks and flavorings in small portions. Mix and scrape well after each addition.
5. Cover, and refrigerate until needed.

Chocolate Filling

(for Croissant)

PASTRY TECHNIQUES:

Melting, Combining

Melting:

1. Prepare the food product to be melted.
2. Place the food product in an appropriate-sized pot over direct heat or over a double boiler.
3. Stir frequently or occasionally, depending on the delicacy of the product, until melted.

OR

1. Place the product on a sheet pan or in a bowl, and place in a low-temperature oven until melted.

Combining:

Bringing together two or more components.

1. Prepare the components to be combined.
2. Add one to the other, using the appropriate mixing method (if needed).

HACCP:

Store at 60° F to 65° F.

YIELD:	3 pounds.	6 pounds.
INGREDIENTS:		
Shortening, all-purpose	1 pound	2 pounds
Chocolate, dark, semi-sweet, chopped	2 pounds	4 pounds

METHOD OF PREPARATION:

1. Gather the equipment and ingredients.
2. Over direct heat, melt the shortening; remove from the heat.
3. Add the chopped dark chocolate to the melted shortening. Stir until dissolved; strain.
4. Store in a covered container in a dry storage area.

CHEF NOTE:

To use, heat the filling slightly to a soft, pliable paste consistency. Place the filling in the center of the croissant triangle, and roll.

German Chocolate Cake Filling

PASTRY TECHNIQUES:

Combining, Boiling, Thickening

Combining:

Bringing together two or more components.

1. Prepare the components to be combined.
2. Add one to the other, using the appropriate mixing method (if needed).

Boiling:

1. Bring the cooking liquid to a rapid boil.
2. Stir the contents.

Thickening:

1. Mix a small amount of sugar with the starches.
2. Create a slurry.
3. Whisk vigorously until thickened and translucent.

HACCP:

Store at 40° F.

HAZARDOUS FOOD:

Egg yolks

YIELD:	6 pounds, 1½ ounces.	12 pounds, 3 ounces.

INGREDIENTS:

Milk, evaporated	1 pound, 8 ounces	3 pounds
Egg yolks, sugared	4½ ounces	9 ounces
Butter, unsalted, melted	1 pound, 8 ounces	3 pounds
Sugar, granulated	1 pound, 5 ounces	2 pounds, 10 ounces
Pecan pieces	12 ounces	1 pound, 8 ounces
Coconut flakes	12 ounces	1 pound, 8 ounces

METHOD OF PREPARATION:

1. Gather the equipment and ingredients.
2. Combine the evaporated milk and egg yolks together.
3. Melt the butter.
4. Add the granulated sugar and melted butter to the milk-yolk mixture.
5. Place in a bowl, and place over a double boiler; cook until the mixture thickens.
6. Add the pecan pieces and coconut flakes to the cooked mixture.
7. Cool.

CHEF NOTE:

Do *not* cook the filling to a caramel color; the filling should be light in color.

Tiramisu Filling

YIELD:	2 pounds, 6 ounces.	4 pounds, 12 ounces.

INGREDIENTS:

Cream cheese	12 ounces	1 pound, 8 ounces
Mascarpone, cheese	12 ounces	1 pound, 8 ounces
Sugar, confectionery, sifted	6 ounces	12 ounces
Cream, heavy, whipped	8 ounces	1 pound

METHOD OF PREPARATION:

1. Gather the equipment and ingredients.
2. Place the cream cheese in a bowl, and soften; scrape the bowl occasionally.
3. Add the mascarpone cheese to the cream cheese, and incorporate thoroughly.
4. Add the confectionery sugar, and combine.
5. Remove from the mixer, and fold in the whipped cream.

PASTRY TECHNIQUES:

Creaming, Combining, Whipping, Folding

Creaming:

1. Soften the fats on low speed.
2. Add the sugar(s) and cream; increase the speed slowly.
3. Add the eggs one at a time; scrape the bowl frequently.

Combining:

Bringing together two or more components.

1. Prepare the components to be combined.
2. Add one to the other, using the appropriate mixing method (if needed).

Whipping:

1. Hold the whip at a 55-degree angle.
2. Create circles, using a circular motion.
3. The circular motion needs to be perpendicular to the bowl.

Folding:

Do steps 1, 2, and 3 in one continuous motion.

1. Run a bowl scraper under the mixture, across the bottom of the bowl.
2. Turn the bowl counterclockwise.
3. Bring the bottom mixture to the top.

HACCP:

Store at 40° F.

HAZARDOUS FOODS:

Cream cheese
Mascarpone cheese
Heavy cream

CHEF NOTE:

Tiramisu is an Italian word that is translated as "pick me up." Often, the tiramisu cheese filling has espresso incorporated into the mix. Gusto Rico or espresso powder also can be used. The filling is served with espresso-soaked lady fingers.

Walnut Strudel Filling

PASTRY TECHNIQUE:

Boiling, Reducing, Combining

Boiling:

1. Bring the cooking liquid to a rapid boil.
2. Stir the contents.

Reducing:

1. Bring the sauce to a boil; then reduce to a simmer.
2. Stir often; reduce to the desired consistency.

Combining:

Bringing together two or more components.

1. Prepare the components to be combined.
2. Add one to the other, using the appropriate mixing method (if needed).

HACCP:

Store at 40° F.

HAZARDOUS FOOD:

Milk

YIELD:	2 pounds, 15¼ ounces.	5 pounds, 14½ ounces.

INGREDIENTS:

Walnuts, crushed	1 pound	2 pounds
Butter, unsalted	4 ounces	8 ounces
Sugar, granulated	6 ounces	12 ounces
Raisins, golden	4 ounces	8 ounces
Lemon zest	¼ each	½ each
Milk, whole	12 ounces	1 pound, 8 ounces
Crumbs, cake, dry	4 ounces	8 ounces
Brandy	1¼ ounces	2½ ounces

METHOD OF PREPARATION:

1. Gather the equipment and ingredients.
2. Place the walnuts, butter, granulated sugar, raisins, lemon zest, and milk in a pot; bring to a boil.
3. Reduce the heat, and simmer to evaporate the liquid.
4. Remove from the heat, and add the cake crumbs and brandy; incorporate well.
5. Cool.
6. Store in a refrigerator; keep covered until used.

Bread Pudding

YIELD: 10 pounds, 1¼ ounces.

INGREDIENTS:

Ingredient	Amount
Butter, unsalted, softened	As needed
Bread, cubed or sliced	2 pounds
Cinnamon, ground	As needed
Nutmeg, ground	As needed
Eggs, whole	2 pounds
Sugar, granulated	1 pound
Salt	¼ ounce
Extract, vanilla	1 ounce
Milk, whole	4 pounds
Cream, heavy	1 pound

METHOD OF PREPARATION:

1. Gather the equipment and ingredients.
2. Butter the hotel pan.
3. Place the cubed or sliced bread in a hotel pan.
4. Sprinkle the bread with cinnamon and nutmeg.
5. Combine and whisk together the eggs, granulated sugar, salt, and vanilla extract.
6. Scald the milk and heavy cream.
7. Temper the egg mixture with the scalded milk and heavy cream.
8. Pour the mixture over the bread.
9. Bake at 350° F in a water bath for 60 minutes, or until set.

PASTRY TECHNIQUES:

Scalding, Slow Baking

Scalding:

1. Heat the liquid on high heat.
2. Do not boil the liquid.

Slow Baking:

1. Use an appropriate baking dish.
2. Use hot water in the pan.
3. Replenish the water when needed.

HACCP:

Store at 40° F.

HAZARDOUS FOODS:

Eggs
Milk
Heavy cream

CHEF NOTES:

1. Use two-day-old white French or Italian bread for more flavor.
2. To enrich pudding, chopped fruit and chocolate pieces can be added.

Indian Pudding

PASTRY TECHNIQUES:
Scalding, Combining

Scalding:
1. Heat the liquid on high heat.
2. Do not boil the liquid.

Combining:
Bringing together two or more components.
1. Prepare the components to be combined.
2. Add one to the other, using the appropriate mixing method (if needed).

HACCP:
Store at 40° F.

HAZARDOUS FOODS:
Milk
Eggs

YIELD:	9 pounds, 3⅛ ounces.	18 pounds, 6¼ ounces.

INGREDIENTS:

Butter, unsalted	As needed	As needed
Milk, whole	6 pounds	12 pounds
Cornmeal	9½ ounces	1 pound, 3 ounces
Butter, unsalted, melted	3 ounces	6 ounces
Molasses	1 pound, 2 ounces	2 pounds, 4 ounces
Salt	¼ ounce	½ ounce
Cinnamon, ground	¼ ounce	½ ounce
Ginger, ground	⅛ ounce	¼ ounce
Eggs, whole	1 pound, 4 ounces	2 pounds, 8 ounces

METHOD OF PREPARATION:

1. Gather the equipment and ingredients.
2. Grease a hotel pan with butter.
3. Scald the milk.
4. Add the cornmeal to the milk.
5. Cook over a double boiler; stir constantly.
6. Combine the butter, molasses, salt, cinnamon, and ginger.
7. Add the eggs to the molasses mixture.
8. Combine the molasses mixture with the cornmeal mixture.
9. Pour into the greased hotel pan.
10. Bake at 350° F in a water bath for 30 to 40 minutes.

Rice Pudding

YIELD: 9 pounds, 12¾ ounces.

INGREDIENTS:

Ingredient	Amount
Butter, unsalted	As needed
Rice, medium-grain, uncooked	1 pound
Sugar, confectionery	As needed
Milk, whole	6 pounds
Extract, vanilla	½ ounce
Salt	¼ ounce
Egg yolks	12 ounces
Sugar, granulated	1 pound
Cream, heavy	1 pound
Cinnamon, ground	As needed

METHOD OF PREPARATION:

1. Gather the equipment and ingredients.
2. Butter a hotel pan.
3. Place the rice in water with a small amount of confectionery sugar.
4. Bring to a boil; blanch for 3 minutes. Drain the rice.
5. Place the blanched rice, milk, vanilla extract, and salt in a pot, and cook for approximately 30 minutes, until tender. Stir occasionally.
6. Combine the egg yolks, granulated sugar, and heavy cream.
7. Temper the egg mixture, and add to the hot milk mixture.
8. Cook for an additional 2 minutes.
9. Remove, and pour into the buttered hotel pan.
10. Sprinkle with cinnamon.
11. Place on a sheet pan, and add water (double-boiler effect).
12. Bake at 350° F for 30 to 40 minutes, or until set.

PASTRY TECHNIQUES:

Boiling, Combining, Tempering, Slow Baking

Boiling:

1. Bring the cooking liquid to a rapid boil.
2. Stir the contents.

Combining:

Bringing together two or more components.

1. Prepare the components to be combined.
2. Add one to the other, using the appropriate mixing method (if needed).

Tempering:

To equalize two extreme temperatures.

1. Whisk the eggs vigorously while ladling hot liquid.

Slow Baking:

1. Use an appropriate baking dish.
2. Use hot water in the pan.
3. Replenish the water when needed.

HAZARDOUS FOODS:

Milk
Egg yolks
Heavy cream

Tapioca Pudding

PASTRY TECHNIQUES:

Combining, Tempering, Whipping, Folding

Combining:

Bringing together two or more components.

1. Prepare the components to be combined.
2. Add one to the other, using the appropriate mixing method (if needed).

Tempering:

To equalize two extreme temperatures.

1. Whisk the eggs vigorously while ladling hot liquid.

Whipping:

1. Hold the whip at a 55-degree angle.
2. Create circles, using a circular motion.
3. The circular motion needs to be perpendicular to the bowl.

Folding:

Do steps 1, 2, and 3 in one continuous motion.

1. Run a bowl scraper under the mixture, across the bottom of the bowl.
2. Turn the bowl counterclockwise.
3. Bring the bottom mixture to the top.

HACCP:

Store at 40° F.

HAZARDOUS FOODS:

Milk
Egg yolks
Heavy cream
Egg whites

YIELD:	8 pounds, 10¼ ounces.	17 pounds, 4¾ ounces.
INGREDIENTS:		
Butter, unsalted	As needed	As needed
Tapioca	4 pounds	8 pounds
Milk, whole	3 pounds	6 pounds
Extract, vanilla	¼ ounce	½ ounce
Salt	Pinch	¼ ounce
Egg yolks	6 ounces	12 ounces
Sugar, granulated	6 ounces	12 ounces
Cream, heavy	8 ounces	1 pound
Cinnamon, ground	As needed	As needed
Egg whites	4 ounces	8 ounces
Sugar, granulated	2 ounces	4 ounces

METHOD OF PREPARATION:

1. Gather the equipment and ingredients.
2. Butter a hotel pan.
3. Place the tapioca, milk, vanilla extract, and salt in a pot; cook until tender.
4. Combine the egg yolks, granulated sugar, heavy cream, and cinnamon.
5. Temper the egg yolk mixture, and add to the milk mixture.
6. Cook for 2 to 3 minutes, stirring constantly; remove from the stove.
7. Place the egg whites in a bowl, whip to a soft peak, and slowly add the granulated sugar to make a meringue.
8. Fold the meringue into the tapioca mixture.
9. Pour into the buttered hotel pan.
10. Refrigerate until set.

Vanilla Pudding

YIELD:	3 pounds, 2 ounces.	6 pounds, 4 ounces.

INGREDIENTS:

Sugar, granulated	8 ounces	1 pound
Milk, whole	2 pounds	4 pounds
Cornstarch	1½ ounces	3 ounces
Egg yolks	4 ounces	8 ounces
Eggs, whole	2 ounces	4 ounces
Butter, unsalted	2 ounces	4 ounces
Extract, vanilla	½ ounce	1 ounce

METHOD OF PREPARATION:

1. Gather the equipment and ingredients.
2. Place half of the sugar and the milk in a pot; bring to a boil.
3. Place the cornstarch and the remaining sugar in a dry bowl; slowly whisk in the egg yolks and whole eggs; combine well.
4. Temper the cornstarch mixture, and add to the boiling milk.
5. Bring the mixture back to a second boil, stirring constantly.
6. Remove from the heat, and stir in the butter and vanilla extract.
7. Cover, and refrigerate.

PASTRY TECHNIQUES:

Boiling, Tempering

Boiling:

1. Bring the cooking liquid to a rapid boil.
2. Stir the contents.

Tempering:

1. Whisk the eggs vigorously while ladling hot liquid.

HACCP:

Store at 40° F.

HAZARDOUS FOODS:

Milk
Egg yolks
Eggs

Basic Pie Dough

PASTRY TECHNIQUES:

Combining

Combining:

Bringing together two or more components.

1. Prepare the components to be combined.
2. Add one to the other, using the appropriate mixing method (if needed).

HACCP:

Store at 40° F.

YIELD:	1 pound, 8¼ ounces.	3 pounds, ½ ounce.
	3, 8-ounce Crusts.	6, 8-ounce Crusts.

INGREDIENTS:

Flour, pastry	12 ounces	1 pound, 8 ounces
Shortening, all-purpose	8 ounces	1 pound
Salt	¼ ounce	½ ounce
Water, ice-cold	4 ounces	8 ounces
Dried milk solids (DMS) (optional)	0 to 1 ounce	0 to 2 ounces
Sugar, granulated (optional)	0 to 1 ounce	0 to 2 ounces

METHOD OF PREPARATION:

1. Gather the equipment and ingredients.
2. Sift the flour to aerate it; remove lumps and impurities.
3. Break the shortening, by hand, into the flour.
4. Dissolve the salt in the cold water.
5. Incorporate the water into the flour until it is sticky. *Do not overwork* the dough.
6. Allow the dough to rest, and chill properly, preferably overnight.

CHEF NOTE:

The DMS and the sugar can be sifted at the beginning with the pastry flour. The process would be continued in the same manner.

PASTRY TECHNIQUE:

Combining

Combining:

Bringing together two or more components.

1. Prepare the components to be combined.
2. Add one to the other, using the appropriate mixing method (if needed).

Graham Cracker Crust

YIELD:	15¾ ounces. 2, 9-inch Pans.	1 pound, 15½ ounces. 4, 9-inch Pans.

INGREDIENTS:		
Butter (unsalted) or margarine	5 ounces	10 ounces
Flour, cake	1 ounce	2 ounces
Crumbs, graham cracker	8 ounces	1 pound
Sugar, confectionery	1¾ ounces	3½ ounces

METHOD OF PREPARATION:

1. Gather the equipment and ingredients.
2. Melt the butter.
3. Combine the cake flour, graham cracker crumbs, and sugar in a bowl.
4. Add the melted butter into the dry ingredients.
5. Combine well.
6. Press into 9-inch pans.
7. Bake at 350° F for 8 to 10 minutes, until firm.
8. Remove from the oven, and cool.

Prebaked Pie Shells

PASTRY TECHNIQUES:

Rolling, Baking

Rolling:

1. Prepare the rolling surface by dusting with the appropriate medium (flour, cornstarch, etc.).
2. Use the appropriate style pin (stick pin or ball bearing pin) to roll the dough to desired thickness; rotate the dough during rolling to prevent sticking.

Baking:

1. Preheat the oven.
2. Position the item appropriately in the oven.
3. Check for appropriate firmness and/or color.

YIELD: 1, 9-inch Shell.

INGREDIENTS:

Pie dough (see page 807)	8 ounces

METHOD OF PREPARATION:

1. Gather the equipment and ingredients.
2. Weigh out the dough into 8-ounce portions.
3. Using a rolling pin, roll out 8 ounces of pie dough, slightly larger around than a 9-inch pie pan.
4. Using a dough docker, dock the entire surface of the dough.
5. Fold the dough in half, and lift up; place over a pie pan.
6. Unfold the dough to fit the pan.
7. Remove the air between the dough and the pan by shaking slightly.
8. Trim the edges.
9. Place another pie pan on top of the dough; turn upside down.
10. Place four pie pans on a sheet pan.
11. Place the sheet pan in the oven; place a second sheet pan, upside down, on top of the pie pans to keep the pie bottoms flat.
12. Bake at 400° F for 10 minutes, or until the edges begin to brown.
13. Remove the top sheet pan, and continue baking until the dough is golden brown.
14. Remove from the oven.
15. Cool.
16. Turn over onto a single pie pan.

Streusel Topping No. 1

PASTRY TECHNIQUES:

Creaming, Combining

Creaming:

1. Soften the fats on low speed.
2. Add the sugar(s) and cream; increase the speed slowly.
3. Add the eggs one at a time; scrape the bowl frequently.
4. Add the dry ingredients in stages.

Combining:

Bringing together two or more components.

1. Prepare the components to be combined.
2. Add one to the other, using the appropriate mixing method (if needed).

YIELD:	1 pound, 8 ounces.	3 pounds.

INGREDIENTS:

Butter (unsalted) or margarine	6 ounces	12 ounces
Sugar, brown	6 ounces	12 ounces
Cinnamon, ground	To taste	To taste
Salt	To taste	To taste
Flour, bread	12 ounces	1 pound, 8 ounces

METHOD OF PREPARATION:

1. Gather the equipment and ingredients.
2. Place the butter and brown sugar in a bowl; cream together with a paddle until light and well blended.
3. Add the cinnamon and salt; combine well.
4. Add the bread flour, and mix until the mixture becomes crumbly.

CHEF NOTE:

Streusel may be used on a variety of pies, bread pudding, or cobblers. Nuts may be added for extra flavor and texture.

Streusel Topping No. 2

PASTRY TECHNIQUE:

Combining

Combining:

Bringing together two or more components.

1. Prepare the components to be combined.
2. Add one to the other, using the appropriate mixing method (if needed).

YIELD:	4 pounds, 11 ounces.	9 pounds, 6 ounces.
INGREDIENTS:		
Sugar, granulated	10 ounces	1 pound, 4 ounces
Sugar, brown	8 ounces	1 pound
Shortening, all-purpose	1 pound	2 pounds
Flour, cake	2 pound, 8 ounces	5 pounds
Extract, vanilla	1 ounce	2 ounces

METHOD OF PREPARATION:

1. Gather the equipment and ingredients.
2. In a large bowl, combine all of the ingredients by hand.

Apple Pie Filling

Canned Apples

YIELD:	2 pounds, 14 ounces.	5 pounds, 12 ounces.
	2, 9-inch Pies.	4, 9-inch Pies.

INGREDIENTS:		
Apples, canned, sliced	1 pound, 10 ounces	3 pounds, 4 ounces
Water, cold	12 ounces	1 pound, 8 ounces
Sugar, granulated	6 ounces	12 ounces
Salt	Pinch	Pinch
Cinnamon, ground	Pinch	Pinch
Nutmeg, ground	Pinch	Pinch
Cornstarch	1½ ounces	3 ounces

METHOD OF PREPARATION:

1. Gather the equipment and ingredients.
2. Place the sliced apples, half of the amount of water, half of the amount of sugar, the salt, cinnamon, and nutmeg in a pot.
3. Bring to a boil.
4. In a dry bowl, place the cornstarch and the remaining amount of sugar; combine.
5. Add the remaining water to the cornstarch mixture slowly to prevent lumps from occurring.
6. When the apple mixture has boiled, stir in the starch mixture slowly to thicken.
7. Continue to cook and stir for 1 minute.
8. Remove from the heat.
9. Pour onto clean sheet pans, and cool.
10. Cover. Store for later use.

PASTRY TECHNIQUES:

Boiling, Thickening

Boiling:

1. Bring the cooking liquid to a rapid boil.
2. Stir the contents.

Thickening:

1. Mix a small amount of sugar with the starches.
2. Create a slurry.
3. Whisk vigorously until thickened and translucent.

HACCP:

Store at 40° F.

CHEF NOTE:

Place the filling in an unbaked pie shell; bake until the crust is golden brown.

Apple Pie Filling

Fresh Apples

YIELD:	4 pounds, 4 ounces. 2, 9-inch Pies.	8 pounds, 8 ounces. 4, 9-inch Pies.

INGREDIENTS:		
Apples, fresh, peeled and sliced	2 pounds, 8 ounces	5 pounds
Water	1 pound, 4 ounces	2 pounds, 8 ounces
Sugar, granulated	6 ounces	12 ounces
Salt	Pinch	Pinch
Cinnamon, ground	To taste	To taste
Cornstarch	2 ounces	4 ounces

METHOD OF PREPARATION:

1. Gather the equipment and ingredients.
2. Combine the apples with two thirds of the water, all of the sugar, salt, and cinnamon; place in a pot.
3. Bring to a boil.
4. Place the cornstarch in a dry bowl; slowly add the other third of water; whisk constantly to prevent lumps from occurring.
5. Once the apple mixture boils, slowly add the cornstarch mixture, and whisk constantly.
6. Bring the mixture back to a boil; boil for approximately 1 minute.
7. Remove from the heat.
8. Cool the filling; store for later use.

PASTRY TECHNIQUES:

Boiling, Thickening

Boiling:

1. Bring the cooking liquid to a rapid boil.
2. Stir the contents.

Thickening:

1. Mix a small amount of sugar with the starches.
2. Create a slurry.
3. Whisk vigorously until thickened and translucent.

HACCP:

Store at 40° F.

CHEF NOTE:

Place the filling in an unbaked pie shell; bake until the crust is golden brown.

Blueberry Pie Filling

Frozen/Canned

PASTRY TECHNIQUES:

Boiling, Thickening

Boiling:

1. Bring the cooking liquid to a rapid boil.
2. Stir the contents.

Thickening:

1. Mix a small amount of sugar with the starches.
2. Create a slurry.
3. Whisk vigorously until thickened and translucent.

HACCP:

Store at 40° F.

YIELD:	1 pound, 12½ ounces. 1, 9-inch Pie.	3 pounds, 9 ounces. 2, 9-inch Pies.

INGREDIENTS:

Ingredient		
Blueberries, canned or frozen	1 pound	2 pounds
Water or juice from blueberries	7 ounces	14 ounces
Sugar, granulated	4½ ounces	9 ounces
Cornstarch	1⅛ ounces	2¼ ounces
Nutmeg	To taste	To taste
Lemon juice	To taste	To taste

METHOD OF PREPARATION:

1. Gather the equipment and ingredients.
2. If the blueberries are packed in water, drain completely; reserve the water. If the blueberries are frozen, defrost them, and drain well.
3. Place three fourths of the water or juice in a pot; add the granulated sugar.
4. Bring the mixture to a boil.
5. Place the cornstarch in a dry bowl; slowly add the remaining water to prevent lumps from forming. Make sure there are no lumps.
6. Add the nutmeg and lemon juice to the cornstarch mixture.
7. When the sugar-water mixture begins to boil, slowly add the cornstarch mixture; whisk constantly, and return to a second boil.
8. Cook until thickened properly, approximately 1 minute. The mixture will turn translucent.
9. Remove from the heat.
10. Fold in the blueberries gently so they do not become crushed.
11. Cool the filling; store for later use.

CHEF NOTE:

Place the filling in an unbaked pie shell; bake until the crust is golden brown.

Cherry Pie Filling

Canned/Frozen/Fresh

PASTRY TECHNIQUES:
Boiling, Thickening

Boiling:
1. Bring the cooking liquid to a rapid boil.
2. Stir the contents.

Thickening:
1. Mix a small amount of sugar with the starches.
2. Create a slurry.
3. Whisk vigorously until thickened and translucent.

HACCP:
Store at 40° F.

YIELD:	1 pound, 13 ounces. 1, 9-inch Pie.	3 pounds, 10 ounces. 2, 9-inch Pies.

INGREDIENTS:

Cherries, dark, canned, frozen or fresh	1 pound	2 pounds
Water or cherry juice	7½ ounces	15 ounces
Sugar, granulated	4½ ounces	9 ounces
Cornstarch	1 ounce	2 ounces
Lemon juice	To taste	To taste
Salt	To taste	To taste

METHOD OF PREPARATION:

1. Gather the equipment and ingredients.
2. If the cherries are packed in water, drain completely; reserve the water. If the cherries are frozen, keep them frozen.
3. Place three fourths of the water or juice and the granulated sugar in a pot.
4. Bring this mixture to a boil.
5. Place the cornstarch in a dry bowl, and slowly add the remaining juice or water; whisk to prevent lumps from forming.
6. Add the lemon juice and salt to the cornstarch mixture.
7. When the sugar-water mixture begins to boil, slowly add the cornstarch mixture; stir constantly, and return to a second boil.
8. Cook until thickened properly and the mixture is translucent, approximately 1 minute.
9. Remove from the heat.
10. Gently fold in the cherries.
11. Cool the filling; store for later use.

CHEF NOTE:
Place the filling in an unbaked pie shell; bake until the crust is golden brown.

Banana Cream Pie

PASTRY TECHNIQUES:
Boiling, Tempering, Thickening

Boiling:
1. Bring the cooking liquid to a rapid boil.
2. Stir the contents.

Tempering:
1. Whisk the eggs vigorously while ladling hot liquid.

Thickening:
1. Mix a small amount of sugar with the starches.
2. Create a slurry.
3. Whisk vigorously until thickened and translucent.

HACCP:
Store at 40° F.

HAZARDOUS FOODS:
Eggs
Heavy cream

YIELD:	1, 9-inch Pie.	2, 9-inch Pies.
INGREDIENTS:		
Pie shell, prebaked (see page 809)	1, 9-inch shell	2, 9-inch shells
Bananas	3 each	6 each
Filling:		
Cornstarch	1½ ounces	3 ounces
Water	1 pound	2 pounds
Eggs, whole	4 ounces	8 ounces
Dried milk solids (DMS)	2 ounces	4 ounces
Sugar, granulated	4 ounces	8 ounces
Salt	⅛ ounce	¼ ounce
Butter, unsalted	1½ ounces	3 ounces
Extract, vanilla	To taste	To taste
Garnish:		
Cream, heavy, whipped	As needed	As needed

METHOD OF PREPARATION:

1. Gather the equipment and ingredients.
2. Prepare the pie shells, as needed.
3. Slice the bananas, and place evenly in the bottom of a pie shell.
4. Place the cornstarch in a dry bowl, and add a small amount of the water to dissolve it. Make sure there are no lumps.
5. Whisk the eggs with the cornstarch mixture; set aside.
6. Place the DMS in a pot; slowly add the remaining water to prevent the mixture from forming lumps.
7. Add the sugar and salt to the pot.
8. Bring this to a boil.
9. Temper the cornstarch mixture, and add to the boiling milk.
10. Bring the mixture to a second boil, and allow it to thicken; whisk constantly.
11. Remove from the heat.
12. Add the butter and vanilla extract.
13. Pour immediately on top of the sliced bananas in a prebaked pie shell.
14. Let the pie cool completely.
15. Garnish with whipped cream, as desired.

Boston Cream Pie

YIELD:	1, 9-inch Pie.	2, 9-inch Pies.
INGREDIENTS:		
Liquid-shortening sponge cake mix, vanilla (see page 758)	10 ounces	1 pound, 4 ounces
Pastry cream, vanilla (see page 788)	8 ounces	1 pound
Chocolate fudge icing (see page 772)	8 ounces	1 pound
Fondant, warmed	As needed	As needed

METHOD OF PREPARATION:

1. Gather the equipment and ingredients.
2. Scale 10 ounces of vanilla sponge cake batter into a greased pie pan.
3. Bake the batter at 350° F, until golden brown, approximately 25 minutes.
4. Let it cool in the pan; remove completely.
5. After the cake is removed from the pan, turn the cake upside down, and cut in half. (The bottom of the cake becomes the top of the cake.)
6. Place the cake on a circle.
7. Fill the center of the cake with the 8 ounces of pastry cream.
8. Spread the 8 ounces of chocolate fudge icing on top of the cake; make it nice and smooth.
9. Use the warm fondant to garnish the top of the cake.

PASTRY TECHNIQUES:

Filling, Covering, Baking

Filling:

1. Cut open the food product.
2. Carefully spread the filling using an icing spatula, or
3. Carefully pipe the filling using a pastry bag.

Covering:

Sealing a product with a layering of another product such as marzipan or rolled fondant.

1. Prepare the product to be covered with a sticky layer of buttercream, food gel, or other medium that is called for in the formula.
2. Roll out the covering material.
3. Cover the product with the covering material.
4. Smooth the covered product with a bowl scraper to remove wrinkles and air bubbles.

Baking:

1. Preheat the oven.
2. Position the item appropriately in the oven.
3. Check for appropriate firmness and/or color.

HACCP:

Store at 40° F.

PASTRY TECHNIQUES:

Dredging, Filling

Dredging:

1. Coat the food product.
2. Sprinkle or toss the product in an appropriate dredging application.

Filling:

1. Cut open the food product.
2. Carefully spread the filling using an icing spatula, or
3. Carefully pipe the filling using a pastry bag.

Dutch Apple Pie

YIELD:	1, 9-inch Pie.	2, 9-inch Pies.

INGREDIENTS:

Pie shell, unbaked (see page 807)	1, 9-inch shell	2, 9-inch shells
Filling:		
Apples	8 to 9 each	16 to 18 each
Sugar, granulated	As needed	As needed
Cinnamon, ground	As needed	As needed
Flour, cake	As needed	As needed
Raisins	4 ounces	8 ounces
Streusel Topping:		
Butter, unsalted	8 ounces	1 pound
Sugar, granulated	5 ounces	10 ounces
Sugar, brown	4 ounces	8 ounces
Salt	Pinch	Pinch
Cinnamon, ground	To taste	To taste
Flour, pastry	15 ounces	1 pound, 14 ounces

METHOD OF PREPARATION:

1. Gather the equipment and ingredients.
2. Prepare an unbaked pie shell.
3. Peel and core the apples.
4. Cut the apples into medium-size slices.
5. Combine the sugar and cinnamon.
6. Dredge the apples in the cinnamon and sugar mixture.
7. Dredge the apples in the flour.
8. Place the apples and raisins in the unbaked pie shell.
9. Sprinkle the topping over the apples.
10. Bake at 400° F to 425° F for 45 minutes, or until done.

Preparation of Topping:

1. Cream together the butter, sugars, salt, and cinnamon.
2. Add the flour.

PASTRY TECHNIQUE:

Combining

Combining:

Bringing together two or more components.

1. Prepare the components to be combined.
2. Add one to the other, using the appropriate mixing method (if needed).

HACCP:

Store at 40° F.

HAZARDOUS FOODS:

Milk
Eggs

Fruit Cobbler with Topping

YIELD:	3 pounds, 5½ ounces.	6 pounds, 11 ounces.
INGREDIENTS:		
Milk, whole	12 ounces	1 pound, 8 ounces
Extract, vanilla	1 ounce	2 ounces
Eggs, whole	8 ounces	1 pound
Sugar, granulated	1 pound	2 pounds
Flour, bread	1 pound	2 pounds
Baking powder	¼ ounce	½ ounce
Cinnamon, ground	¼ ounce	½ ounce
Short dough (see page 840)	As needed	As needed
Fruit filling, as desired	As needed	As needed

METHOD OF PREPARATION:

1. Gather the equipment and ingredients.
2. Roll out the short dough crust, and place in a well-greased pan.
3. Place the fruit filling on the top of the short crust.
4. Sprinkle with topping.
5. Bake at 350° F for approximately 35 minutes.

Preparation of Cobbler Topping:

1. Place the milk, vanilla extract, eggs, and granulated sugar in a bowl with a paddle.
2. Combine well.
3. Sift together the bread flour, baking powder, and cinnamon.
4. Add the dry ingredients in the sugar mixture.
5. Combine well. *Do not overmix.*

Fruit Pie

YIELD:	1, 9-inch Pie.	2, 9-inch Pies.
INGREDIENTS:		
Pie dough (see page 807)	1 pound	2 pounds
Pie filling	1 pound, 12 ounces	3 pounds, 8 ounces
Egg wash (see page 964)	As needed	As needed

METHOD OF PREPARATION:

1. Gather the equipment and ingredients.
2. Divide the dough into two (8-ounce) portions.
3. Roll out one (8-ounce) portion, and place it in a pie pan; be sure that all of the air between the pan and the dough is released.
4. Trim the edges. (Do not dock.)
5. Fill the shell with pie filling.
6. Wet the edges of the pie with egg wash; use a pastry brush.
7. Roll out the remaining 8-ounce portion of pie dough, and place it over the top of the pie; seal the bottom and top layers of dough completely.
8. Trim the edges.
9. Using a pastry brush, brush the top of the pie completely with the egg wash. Do not brush too heavily. (Granulated sugar may be sprinkled on top, if desired.)
10. Cut a hole in the center of the top.
11. Bake the pie at 400° F 45 minutes, or until the crust is golden brown and the filling is bubbling.
12. Cool the pie before serving.

Variations:

1. Cherry Pie Filling
   a. Canned
   b. Frozen
   c. Fresh
2. Blueberry Pie Filling
   a. Canned
   b. Frozen
3. Apple Pie Filling
   a. Canned
   b. Fresh

PASTRY TECHNIQUES:

Rolling, Filling

Rolling:

1. Prepare the rolling surface by dusting with the appropriate medium (flour, cornstarch, etc.).
2. Use the appropriate style pin (stick pin or ball bearing pin) to roll the dough to desired thickness; rotate the dough during rolling to prevent sticking.

Filling:

1. Cut open the food product.
2. Carefully spread the filling using an icing spatula, or
3. Carefully pipe the filling using a pastry bag.

CHEF NOTE:

Before removing pie from oven, make sure that the shell is baked. Top color is not an indication of doneness.

Lemon Meringue Pie

PASTRY TECHNIQUES:

Baking, Boiling, Tempering, Thickening, Whipping

Baking:

1. Preheat the oven.
2. Position the item appropriately in the oven.
3. Check for appropriate firmness and/or color.

Boiling:

1. Bring the cooking liquid to a rapid boil.
2. Stir the contents.

Tempering:

1. Whisk the eggs vigorously while ladling hot liquid.

Thickening:

1. Mix a small amount of sugar with the starches.
2. Create a slurry.
3. Whisk vigorously until thickened and translucent.

Whipping:

1. Hold the whip at a 55-degree angle.
2. Create circles, using a circular motion.
3. The circular motion needs to be perpendicular to the bowl.

HACCP:

Store at 40° F.

HAZARDOUS FOODS:

Eggs
Egg whites

YIELD:	1, 9-inch Pie.	2, 9-inch Pies.
INGREDIENTS:		
Pie shell, prebaked (see page 809)	1, 9-inch shell	2, 9-inch shells
Filling:		
Cornstarch	1¾ ounces	3½ ounces
Water	1 pound	2 pounds
Eggs, whole	4 ounces	8 ounces
Sugar, granulated	6 ounces	12 ounces
Salt	To taste	To taste
Lemon rind	1½ lemons	3 lemons
Lemon juice, fresh	2½ lemons	5 lemons
Butter, unsalted	1 ounce	2 ounces
Meringue:		
Egg whites	2 ½ ounces	5 ounces
Sugar, granulated	3 ounces	6 ounces

METHOD OF PREPARATION:

1. Gather the equipment and ingredients.
2. Prepare the prebaked pie shells, as needed.
3. Place the cornstarch in a dry bowl; slowly add some water into the cornstarch, a little at a time, and whisk until all is dissolved. Make sure no lumps are present.
4. Add the eggs to the cornstarch mixture.
5. Combine the remaining water, sugar, salt, and lemon rind into a pot, and bring to a boil.
6. When boiling, temper the cornstarch mixture; add to the hot mixture.
7. Bring back to a second boil; allow the mixture to thicken.
8. Remove from the heat.
9. Add the lemon juice and butter.
10. Pour the filling into a prebaked pie shell.
11. Allow to cool.
12. Place the egg whites in a bowl, whip, and slowly add the granulated sugar; whip to a soft peak.
13. Garnish the top of the pie with meringue.
14. Bake the pie until the meringue turns brown, or use a propane torch to brown the meringue.

Pecan Pie

PASTRY TECHNIQUE:
Combining

Combining:
Bringing together two or more components.
1. Prepare the components to be combined.
2. Add one to the other, using the appropriate mixing method (if needed).

HACCP:
Store at 40° F.

HAZARDOUS FOOD:
Eggs

YIELD:	1, 9-inch Pie.	2, 9-inch Pies.

INGREDIENTS:

Pie shell, unbaked (see page 807)	1, 9-inch shell	2, 9-inch shells
Filling:		
Eggs, whole	6 ounces	12 ounces
Butter, unsalted	2 ounces	4 ounces
Corn syrup, light	12 ounces	1 pound, 8 ounces
Sugar, granulated	3½ ounces	7 ounces
Bourbon	To taste	To taste
Extract, vanilla	To taste	To taste
Pecan, pieces	5½ ounces	11 ounces

METHOD OF PREPARATION:

1. Gather the equipment and ingredients.
2. Prepare the pie shells, as needed.
3. Place the eggs in bowl, and slightly beat.
4. Melt the butter.
5. Add the melted butter, corn syrup, sugar, bourbon, and vanilla extract into the eggs; mix well using a whisk.
6. Place the pecans into an unbaked pie shell.
7. Pour the liquid mixture on top of the pecans.
8. Bake at 425° F for 15 minutes; reduce the oven temperature to 350° F. Bake for an additional 20 to 30 minutes, or until set but still slightly loose in the center.

Pumpkin Pie

PASTRY TECHNIQUES:
Combining, Filling

Combining:
Bringing together two or more components.
1. Prepare the components to be combined.
2. Add one to the other, using the appropriate mixing method (if needed).

Filling:
1. Cut open the food product.
2. Carefully spread the filling using an icing spatula, or
3. Carefully pipe the filling using a pastry bag.

HACCP:
Store at 40° F.

HAZARDOUS FOOD:
Eggs

YIELD:	1, 9-inch Pie.	2, 9-inch Pies.

INGREDIENTS:

Pie shell, unbaked (see page 807)	1, 9-inch shell	2, 9-inch shells
Filling:		
Milk, evaporated	8 ounces	1 pound
Sugar, brown, packed	5 ounces	10 ounces
Corn syrup	2 ounces	4 ounces
Eggs, whole	2 each	4 each
Pumpkin pie filling, solid pack	12 ounces	1 pound, 8 ounces
Salt	To taste	To taste
Extract, vanilla	To taste	To taste
Cinnamon, ground	To taste	To taste
Cloves, ground	To taste	To taste
Mace, ground	To taste	To taste
Ginger, ground	To taste	To taste
Nutmeg, ground	To taste	To taste
Allspice, ground	To taste	To taste

METHOD OF PREPARATION:

1. Gather the equipment and ingredients.
2. Prepare a 9-inch unbaked pie shell.
3. Combine the evaporated milk, sugar, corn syrup, and eggs.
4. Add the pumpkin pie filling.
5. Combine.
6. Add the flavorings: salt, vanilla extract, cinnamon, clove, mace, ginger, nutmeg, and allspice.
7. Place in an unbaked pie shell.
8. Bake at 350° F or bake at 425° F until the crust begins to brown; reduce the heat to 350° F, until slightly firm, approximately 40 minutes.

Sour Cream Apple Pie

PASTRY TECHNIQUES:
Mixing, Filling

Mixing:
1. Follow the proper mixing procedure: creaming, blending, whipping, or combination.

Filling:
1. Cut open the food product.
2. Carefully spread the filling using an icing spatula, or
3. Carefully pipe the filling using a pastry bag.

HACCP:
Store at 40° F.

HAZARDOUS FOODS:
Sour cream
Eggs

YIELD:	1, 9-inch Pie.	2, 9-inch Pies.

INGREDIENTS:

Pie shell, unbaked (see page 807)	1, 9-inch shell	2, 9-inch shells

Walnut Streusel:

Sugar, granulated	1 ounce	2 ounces
Cinnamon, ground	To taste	To taste
Salt	Pinch	Pinch
Butter, unsalted	1½ ounces	3 ounces
Sugar, brown	1 ounce	2 ounces
Flour, bread	1½ ounces	3 ounces
Walnuts, chopped or pieces	1½ ounces	3 ounces

Filling:

Apples, Granny Smith	1 pound, 8 ounces	3 pounds
Sour cream	5 ½ ounces	11 ounces
Sugar, granulated	2¾ ounces	5½ ounces
Flour, bread	¾ ounce	1½ ounces
Eggs, whole	1 ounce	2 ounces
Extract, vanilla	To taste	To taste
Salt	Pinch	Pinch

METHOD OF PREPARATION:

1. Gather the equipment and ingredients.
2. Prepare an unbaked pie shell with fluted edges.

Preparation of Walnut Streusel:

1. Place the granulated sugar, cinnamon, salt, butter, and brown sugar in bowl.
2. Combine well.
3. Add the bread flour and walnuts.
4. Combine well, and set aside.

Preparation of Filling:

1. Peel and core the apples.
2. Slice the apples into ⅛-inch-thick slices.
3. Combine the sour cream, granulated sugar, bread flour, eggs, vanilla extract, and salt.
4. Combine the apples with the sour cream mixture.

Baking Instructions:

1. Place the apple mixture into an unbaked 9-inch pie shell.
2. Bake at 450° F for 10 minutes.
3. Reduce the oven temperature to 375° F, and continue to bake until the apples are cooked. (The apples will be done when soft to the touch with a fork.)
4. Remove the pie from the oven when it is done.
5. Cover the pie with the streusel topping. (Do not cover the edges.)
6. Return the pie to the oven, and bake at 375° F for an additional 10 minutes.

Strawberry Chiffon Pie

PASTRY TECHNIQUES:

Boiling, Thickening, Blooming, Folding, Whipping

Boiling:

1. Bring the cooking liquid to a rapid boil.
2. Stir the contents.

Thickening:

1. Mix a small amount of sugar with the starches.
2. Create a slurry.
3. Whisk vigorously until thickened and translucent.

Blooming:

Gelatin sheets or leaves:

1. Fan the sheets out.
2. Cover the sheets in liquid.
3. Sheets are bloomed when softened.

Granular gelatin:

1. Sprinkle the gelatin.
2. Gelatin is ready when it is cream of wheat consistency.

Folding:

Do steps 1, 2, and 3 in one continuous motion.

1. Run a bowl scraper under the mixture, across the bottom of the bowl.
2. Turn the bowl counterclockwise.
3. Bring the bottom mixture to the top.

Whipping:

1. Hold the whip at a 55-degree angle.
2. Create circles, using a circular motion.
3. The circular motion needs to be perpendicular to the bowl.

HACCP:

Store at 40° F.

HAZARDOUS FOOD:

Egg whites

YIELD:	1, 9-inch Pie.	2, 9-inch Pies.
INGREDIENTS:		
Pie shell, prebaked (see page 809)	1, 9-inch shell	2, 9-inch shells
Filling:		
Strawberries, frozen and drained *or*	10½ ounces	1 pound, 5 ounces
fresh, halved	8 ounces	1 pound
Sugar, granulated	2½ ounces	5 ounces
Salt	Pinch	Pinch
Cornstarch	⅛ ounce	¼ ounce
Water	½ ounce	1 ounce
Gelatin	⅛ ounce	¼ ounce
Water	¾ ounce	1½ ounces
Lemon juice	⅛ ounce	¼ ounce
Chiffon:		
Egg whites	2½ ounces	5 ounces
Sugar, granulated	2 ounces	4 ounces

METHOD OF PREPARATION:

1. Gather the equipment and ingredients.
2. Prepare a prebaked pie shell, and cool properly.

Filling:

1. Place two thirds of the strawberries, the granulated sugar, and the salt in a pot; bring to a boil.
2. In a dry bowl, place the cornstarch, and slowly add a small amount of water to dissolve it; set aside.
3. In another dry bowl, place the gelatin, and bloom with water; set aside.
4. When the strawberry mixture comes to a boil, vigorously whisk the cornstarch mixture into the boiling strawberries.
5. Bring to a second boil, and let thicken.

6. Remove from the heat.
7. Add the lemon juice.
8. Add the bloomed gelatin to the strawberry mixture; combine well.
9. Cool slightly.
10. Fold in the remaining strawberries.
11. Set aside until cool but not cold.

Chiffon:

1. Place the egg whites in a bowl, and whip to full volume; slowly add the granulated sugar to the egg whites.
2. Fold the meringue into the strawberry filling.
3. Place in a prebaked pie shell.
4. Chill well prior to serving.

Basic Custard

PASTRY TECHNIQUES:

Combining, Slow Baking

Combining:

Bringing together two or more components.

1. Prepare the components to be combined.
2. Add one to the other, using the appropriate mixing method (if needed).

Slow Baking:

1. Use an appropriate baking dish.
2. Use hot water in the pan.
3. Replenish the water when needed.

HACCP:

Store at 40° F.

HAZARDOUS FOODS:

Eggs
Milk

YIELD:	1 pound, 6½ ounces.	2 pounds, 13 ounces.

INGREDIENTS:

Eggs, whole	6 ounces	12 ounces
Sugar, granulated	3¼ ounces	6½ ounces
Milk, whole	13 ounces	1 pound, 10 ounces
Extract, vanilla	¼ ounce	½ ounce

METHOD OF PREPARATION:

1. Gather the equipment and ingredients.
2. Combine all of the ingredients, and strain through a chinois.
3. Pour the custard mixture into custard cups.
4. Place the cups in hotel pans containing hot water.
5. Bake at 325° F, until firm.
6. Cool thoroughly.

Variations:

1. Crème Caramel
2. Crème Brûlée
3. Bread Pudding

CHEF NOTE:

Do not overbake. Overbaking creates steam in the mixture which will cause air bubbles in the custard.

Cheesecake, French Style

YIELD:	2 pounds, 2 ounces. 1, 9-inch Cake	4 pounds, 4 ounces. 2, 9-inch Cakes

INGREDIENTS:		
Cream cheese, room temperature	1 pound	2 pounds
Sugar, granulated	5 ounces	10 ounces
Butter, unsalted, softened	2½ ounces	5 ounces
Flour, bread	1¼ ounces	2½ ounces
Salt	Pinch	Pinch
Extract, vanilla	To taste	To taste
Lemon juice	To taste	To taste
Extract, lemon (optional)	To taste	To taste
Egg yolks	1⅛ ounces	2¼ ounces
Eggs, whole	1⅜ ounces	2¾ ounces
Cream, heavy	6 ounces	12 ounces
Egg whites	1⅜ ounces	2¾ ounces
Butter, additional, unsalted	As needed	As needed
Cake layers, vanilla	1, 9-inch layer, ¼-inch thick	2, 9-inch layers, ¼-inch thick

METHOD OF PREPARATION:

1. Gather the equipment and ingredients.
2. Place the cream cheese, half the amount of granulated sugar, butter, flour, salt, vanilla extract, lemon juice, and lemon extract in a bowl with a paddle.
3. Paddle until smooth; scrape the bowl often.
4. Add the egg yolks and whole eggs; combine, and scrape well.
5. Add the heavy cream; combine, and scrape well.
6. In a separate bowl, whip the egg whites to a medium peak; slowly add the remaining amount of granulated sugar.
7. Add a small amount of the cheese mixture to the egg white mixture.
8. Gently fold the remaining cheese and egg mixtures together.
9. Prepare the 9-inch cake pans by greasing them with butter and lining the bottoms with a vanilla cake layer.

PASTRY TECHNIQUES:

Creaming, Combining, Whipping, Slow Baking

Creaming:

1. Soften the fats on low speed.
2. Add the sugar(s) and cream; increase the speed slowly.
3. Add the eggs one at a time; scrape the bowl frequently.
4. Add the dry ingredients in stages.

Combining:

Bringing together two or more components.

1. Prepare the components to be combined.
2. Add one to the other, using the appropriate mixing method (if needed).

Whipping:

1. Hold the whip at a 55-degree angle.
2. Create circles, using a circular motion.
3. The circular motion needs to be perpendicular to the bowl.

Slow Baking:

1. Use an appropriate baking dish.
2. Use hot water in the pan.
3. Replenish the water when needed.

HACCP:

Store at 40° F.

HAZARDOUS FOODS:

Cream cheese
Egg yolks
Eggs
Heavy cream
Egg whites

CHEF NOTE:

French-style cheesecake can be baked without lining the pans with vanilla layer cake or by lining the pans with cake crumbs.

10. Fill the prepared pans with 2 pounds, 2 ounces of filling.
11. Place the pan in a water bath.
12. Bake at 325° F for 45 minutes, or until firm.
13. Allow to cool properly and refrigerate overnight.

Crème Brûlée

YIELD:	2 pounds, 9 ounces.	5 pounds, 2 ounces.
INGREDIENTS:		
Egg yolks	6 ounces	12 ounces
Sugar, granulated	3 ounces	6 ounces
Extract, vanilla	To taste	To taste
Cream, heavy	2 pounds	4 pounds
Sugar, additional, granulated	As needed	As needed

METHOD OF PREPARATION:

1. Gather the equipment and ingredients.
2. Whisk the egg yolks and sugar together.
3. Add the vanilla extract.
4. Add the heavy cream.
5. Strain the mixture; remove any foam on top.
6. Fill the custard molds.
7. Place the molds in a hot-water bath.
8. Bake at 325° F, until firm.
9. Chill.
10. When the custard is chilled thoroughly, sprinkle some granulated sugar on top of it, and caramelize.

PASTRY TECHNIQUES:

Combining, Slow Baking

Combining:

Bringing together two or more components.

1. Prepare the components to be combined.
2. Add one to the other, using the appropriate mixing method (if needed).

Slow Baking:

1. Use an appropriate baking dish.
2. Use hot water in the pan.
3. Replenish the water when needed.

HACCP:

Store at 40° F.

HAZARDOUS FOODS:

Egg yolks
Heavy cream

CHEF NOTE:

Other flavorings may be added to the mixture, such as praline paste.

Deli Cheesecake

(New York Style)

PASTRY TECHNIQUES:

Creaming, Slow Baking

Creaming:

1. Soften the fats on low speed.
2. Add the sugar(s) and cream; increase the speed slowly.
3. Add the eggs one at a time; scrape the bowl frequently.
4. Add the dry ingredients in stages.

Slow Baking:

1. Use an appropriate baking dish.
2. Use hot water in the pan.
3. Replenish the water when needed.

HACCP:

Store at 40° F.

HAZARDOUS FOODS:

Cream cheese
Heavy cream
Eggs
Sour cream
Half-and-half

YIELD:	2 pounds, 15 ounces. 1, 9-inch Cake	5 pounds, 14 ounces. 2, 9-inch Cakes

INGREDIENTS:

Crust:		
Graham cracker crust (see page 808)	As needed	As needed
Filling:		
Cream cheese, room temperature	1 pound, 8 ounces	3 pounds
Lemon rind, grated	2 each	4 each
Sugar, granulated	8 ounces	1 pound
Cream, heavy	4 ounces	8 ounces
Eggs, whole	8 ounces	1 pound
Sour cream	1 ounce	2 ounces
Half-and-half	2 ounces	4 ounces
Extract, vanilla	To taste	To taste

METHOD OF PREPARATION:

1. Gather the equipment and ingredients.
2. Prepare the cake pan with the graham cracker crust.
3. Place the cream cheese in a bowl with the lemon rind.
4. Paddle until smooth.
5. Add the sugar, and paddle until smooth; scrape often.
6. Add the heavy cream gradually; scrape often.
7. Add the eggs in small amounts, incorporating well after each addition.
8. Scrape well.
9. Add the sour cream, half-and-half, and vanilla extract; scrape well; combine.
10. Pour 2 pounds, 15 ounces of mixture into a prepared cake pan.
11. Place in a water bath.
12. Bake at 375° F for approximately 1½ hours, or until the center is firm.
13. Remove, and cool.

Common Meringue

PASTRY TECHNIQUE:

Whipping

Whipping:

1. Hold the whip at a 55-degree angle.
2. Create circles, using a circular motion.
3. The circular motion needs to be perpendicular to the bowl.

HAZARDOUS FOOD:

Egg whites

YIELD:	1 pound, 8 ounces.	3 pounds.
INGREDIENTS:		
Egg whites	8 ounces	1 pound
Sugar, granulated	1 pound	2 pounds

METHOD OF PREPARATION:

1. Gather the equipment and ingredients.
2. Place the egg whites in bowl, and begin to whip.
3. Slowly add the granulated sugar while whipping the whites.
4. When all of the sugar has been added, whip to the desired consistency.
5. Use immediately.

Italian Meringue

YIELD:	1 pound, 12 ounces.	3 pounds, 8 ounces.

PASTRY TECHNIQUES:

Boiling, Whipping

Boiling:

1. Bring the cooking liquid to a rapid boil.
2. Stir the contents.

Whipping

1. Hold the whip at a 55-degree angle.
2. Create circles, using a circular motion.
3. The circular motion needs to be perpendicular to the bowl.

HAZARDOUS FOOD:

Egg whites

INGREDIENTS:

Sugar, granulated	1 pound	2 pounds
Water	4 ounces	8 ounces
Egg whites	8 ounces	1 pound

METHOD OF PREPARATION:

1. Gather the equipment and ingredients.
2. Place the sugar and water in a pot.
3. Cook to 250° F.
4. Place the egg whites in a bowl, and whip.
5. When the sugar has reached the correct temperature, slowly add the hot sugar in a slow, steady stream into the egg whites.
6. Whip to the desired consistency.

Swiss Meringue

PASTRY TECHNIQUES:

Heating, Whipping

Heating:

1. Prepare the food product according to the formula's instructions.
2. Choose the appropriate method of heating (on the range or stove top, in the oven, etc.)
3. Apply the product to the heat.

Whipping:

1. Hold the whip at a 55-degree angle.
2. Create circles, using a circular motion.
3. The circular motion needs to be perpendicular to the bowl.

HAZARDOUS FOOD:

Egg whites

YIELD:	1 pound, 8 ounces.	3 pounds.

INGREDIENTS:

Egg whites	8 ounces	1 pound
Sugar, granulated	1 pound	2 pounds

METHOD OF PREPARATION:

1. Gather the equipment and ingredients.
2. Place the egg whites and sugar in a bowl.
3. Place over a double boiler; whip constantly until the mixture is warmed to 120° F.
4. Remove from the double boiler, and place on a mixer.
5. Whip at high speed until stiff peaks form.
6. Ready for use.

Beignet Batter

YIELD:	2 pounds, 11 ounces.	5 pounds, 6 ounces.
INGREDIENTS:		
Flour, pastry	13 ounces	1 pound, 10 ounces
Sugar, granulated	2 ounces	4 ounces
Milk, whole	1 pound	2 pounds
Butter, unsalted, chopped	4 ounces	8 ounces
Eggs, whole	8 ounces	1 pound
Oil, vegetable (for frying)	As needed	As needed
Cinnamon and sugar mixture	As needed	As needed

METHOD OF PREPARATION:

1. Gather the equipment and ingredients.
2. Combine the flour and sugar.
3. Place the milk and butter in a pot; bring to a boil.
4. Add the flour mixture all at once to the boiling milk; stir until well mixed and a skin begins to form on the bottom of the pan.
5. Remove from the heat, and place in a bowl with a paddle; mix on low speed until cool.
6. Add in the eggs gradually, and incorporate well.
7. Pipe the mixture into beignets or cruellers; use a pastry bag or scoop with a metal spoon, and place them into the hot oil.
8. Fry in 360° F hot oil, on each side, until a nice golden brown.
9. Remove, and drain on paper towels.
10. Dredge in cinnamon and sugar.
11. Serve.

PASTRY TECHNIQUES:

Combining, Boiling, Piping, Freezing

Combining:

Bringing together two or more components.

1. Prepare the components to be combined.
2. Add one to the other, using the appropriate mixing method (if needed).

Boiling:

1. Bring the cooking liquid to a rapid boil.
2. Stir the contents.

Piping:

With bag:

1. Use a bag with a disposable tip; cut the bag at 45-degree angle.
2. Fill to no more than half full.
3. Burp the bag.

With cone:

1. Cut and fold the piping cone to the appropriate size.
2. Fill the cone with a small amount.
3. Fold the ends to form a triangle.
4. Pipe the desired designs.

Freezing:

1. Prepare the product.
2. Place the product in the freezing cabinet for the appropriate length of time.

HAZARDOUS FOODS:

Milk
Eggs

Pâté à Choux No. 1

Solid Fat-Based

PASTRY TECHNIQUES:

Boiling, Combining, Piping

Boiling:

1. Bring the cooking liquid to a rapid boil.
2. Stir the contents.

Combining:

Bringing together two or more components.

1. Prepare the components to be combined.
2. Add one to the other, using the appropriate mixing method (if needed).

Piping:

With bag:

1. Use a bag with a disposable tip; cut the bag at 45-degree angle.
2. Fill to no more than half full.
3. Burp the bag.

With cone:

1. Cut and fold the piping cone to the appropriate size.
2. Fill the cone with a small amount.
3. Fold the ends to form a triangle.
4. Pipe the desired designs.

HAZARDOUS FOODS:

Milk
Eggs

YIELD:	3 pounds, 3 ounces.	6 pounds, 6 ounces.
INGREDIENTS:		
Butter (unsalted) or shortening (all-purpose)	8 ounces	1 pound
Salt	¼ ounce	½ ounce
Sugar, granulated (optional)	¼ ounce	½ ounce
Water or milk (whole)	1 pound	2 pounds
Flour, bread, sifted	10½ ounces	1 pound, 5 ounces
Eggs, whole	1 pound	2 pounds

METHOD OF PREPARATION:

1. Gather the equipment and ingredients.
2. Place the butter, salt, granulated sugar, and water or milk in a pot.
3. Bring to a boil.
4. Add all of the sifted flour at once.
5. Stir with a wooden spoon for approximately 5 minutes or until the mixture forms a ball that does not stick to the inside of the pot.
6. Cook at this point for an additional 3 minutes.
7. Remove from the heat, and place the mixture in a mixing bowl.
8. Mix on low speed until cooled slightly.
9. Add the eggs gradually; mix at low speed; make sure the eggs are fully incorporated before the next addition.
10. When the eggs are fully incorporated, pipe into the desired shapes on parchment-lined sheet pans.
11. Bake at 400° F to 425° F until brown and dry on the inside.

Variations:

1. Éclairs
2. Swans
3. Profiteroles
4. Paris-Brest
5. Gâteau St. Honore
6. Croquembouche
7. Beignets Soufflé

Pâté à Choux No. 2

Oil-Based

PASTRY TECHNIQUES:

Boiling, Combining, Piping

Boiling:

1. Bring the cooking liquid to a rapid boil.
2. Stir the contents.

Combining:

Bringing together two or more components.

1. Prepare the components to be combined.
2. Add one to the other, using the appropriate mixing method (if needed).

Piping:

With bag:

1. Use a bag with a disposable tip; cut the bag at 45-degree angle.
2. Fill to no more than half full.
3. Burp the bag.

With cone:

1. Cut and fold the piping cone to the appropriate size.
2. Fill the cone with a small amount.
3. Fold the ends to form a triangle.
4. Pipe the desired designs.

HAZARDOUS FOODS:

Eggs
Egg whites

YIELD:	2 pounds, 9½ ounces.	5 pounds, 3¼ ounces.
INGREDIENTS:		
Water	12 ounces	1 pound, 8 ounces
Salt	To taste	To taste
Oil, vegetable	6 ounces	12 ounces
Flour, bread	8½ ounces	1 pound, 1¼ ounces
Eggs, whole	12 ounces	1 pound, 8 ounces
Egg whites	3 ounces	6 ounces

METHOD OF PREPARATION:

1. Gather the equipment and ingredients.
2. Place the water, salt, and oil in a pot.
3. Bring to a boil.
4. Add all of the flour at once.
5. Stir with a wooden spoon for approximately 5 minutes or until the mixture forms a ball that does not stick to the inside of the pot.
6. Remove from the heat, and place in a mixer; cool slightly.
7. Add the whole eggs and egg whites gradually; make sure the eggs are fully incorporated before the next addition.
8. Pipe into the desired shapes on parchment-lined sheet pans.
9. Bake at 400° F to 425° F until golden brown and dry in appearance.

Variations:

1. Éclairs
2. Swans
3. Profiteroles
4. Paris-Brest
5. Gâteau St. Honore
6. Croquembouche
7. Beignets Soufflé

Linzer Dough

PASTRY TECHNIQUES:
Creaming, Combining

Creaming:
1. Soften the fats on low speed.
2. Add the sugar(s) and cream; increase the speed slowly.
3. Add the eggs one at a time; scrape the bowl frequently.
4. Add the dry ingredients in stages.

Combining:
Bringing together two or more components.
1. Prepare the components to be combined.
2. Add one to the other, using the appropriate mixing method (if needed).

HACCP:
Store at 40° F.

HAZARDOUS FOODS:
Eggs
Milk

YIELD:	6 pounds, 10 ounces.	13 pounds, 4 ounces.
INGREDIENTS:		
Hazelnuts, toasted	10½ ounces	1 pound, 5 ounces
Butter, unsalted, chopped	1 pound, 1 ounce	2 pounds, 2 ounces
Sugar, granulated	1 pound, 1 ounce	2 pounds, 2 ounces
Eggs, whole	5 ounces	10 ounces
Flour, pastry	2 pounds	4 pounds
Flour, bread	1 ounce	2 ounces
Cinnamon, ground	½ ounce	1 ounce
Cloves, ground	¼ ounce	½ ounce
Lemon or orange zest	¼ ounce	½ ounce
Crumbs, cake or bread	8 ounces	1 pound
Glucose	14½ ounces	9 ounces
Milk, whole	As needed	As needed

METHOD OF PREPARATION:

1. Gather the equipment and ingredients.
2. Ground the hazelnuts fine in robot coupe or food processor.
3. Cream the butter and the sugar together in a bowl until smooth and creamy; use a paddle.
4. Add the eggs, a little at a time, scraping the bowl down after each addition.
5. Sift the flours, cinnamon, and cloves together, and add to the creamed mixture in two stages.
6. Add the zest and cake crumbs; combine.
7. Add the ground hazelnuts.
8. Mix until incorporated.
9. Add the glucose and milk (if the mixture seems to be too dry). Mix just enough to incorporate; *do not overmix.*
10. Allow the dough to rest for 20 minutes before using.

Short Dough

1-2-3 Cookie Dough

PASTRY TECHNIQUES:
Creaming, Combining

Creaming:
1. Soften the fats on low speed.
2. Add the sugar(s) and cream; increase the speed slowly.
3. Add the eggs one at a time; scrape the bowl frequently.
4. Add the dry ingredients in stages.

Combining:
Bringing together two or more components.
1. Prepare the components to be combined.
2. Add one to the other, using the appropriate mixing method (if needed).

HACCP:
Store at 40° F.

HAZARDOUS FOOD:
Eggs

YIELD:	6 pounds, 6 ounces.	12 pounds, 12 ounces.
INGREDIENTS:		
Sugar, granulated	1 pound	2 pounds
Butter, unsalted, room temperature	2 pounds	4 pounds
Eggs, whole	6 ounces	12 ounces
Flour, cake	3 pounds	6 pounds

METHOD OF PREPARATION:

1. Gather the equipment and ingredients.
2. Place the sugar and butter in a bowl; combine, using a paddle.
3. Add the eggs, and blend well.
4. Add the flour, and mix on the lowest speed until well blended.
5. Chill to firm the dough before use.

Sour Cream Pastry Dough

PASTRY TECHNIQUE:

Blending

Blending:

1. Combine the dry ingredients on low speed.
2. Add the softened fat(s) and liquid(s).
3. Mix the ingredients on low speed.
4. Increase the speed gradually.

HACCP:

Store at 40° F.

HAZARDOUS FOOD:

Sour cream

YIELD:	4 pounds.	8 ounces.

INGREDIENTS:

Flour, pastry	1 pound, 8 ounces	3 pounds
Butter (unsalted) or margarine, room temperature	1 pound, 8 ounces	3 pounds
Sour cream	1 pound	2 pounds

METHOD OF PREPARATION:

1. Gather the equipment and ingredients.
2. Place all of the ingredients in a bowl, and mix together on low speed. *Do not overmix.*
3. Remove the dough from the mixer, and place on a lightly floured sheet pan.
4. Prior to use, refrigerate the dough to chill completely.

CHEF NOTE:

Sour cream pastry dough may be used for all types of desserts, including individual pieces, tart shells, and pie shells.

Steps of Laminated Yeast Dough Preparation

1. Mix according to the straight dough mixing method; mix by hand or machine. Develop only until incorporated.
2. Rest the dough, covered, for 1 to 1½ hours.
3. While the dough rests, soften the roll-in fat until smooth.
4. After the dough has rested, roll out the dough ½-inch to ¾-inch thick to form a rectangle shape.
5. Spot the roll-in fat over two thirds of the length of the dough.
6. Fold the third of dough without fat over the center third.
7. Fold the remaining third of dough on top.
8. Rest the dough for 20 to 30 minutes in the refrigerator.
9. Place the dough on a bench, and turn 90 degrees; roll out the dough again as previously done. Repeat the folding-in procedure again until *three* 3-folds have been achieved. Be sure to brush off any excess dusting flour from between the folds.
10. After the third fold, rest the dough for several hours, or overnight. Cover with plastic wrap to prevent a crust from forming.
11. Make up and bake, as demonstrated by the instructor.

American Danish Dough

PASTRY TECHNIQUE:

Steps of Laminated Yeast Dough Preparation (see page 842)

HAZARDOUS FOOD:

Eggs

YIELD: 11 pounds, 4½ ounces. Baker's Percentage.

INGREDIENTS:

Dough (unlaminated):	9 pounds, 15½ ounces	
Flour, patent	5 pounds	100
Dry milk solids (DMS)	5 ounces	6
Salt	1½ ounces	1.75
Cardamom, ground	½ ounce	0.60
Water	1 pound, 4 ounces	25
Yeast, compressed	5 ounces	6
Extract, vanilla	1½ ounces	2
Sugar, granulated	1 pound	20
Shortening, all-purpose	10 ounces	14
Eggs, whole	1 pound, 4 ounces	25
Roll-in Fat:		13
Margarine	1 pound, 5 ounces	100

METHOD OF PREPARATION:

1. Gather the equipment and ingredients.
2. Follow the steps of laminated yeast dough preparation (see page 842).

PASTRY TECHNIQUE:

Steps of Laminated Yeast Dough Preparation (see page 842)

Croissant Dough No. 1

YIELD:	11 pounds, 13 ounces.	Baker's Percentage.

INGREDIENTS:

Dough:

Ingredient	Amount	Baker's Percentage
Water, cold	3 pounds	60
Yeast, compressed	3 ounces	4
Flour, bread	4 pounds	80
Flour, pastry	1 pound	20
Salt	2 ounces	2
Sugar, granulated	8 ounces	10

Roll-in Fat:

Ingredient	Amount	Baker's Percentage
Butter, unsalted	3 pounds	60

METHOD OF PREPARATION:

1. Gather the equipment and ingredients.
2. Follow the steps of laminated yeast dough preparation (page 842).

Variations:

1. Mini-croissant
2. Regular croissant

PASTRY TECHNIQUE:

Steps of Laminated Yeast Dough Preparation (see page 842)

Croissant Dough No. 2

YIELD:	12 pounds, ½ ounce.	Baker's Percentage.

INGREDIENTS:

Dough (unlaminated):	9 pounds, ½ ounce.	
Flour, bread	5 pounds	100
Dry milk solids (DMS)	5 ounces	6
Salt	1 ounce	1
Sugar, granulated	3 ounces	4
Water, cold	2 pounds, 13½ ounces	57
Yeast, compressed	5 ounces	6
Shortening, all-purpose	5 ounces	6
Roll-in Fat:		33
Butter, unsalted (60° F)	2 pounds, 13½ ounces	95
Flour, bread, sifted	2½ ounces	5

METHOD OF PREPARATION:

1. Gather the equipment and ingredients.
2. Follow the steps of laminated yeast dough preparation (page 842).

Variations:

1. Mini-croissant
2. Regular croissant

Danish Dough No. 1

PASTRY TECHNIQUE:

Steps of Laminated Yeast Dough Preparation (see page 842)

HAZARDOUS FOOD:

Eggs

YIELD: 12 pounds, 1 ounce. Baker's Percentage.

INGREDIENTS:

Dough:

Water, cold	2 pounds	40
Eggs, whole	1 pound	20
Yeast, compressed	3 ounces	4
Flour, bread	4 pounds	80
Flour, pastry	1 pound	20
Salt	2 ounces	2
Sugar, granulated	8 ounces	10
Dried milk solids (DMS)	4 ounces	5

Roll-in Fat:

Butter, unsalted	3 pounds	60

METHOD OF PREPARATION:

1. Follow the steps of laminated yeast dough preparation (see page 842).

Danish Dough No. 2

PASTRY TECHNIQUE:

Steps of Laminated Yeast Dough Preparation (see page 842)

HAZARDOUS FOOD:

Eggs

YIELD:	19 pounds, 5⅞ ounces.	Baker's Percentage.

INGREDIENTS:

Dough (unlaminated):	18 pounds, 15⅞ ounces	
Yeast, compressed	8 ounces	5.30
Water	3 pounds, 8 ounces	37
Shortening, high-ratio	1 pound, 2 ounces	12
Dried milk solids (DMS)	6 ounces	4
Sugar, granulated	1 pound, 14 ounces	20
Salt	2¼ ounces	1.50
Eggs, whole	2 pounds	21.30
Extract, vanilla	1 ounce	0.70
Nutmeg, ground	⅛ ounce	0.08
Cardamom, ground	¼ ounce	0.16
Cinnamon, ground	¼ ounce	0.16
Flour, bread	7 pounds, 8 ounces	80
Flour, cake	1 pound, 14 ounces	20
Roll-in Fat:		
Butter, unsalted	6 pounds	32

METHOD OF PREPARATION:

1. Gather the equipment and ingredients.
2. Follow the steps of laminated yeast dough preparation (see page 842).

Puff Pastry Dough No. 1

PASTRY TECHNIQUES:

Laminating, Rolling

Laminating:

1. Allow a proper time to rest dough.
2. Roll the dough out to a ½-inch to ¾-inch thickness.
3. Evenly spread the fat.
4. Allow a proper time for the dough to rest.
5. Refrigerate for several hours.

Rolling:

1. Prepare the rolling surface by dusting with the appropriate medium (flour, cornstarch, etc.).
2. Use the appropriate style pin (stick pin or ball bearing pin) to roll the dough to desired thickness; rotate the dough during rolling to prevent sticking.

HAZARDOUS FOOD:

Eggs

YIELD:	7 pounds, 8¾ ounces.	15 pounds, 1¼ ounces.

INGREDIENTS:

Flour, bread	2 pounds, 8 ounces	5 pounds
Salt	½ ounce	1 ounce
Butter, unsalted (55° F)	8 ounces	1 pound
Water, ice cold	1 pound, 4¼ ounces	2 pounds, 8½ ounces
Eggs, whole	2 ounces	4 ounces

Roll-in Fat:

Butter (unsalted) or puff pastry shortening	2 pounds, 12 ounces	5 pounds, 9 ounces
Flour, bread	2¾ ounces	5½ ounces
Flour, cake	2¾ ounces	5½ ounces

METHOD OF PREPARATION:

1. Gather the equipment and ingredients.

Methods for Mixing Dough Portion of Puff Pastry:

Sable Method (Bench Method):

1. Place the bread flour and salt in a mound on the table.
2. Cut the butter (which is at 55° F) into cubes.
3. Place the butter cubes on top of the flour, and cut the cubes into the flour using a bench scraper. Stop when the flour resembles sand.
4. Whisk together the water and whole eggs.
5. Make a well in the flour-butter mixture, and pour the water-egg mixture into the well.
6. Starting with the flour on the inside of the well, work your way around, and slowly start to mix the flour into the water mixture.
7. Continue to mix until all of the water is absorbed into the flour.
8. Knead the dough by hand for 3 minutes.
9. Cover with plastic, and let the dough relax for 20 to 30 minutes.

Machine Method:

1. Place the flour and salt in the appropriate-size bowl.
2. Cut the butter (which is at 55° F.) into 1-inch cubes.
3. Mix the cubes into the flour.

4. Using the paddle attachment, cut the butter into the flour until it resembles sand.
5. Replace the paddle with a dough hook.
6. Make a well in the center of the flour.
7. Whisk together the water and eggs.
8. Add the egg-water mixture to the flour.
9. Mix the dough on second speed for 3 to 5 minutes, or until it comes together.
10. Remove the dough from the mixer.
11. Wrap the dough, and let it rest for 20 to 30 minutes.

Roll-in Fat:

1. Place the butter or puff pastry shortening and the flours in the appropriate-sized bowl.
2. Using the paddle attachment, mix the ingredients together on medium speed; incorporate thoroughly.
3. Scrape the bowl well during this process.
4. Shape into a block form, or leave in the bowl.
5. Allow to chill, if needed.

To Assemble the Dough (Spot Method):

1. On a well-floured surface or on a canvas, roll the dough portion into a rectangular shape that is ¾-inch to 1-inch thick.
2. Place the fat over two thirds of the dough, leaving a small border all around the fat.
3. Fold one third of the dough over the portion of dough covered with butter.
4. Fold the remaining third on top. (The dough is laminated.)
5. Roll out the dough, again to a rectangular shape.
6. Mark the dough into thirds.
7. Fold one third of the dough on top of another third.
8. Fold the remaining third on top (complete a 3-fold).
9. Make a vertical mark in the center of the dough, indicating the center.
10. Fold both ends to this mark.
11. Then fold the ends on top of each (complete a 4-fold).
12. Rest all of the dough for approximately 30 minutes.
13. Roll out the dough, again to a rectangular shape.
14. Complete another 3-fold; allow the dough to rest for 30 minutes.
15. Roll out the dough, again to a rectangular shape.
16. Complete another 4-fold; allow the dough to rest for 45 minutes.
17. The dough is ready for use.

To Assemble the Dough (Block Method):

1. On a well-floured surface or on a canvas, roll out the dough into a square shape that is approximately ¾-inch to 1-inch thick. The square should be slightly larger than the block of butter.

2. Place the butter on the square diagonally so that there are four dough triangles showing.
3. Fold each of the dough triangles over the butter so they will meet in the center.
4. Pinch the edges together to seal in the butter.
5. Roll out the dough to a rectangular shape.
6. Mark the dough into thirds.
7. Fold one third of the dough over another third of the dough.
8. Fold the remaining third of the dough (complete a 3-fold).
9. Allow the dough to rest for 30 minutes.
10. Roll out the dough into a rectangular shape.
11. Indicate the middle of the rectangle.
12. Fold each end of the dough over to meet the center of the dough.
13. Fold each side on top of each other (complete a 4-fold).
14. Allow the dough to rest for 30 minutes.
15. Roll out the dough, and repeat the 3-fold procedure.
16. Allow the dough to rest for 30 minutes.
17. Roll out the dough, again into a rectangular shape; repeat the 4-fold procedure.
18. Allow the dough to rest for 45 minutes.
19. The dough is ready for use.

Variations:

1. Patty shells
2. Vol-au-vent
3. Cream horns
4. Cheese sticks
5. Cream slices
6. Turnovers
7. Napoleons
8. Butterflies
9. Palm leaves
10. Millefeuilles
11. Pithiviers

Puff Pastry Dough No. 2

YIELD: 11 pounds, 9 ounces.

INGREDIENTS:

Water, ice cold	2 pounds, 12 ounces
Flour, bread	4 pounds, 6 ounces
Salt	1 ounce

Roll-in Fat:

Butter (unsalted) or puff pastry shortening	4 pounds, 6 ounces

Number of Folds:

4 by 4

METHOD OF PREPARATION:

1. Gather the equipment and ingredients.
2. Place the water, bread flour, and salt in a 20-quart mixing bowl. Mix with a dough hook on low speed for 10 minutes, or until gluten is developed.
3. Remove the dough from the mixer. Place on a sheet pan, cover, and allow to rest in the refrigerator for 15 minutes.
4. Roll the dough into a rectangular shape; place the fat over half of the dough; leave a small border of dough all around the fat. Fold the border over the fat, and incorporate the fat into the dough using a 2-fold method.
5. Give the dough *four* 4-folds (book-folds), as demonstrated by the instructor. Allow the dough to rest in the refrigerator for 15 minutes between each fold.
6. Allow the puff pastry dough 30 minutes of rest time after make-up and before baking.
7. Bake at 425° F, or until the puff pastry is an even golden brown and has a dry appearance. Bake smaller items for approximately 8 to 10 minutes, and bake larger items for approximately 15 to 30 minutes.

Variations:

1. Patty shells
2. Vol-au-vent
3. Cream horns
4. Cheese sticks
5. Cream slices
6. Turnovers
7. Napoleons
8. Butterflies
9. Palm leaves
10. Millefeuilles
11. Pithiviers

PASTRY TECHNIQUES:

Combining, Kneading, Rolling, Laminating

Combining:

Bringing together two or more components.

1. Prepare the components to be combined.
2. Add one to the other, using the appropriate mixing method (if needed).

Kneading:

1. Prepare the kneading surface with the appropriate medium (flour, cornstarch, etc.).
2. Press and form the dough into a mass using soft, determined strokes.
3. Continue kneading until appropriate consistency and/or temperature is achieved.

Rolling:

1. Prepare the rolling surface by dusting with the appropriate medium (flour, cornstarch, etc.).
2. Use the appropriate style pin (stick pin or ball bearing pin) to roll the dough to desired thickness; rotate the dough during rolling to prevent sticking.

Laminating:

1. Allow a proper time to rest dough.
2. Roll the dough out to a ½-inch to ¾-inch thickness.
3. Evenly spread the fat.
4. Allow a proper time for the dough to rest.
5. Refrigerate for several hours.

Puff Pastry Dough: Blitz Method

PASTRY TECHNIQUES:

Combining, Laminating

Combining:

Bringing together two or more components.

1. Prepare the components to be combined.
2. Add one to the other, using the appropriate mixing method (if needed).

Laminating:

1. Allow a proper time to rest dough.
2. Roll the dough out to a ½-inch to ¾-inch thickness.
3. Evenly spread the fat.
4. Allow a proper time for the dough to rest.
5. Refrigerate for several hours.

YIELD: 15 pounds, 1¾ ounces. 30 pounds, 3½ ounces.

INGREDIENTS:		
Water, cold	4 pounds	8 pounds
Flour, bread, sifted	4 pounds, 8 ounces	9 pounds
Flour, pastry, sifted	1 pound, 8 ounces	3 pounds
Salt	¾ ounce	1½ ounces
Cream of tartar	1 ounce	2 ounces
Shortening, puff pastry	5 pounds	10 pounds
Flour, bread or pastry	As needed	As needed

METHOD OF PREPARATION:

1. Gather the equipment and ingredients.
2. Place the cold water in a mixing bowl.
3. Add the bread flour, pastry flour, salt, and cream of tartar to the water.
4. Using a paddle attachment, begin to mix slowly.
5. Break the puff pastry shortening into small pieces, and add slowly to the flour mixture.
6. Mix together until just incorporated. *Do not overmix.*
7. Remove the dough from the mixer, and place it onto flour-dusted sheet pans.
8. Place in a refrigerator to chill and relax for 30 minutes.
9. Remove the pan from the refrigerator, and lay out the dough on a lightly flour-dusted bench.
10. Roll out the dough to approximately the size of a sheet pan or into a rectangle (1-inch thickness).
11. Brush the flour from the dough, and fold one side two thirds of the way to the opposite end.
12. Fold the remaining third over the top of the dough (a 3-fold).
13. Refrigerate the dough again for 20 minutes to relax.
14. Repeat the steps of rolling, folding, and resting three more times.
15. The dough should be ready for use.

Apricot Sauce No. 1

Canned Apricots

PASTRY TECHNIQUES:

Puréeing, Boiling, Combining

Puréeing:

1. Do not overfill the food processor.
2. First pulse the food processor.
3. Turn food processor to maximum to purée food.

Boiling:

1. Bring the cooking liquid to a rapid boil.
2. Stir the contents.

Combining:

Bringing together two or more components.

1. Prepare the components to be combined.
2. Add one to the other, using the appropriate mixing method (if needed).

HACCP:

Store at 40° F.

YIELD:	1 pound.	2 pounds.
INGREDIENTS:		
Apricot, halves, canned, drained	12¼ ounces	1 pound, 8½ ounces
Sugar, granulated	3 ounces	6 ounces
Water	¼ ounce	½ ounce
Brandy, apricot	½ ounce	1 ounce

METHOD OF PREPARATION:

1. Gather the equipment and ingredients.
2. Purée the apricots.
3. Boil the sugar with the water, and cook to a syrup.
4. Add the puréed apricots to the syrup.
5. Add the apricot brandy, to taste.

CHEF NOTE:

Can be served hot or cold.

Apricot Sauce No. 2

Apricot Preserves

PASTRY TECHNIQUES:

Boiling, Thickening, Combining

Boiling:

1. Bring the cooking liquid to a rapid boil.
2. Stir the contents.

Thickening:

1. Mix a small amount of sugar with the starches.
2. Create a slurry.
3. Whisk vigorously until thickened and translucent.

Combining:

Bringing together two or more components.

1. Prepare the components to be combined.
2. Add one to the other, using the appropriate mixing method (if needed).

HACCP:

Store at 40° F.

YIELD:	1 pound, 2 ounces.	2 pounds, 3 ounces.
INGREDIENTS:		
Preserves, apricot	12 ounces	1 pound, 8 ounces
Water	2 ounces	3 ounces
Sugar, granulated	3 ounces	6 ounces
Rum	1 ounce	2 ounces

METHOD OF PREPARATION:

1. Gather the equipment and ingredients.
2. Bring the preserves, water, and sugar to a boil.
3. Purée the mixture.
4. Strain the mixture.
5. Add the rum to the sauce.

CHEF NOTE:

Can be served hot or cold.

Apricot Sauce No. 3

Starch-Thickened

PASTRY TECHNIQUES:

Puréeing, Boiling, Thickening

Puréeing:

1. Do not overfill the food processor.
2. First pulse the food processor.
3. Turn food processor to maximum to purée food.

Boiling:

1. Bring the cooking liquid to a rapid boil.
2. Stir the contents.

Thickening:

1. Mix a small amount of sugar with the starches.
2. Create a slurry.
3. Whisk vigorously until thickened and translucent.

HACCP:

Store at 40° F.

YIELD: 1 pound, 8 ounces. 3 pounds.

INGREDIENTS:		
Apricots or apricot pulp, canned	13½ ounces	1 pound, 11 ounces
Water	8 ounces	1 pound
Sugar, granulated	8 ounces	1 pound
Lemon juice	¼ ounce	½ ounce
Cornstarch	¼ ounce	1½ ounce
Water	2 ounces	4 ounces

METHOD OF PREPARATION:

1. Gather the equipment and ingredients.
2. Purée the apricots and water together.
3. Place the puréed fruit and water in a pot with the sugar and lemon juice.
4. Bring to a boil.
5. Place the cornstarch in a dry bowl, and slowly add a small amount of water to dissolve it; stir well.
6. Add the cornstarch mixture to the boiling apricots.
7. Simmer for 3 minutes; stir continuously.
8. Remove from the heat, and strain.

CHEF NOTE:

Can be served hot or cold.

Blueberry Sauce

YIELD:	1 pound, 6¼ ounces.	2 pounds, 12½ ounces.

INGREDIENTS:

Ingredient		
Blueberries, whole, fresh	1 pound	2 pounds
Water	2 ounces	4 ounces
Sugar, granulated	4 ounces	8 ounces
Arrowroot	¼ ounce	½ ounce

METHOD OF PREPARATION:

1. Gather the equipment and ingredients.
2. Place the blueberries, water, and sugar in a bowl together. Let them stand for 2 hours.
3. Purée the blueberries.
4. Strain.
5. Place the arrowroot in a bowl; slowly add 2 ounces of strained purée; stir well.
6. Place the remaining purée in a pot and bring it to a boil.
7. Allow the purée to reduce by 25%.
8. Add the arrowroot mixture to the purée.
9. Return to a boil; cook to a desired consistency.

PASTRY TECHNIQUES:

Puréeing, Boiling, Thickening

Puréeing:

1. Do not overfill the food processor.
2. First pulse the food processor.
3. Turn food processor to maximum to purée food.

Boiling:

1. Bring the cooking liquid to a rapid boil.
2. Stir the contents.

Thickening:

1. Mix a small amount of sugar with the starches.
2. Create a slurry.
3. Whisk vigorously until thickened and translucent.

HACCP:

Store at 40° F.

CHEF NOTE:

Can be served hot or cold.

Caramel Sauce

YIELD:	1 pound, 4¼ ounces.	2 pounds, 8½ ounces.

INGREDIENTS:		
Sugar, granulated	12 ounces	1 pound, 8 ounces
Butter, unsalted	1½ ounces	3 ounces
Cream, heavy	6¾ ounces	13½ ounces

METHOD OF PREPARATION:

1. Gather the equipment and ingredients.
2. Melt the sugar in a heavy-bottomed saucepan, a small amount at a time, until it becomes caramelized.
3. Stir constantly; add the butter.
4. Add the cream; stir.
5. Store in a refrigerator.

PASTRY TECHNIQUES:

Melting, Combining

Melting:

1. Prepare the food product to be melted.
2. Place the food product in an appropriate sized pot over direct heat or over a double boiler.
3. Stir frequently or occasionally, depending on the delicacy of the product, until melted.

Combining:

Bringing together two or more components.

1. Prepare the components to be combined.
2. Add one to the other, using the appropriate mixing method (if needed).

HACCP:

Store at 40° F.

HAZARDOUS FOOD:

Heavy cream

CHEF NOTE:

Can be served hot or cold.

Chocolate Rum Sauce

PASTRY TECHNIQUES:

Chopping, Melting, Combining

Chopping:

1. Use a sharp knife.
2. Hold the food product properly.
3. Cut with a quick downward motion.

Melting:

1. Prepare the food product to be melted.
2. Place the food product in an appropriate sized pot over direct heat or over a double boiler.
3. Stir frequently or occasionally, depending on the delicacy of the product, until melted.

OR

1. Place the product on a sheet pan or in a bowl, and place in a low-temperature oven until melted.

Combining:

Bringing together two or more components.

1. Prepare the components to be combined.
2. Add one to the other, using the appropriate mixing method (if needed).

HACCP:

Store at 40° F.

CHEF NOTE:

Can be served hot or cold.

YIELD:	15½ ounces.	1 pound, 15 ounces.
INGREDIENTS:		
Chocolate, dark, unsweetened	2 ounces	4 ounces
Sugar, granulated	7½ ounces	15 ounces
Cream of tartar	Pinch	Pinch
Milk, evaporated	3 ounces	6 ounces
Rum, dark	3 ounces	6 ounces
Extract, vanilla	To taste	To taste
Salt	Pinch	Pinch

METHOD OF PREPARATION:

1. Gather the equipment and ingredients.
2. Chop the chocolate, and melt over a double boiler.
3. Add the sugar and cream of tartar to the chocolate; stir.
4. Add the milk and dark rum to the chocolate; stir.
5. Add the vanilla extract and salt to the mixture; stir.

PASTRY TECHNIQUES:
Combining, Chopping

Combining:
Bringing together two or more components.
1. Prepare the components to be combined.
2. Add one to the other, using the appropriate mixing method (if needed).

Chopping:
1. Use a sharp knife.
2. Hold the food product properly.
3. Cut with a quick downward motion.

HACCP:
Store at 40° F.

Chocolate Sauce

YIELD:	1 pound, 6 ounces.	2 pounds, 12 ounces.
INGREDIENTS:		
Cocoa powder	4 ounces	8 ounces
Water	8 ounces	1 pound
Sugar, granulated	8 ounces	1 pound
Couverture, dark, semi-sweet	2 ounces	4 ounces

METHOD OF PREPARATION:

1. Gather the equipment and ingredients.
2. Place the cocoa powder in a dry bowl, and slowly add a small amount of water to make a smooth paste.
3. Chop the couverture.
4. Place the sugar and remaining water in a pot; heat over low heat.
5. Stir in the cocoa paste mixture.
6. Bring to a boil; stir constantly.
7. Remove from the heat, and add the chopped couverture; stir constantly until blended.
8. Pass the sauce through a fine sieve.

CHEF NOTE:
Can be served hot or cold.

Chocolate Sauce: Quick Method

YIELD:	1 pound.	2 pounds.

INGREDIENTS:

Chocolate, dark, semi-sweet	11 ounces	1 pound, 6 ounces
Water	5 ounces	10 ounces

METHOD OF PREPARATION:

1. Gather the equipment and ingredients.
2. Chop the chocolate.
3. Melt over a double boiler.
4. Add the water slowly to the melted chocolate.

PASTRY TECHNIQUE:

Chopping, Melting, Combining

Chopping:

1. Use a sharp knife.
2. Hold the food product properly.
3. Cut with a quick downward motion.

Melting:

1. Prepare the food product to be melted.
2. Place the food product in an appropriate sized pot over direct heat or over a double boiler.
3. Stir frequently or occasionally, depending on the delicacy of the product, until melted.

OR

1. Place the product on a sheet pan or in a bowl, and place in a low-temperature oven until melted.

Combining:

Bringing together two or more components.

1. Prepare the components to be combined.
2. Add one to the other, using the appropriate mixing method (if needed).

HACCP:

Store at 40° F.

CHEF NOTE:

Can be served hot or cold.

Clear Fruit Sauce

YIELD:	1 pound, ¾ ounce.	2 pounds, 1½ ounces.
INGREDIENTS:		
Cornstarch	½ ounce	1 ounce
Fruit juice, sweetened	8 ounces	1 pound
Water	6 ounces	12 ounces
Wine, red	2 ounces	4 ounces
Lemon juice	¼ ounce	½ ounce

METHOD OF PREPARATION:

1. Gather the equipment and ingredients.
2. Place the cornstarch in a dry bowl, and slowly add a small amount of water to dissolve it.
3. Place the fruit juice, water, wine, and lemon juice in a pot; heat.
4. Simmer for approximately 3 minutes; stir occasionally.
5. Stir the cornstarch into the simmering fruit juice mixture; whip constantly to prevent burning and lumps.
6. Bring to a boil.
7. Let cool.

PASTRY TECHNIQUES:

Combining, Thickening, Boiling

Combining:

Bringing together two or more components.

1. Prepare the components to be combined.
2. Add one to the other, using the appropriate mixing method (if needed).

Thickening:

1. Mix a small amount of sugar with the starches.
2. Create a slurry.
3. Whisk vigorously until thickened and translucent.

Boiling:

1. Bring the cooking liquid to a rapid boil.
2. Stir the contents.

HACCP:

Store at 40° F.

CHEF NOTES:

1. Cherry juice, raspberry juice, or strawberry juice may be used.
2. Can be served hot or cold.

PASTRY TECHNIQUE:

Combining

Combining:

Bringing together two or more components.

1. Prepare the components to be combined.
2. Add one to the other, using the appropriate mixing method (if needed).

HACCP:

Store at 40° F.

Coulis

Fruit Sauce, Uncooked

YIELD:	1 pound.	1 pound, 8 ounces.
INGREDIENTS:		
Fruit, fresh, frozen, or canned	12 ounces	1 pound, 8 ounces
Lemon juice	½ lemon	1 lemon
Simple syrup (see page 966)	4 ounces	8 ounces

METHOD OF PREPARATION:

1. Gather the equipment and ingredients.
2. Prepare the fruit; clean and hull; thaw or drain, as needed.
3. Purée the fruit with lemon juice.
4. Add enough simple syrup to thin to the desired consistency.
5. Strain, if necessary.
6. Store in a refrigerator.

Variations:

1. Kiwi
2. Mango
3. Peach
4. Pineapple
5. Raspberry
6. Strawberry

Cream Sauce

YIELD:	1 pound, 5½ ounces.	2 pounds, 11 ounces.
INGREDIENTS:		
Sugar, granulated	10½ ounces	1 pound, 5 ounces
Salt	To taste	To taste
Cream, heavy	8 ounces	1 pound
Butter, unsalted, chopped	3 ounces	6 ounces
Flavorings/compounds, zest	To taste	To taste

METHOD OF PREPARATION:

1. Gather the equipment and ingredients.
2. Place all of ingredients in a sauce pan; heat on medium-high.
3. Bring the mixture to a boil; reduce the heat.
4. Continue cooking until the sauce is reduced to a desired consistency, approximately 5 minutes. (The sauce should be slightly thick.)
5. Remove from the heat.
6. Add the compounds or other flavorings, if desired.
7. Store in a refrigerator.

PASTRY TECHNIQUES:

Combining, Boiling, Reducing

Combining:

Bringing together two or more components.

1. Prepare the components to be combined.
2. Add one to the other, using the appropriate mixing method (if needed).

Boiling:

1. Bring the cooking liquid to a rapid boil.
2. Stir the contents.

Reducing:

1. Bring the sauce to a boil; then reduce to a simmer.
2. Stir often; reduce to the desired consistency.

HACCP:

Store at 40° F.

HAZARDOUS FOOD:

Heavy cream

CHEF NOTE:

Can be served hot or cold.

English Sauce

(Sauce Anglaise)

PASTRY TECHNIQUES:

Combining, Boiling, Tempering, Thickening

Combining:

Bringing together two or more components.

1. Prepare the components to be combined.
2. Add one to the other, using the appropriate mixing method (if needed).

Boiling:

1. Bring the cooking liquid to a rapid boil.
2. Stir the contents.

Tempering:

1. Whisk the eggs vigorously while ladling hot liquid.

Thickening:

1. Mix a small amount of sugar with the starches.
2. Create a slurry.
3. Whisk vigorously until thickened and translucent.

HACCP:

Store at 40° F.

HAZARDOUS FOODS:

Milk
Heavy cream
Egg yolks

YIELD:	12¼ ounces.	1 pound, 8½ ounces.
INGREDIENTS:		
Milk, whole	4 ounces	8 ounces
Cream, heavy	4 ounces	8 ounces
Sugar, granulated	2 ounces	4 ounces
Salt	Pinch	Pinch
Egg yolks	2¼ ounces	4½ ounces
Extract, vanilla	To taste	To taste

METHOD OF PREPARATION:

1. Gather the equipment and ingredients.
2. Place the milk, heavy cream, granulated sugar, and salt in a pot; bring to a boil.
3. Place the egg yolks and vanilla extract in a bowl; whip slightly.
4. Temper the egg yolk mixture with the boiling milk.
5. Cook the mixture to 175° F, and stir constantly. The mixture should be able to coat the back of a spoon.
7. When the mixture thickens, remove it from the heat, and put it in an ice bath to stop the cooking process. Continue to stir frequently to avoid skin creation.
8. Strain, if desired.
9. Refrigerate until use.

Hard Sauce

YIELD:	10 ounces.	1 pound, 4 ounces.
INGREDIENTS:		
Butter, unsalted, softened	5 ounces	10 ounces
Sugar, confectionery	3 ounces	6 ounces
Rum or brandy	1 ounce	2 ounces
Lemon juice	1 ounce	2 ounces

METHOD OF PREPARATION:

1. Gather the equipment and ingredients.
2. Cream the butter until soft and light.
3. Add the confectionery sugar to the butter.
4. Add the liquor and lemon juice to incorporate thoroughly.
5. Store in a refrigerator.

PASTRY TECHNIQUE:

Creaming

Creaming:

1. Soften the fats on low speed.
2. Add the sugar(s) and cream; increase the speed slowly.
3. Add the eggs one at a time; scrape the bowl frequently.
4. Add the dry ingredients in stages.

HACCP:

Store at 40° F.

CHEF NOTE:

Hard sauce is generally served as an accompaniment to bread pudding.

Lemon Custard Sauce

PASTRY TECHNIQUES:

Combining, Boiling, Tempering, Thickening

Combining:

Bringing together two or more components.

1. Prepare the components to be combined.
2. Add one to the other, using the appropriate mixing method (if needed).

Boiling:

1. Bring the cooking liquid to a rapid boil.
2. Stir the contents.

Tempering:

1. Whisk the eggs vigorously while ladling hot liquid.

Thickening:

1. Mix a small amount of sugar with the starches.
2. Create a slurry.
3. Whisk vigorously until thickened and translucent.

HACCP:

Store at 40° F.

HAZARDOUS FOOD:

Egg yolks

YIELD:	15¼ ounces.	1 pound, 15½ ounces.

INGREDIENTS:

Ingredient		
Cornstarch	½ ounce	1 ounce
Water	8 ounces	1 pound
Sugar, granulated	4 ounces	8 ounces
Lemon juice	¾ ounce	1½ ounces
Egg yolks	2 ounces	4 ounces
Lemon rind	1 lemon	2 lemons

METHOD OF PREPARATION:

1. Gather the equipment and ingredients.
2. Place the cornstarch in a dry bowl, and slowly add a small amount of water to dissolve.
3. Add half of the sugar, the lemon juice, and the egg yolks into the cornstarch mixture; combine well, and set aside.
4. Place the remaining water, sugar, and lemon rind in a pot; bring to a boil.
5. Temper the cornstarch mixture; add to the boiling liquid.
6. Boil the liquid for approximately 1 minute; stir constantly.
7. Remove from the heat.
8. Cool.
9. Store in a refrigerator.

Melba Sauce

PASTRY TECHNIQUES:

Combining, Boiling

Combining:

Bringing together two or more components.

1. Prepare the components to be combined.
2. Add one to the other, using the appropriate mixing method (if needed).

Boiling:

1. Bring the cooking liquid to a rapid boil.
2. Stir the contents.

HACCP:

Store at 40° F.

YIELD:	12¼ ounces.	1 pound, 8½ ounces.

INGREDIENTS:

Preserves, raspberry	8 ounces	1 pound
Simple syrup (see page 966)	4 ounces	8 ounces
Salt	Pinch	Pinch
Extract, vanilla	To taste	To taste
Lemon juice	¼ ounce	½ ounce

METHOD OF PREPARATION:

1. Gather the equipment and ingredients.
2. Place the raspberry preserves, simple syrup, and salt in a pot.
3. Bring to a boil; boil for approximately 1 minute.
4. Remove from the heat, and strain.
5. Add the vanilla extract and lemon juice.
6. Cool.
7. Store in a refrigerator.

Orange Custard Sauce

YIELD:	15½ ounces.	1 pound, 15 ounces.

INGREDIENTS:		
Water	8 ounces	1 pound
Sugar, granulated	4 ounces	8 ounces
Orange rind	1 each	2 each
Cornstarch	½ ounce	1 ounce
Orange juice	1 ounce	2 ounces
Egg yolks	2 ounces	4 ounces

METHOD OF PREPARATION:

1. Gather the equipment and ingredients.
2. Place the water, half of the granulated sugar, and orange rind in a pot; bring to a boil.
3. Place the cornstarch in a dry bowl, and slowly add a small amount of the orange juice to dissolve it.
4. Add in the remaining orange juice, sugar, and egg yolks to the cornstarch mixture; combine well.
5. Temper the cornstarch mixture, and add to the boiling liquid.
6. Boil the liquid for approximately 1 minute; stir constantly.
7. Remove from the heat.
8. Cool.
9. Store in a refrigerator.

PASTRY TECHNIQUES:

Boiling, Tempering, Thickening

Boiling:

1. Bring the cooking liquid to a rapid boil.
2. Stir the contents.

Tempering:

1. Whisk the eggs vigorously while ladling hot liquid.

Thickening:

1. Mix a small amount of sugar with the starches.
2. Create a slurry.
3. Whisk vigorously until thickened and translucent.

HACCP:

Store at 40° F.

HAZARDOUS FOOD:

Egg yolks

Orange Sauce

YIELD:	1 pound, 7 ounces.	2 pounds, 14 ounces.

INGREDIENTS:

Cornstarch or arrowroot	½ ounce	1 ounce
Orange juice	1 pound, 3 ounces	2 pounds, 6 ounces
Sugar, granulated	1½ ounces	3 ounces
Grand Marnier	2 ounces	4 ounces

METHOD OF PREPARATION:

1. Gather the equipment and ingredients.
2. Place the cornstarch or arrowroot in a dry bowl, and add a small amount of orange juice to dissolve it.
3. Place the remaining orange juice and sugar in a pot; place over medium heat.
4. Bring the mixture to a boil.
5. Add the cornstarch or arrowroot to the boiling orange juice; stir gently.
6. Bring the orange juice back to a boil; continue to cook for 1 minute.
7. Remove from the heat.
8. Allow the sauce to cool.
9. Strain the sauce through a fine sieve.
10. Add the Grand Marnier.
11. Store in a refrigerator.

PASTRY TECHNIQUES:

Combining, Boiling, Thickening

Combining:

Bringing together two or more components.

1. Prepare the components to be combined.
2. Add one to the other, using the appropriate mixing method (if needed).

Boiling:

1. Bring the cooking liquid to a rapid boil.
2. Stir the contents.

Thickening:

1. Mix a small amount of sugar with the starches.
2. Create a slurry.
3. Whisk vigorously until thickened and translucent.

HACCP:

Store at 40° F.

Raisin Rum Sauce

PASTRY TECHNIQUES:
Combining, Simmering

Combining:
Bringing together two or more components.
1. Prepare the components to be combined.
2. Add one to the other, using the appropriate mixing method (if needed).

Simmering:
1. Place the prepared product in an appropriate-sized pot.
3. Bring the product to a boil, then reduce the heat to allow the product to barely boil.
4. Cook until desired doneness is achieved.

HACCP:
Store at 40° F.

YIELD:	1 pound, 2 ounces.	2 pounds, 4 ounces.

INGREDIENTS:		
Water	4 ounces	8 ounces
Sugar, granulated	7 ounces	14 ounces
Orange juice	⅛ ounce	¼ ounce
Cloves, whole	2 each	2 each
Raisins, seedless	3 ounces	6 ounces
Rum	4 ounces	8 ounces

METHOD OF PREPARATION:
1. Gather the equipment and ingredients.
2. Place the water, sugar, orange juice, and cloves in a pot; bring to a boil.
3. Simmer for 10 minutes.
4. Remove the cloves, and add the raisins.
5. Simmer for an additional 5 minutes.
6. Remove the sauce from the heat.
7. Add the rum.
8. Store in a refrigerator.

CHEF NOTE:
Can be served hot or cold.

Raspberry Sauce

YIELD:	9 ounces.	1 pound, 2 ounces.

INGREDIENTS:		
Raspberries, whole, fresh, or frozen	8 ounces	1 pound
Water	1 ounce	2 ounces
Sugar, granulated	2 ounces	4 ounces
Arrowroot	⅛ ounce	¼ ounce

METHOD OF PREPARATION:

1. Gather the equipment and ingredients.
2. Place the raspberries, water, and sugar in a bowl. Let stand for 2 hours.
3. Purée all of the ingredients.
4. Strain.
5. Place the arrowroot in a bowl, and slowly add 2 ounces of strained purée; stir well.
6. Place the remaining purée in a pot, and bring to a boil.
7. The purée will reduce by 25%.
8. Add the arrowroot mixture to the purée.
9. Return to a boil; cook to a desired consistency.
10. Store in a refrigerator.

PASTRY TECHNIQUES:

Puréeing, Boiling, Thickening

Puréeing:

1. Do not overfill the food processor.
2. First pulse the food processor.
3. Turn food processor to maximum to purée food.

Boiling:

1. Bring the cooking liquid to a rapid boil.
2. Stir the contents.

Thickening:

1. Mix a small amount of sugar with the starches.
2. Create a slurry.
3. Whisk vigorously until thickened and translucent.

HACCP:

Store at 40° F.

CHEF NOTE:

Serve cold.

Sabayon à l'Marsala

PASTRY TECHNIQUE:
Whipping

Whipping:
1. Hold the whip at a 55-degree angle.
2. Create circles, using a circular motion.
3. The circular motion needs to be perpendicular to the bowl.

HACCP:
Store at 40° F.

HAZARDOUS FOOD:
Egg yolks

YIELD:	10 ounces.	1 pound, 4 ounces.

INGREDIENTS:

Egg yolks	4½ ounces	9 ounces
Sugar, granulated	1 ounce	2 ounces
Wine, Marsala	4½ ounces	9 ounces

METHOD OF PREPARATION:

1. Gather the equipment and ingredients.
2. Place the egg yolks and sugar in a bowl; place over a double boiler.
3. Whip constantly until at least 140° F to form a thick, fluffy mixture.
4. When some volume is reached, add the Marsala slowly; continue to whip. Be careful to keep the sides clean so that the egg yolks will not cook along the sides of the bowl.
5. When the mixture has reached a ribbon-like texture, remove from the heat.
6. Serve warm.
7. Store in a refrigerator.

Serving Suggestions:

1. Pour sabayon over fruit.
2. Serve with lady fingers or cubed sponge cake.

Variations:

Supplement or substitute Marsala with a sweet white wine, Grand Marnier, or Kirschwasser.

CHEF NOTE:
The name will change according to the flavoring liquid used.

Sabayon à l'Orange

PASTRY TECHNIQUE:

Whipping

Whipping:

1. Hold the whip at a 55-degree angle.
2. Create circles, using a circular motion.
3. The circular motion needs to be perpendicular to the bowl.

HACCP:

Store at 40° F.

HAZARDOUS FOOD:

Egg yolks

YIELD:	14¾ ounces.	1 pound, 13½ ounces.
INGREDIENTS:		
Egg yolks	3¾ ounces	7½ ounces
Sugar, granulated	3 ounces	6 ounces
Wine, sweet	4 ounces	8 ounces
Lemon juice	2 ounces	4 ounces
Orange juice	2 ounces	4 ounces

METHOD OF PREPARATION:

1. Gather the equipment and ingredients.
2. Place all of the ingredients in a bowl.
3. Place over a double boiler; whip constantly until it reaches at least 140° F, to form a thick fluffy mixture.
4. Serve warm.
5. Store in a refrigerator.

Almond Lace Cookies

(Florentines)

PASTRY TECHNIQUES:

Combining, Rolling

Combining:

Bringing together two or more components.

1. Prepare the components to be combined.
2. Add one to the other, using the appropriate mixing method (if needed).

Rolling:

1. Prepare the rolling surface by dusting with the appropriate medium (flour, cornstarch, etc.).
2. Use the appropriate style pin (stick pin or ball bearing pin) to roll the dough to desired thickness; rotate the dough during rolling to prevent sticking.

HACCP:

Store unbaked dough at 40° F.

YIELD: 2 pounds, 14¼ ounces. 5 pounds, 12½ ounces.

INGREDIENTS:		
Flour, pastry	9 ounces	1 pound, 2 ounces
Sugar, granulated	9 ounces	1 pound, 2 ounces
Butter, unsalted, softened	9 ounces	1 pound, 2 ounces
Corn syrup	10¼ ounces	1 pound, 4½ ounces
Almonds, blanched, sliced, and crushed	9 ounces	1 pound, 2 ounces

METHOD OF PREPARATION:

1. Gather the equipment and ingredients.
2. Place the flour, sugar, and slightly softened butter in a bowl, and place on a mixer.
3. Using a paddle, blend together.
4. Add the corn syrup; mix until creamy.
5. Add the crushed almonds, and mix until the nuts are evenly dispersed.
6. Remove from the mixer.
7. Place 1 pound of dough on parchment paper lengthwise, and roll into a log shape.
8. Refrigerate or freeze until firm.
9. When the dough is firm, slice the log into ⅛-inch slices; place on a greased sheet pan.
10. Bake at 350° F for approximately 5 to 7 minutes, or until golden brown.
11. Remove from the oven, and shape as desired.

Almond Macaroons

YIELD:	2 pounds, 7½ ounces.	4 pounds, 15 ounces.

INGREDIENTS:

Ingredient		
Almond meal	11½ ounces	1 pound, 7 ounces
Sugar, confectionery	1 pound, 2½ ounces	2 pounds, 5 ounces
Egg whites	9 ounces	1 pound, 2 ounces
Sugar, granulated	½ ounce	1 ounce

METHOD OF PREPARATION:

1. Gather the equipment and ingredients.
2. Sift together the almond meal and confectionery sugar.
3. Place the egg whites in a bowl; whip to a medium peak; slowly add the granulated sugar to create a meringue.
4. Fold the dry ingredients into the meringue.
5. Pipe into desired shapes and sizes on parchment-lined sheet pans.
6. Bake at 400° F on double sheet pans for approximately 12 to 15 minutes, or until the edges are golden brown.

PASTRY TECHNIQUES:

Whipping, Combining, Piping

Whipping:

1. Hold the whip at a 55-degree angle.
2. Create circles, using a circular motion.
3. The circular motion needs to be perpendicular to the bowl.

Combining:

Bringing together two or more components.

1. Prepare the components to be combined.
2. Add one to the other, using the appropriate mixing method (if needed).

Piping:

With bag:

1. Use a bag with a disposable tip; cut the bag at 45-degree angle.
2. Fill to no more than half full.
3. Burp the bag.

With cone:

1. Cut and fold the piping cone to the appropriate size.
2. Fill the cone with a small amount.
3. Fold the ends to form a triangle.
4. Pipe the desired designs.

HAZARDOUS FOOD:

Egg whites

Almond Ring Cookies

YIELD:	1 pound, 14 ounces. 25 Cookies	3 pounds, 12 ounces. 50 Cookies

INGREDIENTS:

Short dough, (see page 840)	12 ounces	1 pound, 8 ounces
Macaroons, (see page 875)	1 pound, 2 ounces	2 pounds, 4 ounces
Preserves, raspberry	As needed	As needed

METHOD OF PREPARATION:

1. Gather the equipment and ingredients.
2. Roll out the short dough to ⅛-inch thickness.
3. Cut out small, round discs from the short dough; place on a sheet pan.
4. Place the macaroon mix in a pastry bag; use a medium-star tip.
5. Pipe the macaroon mix around the edge of the cookie.
6. Fill the center with raspberry preserves.
7. Bake at 350° F for approximately 15 minutes, or until golden brown.
8. Cool.

PASTRY TECHNIQUES:

Rolling, Piping, Filling

Rolling:

1. Prepare the rolling surface by dusting with the appropriate medium (flour, cornstarch, etc.).
2. Use the appropriate style pin (stick pin or ball bearing pin) to roll the dough to desired thickness; rotate the dough during rolling to prevent sticking.

Piping:

With bag:
1. Use a bag with a disposable tip; cut the bag at 45-degree angle.
2. Fill to no more than half full.
3. Burp the bag.

With cone:
1. Cut and fold the piping cone to the appropriate size.
2. Fill the cone with a small amount.
3. Fold the ends to form a triangle.
4. Pipe the desired designs.

Filling:

1. Cut open the food product.
2. Carefully spread the filling using an icing spatula, or
3. Carefully pipe the filling using a pastry bag.

CHEF NOTE:

To add a shiny finish to these cookies, brush the macaroon portion of the cooled cookies with heated apricot glaze.

Brownies, Cake

PASTRY TECHNIQUES:

Combining, Spreading

Combining:

Bringing together two or more components.

1. Prepare the components to be combined.
2. Add one to the other, using the appropriate mixing method (if needed).

Spreading:

1. Using an icing spatula or off-set spatula, smooth the icing or other spreading medium over the surface area.

HAZARDOUS FOOD:

Eggs

YIELD:	4 pounds. ½ Sheet Pan	8 pounds. 1 Sheet Pan

INGREDIENTS:

Batter:

Cocoa powder, sifted	3¾ ounces	7½ ounces
Flour, cake, sifted	7½ ounces	15 ounces
Flour, pastry, sifted	4 ounces	8 ounces
Salt	¼ ounce	½ ounce
Baking soda, sifted	⅛ ounce	¼ ounce
Shortening, high-ratio	11½ ounces	1 pound, 7 ounces
Sugar, granulated	7½ ounces	15 ounces
Sugar, brown	7½ ounces	15 ounces
Corn syrup	4½ ounces	9 ounces
Eggs, whole	9 ounces	1 pound, 2 ounces
Water	1½ ounces	3 ounces
Walnuts, pieces, toasted (optional)	7½ ounces	15 ounces

Garnish:

Simple syrup (see page 966	4 ounces	8 ounces
Chocolate fudge icing (see page 772)	2 pounds, 3¾ ounces	4 pounds, 7½ ounces

METHOD OF PREPARATION:

1. Gather the equipment and ingredients.
2. Sift all of the dry ingredients together: cocoa powder, cake flour, pastry flour, salt, and baking soda.
3. Place all of the ingredients in a bowl.
4. Blend well on low speed to combine.
5. Mix for 2 minutes on medium speed.
6. Scale the batter, and pour onto greased, paper-lined sheet pans.
7. Spread the batter evenly with a spatula. Clean the pan edge.
8. Bake at 375° F for approximately 25 to 30 minutes, or until firm but not dry. *Do not overbake.*
9. Remove from the oven; cool completely.
10. Coat the top of the brownies with simple syrup.
11. Cover with fudge icing.
12. Cut a sheet pan of brownies with a sharp knife into the desired sizes.

Brownies, Chocolate Fudge

PASTRY TECHNIQUES:

Creaming, Combining, Spreading

Creaming:

1. Soften the fats on low speed.
2. Add the sugar(s) and cream; increase the speed slowly.
3. Add the eggs one at a time; scrape the bowl frequently.
4. Add the dry ingredients in stages.

Combining:

Bringing together two or more components.

1. Prepare the components to be combined.
2. Add one to the other, using the appropriate mixing method (if needed).

Spreading:

1. Using an icing spatula or off-set spatula, smooth the icing or other spreading medium over the surface area.

HAZARDOUS FOOD:

Eggs

YIELD:	5 pounds.	10 pounds.
	½ Sheet Pan	1 Sheet Pan

INGREDIENTS:

Batter:		
Sugar, granulated	1 pound, 8 ounces	3 pounds
Shortening, all-purpose	12 ounces	1 pound, 8 ounces
Corn syrup	4 ounces	8 ounces
Cocoa powder, sifted	4 ounces	8 ounces
Extract, vanilla	¼ ounce	½ ounce
Eggs, whole	12 ounces	1 pound, 8 ounces
Flour, pastry	1 pound	2 pounds
Walnuts, pieces, toasted (optional)	8 ounces	1 pound
Garnish:		
Chocolate fudge icing (see page 772)	2 pounds, 3¾ ounces	4 pounds, 7½ ounces

METHOD OF PREPARATION:

1. Gather the equipment and ingredients.
2. Place the sugar, shortening, corn syrup, cocoa powder, and vanilla extract in a bowl.
3. Cream well on low speed.
4. Add the whole eggs slowly; scrape the bowl.
5. Add the pastry flour, and blend.
6. Add the walnuts, and incorporate.
7. Scrape the bowl; mix until blended.
8. Scale the batter, and pour into greased, paper-lined sheet pans.
9. Spread the batter evenly with spatula. Clean the pan edge.
10. Bake at 375° F for approximately 25 to 30 minutes, or until firm but not dry. *Do not overbake.*
11. Remove from the oven; cool completely.
12. Cover with fudge icing.
13. Cut a sheet pan of brownies with a sharp knife into the desired sizes.

Butter Cookies

YIELD:	1 pound, 6½ ounces.	7 pounds, 4 ounces.
	3 Dozen	12 Dozen

INGREDIENTS:

Ingredient		
Sugar, granulated	3¾ ounces	1 pound, 4 ounces
Butter, unsalted (70° F)	7½ ounces	2 pounds, 8 ounces
Egg whites	1½ ounces	8 ounces
Flour, bread, sifted	9¾ ounces	3 pounds

METHOD OF PREPARATION:

1. Gather the equipment and ingredients.
2. Place the granulated sugar and butter in a bowl; cream until light and fluffy.
3. Add the egg whites slowly; continue to cream the mixture.
4. Scrape the bowl; continue mixing.
5. Add the sifted bread flour, and blend; scrape the bowl; continue mixing until smooth.
6. Place the mixture into a pastry bag; use a medium-star tip.
7. Pipe out the desired shapes on a parchment-lined sheet pan.
8. Bake at 375° F for approximately 10 to 12 minutes, or until light golden brown.

PASTRY TECHNIQUES:

Creaming, Combining, Piping

Creaming:

1. Soften the fats on low speed.
2. Add the sugar(s) and cream; increase the speed slowly.
3. Add the eggs one at a time; scrape the bowl frequently.
4. Add the dry ingredients in stages.

Combining:

Bringing together two or more components.

1. Prepare the components to be combined.
2. Add one to the other, using the appropriate mixing method (if needed).

Piping:

With bag:

1. Use a bag with a disposable tip; cut the bag at 45-degree angle.
2. Fill to no more than half full.
3. Burp the bag.

With cone:

1. Cut and fold the piping cone to the appropriate size.
2. Fill the cone with a small amount.
3. Fold the ends to form a triangle.
4. Pipe the desired designs.

HAZARDOUS FOOD:

Egg whites

CHEF NOTE:

The batter may be colored, if desired.

Chocolate Chip Cookies

PASTRY TECHNIQUES:

Creaming, Combining, Portioning

Creaming:

1. Soften the fats on low speed.
2. Add the sugar(s) and cream; increase the speed slowly.
3. Add the eggs one at a time; scrape the bowl frequently.
4. Add the dry ingredients in stages.

Combining:

Bringing together two or more components.

1. Prepare the components to be combined.
2. Add one to the other, using the appropriate mixing method (if needed).

Portioning:

1. Mark the product for portioning, using a ruler, if necessary.
3. Cut, spoon, or scoop the product with the appropriate-sized utensil.

HAZARDOUS FOOD:

Eggs

YIELD:	2 pounds. 2 Dozen	10 pounds. 10 Dozen

INGREDIENTS:

Butter, unsalted	2¾ ounces	13¾ ounces
Margarine	2¾ ounces	13¾ ounces
Sugar, granulated	4 ounces	1 pound, 4 ounces
Sugar, brown	3 ounces	15 ounces
Salt	⅛ ounce	⅝ ounce
Baking soda	⅛ ounce	⅝ ounce
Flour, pastry	7½ ounces	2 pounds, 5½ ounces
Eggs, whole	2¼ ounces	11¼ ounces
Extract, vanilla	To taste	To taste
Chips, dark or white	9 ounces	2 pounds, 13 ounces
Pecans or walnuts, chopped (optional)	½ ounce	2½ ounces

METHOD OF PREPARATION:

1. Gather the equipment and ingredients.
2. Place the butter, margarine, granulated sugar, and brown sugar in a bowl with a paddle; cream.
3. Sift the dry ingredients together: salt, baking soda, and pastry flour.
4. Add the eggs and vanilla extract into the butter mixture.
5. Add the sifted dry ingredients; scrape well.
6. Add the chocolate chips and nuts; combine.
7. Portion the dough into 1½-ounce pieces; place on a parchment-lined sheet pan.
8. Bake at 375° F for 8 to 10 minutes, or until the cookies are golden brown.

Coconut Macaroons

PASTRY TECHNIQUES:

Heating, Whipping, Folding, Piping

Heating:

1. Prepare the food product according to the formula's instructions.
2. Choose the appropriate method of heating (on the range or stove top, in the oven, etc.)
3. Apply the product to the heat.

Whipping:

1. Hold the whip at a 55-degree angle.
2. Create circles, using a circular motion.
3. The circular motion needs to be perpendicular to the bowl.

Folding:

Do steps 1, 2, and 3 in one continuous motion.

1. Run a bowl scraper under the mixture, across the bottom of the bowl.
2. Turn the bowl counterclockwise.
3. Bring the bottom mixture to the top.

Piping:

With bag:

1. Use a bag with a disposable tip; cut the bag at 45-degree angle.
2. Fill to no more than half full.
3. Burp the bag.

With cone:

1. Cut and fold the piping cone to the appropriate size.
2. Fill the cone with a small amount.
3. Fold the ends to form a triangle.
4. Pipe the desired designs.

HAZARDOUS FOOD:

Egg whites

YIELD:	2 pounds, 13 ounces.	5 pounds, 10 ounces.

INGREDIENTS:

Ingredient		
Egg whites	8 ounces	1 pound
Almond meal	4½ ounces	9 ounces
Sugar, granulated	1 pound, 4 ounces	2 pounds, 8 ounces
Macaroon coconut	10 ounces	1 pound, 4 ounces
Flour, pastry	2½ ounces	5 ounces

METHOD OF PREPARATION:

1. Gather the equipment and ingredients.
2. Place the egg whites in a bowl over a double boiler; heat until they reach 120° F, whipping constantly.
3. Remove from the heat.
4. Combine the almond meal, granulated sugar, macaroon coconut, and pastry flour.
5. Combine the dry ingredients into the heated egg whites.
6. Return the mixture back to heat. Heat until the mixture reaches 120° F.
7. Rest covered for 5 minutes.
8. Pipe into 1-inch rounds, using no tubes.
9. Bake on double pans at 375° F for approximately 12 to 15 minutes, or until golden brown.

Florentines

Traditional

PASTRY TECHNIQUES:

Boiling, Combining, Portioning

Boiling:

1. Bring the cooking liquid to a rapid boil.
2. Stir the contents.

Combining:

Bringing together two or more components.

1. Prepare the components to be combined.
2. Add one to the other, using the appropriate mixing method (if needed).

Portioning:

1. Mark the product for portioning, using a ruler, if necessary.
3. Cut, spoon, or scoop the product with the appropriate-sized utensil.

HACCP:

Store unbaked dough at 40° F.

HAZARDOUS FOOD:

Heavy cream

YIELD:	2 pounds, 2 ounces.	4 pounds, 4 ounces.
INGREDIENTS:		
Couverture, dark, semisweet	As needed	As needed
Dough:		
Butter, unsalted	8 ounces	1 pound
Sugar, granulated	8 ounces	1 pound
Honey	3 ounces	6 ounces
Cream, heavy	3 ounces	6 ounces
Almonds, blanched, sliced, and toasted	3 ounces	6 ounces
Almonds, blanched, chopped, and toasted	2 ounces	4 ounces
Walnut pieces	2 ounces	4 ounces
Fruit, candied	5 ounces	10 ounces

METHOD OF PREPARATION:

1. Gather the equipment and ingredients.
2. Place the butter, sugar, honey, and heavy cream in a pot; bring to a boil, and heat to 240° F.
3. Add the sliced almonds, chopped almonds, walnut pieces, and candied fruit; return to a boil.
4. Cool.
5. Portion ½-ounce pieces.
6. Flatten the pieces on greased, 3-inch florentine pan, or place the dough on parchment-lined sheet pans.
7. Bake at 350° F for approximately 8 minutes, or until golden brown.
8. Cool.
9. Coat the underside of the florentine with melted couverture, and mark with a decorating comb.

CHEF NOTE:

Honey can be replaced with a honey-glucose mixture.

French Macaroons

YIELD:	2 pounds, 10½ ounces.	5 pounds, 5 ounces.

INGREDIENTS:

Ingredient		
Egg whites	9 ounces	1 pound, 2 ounces
Sugar, granulated	1½ ounces	3 ounces
Cream of tartar	Pinch	Pinch
Almond meal	13½ ounces	1 pound, 11 ounces
Sugar, confectionery	1 pound, 2½ ounces	2 pounds, 5 ounces

METHOD OF PREPARATION:

1. Gather the equipment and ingredients.
2. Place the egg whites in a bowl with a whip. Whip to a soft peak, slowly adding the granulated sugar and cream of tartar to make a meringue.
3. Sift together the almond meal and confectionery sugar.
4. Fold the sifted ingredients carefully into the meringue.
5. Using a no. 3 plain tube, pipe immediately into ¾-inch rounds.
6. Bake on double pans at 400° F for approximately 10 to 12 minutes, or until golden on the bottom of the macaroon.

PASTRY TECHNIQUES:

Whipping, Folding, Piping

Whipping:

1. Hold the whip at a 55-degree angle.
2. Create circles, using a circular motion.
3. The circular motion needs to be perpendicular to the bowl.

Folding:

Do steps 1, 2, and 3 in one continuous motion.

1. Run a bowl scraper under the mixture, across the bottom of the bowl.
2. Turn the bowl counterclockwise.
3. Bring the bottom mixture to the top.

Piping:

With bag:

1. Use a bag with a disposable tip; cut the bag at 45-degree angle.
2. Fill to no more than half full.
3. Burp the bag.

With cone:

1. Cut and fold the piping cone to the appropriate size.
2. Fill the cone with a small amount.
3. Fold the ends to form a triangle.
4. Pipe the desired designs.

HAZARDOUS FOOD:

Egg whites

Hazelnut Biscotti

YIELD:	2 pounds, 4⅛ ounces. 30 Biscuits	4 pounds, 8¼ ounces. 60 Biscuits

INGREDIENTS:

Hazelnuts	4 ounces	8 ounces
Butter, unsalted	4 ounces	8 ounces
Sugar, granulated	5 ounces	10 ounces
Lemon zest	½ ounce	1 ounce
Eggs, whole	6 ounces	12 ounces
Extract, vanilla	To taste	To taste
Flour, pastry, sifted	1 pound	2 pounds
Baking powder	½ ounce	1 ounce
Salt	⅛ ounce	¼ ounce

METHOD OF PREPARATION:

1. Gather the equipment and ingredients.
2. Toast the hazelnuts, chop coarsely, and set aside.
3. Place the butter, granulated sugar, and lemon zest in a bowl; cream together until light and fluffy. Scrape the bowl well.
4. Add in the eggs and vanilla extract gradually; incorporate well.
5. Sift together the dry ingredients: pastry flour, baking powder, and salt.
6. Add the sifted dry ingredients to the butter mixture in three stages.
7. Stir in the chopped hazelnuts.
8. Divide the dough into three equal portions.
9. Roll each portion into a 3½-inch-diameter log.
10. Place the logs on parchment-lined sheet pans.
11. Press the dough down to increase the diameter, and create an arch shape.
12. Bake at 350° F for approximately 20 minutes.
13. Remove from the oven, and, with a sharp knife, cut the logs into ¾-inch slices.
14. Place the slices on sheet pans, and continue baking until they are dry and lightly browned on the sides.
15. Remove from the oven, and transfer to sheet pans to cool.

PASTRY TECHNIQUES:

Creaming, Combining, Rolling, Baking, Slicing

Creaming:

1. Soften the fats on low speed.
2. Add the sugar(s) and cream; increase the speed slowly.
3. Add the eggs one at a time; scrape the bowl frequently.
4. Add the dry ingredients in stages.

Combining:

Bringing together two or more components.

1. Prepare the components to be combined.
2. Add one to the other, using the appropriate mixing method (if needed).

Rolling:

1. Prepare the rolling surface by dusting with the appropriate medium (flour, cornstarch, etc.).
2. Use the appropriate style pin (stick pin or ball bearing pin) to roll the dough to desired thickness; rotate the dough during rolling to prevent sticking.

Baking:

1. Preheat the oven.
2. Position the item appropriately in the oven.
3. Check for appropriate firmness and/or color.

Slicing:

1. Prepare the product for cutting; clean and clear the work area.
2. Slice the product using the "claw" grasp or the rocking motion.

HAZARDOUS FOOD:

Eggs

CHEF NOTE:

If preferred, the ends can be dipped into chocolate; make sure the biscotti are cool before dipping.

Hippen Paste No. 1

Almond Wafer Paste

PASTRY TECHNIQUES:

Combining, Spreading

Combining:

Bringing together two or more components.

1. Prepare the components to be combined.
2. Add one to the other, using the appropriate mixing method (if needed).

Spreading:

1. Using an icing spatula or off-set spatula, smooth the icing or other spreading medium over the surface area.

HACCP:

Store unbaked paste at 40° F.

HAZARDOUS FOODS:

Egg whites
Milk

YIELD:	3 pounds, 14 ounces.	7 pounds, 12 ounces.
INGREDIENTS:		
Almond paste	1 pound, 5 ounces	2 pounds, 10 ounces
Sugar, confectionery	14 ounces	1 pound, 12 ounces
Flour, bread	7 ounces	14 ounces
Cinnamon, ground (optional)	To taste	To taste
Egg whites	1 pound, 1 ounce	2 pounds, 2 ounces
Milk, whole	3 ounces	6 ounces
Extract, vanilla	To taste	To taste

METHOD OF PREPARATION:

1. Gather the equipment and ingredients.
2. Place the almond paste, sugar, bread flour, and cinnamon (optional) in a bowl; paddle until crumbly.
3. Combine the egg whites, milk, and vanilla extract.
4. Slowly add to the almond paste mixture. This must be done slowly, or the mixture will become lumpy and will have to be strained.
5. Store in a container.

To Make Cookies:

1. Grease the sheet pan well.
2. Make a pattern form, or use a stencil.
3. Spread the batter evenly over the stencil.
4. Remove the stencil.
5. Bake at 400° F for about 5 to 7 minutes, or until golden brown.
6. Remove from the pan, and shape accordingly while still warm.

Hippen Paste No. 2

Almond Wafer Paste

PASTRY TECHNIQUES:

Combining, Spreading

Combining:

Bringing together two or more components.

1. Prepare the components to be combined.
2. Add one to the other, using the appropriate mixing method (if needed).

Spreading:

1. Using an icing spatula or off-set spatula, smooth the icing or other spreading medium over the surface area.

HACCP:

Store unbaked paste at 40° F.

HAZARDOUS FOODS:

Egg whites
Milk

YIELD:	15 ounces.	3 pounds, 8 ounces.
INGREDIENTS:		
Almond paste	7 ounces	1 pound, 12 ounces
Sugar, granulated	2 ounces	7 ounces
Flour, bread	2 ounces	7 ounces
Egg whites	2 ounces	7 ounces
Milk, whole	2 ounces	7 ounces
Extract, vanilla	To taste	To taste

METHOD OF PREPARATION:

1. Gather the equipment and ingredients.
2. Place the almond paste, sugar, and flour in a bowl, and paddle until crumbly.
3. Combine the egg whites, milk, and vanilla extract.
4. Slowly add the liquid mixture to the almond paste mixture. This must be done slowly to prevent lumps from forming in the batter.
5. Mix well.
6. Store in a container.

To Make Cookies:

1. Grease the sheet pans well.
2. Make a pattern form, or use a stencil.
3. Spread the batter evenly over the stencil.
4. Remove the stencil.
5. Bake at 400° F for 5 to 7 minutes, or until light, golden brown.
6. Remove from the pan.
7. Shape accordingly while still warm.

Ice Box Cookies

PASTRY TECHNIQUES:
Creaming, Rolling

Creaming:
1. Soften the fats on low speed.
2. Add the sugar(s) and cream; increase the speed slowly.
3. Add the eggs one at a time; scrape the bowl frequently.
4. Add the dry ingredients in stages.

Rolling:
1. Prepare the rolling surface by dusting with the appropriate medium (flour, cornstarch, etc.).
2. Use the appropriate style pin (stick pin or ball bearing pin) to roll the dough to desired thickness; rotate the dough during rolling to prevent sticking.

HACCP:
Store unbaked dough at 40° F.

HAZARDOUS FOODS:
Egg yolks
Eggs
Milk

YIELD:			
	Vanilla	3 pounds, 1½ ounces.	6 pounds, 3 ounces.
	Chocolate	3 pounds, 2¾ ounces	6 pounds, 5½ ounces

INGREDIENTS:

Vanilla Portion:

Sugar, confectionery	8 ounces	1 pound
Butter, unsalted, softened	8 ounces	1 pound
Shortening, all-purpose	8 ounces	1 pound
Egg yolks	1½ ounces	3 ounces
Salt	Pinch	Pinch
Lemon rind, grated	1 each	2 each
Flour, cake	12 ounces	1 pound, 8 ounces
Flour, bread	12 ounces	1 pound, 8 ounces

Chocolate Portion:

Sugar, confectionery	10 ounces	1 pound, 4 ounces
Butter, unsalted	5 ounces	10 ounces
Shortening, all-purpose	5 ounces	10 ounces
Eggs, whole	5 ounces	10 ounces
Salt	Pinch	Pinch
Milk, whole	2 ounces	4 ounces
Baking soda	Pinch	Pinch
Flour, cake	10 ounces	1 pound, 4 ounces
Flour, bread	10 ounces	1 pound, 4 ounces
Cocoa powder	3¾ ounces	7½ ounces

METHOD OF PREPARATION:

Gather the equipment and ingredients.

Preparation of Vanilla Portion:

1. Place the sugar, butter, and shortening in a bowl; cream until smooth.
2. Add the eggs, salt, and flavorings; combine well.
3. Sift the dry ingredients; add to the creamed mixture.
4. Mix slowly until a smooth paste, but *do not overmix*.
5. Refrigerate until cold.

Preparation of Chocolate Portion:

1. Place the sugar, butter, and shortening in a bowl; cream until smooth.
2. Add the eggs, salt, and milk; combine well.
3. Sift the dry ingredients; add to the creamed mixture.
4. Mix slowly until a smooth paste, but *do not overmix.*
5. Refrigerate until cold.

Construction of Cookies:

1. Roll and shape.
2. Place on parchment-lined sheet pans.
3. Bake at 365° F for approximately 10 to 12 minutes, or until golden brown.

Ladyfingers No. 1

YIELD:	1 pound, 3½ ounces.	2 pounds, 7 ounces.

INGREDIENTS:

Egg yolks	4 ounces	8 ounces
Sugar, granulated	6 ounces	12 ounces
Egg whites	5 ounces	10 ounces
Cream of tartar	Pinch	Pinch
Lemon juice	½ ounce	1 ounce
Flour, cake, sifted	4 ounces	8 ounces
Sugar, confectionery (optional)	As needed	As needed

METHOD OF PREPARATION:

1. Gather the equipment and ingredients.
2. Place the egg yolks and 3 ounces of granulated sugar in bowl; whip until thick and light in color.
3. In another bowl, place the egg whites; add a pinch of cream of tartar, and whip until firm; gradually add 3 more ounces of granulated sugar.
4. Add the lemon juice to the yolks, and incorporate.
5. Fold the yolks into the whipped egg whites.
6. Fold the sifted cake flour gently into the mixture.
7. Using a plain tube, pipe the batter immediately onto a parchment-lined sheet pan.
8. Bake at 425° F for 8 minutes, or until golden brown.

PASTRY TECHNIQUES:

Whipping, Folding, Piping

Whipping:

1. Hold the whip at a 55-degree angle.
2. Create circles, using a circular motion.
3. The circular motion needs to be perpendicular to the bowl.

Folding:

Do steps 1, 2, and 3 in one continuous motion.

1. Run a bowl scraper under the mixture, across the bottom of the bowl.
2. Turn the bowl counterclockwise.
3. Bring the bottom mixture to the top.

Piping:

With bag:

1. Use a bag with a disposable tip; cut the bag at 45-degree angle.
2. Fill to no more than half full.
3. Burp the bag.

With cone:

1. Cut and fold the piping cone to the appropriate size.
2. Fill the cone with a small amount.
3. Fold the ends to form a triangle.
4. Pipe the desired designs.

HAZARDOUS FOODS:

Egg yolks
Egg whites

CHEF NOTE:

If a crispy surface is desired, sift confectionery sugar on the top.

Ladyfingers No. 2

YIELD:	1 pound, 9¾ ounces.	3 pounds, 3½ ounces.

INGREDIENTS:

Ingredient		
Egg whites	7 ounces	14 ounces
Sugar, granulated	6½ ounces	13 ounces
Egg yolks	4¾ ounces	9½ ounces
Flour, pastry, sifted	5 ounces	10 ounces
Cornstarch, sifted	2½ ounces	5 ounces
Sugar, confectionery	As needed	As needed

METHOD OF PREPARATION:

1. Gather the equipment and ingredients.
2. Place the egg whites in a bowl, and whip to medium-stiff peak; slowly add in the granulated sugar to make a meringue.
3. Place the egg yolks in a bowl, and whip to full volume.
4. Sift together the pastry flour and cornstarch.
5. Gently fold the meringue into the whipped egg yolks.
6. Gently fold the sifted dry ingredients into the egg mixture.
7. Pipe immediately using a medium-straight tip; pipe the mixture into finger-shaped cookies on a parchment-lined sheet pan.
8. Sprinkle the cookies with confectionery sugar.
9. Bake at 350° F for 5 minutes, or until just set.

PASTRY TECHNIQUES:

Whipping, Folding, Piping

Whipping:

1. Hold the whip at a 55-degree angle.
2. Create circles, using a circular motion.
3. The circular motion needs to be perpendicular to the bowl.

Folding:

Do steps 1, 2, and 3 in one continuous motion.

1. Run a bowl scraper under the mixture, across the bottom of the bowl.
2. Turn the bowl counterclockwise.
3. Bring the bottom mixture to the top.

Piping:

With bag:

1. Use a bag with a disposable tip; cut the bag at 45-degree angle.
2. Fill to no more than half full.
3. Burp the bag.

With cone:

1. Cut and fold the piping cone to the appropriate size.
2. Fill the cone with a small amount.
3. Fold the ends to form a triangle.
4. Pipe the desired designs.

HAZARDOUS FOODS:

Egg whites
Egg yolks

Oatmeal Chocolate Chip Cookies

PASTRY TECHNIQUES:

Creaming, Combining, Portioning

Creaming:

1. Soften the fats on low speed.
2. Add the sugar(s) and cream; increase the speed slowly.
3. Add the eggs one at a time; scrape the bowl frequently.
4. Add the dry ingredients in stages.

Combining:

Bringing together two or more components.

1. Prepare the components to be combined.
2. Add one to the other, using the appropriate mixing method (if needed).

Portioning:

1. Mark the product for portioning, using a ruler, if necessary.
3. Cut, spoon, or scoop the product with the appropriate-sized utensil.

HAZARDOUS FOOD:

Eggs

YIELD:	2 pounds, 1½ ounces.	8 pounds, 6½ ounces.

INGREDIENTS:

Shortening, all-purpose	2 ounces	8 ounces
Butter, unsalted	2 ounces	8 ounces
Sugar, brown	3½ ounces	14 ounces
Sugar, granulated	3¾ ounces	15 ounces
Eggs, whole	2 ounces	8 ounces
Extract, vanilla	To taste	To taste
Coconut, shredded	2 ounces	8¼ ounces
Flour, pastry	3 ounces	12½ ounces
Salt	Pinch	¼ ounce
Baking soda	Pinch	¼ ounce
Cinnamon, ground	Pinch	¼ ounce
Oats, old-fashioned	6¼ ounces	1 pound, 8 ounces
Raisins (optional)	3 ounces	12 ounces
Nuts (optional)	3 ounces	12 ounces
Chips, chocolate or butterscotch	3 ounces	12 ounces

METHOD OF PREPARATION:

1. Gather the equipment and ingredients.
2. Paddle together the shortening, butter, brown sugar, and granulated sugar until light and fluffy.
3. In two stages, add the eggs and vanilla extract; mix well.
4. Add the shredded coconut.
5. Add the sifted dry ingredients: pastry flour, salt, baking soda, and cinnamon; mix only until blended.
6. Fold in the oats by hand.
7. Add the raisins, nuts, and chips, if desired.
8. Scale 1½-ounce portions, and place on parchment-lined sheet pans.
9. Bake at 360° F for 10 to 12 minutes, or until golden brown.

Oatmeal Cookies

PASTRY TECHNIQUES:

Creaming, Combining, Rolling

Creaming:

1. Soften the fats on low speed.
2. Add the sugar(s) and cream; increase the speed slowly.
3. Add the eggs one at a time; scrape the bowl frequently.
4. Add the dry ingredients in stages.

Combining:

Bringing together two or more components.

1. Prepare the components to be combined.
2. Add one to the other, using the appropriate mixing method (if needed).

Rolling:

1. Prepare the rolling surface by dusting with the appropriate medium (flour, cornstarch, etc.).
2. Use the appropriate style pin (stick pin or ball bearing pin) to roll the dough to desired thickness; rotate the dough during rolling to prevent sticking.

HAZARDOUS FOOD:

Eggs

YIELD:	1 pound, 10 ounces. 3¼ Rolls	1 pound, 10 ounces. 6½ Rolls
INGREDIENTS:		
Sugar, granulated	10 ounces	1 pound, 4 ounces
Sugar, brown	1 pound	2 pounds
Shortening, all-purpose	14 ounces	1 pound, 12 ounces
Nulomoline (invert sugar)	2 ounces	4 ounces
Salt	½ ounce	1 ounce
Baking soda	¼ ounce	½ ounce
Extract, vanilla	To taste	To taste
Eggs, whole	8 ounces	1 pound
Oats, old-fashioned	1 pound, 2 ounces	2 pounds, 4 ounces
Flour, pastry	13 ounces	1 pound, 10 ounces
Nuts, chopped	4 ounces	8 ounces

METHOD OF PREPARATION:

1. Gather the equipment and ingredients.
2. Place the granulated sugar, brown sugar, shortening, nulomoline, salt, and baking soda in a bowl with a paddle. Paddle until well blended.
3. Combine the vanilla extract and eggs.
4. Add the egg mixture slowly to the sugar-shortening mixture; scrape the bowl well.
5. Add the oatmeal, pastry flour, and nuts; mix until blended together.
6. Scale 1 pound, 10-ounce pieces of mixture; roll, and place into even logs.
7. Place the rolls on sheets pans, and refrigerate to chill.
8. To bake, remove the parchment paper from the logs, and cut to a desired thickness.
9. Bake at 375° F for 8 to 10 minutes, or until golden brown.

Peanut Butter Cookies

PASTRY TECHNIQUES:

Creaming, Combining, Rolling

Creaming:

1. Soften the fats on low speed.
2. Add the sugar(s) and cream; increase the speed slowly.
3. Add the eggs one at a time; scrape the bowl frequently.
4. Add the dry ingredients in stages.

Combining:

Bringing together two or more components.

1. Prepare the components to be combined.
2. Add one to the other, using the appropriate mixing method (if needed).

Rolling:

1. Prepare the rolling surface by dusting with the appropriate medium (flour, cornstarch, etc.).
2. Use the appropriate style pin (stick pin or ball bearing pin) to roll the dough to desired thickness; rotate the dough during rolling to prevent sticking.

HAZARDOUS FOODS:

Eggs
Milk

YIELD:	4 pounds, 6⅛ ounces.	8 pounds, 12½ ounces.
INGREDIENTS:		
Shortening, all-purpose	7½ ounces	15 ounces
Peanut butter	12 ounces	1 pound, 8 ounces
Sugar, granulated	12 ounces	1 pound, 8 ounces
Salt	⅓ ounce	⅔ ounce
Extract, vanilla	To taste	To taste
Baking soda, sifted	⅛ ounce	¼ ounce
Eggs, whole	3½ ounces	7 ounces
Milk, whole	12 ounces	1 pound, 8 ounces
Flour, pastry	1 pound, 6 ounces	2 pounds, 12 ounces
Baking powder	¾ ounce	1½ ounces

METHOD OF PREPARATION:

1. Gather the equipment and ingredients together.
2. Place the shortening, peanut butter, granulated sugar, salt, vanilla extract, and baking soda in a bowl with a paddle; cream together for about 3 minutes.
3. Add the whole eggs, and mix well; scrape the bowl.
4. Add the milk, and mix well; scrape the bowl.
5. Sift together the pastry flour and baking powder; add to the mixture, and mix well.
6. Scale 1 pound, 2 ounces of dough onto parchment paper; roll into an even log approximately 12 inches in length.
7. Chill until firm.
8. To bake, remove the paper from the log, and cut into the desired thickness.
9. Place on parchment-lined sheet pans.
10. Bake at 375° F for 8 to 10 minutes, or until light golden brown.

Pecan Diamonds

YIELD:	½ Sheet Pan.	1 Sheet Pan.

INGREDIENTS:

Short dough (see page 840)	1 pound, 12 ounces	3 pounds, 8 ounces

Filling:

Butter, unsalted	1 pound	2 pounds
Sugar, granulated	4 ounces	8 ounces
Honey	12 ounces	1 pound, 8 ounces
Sugar, brown	1 pound	2 pounds
Pecan pieces	2 pounds	4 pounds
Cream, heavy	4 ounces	8 ounces

METHOD OF PREPARATION:

Gather the equipment and ingredients.

To Prepare Pan:

1. Roll out the dough, and line the sheet pan with dough.
2. Bake the dough at 425° F for 5 minutes, or until prebaked.

Filling:

1. In a pan, combine the butter, granulated sugar, honey, and brown sugar.
2. Boil for 3 minutes, or to 257° F.
3. Stir in the pecan pieces and heavy cream.
4. Pour into the prepared pan.
5. Bake at 350° F for 20 minutes, or until the mixture bubbles and the top is semi-firm.
6. Cool.
7. Cut into diamond shapes.

PASTRY TECHNIQUES:

Rolling, Boiling, Combining, Pouring

Rolling:

1. Prepare the rolling surface by dusting with the appropriate medium (flour, cornstarch, etc.).
2. Use the appropriate style pin (stick pin or ball bearing pin) to roll the dough to desired thickness; rotate the dough during rolling to prevent sticking.

Boiling:

1. Bring the cooking liquid to a rapid boil.
2. Stir the contents.

Combining:

Bringing together two or more components.

1. Prepare the components to be combined.
2. Add one to the other, using the appropriate mixing method (if needed).

Pouring:

1. Place the product in an appropriate container for pouring: a pitcher or large ladle.
2. Pour the product into desired containers or over another product.

HAZARDOUS FOOD:

Heavy cream

Tuiles

PASTRY TECHNIQUES:

Combining, Spreading

Combining:

Bringing together two or more components.

1. Prepare the components to be combined.
2. Add one to the other, using the appropriate mixing method (if needed).

Spreading:

1. Using an icing spatula or off-set spatula, smooth the icing or other spreading medium over the surface area.

HACCP:

Store unbaked batter at 40° F.

HAZARDOUS FOODS:

Eggs
Egg whites

YIELD:	1 pound, 3½ ounces.	2 pounds, 7 ounces.

INGREDIENTS:

Sugar, confectionery	4½ ounces	9 ounces
Flour, pastry	2 ounces	4 ounces
Eggs, whole	4 ounces	8 ounces
Egg whites	2 ounces	4 ounces
Extract, vanilla	To taste	To taste
Butter, unsalted, softened	1½ ounces	3 ounces
Almonds, blanched and sliced	5½ ounces	11 ounces
Butter, unsalted	As needed	As needed

METHOD OF PREPARATION:

1. Gather the equipment and ingredients.
2. Sift together the confectionery sugar and pastry flour.
3. Whisk together the whole eggs, egg whites, and vanilla extract.
4. Place the dry ingredients into a bowl, and slowly whisk in the liquids; be careful to prevent lumps from forming.
5. Add the softened butter.
6. Fold in the sliced almonds.
7. Allow the dough to rest for 1 hour.
8. Butter the bottom of a sheet pan, and freeze.
9. Spoon the rested dough onto the inverted frozen sheet pan, and press out with a fork dipped in water until the dough is 2½ inches in diameter and as thin as a sliced almond.
10. Bake at 375° F for 5 minutes, or until golden brown.
11. Form the warm cookies into a cup shape over the end of a 1½-inch-diameter dowel.

Tulip or Cigarette Paste

PASTRY TECHNIQUES:

Combining, Spreading

Combining:

Bringing together two or more components.

1. Prepare the components to be combined.
2. Add one to the other, using the appropriate mixing method (if needed).

Spreading:

1. Using an icing spatula or off-set spatula, smooth the icing or other spreading medium over the surface area.

Stuffing:

1. Place the stuffing inside the cavity of the food product using a gloved hand or a piping bag.
2. Be sure to fill the cavity completely.

HACCP:

Store unbaked batter at 40° F.

HAZARDOUS FOOD:

Egg whites

YIELD:	1 pound, 2 ounces.	2 pounds, 4 ounces.

INGREDIENTS:		
Butter, unsalted	3½ ounces	7 ounces
Sugar, confectionery	6 ounces	12 ounces
Flour, bread	5 ounces	10 ounces
Egg whites	3½ ounces	7 ounces
Butter, additional, unsalted	As needed	As needed

METHOD OF PREPARATION:

1. Gather the equipment and ingredients.
2. Place the butter, sugar, and bread flour in a bowl; paddle until smooth.
3. Add the egg whites slowly to avoid creating lumps; mix until smooth.
4. Butter a sheet pan, and place in the freezer for 10 minutes.
5. Make a pattern form, or use a stencil.
6. Place the stencil on the cold sheet pan, and spread the batter evenly over the stencil.
7. Bake at 350° F for 5 minutes, or until very light brown around the edges.
8. Remove from the pan immediately; mold into shape while hot, as desired.

The Ten Steps of Bread Making

There are basically 10 steps in the baking process when using a straight dough method. If a sponge or pre-dough method is used, these 10 steps are usually followed from the mixing stage (step 2) *after* the mixing and fermentation of the sponge. Although formulas vary, if the 10 steps to baking bread are understood, they can be used as a guideline for making any yeast-raised product.

- **Step 1: Scaling the Ingredients**
  The exact scaling of ingredients is an integral part of the baking process. If ingredients have not been scaled properly, an inconsistent and often inferior product will result. The proper use of a baker's scale can ensure that amounts of ingredients conform to a given formula.

  Eggs, oil, liquid shortening, malt, honey, molasses, and other heavy liquids with a viscosity denser than milk or water should be weighed on the baker's scale. Water and milk can be measured with liquid measures: 1 pint equals 1 pound.
- **Step 2: Mixing of Yeast-Raised Dough**
  Mixing accomplishes two major objectives: It evenly and thoroughly distributes ingredients, and it allows for maximum development of the gluten. The procedure for mixing doughs can vary with the type of yeast used. The method that follows uses compressed, fresh yeast. If instant yeast is used, it must be added just after the liquids have absorbed into the flour. For a straight dough, the water and fresh yeast are mixed together.

  Over-mixing will give a let-down, or complete breakdown, of the dough. The dough will be warm and sticky and will come apart easily. There are four stages in the mixing process:

1. **Pick-up**
   A low speed is used to mix the flour and dry ingredients together; then the water and yeast are added. Any other wet ingredients are usually added to the water and yeast just before they are added to the dry ingredients. All of the water should be added before the kneader is accelerated to medium speed. Any solid fats that are used in the formula are added last; oils are usually added at the beginning of the mixing process, just after the water and yeast.
2. **Clean-up**
   During the clean-up stage, the ingredients come together, and the bottom of the bowl can be seen clearly. At this point, the kneader is accelerated to medium speed.
3. **Development**
   This is the longest stage in the mixing process. It is called the development stage because the gluten is being developed. Oxygen is being absorbed into the dough. The starch will come to the surface as gluten is forming, and the dough will be whiter. The dough will tear easily, and the color will be uneven.
4. **Final clear**
   The final clear stage occurs when the proper development of gluten has been obtained; when a small piece of dough is cut from the mass, it is stretched to a thinness that light will shine clearly through. The dough also can be stretched a few times without tearing. Once the dough reaches this stage, it is properly developed and can be removed from the mixing bowl.

   The dough should be lightly coated with oil before floor fermentation to prevent the dough from sticking to its container (usually a bulk proof box or trough) and from developing a skin. If dough improvers, conditioners, or instant yeast are used, bulk fermentation (step 3) is not necessary. If conditioners are used, bulk fermentation is eliminated, and the dough may be divided immediately.

- **Step 3: Bulk or Floor Fermentation**
  During the fermentation process, yeast cells act on sugary agents in the dough and produce carbon dioxide (CO_2) gas and alcohol. The gluten structure, formed by mixing, will contain the CO_2 gas produced by the yeast. The gluten is conditioned and mellowed and becomes more elastic as a result of the effect of the alcohol given off by the yeast and the lower acidity in the dough. The important factors for good fermentation are
  - Properly developed dough
  - Proper humidity of between 75% and 85%. The dough should be kept covered and away from drafts. Bulk fermentation is done in a proof box where the humidity can be controlled and observed.
  - A dough temperature between 78° F and 82° F, for optimum fermentation.

  Fermentation must be regulated throughout to ensure proper flavor and conditioning.

SIDEBAR:

In many bake shops, fermentation time is cut considerably with the use of dough conditioners; these chemically conditioned additives eliminate the need for bulk fermentation and save on labor costs. Unfortunately, the rich yeast flavor of bulk-fermented breads is lost in this type of product. In addition, the texture and crumb of a bread or roll made with conditioners is not as high in quality as that of bulk-fermented bread.

- **Step 4: Dividing the Dough and Punch Down**
  When bulk fermentation has caused the dough to rise (in most cases, close to double in volume or 50% fermentation), it is tested for punch down—the fingers are inserted into the dough to the knuckles, and the dough is observed to see if the finger marks leave a slight indentation and then close very slowly—if so, the dough is ready for punch down. Punch down is done by folding the sides of the dough into the middle and turning the dough over. Punching keeps the dough at an even temperature by turning the dough inside out; it also releases some of the carbon dioxide, which, if allowed to stay concentrated, will eventually restrict fermentation; it introduces fresh oxygen to the dough to help with fermentation, and it helps develop gluten.

  Dividing the dough should be done as quickly and accurately as possible. The dough continues to ferment during dividing, and delay will cause over-fermentation of the last unit scaled. The dough is divided using a bench scraper and is weighed to the desired weight, either for a dough press or for individual loaves. Leftover dough from inaccurate scaling should be divided evenly among the number of large pieces and tucked under. Smaller pieces will ferment faster and must be incorporated properly.
- **Step 5: Rounding or Folding Over**
  Rounding or folding the dough over are ways for the newly divided and scaled dough to be shaped into better condition for the further shaping that will come after the second fermentation. Rounding also prevents too much gas loss by providing the dough with a thin skin that entraps the gas. Folding over allows old gases to escape and new gases to be created. When rounding or folding over, the pieces of dough should be kept in order—the first piece rounded or folded should be the first piece shaped after having been bench rested, so that all of the rounded pieces of dough will receive the proper fermentation time.
- **Step 6: Bench Rest/Intermediate Proof**
  This is a short resting period in which the rounded or folded pieces of dough are relaxed. Fermentation continues, and the pieces become gassy; the gluten relaxes in this step to make future handling of the dough easier. The dough must be covered to keep in moisture and prevent crust formation; usually the dough is put back in the proof box, even for this short period, to ensure that the proper humidity and temperature are maintained. Sometimes it is left in the dough troughs; the troughs cover each other as they are stacked and prevent crust formation. It is during this rest period that the dough will begin to feel lighter and softer, and more yeast flavor will develop.
- **Step 7: Shaping**
  This is the step in which the properly rested dough is formed into the desired shape. Shaping should be

done quickly, with as little dusting flour as possible, to prevent the dough from drying. If a seam is a part of the finished product, it should be placed on the bottom; a proper seam should be straight and tight, because the seam is the weakest part of any shaped product and could open during baking. Rolls, braids, and other rustic loaves are all shaped at this stage. The first piece of dough that was rounded or folded over is the first piece of dough to be shaped.

- **Step 8: Panning**
  At this stage, the proper pan is selected and prepared accordingly, depending on the type of bread being made: pullman loaves, French baguettes, soft rolls, and so on. (Bread formulas will indicate the proper mise en place for panning.) All washes and appropriate garnishes should be applied after panning. Generally, lean doughs are placed on perforated pans with cornmeal and are water-washed; soft or medium doughs are placed on sheet pans with parchment or in lightly greased loaf pans and are egg-washed.
- **Step 9: Final Proof**
  This stage should achieve maximum fermentation prior to the baking of the product. Final proofing should occur in a proof box, which produces a warm and humid environment. The proof box should be maintained at a temperature of 90° F to 95° F, with a humidity of 80% to 90%. In most cases, a properly proofed product should be nearly doubled in volume and light and airy to the touch, and will close around a finger indentation slowly without collapsing.

  Proofing time is affected by the type of dough and the size and shape of the product. Sweet doughs should be proofed only partially before they are baked to prevent the richness of the dough from exerting extra weight on the gluten strands and compromising the final structure of the item.

  Some items, particularly hard or lean doughs, are stippled (cut) either before, during, or after final proof. *Stippling* is the term used for the characteristic and decorative cuts made on the top of certain types of bread. Stippling improves the look of the product by making it more interesting visually; it also lightens the product by allowing the crumb to move upward and expand during baking. If done correctly, stippling cuts will look like a scar on the surface of the product.

  For most breads, stippling is done during the last 20% of the final proof, or when the loaf has three fourths of its maximum volume. The loaf will then expand more by moving up. A razor is better than a serrated knife for stippling, because the knife tends to drag across the dough. If the dough is very wet, using light rye flour will help the knife cut cleanly. In either case, the blade should never be dragged across the dough; a clear, sharp cut is desired. The following describes the method for stippling:
  - Hold the blade almost parallel to the length of the dough.
  - Cut just under the surface of the dough, not a deep, straight cut.
  - The cuts must overlap by one third the length of each cut.
  - All cuts must be the same length.
  - The cuts must cover the full length of the dough.
  - Use both hands, one to steady the product and the other to cut.
- **Step 10: Baking**
  Baking changes and transforms the dough into an appetizing, desirable product. Baking times and temperatures are determined by the type of dough, the size of the unit, the richness of the unit, the crust color desired, and, often, the weather. Generally, the leaner the dough, the higher the oven temperature and the shorter the baking time; richer doughs usually require lower oven temperatures and longer baking periods. As the product is baked, the internal temperature rises, and four major changes take place, in the following order:
  - **Natural Oven Spring**
    This is a sudden rise and expansion due to the last effort of the yeast, reacting to the heat of the oven, to ferment before it is killed at 138° F to 140° F and the rapid expansion of the carbon dioxide created during the proofing process. This process occurs during the first 5 minutes or so of baking. Some reasons for the failure of this reaction include the following: There was too much salt in the dough;

the product was over-proofed; not enough yeast was added during mixing. The dough is very soft at this stage and will collapse if touched.

- **Starch Gelatinization**
  At 130° F., the starch granules will begin to swell from a transfer of moisture from other ingredients; as they swell, they will become fixed in the gluten structure. At 150° F, the starches will gelatinize and become the chief structure of the dough, rather than the gluten, which begins to dry out and coagulate at 165° F and continues to coagulate until the product is finished baking. The structure will begin to finalize in the crumb.
- **Protein Coagulation**
  The crust will be formed at 165° F due to the exposed starch and sugar at the surface of the dough. This finalizes the structure of the product. At this point, the product will begin to appear done, but the taste will still be heavy with alcohol because evaporation of alcohol has not yet taken place.
- **Alcohol Evaporation**
  The alcohol given off by the yeast as a by-product will be burned off from the product at 176° F. The finished bread will have an approximate internal temperature of 220° F.
- **Baking with Steam**
  Some breads, such as French and Italian loaves, require a thin, hard, crispy crust. Most professional bakers accomplish this through the use of steam in the oven during baking. Steam provides a moist environment that prevents a crust from forming too soon; the sugar present in the dough mixes with the moisture on the surface of the product and caramelizes. When the steam is released by the opening of the oven's damper, a crispy crust is formed. All ovens have different steam pressures, and timing must be adjusted to each situation, as well as how many seconds of steam will actually be applied. Older ovens and deck ovens may need to be loaded with steam prior to placing the product in the oven. Steam will produce a few major effects on a product: (1) The moisture helps improve natural oven spring by creating a proof box–type environment. This keeps the product soft and creates a glossy shine. (2) Steam prevents cracking of the crust during baking. (3) Starches and sugars mix with the steam, caramelizing to create crust color. The damper is then opened to let the steam out and prevent over-browning. In older products (old dough), the yeast will have digested a large amount of sugar, and the product will have less color.

PASTRY TECHNIQUE:

The Ten Steps of Bread Making (see page 897)

Basic Lean Dough

(Hard)

YIELD: 24 pounds, 15½ ounces. Baker's Percentage.

INGREDIENTS:		
Flour, high-gluten, sifted	15 pounds	100
Water	9 pounds	60
Yeast, compressed	4½ ounces	1.8
Salt	4½ ounces	1.8
Dough conditioner (optional)	6½ ounces	2

METHOD OF PREPARATION:

1. Gather the equipment and ingredients.
2. Follow the 10 steps of bread making (page 897).

Variations (items that can be produced from basic lean dough):

1. French breads: baguettes, batard
2. Italian breads
3. Crusty rolls
4. Pizza shells

Biga

(White Starter)

PASTRY TECHNIQUE:
Mixing

Mixing:
1. Follow the proper mixing procedure: creaming, blending, whipping, or combination.

HACCP:
Store for 12 hours at 60° F to 65° F, or refrigerate for up to 7 days.

YIELD: 16 pounds, 1¼ ounces. Baker's Percentage.

INGREDIENTS:		
Yeast, compressed	1¼ ounces	0.86
Water	7 pounds	77.80
Flour, high-gluten	9 pounds	100

METHOD OF PREPARATION:

1. Gather the equipment and ingredients.
2. Dissolve the yeast in water.
3. Add the flour.
4. Mix for 2 minutes on low speed.
5. Mix for 2 minutes on medium speed.
6. Place in bowl and leave for 12 hours at room temperature or refrigerate for up to 7 days.

Brioche

PASTRY TECHNIQUE:

The Ten Steps of Bread Making (see page 897)

HAZARDOUS FOODS:

Milk
Eggs

YIELD:	10 pounds, 5½ ounces.	Baker's Percentage.

INGREDIENTS:		
Yeast, compressed	5 ounces	6.9
Milk, whole	1 pound	21
Eggs, whole	2 pounds	44
Flour, pastry	1 pound, 2 ounces	25
Flour, bread	3 pounds, 6 ounces	75
Sugar, granulated	5 ounces	6.9
Salt	½ ounce	0.7
Butter, unsalted, soft	2 pounds, 4 ounces	50

METHOD OF PREPARATION:

1. Gather the equipment and ingredients.
2. Dissolve the yeast in the milk and eggs.
3. Add all of the dry ingredients; mix on medium speed for 5 minutes.
4. Slice the butter into ½-inch pieces; incorporate into dough on medium speed for 2 minutes.
5. Refrigerate overnight on a floured surface. Cover with a damp cloth, and seal with a plastic bag.
6. On the next day, remove the dough from the refrigerator.
7. Scale.
8. Mold.
9. Proof.
10. Egg-wash.
11. Bake at 375° F for approximately 20 minutes, or until brown on all sides.

Challah

PASTRY TECHNIQUE:

The Ten Steps of Bread Making (see page 897)

HAZARDOUS FOOD:

Eggs

YIELD: 7 pounds, 11½ ounces. Baker's Percentage.

INGREDIENTS:

Ingredient	Amount	Baker's Percentage
Water	2 pounds	50
Yeast, compressed	4 ounces	6.25
Sugar, granulated	6 ounces	9.38
Salt	1½ ounces	2.30
Oil, vegetable	6 ounces	9.38
Eggs, whole	10 ounces	56
Flour, high-gluten	4 pounds	100

METHOD OF PREPARATION:

1. Gather the equipment and ingredients.
2. Follow the 10 steps of bread making (page 897).

Cinnamon-Raisin Bread

PASTRY TECHNIQUE:

The Ten Steps of Bread Making (see page 897)

HAZARDOUS FOOD:

Eggs

YIELD:	19 pounds, 14 ounces.	Baker's Percentage.

INGREDIENTS:

Flour, high-gluten, sifted	10 pounds	100
Sugar, granulated	13 ounces	8
Salt	3 ounces	2
Water	3 pounds, 14 ounces	39
Yeast, compressed	12½ ounces	7.75
Eggs, whole	1 pound	10
Shortening, high-ratio	10 ounces	6
Raisins	2 pounds, 8 ounces	25
Cinnamon, ground	1½ ounces	1
Dough conditioner (optional)	3½ ounces	1–2

METHOD OF PREPARATION:

1. Gather the equipment and ingredients.
2. Follow the 10 steps of bread making (page 897).

Variations:

1. Pullman loaves
2. Rolls

BAKING

French Bread

PASTRY TECHNIQUE:

The Ten Steps of Bread Making (see page 897)

YIELD: 30 pounds, 10 ounces. Baker's Percentage.

INGREDIENTS:

Sponge:

Water	6 pounds	31.6
Yeast, compressed	6 ounces	2
Flour, high-gluten	6 pounds	31.6
Malt powder	3 ounces	1

Dough:

Water	4 pounds, 8 ounces	23.7
Flour, high-gluten	13 pounds	68.4
Salt	6 ounces	2
Dough conditioner (optional)	3 ounces	1

METHOD OF PREPARATION:

1. Gather the equipment and ingredients.
2. Follow the 10 steps of bread making (page 897).

PASTRY TECHNIQUE:

The Ten Steps of Bread Making (see page 897)

Light Rye

YIELD:	17 pounds, 6½ ounces.	Baker's Percentage.

INGREDIENTS:

Flour, first clear, sifted	7 pounds	70
Flour, light rye, sifted	3 pounds	30
Salt	3 ounces	2
Water (variable)	5 pounds, 6 ounces	54
Molasses	8 ounces	5
Yeast, compressed	10 ounces	6
Shortening, all-purpose	10 ounces	6
Seeds, caraway (optional)	1½ ounces	1
Rye sour starter	1 pound	10

METHOD OF PREPARATION:

1. Gather the equipment and ingredients.
2. Follow the 10 steps of bread making (page 897).

Variations:

1. Pullman loaves
2. Rolls
3. Boulé loaves
4. Oblong loaves

PASTRY TECHNIQUE:

The Ten Steps of Bread Making (see page 897)

Medium Wheat Dough

YIELD:	17 pounds, 8½ ounces.	Baker's Percentage.

INGREDIENTS:

Flour, first clear, sifted	6 pounds, 11 ounces	67
Flour, whole wheat, *not sifted*	3 pounds, 5 ounces	33
Salt	3 ounces	2
Dry milk solids (DMS)	10 ounces	6.25
Water (variable)	5 pounds, 8 ounces	55
Yeast, compressed	9½ ounces	6
Shortening, all-purpose	10 ounces	6.25

METHOD OF PREPARATION:

1. Gather the equipment and ingredients.
2. Follow the 10 steps of bread making (page 897).

Variations:

1. Pullman bread
2. Rolls

Pecan Roll Smear

(Pan Preparation for Pecan Rolls)

PASTRY TECHNIQUE:

Combining

Combining:

Bringing together two or more components.

1. Prepare the components to be combined.
2. Add one to the other, using the appropriate mixing method (if needed).

YIELD:	4 pounds.	8 pounds.
INGREDIENTS:		
Sugar, brown	2 pounds	4 pounds
Margarine	1 pound, 2 ounces	2 pounds, 4 ounces
Salt	½ ounce	1 ounce
Syrup, corn	12 ounces	1 pound, 8 ounces
Extract, rum	¾ ounce	1½ ounces
Extract, vanilla	¾ ounce	1½ ounces
Pecan, halves	As needed	As needed

METHOD OF PREPARATION:

1. Gather the equipment and ingredients.
2. Place all of the ingredients in a bowl with a paddle, and combine together until a smooth, spreading consistency is achieved. If too dry, moisten with a small amount of water.
3. Using a spoon, deposit 1 ounce in the bottom of each cup of a heavily greased muffin tin.
4. Place three or four pecan halves on top of the mixture in each cup.

CHEF NOTE:

This preparation is an excellent base for pineapple upside-down cake.

Pre-ferment

YIELD:	5 pounds, 2 ounces.	Baker's Percentage.
INGREDIENTS:		
Flour, high-gluten	3 pounds	100
Water	2 pounds	66
Salt	1 ounce	2
Yeast, compressed	1 ounce	2

METHOD OF PREPARATION:

1. Gather the equipment and ingredients.
2. Use the straight mixing method.
3. Ferment for 2 to 5 hours, or for up to 7 days in a refrigerator.

PASTRY TECHNIQUE:

Mixing

Mixing:

1. Follow the proper mixing procedure: creaming, blending, whipping, or combination.

HACCP:

Ferment for 2 to 5 hours, or up to 7 days in a refrigerator.

Pre-ferment Rye

YIELD:	4 pounds, 9 ounces.	Baker's Percentage.

INGREDIENTS:

Water	2 pounds	80
Flour, rye, white or dark	2 pounds, 8 ounces	100
Yeast, compressed	1 ounce	2.5
Onion	1 each	

METHOD OF PREPARATION:

1. Gather the equipment and ingredients.
2. Combine all of the ingredients.
3. After all of the ingredients are combined, bury the onion in the mixture.
4. Let the dough rest for 5 hours.

PASTRY TECHNIQUE:

Combining

Combining:

Bringing together two or more components.

1. Prepare the components to be combined.
2. Add one to the other, using the appropriate mixing method (if needed).

HACCP:

Ferment for 2 to 5 hours, or up to 7 days in a refrigerator.

CHEF NOTE:

If kept for 24 hours before using, the onions must be removed.

Rustic Wheat Dough

PASTRY TECHNIQUES:

Soaking, The Ten Steps of Bread Making (see page 897)

Soaking:

1. Place the item(s) to be soaked in a large bowl or appropriate container.
2. Pour water or other liquid over the items to be soaked.
3. Allow to sit until desired saturation or softening is achieved.

YIELD: 17 pounds, 12½ ounces. Baker's Percentage.

INGREDIENTS:

Ingredient	Amount	Baker's Percentage
Wheat flakes or cracked wheat	3 pounds, 5 ounces	33
Water (variable)	5 pounds	50
Salt	5 ounces	3

Dough:

Ingredient	Amount	Baker's Percentage
Flour, first clear, sifted	6 pounds, 11 ounces	67
Water	8 ounces	5
Molasses	11 ounces	7
Yeast, compressed	9½ ounces	6
Shortening, all-purpose	11 ounces	7

METHOD OF PREPARATION:

1. Gather the equipment and ingredients.
2. Place the wheat flakes, water, and salt in a bowl. Allow to soak for 90 minutes to 3 hours (ideally, overnight).
3. Follow the 10 steps of bread making (page 897).

Variations:

1. Pullman loaves
2. Rolls

Soft Rolls

PASTRY TECHNIQUE:

The Ten Steps of Bread Making (see page 897)

YIELD: 26 pounds, 15 ounces. Baker's Percentage.

INGREDIENTS:		
Water	9 pounds	64
Dry milk solids (DMS)	1 pound	7
Sugar, granulated	1 pound	7
Yeast, compressed	8 ounces	3.5
Flour, bread	14 pounds	100
Salt	4½ ounces	2
Shortening, all-purpose	1 pound	7
Dough conditioner (optional)	2½ ounces	1

METHOD OF PREPARATION:

1. Gather the equipment and ingredients.
2. Follow the 10 steps of bread making (page 897).

Variations:

1. Rolls
2. Pecan rolls
3. Cinnamon rolls
4. Coffee cakes

Swedish Rye Bread

PASTRY TECHNIQUES:

Boiling, The Ten Steps of Bread Making (see page 897)

Boiling:

1. Bring the cooking liquid to a rapid boil.
2. Stir the contents.

YIELD:	29 pounds, 5 ounces.	Baker's Percentage.

INGREDIENTS:

Ingredient	Amount	Baker's Percentage
Water	2 pounds	13
Molasses	1 pound, 8 ounces	10
Orange peels, dried	8 ounces	3
Anise, ground	3 ounces	1.25
Fennel, ground	2 ounces	0.80
Water	4 pounds	27
Yeast, compressed	2 ounces	0.80
Flour, rye	2 pounds, 8 ounces	17
Water	2 pounds	13
Yeast, compressed	10 ounces	4
Salt	4 ounces	1.60
Shortening, all-purpose	1 pound	6
Sugar, brown	2 pounds	13
Flour, bread	12 pounds, 8 ounces	83

METHOD OF PREPARATION:

1. *The day before:* Place the water (2 pounds), molasses, dried orange peels, ground anise, and ground fennel in a pot, and bring to a boil. Allow to stand overnight.
2. Gather the equipment and ingredients.
3. Mix additional water (4 pounds), yeast (2 ounces), and rye flour together; allow to rest 1½ hours, or until double in volume.
4. Place the remaining ingredients in a bowl; add the sponge and the boiled mixture.
5. Mix on medium speed for about 6 to 7 minutes, or until the dough clears the bowl.
6. Follow any steps of the 10 steps of bread making that remain (page 897).

Variation:

Round loaves

Sweet Dough

PASTRY TECHNIQUE:

The Ten Steps of Bread Making (see page 897)

HAZARDOUS FOOD:

Eggs

YIELD: 18 pounds, 9½ ounces. Baker's Percentage.

INGREDIENTS:		
Sugar, granulated	1 pound, 8 ounces	17
Dry milk solids (DMS)	8 ounces	5.50
Salt	2½ ounces	1.70
Shortening, all-purpose	1 pound, 8 ounces	17
Eggs, whole	1 pound	11
Water	4 pounds	44.40
Yeast, compressed	14 ounces	1.0
Color, egg shade	5 drops	0.10
Compound or flavor, lemon	1 ounce	0.10
Flour, bread	9 pounds	100

METHOD OF PREPARATION:

1. Gather the equipment and ingredients.
2. Follow the 10 steps of bread making (page 897).

Vienna Bread

PASTRY TECHNIQUE:
The Ten Steps of Bread Making (see page 897)

HAZARDOUS FOOD:
Eggs

YIELD: 16 pounds, 11¼ ounces. Baker's Percentage.

INGREDIENTS:		
Flour, high-gluten, sifted	10 pounds	100
Water (variable)	5 pounds, 2 ounces	51
Oil, vegetable	4 ounces	2.50
Eggs, whole, or egg whites	4 ounces	2.50
Yeast, compressed	8 ounces	5
Sugar, sifted	6 ounces	3.70
Salt	3¼ ounces	2
Dough conditioner (optional)	2 ounces	1.25
Malt syrup (optional)	2 ounces	1.25

METHOD OF PREPARATION:

1. Gather the equipment and ingredients.
2. Follow the 10 steps of bread making (page 897).

Variations:

1. Braid, three-strand
2. Loaves
3. Rolls

Apple Date Muffins

PASTRY TECHNIQUE:

Creaming

Creaming:

1. Soften the fats on low speed.
2. Add the sugar(s) and cream; increase the speed slowly.
3. Add the eggs one at a time; scrape the bowl frequently.
4. Add the dry ingredients in stages.

HAZARDOUS FOOD:

Eggs

YIELD:	8 pounds, 2⅜ ounces.	16 pounds, 4¾ ounces.
INGREDIENTS:		
Butter, unsalted	1 pound	2 pounds
Sugar, brown	1 pound, 4 ounces	2 pounds, 8 ounces
Eggs, whole	7 ounces	14 ounces
Flour, all-purpose	1 pound, 4 ounces	2 pounds, 8 ounces
Baking powder	⅝ ounce	1¼ ounces
Cinnamon, ground	¼ ounce	½ ounce
Salt	¼ ounce	½ ounce
Applesauce	1 pound, 2 ounces	2 pounds, 4 ounces
Dates, chopped	1 pound, 8 ounces	3 pounds
Oats, rolled, regular	1 pound	2 pounds

METHOD OF PREPARATION:

1. Gather the equipment and ingredients.
2. Place the butter and brown sugar in a bowl with a paddle; cream together.
3. Add the eggs gradually until blended.
4. Sift the dry ingredients together: all-purpose flour, baking powder, cinnamon, and salt.
5. Slowly add the dry ingredients to the sugar-egg mixture alternately with applesauce until blended. *Do not overmix.*
6. Slowly add the dates and oats until blended.
7. Place the batter in well-greased muffin tins.
8. Bake at 375° F for 25 minutes, or until a muffin springs back when lightly touched.

Apple Walnut Bread

PASTRY TECHNIQUE:

Blending

Blending:

1. Combine the dry ingredients on low speed.
2. Add the softened fat(s) and liquid(s).
3. Mix the ingredients on low speed.
4. Increase the speed gradually.

HAZARDOUS FOOD:

Eggs

YIELD:	18 pounds, 6½ ounces.	36 pounds, 13 ounces.
INGREDIENTS:		
Flour, pastry	5 pounds	10 pounds
Sugar, granulated	3 pounds, 5 ounces	6 pounds, 10 ounces
Baking powder	4 ounces	8 ounces
Salt	1 ounce	2 ounces
Cinnamon, ground	½ ounce	1 ounce
Eggs, whole	1 pound	2 pounds
Applesauce	5 pound, 8 ounces	11 pounds
Butter, unsalted, melted	8 ounces	1 pound
Walnuts, chopped	2 pounds, 12 ounces	5 pounds, 8 ounces

METHOD OF PREPARATION:

1. Gather the equipment and ingredients.
2. Sift together all of the dry ingredients: pastry flour, granulated sugar, baking powder, salt, and cinnamon.
3. Place the dry ingredients in a bowl.
4. Combine the eggs and applesauce.
5. Add the liquid ingredients to the dry ingredients. Using a paddle attachment, mix only enough to incorporate; do not remove.
6. Add the melted butter.
7. Add the walnuts; combine.
8. Scale at 1 pound, 6 ounces for small loaves or 2 pounds, 8 ounces for large loaves. Pans should be well greased.
9. Bake at 350° F for 45 to 55 minutes, or until loaves are golden brown and firm in center.

Baking Powder Biscuits

PASTRY TECHNIQUES:

Rubbing, Combining, Rolling

Rubbing:

1. Use a pastry cutter to keep the fat in large pieces.
2. Add the liquid in stages.

Combining:

Bringing together two or more components.

1. Prepare the components to be combined.
2. Add one to the other, using the appropriate mixing method (if needed).

Rolling:

1. Prepare the rolling surface by dusting with the appropriate medium (flour, cornstarch, etc.).
2. Use the appropriate style pin (stick pin or ball bearing pin) to roll the dough to desired thickness; rotate the dough during rolling to prevent sticking.

HAZARDOUS FOOD:

Eggs

YIELD:	2 pounds, 11¼ ounces.	5 pounds, 6½ ounces.
INGREDIENTS:		
Flour, bread	10 ounces	1 pound, 4 ounces
Flour, cake	10 ounces	1 pound, 4 ounces
Sugar, granulated	1½ ounces	3 ounces
Dry milk solids (DMS)	1½ ounces	3 ounces
Baking powder	1¼ ounces	2½ ounces
Salt	¼ ounce	½ ounce
Shortening, all-purpose	5 ounces	10 ounces
Water, cold	11¾ ounces	1 pound, 7½ ounces
Eggs, whole	2 ounces	4 ounces
Flour, bread, additional	As needed	As needed

METHOD OF PREPARATION:

1. Gather the equipment and ingredients.
2. Sift the dry ingredients together: bread flour, cake flour, granulated sugar, DMS, baking powder, and salt; place in a bowl.
3. Add the shortening. Using your hands, mix the dough until the shortening is broken into pieces about the size of peas.
4. Mix the water and eggs together.
5. Add the liquid to the mixture, and mix lightly. *Do not overmix* or *overwork* this dough.
6. Place the dough on a bench on a thin layer of bread flour.
7. Using a rolling pin, bench pin, or your hands, roll or press the dough to a 1-inch thickness.
8. With a standard biscuit cutter, cut out biscuits close together to avoid leaving excess scraps.
9. Place the biscuits 1-inch apart on parchment-lined sheet pans. If desired, allow the dough to relax for 10 to 15 minutes before baking.
10. Bake at 425° F for 10 to 15 minutes, or until golden brown.

Banana Nut Bread No. 1

PASTRY TECHNIQUES:

Creaming, Combining

Creaming:

1. Soften the fats on low speed.
2. Add the sugar(s) and cream; increase the speed slowly.
3. Add the eggs one at a time; scrape the bowl frequently.
4. Add the dry ingredients in stages.

Combining:

Bringing together two or more components.

1. Prepare the components to be combined.
2. Add one to the other, using the appropriate mixing method (if needed).

HAZARDOUS FOOD:

Eggs

YIELD:	6 pounds, 3⅝ ounces.	12 pounds, 7¼ ounces.
INGREDIENTS:		
Sugar, granulated	1 pound, 4 ounces	2 pounds, 8 ounces
Shortening, high-ratio	6 ounces	12 ounces
Baking soda, sifted	½ ounce	1 ounce
Lemon powder	½ ounce	1 ounce
Salt	⅛ ounce	¼ ounce
Bananas, fresh or canned, mashed	8 ounces	1 pound
Eggs, whole	2 ounces	4 ounces
Water, cold	1 pound, 8 ounces	3 pounds
Flour, bread, sifted	1 pound	2 pounds
Flour, cake, sifted	1 pound	2 pounds
Baking powder, sifted	½ ounce	1 ounce
Nuts, finely chopped	4 ounces	8 ounces
Compound, banana	2 ounces	4 ounces

METHOD OF PREPARATION:

1. Gather the equipment and ingredients.
2. Place the granulated sugar, shortening, baking soda, lemon powder, and salt in a bowl; paddle for 2 minutes.
3. Add the bananas and eggs to the mixture in the bowl; cream for an additional 1 minute.
4. Add one third the amount of the water, and mix at low speed.
5. Sift together the flours and baking powder.
6. Add the sifted ingredients to the mixture in two stages. Mix at low speed.
7. Add one third of the water, and mix only until everything is incorporated; *do not overmix*.
8. Scrape the bowl well.
9. Add the chopped nuts and banana compound.
10. Add the remaining third of water; mix until incorporated.
11. Scale at 1 pound, 6 ounces for small loaves or 2 pounds, 8 ounces for large loaf pans.
12. Bake at 375° F until the loaves are light brown overall and firm in the center.
13. Cool.
14. Remove from the pans.

Banana Nut Bread No. 2

PASTRY TECHNIQUE:

Blending

Blending:

1. Combine the dry ingredients on low speed.
2. Add the softened fat(s) and liquid(s).
3. Mix the ingredients on low speed.
4. Increase the speed gradually.

HAZARDOUS FOOD:

Eggs

YIELD:	17 pounds, 4½ ounces.	34 pounds, 8½ ounces.
INGREDIENTS:		
Flour, bread	1 pound	2 pounds
Flour, pastry	4 pounds	8 pounds
Sugar, granulated	2 pounds	4 pounds
Baking powder	4 ounces	8 ounces
Baking soda	¾ ounce	1½ ounces
Salt	1 ounce	2 ounces
Bananas, fresh or canned, mashed	5 pounds	10 pounds
Eggs, whole	2 pounds	4 pounds
Oil, vegetable	1 pound, 10½ ounces	3 pounds, 5 ounces
Nuts, chopped (optional)	1 pound, 4 ounces	2 pounds, 8 ounces

METHOD OF PREPARATION:

1. Gather the equipment and ingredients.
2. Sift together all the dry ingredients: bread flour, pastry flour, granulated sugar, baking powder, baking soda, and salt.
3. Place the dry ingredients in a bowl with a paddle.
4. Add the bananas, and combine.
5. Combine the eggs, oil, and nuts (optional).
6. Add the liquid ingredients to the mixture; combine. *Do not overmix.*
7. Scale at 1 pound, 6 ounces for small loaves or 2 pounds, 8 ounces for large loaves. Pans should be lightly greased.
8. Bake at 350° F for 35 to 40 minutes, or until golden brown.

Basic Muffins

Oil Mix

PASTRY TECHNIQUE:

Blending

Blending:

1. Combine the dry ingredients on low speed.
2. Add the softened fat(s) and liquid(s).
3. Mix the ingredients on low speed.
4. Increase the speed gradually.

HAZARDOUS FOOD:

Eggs

YIELD:	6 Dozen Muffins.	12 Dozen Muffins.
INGREDIENTS:		
Sugar, granulated	2 pounds, 8 ounces	5 pounds
Salt	¾ ounce	1½ ounces
Eggs, whole	1 pound	2 pounds
Oil, vegetable	1 pound, 8 ounces	3 pounds
Water	2 pounds, 8 ounces	5 pounds
Flour, high-gluten, sifted	4 pounds, 12 ounces	9 pounds, 8 ounces
Baking powder, sifted	3¼ ounces	6½ ounces
Dry milk solids (DMS)	8 ounces	1 pound
Compound, lemon	2 ounces	4 ounces
Compound, orange	2 ounces	4 ounces
Fruit (optional)	See variations	See variations

METHOD OF PREPARATION:

1. Gather the equipment and ingredients.
2. Put the sugar and salt in a bowl.
3. Using the paddle to mix, add the eggs slowly, and blend well.
4. Add the oil on low speed.
5. Add the water on low speed.
6. Sift all of the dry ingredients together.
7. Add the dry ingredients; mix just enough to incorporate; this batter should be rough, with small lumps.
8. If adding fruit to the mix, dredge it first with flour to prevent it from bleeding color into the batter. Fold in the fruit.
9. Drop 3 ounces of batter into greased muffin tins or to three-fourths full.
10. Bake at 400° F for 20 minutes, or until light brown overall and firm in the center.

Variations:

1. *Blueberry Muffins* (with frozen blueberries):
   a. ¾ quart (6 dozen muffins)
   b. 1½ quart (12 dozen muffins)
2. *Cranberry Muffins* (with frozen cranberries):
   a. 1 pound (6 dozen muffins)
   b. 2 pounds (12 dozen muffins)
3. *Spiced Apple Muffins:*
   a. Place an apple slice in each cup.
   b. Sprinkle the tops with cinnamon-sugar mixture.

Buttermilk Biscuits

PASTRY TECHNIQUES:

Combining, Rolling

Combining:

Bringing together two or more components.

1. Prepare the components to be combined.
2. Add one to the other, using the appropriate mixing method (if needed).

Rolling:

1. Prepare the rolling surface by dusting with the appropriate medium (flour, cornstarch, etc.).
2. Use the appropriate style pin (stick pin or ball bearing pin) to roll the dough to desired thickness; rotate the dough during rolling to prevent sticking.

HAZARDOUS FOODS:

Eggs
Buttermilk

YIELD:	6 pounds, 14½ ounces. 4 Dozen	13 pounds, 13 ounces. 8 Dozen
INGREDIENTS:		
Sugar, granulated	8 ounces	1 pound
Salt	½ ounce	1 ounce
Dry milk solids (DMS)	4 ounces	8 ounces
Baking powder, sifted	3 ounces	6 ounces
Flour, bread, sifted	1 pound, 10 ounces	3 pounds, 4 ounces
Flour, pastry, sifted	1 pound, 10 ounces	3 pounds, 4 ounces
Shortening, all-purpose	5 ounces	10 ounces
Margarine or butter, unsalted	6 ounces	12 ounces
Eggs, whole	5 ounces	10 ounces
Extract, vanilla	1 ounce	2 ounces
Water	10 ounces	1 pound, 4 ounces
Buttermilk	1 pound	2 pounds
Flour, pastry or flour	As needed	As needed
Egg wash (see page 964)	As needed	As needed

METHOD OF PREPARATION:

1. Gather the equipment and ingredients.
2. Sift all the dry ingredients together: granulated sugar, salt, DMS, baking powder, bread flour, and pastry flour; place in a dry bowl.
3. Add the fats to the dry ingredients, and mix until crumbly in texture.
4. Combine all of the liquid ingredients together: eggs, vanilla extract, water, and buttermilk.
5. Add the liquid ingredients to the mixture, and mix only until blended.
6. Place the dough on a floured surface, and cover with parchment.
7. Allow to relax for 15 to 20 minutes at room temperature.
8. Roll out the dough to a 1-inch thickness.
9. Cut using a standard biscuit cutter.
10. Place on parchment-lined sheet pans.
11. Egg-wash the tops.
12. Bake at 360° F until raised and the tops are golden brown.

Buttermilk Scones

PASTRY TECHNIQUES:

Combining, Rolling

Combining:

Bringing together two or more components.

1. Prepare the components to be combined.
2. Add one to the other, using the appropriate mixing method (if needed).

Rolling:

1. Prepare the rolling surface by dusting with the appropriate medium (flour, cornstarch, etc.).
2. Use the appropriate style pin (stick pin or ball bearing pin) to roll the dough to desired thickness; rotate the dough during rolling to prevent sticking.

HAZARDOUS FOOD:

Buttermilk

YIELD:	5 pounds, 6⅛ ounces.	10 pounds, 12¼ ounces.
INGREDIENTS:		
Flour, bread	1 pound, 14 ounces	3 pounds, 12 ounces
Flour, cake	10 ounces	1 pound, 4 ounces
Salt	¼ ounce	½ ounce
Sugar, granulated	5¼ ounces	10½ ounces
Baking powder	1⅞ ounces	3¾ ounces
Butter, unsalted	10 ounces	1 pound, 4 ounces
Buttermilk	1 pound, 4 ounces	2 pounds, 8 ounces
Currants	8¾ ounces	1 pound, 1½ ounces
Flour, bread or cake	As needed	As needed
Egg wash (see page 964)	As needed	As needed

METHOD OF PREPARATION:

1. Gather the equipment and ingredients.
2. Sift all of the dry ingredients together: bread flour, cake flour, salt, granulated sugar, and baking powder.
3. Place the dry ingredients in a bowl.
4. Add the butter. Using your hands, mix the dough until the butter is broken into pieces about the size of peas.
5. Add the buttermilk to the mixture, and mix lightly.
6. Add the currants; *do not overmix* or *overwork* this dough.
7. Place the dough on a floured surface; allow it to rest for 15 minutes.
8. Using a rolling pin, bench pin, or your hands, roll or press the dough to a 1-inch thickness.
9. With a standard biscuit cutter, cut out biscuits close together to avoid leaving excess scraps.
10. Place the biscuits 1-inch apart on parchment-lined sheet pans.
11. Egg-wash the tops.
12. Bake at 425° F for 10 to 15 minutes, or until golden brown.

Carrot Apple Muffins

PASTRY TECHNIQUE:

Blending

Blending:

1. Combine the dry ingredients on low speed.
2. Add the softened fat(s) and liquid(s).
3. Mix the ingredients on low speed.
4. Increase the speed gradually.

HAZARDOUS FOOD:

Eggs

YIELD:	6 pounds, 11¾ ounces.	13 pounds, 7½ ounces.
INGREDIENTS:		
Flour, pastry, sifted	12 ounces	1 pound, 8 ounces
Flour, bread, sifted	5 ½ ounces	11 ounces
Sugar, granulated	1 pound, 4 ounces	2 pounds, 8 ounces
Salt	⅜ ounce	¾ ounce
Baking soda	¾ ounce	1½ ounces
Cinnamon, ground	¾ ounce	1½ ounces
Eggs, whole	9 ounces	1 pound, 2 ounces
Oil, vegetable	14 ounces	1 pound, 12 ounces
Extract, vanilla	½ ounce	1 ounce
Raisins, seedless	6 ounces	12 ounces
Coconut, shredded	3 ounces	6 ounces
Apples, shredded	12 ounces	1 pound, 8 ounces
Carrots, shredded	1 pound, 4 ounces	2 pounds, 8 ounces
Walnuts, chopped	4 ounces	8 ounces

METHOD OF PREPARATION:

1. Gather the equipment and ingredients.
2. Sift together all of the dry ingredients: pastry flour, bread flour, granulated sugar, salt, baking soda, and cinnamon.
3. Place the dry ingredients in a bowl.
4. Combine all of the liquid ingredients: eggs, salad oil, and vanilla extract.
5. Add the liquid ingredients to the dry ingredients using a paddle attachment. Mix only enough to incorporate; *do not overmix*.
6. Add the raisins and coconut; mix well.
7. Add the shredded apples, carrots, and walnuts; combine.
8. Drop into greased muffin tins to three-fourths full.
9. Bake at 360° F for 25 to 30 minutes, or until they are golden brown overall and firm in the center.

Corn Bread

PASTRY TECHNIQUE:

Blending

Blending:

1. Combine the dry ingredients on low speed.
2. Add the softened fat(s) and liquid(s).
3. Mix the ingredients on low speed.
4. Increase the speed gradually.

HAZARDOUS FOOD:

Eggs

YIELD:	4 pounds, 10⅞ ounces.	9 pounds, 5¾ ounces.
	½ sheet pan	1 sheet pan

INGREDIENTS:

Flour, bread, sifted	14 ounces	1 pound, 12 ounces
Flour, pastry, sifted	6 ounces	12 ounces
Baking powder	1⅜ ounces	2¾ ounces
Salt	½ ounce	1 ounce
Dry milk solids (DMS)	3 ounces	6 ounces
Cornmeal	8 ounces	1 pound
Sugar, granulated	13 ounces	1 pound, 10 ounces
Water	15 ounces	1 pound, 14 ounces
Eggs, whole	8 ounces	1 pound
Oil, vegetable	6 ounces	12 ounces

METHOD OF PREPARATION:

1. Gather the equipment and ingredients.
2. Sift the dry ingredients together: bread flour, pastry flour, baking powder, salt, DMS, cornmeal, and granulated sugar.
3. Place all of the dry ingredients in bowl.
4. Mix the water and eggs together.
5. Add the water-egg mixture to the dry ingredients only until combined.
6. Add the oil, and mix just enough to mix together.
7. Scale out the batter into lightly greased sheet pans.
8. Bake at 370° F for approximately 22 to 25 minutes.

CHEF NOTE:

This batter also can be used to make corn muffins.

Cranberry Bread

YIELD:	4, 1-pound Loaves.	8, 1-pound Loaves.
INGREDIENTS:		
Sugar, granulated	11½ ounces	1 pound, 7 ounces
Shortening, high-ratio	3 ounces	6 ounces
Baking soda, sifted	¼ ounce	½ ounce
Salt	¼ ounce	½ ounce
Cinnamon, ground	¼ ounce	½ ounce
Eggs, whole	3½ ounces	7 ounces
Flour, bread	15 ounces	1 pound, 14 ounces
Flour, cake	4½ ounces	9 ounces
Baking powder	½ ounce	1 ounce
Water	13½ ounces	1 pound, 11 ounces
Cranberries	8 ounces	1 pound
Oranges, peeled, rind and juice	1½ each	3 each
Pecans, chopped	3 ounces	6 ounces
Compound, orange	1 ounce	2 ounces
Compound, lemon	1 ounce	2 ounces

METHOD OF PREPARATION:

1. Gather the equipment and ingredients.
2. Place the granulated sugar, shortening, baking soda, salt, and cinnamon in a bowl; cream with a paddle for 3 minutes on medium speed.
3. Add the whole eggs gradually; mix for 3 minutes at medium speed.
4. Sift together the bread flour, cake flour, and baking powder.
5. Alternately add the dry ingredients and water into the cream mixture.
6. Mix on slow speed for 3 minutes, and scrape the bowl well.
7. Coarsely grind the cranberries, and drain the juice.
8. Grate the orange rind, and extract the juice; reserve.
9. Add the cranberries, orange rind, orange juice, nuts, orange compound, and lemon compound to the mixture; mix only until incorporated.
10. Scale 1 pound of batter into greased loaf pans.
11. Place the pans on sheet pans, and bake at 380° F until light brown overall and firm in the center.
12. Cool, and remove from the loaf pans.

PASTRY TECHNIQUES:

Creaming, Combining

Creaming:

1. Soften the fats on low speed.
2. Add the sugar(s) and cream; increase the speed slowly.
3. Add the eggs one at a time; scrape the bowl frequently.
4. Add the dry ingredients in stages.

Combining:

Bringing together two or more components.

1. Prepare the components to be combined.
2. Add one to the other, using the appropriate mixing method (if needed).

HAZARDOUS FOOD:

Eggs

CHEF NOTE:

Other types of fruit, such as dates, raisins, blueberries, or raspberries, can be used in place of cranberries and oranges.

Irish Soda Bread

PASTRY TECHNIQUE:

Combining, Shaping

Combining:

Bringing together two or more components.

1. Prepare the components to be combined.
2. Add one to the other, using the appropriate mixing method (if needed).

Shaping:

1. Prepare the medium to be shaped.
2. Prepare the surface area for shaping.
3. Mold medium into desired shapes according to the instructor's directions.

HAZARDOUS FOOD:

Buttermilk

YIELD: 24 pounds, 11 ounces. Baker's Percentage.

INGREDIENTS:

Ingredient	Amount	Baker's %
Flour, bread	7 pounds	70
Flour, whole wheat	3 pounds	30
Salt	4 ounces	2.50
Caraway	4 ounces	2.50
Baking powder	2 ounces	1.25
Baking soda	1 ounce	0.62
Butter, unsalted	1 pound	10
Buttermilk	8 pounds	80
Currants	5 pounds	50
Flour, white rye	As needed	

METHOD OF PREPARATION:

1. Gather the equipment and ingredients.
2. Place all of the dry ingredients on a surface, and cut butter into the flour until a mealy consistency is achieved.
3. Add the buttermilk until incorporated.
4. Add the currants, and blend slightly.
5. Scale, as instructed. Dust with white rye flour; score a deep X into the top of each loaf.
6. Bake at 425° F until lightly browned.

Lemon–Poppy Seed Muffins

PASTRY TECHNIQUE:

Creaming

Creaming:

1. Soften the fats on low speed.
2. Add the sugar(s) and cream; increase the speed slowly.
3. Add the eggs one at a time; scrape the bowl frequently.
4. Add the dry ingredients in stages.

HAZARDOUS FOODS:

Eggs
Milk

YIELD:	9 pounds, 3⅞ ounces.	18 pounds, 6¾ ounces.

INGREDIENTS:

Ingredient		
Sugar, granulated	1 pound, 13 ounces	3 pounds, 10 ounces
Butter, unsalted, softened	11½ ounces	1 pound, 7 ounces
Shortening, high-ratio	7 ounces	14 ounces
Salt	⅝ ounce	1¼ ounces
Eggs, whole	1 pound, 2 ounces	2 pounds, 4 ounces
Seeds, poppy	2 ounces	4 ounces
Flour, bread	1 pound, 8 ounces	3 pounds
Flour, pastry	1 pound	2 pounds
Baking powder	2 ounces	4 ounces
Milk, whole	1 pound, 15 ounces	3 pounds, 14 ounces
Lemon juice	2¼ ounces	4½ ounces
Lemon rind	4½ ounces	9 ounces

METHOD OF PREPARATION:

1. Gather the equipment and ingredients.
2. Place the granulated sugar, softened butter, shortening, and salt in bowl with a paddle, and cream together.
3. Add the eggs gradually until blended.
4. Add the poppy seeds to the mixture; combine.
5. Sift the dry ingredients together: bread flour, pastry flour, and baking powder.
6. Combine the milk, lemon juice, and lemon rind.
7. Slowly add the dry ingredients to the sugar-egg mixture alternately with the milk mixture until blended. *Do not overmix.*
8. Drop into greased muffin tins to three-fourths full.
9. Bake at 375° F for about 20 minutes, or until the muffins spring back when lightly touched.

Quick Coffee Cake

PASTRY TECHNIQUE:

Blending

Blending:

1. Combine the dry ingredients on low speed.
2. Add the softened fat(s) and liquid(s).
3. Mix the ingredients on low speed.
4. Increase the speed gradually.

HAZARDOUS FOOD:

Eggs

YIELD:	6 pounds. 1 Sheet Pan	12 pounds. 2 Sheet Pans
INGREDIENTS:		
Eggs, whole	10 ounces	1 pound, 4 ounces
Oil, vegetable	12 ounces	1 pound, 8 ounces
Extract, vanilla	To taste	To taste
Water	1 pound, 8 ounces	3 pounds
Flour, pastry, sifted	1 pound, 12 ounces	3 pounds, 8 ounces
Baking powder	1¼ ounces	2½ ounces
Dry milk solids (DMS)	3 ounces	6 ounces
Salt	½ ounce	1 ounce
Sugar, granulated	1 pound, 8 ounces	3 pounds

METHOD OF PREPARATION:

1. Gather the equipment and ingredients.
2. Place the whole eggs, oil, vanilla extract, and water in a bowl; combine well.
3. Sift together the pastry flour, baking powder, DMS, salt, and sugar.
4. Add the sifted dry ingredients to the liquids, and mix lightly.
5. Scale into parchment-lined sheet pans.
6. Spread evenly.
7. Bake at 375° F until light golden brown.

Variations:

1. *Streusel Topping:*
   a. If used, sprinkle on the top of the cake.
   b. Bake.
2. *Fruit Topping:*
   a. Remove the finished cake from the oven, and allow it to cool before topping.
   b. Top with desired fruit, and add a layer of streusel.
   c. Return to the oven to brown the streusel.
   d. Use double sheet pans when browning the streusel to prevent excessive browning of the bottom of the cake.

Zucchini-Raisin-Walnut Quick Bread

PASTRY TECHNIQUE:

Blending

Blending:

1. Combine the dry ingredients on low speed.
2. Add the softened fat(s) and liquid(s).
3. Mix the ingredients on low speed.
4. Increase the speed gradually.

HAZARDOUS FOOD:

Eggs

YIELD:	14 pounds, 8 ounces.	29 pounds.

INGREDIENTS:

Zucchini	4 pounds	8 pounds
Sugar, granulated	2 pounds	4 pounds
Sugar, brown	1 pound, 4 ounces	2 pounds, 8 ounces
Oil, vegetable	1 pound, 4 ounces	2 pounds, 8 ounces
Eggs, whole	1 pound, 4 ounces	2 pounds, 8 ounces
Salt	½ ounce	1 ounce
Dry milk solids (DMS)	1½ ounces	3 ounces
Baking soda	½ ounce	1 ounce
Baking powder	1 ounce	2 ounces
Cinnamon, ground	¼ ounce	½ ounce
Nutmeg, ground	¼ ounce	½ ounce
Flour, bread	2 pounds, 6 ounces	4 pounds, 12 ounces
Flour, cake	8 ounces	1 pound
Walnuts, ground	14 ounces	1 pound, 12 ounces
Raisins	12 ounces	1 pound, 8 ounces

METHOD OF PREPARATION:

1. Gather the equipment and ingredients.
2. Wash and split zucchini in lengths, deseed, grate, and keep as dry as possible.
3. Place the granulated sugar, brown sugar, vegetable oil, and eggs in a bowl with a paddle, and combine.
4. Add the grated zucchini.
5. Sift all of the dry ingredients: salt, DMS, baking soda, baking powder, cinnamon, nutmeg, bread flour, and cake flour. Blend into the zucchini mixture just until moist.
6. Add the walnuts and raisins just until combined; *do not overmix*.
7. Place in pans, and run a trowel down the center before baking.
8. Bake at 375° F to 400° F for approximately 30 minutes, or until the bread springs back when lightly touched.

CHEF NOTE:

Zucchini should be washed and patted dry. Cut off ends and split in the length.

Chocolate Soufflé

YIELD: 2 pounds, 1¾ ounces.

INGREDIENTS:	
Butter, unsalted	As needed
Sugar, granulated	As needed
Egg yolks	5 ounces
Sugar, granulated	5 ounces
Salt	Pinch
Chocolate, dark, bittersweet, or semi-sweet	8 ounces
Chocolate, dark, unsweetened	2 ounces
Cream, heavy	2 ounces
Espresso, instant	¼ ounce
Rum or brandy	1½ ounces
Egg whites	10 ounces

METHOD OF PREPARATION:

1. Gather the equipment and ingredients.
2. Butter the soufflé dishes, sprinkle with granulated sugar, as needed, and set aside.
3. Ribbon the egg yolks, sugar, and salt.
4. Melt both chocolates together slowly over a double boiler.
5. Stir the melted chocolates into the egg yolk–sugar mixture.
6. Stir in the heavy cream, instant espresso, and liquor.
7. Whip the egg whites to a medium peak.
8. Carefully fold the egg whites into the chocolate mixture.
9. Pour the batter into prepared soufflé dishes immediately.
10. Bake at 400° F in a deck oven for 30 minutes, or until done. (*Do not bake this soufflé in a convection oven.*)
11. Serve immediately.

PASTRY TECHNIQUES:

Melting, Folding, Whipping, Ribboning

Melting:

1. Prepare the food product to be melted.
2. Place the food product in an appropriate sized pot over direct heat or over a double boiler.
3. Stir frequently or occasionally, depending on the delicacy of the product, until melted.

OR

1. Place the product on a sheet pan or in a bowl, and place in a low-temperature oven until melted.

Folding:

Do steps 1, 2, and 3 in one continuous motion.

1. Run a bowl scraper under the mixture, across the bottom of the bowl.
2. Turn the bowl counterclockwise.
3. Bring the bottom mixture to the top.

Whipping:

1. Hold the whip at a 55-degree angle.
2. Create circles, using a circular motion.
3. The circular motion needs to be perpendicular to the bowl.

Ribboning:

1. Use a high speed on the mixer.
2. Do not overwhip the egg yolks.

HAZARDOUS FOODS:

Egg yolks
Heavy cream
Egg whites

CHEF NOTE:

The 2 ounces of unsweetened chocolate can be replaced with 1¾ ounces of cocoa powder and 1 ounce of oil.

Frozen Soufflé No. 1

(Soufflé Glacé)

YIELD:	1 pound, 8 ounces.	3 pounds.
INGREDIENTS:		
Egg yolks	2 ounces	4 ounces
Eggs, whole	2 ounces	4 ounces
Sugar, granulated	4 ounces	8 ounces
Cream, heavy	1 pound	2 pounds
Flavorings	To taste	To taste

METHOD OF PREPARATION:

1. Gather the equipment and ingredients.
2. Place the egg yolks, whole eggs, and granulated sugar in a bowl, and place over a double boiler; heat to 145° F, whipping constantly.
3. Remove from the heat, and whip to a medium peak; cool.
4. Whip the heavy cream to a soft peak.
5. Fold the cooled egg mixture into the whipped cream.
6. Flavor, as desired.
7. Pipe into desired containers.
8. Freeze.

PASTRY TECHNIQUES:

Heating, Whipping, Folding, Piping

Heating:

1. Prepare the food product according to the formula's instructions.
2. Choose the appropriate method of heating (on the range or stove top, in the oven, etc.)
3. Apply the product to the heat.

Whipping:

1. Hold the whip at a 55-degree angle.
2. Create circles, using a circular motion.
3. The circular motion needs to be perpendicular to the bowl.

Folding:

Do steps 1, 2, and 3 in one continuous motion.

1. Run a bowl scraper under the mixture, across the bottom of the bowl.
2. Turn the bowl counterclockwise.
3. Bring the bottom mixture to the top.

Piping:

With bag:

1. Use a bag with a disposable tip; cut the bag at 45-degree angle.
2. Fill to no more than half full.
3. Burp the bag.

With cone:

1. Cut and fold the piping cone to the appropriate size.
2. Fill the cone with a small amount.
3. Fold the ends to form a triangle.
4. Pipe the desired designs.

HACCP:

Store at 0° F

HAZARDOUS FOODS:

Egg yolks
Eggs
Heavy cream

CHEF NOTES:

1. Flavorings may be added according to the name of the frozen soufflé (e.g., Soufflé Glacé Grand Marnier.)
2. Soufflé cups can be used; prior to piping in the mixture, place a paper collar around the cups.

Frozen Soufflé No. 2

(Pâté à Bombe Method)

PASTRY TECHNIQUES:
Whipping, Cooking, Folding

Whipping:
1. Hold the whip at a 55-degree angle.
2. Create circles, using a circular motion.
3. The circular motion needs to be perpendicular to the bowl.

Cooking:
Preparing food through the use of various sources of heating.
1. Choose the appropriate heat application: baking, boiling, simmering, etc.
2. Prepare the formula according to instructions.
3. Cook according to instructions.

Folding:
Do steps 1, 2, and 3 in one continuous motion.
1. Run a bowl scraper under the mixture, across the bottom of the bowl.
2. Turn the bowl counterclockwise.
3. Bring the bottom mixture to the top.

HACCP:
Store at 0° F.

HAZARDOUS FOODS:
Egg yolks
Heavy cream

YIELD: 2 pounds, 5½ ounces. | 4 pounds, 11 ounces.

INGREDIENTS:		
Egg yolks	7 ounces	14 ounces
Sugar, granulated	7½ ounces	15 ounces
Water	3 ounces	6 ounces
Cream, heavy, whipped	1 pound, 4 ounces	2 pounds, 8 ounces
Flavorings	To taste	To taste

METHOD OF PREPARATION:

1. Gather the equipment and ingredients.
2. Place the egg yolks in a bowl, and whip to full volume.
3. Place the sugar and water in a pot, and cook to 238° F.
4. Add the cooked sugar to the whipped egg yolks; mix on medium speed.
5. Whip the heavy cream to a soft peak.
6. Fold the whipped cream into the egg yolk–sugar mixture.
7. Flavor, as desired.
8. Pipe into soufflé cups. Prior to piping in the mixture, place a paper collar around each cup.
9. Freeze.

Soufflé

YIELD:	1 pound, 6¾ ounces.	2 pounds, 13½ ounces.
INGREDIENTS:		
Butter, unsalted	As needed	As needed
Sugar, granulated	As needed	As needed
Milk, whole	4¼ ounces	8½ ounces
Sugar, granulated	2 ounces	4 ounces
Milk, whole	4¼ ounces	8½ ounces
Flour, cake	1¾ ounces	3½ ounces
Egg yolks	1½ ounces	3 ounces
Liqueur (see chef note 1)	1¾ ounces	3½ ounces
Egg whites	6¾ ounces	13½ ounces
Sugar, granulated	½ ounce	1 ounce
Cream of tartar	Pinch	Pinch

METHOD OF PREPARATION:

1. Gather the equipment and ingredients.
2. Butter and sugar the soufflé cups; be sure the top edge is clean to get maximum rise.
3. Combine the milk and sugar; bring to a simmer.
4. Make a paste of the milk and cake flour.
5. Combine the paste with the simmering milk, and cook for 2 minutes, or until the mixture thickens.
6. Remove from the heat, and cool.
7. Whisk the egg yolks and liqueur together.
8. Add the egg yolk mixture to the milk mixture.
9. Place the egg whites in a bowl, and whip to a medium-soft peak; gradually add the granulated sugar and cream of tartar to make a meringue.
10. Combine the meringue and the milk-egg mixture carefully.
11. Place in prepared soufflé cups about three fourths of the way full.
12. Bake at 375° F for 8 to 10 minutes, or until the soufflé is firm but not dry.

PASTRY TECHNIQUES:
Combining, Whipping, Folding

Combining:
Bringing together two or more components.
1. Prepare the components to be combined.
2. Add one to the other, using the appropriate mixing method (if needed).

Whipping:
1. Hold the whip at a 55-degree angle.
2. Create circles, using a circular motion.
3. The circular motion needs to be perpendicular to the bowl.

Folding:
Do steps 1, 2, and 3 in one continuous motion.
1. Run a bowl scraper under the mixture, across the bottom of the bowl.
2. Turn the bowl counterclockwise.
3. Bring the bottom mixture to the top.

HACCP:
Store at 40° F.

HAZARDOUS FOODS:
Milk
Egg yolks
Egg whites

CHEF NOTES:
1. Suggestions for liqueur include Grand Marnier, Kahlua, or Triple Sec.
2. This mixture can be prepared in advance; however, the meringue can be whipped up at the last moment before baking and folded into mixture.

Crêpe Batter No. 1

PASTRY TECHNIQUE:
Combining

Combining:
Bringing together two or more components.
1. Prepare the components to be combined.
2. Add one to the other, using the appropriate mixing method (if needed).

HACCP:
Store at 40° F.

HAZARDOUS FOODS:
Milk
Eggs

YIELD:	2 pounds, 4 ounces.	4 pounds, 8 ounces.

INGREDIENTS:

Butter, unsalted	¾ ounce	1½ ounces
Milk, whole	1 pound, 2 ounces	2 pounds, 4 ounces
Eggs, whole	6 ounces	12 ounces
Rum	½ ounce	1 ounce
Oil, vegetable	1¾ ounces	3½ ounces
Sugar, granulated	1¾ ounces	3½ ounces
Salt	Pinch	⅛ ounce
Flour, bread	7 ounces	14 ounces
Orange, zest only	1 each	2 each

METHOD OF PREPARATION:

1. Gather the equipment and ingredients.
2. Melt the butter, and set aside.
3. Combine the milk, eggs, rum, and oil.
4. Add the melted butter into the milk mixture.
5. Place all of the dry ingredients into a bowl: sugar, salt, flour.
6. Using a whip, slowly add the liquid mixture into the dry mixture. Add zest.
7. Let the mixture rest for 30 minutes.
8. To prepare crêpes, use a non-stick crepe pan.

Crêpe Batter No. 2

YIELD:	1 pound, 6 ounces.	2 pounds, 12 ounces.
INGREDIENTS:		
Eggs, whole	4 ounces	8 ounces
Sugar, confectionery	¾ ounce	1½ ounces
Salt	To taste	To taste
Milk, whole	10½ ounces	1 pound, 5 ounces
Oil, vegetable	½ ounce	1 ounce
Flour, pastry	6 ounces	12 ounces
Extract, vanilla	To taste	To taste
Lemon, rind only	½ each	1 each
Orange, rind only	½ each	1 each

METHOD OF PREPARATION:

1. Gather the equipment and ingredients.
2. Combine the eggs, sugar, salt, milk, and oil. Mix until the liquid is smooth.
3. Place the pastry flour in a dry bowl, and slowly add the liquid ingredients to it. (Strain the mixture.)
4. Add the vanilla extract, lemon rind, and orange rind.
5. Let the mixture rest for 30 minutes.
6. To prepare crêpes, use a non-stick crepe pan.

PASTRY TECHNIQUE:

Combining

Combining:

Bringing together two or more components.

1. Prepare the components to be combined.
2. Add one to the other, using the appropriate mixing method (if needed).

HACCP:

Store at 40° F.

HAZARDOUS FOODS:

Eggs
Milk

CHEF NOTES:

1. If a thicker batter is desired, add more flour.
2. If a thinner batter is desired, add more liquid.
3. Milk can be partially replaced with club soda in a 1:1 ratio, which will result in a lighter product.

Crêpe Batter No. 3

PASTRY TECHNIQUE:

Combining

Combining:

Bringing together two or more components.
1. Prepare the components to be combined.
2. Add one to the other, using the appropriate mixing method (if needed).

HACCP:

Store at 40° F

HAZARDOUS FOODS:

Milk
Eggs

YIELD:	1 pound, 4½ ounces.	2 pounds, 9 ounces.
INGREDIENTS:		
Flour, pastry	3½ ounces	7 ounces
Sugar, granulated	⅞ ounce	1¾ ounces
Salt	Pinch	¼ ounce
Milk, whole	9 ounces	1 pound, 2 ounces
Eggs, whole	6 ounces	12 ounces
Oil, vegetable	1 ounce	2 ounces
Extract, vanilla	To taste	To taste

METHOD OF PREPARATION:

1. Gather the equipment and ingredients.
2. Place the pastry flour, granulated sugar, and salt in a bowl.
3. Slowly add the milk into the dry ingredients. (Strain the mixture.)
4. Add the eggs, oil, and vanilla extract.
5. Combine well.
6. Let the mixture rest for 1 hour.
7. To prepare crêpes, use a non-stick omelette pan.

PASTRY TECHNIQUES:

Combining

Combining:

Bringing together two or more components.

1. Prepare the components to be combined.
2. Add one to the other, using the appropriate mixing method (if needed).

Crêpes Georgette

YIELD:	12 Servings.	24 Servings.

INGREDIENTS:

Crêpes (see page 936)	24 each	48 each
English sauce (see page 864)	As needed	As needed
Melba sauce (see page 867)	As needed	As needed
Filling:		
Walnuts, finely chopped and toasted	10 ounces	1 pound, 4 ounces
Sugar, granulated	6 ounces	12 ounces
Cinnamon, ground	¼ ounce	½ ounce

METHOD OF PREPARATION:

1. Gather the equipment and ingredients.
2. Combine the walnuts, sugar, and cinnamon.
3. Place a small spoonful of the walnut mixture in the center of each crêpe.
4. Roll or bundle the crêpe.
5. Place the crêpe on a dessert plate, and garnish with English sauce and melba sauce.

Crêpes Suzette

YIELD:	6 Servings.	12 Servings.
INGREDIENTS:		
Oranges, zest only	3 each	6 each
Lemons, zest only	3 each	6 each
Sugar, granulated	2⅓ ounces	5¼ ounces
Butter, unsalted	3 ounces	6 ounces
Orange juice	12 ounces	1 pound, 8 ounces
Grand Marnier	6 ounces	12 ounces
Crêpes (see page 936)	12 each	24 each
Cognac	3 ounces	6 ounces

METHOD OF PREPARATION:

1. Gather the equipment and ingredients.
2. Zest all of the oranges and lemons.
3. Heat a suzette pan, sprinkle sugar over the pan, add the butter, and mix until all of the sugar is dissolved.
4. While the sugar is dissolving, add the zest from the oranges and lemons.
5. Add the orange juice.
6. Remove the pan from the flame, and add the Grand Marnier; *do not flame.*
7. Return the pan to the heat and dip the crêpes into the sauce one at a time. Fold the crêpes into quarters.
8. Move the crêpes to the side of the pan.
9. When all crêpes are folded, remove the pan from the flame, and add the cognac.
10. Return to flame, and ignite.
11. Remove the crêpes from the pan; place on a heated platter.
12. Reduce the sauce, and serve; spoon the sauce over the crêpes.

PASTRY TECHNIQUES:

Combining, Folding

Combining:

Bringing together two or more components.

1. Prepare the components to be combined.
2. Add one to the other, using the appropriate mixing method (if needed).

Folding:

Do steps 1, 2, and 3 in one continuous motion.

1. Run a bowl scraper under the mixture, across the bottom of the bowl.
2. Turn the bowl counterclockwise.
3. Bring the bottom mixture to the top.

CHEF NOTES:

The following are variations that can be used:

1. For a more syrupy sauce, caramelize the sugar before adding the butter.
2. If no fresh oranges are available, orange preserves or orange marmalade can be used as a substitute.
3. If marmalade is used, use less sugar.

American Ice Cream

Vanilla

PASTRY TECHNIQUES:

Combining, Heating, Freezing

Combining:

Bringing together two or more components.

1. Prepare the components to be combined.
2. Add one to the other, using the appropriate mixing method (if needed).

Heating:

1. Prepare the food product according to the formula's instructions.
2. Choose the appropriate method of heating (on the range or stove top, in the oven, etc.)
3. Apply the product to the heat.

Freezing:

1. Prepare the product.
2. Place the product in the freezing cabinet for the appropriate length of time.

HACCP:

Store at 0° F.

HAZARDOUS FOODS:

Heavy cream
Milk

YIELD: 2 pounds, 10 ounces. 5 pounds, 4 ounces.

INGREDIENTS:		
Cream, heavy	10½ ounces	1 pound, 5 ounces
Milk, whole	1 pound, 5 ounces	2 pounds, 10 ounces
Sugar, granulated	10½ ounces	1 pound, 5 ounces
Extract, vanilla	To taste	To taste
Salt	Pinch	Pinch

METHOD OF PREPARATION:

1. Gather the equipment and ingredients.
2. Combine all of the ingredients.
3. Heat to 180° F.
4. Cool immediately.
5. Freeze in an ice cream machine.

French Vanilla Ice Cream

YIELD:	3 pounds, 1½ ounces.	6 pounds, 3 ounces.

INGREDIENTS:

Ingredient		
Milk, whole	1 pound	2 pounds
Cream, heavy	1 pound	2 pounds
Salt	Pinch	Pinch
Sugar, granulated	9 ounces	1 pound, 2 ounces
Egg yolks	8 ounces	1 pound
Vanilla bean, *or*	½ each	1 each
Vanilla extract	½ ounce	1 ounce

METHOD OF PREPARATION:

1. Gather the equipment and ingredients.
2. Place the milk, heavy cream, salt, and granulated sugar in a pot, and bring to a boil.
3. Place the egg yolks and vanilla in a bowl; whisk together.
4. Temper the egg yolk mixture, and add to the boiling milk.
5. Place over the double boiler, and heat to 175° F, or until the mixture coats the back of a spoon.
6. Cool immediately over an ice bath.
7. Freeze in an ice cream machine.

PASTRY TECHNIQUES:

Combining, Boiling, Tempering, Freezing

Combining:

Bringing together two or more components.

1. Prepare the components to be combined.
2. Add one to the other, using the appropriate mixing method (if needed).

Boiling:

1. Bring the cooking liquid to a rapid boil.
2. Stir the contents.

Tempering:

1. Whisk the eggs vigorously while ladling hot liquid.

Freezing:

1. Prepare the product.
2. Place the product in the freezing cabinet for the appropriate length of time.

HACCP:

Store at 0° F.

HAZARDOUS FOODS:

Milk
Heavy cream
Egg yolks

CHEF NOTE:

This mixture is best if left to sit in the refrigerator overnight for freezing the next day.

Frozen Yogurt

YIELD:	2 pounds, 6 ounces.	4 pounds, 12 ounces.
INGREDIENTS:		
Yogurt, plain	8 ounces	1 pound
Milk, whole	1 pound, 8 ounces	3 pounds
Sugar, granulated	6 ounces	12 ounces
Extract, vanilla	To taste	To taste
Egg shade, liquid color (optional)	As needed	As needed

METHOD OF PREPARATION:

1. Gather the equipment and ingredients.
2. Combine all of the ingredients.
3. Strain the mixture.
4. Freeze in an ice cream machine.

IF USING CORNSTARCH:

Yield:

2 ounces of cornstarch for 2 pounds, 6 ounces of batter.
4 ounces of cornstarch for 4 pounds, 12 ounces of batter.

Method of Preparation:

1. Place the cornstarch in a bowl, and add one third of the milk to the cornstarch; slowly stir with a whip until smooth and well mixed.
2. Cook the milk-starch mixture over medium heat; stir constantly with a whip until the mixture comes to a rolling boil and no starch taste remains.
3. Add all of the remaining ingredients, including the cooked milk-starch mixture. Stir to remove any lumps.
4. Strain.
5. Freeze in an ice cream machine.

PASTRY TECHNIQUES:

Combining, Freezing

Combining:

Bringing together two or more components.

1. Prepare the components to be combined.
2. Add one to the other, using the appropriate mixing method (if needed).

Freezing:

1. Prepare the product.
2. Place the product in the freezing cabinet for the appropriate length of time.

HACCP:

Store at 0° F.

HAZARDOUS FOODS:

Yogurt
Milk

CHEF NOTES:

Cornstarch can be used to prevent this product from becoming icy, if stored in the freezer overnight. If available, an instant starch or modified starch could be substituted and will produce a more stable frozen yogurt than when using cornstarch.

Ice Cream Flavors

PASTRY TECHNIQUE:

Combining

Combining:

Bringing together two or more components.

1. Prepare the components to be combined.
2. Add one to the other, using the appropriate mixing method (if needed).

HACCP:

Store at 0° F.

INGREDIENTS:

	YIELD:
Ice cream, American or French (see pages 941 and 942)	1 quart

Flavorings:

Chocolate:

Chocolate, dark, semi-sweet, melted	6 ounces

Banana:

Banana purée	1 pound
Lemon juice	½ each

Strawberry:

Strawberries, chopped and cleaned	3 pounds
Strawberry compound	1 ounce

Rum and Raisin:

Rum	To taste
Raisins, macerated in rum	3 ounces

Chocolate Chip:

Chocolate chips	3 ounces

Praline:

Praline paste	5 ounces

Nougat:

Praline paste	2 ounces
Nougat, ground	5 ounces

Eggnog:

Nutmeg, ground	⅛ ounce
Rum or brandy	To taste

Mocha/Coffee:

Instant coffee, *or*	1 ounce
Coffee paste	2 ounces

Almond:

Almond paste	3 ounces
Almonds, ground, toasted	2 ounces

METHOD OF PREPARATION:

1. Gather the equipment and ingredients.
2. Prepare the ice cream, and freeze as instructed.

Lemon Sorbet

PASTRY TECHNIQUES:

Combining, Whipping, Freezing

Combining:

Bringing together two or more components.

1. Prepare the components to be combined.
2. Add one to the other, using the appropriate mixing method (if needed).

Whipping:

1. Hold the whip at a 55-degree angle.
2. Create circles, using a circular motion.
3. The circular motion needs to be perpendicular to the bowl.

Freezing:

1. Prepare the product.
2. Place the product in the freezing cabinet for the appropriate length of time.

HACCP:

Store at 0° F.

HAZARDOUS FOOD:

Egg whites

YIELD: 3 pounds, 1½ ounces. 6 pounds, 3 ounces.

INGREDIENTS:

Sugar, granulated	1 pound	2 pounds
Water	1 pound, 10½ ounces	3 pounds, 5 ounces
Lemon juice	5 ounces	10 ounces
Salt	Pinch	Pinch
Lemons, rind, freshly grated	1½ each	3 each
Egg whites	2 ounces	4 ounces

METHOD OF PREPARATION:

1. Gather the equipment and ingredients.
2. Place the sugar, water, lemon juice, salt, and grated lemon rind in a container.
3. Let the mixture stand for 1 hour, or until all of the sugar has dissolved.
4. Strain through a very fine sieve.
5. Beat the egg whites to the froth stage.
6. Add the egg whites to the sugar mixture; combine.
7. Freeze in an ice cream machine.

Sorbet

PASTRY TECHNIQUES:

Combining, Reducing, Freezing

Combining:

Bringing together two or more components.

1. Prepare the components to be combined.
2. Add one to the other, using the appropriate mixing method (if needed).

Reducing:

1. Bring the sauce to a boil; then reduce to a simmer.
2. Stir often; reduce to the desired consistency.

Freezing:

1. Prepare the product.
2. Place the product in the freezing cabinet for the appropriate length of time.

HACCP:

Store at 0° F.

YIELD:	2 pounds.	4 pounds.
INGREDIENTS:		
Lemon, rind, juice	1 each	2 each
Simple syrup (see page 966)	1 pound	2 pounds
Purée, fruit	1 pound	2 pounds

METHOD OF PREPARATION:

1. Gather the equipment and ingredients.
2. Add the lemon rind to the simple syrup when preparing it.
3. Reduce the simple syrup so it will read 18° on the Baumé scale.
4. Combine the simple syrup, lemon juice, and fruit purée.
5. Strain, if necessary.
6. Freeze in an ice cream machine.

Variations:

The following purées can be used:

1. Raspberry purée
2. Apricot purée
3. Peach purée
4. Strawberry purée
5. Pineapple purée

Wine Sorbet

YIELD:	2 pounds, 9 ounces.	5 pounds, 2 ounces.
INGREDIENTS:		
Water	1 pound	2 pounds
Sugar, granulated	6 ounces	12 ounces
Honey or glucose	2 ounces	4 ounces
Orange, zest and juice	½ each	1 each
Lemon, zest and juice	½ each	1 each
Wine, red, white, rosé, champagne, or apple	1 pound	2 pounds
Egg whites, beaten	1 ounce	2 ounces

METHOD OF PREPARATION:

1. Gather the equipment and ingredients.
2. Place the water, sugar, and honey or glucose in a pot; bring to a boil. Remove from heat after the sugar has been dissolved totally.
3. Add the orange and lemon zests and juices into the hot syrup.
4. Cool the syrup.
5. Strain the syrup, and add the wine.
6. Freeze the sorbet in an ice cream machine until two-thirds frozen.
7. Whip the egg whites to a soft peak.
8. Add the whipped egg whites to the sorbet.
9. Continue to freeze.

PASTRY TECHNIQUES:

Combining, Boiling, Freezing, Whipping

Combining:

Bringing together two or more components.

1. Prepare the components to be combined.
2. Add one to the other, using the appropriate mixing method (if needed).

Boiling:

1. Bring the cooking liquid to a rapid boil.
2. Stir the contents.

Freezing:

1. Prepare the product.
2. Place the product in the freezing cabinet for the appropriate length of time.

Whipping:

1. Hold the whip at a 55-degree angle.
2. Create circles, using a circular motion.
3. The circular motion needs to be perpendicular to the bowl.

HACCP:

Store at 0° F.

HAZARDOUS FOODS:

Egg whites

CHEF NOTE:

When served, additional wine or champagne can be poured over the sorbet. For more intense flavor and/or color, more wine can be substituted for the water.

PASTRY TECHNIQUES:

Mixing, Piping, Coating, Dipping, Tempering, Spiking

Mixing:

1. Follow the proper mixing procedure: creaming, blending, whipping, or combination.

Piping:

With bag:
1. Use a bag with a disposable tip; cut the bag at 45-degree angle.
2. Fill to no more than half full.
3. Burp the bag.

With cone:
1. Cut and fold the piping cone to the appropriate size.
2. Fill the cone with a small amount.
3. Fold the ends to form a triangle.
4. Pipe the desired designs.

Coating:

1. Use a coating screen, with a sheet pan underneath.
2. Ensure that the product is the correct temperature.
3. Coat the product; use an appropriately-sized utensil.

Dipping:

1. Prepare the product to the proper dipping temperature.
2. Carefully submerge the product.
3. Dry on parchment paper or a screen.

Tempering:

1. Whisk the eggs vigorously while ladling hot liquid.

Spiking:

1. Use chocolate that is appropriately tempered.
2. Drag the truffles across a screen.
3. Use three to four clean swift strokes.

Dark Truffles

YIELD: 3 pounds, 3 ounces. 200 Truffles.

INGREDIENTS:	
Couverture, dark, semi-sweet	2 pounds
Cream, heavy	1 pound
Butter, unsalted	3 ounces
Sugar, confectionery (rolling)	As needed
Couverture, additional dark, semi-sweet, tempered (for dipping)	As needed

METHOD OF PREPARATION:

1. Gather the equipment and ingredients.
2. Chop the couverture into small pieces.
3. Scald the heavy cream, and remove from the heat.
4. Add the chopped couverture and butter into the scalded cream.
5. Whisk until smooth.
6. Wrap well, and allow to cool to room temperature.
7. As soon as the ganache mixture is cooled and set, transfer to a mixing bowl; mix, using a rubber spatula, until light in color and texture.
8. Using a pastry bag and a round tip, immediately pipe out ¾-inch round dots of ganache mixture onto parchment-lined sheet pans.
9. When the piping is complete, place the truffles in the refrigerator; allow them to set to marzipan consistency (not fully hardened).
10. Powder your hands with confectionery sugar to reduce sticking; roll each truffle in your hand until rounded.
11. Return the truffles to the parchment-lined sheet pan, and return to the refrigerator; allow them to set.
12. Melt the dark couverture, and temper.
13. Remove the truffles from the refrigerator when fully set.
14. Place a small amount of tempered couverture in one hand, pick up a truffle with the other hand, and roll the truffle gently between the palms of your hands to coat evenly.

(continued)

HACCP:

Store at 60° F to 65° F.

HAZARDOUS FOOD:

Heavy cream

CHEF NOTES:

1. After piping the ganache, do *not* leave the truffles in the refrigerator too long, or they will be too hard to roll.
2. Do *not* overspike, or the chocolate will appear dull.
3. Do *not* spike the truffles too soon, or the texture will melt.
4. Use gloves (latex) to roll the truffle in your hands.

15. Return the truffles to the sheet pans, and allow to set at room temperature.
16. Use a dipping fork to dip each truffle in tempered dark couverture.
17. Immediately place the truffle on a wire rack.
18. Using two forks, roll the truffle three to four times on the screen; allow the chocolate to develop a spiked texture.
19. Using two dipping forks, place the truffles back on parchment paper. (Do *not* allow the truffles to set on the screen.)

Milk Truffles

YIELD: 3 pounds, 11 ounces. 230 Truffles

INGREDIENTS:

Couverture, milk	2 pounds, 8 ounces
Cream, heavy	1 pound
Butter, unsalted	3 ounces
Sugar, confectionery (rolling)	As needed
Couverture, additional milk, tempered (for dipping)	As needed

METHOD OF PREPARATION:

1. Gather the equipment and ingredients.
2. Chop the couverture into small pieces.
3. Scald the heavy cream, and remove from the heat.
4. Add the chopped couverture and butter into the scalded cream.
5. Whisk until smooth.
6. Wrap well, and allow to cool to room temperature.
7. As soon as the ganache mixture is cooled and set, transfer to a mixing bowl, and mix, using a rubber spatula, until light in color and texture.
8. Using a pastry bag and a round tip, immediately pipe out ¾-inch round dots of ganache mixture onto parchment-lined sheet pans.
9. When the piping is complete, place the truffles in the refrigerator; allow them to set to marzipan consistency (not fully hardened).
10. Powder your hands with confectionery sugar to reduce sticking, and roll each truffle in your hand until rounded.
11. Return the truffles to the parchment-lined sheet pan, and return to the refrigerator; allow them to set.
12. Melt the milk couverture, and temper.
13. Remove the truffles from the refrigerator when fully set.
14. Place a small amount of tempered couverture in one hand, pick up a truffle with the other hand, and roll the truffle gently between the palms of your hands to coat evenly.

(continued)

PASTRY TECHNIQUES:

Mixing, Piping, Coating, Dipping, Tempering, Spiking

Mixing:

1. Follow the proper mixing procedure: creaming, blending, whipping, or combination.

Piping:

With bag:
1. Use a bag with a disposable tip; cut the bag at 45-degree angle.
2. Fill to no more than half full.
3. Burp the bag.

With cone:
1. Cut and fold the piping cone to the appropriate size.
2. Fill the cone with a small amount.
3. Fold the ends to form a triangle.
4. Pipe the desired designs.

Coating:

1. Use a coating screen, with a sheet pan underneath.
2. Ensure that the product is the correct temperature.
3. Coat the product; use an appropriately-sized utensil.

Dipping:

1. Prepare the product to the proper dipping temperature.
2. Carefully submerge the product.
3. Dry on parchment paper or a screen.

Tempering:

1. Whisk the eggs vigorously while ladling hot liquid.

Spiking:

1. Use chocolate that is appropriately tempered.
2. Drag the truffles across a screen.
3. Use three to four clean swift strokes.

HACCP:

Store at 60° F to 65° F.

HAZARDOUS FOOD:

Heavy cream

CHEF NOTES:

1. After piping the ganache, do *not* leave the truffles in the refrigerator too long, or they will be too hard to roll.
2. Do *not* overspike, or the chocolate will appear dull.
3. Do *not* spike the truffles too soon, or the texture will melt.
4. Use gloves (latex) to roll the truffles in your hand.

15. Return the truffles to the sheet pans, and allow them to set at room temperature.

16. Using a dipping fork, dip each truffle in tempered milk couverture.

17. Immediately place the truffle on a wire rack.

18. Using two forks, roll the truffle three to four times on the screen; allow the chocolate to develop a spiked texture.

19. Using two dipping forks, place the truffles back on the parchment paper. (Do *not* allow the truffles to set on the screen.)

White Truffles

YIELD: 4 pounds, 6 ounces. 270 Truffles

INGREDIENTS:	
Couverture, white	3 pounds, 3 ounces
Cream, heavy	1 pound
Liqueur, Kirschwasser	3 ounces
Sugar, confectionery (rolling)	1 pound
Couverture, additional white (for dipping)	As needed

METHOD OF PREPARATION:

1. Gather the equipment and ingredients.
2. Chop the couverture into small pieces.
3. Scald the heavy cream, and remove from the heat.
4. Add the chopped couverture into the scalded cream.
5. Add the Kirschwasser, and whisk lightly until smooth.
6. Wrap well, and allow to cool at room temperature.
7. As soon as the ganache mixture is cooled and set, transfer to a mixing bowl, and mix, using a rubber spatula, until light in color.
8. Using a pastry bag with a round tip, immediately pipe out ¾-inch round dots of ganache mixture onto parchment-lined sheet pans.
9. When the piping is complete, place the truffles in the refrigerator; allow them to set to marzipan consistency (not fully hardened).
10. Powder your hands with confectionery sugar to reduce sticking, and roll each truffle in your hands until rounded.
11. Return the truffles to the parchment-lined sheet pan, and return to the refrigerator; allow them to set.
12. Melt the white couverture, and temper.
13. Remove the truffles from the refrigerator when fully set.

(continued)

PASTRY TECHNIQUES:

Mixing, Piping, Coating, Dipping, Tempering, Spiking

Mixing:

1. Follow the proper mixing procedure: creaming, blending, whipping, or combination.

Piping:

With bag:
1. Use a bag with a disposable tip; cut the bag at 45-degree angle.
2. Fill to no more than half full.
3. Burp the bag.

With cone:
1. Cut and fold the piping cone to the appropriate size.
2. Fill the cone with a small amount.
3. Fold the ends to form a triangle.
4. Pipe the desired designs.

Coating:

1. Use a coating screen, with a sheet pan underneath.
2. Ensure that the product is the correct temperature.
3. Coat the product, use an appropriately sized utensil.

Dipping:

1. Prepare the product to the proper dipping temperature.
2. Carefully submerge the product.
3. Dry on parchment paper or a screen.

Tempering:

1. Whisk the eggs vigorously while ladling hot liquid.

Spiking:

1. Use chocolate that is appropriately tempered.
2. Drag the truffles across a screen.
3. Use three to four clean swift strokes.

HACCP:

Store at 60° F to 65° F.

HAZARDOUS FOOD:

Heavy cream

CHEF NOTES:

1. After piping the ganache, do *not* leave the truffles in the refrigerator too long, or they will be too hard to roll.
2. Do *not* overspike, or the chocolate will appear dull.
3. Do *not* spike the truffles too soon, or the texture will melt.
4. Use gloves (latex) to roll the truffle in your hands.

14. Place a small amount of tempered couverture in one hand, pick up a truffle with the other hand, and roll the truffle gently between the palms of your hands to coat evenly.
15. Return the truffles to the sheet pans, and allow them to set at room temperature.
16. Using a dipping fork, dip each truffle in white tempered couverture.
17. Immediately place the truffle on a wire rack.
18. Using two forks, roll the truffle three to four times on the screen, allowing the chocolate to develop a spiked texture.
19. Using two dipping forks, place the truffles back on the parchment paper. (Do *not* allow the truffles to set on the screen.)

Ganache No. 1

PASTRY TECHNIQUES:

Chopping, Scalding, Combining

Chopping:

1. Use a sharp knife.
2. Hold the food product properly.
3. Cut with a quick downward motion.

Scalding:

1. Heat the liquid on high heat.
2. Do not boil the liquid.

Combining:

Bringing together two or more components.

1. Prepare the components to be combined.
2. Add one to the other, using the appropriate mixing method (if needed).

HACCP:

Store at 40° F.

HAZARDOUS FOOD:

Heavy cream

YIELD:	SOFT 2 pounds, 12 ounces.	SEMI-HARD 2 pounds.	HARD 1 pound, 12 ounces.
INGREDIENTS:			
Cream, heavy	1 pound, 12 ounces	1 pound	8 ounces
Butter, unsalted	As needed	As needed	4 ounces
Couverture, dark, semi-sweet	1 pound	1 pound	1 pound

METHOD OF PREPARATION:

1. Gather the equipment and ingredients.
2. Place the heavy cream and butter, if necessary, in a saucepan and scald.
3. Chop the dark couverture very fine.
4. Add the couverture to the scalded cream.
5. Stir until smooth and the consistency is uniform.
6. Cool to room temperature.

CHEF NOTE:

For best results, let the ganache sit for at least 24 hours before use. Butter also may be added to either soft or semi-hard ganache to enhance its texture.

Ganache No. 2

YIELD:	1 pound, 14 ounces.	3 pounds, 12 ounces.

INGREDIENTS:

Couverture, dark, semi-sweet	1 pound, 2 ounces	2 pounds, 4 ounces
Cream, heavy	8 ounces	1 pound
Butter, unsalted	4 ounces	8 ounces

METHOD OF PREPARATION:

1. Gather the equipment and ingredients.
2. Place the heavy cream and butter in a saucepan, and scald.
3. Chop the dark couverture very fine.
4. Add couverture to scalded cream.
5. Stir until smooth and uniform in consistency.
6. Cool to room temperature.

PASTRY TECHNIQUES:

Chopping, Scalding, Combining

Chopping:

1. Use a sharp knife.
2. Hold the food product properly.
3. Cut with a quick downward motion.

Scalding:

1. Heat the liquid on high heat.
2. Do not boil the liquid.

Combining:

Bringing together two or more components.

1. Prepare the components to be combined.
2. Add one to the other, using the appropriate mixing method (if needed).

HACCP:

Store at 40° F.

HAZARDOUS FOOD:

Heavy cream

CHEF NOTE:

For best results, let the ganache sit for at least 24 hours before use.

Apple Strudel

PASTRY TECHNIQUES:

Combining, Kneading, Filling, Rolling

Combining:

Bringing together two or more components.

1. Prepare the components to be combined.
2. Add one to the other, using the appropriate mixing method (if needed).

Kneading:

1. Prepare the kneeding surface with the appropriate medium (flour, cornstarch, etc.).
2. Press and form the dough into a mass using soft, determined strokes.
3. Continue kneading until appropriate consistency and/or temperature is achieved.

Filling:

1. Cut open the food product.
2. Carefully spread the filling using an icing spatula, or
3. Carefully pipe the filling using a pastry bag.

Rolling:

1. Prepare the rolling surface by dusting with the appropriate medium (flour, cornstarch, etc.).
2. Use the appropriate style pin (stick pin or ball bearing pin) to roll the dough to desired thickness; rotate the dough during rolling to prevent sticking.

HAZARDOUS FOOD:

Eggs

YIELD:	3 pounds, 13¾ ounces.	13 pounds, 11½ ounces.

INGREDIENTS:

Oil	As needed	As needed
Flour, bread	As needed	As needed
Butter, unsalted, melted	3 ounces	12 ounces
Crumbs, cake or bread	4 ounces	1 pound
Sugar, confectionery	As needed	As needed

Dough:

Flour, bread	10 ounces	2 pounds, 8 ounces
Butter, unsalted, melted (or oil)	2 ounces	8 ounces
Salt	Pinch	½ ounce
Eggs, whole	2 ounces	8 ounces
Vinegar, white	Pinch	Pinch
Water, lukewarm	3½ ounces	14 ounces

Filling:

Apples, Granny Smith	1 pound, 8 ounces	6 pounds
Cinnamon, ground	¼ ounce	1 ounce
Sugar, granulated	4 ounces	1 pound
Raisins, seedless	3 ounces	12 ounces
Walnuts, chopped	3 ounces	12 ounces

METHOD OF PREPARATION:

Gather the equipment and ingredients.

Preparation of Dough:

1. Place the bread flour, melted butter or oil, salt, eggs, vinegar, and water in a bowl with a dough hook.
2. Mix on high speed for 7 to 9 minutes.
3. Remove, and brush the dough with oil, as needed; place in a covered bowl in a warm place for 1 hour.

Preparation of Filling:

1. Peel, core, and slice or cube the apples.
2. Combine the apples with the cinnamon, sugar, raisins, and walnuts.

Preparation of Rolling Strudel:

1. Cover a table with a pastry cloth, and dust lightly with flour.
2. Have the melted butter, cake crumbs, and rested dough ready.
3. Stretch the dough evenly into a rectangular shape, and pin out, as much as possible.
4. Pull out the edges, using the back of your hands.
5. Using the floured backs of your hands, stretch the dough uniformly.
6. Continue to stretch toward the opposite corners and the sides until the table is completely covered.
7. Trim the excess dough with a pastry wheel.
8. Sprinkle the dough with butter.
9. Sprinkle the cake crumbs over one edge of the dough to cover an area approximately 1 to 1½ feet wide.
10. Place the fruit filling on top of the crumbs.
11. Roll up, using the table cloth to hold and guide the dough.
12. Transfer to a sheet pan; brush the roll with butter.
13. Bake at 400° F for 20 minutes, or until golden brown.
14. Serve fresh and warm, dusted with confectionery sugar.

Variations:

Use the same dough, but substitute any of the following for filling:

1. Cheese filling
2. Poppy seed filling
3. Walnut filling

Baked Apples

PASTRY TECHNIQUES:
Combining, Filling

Combining:
Bringing together two or more components.
1. Prepare the components to be combined.
2. Add one to the other, using the appropriate mixing method (if needed).

Filling:
1. Cut open the food product.
2. Carefully spread the filling using an icing spatula, or
3. Carefully pipe the filling using a pastry bag.

YIELD:	5 Portions.	10 Portions.

INGREDIENTS:

Apples	5 each	10 each
Lemon juice	As needed	As needed
Water	As needed	As needed
Sugar, granulated	1 ounce	2 ounces
Walnuts, chopped	1 ounce	2 ounces
Raisins, seedless	2 ounces	4 ounces
Cinnamon, ground	⅛ ounce	¼ ounce
Butter, unsalted	2 ounces	4 ounces

Syrup:

Sugar, granulated	12 ounces	1 pound, 8 ounces
Wine, sweet white	4 ounces	8 ounces
Lemon juice	½ ounce	1 ounce

METHOD OF PREPARATION:

1. Gather the equipment and ingredients.
2. Wash, peel, and core the apples. (Place the apples in lemon water after they have been peeled.)
3. Combine the sugar, walnuts, raisins, and cinnamon.
4. Place apples on a half sheet pan.
5. Place the mixture in the center of the apples.
6. Melt the butter, and pour over the apples.
7. Bake at 375° F until done.
8. Place the granulated sugar, white wine, and lemon juice in a pot; boil to a clear syrup.
9. Place a baked apple on preheated dessert plate, and coat with the syrup. Serve hot.

CHEF NOTE:
Remove apples from lemon water and let drip dry.

Baklava

YIELD:	½ Sheet Pan.	1 Sheet Pan.
INGREDIENTS:		
Butter, unsalted	10 ounces	1 pound, 4 ounces
Dough, phyllo	15 leaves	30 leaves
Walnuts, finely chopped	10 ounces	1 pound, 4 ounces
Nuts, pistachio, chopped	5 ounces	10 ounces
Honey	8 ounces	1 pound
Sugar, granulated	4 ounces	8 ounces
Water	3 ounces	6 ounces
Cinnamon sticks	1 each	2 each

METHOD OF PREPARATION:

1. Gather the equipment and ingredients.
2. Melt the butter, and brush on a half sheet pan, or hotel pan.
3. Place one layer of dough on the bottom of the pan, and drizzle with butter.
4. Repeat the procedure with four additional layers of dough.
5. On the fifth layer, sprinkle half of the nuts to completely cover the surface of the dough.
6. Place another five layers of dough on top of the nuts, drizzling butter between each layer.
7. Sprinkle the remaining nuts over the surface of the dough; cover it completely.
8. Place another five layers of dough on top of the nuts, drizzling butter between each layer.
9. Slice the dough into diamond shapes, using a very sharp knife. Be certain to slice completely through the layers, but do not remove any layers of dough.
10. Bake at 420° F until the pastry is golden brown and crispy.
11. Combine the honey, sugar, water, and cinnamon stick in a saucepan.
12. Bring to a boil, and cook for 1 minute, or until a syrupy consistency is reached.
13. Remove the baklava from the oven and, while it is still hot, pour the hot honey syrup over it.
14. Set aside to cool; then separate the diamond-shaped pieces.

PASTRY TECHNIQUES:

Combining, Filling

Combining:

Bringing together two or more components.

1. Prepare the components to be combined.
2. Add one to the other, using the appropriate mixing method (if needed).

Filling:

1. Cut open the food product.
2. Carefully spread the filling using an icing spatula, or
3. Carefully pipe the filling using a pastry bag.

HACCP:

Store at 60° F to 65° F.

CHEF NOTES:

1. Baklava should sit overnight before serving to allow the honey syrup to soak into the layers and provide better flavor.
2. Do not place in refrigerator.

PASTRY TECHNIQUES:

Heating, Coating

Heating:

1. Prepare the food product according to the formula's instructions.
2. Choose the appropriate method of heating (on the range or stove top, in the oven, etc.)
3. Apply the product to the heat.

Coating:

1. Use a coating screen, with a sheet pan underneath.
2. Ensure that the product is the correct temperature.
3. Coat the product; use an appropriately-sized utensil.

Petit Four Glacé

YIELD: 150 Each.

INGREDIENTS:	
Apricot, coating	10 ounces
Frangipane (see pages 753 and 754)	½ sheet
Sugar, granulated	As needed
Marzipan (see page 965)	2 pounds
Fondant	7 pounds, 8 ounces
Simple syrup (see page 966)	As needed

METHOD OF PREPARATION:

Gather the equipment and ingredients.

Building:

1. Bring to a boil the apricot coating, and set aside; keep warm.
2. Sprinkle a half sheet of frangipane (18 inches by 12 inches, 1-inch thick) with sugar. Place a piece of sheet paper on top, and cover with a cardboard sheet. Flip upside down to unmold.
3. Trim the edges to square up.
4. Mark and divide into five 3¼-inch rectangles.
5. Place each on a 3¾-inch by 11½-inch piece of cardboard.
6. Level each piece.
7. Brush the surface of the frangipane with approximately 2 ounces of the apricot coating that was previously brought to a boil.
8. Roll out the marzipan to ⅛-inch thick and slightly wider and longer than the cake.
9. Roll up on the pin, unroll onto the top of the cake, and pin-level.
10. Place a cardboard on the marzipan, and flip over.
11. Carefully trim excess marzipan from the edge of the cake, using a French knife; avoid dragging the knife. (Reserve clean marzipan for reuse.)
12. Measure a 1-inch bar along the length of the cake, and cut, using the French knife. Keep the blade straight; avoid dragging. Cut 1-inch squares from the bar.
13. You should end up with thirty 1-inch by 1-inch pieces.
14. Place the pieces on a glazing screen, leaving 1½ inches of space around each piece; keep in neat rows.

Preparing Fondant:

1. Remove 1 pound, 8 ounces of fondant from the pail of prepared fondant (30 petit fours).
2. Warm to 98° F to 100° F. Either use a water bath and stir, testing often to

determine the temperature, or use a saucepan over low heat; stir with hand, on and off the fire.

3. Flavor and color as desired; use compounds, liqueurs, nuts, butters, and so forth.
4. Adjust the consistency, using 100° F simple syrup until the fondant evenly coats a test piece.
5. The fondant should coat a piece thinly over the entire surface and leave no "foot" at the base.

Glazing (Spoon Method):

1. Place a screen on a clean plastic sheet pan.
2. Using a solid stainless spoon, pour the glaze over an individual piece to coat.
3. Pour from the side of the spoon, slowly and very close to the top of the piece.
4. When you reach the back, stop; allow the fondant to fall down and cover the back side.
5. Twist the spoon to catch drips, and remove.

Note: If necessary, additional methods can be added.

Finishing:

1. Decorate by using the traditional decoration, as instructed.
2. Loosen the petit fours from the screen by using a paring knife dipped in hot water.
3. Dip your fingers in cold water before lifting each piece, to avoid sticking.

Note: Fondant that flows through the screen onto the plastic sheet should be reused, as needed. Pick out any cake crumbs that may have fallen in. Fondant can be warmed and strained, if necessary.

Tart Tatin

YIELD: 1, 8-inch Tart.

INGREDIENTS:

Apples, Granny Smith	2 to 3 each
Butter, unsalted, softened	1 ounce
Sugar, granulated	2 ounces
Puff pastry (see page 848)	1, 8-inch circle, ¼-inch thick

METHOD OF PREPARATION:

1. Gather the equipment and ingredients.
2. Preheat the oven to 425° F.
3. Peel, core, and cut the apples into quarters.
4. In an 8-inch teflon-coated sauté pan, spread the softened butter on the bottom.
5. Sprinkle the granulated sugar over the softened butter.
6. Place each apple quarter on its side in the pan, the thinner portion of the apple being placed toward the center of pan.
7. Place the pan over medium heat, and cook the sugar, butter, and apples until the sugar begins to caramelize and the apples to soften.
8. When the sugar has caramelized, remove the pan from the heat, and lay the puff pastry circle over the top of the apples.
9. Turn the oven temperature to 400° F. Immediately place the pan into the oven, and bake for approximately 15 to 20 minutes; remove from the oven when the puff pastry is baked completely.
10. After removing the tart from the oven, invert it immediately onto a preheated serving plate.

PASTRY TECHNIQUES:

Rolling, Peeling, Caramelizing

Rolling:

1. Prepare the rolling surface by dusting with the appropriate medium (flour, cornstarch, etc.).
2. Use the appropriate style pin (stick pin or ball bearing pin) to roll the dough to desired thickness; rotate the dough during rolling to prevent sticking.

Peeling:

1. Use a clean paring knife or peeler.
2. Do not peel over an unsanitary surface.

Caramelizing:

Wet method:

1. Use an extremely clean pot.
2. Place the sugar and water on high heat.
3. Never stir the mixture once the sugar begins to dissolve.
4. Once caramelized, shock in ice water.

CHEF NOTES:

1. Pears and mangos are great substitutes for the apples.
2. Use a sauté pan with a heat-resistant handle.

Egg Wash

YIELD:	6 ounces.	12 ounces.
INGREDIENTS:		
Eggs, whole	3 ounces	6 ounces
Milk, whole or water	3 ounces	6 ounces

METHOD OF PREPARATION:

1. Gather the equipment and ingredients.
2. Combine the ingredients; use a whisk to make a smooth mixture.
3. When using an egg wash, apply it with a pastry brush.

PASTRY TECHNIQUE:

Combining

Combining:

Bringing together two or more components.

1. Prepare the components to be combined.
2. Add one to the other, using the appropriate mixing method (if needed).

HACCP:

Store at 40° F.

HAZARDOUS FOODS:

Eggs
Milk

CHEF NOTES:

1. If additional browning is required, milk can be used instead of water.
2. If additional browning is required, sugar can be added.
3. If used for savory items, salt may be added.

Marzipan

YIELD:	2 pounds, 8 ounces.	5 pounds.
INGREDIENTS:		
Sugar, confectionery	1 pound	2 pounds
Almond paste	1 pound	2 pounds
Fondant	4 ounces	8 ounces
Glucose	4 ounces	8 ounces

METHOD OF PREPARATION:

1. Gather the equipment and ingredients.
2. Place the ingredients in a mixing bowl in this order: confectionery sugar, almond paste, fondant, and glucose.
3. Mix with a paddle until the mixture is combined.
4. Remove from the mixer.
5. Knead the mixture, working it into one piece.
6. Place it in a plastic bag to avoid its becoming dry.

PASTRY TECHNIQUES:

Mixing, Kneading

Mixing:

1. Follow the proper mixing procedure: creaming, blending, whipping, or combination.

Kneading:

1. Prepare the kneeding surface with the appropriate medium (flour, cornstarch, etc.).
2. Press and form the dough into a mass using soft, determined strokes.
3. Continue kneading until appropriate consistency and/or temperature is achieved.

HACCP:

Store at 60° F to 65° F.

CHEF NOTES:

1. Instead of using a mixer to combine ingredients, your hands may be used.
2. If an almond paste, which contains a high percentage of glucose, is used, reduce or omit the glucose listed in the formula to obtain the desired consistency.

Simple Syrup

YIELD: 16 pounds.

PASTRY TECHNIQUES:
Combining, Boiling

Combining:
Bringing together two or more components.
1. Prepare the components to be combined.
2. Add one to the other, using the appropriate mixing method (if needed).

Boiling:
1. Bring the cooking liquid to a rapid boil.
2. Stir the contents.

HACCP:
Store at 40° F.

INGREDIENTS:

Sugar, granulated	8 pounds
Water	8 pounds
Lemon, sliced	1 each

METHOD OF PREPARATION:

1. Gather the equipment and ingredients.
2. Place the sugar and water in a pot, and bring to a boil.
3. Remove from the heat, and add the lemon slices. (Another form of acid may be substituted for the sliced lemons.)
4. Cool.
5. Store in a refrigerator.

Variations:

Combine all of the ingredients for each flavored simple syrup.

Kahlua:	1 torte
Kahlua	½ ounce
Simple syrup	2½ ounces

Raspberry/chocolate:	1 torte
Raspberry compound	¼ ounce
Simple syrup	2½ ounces

Mandarin orange:	1 torte
Orange compound	¼ ounce
Simple syrup	2½ ounces

Pistachio:	1 torte
Pistachio compound	¼ ounce
Simple syrup	2½ ounces
Lemon juice	¼ ounce

Kirschwasser:	1 torte
Kirschwasser	½ ounce
Simple syrup	1½ ounces
Cherry syrup	1 ounce

Coconut:	1 torte
Coconut liqueur	½ ounce
Simple syrup	2½ ounces

Rum or brandy:	1 torte
Rum or brandy	½ ounce
Simple syrup	2½ ounces

Poached Fruit

PASTRY TECHNIQUE:

Poaching

Poaching:

1. Bring the liquid to a boil; then reduce to a simmer.
2. Submerge and anchor the product.
3. Do not overcook the product.

YIELD:

INGREDIENTS:

Fruit, desired	As needed
Liquid, basic (for poaching) (see page 969)	To cover fruit

METHOD OF PREPARATION:

1. Gather the equipment and ingredients.
2. Peel and core the desired fruit.
3. Place the poaching liquid in a saucepan, deep enough to cover the fruit.
4. Place the fruit in the liquid, and weigh it down so it does not rise to the surface.
5. Simmer, *do not boil,* over a low flame until the fruit is tender but firm enough to hold its shape and size.

CHEF NOTES:

1. The liquid should be at about 190° F and should not move; it, however, should have steam rising from the surface.
2. For variation, fruit can be poached in the oven at 325° F to 350° F.

PASTRY TECHNIQUE:

Combining

Combining:

Bringing together two or more components.

1. Prepare the components to be combined.
2. Add one to the other, using the appropriate mixing method (if needed).

Poaching Liquid, Basic

YIELD:	2 pounds, 8 ounces.	5 pounds.
INGREDIENTS:		
Wine, red or white	1 pound	2 pounds
Water	1 pound	2 pounds
Sugar, granulated	8 ounces	1 pound
Cinnamon	½ stick	1 stick
Cloves	1 to 2 each	2 to 4 each

METHOD OF PREPARATION:

1. Gather the equipment and ingredients.
2. Place all of the ingredients in a pot.
3. Bring to a simmer, and poach the fruit, as desired.

Index